Theoretical and Mathematical Physics

The series founded in 1975 and formerly (until 2005) entitled *Texts and Monographs in Physics* (TMP) publishes high-level monographs in theoretical and mathematical physics. The change of title to *Theoretical and Mathematical Physics* (TMP) signals that the series is a suitable publication platform for both the mathematical and the theoretical physicist. The wider scope of the series is reflected by the composition of the editorial board, comprising both physicists and mathematicians.

The books, written in a didactic style and containing a certain amount of elementary background material, bridge the gap between advanced textbooks and research monographs. They can thus serve as basis for advanced studies, not only for lectures and seminars at graduate level, but also for scientists entering a field of research.

Howard J. Carmichael

Statistical Methods in Quantum Optics 2

Non-Classical Fields

With 89 Figures

Professor Howard J. Carmichael
University of Auckland
Physics Department
Private Bag 92019, Auckland, New Zealand
E-Mail: h.carmichael@auckland.ac.nz

Originally published with the title "Texts and Monographs in Physics" with ISSN 0172-5998
ISBN 978-3-642-09041-7 ISSN 1864-5879
ISBN 978-3-540-71320-3 (eBook) Springer Berlin Heidelberg New York

Springer is a part of Springer Science+Business Media
springer.com

Softcover reprint of the hardcover 1st edition 2008

Cover: eStudio Calamar S.L., F. Steinen-Broo, Girona, Spain

For Marybeth

Preface

I must admit with regret, and not a little embarrassment, that eight years have passed since I sat down to write a preface for the first volume of this book. A great deal has changed for the community of quantum opticians in the meantime. The interests of some have been turned to the fascinating properties of degenerate quantum gases where a number of analogies with quantum optics are to be found. Then there is the quantum information revolution: a whole new language to be learned, built around John Bell's reading of the Bohr–Einstein debate and venerable words like entanglement, launched in a new direction, with the goals of achieving an unbreakable code and a new paradigm for computation—a quantum-mechanical one. Considering the passage of time and what has occurred, I can only trust it will not disappoint to announce that this second volume of *Statistical Methods* has not been diverted in either direction—or, perhaps, rather closer to the truth, it could not: a path was already set in the preface to Volume 1, and this is the path I have followed in preparing Chaps. 9 through 19 of Volume 2.

The subtitle, *Nonclassical Fields*, is perhaps not as accurate as it might be as a summary of content; or to put it another way, if my aim from the start had been to write a book on this topic, parts of that book would differ significantly from what follows here. Possibly the most important thing missing, and something that should be said, is that there are two quite distinct paths to a definition of nonclassicality in quantum optics. The first is grounded in the existence, or otherwise, of a nonsingular and positive Glauber–Sudarshan P function. The physical grounding is in the treatment of optical measurements, specifically the photoelectric effect: for a given optical field, can the photoelectron counting statistics, including all correlations, be reproduced by a Poisson process of photoelectron generation driven by a classical light intensity, allowed most generally to be stochastic? Viewed at a more informal level, the question asks whether or not the infamous proposal of Bohr, Kramers, and Slater for the interaction of classical light and quantized atoms can be upheld in the presence of the observable photoelectron counting statistics.

This criterion for nonclassicality is likely to be the one offered up by most quantum opticians when pressed for a definition. There is, however, a second. It is an outgrowth of John Bell's work and does not speak directly about measurements of any sort. At issue are the variances and covariances of a set of quantum mechanical observables—the quadrature amplitudes occupying this or that optical mode: can these quantities all be computed from a classical probability distribution, admitting hidden variables but no nonlocal connections between the values they take? Generally speaking, but not always, variances and covariances computed from an entangled state within quantum mechanics cannot be recovered from a classical distribution. Squeezed light provides a notable counterexample; for it, a positive definite Wigner function serves as the required classical probability distribution. Thus, by the second Bell-based criterion, squeezed light is not nonclassical. (Though in a perverse reversal of Bell's argument, the entangled character of the two-mode squeezed state is often seen to trump this observation.) Squeezed light is of course nonclassical by the former P function criterion.

In this volume squeezed light is nonclassical. The "Nonclassical" of the subtitle is to be read in the P function sense. Starting with two chapters on squeezing in the degenerate parametric oscillator, the volume continues on with the theme taken up in Volume 1 of "methods developed in quantum optics for analyzing quantum fluctuations in terms of a visualizable evolution over time." These are the methods of the quantum–classical correspondence: the phase-space representations, which when applied to an operator master equation yield a Fokker–Planck equation, albeit, in many cases, only after a system size expansion of the full equation of motion is made—i.e., only when the quantum noise is sufficiently small. Applied to the degenerate parametric oscillator, the methods fail, though the positive P representation of Drummond and Gardiner does manage to resurrect "a visualizable evolution over time"—qualified, however, by serious difficulties of a new kind.

Chapters 9 and 10 deal with squeezing, the degenerate parametric oscillator, and how squeezed light generation causes the standard phase-space methods to fail. Chapter 11 then develops the positive P approach, while Chap. 12 uncovers the problems it encounters when the system size expansion no longer holds.

Problems with the positive P representation aside, much of the appeal of the phase-space approach is lost when the system size expansion fails. Its very premise is a classical dynamic plus quantum fluctuation "fuzz," the "fuzz" a perturbation by definition; "fluctuation" is defined in a classical sense from the very beginning. While the positive P representation escapes this background to some extent, it also retreats from all but a formal connection with the physics—as a generator of quantum averages—and any resolution of its difficulties can only deepen that retreat.

On this score it is worthwhile to recall my appeal in the preface to Volume 1: "Nothing in the Schrödinger equation fluctuates. What then *is* a quantum fluctuation?" A classically inspired method for computing quantum averages is unlikely to illuminate this question. The seven chapters from Chap. 13 to Chap. 19 work towards an outlook that possibly can.

The context for the development is provided by cavity QED, which is explored in Chaps. 13–16. Its defining conditions of strong dipole coupling between a resonant atomic transition and an optical cavity mode are essentially the same—for single atoms—as those defining a small system size, such that the system size expansion fails, and experiments have reached a remarkable level of sophistication, a level hardly imagined as researchers set out to realize strong dipole coupling some 20 years ago.

My attempt to illuminate the "What *is* a quantum fluctuation?" question occupies Chaps. 17–19. Here quantum trajectory theory is developed. The approach, at bottom, is conventional, recalling observations that have been made about the meaning of quantum mechanics since the time of Niels Bohr. Certainly nothing fluctuates in the Schrödinger equation; indeed, the Schrödinger equation describes no realized happenings of any sort—no realized events; it governs the time evolution of probabilities of events. To actually *realize* events, the probabilities must be put into action, to play out as a stochastic process. But here is the sticking point: the playing out is not unique, not only in the trivial sense that the throwing of a die yields different answers on every throw, but because the very shape of the die is not uniquely defined from within the Schrödinger equation itself. It is we the commentators who chose a shape through the question we chose to ask—or so it might appear, though in practice it is not so much a matter of commentators and their questions, but a subdivision of the physical world into a subsystem acting and one acted (irreversibly) upon. With only the "acting" subsystem defined, there are, of course, many possibilities for the subsystem "acted (irreversibly) upon" and such a division is not unique.

The many years that have passed since I began writing this book have left me indebted to numerous people, for their support and encouragement, and for the detection of many of those irritating errors that inevitably seem to make it into the typeset text. I thank both the University of Oregon and the University of Auckland for support during periods of concentrated work on the book. I am also indebted to the Alexander Humboldt Foundation, my German sponsor, Wolfgang Schleich, and his tireless wife Kathy, for their support during a year spent in Ulm; the visit allowed me to restart a project that had languished for quite some time. Then the patience of the editorial office of Prof. Wolf Beiglböck at Springer can only be wondered at. Finally, there are my students in Auckland, Mile Gu, Andy Chia, Changsuk Noh, Rob Fisher, and Felipe Dimer de Oliveira, who provided indispensable service by reading parts of the text and detecting so many of those irritating errors, and my special thanks go to Hyunchul Nha who, as my postdoc, made numerous contributions that enabled me to improve what is written.

I must add that work on the book has stolen many hours away from my wife Marybeth. My principal debt is to her. We can both now be happy that this one cause, at least, of stolen hours is at a close.

Auckland
January 2007

Howard Carmichael

Contents

9 The Degenerate Parametric Oscillator I: Squeezed States .. 1
9.1 Introduction ... 1
9.2 Degenerate Parametric Amplification and Squeezed States 4
9.2.1 Degenerate Parametric Amplification Without Pump Depletion ... 4
9.2.2 Quantum Fluctuations and Squeezed States ... 8
9.2.3 The Degenerate Parametric Oscillator ... 14
9.2.4 Master Equation for the Degenerate Parametric Oscillator ... 20
9.2.5 Cavity Output Fields ... 27
9.3 The Spectrum of Squeezing ... 31
9.3.1 Intracavity Field Fluctuations ... 32
9.3.2 Definition of the Spectrum of Squeezing ... 38
9.3.3 Homodyne Detection: The Source-Field Spectrum of Squeezing ... 40
9.3.4 The Source-Field Spectrum of Squeezing with Unit Efficiency ... 44
9.3.5 Free-Field Contributions ... 47
9.3.6 Vacuum Fluctuations ... 49
9.3.7 Squeezing in the Wigner Representation: A Comment on Interpretation ... 54

10 The Degenerate Parametric Oscillator II: Phase-Space Analysis in the Small-Noise Limit ... 61
10.1 Phase-Space Formalism for the Degenerate Parametric Oscillator ... 61
10.1.1 Phase-Space Equation of Motion in the P Representation ... 62
10.1.2 Phase-Space Equations of Motion in the Q and Wigner Representations ... 66
10.2 Squeezing: Quantum Fluctuations in the Small-Noise Limit ... 71

10.2.1 System Size Expansion Far from Threshold 71
10.2.2 Quantum Fluctuations Below Threshold 74
10.2.3 Quantum Fluctuations Above Threshold 83
10.2.4 Quantum Fluctuations at Threshold 86

11 The Positive *P* Representation 95
11.1 The Positive P Representation 96
11.1.1 The Characteristic Function and Associated Distribution 98
11.1.2 Fokker–Planck Equation for the Degenerate Parametric Oscillator 103
11.1.3 Linear Theory of Quantum Fluctuations 112
11.2 Miscellaneous Topics 117
11.2.1 Alternative Approaches to the Linear Theory of Quantum Fluctuations 117
11.2.2 Dynamical Stability of the Classical Phase Space 124
11.2.3 Preservation of Conjugacy for Stochastic Averages 127

12 The Degenerate Parametric Oscillator III: Phase-Space Analysis Outside the Small-Noise Limit 133
12.1 The Degenerate Parametric Oscillator with Adiabatic Elimination of the Pump 134
12.1.1 Adiabatic Elimination in the Stochastic Differential Equations 135
12.1.2 A Note About Superoperators 138
12.1.3 Adiabatic Elimination in the Master Equation 141
12.1.4 Numerical Simulation of the Stochastic Differential Equations 145
12.1.5 Deterministic Dynamics in the Extended Phase Space 151
12.1.6 Steady-State Solution for the Positive P Distribution 156
12.1.7 Quantum Fluctuations and System Size 160
12.1.8 Quantum Dynamics Beyond Classical Trajectories plus "Fuzz" 169
12.1.9 Higher-Order Corrections to the Spectrum of Squeezing at Threshold 178
12.2 Difficulties with the Positive P Representation 181
12.2.1 Technical Difficulties: Two-Photon Damping 182
12.2.2 Physical Interpretation: The Anharmonic Oscillator 189

13 Cavity QED I: Simple Calculations 195
13.1 System Size and Coupling Strength 196
13.2 Cavity QED in the Perturbative Limit 198
13.2.1 Cavity-Enhanced Spontaneous Emission 202
13.2.2 Cavity-Enhanced Resonance Fluorescence 210

13.2.3 Forwards Photon Scattering in the Weak-Excitation Limit 217
13.2.4 A One-Atom "Laser" 222
13.3 Nonperturbative Cavity QED 231
13.3.1 Spontaneous Emission from a Coupled Atom and Cavity 232
13.3.2 Vacuum Rabi Splitting 239
13.3.3 Vacuum Rabi Resonances in the Two-State Approximation 243

14 Many Atoms in a Cavity: Macroscopic Theory 247
14.1 Optical Bistability: Steady-State Transmission of a Nonlinear Fabry–Perot 247
14.2 The Mean-Field Limit for a Homogeneously Broadened Two-Level Medium 253
14.2.1 Steady State 254
14.2.2 Maxwell–Bloch Equations 263
14.2.3 Stability of the Steady State 268
14.3 Relationship Between Macroscopic and Microscopic Variables . 271
14.4 Cavity QED with Many Atoms 275
14.4.1 Weak-Probe Transmission Spectra 276
14.4.2 A Comment on Spatial Effects 280

15 Many Atoms in a Cavity II: Quantum Fluctuations in the Small-Noise Limit 285
15.1 Microscopic Model 286
15.1.1 Master Equation for Optical Bistability 286
15.1.2 Fokker–Planck Equation in the P Representation 287
15.1.3 Fokker–Planck Equation in the Q Representation 289
15.1.4 Fokker–Planck Equation in the Wigner Representation . 292
15.2 Linear Theory of Quantum Fluctuations 298
15.2.1 System Size Expansion for Optical Bistability 299
15.2.2 Linearization About the Steady State 306
15.2.3 Covariance Matrix for Absorptive Bistability 312
15.2.4 Atom–Atom Correlations 317
15.2.5 A Comment on Measures of Squeezing 320
15.2.6 Spectrum of the Transmitted Light in the Weak-Excitation Limit 322
15.2.7 Forwards Photon Scattering in the Weak-Excitation Limit 330

16 Cavity QED II: Quantum Fluctuations 335
16.1 Density Matrix Expansion for the Weak-Excitation Limit 335
16.1.1 Pure-State Factorization of the Density Operator for One Atom 336

16.1.2 Pure-State Factorization of the Density Operator for Many Atoms . . . 339
16.1.3 Forwards Photon Scattering for N Atoms in a Cavity . . 345
16.1.4 Corrections to the Small-Noise Approximation . . . 350
16.1.5 Antibunching of Fluorescence for One Atom in a Cavity . . . 352
16.1.6 Spectra of Squeezing in the Weak-Excitation Limit . . . 357
16.2 Spatial Effects . . . 360
16.3 Beyond Classical Trajectories plus "Fuzz": Spontaneous Dressed-State Polarization . . . 368
16.3.1 Maxwell–Bloch Equations for "Zero System Size" . . . 369
16.3.2 Dressed Jaynes–Cummings Eigenstates . . . 376
16.3.3 Secular Approximation in the Basis of Dressed Jaynes–Cummings Eigenstates . . . 383
16.3.4 Spectrum of the Transmitted Light in the Strong-Coupling and Weak-Excitation Limits . . . 385
16.3.5 The $\sqrt{n}$ Anharmonic Oscillator . . . 391
16.3.6 Quantum Fluctuations for Strong Excitation . . . 395

17 Quantum Trajectories I: Background and Interpretation . . . 401
17.1 Density Operators and Scattering Records . . . 403
17.2 Generalizing the Bohr–Einstein Quantum Jump . . . 410
17.2.1 The Einstein Stochastic Process . . . 410
17.2.2 Conditional Evolution: Trajectories for Known Initial States . . . 412
17.2.3 Conditional Evolution: Trajectories for "Blind" Realizations . . . 415
17.2.4 The Master Equation . . . 420
17.2.5 Quantum Jumps in the Presence of Coherence . . . 422
17.3 Miscellaneous Observations . . . 426
17.3.1 Time Evolution Under Null Measurements . . . 426
17.3.2 Conditional States and Nonlinearity . . . 429
17.3.3 Record Probabilities and Norms . . . 430
17.3.4 Monte Carlo Simulations . . . 431
17.3.5 The Waiting-Time Distribution . . . 433

18 Quantum Trajectories II: The Degenerate Parametric Oscillator . . . 437
18.1 Scattering Records and Photoelectron Counting . . . 437
18.1.1 Record Probabilities . . . 438
18.1.2 Factorization for Pure Initial States . . . 446
18.2 Unraveling the Density Operator . . . 447
18.2.1 Photoelectron Counting Records . . . 447
18.2.2 Homodyne-Current Records . . . 451
18.2.3 Heterodyne-Current Records . . . 460

18.3 Physical Interpretation . . . 466
18.3.1 Systems, Environments, and Complementarity . . . 466
18.3.2 Modeling Projective Measurements . . . 473

19 Quantum Trajectories III: More Examples . . . 479
19.1 Photon Scattering in the Weak-Excitation Limit . . . 479
19.2 Unraveling the Density Operator: Cascaded Systems . . . 484
19.2.1 System–Reservoir Interaction Hamiltonian . . . 486
19.2.2 Reservoir Field . . . 487
19.2.3 The Cascaded Systems Master Equation . . . 488
19.2.4 Photoelectron Counting Unraveling . . . 491
19.2.5 Coherent Driving Fields . . . 492
19.2.6 Symmetric Irreversible Coupling . . . 497
19.3 Optical Spectra . . . 499
19.3.1 Optical Spectrum Using a Scanning Interferometer . . . 500
19.3.2 Spontaneous Emission from a Driven Excited-State Doublet . . . 506
19.3.3 Optical Spectrum Using Heterodyne Detection . . . 510

References . . . 515

Index . . . 527

9

The Degenerate Parametric Oscillator I: Squeezed States

9.1 Introduction

Volume 1 of this book introduced the formalism of open systems in quantum optics as it existed in the early 1980s, two decades after the invention of the laser. In it we met the operator master equation, the quantum regression formula, and the phase-space methods, based on the P, Q, and Wigner representations, that provide "classical" visualizations of the fluctuations of the radiation field—the so-called quantum–classical correspondence. We also met Fokker–Planck equations and stochastic differential equations, and the various methods of classical statistical physics that help with the analysis within the phase-space representations. The application of all these things was illustrated by two classic examples: resonance fluorescence and the single-mode laser.

We begin Volume 2 by recalling some elementary aspects of laser theory, which introduce us to the theme that will carry us forward into a discussion of nonclassical fields. The quantum theory of the laser illustrates how an operator master equation, the quantum–classical correspondence, and the methods of classical statistical physics can be used to solve a nontrivial problem of some practical importance. There is, however, something missing in this example; we alluded to this at the end of Sect. 7.1.3. The laser is essentially a classical device. Thus, we seem to lose the distinction between quantum statistics and classical statistics in passing from the Hamiltonian of Sect. 7.2.1 to the classical stochastic description of the laser developed in Chap. 8. That is not to say, of course, that laser action has nothing to do with quantum mechanics. Stimulated emission gain results from an inverted population in *quantized* atomic states, and the fluctuations of the laser field, characterized by n_{spon}, are certainly quantum fluctuations; they arise from the intrinsic probabilistic character of quantum mechanics, not from any externally imposed statistical assumption. On the other hand, these fluctuations can be described by a *classical* stochastic process. The laser operates as a classical nonlinear oscillator with additive noise. Surely, in general, we would not expect a quantum field to be describable in the language of classical statistics, where the fundamental quantities are probabilities. The fundamental quantities in quantum mechanics are probability amplitudes. Why does our theory of the laser not recognize the distinction?

In fact the language of the laser theory developed in Chap. 8 *is*, in a sense, quantum mechanical, but in a limited sense only. After all, the Fokker–Planck equations (8.61a), (8.86a), (8.121a), and (8.131a) determine the Glauber–Sudarshan P representation for a quantum-mechanical density operator; they evolve a quantum state. The important qualification is that the laser example does not exploit the full flexibility of the P representation and consequently avoids the need for probability amplitudes. We deduce this, on the one hand, from the fact that the P distribution need not satisfy all the requirements of a classical probability density—it may take on negative values, or be more singular than a δ-function, as it is for a Fock state (Eqs. 3.31, 3.40, and 3.41)—yet for the laser it does. Thus, in quantum-mechanical language, the state of the laser field (Eq. 3.15) is a statistical mixture of coherent states. To define it, we need only the probabilities that enter this mixture. There is no call for probability amplitudes. Wherever amplitude (phase) information is needed, it is provided by the phase-space variable α, the complex amplitude of the laser field.

A second issue, beyond that of merely representing states, is the issue of dynamics. We cannot generally expect the state of a quantum system to evolve over time according to the rules governing the time evolution of a classical stochastic process. Our experience with the phase-space representation for two-level atoms (Chap. 6) provides an example showing that there is no guarantee that the quantum–classical correspondence will identify a Fokker–Planck equation with a given master equation. In the laser case, however, it does just this.

The two mentioned properties—a positive, nonsingular P distribution, and the Fokker–Planck form for the phase-space equation of motion—are not entirely independent. It is as well, however, to note them both. What if, for example, we substitute the Q representation for the P representation? The Q distribution is positive and nonsingular by definition, but it need not satisfy a Fokker–Planck equation, as we will see very shortly (Sect. 10.1.2).

Indeed, there is a choice to be made between different phase-space representations (Chap. 4). As a result, there is apparently some ambiguity in asserting that an optical field, like the field emitted by a laser, is essentially classical because it is described within a particular phase-space representation by a classical stochastic process. Such a description may exist in one representation and not in another. The P representation is special, though, as it is this representation, not the Q or Wigner representation, that gives the normal-ordered, time-ordered averages (Sect. 4.3.3) that enter the quantum theory of photodetection and coherence [9.1, 9.2]. Because of this, any field that is described by a classical stochastic process within the P representation can be described, equivalently, within the framework of classical statistical optics and the semiclassical theory of photoelectric detection [9.3]. On this basis, such fields are designated *classical fields.* Those that remain are *nonclassical fields*—i.e., nonclassical fields are those that are not described by a classical stochastic process within the Glauber–Sudarshan P representation.

Volume 2 deals with nonclassical fields. Here we explore the statistical methods developed in the last two and half decades to treat fluctuations of nonclassical fields. In order to do this we must add to the repertoire of quantum-statistical methods learned in Volume 1. The focus initially stays with the quantum–classical correspondence and the associated phase-space representations. Various questions arise: in what ways does the quantum–classical correspondence break down? Can the breakdowns be mended? What physical conditions result in breakdowns? These questions lead us to the topics of squeezing, the positive P representation, and cavity QED. Ultimately we will find that a different way of connecting the quantum to the classical is called for, something that brings us closer to the foundations of quantum mechanics and the long-standing debate over quantum measurements. The connection is provided by the method of quantum trajectories. This is the last of the topics dealt with in Volume 2.

As in Volume 1, formalism is developed around specific examples. We start by taking a close look at the degenerate parametric oscillator. This device is closely related to the laser. It is a self-sustained oscillator whose operation depends on a nonlinear optical interaction taking place inside a cavity. Like the laser, it exhibits threshold behavior as an adjustable parameter controlling the energy supply is increased. In this instance, the oscillation is sustained by the process of parametric amplification (frequency down conversion). Thus, the pump is a coherent field of frequency 2ω which pumps a cavity mode of frequency ω via an interaction in a $\chi^{(2)}$ medium. Below threshold, the output at frequency ω is a fluctuating field with zero mean amplitude, somewhat analogous to the chaotic output of a laser below threshold. Above threshold, the output field has nonzero mean amplitude and exhibits small fluctuations, again similar to the output from a laser above threshold. Unlike the laser, however, the degenerate parametric oscillator produces a field of definite phase, the phase reference being provided by the coherent pump. There is then no phase diffusion above threshold. Instead, the output field "chooses" between stable amplitudes 180° out of phase with one another.

The most interesting difference from the laser from our point of view is in the quantum fluctuations. The fluctuations are nonclassical, they are squeezed, both below and above threshold. Experiments with parametric oscillators have demonstrated substantial amounts of squeezing [9.4, 9.5, 9.6, 9.7]. We are not primarily interested in the squeezed states themselves, though; our interest is the comparison between the quantum-statistical treatment of an essentially classical device—the laser—and a device that produces nonclassical light. We therefore touch only indirectly on the properties of squeezed states. There are a number of works available which adequately review this topic [9.8, 9.9]. For an introduction to parametric processes, the book by Yariv is a suitable place to start [9.10], and Bloembergen's volume on nonlinear optics [9.11] is a good source of references for further reading on this subject.

9.2 Degenerate Parametric Amplification and Squeezed States

There is nothing new about states of the electromagnetic field that have a singular or non-positive-definite P distribution. The P representation was first introduced to represent statistical mixtures of coherent states, thus allowing traditional statistical optics to be translated into the language of quantum mechanics. Application of the representation to states outside this class was also considered, however, particularly by Sudarshan and Klauder [9.12, 9.13]. In particular, Mollow and Glauber published a work in 1967 that is closely related to our interests in this section [9.14]. They analyzed the process of parametric amplification and derived a number of results concerning the existence, or otherwise, of a positive, nonsingular P distribution. Parametric amplification provides the gain for a parametric oscillator and is thus intimately connected with the generation of squeezed states. Let us therefore begin with some background on degenerate parametric amplification and its relationship to squeezed states.

9.2.1 Degenerate Parametric Amplification Without Pump Depletion

First we review the classical theory of degenerate parametric amplification without pump depletion. Consider the lossless standing-wave cavity illustrated in Fig. 9.1. The cavity supports two resonant electromagnetic field modes that couple to one another through the $\chi^{(2)}_{yxy} = \chi^{(2)}_{yyx} = \chi^{(2)}_{xyy}$ component of the nonlinear susceptibility tensor of an intracavity crystal ($LiNbO_3$, for example, with the optic axis aligned in the x direction). These fields, the subharmonic and pump, are expanded as

$$\boldsymbol{E}(z,t) = \hat{e}_y \mathcal{E}(t) A(z) \cos[\Phi(z) + \phi] e^{-i\omega_C t} + \text{c.c.}, \tag{9.1a}$$

$$\boldsymbol{E}_p(z,t) = \hat{e}_x \mathcal{E}_p A(z) \cos[2\Phi(z) + \phi_p] e^{-i2\omega_C t} + \text{c.c.}, \tag{9.1b}$$

with

$$A(z) \equiv 1 + \left(1/\sqrt{n} - 1\right)[\theta(z) - \theta(z-\ell)], \tag{9.2a}$$

$$\Phi(z) \equiv (\omega_C/c)z + (n-1)(\omega_C/c)\int_0^z dz'[\theta(z') - \theta(z'-\ell)], \tag{9.2b}$$

where $\theta(\xi) \equiv 0, 1$ for $\xi < 0$ and $\xi \geq 0$, respectively, $\mathcal{E}(t)$ and $\mathcal{E}_p$ are complex mode amplitudes, ϕ and ϕ_p are constants which set the phases of the standing waves inside the crystal, $\hat{e}_x$ and $\hat{e}_y$ are unit polarization vectors, and n is

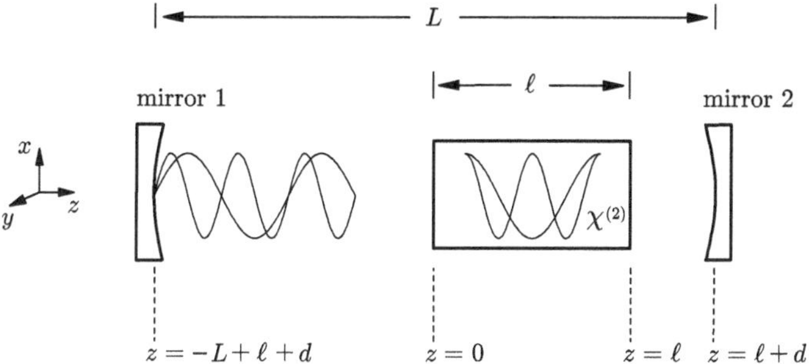

Fig. 9.1. Cavity geometry for a standing-wave degenerate parametric amplifier. The relative phases of the pump and subharmonic standing-wave mode functions are shown for maximum coupling inside the crystal and are determined by reflection boundary conditions at the mirrors. The incompatible standing-wave patterns are matched by dispersive elements placed inside the cavity (see Note 9.3)

the crystal refractive index. We assume perfect phase matching, and small parametric gain so that the forwards and backwards field amplitudes may be taken to be equal. In the undepleted pump approximation, we take $\mathcal{E}_p$ to be constant and seek an equation of motion for the amplitude $\mathcal{E}(t)$.

Note 9.1. We have assumed that all of the power incident upon a face of the crystal propagates into the crystal—i.e., no reflection. Thus, the field amplitudes in vacuum and inside the crystal are matched by setting $\frac{1}{2}\epsilon_0 c|\mathcal{E}_{\text{vac}}|^2 = \frac{1}{2}n\epsilon_0 c|\mathcal{E}_{\text{crystal}}|^2$ and $\frac{1}{2}\epsilon_0 c|\mathcal{E}_p^{\text{vac}}|^2 = \frac{1}{2}n\epsilon_0 c|\mathcal{E}_p^{\text{crystal}}|^2$.

The nonlinear interaction between the two electromagnetic field modes inside the crystal generates a polarization

$$\begin{aligned}\boldsymbol{P}_{\omega_C}(z,t) = \hat{e}_y(2\epsilon_0\chi^{(2)}/n)\mathcal{E}^*(t)\mathcal{E}_p\cos[\Phi(z)+\phi]\cos[2\Phi(z)+\phi_p]e^{-i\omega_C t}\\ + \text{c.c.},\end{aligned} \tag{9.3}$$

oscillating at the frequency ω_C, where $\chi^{(2)} \equiv \chi^{(2)}_{yxy} = \chi^{(2)}_{yyx} = \chi^{(2)}_{xyy}$. More precisely, we identify the polarization components that radiate forward- and backward-traveling subharmonic waves by expanding the product of cosines in (9.3) as a sum of exponentials; thus, there are forward- and backward-traveling electric fields

$$\boldsymbol{E}_f(z,t) = \hat{e}_y\tfrac{1}{2}\mathcal{E}(t)A(z)e^{-i[\omega_C t-\Phi(z)-\phi]} + \text{c.c.}, \tag{9.4a}$$

and

$$\boldsymbol{E}_b(z,t) = \hat{e}_y\tfrac{1}{2}\mathcal{E}(t)A(z)e^{-i[\omega_C t+\Phi(z)+\phi]} + \text{c.c.} \tag{9.4b}$$

radiated, respectively, by the polarization components ($0 < z < \ell$)

$$\boldsymbol{P}^f_{\omega_C}(z,t) = \hat{e}_y\tfrac{1}{2}\mathcal{P}_f(t)e^{-i[\omega_C(t-z/c)-\phi]} + \text{c.c.}, \tag{9.5a}$$

$$\mathcal{P}_f(t) \equiv (\epsilon_0\chi^{(2)}/n)\mathcal{E}^*(t)\mathcal{E}_p e^{i(\phi_p-2\phi)}, \tag{9.5b}$$

and

$$\boldsymbol{P}^b_{\omega_C}(z,t) = \hat{e}_y \tfrac{1}{2}\mathcal{P}_b(t) e^{-i[\omega_C(t+z/c)+\phi]} + \text{c.c.}, \tag{9.6a}$$

$$\mathcal{P}_b(t) \equiv (\epsilon_0 \chi^{(2)}/n)\mathcal{E}^*(t)\mathcal{E}_p e^{-i(\phi_p - 2\phi)}. \tag{9.6b}$$

When the parametric gain is small, the subharmonic field must make many round trips in the cavity before its amplitude changes significantly. The rate of change can be obtained from the ratio of the change $\Delta\mathcal{E}$ on a single round trip and the cavity round-trip time

$$t_C = 2\bar{L}/c, \qquad \bar{L} \equiv L + (n-1)\ell; \tag{9.7}$$

L is the cavity length and ℓ is the length of the crystal. By following the forward-traveling field once around the cavity, starting at $z = 0$, we find

$$\begin{aligned}\mathcal{E} + \Delta\mathcal{E} = &\left\{\left(\frac{1}{\sqrt{n}}\mathcal{E} + i\frac{\omega_C \ell}{2\epsilon_0 c n}\mathcal{P}_f\right)\sqrt{n} e^{i\phi_R} e^{2i[\Phi(\ell+d)+\phi]}\frac{1}{\sqrt{n}} + i\frac{\omega_C \ell}{2\epsilon_0 c n}\mathcal{P}_b\right\} \\ &\times \sqrt{n} e^{i\phi_R} e^{-2i[\Phi(-L+\ell+d)+\phi]},\end{aligned} \tag{9.8}$$

where the terms $i(\omega_C\ell/2\epsilon_0 cn)\mathcal{P}_f$ and $i(\omega_C\ell/2\epsilon_0 cn)\mathcal{P}_b$ are increments to the field amplitude arising from its propagation through the crystal in the forwards and backwards directions, respectively; factors $1/\sqrt{n}$ and $\sqrt{n}$ transform field amplitudes into and out of the crystal, ϕ_R is a phase change due to reflection at the mirrors, and phase changes $2[\Phi(\ell+d)+\phi]$ and $-2[\Phi(-L+\ell+d)+\phi]$ are required by boundary conditions at the mirrors. Substituting (9.5b) and (9.6b) into (9.8), and using the resonance condition

$$2[\phi_R + \Phi(\ell+d) - \Phi(-L+\ell+d)] = N2\pi, \qquad N \text{ an integer}, \tag{9.9}$$

and boundary condition at $z = \ell + d$,

$$\phi_R + 2[\Phi(\ell+d)+\phi] = M2\pi, \qquad M \text{ an integer}, \tag{9.10}$$

we arrive at the desired *equation of motion for the subharmonic field amplitude:*

$$\dot{\mathcal{E}} = \frac{\Delta\mathcal{E}}{2\bar{L}/c} = K\mathcal{E}^*, \tag{9.11}$$

where

$$K \equiv i\frac{\omega_C \ell \chi^{(2)}}{2n^{3/2}\bar{L}}\mathcal{E}_p \cos(\phi_p - 2\phi). \tag{9.12}$$

Note 9.2. The fields radiated by the polarization components (9.5) and (9.6) are calculated in the slowly-varying-amplitude approximation. The approximation is defined as follows. Inside the crystal the field

$$\begin{aligned}\boldsymbol{E}(z,t) = \hat{e}_y\left[\tfrac{1}{2}\mathcal{E}_f(z,t)e^{i[n(\omega_C/c)z+\phi]} + \tfrac{1}{2}\mathcal{E}_b(z,t)e^{-i[n(\omega_C/c)z+\phi]}\right]e^{-i\omega_C t} \\ + \text{c.c.}\end{aligned} \tag{9.13}$$

satisfies Maxwell's equation

$$\left(\frac{\partial^2}{\partial z^2} - \frac{n^2}{c^2}\frac{\partial^2}{\partial t^2}\right)\boldsymbol{E} = \frac{1}{\epsilon_0 c^2}\frac{\partial^2}{\partial t^2}\left(\boldsymbol{P}^f_{\omega_C} + \boldsymbol{P}^b_{\omega_C}\right). \tag{9.14}$$

In the *slowly-varying-amplitude approximation* we assume

$$\left.\begin{aligned} \frac{\partial^2\mathcal{E}_{f,b}}{\partial z^2} &\ll n(\omega_C/c)\frac{\partial\mathcal{E}_{f,b}}{\partial z} \ll [n(\omega_C/c)]^2\mathcal{E}_{f,b}, \\ \frac{\partial^2\mathcal{E}_{f,b}}{\partial t^2} &\ll \omega_C\frac{\partial\mathcal{E}_{f,b}}{\partial t} \ll \omega_C^2\mathcal{E}_{f,b}, \\ \frac{\partial^2\mathcal{P}_{f,b}}{\partial t^2} &\ll \omega_C\frac{\partial\mathcal{P}_{f,b}}{\partial t} \ll \omega_C^2\mathcal{P}_{f,b}. \end{aligned}\right\} \tag{9.15}$$

Then neglecting small terms in (9.14) yields the *Maxwell equations for slowly-varying amplitudes*

$$\frac{\partial\mathcal{E}_f}{\partial z} + \frac{n}{c}\frac{\partial\mathcal{E}_f}{\partial t} = i\frac{\omega_C}{2\epsilon_0 cn}\mathcal{P}_f, \tag{9.16a}$$

$$-\frac{\partial\mathcal{E}_b}{\partial z} + \frac{n}{c}\frac{\partial\mathcal{E}_b}{\partial t} = i\frac{\omega_C}{2\epsilon_0 cn}\mathcal{P}_b. \tag{9.16b}$$

The field increments appearing in (9.8) are obtained by integrating (9.16a) and (9.16b) through the crystal (in a retarded frame) with the right-hand sides of the equations treated as constants.

Note 9.3. The coupling constant defined by (9.12) depends on the relative phases of the pump and subharmonic field standing waves. The boundary condition satisfied by the subharmonic field at $z = \ell + d$ is given by (9.10), while the pump field must satisfy the boundary condition

$$\phi_R + 2[2\Phi(\ell + d) + \phi_p] = M'2\pi, \qquad M' \text{ an integer.} \tag{9.17}$$

From (9.10) and (9.17), we find

$$|\cos(\phi_p - 2\phi)| = \left|\cos\left(\tfrac{1}{2}\phi_R\right)\right|. \tag{9.18}$$

It follows that the phase change $\phi_R = \pi$ required to produce nodes at the mirrors sets the coupling constant (9.12) to zero. The coupling constant vanishes because the field increments produced by the parametric gain during forwards and backwards propagation through the crystal add out-of-phase; alternatively, the spatial average of the mode function overlap in the crystal, the integral of the product $\cos[n(\omega_C/c)z+\phi]\cos[2n(\omega_C/c)z+\phi_p]$, vanishes. Wu and Kimble have studied these effects in some detail [9.15]. To achieve strong mode coupling, additional dispersive elements must be used in the cavity or at the mirrors.

The solution to (9.11) is best expressed in terms of the two quadrature phase amplitudes of the subharmonic field. For an arbitrary phase θ, the *quadrature phase amplitudes* $\mathcal{E}_\theta(t)$ and $\mathcal{E}_{\theta+\pi/2}(t)$ are defined by writing (9.1a) as the sum of oscillating cosine and sine functions:

$$\begin{aligned}&\boldsymbol{E}(z,t)\\&=\hat{e}_y 2A(z)\cos[\Phi(z)+\phi]\big[\mathcal{E}_\theta(t)\cos(\omega_C t-\theta)+\mathcal{E}_{\theta+\pi/2}(t)\sin(\omega_C t-\theta)\big],\end{aligned}\tag{9.19}$$

with

$$\mathcal{E}_\theta \equiv \tfrac{1}{2}\big(\mathcal{E}e^{-i\theta}+\mathcal{E}^* e^{i\theta}\big).\tag{9.20}$$

Equation 9.11 is equivalent to the pair of equations for quadrature phase amplitudes

$$\dot{\mathcal{X}}=|K|\mathcal{X}, \qquad \dot{\mathcal{Y}}=-|K|\mathcal{Y},\tag{9.21}$$

where $\mathcal{X}\equiv\mathcal{E}_\theta$, $\mathcal{Y}\equiv\mathcal{E}_{\theta+\pi/2}$, with $\theta=\frac{1}{2}\arg(K)$. The straightforward solution of these equations gives

$$\mathcal{X}(t)=e^{|K|t}\mathcal{X}(0), \qquad \mathcal{Y}(t)=e^{-|K|t}\mathcal{Y}(0).\tag{9.22}$$

Thus, the degenerate parametric amplifier is a phase-sensitive amplifier; with the appropriate choice of phase, one quadrature phase amplitude of the subharmonic field is amplified while the other is deamplified.

9.2.2 Quantum Fluctuations and Squeezed States

In quantum-mechanical language the energy exchange in degenerate parametric amplification results from an interaction that annihilates one pump photon, of frequency $2\omega_C$, and creates a pair of subharmonic photons of frequency ω_C. The conjugate interaction accounts for the process of second harmonic generation. In the *undepleted pump approximation*, or *parametric approximation*, the pump mode is assumed to be highly populated and its loss or gain of photons is assumed to be negligible. Then the pump may be treated as a classical field of constant amplitude, and the Hamiltonian describing the creation and annihilation of subharmonic photons is

$$H\equiv\hbar\omega_C a^\dagger a+i\hbar\tfrac{1}{2}\big(Ke^{-i2\omega_C t}a^{\dagger 2}-K^* e^{i2\omega_C t}a^2\big),\tag{9.23}$$

where K is the coupling constant defined in (9.12). From (9.23) we obtain the Heisenberg equations of motion

$$\dot{\tilde{a}}=K\tilde{a}^\dagger, \qquad \dot{\tilde{a}}^\dagger=K^*\tilde{a},\tag{9.24}$$

where the annihilation and creation operators $\tilde{a}$ and $\tilde{a}^\dagger$ are defined in a frame rotating at frequency ω_C. The Heisenberg equations (9.24) are quantized versions of the classical equation (9.11). To complete the translation of our degenerate parametric amplifier model into quantum-mechanical language we replace the classical field (9.1a) by the field operator

$$\hat{\boldsymbol{E}}(z,t) = i\hat{e}_y\sqrt{\frac{\hbar\omega_C}{\epsilon_0 A\bar{L}}}A(z)\cos[\Phi(z)+\phi]\left(\tilde{a}(t)e^{-i\omega_C t} - \tilde{a}^\dagger(t)e^{i\omega_C t}\right), \quad (9.25)$$

where A is the cross-sectional area of the field mode [not to be confused with the function $A(z)$]. In parallel with (9.19) and (9.20), we introduce operator quadrature phase amplitudes, or *quadrature phase operators*

$$\hat{A}_\theta \equiv \tfrac{1}{2}\left(\tilde{a}e^{-i\theta} + \tilde{a}^\dagger e^{i\theta}\right). \quad (9.26)$$

Within the quantum theory, amplification and deamplification of quadrature phase amplitudes occurs in much the same way as it does in the classical treatment. Writing the Heisenberg equations of motion (9.24) in the form

$$\dot{\hat{X}} = |K|\hat{X}, \qquad \dot{\hat{Y}} = -|K|\hat{Y}, \quad (9.27)$$

with $\hat{X} \equiv \hat{A}_\theta$, $\hat{Y} \equiv \hat{A}_{\theta+\pi/2}$, for $\theta = \frac{1}{2}\arg(K)$, we obtain solutions

$$\hat{X}(t) = e^{|K|t}\hat{X}(0), \qquad \hat{Y}(t) = e^{-|K|t}\hat{Y}(0), \quad (9.28)$$

analogous to those given in (9.22). So far everything looks just the same. There is one important difference, however. The quadrature phase amplitudes are now operators, and they therefore exhibit fluctuations. We characterize these fluctuations by the standard deviation

$$\Delta A_\theta \equiv \sqrt{\left\langle\left(\hat{A}_\theta - \langle\hat{A}_\theta\rangle\right)^2\right\rangle}. \quad (9.29)$$

The size of the fluctuations will depend on the state of the field. In principle, the fluctuation of any particular quadrature phase amplitude may be arbitrarily small. According to the Heisenberg uncertainty principle, however, for any phase θ the uncertainty product $\Delta A_\theta \Delta A_{\theta+\pi/2}$ is bounded below by

$$\Delta A_\theta \Delta A_{\theta+\pi/2} \geq \tfrac{1}{2}|\langle\hat{C}\rangle|, \quad (9.30)$$

where $\hat{C}$ is the commutator

$$\begin{aligned}\hat{C} &= \left[\hat{A}_\theta, \hat{A}_{\theta+\pi/2}\right] \\ &= -\tfrac{1}{4}i\left[\left(\tilde{a}e^{-i\theta} + \tilde{a}^\dagger e^{i\theta}\right), \left(\tilde{a}e^{-i\theta} - \tilde{a}^\dagger e^{i\theta}\right)\right] \\ &= \tfrac{1}{4}i\left\{\left[\tilde{a}, \tilde{a}^\dagger\right] - \left[\tilde{a}^\dagger, \tilde{a}\right]\right\} \\ &= \tfrac{1}{2}i. \end{aligned} \quad (9.31)$$

Thus, in contrast to the classical theory, quantum mechanics imposes an *uncertainty relation for quadrature phase amplitudes*

$$\Delta A_\theta \Delta A_{\theta+\pi/2} \geq \tfrac{1}{4}. \tag{9.32}$$

This uncertainty relation for quadrature phase amplitudes is formally equivalent to the uncertainty relation for position and momentum operators of a mechanical oscillator. We may recast (9.32) with the definitions

$$\hat{q}_\theta \equiv \sqrt{\frac{2\hbar}{m\omega_C}}\hat{A}_\theta, \qquad \hat{p}_\theta \equiv \sqrt{2\hbar m\omega_C}\hat{A}_{\theta+\pi/2}, \tag{9.33}$$

to obtain the familiar result

$$\Delta q_\theta \Delta p_\theta \geq \tfrac{1}{2}\hbar; \tag{9.34}$$

$\hat{q}_\theta$ and $\hat{p}_\theta$ are related to the position operator $\hat{q}$ and momentum operator $\hat{p}$ of a mechanical oscillator by a rotation of coordinates through the angle $\omega_C t - \theta$ [use (9.26) and (3.5)].

We saw in Chap. 3 that coherent states of the harmonic oscillator are minimum uncertainty states, satisfying the uncertainty relation (9.34) with the equality $\Delta q \Delta p = \frac{1}{2}\hbar$ (Eqs. 3.4–3.6). Also, for coherent states the uncertainties in the scaled operators $\sqrt{2m\omega/\hbar}\hat{q}$ and $\sqrt{2/\hbar m\omega}\hat{p}$ are equal to one another (Eqs. 3.6). With the help of (9.33), these results are transfered to the quadrature phase amplitudes. Thus, a freely evolving field mode prepared in a coherent state satisfies

$$\Delta A_\theta \Delta A_{\theta+\pi/2} = \tfrac{1}{4}, \qquad \Delta A_\theta = \Delta A_{\theta+\pi/2} = \tfrac{1}{2}. \tag{9.35}$$

Finally, under free evolution, an initial coherent state remains as a coherent state for all times (Eq. 3.3). Taken together, these observations provide the picture of quantum fluctuations for a free field mode in a coherent state illustrated in Fig. 9.2a.

What happens to the quantum fluctuations when a mode prepared in a coherent state undergoes degenerate parametric amplification? Of course, the mean quadrature phase amplitudes $\langle\hat{X}\rangle$ and $\langle\hat{Y}\rangle$ are amplified and deamplified just like $\mathcal{X}$ and $\mathcal{Y}$ in the classical treatment. What, though, about the standard deviations? Equations 9.28, 9.29, and 9.35 provide the simple answer: from these equations,

$$\Delta X(t) = e^{|K|t}\Delta X(0) = e^{|K|t}\tfrac{1}{2}, \tag{9.36a}$$

$$\Delta Y(t) = e^{-|K|t}\Delta Y(0) = e^{-|K|t}\tfrac{1}{2}. \tag{9.36b}$$

Thus, the fluctuations in quadrature phase amplitudes [with $\theta = \frac{1}{2}\arg(K)$] are amplified and deamplified in the same way as the means; as a result they continue to satisfy the minimum uncertainty condition – $\Delta X(t)\Delta Y(t) = \frac{1}{4}$.

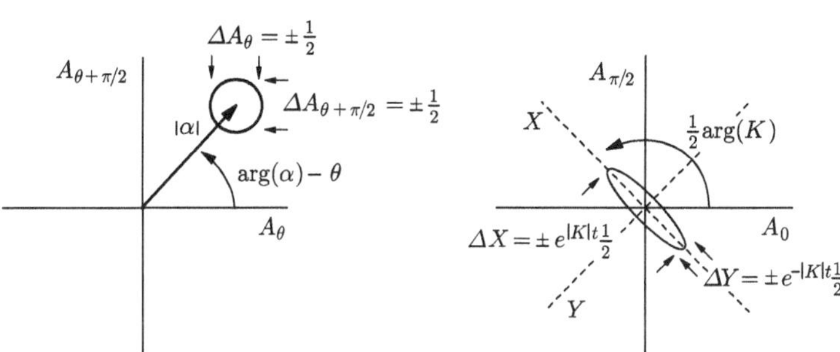

Fig. 9.2. Phase-space picture of the quantum fluctuations of (**a**) a freely evolving field mode prepared in the coherent state $|\alpha\rangle$, and (**b**) the subharmonic field for degenerate parametric amplification of the vacuum state. Fluctuations of the field amplitude are concentrated in the regions within the circle, part (**a**), and the ellipse, part (**b**). For a rigorous interpretation, A_θ should be defined in terms of the complex argument α of the Wigner distribution (see Eq. 9.47b). Note that the figure adopts a rotating frame; in a nonrotating frame, the circle and the ellipse rotate clockwise about the origin at frequency ω_C, and the ordinate is proportional to the oscillating electric field

This process is illustrated for an initial vacuum state in Fig. 9.2b. More generally, for an initial coherent state $|\alpha\rangle$, the ellipse in Fig. 9.2b is displaced from the origin to the point (X, Y), with $X = \exp\big(|K|t\big)|\alpha|\cos\big[\arg(\alpha) - \frac{1}{2}\arg(K)\big]$ and $Y = \exp\big(-|K|t\big)|\alpha|\sin\big[\arg(\alpha) - \frac{1}{2}\arg(K)\big]$.

Let us now shift our point of view from the Heisenberg to the Schrödinger picture. In the Schrödinger picture, degenerate parametric amplification changes a coherent state of the subharmonic field into a squeezed coherent state. Using Hamiltonian (9.23), the Schrödinger equation in the interaction picture is given by

$$\frac{d|\tilde{\psi}(t)\rangle}{dt} = \tfrac{1}{2}\big(Ka^{\dagger 2} - K^* a^2\big)|\tilde{\psi}(t)\rangle, \tag{9.37}$$

where the state in the interaction picture is $|\tilde{\psi}(t)\rangle = e^{i\omega_C a^\dagger a t}|\psi(t)\rangle$, and we have used the result $e^{i\omega_C a^\dagger a t} a e^{-i\omega_C a^\dagger a t} = a e^{-i\omega_C t}$ (Eq. 1.40). The formal solution for an initial coherent state $|\alpha_0\rangle$ is

$$\begin{aligned}
|\psi(t)\rangle &= e^{-i\omega_C a^\dagger a t}\exp\big[\tfrac{1}{2}\big(Ka^{\dagger 2} - K^* a^2\big)t\big]|\alpha_0\rangle \\
&= e^{-i\omega_C a^\dagger a t} S\big(-Kt\big)|\alpha_0\rangle \\
&= S\big(-e^{-i2\omega_C t}Kt\big)|e^{-i\omega_C t}\alpha_0\rangle,
\end{aligned} \tag{9.38}$$

where we introduce the unitary operator

$$S(\xi) \equiv \exp\big[\tfrac{1}{2}(\xi^* a^2 - \xi a^{\dagger 2})\big]. \tag{9.39}$$

The operator $S(\xi)$, $\xi = re^{i2\theta}$, is known as the *squeeze operator*, with r the *squeeze parameter*. It is the generator of the so-called *squeezed coherent states*, which we define by

$$|\alpha, \xi\rangle \equiv D(\alpha)S(\xi)|0\rangle. \tag{9.40}$$

Starting from the vacuum state $|0\rangle$, acting with the squeeze operator produces a *squeezed vacuum state* with quadrature fluctuations $\Delta A_\theta = \frac{1}{2}e^{-r}$ and $\Delta A_{\theta+\pi/2} = \frac{1}{2}e^{r}$, where $\hat{A}_\theta$ and $\hat{A}_{\theta+\pi/2}$ are defined by (9.26) with $\tilde{a} \to a$. Acting with the *displacement operator*,

$$D(\alpha) \equiv \exp\big(\alpha a^\dagger - \alpha^* a\big), \tag{9.41}$$

adds the coherent amplitude α.

Note 9.4. The displacement operator $D(\alpha)$ is derived by writing $|0\rangle = e^{\alpha^* a}|0\rangle$ on the right-hand side of (3.10) and simplifying with the help of the Baker–Hausdorff theorem (4.8).

In the expression (9.38) for $|\psi(t)\rangle$, the squeeze operator rescales the initial coherent amplitude as well as the fluctuations. Taking this into account, we may write the state as the squeezed coherent state

$$|\psi(t)\rangle = |\alpha(t), \xi(t)\rangle, \tag{9.42}$$

where $[\theta = -\omega_C t + \frac{1}{2}\arg(K)]$

$$\begin{aligned}\alpha(t) &= e^{i\theta}\Big\{e^{|K|t}\tfrac{1}{2}\big[\alpha_0 e^{-i(\omega_C t+\theta)} + \alpha_0^* e^{i(\omega_C t+\theta)}\big] \\ &\qquad + e^{-|K|t}\tfrac{1}{2}\big[\alpha_0 e^{-i(\omega_C t+\theta)} - \alpha_0^* e^{i(\omega_C t+\theta)}\big]\Big\} \\ &= e^{-i\omega_C t}\big[\alpha_0 \cosh\big(|K|t\big) + \alpha_0^* e^{i\arg(K)}\sinh\big(|K|t\big)\big],\end{aligned} \tag{9.43a}$$

and

$$\xi(t) = -e^{-i2\omega_C t}Kt. \tag{9.43b}$$

Note 9.5. The common use of the term *squeezed vacuum state* can be a little misleading. The state $S(\xi)|0\rangle$ differs substantially from the vacuum; it is not simply a vacuum state with redistributed fluctuations. The vacuum fluctuations have been amplified, and the *amplified* vacuum fluctuations contain energy and can do work. While the mean photon number in the vacuum is zero, the squeezed vacuum state $S(-e^{-i2\omega_C t}Kt)|0\rangle$ has nonzero mean photon number

$$\begin{aligned}\langle (a^\dagger a)(t)\rangle &= \langle\big(\hat{A}_\theta(t) - i\hat{A}_{\theta+\pi/2}(t)\big)\big(\hat{A}_\theta(t) + i\hat{A}_{\theta+\pi/2}(t)\big)\rangle \\ &= \langle \hat{A}_\theta^2(t)\rangle + \langle \hat{A}_{\theta+\pi/2}^2(t)\rangle + i\langle[\hat{A}_\theta, \hat{A}_{\theta+\pi/2}](t)\rangle \\ &= \big(\Delta A_\theta(t)\big)^2 + \big(\Delta A_{\theta+\pi/2}(t)\big)^2 - \tfrac{1}{2} \\ &= \tfrac{1}{2}\big[\cosh\big(2|K|t\big) - 1\big],\end{aligned} \tag{9.44}$$

where we have made use of (9.26), (9.29), (9.31), and (9.36).

Exercise 9.1. Squeezed states are nonclassical states. They do not possess a positive, nonsingular P distribution. The Q and Wigner distributions do exist, though, as positive-definite functions. Derive the Q and Wigner distributions for a squeezed vacuum state. First show that

$$S^\dagger(\xi)aS(\xi) = a\cosh r - e^{i2\theta}a^\dagger \sinh r, \tag{9.45a}$$

and

$$S^\dagger(\xi)a^\dagger S(\xi) = a^\dagger \cosh r - e^{-i2\theta}a\sinh r. \tag{9.45b}$$

Use these results to obtain the characteristic functions

$$\begin{aligned}&\chi_A(\mu+i\nu,\mu-i\nu)\\ &= \exp\left[-\tfrac{1}{2}(1+e^{-2r})(\mu\cos\theta-\nu\sin\theta)^2-\tfrac{1}{2}(1+e^{2r})(\mu\sin\theta+\nu\cos\theta)^2\right],\end{aligned} \tag{9.46a}$$

and

$$\begin{aligned}&\chi_S(\mu+i\nu,\mu-i\nu)\\ &= \exp\left[-\tfrac{1}{2}e^{-2r}(\mu\cos\theta-\nu\sin\theta)^2-\tfrac{1}{2}e^{2r}(\mu\sin\theta+\nu\cos\theta)^2\right];\end{aligned} \tag{9.46b}$$

hence obtain the Q *distribution for a squeezed vacuum state*,

$$\begin{aligned}Q(x+iy,x-iy) &= \sqrt{\frac{2}{\pi(1+e^{-2r})}}\exp\left[-\frac{1}{2}\frac{(x\cos\theta+y\sin\theta)^2}{(1+e^{-2r})/4}\right]\\ &\quad\times\sqrt{\frac{2}{\pi(1+e^{2r})}}\exp\left[-\frac{1}{2}\frac{(-x\sin\theta+y\cos\theta)^2}{(1+e^{2r})/4}\right],\end{aligned} \tag{9.47a}$$

and the *Wigner distribution for a squeezed vacuum state*

$$\begin{aligned}W(x+iy,x-iy) &= \sqrt{\frac{2}{\pi e^{-2r}}}\exp\left[-\frac{1}{2}\frac{(x\cos\theta+y\sin\theta)^2}{e^{-2r}/4}\right]\\ &\quad\times\sqrt{\frac{2}{\pi e^{2r}}}\exp\left[-\frac{1}{2}\frac{(-x\sin\theta+y\cos\theta)^2}{e^{2r}/4}\right].\end{aligned} \tag{9.47b}$$

Note the larger variance of the Q distribution compared to the Wigner distribution, a repetition of the comparison for a thermal state (Eqs. 4.21 and 4.41). Note also that the variance of $\hat{A}_\theta$ involves the symmetrically-ordered product $\frac{1}{2}(a^\dagger a + aa^\dagger)$, which is why the Gaussian widths of the Wigner distribution match the standard deviations, ΔA_θ and $\Delta A_{\theta+\pi/2}$, displayed in Fig. 9.2 (see Sect. 4.1.4). Determine the normal-ordered characteristic function $\chi_N(\mu+i\nu,\mu-i\nu)$. Where does the problem arise in the derivation of the P distribution for a squeezed vacuum state?

9.2.3 The Degenerate Parametric Oscillator

The parametric amplifier model we have just studied will produce arbitrarily large amounts of squeezing no matter how small we make the coupling constant $|K|$. For smaller $|K|$, we simply need to wait longer to achieve the same level of squeezing. This situation is of course not physical. In practice, a parametric amplifier is normally operated as a traveling-wave device, without a cavity, and the interaction time is limited to the propagation time $n\ell/c$ through the nonlinear crystal. The squeezing obtained in this manner is very small unless a pulsed pump of high intensity or a very long (unreasonably so) crystal is used. This limitation led to the use of cavities in early squeezing experiments. Considering such experiments, our treatment of squeezing for a single cavity mode is appropriate, but then the model of Sect. 9.2.1 is incomplete. Specifically, this model does not include any mechanism for getting the pump field into the cavity and the subharmonic field out. A more realistic model must account for the injection of the pump field and extraction of the subharmonic field through mirrors of nonunit reflectivity. The nonunit reflectivity introduces losses for the intracavity fields, and there may be losses from absorption in the crystal as well. With these losses included, the parametric gain on a single round trip must exceed the round-trip loss for amplification to take place – the parametric amplifier becomes a parametric oscillator.

Note 9.6. From the expression (9.12) for the coupling constant K, the single-pass gain coefficient $|K|t$ for a crystal of length ℓ is

$$\begin{aligned} |K|(\bar{L}/c) &= \frac{\omega_C \ell \chi^{(2)}}{2n^{3/2}c}|\mathcal{E}_p| \\ &= \frac{\omega_C \ell \chi^{(2)}}{2n^{3/2}c}\sqrt{\frac{2\times 10^4}{\epsilon_0 c}P_p}, \end{aligned} \tag{9.48}$$

where P_p is the (forward-propagating) pump power in $\mathrm{W/cm^2}$. Numbers for $\mathrm{LiNbO_3}$ provide an estimate of the single-pass gain and squeezing: for $\omega_C = 2\pi c/\lambda_C$, $\lambda_C = 10^{-6}\mathrm{m}$, $\chi^{(2)} = 1.2\times 10^{-11}\mathrm{m/V}$ (see [9.16], with $\chi^{(2)} = 2d_{31}/\epsilon_0$), $n = 2.25$, and $\ell = 10^{-2}\mathrm{m}$, we obtain

$$|K|(\bar{L}/c) = 3.0\times 10^{-4}\sqrt{P_p(\mathrm{W/cm^2})}. \tag{9.49}$$

Thus, for a single pass through a 1-cm crystal, a pump power as large as $\sim 10\mathrm{MW/cm^2}$ is required to achieve a squeeze factor of unity ($r = |K|t = 1$). Note the useful relationship [9.17]

$$\chi^{(2)}_{\mathrm{SI}} = 4\pi\sqrt{40\pi\epsilon_0}\chi^{(2)}_{\mathrm{esu}} \tag{9.50}$$

for converting nonlinear susceptibilities from electrostatic to S.I. units.

We aim to convert our degenerate parametric amplifier model into a model for the degenerate parametric oscillator. To do so we must add losses. Let us first add the losses in the classical theory, then introduce a quantum-mechanical master equation to extend the result into the quantum domain.

At this point we must also allow for depletion of the pump. If we do not do this, once the round-trip gain exceeds the round-trip loss the intensity of the subharmonic field will grow without bound, as it does in the expression (9.44) for the subharmonic mode photon number. As with the laser above threshold, the parametric oscillator relies on nonlinear gain reduction to establish a balance between gain and loss. For the laser the nonlinearity comes from saturation of the lasing transition; its origin for the parametric oscillator is pump depletion.

To derive the loss term to be added to (9.11), we must rewrite (9.8) so that it includes the effects of nonunit reflection coefficients, R_1 and R_2, and absorption in the crystal, with absorption coefficient α. We introduce the absorption by adding the linear polarization $-\frac{1}{2}\alpha\mathcal{E}$ to the right-hand sides of (9.16a) and (9.16b). Equation 9.8 then reads (for perfect resonance)

$$\begin{aligned}\mathcal{E}+\Delta\mathcal{E} &= \left\{\left(\frac{1}{\sqrt{n}}\mathcal{E}-\frac{\alpha\ell}{2}\mathcal{E}+i\frac{\omega_C\ell}{2\epsilon_0 cn}\mathcal{P}_f\right)\sqrt{n}\sqrt{R_2}e^{i\phi_R}e^{2i[\Phi(\ell+d)+\phi]}\frac{1}{\sqrt{n}}\right.\\ &\qquad \left.-\frac{\alpha\ell}{2}\mathcal{E}+i\frac{\omega_C\ell}{2\epsilon_0 cn}\mathcal{P}_b\right\}\sqrt{n}\sqrt{R_1}e^{i\phi_R}e^{-2i[\Phi(-L+\ell+d)+\phi]}\\ &= \left[1-\alpha\ell-\tfrac{1}{2}(T_1+T_2)\right]\mathcal{E}+i\frac{\omega_C\ell}{2\epsilon_0 cn^{1/2}}\mathcal{P}_f+i\frac{\omega_C\ell}{2\epsilon_0 cn^{1/2}}\mathcal{P}_b; \end{aligned} \tag{9.51}$$

we have written $\sqrt{R_1}=\sqrt{1-T_1}\approx 1-\frac{1}{2}T_1$ and $\sqrt{R_2}=\sqrt{1-T_2}\approx 1-\frac{1}{2}T_2$, where $T_1 \ll 1$ and $T_2 \ll 1$ are the mirror transmission coefficients. Dividing this expression for $\Delta\mathcal{E}$ by the cavity round-trip time $2\bar{L}/c$, and substituting $\mathcal{P}_f$ and $\mathcal{P}_b$ from (9.5b) and (9.6b), we obtain the *equation of motion for the subharmonic field amplitude including cavity and crystal loss*:

$$\dot{\mathcal{E}} = -\kappa\mathcal{E}+iK'\mathcal{E}^*\mathcal{E}_p, \tag{9.52}$$

where

$$K' \equiv \frac{\omega_C\ell\chi^{(2)}}{2n^{3/2}\bar{L}}\cos(\phi_p-2\phi), \tag{9.53}$$

and

$$\kappa = \frac{\left[\alpha\ell+\frac{1}{2}(T_1+T_2)\right]}{2\bar{L}/c} \tag{9.54}$$

is the subharmonic mode damping rate.

With pump depletion taken into account the pump field amplitude $\mathcal{E}_p$ is no longer to be treated as a constant. We must derive an equation of motion for it too. The derivation proceeds in a similar manner, with the one difference that injection of the external pump field into the cavity must be included. We take the external field,

$$\boldsymbol{E}_0(z,t) = \hat{e}_x \tfrac{1}{2}\mathcal{E}_0 e^{-i[2\omega_C(t-z/c)-\phi_0]} + \text{c.c.}, \tag{9.55}$$

$z \leq -L + \ell + d$, to be incident from the left of the cavity. To account for its injection, we decompose $E_p(z,t)$ into forward- and backward-propagating components,

$$\boldsymbol{E}_p^f(z,t) = \hat{e}_x \tfrac{1}{2}\mathcal{E}_p^f(t)A(z)e^{-i[2\omega_C t - 2\Phi(z) - \phi_p]} + \text{c.c.} \tag{9.56a}$$

and

$$\boldsymbol{E}_p^b(z,t) = \hat{e}_x \tfrac{1}{2}\mathcal{E}_p^b(t)A(z)e^{-i[2\omega_C t + 2\Phi(z) + \phi_p]} + \text{c.c.}, \tag{9.56b}$$

and impose the boundary condition at $z = -L + \ell + d$

$$\mathcal{E}_p^f = \sqrt{R_p^1}e^{i\phi_R}\mathcal{E}_p^b e^{-2i[2\Phi(-L+\ell+d)+\phi_p]} + \sqrt{T_p^1}e^{i\phi_T}\mathcal{E}_0 e^{i(\phi_0-\phi_p)}, \tag{9.57}$$

where ϕ_T is the phase change upon transmission at the mirrors. The remaining details of the derivation are left as an exercise.

Exercise 9.2. Follow the approach leading to (9.11) and (9.52) to obtain the *equation of motion for the pump field amplitude with cavity loss and injected external field*

$$\dot{\mathcal{E}}_p = -\kappa_p \mathcal{E}_p + iK'\mathcal{E}^2 + e^{i(\phi_T+\phi_0-\phi_p)}(c/2\bar{L})\sqrt{T_p^1}\mathcal{E}_0, \tag{9.58}$$

where the pump mode damping rate is

$$\kappa_p = \frac{\left[\alpha_p \ell + \frac{1}{2}(T_p^1 + T_p^2)\right]}{2\bar{L}/c}; \tag{9.59}$$

T_p^1, T_p^2, and α_p are the mirror transmission coefficients and the absorption coefficient for the pump field in the crystal.

The equations of motion (9.52) and (9.58) can be simplified by introducing scaled field amplitudes

$$\bar{\mathcal{E}} \equiv \sqrt{\xi}(K'/\kappa)e^{-i\frac{1}{2}(\psi+\pi/2)}\mathcal{E}, \tag{9.60a}$$

and

$$\bar{\mathcal{E}}_p \equiv (K'/\kappa)e^{-i\psi}\mathcal{E}_p, \tag{9.60b}$$

where we have defined the ratio of damping rates

$$\xi \equiv \frac{\kappa}{\kappa_p}, \tag{9.61}$$

and the phase

$$\psi \equiv \phi_T + \phi_0 - \phi_p + \arg(\mathcal{E}_0). \tag{9.62}$$

Then (9.52) and (9.58) become a set of *coupled equations of motion in scaled variables for the degenerate parametric oscillator*:

$$\kappa^{-1}\dot{\bar{\mathcal{E}}} = -\bar{\mathcal{E}} + \bar{\mathcal{E}}^*\bar{\mathcal{E}}_p, \tag{9.63a}$$

$$\kappa_p^{-1}\dot{\bar{\mathcal{E}}}_p = -\bar{\mathcal{E}}_p - \bar{\mathcal{E}}^2 + \lambda. \tag{9.63b}$$

With the introduced scaling, λ is a real (and positive) dimensionless *pump parameter*:

$$\begin{aligned}\lambda &\equiv \frac{K'}{\kappa\kappa_p}(c/2\bar{L})\sqrt{T_p^1}|\mathcal{E}_0| \\ &= \frac{\omega_C \ell \chi^{(2)}}{n^{3/2}c}\frac{\sqrt{T_p^1}}{\left[\alpha L + \frac{1}{2}(T_1+T_2)\right]\left[\alpha_p L + \frac{1}{2}(T_p^1+T_p^2)\right]}|\mathcal{E}_0|\cos(\phi_p - 2\phi) \\ &= \frac{\mathcal{F}}{\pi}\frac{\mathcal{F}_p}{\pi}\frac{\omega_C \ell \chi^{(2)}}{n^{3/2}c}\sqrt{T_p^1}|\mathcal{E}_0|\cos(\phi_p - 2\phi),\end{aligned} \tag{9.64}$$

where $\mathcal{F}$ and $\mathcal{F}_p$ are cavity finesses for the subharmonic and pump modes, respectively.

The degenerate parametric amplifier model discussed in Sect. 9.2.1 is contained within this more general model. To recover the equation of motion (9.11), we assume that $|\bar{\mathcal{E}}_p| \sim \lambda \gg |\bar{\mathcal{E}}|^2$ in (9.63b) and $\kappa_p \gg \kappa$. Then, for times that are long compared to the cavity filling time for the pump field, κ_p^{-1}, but short compared to the decay time κ^{-1} for the subharmonic, we may replace (9.63a) by

$$\dot{\bar{\mathcal{E}}} = (\kappa\lambda)\bar{\mathcal{E}}^*. \tag{9.65}$$

In the present notation, $\kappa\lambda = K'(c/2\bar{L})\sqrt{T_p^1}|\mathcal{E}_0|/\kappa_p$ corresponds to the coupling constant $|K|$ in (9.11), where the factor $(c/2\bar{L})\sqrt{T_p^1}|\mathcal{E}_0|/\kappa_p$ is the pump field amplitude inside the cavity (without depletion). The phase $\psi + \pi/2$, which is the phase of K in (9.11), has been absorbed into the definition of the scaled field amplitude $\bar{\mathcal{E}}$ (Eq. 9.60a).

Note 9.7. It is interesting to estimate the squeeze parameter r that can be achieved using intracavity parametric amplification and compare it with the result (9.49) obtained for a single-pass amplifier. Taking the interaction time to be the cavity lifetime yields the result $r = \kappa\lambda t = \kappa\lambda\kappa^{-1} = \lambda$. Then adopting the parameters used to obtain (9.49), we have

$$\begin{aligned}\lambda &= \frac{\mathcal{F}}{\pi}\frac{\mathcal{F}_p}{\pi}\frac{\omega_C \ell \chi^{(2)}}{n^{3/2}c}\sqrt{T_p^1}\sqrt{\frac{2\times 10^4}{\epsilon_0 c}P_0} \\ &= \frac{\mathcal{F}}{\pi}\frac{\mathcal{F}_p}{\pi}\sqrt{T_p^1}\times 6.0\times 10^{-4}\sqrt{P_0(\mathrm{W/cm^2})}.\end{aligned} \tag{9.66}$$

The product of the finesses decreases the external pump power P_0 required to reach $r = \lambda = 1$ by orders of magnitude.

In the long time limit, the parametric oscillator equations (9.63a) and (9.63b) behave very differently to the parametric amplifier equation (9.65). In place of the unbounded growth of the subharmonic field, these equations describe evolution to a steady state, with field amplitudes $\bar{\mathcal{E}}_{\rm ss}$ and $\bar{\mathcal{E}}_p^{\rm ss}$ satisfying

$$-\bar{\mathcal{E}}_{\rm ss} + \bar{\mathcal{E}}_{\rm ss}^* \bar{\mathcal{E}}_p^{\rm ss} = 0, \tag{9.67a}$$

$$-\bar{\mathcal{E}}_p^{\rm ss} - \bar{\mathcal{E}}_{\rm ss}^2 + \lambda = 0. \tag{9.67b}$$

The *first steady-state solution to the degenerate parametric oscillator equations of motion* is the trivial solution

$$\bar{\mathcal{E}}_{\rm ss} = 0, \qquad \bar{\mathcal{E}}_p^{\rm ss} = \lambda. \tag{9.68}$$

There are also other possibilities, however, which we uncover in the following way. First multiply (9.67b) by $\bar{\mathcal{E}}_{\rm ss}^*$ and substitute

$$\bar{\mathcal{E}}_{\rm ss}^* \bar{\mathcal{E}}_p^{\rm ss} = \bar{\mathcal{E}}_{\rm ss}(9.69) \tag{9.69}$$

from (9.67a), to obtain

$$-\bar{\mathcal{E}}_{\rm ss} + (\lambda - \bar{\mathcal{E}}_{\rm ss}^2)\bar{\mathcal{E}}_{\rm ss}^* = 0. \tag{9.70}$$

Then take the conjugate of this equation, multiply by $(\lambda - \bar{\mathcal{E}}_{\rm ss}^2)$, and substitute $(\lambda - \bar{\mathcal{E}}_{\rm ss}^2)\bar{\mathcal{E}}_{\rm ss}^* = \bar{\mathcal{E}}_{\rm ss}$, using (9.70) once again. This leads us to an autonomous equation for the complex amplitude $\bar{\mathcal{E}}_{\rm ss}$:

$$\bar{\mathcal{E}}_{\rm ss}\left(|\lambda - \bar{\mathcal{E}}_{\rm ss}^2|^2 - 1\right) = 0. \tag{9.71}$$

Thus, in addition to the trivial solution (9.68), there are four solutions to (9.67) satisfying

$$|\lambda - \bar{\mathcal{E}}_{\rm ss}^2| = 1. \tag{9.72}$$

The pair satisfying $\lambda - \bar{\mathcal{E}}_{\rm ss}^2 = -1$ is spurious, since this relation leads to an inconsistency: working from (9.70) we find

$$\lambda - \bar{\mathcal{E}}_{\rm ss}^2 = -1 \quad \Rightarrow \quad \bar{\mathcal{E}}_{\rm ss} = -\bar{\mathcal{E}}_{\rm ss}^* \quad \Rightarrow \quad \bar{\mathcal{E}}_{\rm ss} = \pm i|\bar{\mathcal{E}}_{\rm ss}|; \tag{9.73a}$$

hence

$$\lambda - \bar{\mathcal{E}}_{\rm ss}^2 = -1 \quad \Rightarrow \quad \lambda + |\bar{\mathcal{E}}_{\rm ss}|^2 = -1 \quad \Rightarrow \quad |\bar{\mathcal{E}}_{\rm ss}|^2 = -(\lambda + 1). \tag{9.73b}$$

The final result is a contradiction because λ is positive by definition. We therefore require (9.72) to be solved by $\lambda - \bar{\mathcal{E}}_{\rm ss}^2 = +1$, which gives the *second pair of steady-state solutions to the degenerate parametric oscillator equations of motion*

$$\bar{\mathcal{E}}_{\rm ss} = \pm\sqrt{\lambda - 1}, \qquad \bar{\mathcal{E}}_p^{\rm ss} = 1. \tag{9.74}$$

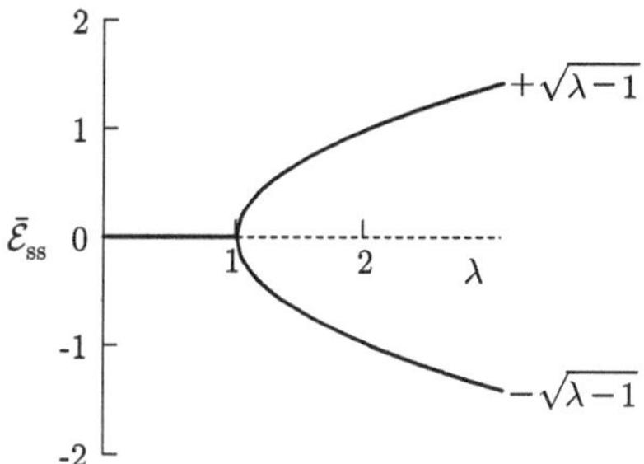

Fig. 9.3. Threshold behavior of the degenerate parametric oscillator. The steady state shown by the *broken line* is unstable

These solutions are admissible for $\lambda \geq 1$. In summary, from (9.68) and (9.74), the degenerate parametric oscillator shows the threshold behavior displayed in Fig. 9.3.

Below threshold ($\lambda < 1$): There is a single acceptable steady-state solution, the trivial solution (9.68):

$$\bar{\mathcal{E}}_< \equiv \bar{\mathcal{E}}_{\rm ss}^< = \sqrt{\xi}(K'/\kappa)e^{-i\frac{1}{2}(\psi+\pi/2)}\mathcal{E}_{\rm ss}^< = 0, \tag{9.75a}$$

$$\bar{\mathcal{E}}_p^< \equiv \bar{\mathcal{E}}_{p,<}^{\rm ss} = (K'/\kappa)e^{-i\psi}\mathcal{E}_{p,<}^{\rm ss} = \lambda. \tag{9.75b}$$

Below threshold, the round-trip gain

$$K'|\mathcal{E}_p^<|(2\bar{L}/c) = \kappa|\bar{\mathcal{E}}_p^<|(2\bar{L}/c) = \kappa\lambda(2\bar{L}/c)$$

is less than the round-trip loss, $\kappa 2\bar{L}/c$.

At threshold ($\lambda = 1$): There is a single three-fold degenerate solution:

$$\bar{\mathcal{E}}_{\rm thr} \equiv \bar{\mathcal{E}}_{\rm ss}^{\rm thr} = \sqrt{\xi}(K'/\kappa)e^{-i\frac{1}{2}(\psi+\pi/2)}\mathcal{E}_{\rm ss}^{\rm thr} = 0, \tag{9.76a}$$

$$\bar{\mathcal{E}}_p^{\rm thr} \equiv \bar{\mathcal{E}}_{p,{\rm thr}}^{\rm ss} = (K'/\kappa)e^{-i\psi}\mathcal{E}_{p,{\rm thr}}^{\rm ss} = 1; \tag{9.76b}$$

the round-trip gain is equal to the round-trip loss.

Above threshold ($\lambda > 1$): There are three acceptable steady-state solutions. The trivial solution (9.68) still exists, but is unstable above threshold, and there exists the pair of stable solutions (9.74):

$$\bar{\mathcal{E}}_> \equiv \bar{\mathcal{E}}_{\rm ss}^> = \sqrt{\xi}(K'/\kappa)e^{-i\frac{1}{2}(\psi+\pi/2)}\mathcal{E}_{\rm ss}^> = \pm\sqrt{\lambda-1}, \tag{9.77a}$$

$$\bar{\mathcal{E}}_p^> \equiv \bar{\mathcal{E}}_{p,>}^{\rm ss} = (K'/\kappa)e^{-i\psi}\mathcal{E}_{p,>}^{\rm ss} = 1. \tag{9.77b}$$

Above threshold, the *undepleted* round-trip gain—determined by the below-threshold solution for the pump amplitude, $\bar{\mathcal{E}}_p^{\rm ss} = \lambda$—exceeds the round-trip loss. This high gain value is unstable, though. The subharmonic field must be amplified to reach its steady-state value $\bar{\mathcal{E}}_{\rm ss}^> = \pm\sqrt{\lambda-1}$; in response, the pump field is depleted, which reduces the round-trip gain. Steady state is reached when the pump field amplitude has been reduced to its threshold value $\bar{\mathcal{E}}^{\rm ss}_{p,>} = \bar{\mathcal{E}}^{\rm ss}_{p,\rm thr} = 1$. This process of *pump depletion* is analogous to inversion clamping in the laser (Eq. 7.22).

Exercise 9.3. Investigate the stability of steady states (9.68) and (9.74). Show that for steady state (9.68) the decay of small perturbations is governed by the eigenvalues

$$\left.\begin{array}{ll} \Lambda_1 = -\kappa(1-\lambda), & \text{real part of } \bar{\mathcal{E}}, \\ \Lambda_2 = -\kappa(1+\lambda), & \text{imaginary part of } \bar{\mathcal{E}}, \\ \Lambda_3 = \Lambda_4 = -\kappa_p, & \text{real and imaginary parts of } \bar{\mathcal{E}}_p. \end{array}\right\} \tag{9.78a}$$

It follows that above threshold ($\lambda > 1$) this steady state is unstable to perturbations in the real part of $\bar{\mathcal{E}}$ (the amplified quadrature phase amplitude). Show that steady states (9.74) are stable above threshold, with the decay of small perturbations governed by eigenvalues

$$\left.\begin{array}{ll} \Lambda_{1,2} = -\frac{1}{2}\kappa_p \pm \frac{1}{2}\sqrt{\kappa_p^2 - 8\kappa\kappa_p(\lambda-1)}, & \text{coupled real parts} \\ \qquad \text{of } \bar{\mathcal{E}} \text{ and } \bar{\mathcal{E}}_p, & \\ \Lambda_{3,4} = -\left(\kappa + \frac{1}{2}\kappa_p\right) \pm \sqrt{\left(\kappa + \frac{1}{2}\kappa_p\right)^2 - 2\kappa\kappa_p\lambda}, & \text{coupled imaginary} \\ \qquad \text{parts of } \bar{\mathcal{E}} \text{ and } \bar{\mathcal{E}}_p. & \end{array}\right\} \tag{9.78b}$$

9.2.4 Master Equation for the Degenerate Parametric Oscillator

The quantum-mechanical model for the degenerate parametric oscillator is illustrated schematically in Fig. 9.4. The system Hamiltonian, H_S, describes the intracavity interaction between the subharmonic and pump, and the driving

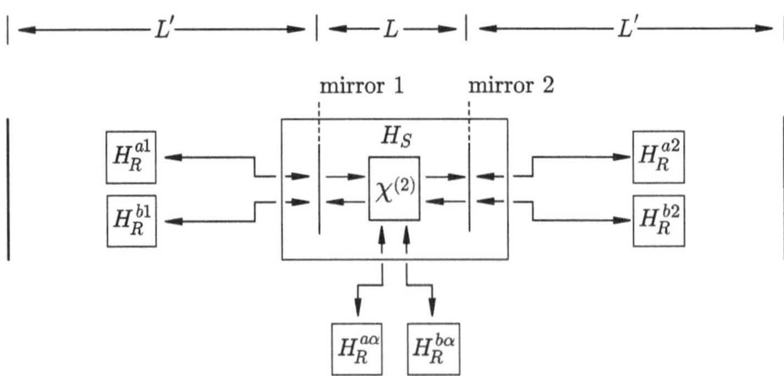

Fig. 9.4. Schematic diagram of the quantum-mechanical model for the degenerate parametric oscillator, with system Hamiltonian H_S (Eq. 9.79) and reservoir Hamiltonian H_R (Eqs. 9.85 and 9.86)

of the pump mode. It is given by

$$\begin{aligned} H_S &= H_a + H_b + H_{ab} + H_{\text{drive}} \\ &\equiv \hbar\omega_C a^\dagger a + \hbar 2\omega_C b^\dagger b \\ &\quad + i\hbar(g/2)(a^{\dagger 2} b - a^2 b^\dagger) + \hbar\big(\bar{\mathcal{E}}_0 e^{-i2\omega_C t} b^\dagger + \bar{\mathcal{E}}_0^* e^{i2\omega_C t} b\big), \end{aligned} \tag{9.79}$$

with coupling constant

$$g \equiv \sqrt{\frac{\hbar 2\omega_C}{\epsilon_0 A\bar{L}}} K' = \sqrt{\frac{\hbar 2\omega_C}{\epsilon_0 A\bar{L}}} \frac{\omega_C \ell \chi^{(2)}}{2n^{3/2}\bar{L}} \cos(\phi_p - 2\phi), \tag{9.80}$$

and effective driving field amplitude

$$\bar{\mathcal{E}}_0 \equiv e^{i(\phi_T + \phi_0 - \phi_p)} \sqrt{\frac{\epsilon_0 A\bar{L}}{\hbar 2\omega_C}} (c/2\bar{L}) \sqrt{T_p^1} \mathcal{E}_0, \tag{9.81}$$

where a and $a^\dagger$ (b and $b^\dagger$) are annihilation and creation operators for the subharmonic (pump) mode, respectively. The quantized field expansions, corresponding to (9.1a) and (9.1b), are

$$\hat{\boldsymbol{E}}(z,t) = i\hat{e}_y \sqrt{\frac{\hbar\omega_C}{\epsilon_0 A\bar{L}}} A(z) \cos[\Phi(z) + \phi][\tilde{a}(t)e^{-i\omega_C t} - \tilde{a}^\dagger(t)e^{i\omega_C t}], \tag{9.82a}$$

and

$$\hat{\boldsymbol{E}}_p(z,t) = i\hat{e}_x \sqrt{\frac{\hbar 2\omega_C}{\epsilon_0 A\bar{L}}} A(z) \cos[2\Phi(z) + \phi_p][\tilde{b}(t)e^{-i2\omega_C t} - \tilde{b}^\dagger(t)e^{i2\omega_C t}], \tag{9.82b}$$

where we introduce annihilation and creation operators in a rotating frame:

$$\begin{aligned} \tilde{a}(t) &\equiv e^{i\omega_C t} a(t), \qquad & \tilde{a}^\dagger(t) &\equiv e^{-i\omega_C t} a^\dagger(t), \\ \tilde{b}(t) &\equiv e^{i2\omega_C t} b(t), \qquad & \tilde{b}^\dagger(t) &\equiv e^{-i2\omega_C t} b^\dagger(t). \end{aligned} \tag{9.83a}$$

Exercise 9.4. Show that Hamiltonian (9.79) gives the Heisenberg equations of motion

$$\dot{a} = -i\omega_C a + g a^\dagger b, \tag{9.84a}$$

$$\dot{b} = -i2\omega_C b - (g/2)a^2 - i\bar{\mathcal{E}}_0 e^{-i2\omega_C t}. \tag{9.84b}$$

Verify that the factorized mean-value equations obtained from these equations reproduce the classical equations of motion without damping terms, (9.52) and (9.58).

Each mode interacts with three reservoirs, which account for the transmission losses at the two mirrors and the absorption loss in the crystal. Although, formally, the three reservoir interactions could be combined into a single term, it is better to keep them separate; their physical origins are distinct, and the differences become important when we consider the cavity output fields in Sect. 9.2.5. The reservoir Hamiltonian is then written as

$$H_R = H_R^{a1} + H_R^{a2} + H_R^{a\alpha} + H_R^{b1} + H_R^{b2} + H_R^{b\alpha}, \tag{9.85}$$

where ($\mu = 1, 2, \alpha$)

$$H_R^{a\mu} \equiv \sum_j \hbar\omega_j r_{a\mu j}^\dagger r_{a\mu j}, \tag{9.86a}$$

$$H_R^{b\mu} \equiv \sum_j \hbar\omega_j r_{b\mu j}^\dagger r_{b\mu j}, \tag{9.86b}$$

and for the interaction between system and reservoir we write

$$H_{SR} = H_{SR}^{a1} + H_{SR}^{a2} + H_{SR}^{a\alpha} + H_{SR}^{b1} + H_{SR}^{b2} + H_{SR}^{b\alpha}, \tag{9.87}$$

where ($\mu = 1, 2, \alpha$)

$$H_{SR}^{a\mu} \equiv \hbar(a\Gamma_{a\mu}^\dagger + a^\dagger \Gamma_{a\mu}), \tag{9.88a}$$

$$H_{SR}^{b\mu} \equiv \hbar(b\Gamma_{b\mu}^\dagger + b^\dagger \Gamma_{b\mu}), \tag{9.88b}$$

with

$$\begin{aligned} \Gamma_{a\mu}^\dagger &\equiv \sum_j \kappa_{a\mu j}^* r_{a\mu j}^\dagger, \qquad & \Gamma_{a\mu} &\equiv \sum_j \kappa_{a\mu j} r_{a\mu j}, \\ \Gamma_{b\mu}^\dagger &\equiv \sum_j \kappa_{b\mu j}^* r_{b\mu j}^\dagger, \qquad & \Gamma_{b\mu} &\equiv \sum_j \kappa_{b\mu j} r_{b\mu j}. \end{aligned} \tag{9.89a}$$

The individual Hamiltonians H_R^{a1}, H_R^{a2} and H_{SR}^{a1}, H_{SR}^{a2}, on the one hand, and H_R^{b1}, H_R^{b2} and H_{SR}^{b1}, H_{SR}^{b2}, on the other, account for mirror transmission losses for the subharmonic and pump modes, respectively; similarly, $H_R^{a\alpha}$ and $H_{SR}^{a\alpha}$, and $H_R^{b\alpha}$ and $H_{SR}^{b\alpha}$ account for absorption losses for the subharmonic and pump. Each reservoir comprises a collection of harmonic oscillators. The reservoir oscillators interacting with the subharmonic mode have annihilation and creation operators $r_{a\mu j}$ and $r_{a\mu j}^\dagger$, $\mu = 1, 2, \alpha$, while those interacting with the pump mode have annihilation and creation operators $r_{b\mu j}$ and $r_{b\mu j}^\dagger$. For the moment we leave all coupling constants, $\kappa_{a\mu j}$ and $\kappa_{b\mu j}$, unspecified.

The derivation of the master equation now runs parallel to the derivation of the master equation for the laser in Sect. 7.2.2. Adopting the notation of Chap. 1, there are four system operators

$$\{s_i\} \equiv \left(a, a^\dagger; b, b^\dagger\right), \tag{9.90a}$$

which couple, respectively, to the four reservoir operators

$$\{\Gamma_i\} \equiv \left(\Gamma_{a1}^\dagger + \Gamma_{a2}^\dagger + \Gamma_{a\alpha}^\dagger, \Gamma_{a1} + \Gamma_{a2} + \Gamma_{a\alpha}; \Gamma_{b1}^\dagger + \Gamma_{b2}^\dagger + \Gamma_{b\alpha}^\dagger, \Gamma_{b1} + \Gamma_{b2} + \Gamma_{b\alpha}\right), \tag{9.90b}$$

We have grouped the interactions associated with each system operator together; thus each Γ_i is a sum of three terms. The individual reservoirs are taken to be in thermal equilibrium. We allow, however, for different reservoir temperatures, T_μ, $\mu = 1, 2, \alpha$. The complete reservoir density operator is given by

$$R_0 = R_0^{a1} R_0^{a2} R_0^{a\alpha} R_0^{b1} R_0^{b2} R_0^{b\alpha}, \tag{9.91}$$

with

$$R_0^{a\mu} = \prod_j \exp\left(-\hbar\omega_j r_{a\mu j}^\dagger r_{a\mu j}/k_B T_\mu\right)\left[1 - \exp\left(-\hbar\omega_j/k_B T_\mu\right)\right], \tag{9.92a}$$

$$R_0^{b\mu} = \prod_j \exp\left(-\hbar\omega_j r_{b\mu j}^\dagger r_{b\mu j}/k_B T_\mu\right)\left[1 - \exp\left(-\hbar\omega_j/k_B T_\mu\right)\right]. \tag{9.92b}$$

With this reservoir state, the mean values $\langle \Gamma_i(t)\rangle_{R_0}$ are all zero, and the only nonvanishing reservoir correlations, $\langle \Gamma_i(t)\Gamma_j(t')\rangle_{R_0}$, are between the pairs of reservoir operators $\Gamma_1 \equiv \Gamma_{a1}^\dagger + \Gamma_{a2}^\dagger + \Gamma_{a\alpha}^\dagger$ and $\Gamma_2 \equiv \Gamma_{a1} + \Gamma_{a2} + \Gamma_{a\alpha}$, and $\Gamma_3 \equiv \Gamma_{b1}^\dagger + \Gamma_{b2}^\dagger + \Gamma_{b\alpha}^\dagger$ and $\Gamma_4 \equiv \Gamma_{b1} + \Gamma_{b2} + \Gamma_{b\alpha}$. The master equation in the interaction picture is

$$\dot{\tilde{\rho}} = \frac{1}{i\hbar}[H_{ab} + \tilde{H}_{\text{drive}}, \tilde{\rho}] + \left(\dot{\tilde{\rho}}\right)_a + \left(\dot{\tilde{\rho}}\right)_b, \tag{9.93}$$

where

$$\tilde{H}_{\text{drive}} \equiv e^{(i/\hbar)(H_a+H_b)t} H_{\text{drive}} e^{-(i/\hbar)(H_a+H_b)t} = \hbar(\bar{\mathcal{E}}_0 b^\dagger + \bar{\mathcal{E}}_0^* b),$$

and $\left(\dot{\tilde{\rho}}\right)_a$ and $\left(\dot{\tilde{\rho}}\right)_b$ each take the form (1.34), with definitions

$$\left(\tilde{s}_1, \tilde{s}_2\right)_a \equiv \left(\tilde{a}, \tilde{a}^\dagger\right), \tag{9.94a}$$

$$\left(\tilde{\Gamma}_1, \tilde{\Gamma}_2\right)_a \equiv \left(\tilde{\Gamma}_{a1}^\dagger + \tilde{\Gamma}_{a2}^\dagger + \tilde{\Gamma}_{a\alpha}^\dagger, \tilde{\Gamma}_{a1} + \tilde{\Gamma}_{a2} + \tilde{\Gamma}_{a\alpha}\right), \tag{9.94b}$$

$$\left(\tilde{s}_1, \tilde{s}_2\right)_b \equiv \left(\tilde{b}, \tilde{b}^\dagger\right), \tag{9.94c}$$

$$\left(\tilde{\Gamma}_1, \tilde{\Gamma}_2\right)_b \equiv \left(\tilde{\Gamma}_{b1}^\dagger + \tilde{\Gamma}_{b2}^\dagger + \tilde{\Gamma}_{b\alpha}^\dagger, \tilde{\Gamma}_{b1} + \tilde{\Gamma}_{b2} + \tilde{\Gamma}_{b\alpha}\right). \tag{9.94d}$$

We neglect the effect of the internal interactions H_{ab} and H_{drive} on the reservoir interactions [see Sect. 2.3.2 and the discussion below (7.92f)]; thus, the interaction picture is defined with $H_a + H_b$ in place of H_S:

$$\{\tilde{s}_i\} \equiv e^{(i/\hbar)(H_a+H_b)t}\{s_i\}e^{-(i/\hbar)(H_a+H_b)t}, \tag{9.95a}$$

$$\{\tilde{\Gamma}_i\} \equiv e^{(i/\hbar)H_R t}\{\Gamma_i\}e^{-(i/\hbar)H_R t}. \tag{9.95b}$$

and

$$\tilde{\rho} \equiv e^{(i/\hbar)(H_a+H_b)t}\rho e^{-(i/\hbar)(H_a+H_b)t}. \tag{9.96}$$

Note 9.8. At each mirror the transmission losses for the two cavity modes arise from interactions with the same electromagnetic field; i.e., both r_{a1j} ($r_{a1j}^\dagger$) and r_{b1j} ($r_{b1j}^\dagger$) are annihilation (creation) operators of the electromagnetic field to the left of the cavity, and r_{a2j} ($r_{a2j}^\dagger$) and r_{b2j} ($r_{b2j}^\dagger$) are both annihilation (creation) operators of the electromagnetic field to the right of the cavity. It might appear that correlations would exist between the operator pairs $(\Gamma_{a1}^\dagger, \Gamma_{a1})$ and $(\Gamma_{b1}^\dagger, \Gamma_{b1})$, and also between $(\Gamma_{a2}^\dagger, \Gamma_{a2})$ and $(\Gamma_{b2}^\dagger, \Gamma_{b2})$. This is not in fact the case, since the cavity modes are polarized in orthogonal directions; hence the reservoir modes with which they interact are distinct. Even if the polarizations were the same there would be no significant correlations, as the subharmonic mode interacts most strongly with reservoir modes in the vicinity of ω_C, while the pump interacts most strongly with reservoir modes in the vicinity of $2\omega_C$.

Explicit evaluation of the last two terms on the right-hand side of (9.93) involves essentially the same calculation as the one carried out in Sect. 1.4.1. The one difference is that the reservoir operators (9.94b) and (9.94d) are broken into three pieces. We note, however, that although three reservoirs couple to each cavity mode, the three reservoirs are statistically independent; hence, every nonvanishing reservoir correlation function is generalized to a sum of three terms, each term taking the form (1.45) or (1.46). Using (1.73), we may pass directly from (9.93) to the *master equation for the degenerate parametric oscillator*:

$$\begin{aligned}\dot{\rho} = & -i\omega_C[a^\dagger a, \rho] - i2\omega_C[b^\dagger b, \rho] \\ & + (g/2)[a^{\dagger 2}b - a^2 b^\dagger, \rho] - i[\bar{\mathcal{E}}_0 e^{-i2\omega_C t} b^\dagger + \bar{\mathcal{E}}_0^* e^{i2\omega_C t} b, \rho] \\ & + \kappa(2a\rho a^\dagger - a^\dagger a\rho - \rho a^\dagger a) + \kappa_p(2b\rho b^\dagger - b^\dagger b\rho - \rho b^\dagger b) \\ & + 2\kappa\bar{n}(a\rho a^\dagger + a^\dagger \rho a - a^\dagger a\rho - \rho a a^\dagger) \\ & + 2\kappa_p\bar{n}_p(b\rho b^\dagger + b^\dagger \rho b - b^\dagger b\rho - \rho b b^\dagger),\end{aligned} \tag{9.97}$$

where

$$\kappa \equiv \tfrac{1}{2}(\gamma_{a1} + \gamma_{a2} + \gamma_{a\alpha}), \qquad \kappa_p \equiv \tfrac{1}{2}(\gamma_{b1} + \gamma_{b2} + \gamma_{b\alpha}), \tag{9.98}$$

with $(\mu = 1, 2, \alpha)$

$$\gamma_{a\mu} \equiv 2\pi g_{a\mu}(\omega_C)|\kappa_{a\mu}(\omega_C)|^2, \tag{9.99a}$$

$$\gamma_{b\mu} \equiv 2\pi g_{b\mu}(\omega_C)|\kappa_{b\mu}(\omega_C)|^2, \tag{9.99b}$$

where $g_{a\mu}(\omega)$ and $g_{b\mu}(\omega)$ are reservoir densities of states, and $\bar{n}$ and $\bar{n}_p$ are weighted thermal photon number averages:

$$\bar{n} \equiv \frac{\gamma_{a1}\bar{n}_{a1} + \gamma_{a2}\bar{n}_{a2} + \gamma_{a\alpha}\bar{n}_{a\alpha}}{\gamma_{a1} + \gamma_{a2} + \gamma_{a\alpha}}, \tag{9.100a}$$

$$\bar{n}_p \equiv \frac{\gamma_{b1}\bar{n}_{b1} + \gamma_{b2}\bar{n}_{b2} + \gamma_{b\alpha}\bar{n}_{b\alpha}}{\gamma_{b1} + \gamma_{b2} + \gamma_{b\alpha}}, \tag{9.100b}$$

with ($\mu = 1, 2, \alpha$)

$$\bar{n}_{a\mu} = \left(e^{\hbar\omega_C/k_B T_\mu} - 1\right)^{-1}, \qquad \bar{n}_{b\mu} = \left(e^{\hbar 2\omega_C/k_B T_\mu} - 1\right)^{-1}. \tag{9.101a}$$

Exercise 9.5. In (9.97) the pump mode is driven by the classical field $\bar{\mathcal{E}}_0 e^{-i2\omega_C t}$. Setting $g = 0$, the reduced density operator for the pump mode obeys the *master equation for a driven damped harmonic oscillator*:

$$\begin{aligned}\dot{\rho}_p = &-i2\omega_C[b^\dagger b, \rho_p] - i[\bar{\mathcal{E}}_0 e^{-i2\omega_C t} b^\dagger + \bar{\mathcal{E}}_0^* e^{i2\omega_C t} b, \rho_p] \\ &+ \kappa_p(2b\rho_p b^\dagger - b^\dagger b\rho_p - \rho_p b^\dagger b) \\ &+ 2\kappa_p \bar{n}_p(b\rho_p b^\dagger + b^\dagger \rho_p b - b^\dagger b\rho_p - \rho_p b b^\dagger).\end{aligned} \tag{9.102}$$

Neglect thermal fluctuations ($\bar{n}_p = 0$) and show that if the initial state is the vacuum, $\rho_p(0) = |0\rangle\langle 0|$, the solution to (9.102) is

$$\rho_p(t) = \big|-i(\bar{\mathcal{E}}_0/\kappa_p)e^{-i2\omega_C t}(1 - e^{-\kappa_p t})\big\rangle\big\langle -i(\bar{\mathcal{E}}_0/\kappa_p)e^{-i2\omega_C t}(1 - e^{-\kappa_p t})\big|, \tag{9.103}$$

where $|\beta(t)\rangle$, $\beta(t) = -i(\bar{\mathcal{E}}_0/\kappa_p)(1 - e^{-\kappa_p t})$, denotes a coherent state. Thus, the pump mode is excited to a coherent state with time-dependent amplitude $\beta(t)$. The amplitude grows exponentially to reach its steady-state value $\beta(\infty) = -i(\bar{\mathcal{E}}_0/\kappa_p)$.

Note 9.9. The interaction term H_{drive} in Hamiltonian (9.79) treats the driving field as a c-number. We have now seen two justifications for this: (i) H_{drive}, as written, yields the desired source term in the equation of motion for the mean pump field amplitude (Exercise 9.4); (ii) in the absence of thermal fluctuations, H_{drive} excites the quantized pump mode to the coherent state $|\beta(t)\rangle$ (Eq. 9.103). For a more convincing justification, we can arrive at the interaction Hamiltonian H_{drive} by associating the classical field $\bar{\mathcal{E}}_0 e^{-i2\omega_C t}$ with the coherent state excitation of a single reservoir mode. Let us write the annihilation and creation operators of this mode as

$$r_0 \equiv r_{b1j}\Big|_{\omega_j = 2\omega_C}, \qquad r_0^\dagger \equiv r_{b1j}^\dagger\Big|_{\omega_j = 2\omega_C}. \tag{9.104}$$

We then consider this mode to be prepared in the coherent state $|\beta_0\rangle\langle\beta_0|$ and remove the term $H_{\text{drive}} = \hbar(\bar{\mathcal{E}}_i e^{-i2\omega_C t} b^\dagger + \bar{\mathcal{E}}_0^* e^{i2\omega_C t} b)$ from the system Hamiltonian H_S. Following through with the derivation of the master equation as before, we will encounter the difficulty that the reservoir operators $\tilde{\Gamma}_{b1}$ and $\tilde{\Gamma}_{b1}^\dagger$ have nonzero means. The same difficulty arose in the treatment of phase destroying processes in Sect. 2.2.4. The way around the difficulty is described there. We must move the interaction with the *mean* reservoir field from H_{SR}^{b1} to the system Hamiltonian H_S. This reintroduces H_{drive}. Explicitly, we define the new reservoir operator

$$\Gamma_{b1} \equiv \sum_{\omega_j \neq 2\omega_C} \kappa_{b1j} r_{b1j} + \kappa_0 (r_0 - \beta_0 e^{-i2\omega_C t}), \tag{9.105a}$$

where

$$\kappa_0 \equiv \kappa_{b1j}\Big|_{\omega_j = 2\omega_C}, \tag{9.105b}$$

and add

$$H_{\text{drive}} \equiv \hbar(\kappa_0 \beta_0 e^{-i2\omega_C t} b^\dagger + \kappa_0^* \beta_0^* e^{i2\omega_C t} b) \tag{9.106}$$

to H_S. Now the reservoir operators $\tilde{\Gamma}_{b1}$ and $\tilde{\Gamma}_{b1}^\dagger$ have zero means and the derivation of the master equation can go ahead as before. The amplitude $\bar{\mathcal{E}}_0$ of the external field is identified with $\kappa_0 \beta_0$.

Note 9.10. It should be noted that although this way of doing things might be more satisfactory than merely asserting that the external field is classical, it does, nevertheless, slip in an assumption that forces the driving field to act in a classical way. Recall that the master equation treatment assumes weak coupling between the system S and the reservoir R. Under this assumption, changes in R due to its interaction with S are neglected. Thus, the coherent state $|\beta_0\rangle$—as well as the amplitude $\bar{\mathcal{E}}_0 \equiv \kappa_0\beta_0$—is assumed to remain unaltered by the interaction between S and R. The assumption asserts a priori that the driving field will act as a constant-amplitude external field. To see a little more of what the weak coupling assumption entails, we observe that as the reservoir size (L' in Fig. 9.4) is taken to infinity, weak coupling requires

$$|\kappa_0|\kappa_p^{-1} \to 0; \tag{9.107}$$

κ_p^{-1} characterizes the typical timescale of the interaction between S and R. Now consider what this means for the photon number in the reservoir mode r_0 compared with that in the driven cavity mode b. With $\bar{\mathcal{E}}_0 \equiv \kappa_0\beta_0$, we may write

$$\langle r_0^\dagger r_0 \rangle = |\beta_0|^2 = |\bar{\mathcal{E}}_0/\kappa_0|^2 = \big(\kappa_p/|\kappa_0|\big)^2 \big(|\bar{\mathcal{E}}_0|/\kappa_p\big)^2 = \big(\kappa_p/|\kappa_0|\big)^2 \langle b^\dagger b\rangle_{\text{ss}}, \tag{9.108}$$

where $\langle b^\dagger b\rangle = (|\bar{\mathcal{E}}_0|/\kappa_p)^2$ is the steady-state photon number in the coherent state (9.103). According to (9.107), $\langle r_0^\dagger r_0\rangle/\langle b^\dagger b\rangle_{\text{ss}}$ goes to infinity with the reservoir size. This ensures that the exchange of photons between S and R can be neglected so far as the photon number in reservoir mode r_0 is concerned. To see explicitly how the reservoir size enters the picture, from (9.99b) we may write

$$|\kappa_0| = \sqrt{\gamma_{b1}/2\pi g_{b1}(2\omega_C)} = \sqrt{c/2L'}\sqrt{\gamma_{b1}}, \tag{9.109}$$

where $g_{b1}(\omega) = L'/\pi c$ is the density of modes in a standing-wave cavity of length L' (Fig. 9.4). Then

$$|\kappa_0|\kappa_p^{-1} = \sqrt{\frac{\kappa_p^{-1}}{2L'/c}}\sqrt{\gamma_{b1}/\kappa_p}. \tag{9.110}$$

We find, under the first square root, the ratio of a system timescale, κ_p^{-1}, and the round-trip time $2L'/c$ for photons in the reservoir. The ratio goes to zero as L' goes to infinity. In this weak-coupling limit, photons removed from the reservoir do not deplete the photon flux impinging upon S, and photons emitted into the reservoir never return to act back upon S. In summary, use of a c-number external field in interaction Hamiltonian (9.106) assumes the limit $\kappa_p^{-1}|\kappa_0| \to 0$, $\beta_0 \to \infty$, with $\kappa_p^{-1}\bar{\mathcal{E}}_0 = \kappa_p^{-1}(\kappa_0\beta_0)$ finite; or more specifically $L' \to \infty$ (infinite reservoir mode volume), $|\beta_0|^2 \to \infty$ (infinite reservoir mode photon number), with $(2L'/c)^{-1}|\beta_0|^2$ finite (finite reservoir mode photon flux).

9.2.5 Cavity Output Fields

Master equation 9.97 provides us with a quantized model for the degenerate parametric oscillator. The quantum-statistical properties of the intracavity fields can be calculated directly from this equation. As we noted for the laser, however, the intracavity fields are not, in fact, the fields that are directly measured (Sect. 7.3). The measured fields are those carried by the modes of the reservoirs that describe the mirror transmission losses. We used a traveling-wave model for the laser output field (Fig. 7.8). For the standing-wave geometry illustrated in Fig. 9.4, the reservoir modes are defined in external standing-wave cavities of length L'. There are four input–output fields, whose positive frequency components are specified as follows:

(i) for $z < -L + \ell + d$,

$$\begin{aligned}&\hat{\boldsymbol{E}}^{(+)}(z,t)\\&= i\hat{\boldsymbol{e}}_y\sum_j\sqrt{\frac{\hbar\omega_j}{\epsilon_0 AL'}}\frac{1}{2i}\left[e^{i(\omega_j/c)(z+L-\ell-d)} + e^{i\phi_R}e^{-i(\omega_j/c)(z+L-\ell-d)}\right]r_{a1j}(t),\end{aligned} \tag{9.111a}$$

$$\begin{aligned}&\hat{\boldsymbol{E}}_p^{(+)}(z,t)\\&= i\hat{\boldsymbol{e}}_x\sum_j\sqrt{\frac{\hbar\omega_j}{\epsilon_0 AL'}}\frac{1}{2i}\left[e^{i(\omega_j/c)(z+L-\ell-d)} + e^{i\phi_R}e^{-i(\omega_j/c)(z+L-\ell-d)}\right]r_{b1j}(t);\end{aligned} \tag{9.111b}$$

(ii) for $z > \ell + d$,

$$\hat{\boldsymbol{E}}^{(+)}(z,t) = i\hat{\boldsymbol{e}}_y \sum_j \sqrt{\frac{\hbar\omega_j}{\epsilon_0 AL'}} \frac{1}{2i} \left[e^{i\phi_R} e^{i(\omega_j/c)(z-\ell-d)} + e^{-i(\omega_j/c)(z-\ell-d)} \right] r_{a2j}(t), \tag{9.112a}$$

$$\hat{\boldsymbol{E}}_p^{(+)}(z,t) = i\hat{\boldsymbol{e}}_x \sum_j \sqrt{\frac{\hbar\omega_j}{\epsilon_0 AL'}} \frac{1}{2i} \left[e^{i\phi_R} e^{i(\omega_j/c)(z-\ell-d)} + e^{-i(\omega_j/c)(z-\ell-d)} \right] r_{b2j}(t). \tag{9.112b}$$

Each of these fields may be written as the sum of a free field and a source field, with the source field derived using the method of Sect. 7.3.1. We write

$$\hat{\boldsymbol{E}}^{(+)}(z,t) = \hat{\boldsymbol{E}}_f^{(+)}(z,t) + \hat{\boldsymbol{E}}_s^{(+)}(z,t), \tag{9.113a}$$

$$\hat{\boldsymbol{E}}_p^{(+)}(z,t) = \hat{\boldsymbol{E}}_{pf}^{(+)}(z,t) + \hat{\boldsymbol{E}}_{ps}^{(+)}(z,t), \tag{9.113b}$$

where the free fields mimic the expansions just given:

(i) for $z < -L + \ell + d$,

$$\hat{\boldsymbol{E}}_f^{(+)}(z,t) = i\hat{\boldsymbol{e}}_y \sum_j \sqrt{\frac{\hbar\omega_j}{\epsilon_0 AL'}} \times \frac{1}{2i} \left[e^{i(\omega_j/c)(z+L-\ell-d)} + e^{i\phi_R} e^{-i(\omega_j/c)(z+L-\ell-d)} \right] r_{a1j}(0) e^{-i\omega_j t}, \tag{9.114a}$$

$$\hat{\boldsymbol{E}}_{pf}^{(+)}(z,t) = i\hat{\boldsymbol{e}}_x \sum_j \sqrt{\frac{\hbar\omega_j}{\epsilon_0 AL'}} \times \frac{1}{2i} \left[e^{i(\omega_j/c)(z+L-\ell-d)} + e^{i\phi_R} e^{-i(\omega_j/c)(z+L-\ell-d)} \right] r_{b1j}(0) e^{-i\omega_j t}; \tag{9.114b}$$

(ii) for $z > \ell + d$,

$$\hat{\boldsymbol{E}}_f^{(+)}(z,t) = i\hat{\boldsymbol{e}}_y \sum_j \sqrt{\frac{\hbar\omega_j}{\epsilon_0 AL'}} \frac{1}{2i} \left[e^{i\phi_R} e^{i(\omega_j/c)(z-\ell-d)} + e^{-i(\omega_j/c)(z-\ell-d)} \right] r_{a2j}(0) e^{-i\omega_j t}, \tag{9.115a}$$

$$\hat{\boldsymbol{E}}_{pf}^{(+)}(z,t)$$
$$= i\hat{\boldsymbol{e}}_x \sum_j \sqrt{\frac{\hbar\omega_j}{\epsilon_0 AL'}} \frac{1}{2i} \left[e^{i\phi_R} e^{i(\omega_j/c)(z-\ell-d)} + e^{-i(\omega_j/c)(z-\ell-d)} \right] r_{b2j}(0) e^{-i\omega_j t}. \tag{9.115b}$$

The derivation of the source fields differs slightly from that in Sect. 7.3.1 due to the standing-wave form of (9.111) and (9.112) (see Note 7.10). The details are left as an exercise.

Exercise 9.6. Show that the source fields at the outputs of the degenerate parametric oscillator take the following form:

(i) for $z < -L + \ell + d$,

$$\hat{\boldsymbol{E}}_s^{(+)}(z,t) = i\hat{\boldsymbol{e}}_y \sqrt{\frac{\hbar\omega_C}{2\epsilon_0 Ac}}\, e^{i[\phi_T - \Phi(-L+\ell+d)-\phi]} \sqrt{\gamma_{a1}}\, a(t'), \tag{9.116a}$$

$$\hat{\boldsymbol{E}}_{ps}^{(+)}(z,t) = i\hat{\boldsymbol{e}}_x \sqrt{\frac{\hbar\omega_C}{\epsilon_0 Ac}}\, e^{i[\phi_T - 2\Phi(-L+\ell+d)-\phi_p]} \sqrt{\gamma_{b1}}\, b(t'), \tag{9.116b}$$

with retarded time $ct' = ct + (z + L - \ell - d)$;

(ii) for $z > \ell + d$,

$$\hat{\boldsymbol{E}}_s^{(+)}(z,t) = i\hat{\boldsymbol{e}}_y \sqrt{\frac{\hbar\omega_C}{2\epsilon_0 Ac}}\, e^{i[\phi_T + \Phi(\ell+d)+\phi]} \sqrt{\gamma_{a2}}\, a(t'), \tag{9.117a}$$

$$\hat{\boldsymbol{E}}_{ps}^{(+)}(z,t) = i\hat{\boldsymbol{e}}_x \sqrt{\frac{\hbar\omega_C}{\epsilon_0 Ac}}\, e^{i[\phi_T + 2\Phi(\ell+d)+\phi_p]} \sqrt{\gamma_{b2}}\, b(t'), \tag{9.117b}$$

with retarded time $ct' = ct - (z - \ell - d)$.

When deriving these results it is necessary to show that the reservoir coupling coefficients $\kappa_{a\mu}(\omega_C)$ and $\kappa_{b\mu}(\omega_C)$, $\mu = 1, 2$, are related to parameters of the cavity through [compare (7.112)]

$$-\sqrt{\frac{2L'}{c}}\kappa_{a1}^*(\omega_C)e^{i\phi_R} = \sqrt{\frac{c}{2\bar{L}}}\sqrt{T_1}e^{i\phi_T}e^{-i[\Phi(-L+\ell+d)+\phi]}, \tag{9.118a}$$

$$-\sqrt{\frac{2L'}{c}}\kappa_{b1}^*(\omega_C)e^{i\phi_R} = \sqrt{\frac{c}{2\bar{L}}}\sqrt{T_p^1}e^{i\phi_T}e^{-i[2\Phi(-L+\ell+d)+\phi_p]}, \tag{9.118b}$$

for mirror 1, and

$$-\sqrt{\frac{2L'}{c}}\kappa_{a2}^*(\omega_C)e^{i\phi_R} = \sqrt{\frac{c}{2\bar{L}}}\sqrt{T_2}e^{i\phi_T}e^{i[\Phi(\ell+d)+\phi]}, \tag{9.119a}$$

$$-\sqrt{\frac{2L'}{c}}\kappa_{b2}^*(\omega_C)e^{i\phi_R} = \sqrt{\frac{c}{2\bar{L}}}\sqrt{T_p^2}e^{i\phi_T}e^{i[2\Phi(\ell+d)+\phi_p]}, \tag{9.119b}$$

for mirror 2. From these expressions and (9.99), we obtain damping rates $\gamma_{a\mu} = T_\mu c/2\bar{L}$, $\gamma_{b\mu} = T_p^\mu c/2\bar{L}$, $\mu = 1, 2$, which agree with those expected from the classical analysis of cavity damping [use (9.98), (9.54), and (9.59)].

The output fields are the components of (9.113a) and (9.113b) that propagate away from the cavity. They are conveniently expressed in photon flux units, as in (7.122) and (7.123a):

(i) the *left-propagating output fields of the degenerate parametric oscillator* are given by ($z < -L + \ell + d$)

$$\hat{\mathcal{E}}_{\leftarrow}(z,t) = \sqrt{c/2L'}r_{a1f}(t') + \sqrt{\gamma_{a1}}a(t'), \tag{9.120a}$$

$$\hat{\mathcal{E}}_{p\leftarrow}(z,t) = \sqrt{c/2L'}r_{b1f}(t') + \sqrt{\gamma_{b1}}b(t'), \tag{9.120b}$$

with

$$r_{a1f}(t') \equiv -ie^{i(\phi_R-\phi_T)}e^{i[\Phi(-L+\ell+d)+\phi]}\sum_j\sqrt{\frac{\omega_j}{\omega_C}}r_{a1j}(0)e^{-i\omega_jt'}, \tag{9.121a}$$

$$r_{b1f}(t') \equiv -ie^{i(\phi_R-\phi_T)}e^{i[2\Phi(-L+\ell+d)+\phi_p]}\sum_j\sqrt{\frac{\omega_j}{2\omega_C}}r_{b1j}(0)e^{-i\omega_jt'}, \tag{9.121b}$$

and retarded time $ct' = ct + (z + L - \ell - d)$;

(ii) the *right-propagating output fields of the degenerate parametric oscillator* are given by ($z > \ell + d$)

$$\hat{\mathcal{E}}_{\rightarrow}(z,t) = \sqrt{c/2L'}r_{a2f}(t') + \sqrt{\gamma_{a2}}a(t'), \tag{9.122a}$$

$$\hat{\mathcal{E}}_{p\rightarrow}(z,t) = \sqrt{c/2L'}r_{b2f}(t') + \sqrt{\gamma_{b2}}b(t'), \tag{9.122b}$$

with

$$r_{a2f}(t') \equiv -ie^{i(\phi_R-\phi_T)}e^{-i[\Phi(\ell+d)+\phi]}\sum_j\sqrt{\frac{\omega_j}{\omega_C}}r_{a2j}(0)e^{-i\omega_jt'}, \tag{9.123a}$$

$$r_{b2f}(t') \equiv -ie^{i(\phi_R-\phi_T)}e^{-i[2\Phi(\ell+d)+\phi_p]}\sum_j\sqrt{\frac{\omega_j}{2\omega_C}}r_{b2j}(0)e^{-i\omega_jt'}, \tag{9.123b}$$

and retarded time $ct' = ct - (z - \ell - d)$.

Finally, we recall that the free field and source field are correlated at each output (Sect. 7.3.3). The correlations for a thermal reservoir state are given by (7.133a) and (7.133b). For the subharmonic mode, we have ($\mu = 1, 2$)

$$\sqrt{c/2L'}\sqrt{\gamma_{a\mu}}\langle\hat{O}(t+\tau)r_{a\mu f}(t)\rangle = \begin{cases} 0 & \tau < 0 \\ \frac{1}{2}\gamma_{a\mu}\bar{n}_{a\mu}\langle[\hat{O}(t+\tau), a(t)]\rangle & \tau = 0 \\ \gamma_{a\mu}\bar{n}_{a\mu}\langle[\hat{O}(t+\tau), a(t)]\rangle & \tau > 0, \end{cases} \tag{9.124a}$$

$$\sqrt{c/2L'}\sqrt{\gamma_{a\mu}}\langle r_{a\mu f}(t)\hat{O}(t+\tau)\rangle$$
$$= \begin{cases} 0 & \tau < 0 \\ \frac{1}{2}\gamma_{a\mu}(\bar{n}_{a\mu}+1)\langle[\hat{O}(t+\tau), a(t)]\rangle & \tau = 0 \\ \gamma_{a\mu}(\bar{n}_{a\mu}+1)\langle[\hat{O}(t+\tau), a(t)]\rangle & \tau > 0, \end{cases} \quad (9.124b)$$

and for the pump mode

$$\sqrt{c/2L'}\sqrt{\gamma_{b\mu}}\langle \hat{O}(t+\tau) r_{b\mu f}(t)\rangle$$
$$= \begin{cases} 0 & \tau < 0 \\ \frac{1}{2}\gamma_{b\mu}\bar{n}_{b\mu}\langle[\hat{O}(t+\tau), b(t)]\rangle & \tau = 0 \\ \gamma_{b\mu}\bar{n}_{b\mu}\langle[\hat{O}(t+\tau), b(t)]\rangle & \tau > 0, \end{cases} \quad (9.125a)$$

$$\sqrt{c/2L'}\sqrt{\gamma_{b\mu}}\langle r_{b\mu f}(t)\hat{O}(t+\tau)\rangle$$
$$= \begin{cases} 0 & \tau < 0 \\ \frac{1}{2}\gamma_{b\mu}(\bar{n}_{b\mu}+1)\langle[\hat{O}(t+\tau), b(t)]\rangle & \tau = 0 \\ \gamma_{b\mu}(\bar{n}_{b\mu}+1)\langle[\hat{O}(t+\tau), b(t)]\rangle & \tau > 0. \end{cases} \quad (9.125b)$$

9.3 The Spectrum of Squeezing

Extracting a prediction of squeezing from the master equation for the degenerate parametric oscillator is not quite as straightforward as solving a simple Heisenberg equation to demonstrate squeezing in parametric amplication (Eqs. 9.27–9.36). The analysis of fluctuations for the model of Sect. 9.2.4 will be tackled in Chap. 10, where we convert the master equation into a Fokker–Planck equation using the quantum–classical correspondence. The emphasis there will not be on the squeezing per se, however, but on how its presence changes the way in which the quantum–classical correspondence works. In fact, if our aim is to understand squeezing, the master equation is not the best place to start. Squeezing, more so than other phenomena in quantum optics, is very much tied up with the measurement used to observed it. It is desirable to first analyze this measurement to learn *what* one should calculate—using the master equation, the quantum–classical correspondence, or any other method.

It turns out that what one should calculate is the so-called spectrum of squeezing, or more specifically, the quantity we define as the *source-field spectrum of squeezing.* The spectrum of squeezing is a recent invention. It is less familiar than the optical spectrum, for example, and its definition raises tricky issues of interpretation not encountered by the latter—talk of vacuum fluctuations etc... What then is the measurement scheme underlying its definition? This is the question taken up in the rest of this chapter.

Note 9.11. Perhaps some qualification is called for. At a fundamental level there is no difference between using a master equation to calculate the squeezing properties of an optical source and using it to calculate something more familiar, such as the optical spectrum or the intensity correlation function. Squeezing appears to be different, but what is easily forgotten is that in *all* cases when performing a quantum-mechanical analysis, the full program involves the same *two* steps. First, the measurement scheme that gives definition to the property of interest must be outlined and analyzed. This step is needed to tell us what to calculate—generally some field correlation function or combination of correlation functions. Only after this step has been taken can the master equation be brought into play, along with the quantum regression formula, to make the calculation. With regard to intensity correlations, for example, the analysis of the appropriate measurement scheme was carried out many years ago by Glauber [9.1]. We are so familiar with the result that there is a tendency to think that the quantity to be calculated, the correlation function $G^{(2)}(\tau)$ (see Sect. 2.3.5), is self-evident. Of course it is not, since a particular ordering of operators is involved, dictated by the fact that the appropriate measurement scheme counts the outgoing particles (photons) of a scattered field.

9.3.1 Intracavity Field Fluctuations

We are led to consider a *spectrum* of squeezing because the subharmonic field is carried by a quasimode—not by a genuine single mode, but by a cavity mode that has a linewidth. We expect fluctuations of the cavity mode amplitude to be distributed over a range of frequencies falling within the cavity linewidth. How should the spectrum of these fluctuations—the spectrum of squeezing—be defined? Perhaps we can define it by decomposing the quadrature phase operators, $\hat{X}$ and $\hat{Y}$, into Fourier components. In fact, this guess is not going to work. It is useful to pursue it as an exercise, though, since the decomposition does teach us something about intracavity field fluctuations. It also helps us appreciate why we need to make a thorough analysis of the measurement procedure used to observe squeezing.

We will consider the degenerate parametric oscillator operating below threshold. To see how the Fourier decomposition should be performed, we complement our master equation treatment with one based on a quantum Langevin equation for the operator a. This approach is closely related to the work on multimode squeezing by Caves and Schumaker [9.18] and Gardiner and coworkers [9.19,9.20]. The required Langevin equation is obtained by generalizing the derivation of (7.130). The explicit derivation is left as an exercise.

Exercise 9.7. Adopt the system Hamiltonian

$$H_S = \hbar\omega_C a^\dagger a + \hbar(g\bar{\mathcal{E}}_0/2\kappa_p)e^{-2i\omega_C t}a^{\dagger 2} + \text{H.c.},$$

which is obtained from the two-mode Hamiltonian (9.79) by replacing the pump mode annihilation operator b by the amplitude of the coherent state

(9.103). Follow the method yielding (7.130) to derive the *quantum Langevin equation for the degenerate parametric oscillator below threshold*:

$$\dot{a} = -(\kappa + i\omega_C)a - ie^{i\psi}\kappa\lambda e^{-2i\omega_C t}a^\dagger - \sqrt{c/2L'}\sqrt{2\kappa}r_{af}, \tag{9.126}$$

where ψ is defined as in (9.62) and

$$r_{af} \equiv (1/\sqrt{2\kappa})\big(\sqrt{\gamma_{a1}}r_{a1f} + \sqrt{\gamma_{a2}}r_{a2f} + \sqrt{\gamma_{a\alpha}}r_{a\alpha f}\big), \tag{9.127}$$

where r_{a1f} and r_{a2f} are the free-field operators (9.121a) and (9.123a), respectively, and

$$r_{a\alpha f}(t) = e^{i\phi_\alpha}\sum_j \sqrt{\frac{\omega_j}{\omega_C}}r_{a\alpha j}(0)e^{-i\omega_j t}, \tag{9.128}$$

with $\phi_\alpha \equiv \arg[\kappa_{a\alpha}(\omega_C)] + \pi/2$. Note that the density of modes for the reservoir that accounts for absorption in the crystal (or mirrors) has been written as $L'/\pi c$—i.e., equal to the density of modes for the reservoirs that carry the cavity outputs. It is not necessary to do this, but it is a convenient way to simplify the equations. Final results do not depend on the reservoir mode density in any case, only on the decay rate $\gamma_{a\alpha} = \alpha\ell c/2\bar{L}$ (Eq. 9.99a).

If we now Fourier-decompose the subharmonic mode annihilation and creation operators, writing

$$a = \sum_\omega \hat{\alpha}_\omega e^{-i(\omega_C+\omega)t}, \qquad a^\dagger = \sum_\omega \hat{\alpha}_\omega^\dagger e^{i(\omega_C+\omega)t}, \tag{9.129}$$

from the quantum Langevin equation (9.126), we find that the Fourier components couple in pairs. In the long-time limit, components oscillating at frequencies displaced symmetrically from the cavity resonance—the frequencies $\omega_C + \omega$ and $\omega_C - \omega$—couple to one another through the steady-state equations

$$0 = -(\kappa - i\omega)\hat{\alpha}_\omega - ie^{i\psi}\kappa\lambda\hat{\alpha}_{-\omega}^\dagger - \sqrt{c/2L'}\sqrt{2\kappa}\hat{f}_\omega, \tag{9.130a}$$

$$0 = -(\kappa - i\omega)\hat{\alpha}_{-\omega}^\dagger + ie^{-i\psi}\kappa\lambda\hat{\alpha}_\omega - \sqrt{c/2L'}\sqrt{2\kappa}\hat{f}_{-\omega}^\dagger, \tag{9.130b}$$

where $\hat{f}_\omega$ is the free-field operator

$$\hat{f}_\omega \equiv \frac{1}{\sqrt{2\kappa}}\Big[ie^{i(\phi_R-\phi_T)}\Big(e^{i[\Phi(-L+\ell+d)+\phi]}\sqrt{\gamma_{a1}}r_{a1j}(0) + e^{-i[\Phi(\ell+d)+\phi]}\sqrt{\gamma_{a2}}r_{a2j}(0)\Big) + e^{i\phi_\alpha}\sqrt{\gamma_{a\alpha}}r_{a\alpha j}(0)\Big]\Big|_{\omega_j=\omega_C+\omega}, \tag{9.131}$$

and $\hat{f}_\omega$ and $\hat{f}_{\omega'}$ obey the boson commutation relations

$$[\hat{f}_\omega, \hat{f}_{\omega'}] = [\hat{f}_\omega^\dagger, \hat{f}_{\omega'}^\dagger] = 0, \qquad [\hat{f}_\omega, \hat{f}_{\omega'}^\dagger] = \delta_{\omega,\omega'}. \tag{9.132}$$

In writing (9.131), we have made the approximation $\omega_j/\omega_C \to 1$, as was done in the derivation of (7.129).

For the thermal reservoir states (9.92a), fluctuations of the free fields exhibit correlations

$$\langle \hat{f}_\omega \hat{f}_{\omega'} \rangle = \langle \hat{f}^\dagger_\omega \hat{f}^\dagger_{\omega'} \rangle = 0, \tag{9.133a}$$

$$\langle \hat{f}^\dagger_\omega \hat{f}_{\omega'} \rangle = \bar{n}\delta_{\omega,\omega'}, \tag{9.133b}$$

$$\langle \hat{f}_\omega \hat{f}^\dagger_{\omega'} \rangle = (\bar{n}+1)\delta_{\omega,\omega'}, \tag{9.133c}$$

where we have made use of (9.98) and (9.100a). To determine the correlations between Fourier components of the subharmonic field, we solve (9.130a) and (9.130b) for

$$\hat{\alpha}_\omega = -\sqrt{\frac{c}{2L'}}\sqrt{2\kappa}\frac{(\kappa - i\omega)\hat{f}_\omega - ie^{i\psi}\kappa\lambda\hat{f}^\dagger_{-\omega}}{(\kappa - i\omega)^2 - \kappa^2\lambda^2}, \tag{9.134a}$$

and

$$\hat{\alpha}^\dagger_{-\omega} = -\sqrt{\frac{c}{2L'}}\sqrt{2\kappa}\frac{(\kappa - i\omega)\hat{f}^\dagger_{-\omega} + ie^{-i\psi}\kappa\lambda\hat{f}_\omega}{(\kappa - i\omega)^2 - \kappa^2\lambda^2}, \tag{9.134b}$$

thus expressing the intracavity Fourier components in terms of their reservoir inputs.

Note 9.12. Steady-state solutions (9.134a) and (9.134b) preserve the commutation relations for a and $a^\dagger$, as they should. Converting frequency sums to integrations, with reservoir mode density $L'/\pi c$, from (9.129), (9.132), (9.134a), and (9.134b), we find

$$\begin{aligned}
[a, a^\dagger] &= \int_{-\infty}^{\infty} d\omega (L'/\pi c)[\hat{\alpha}_\omega, \hat{\alpha}^\dagger_\omega] \\
&= \frac{\kappa}{\pi}\int_{-\infty}^{\infty} d\omega \frac{|\kappa - i\omega|^2 - \kappa^2\lambda^2}{|(\kappa - i\omega)^2 - \kappa^2\lambda^2|^2} \\
&= \frac{\kappa}{\pi}\int_{-\infty}^{\infty} d\omega \frac{\kappa^2(1-\lambda)(1+\lambda) + \omega^2}{[\kappa^2(1-\lambda)^2 + \omega^2][\kappa^2(1+\lambda)^2 + \omega^2]} \\
&= \int_{-\infty}^{\infty} d\omega \frac{1}{2}\left\{\frac{\kappa(1-\lambda)/\pi}{[\kappa(1-\lambda)]^2 + \omega^2} + \frac{\kappa(1+\lambda)/\pi}{[\kappa(1+\lambda)]^2 + \omega^2}\right\} \\
&= 1.
\end{aligned} \tag{9.135}$$

Our aim is to decompose the quadrature phase operators $\hat{X}$ and $\hat{Y}$ into Fourier components, not simply the annihilation and creation operators. As a first step we write $\hat{X}$ and $\hat{Y}$ in terms of quadrature phase operators for individual Fourier components of a and $a^\dagger$. We define $\hat{x}_\omega$ and $\hat{y}_\omega$ by

$$\hat{x}_\omega \equiv \tfrac{1}{2}\left[\hat{\alpha}_\omega e^{-i\frac{1}{2}(\psi - \pi/2)} + \hat{\alpha}^\dagger_\omega e^{i\frac{1}{2}(\psi - \pi/2)}\right], \tag{9.136a}$$

$$\hat{y}_\omega \equiv \tfrac{1}{2}\left[\hat{\alpha}_\omega e^{-i\frac{1}{2}(\psi + \pi/2)} + \hat{\alpha}^\dagger_\omega e^{i\frac{1}{2}(\psi + \pi/2)}\right], \tag{9.136b}$$

and write

$$\begin{aligned}\hat{X} &= \tfrac{1}{2}\left[\tilde{a}e^{-i\frac{1}{2}(\psi-\pi/2)} + \tilde{a}^\dagger e^{i\frac{1}{2}(\psi-\pi/2)}\right] \\ &= \frac{1}{2}\left[\left(\sum_\omega \hat{\alpha}_\omega e^{-i\omega t}\right)e^{-i\frac{1}{2}(\psi-\pi/2)} + \left(\sum_\omega \hat{\alpha}^\dagger_\omega e^{i\omega t}\right)e^{i\frac{1}{2}(\psi-\pi/2)}\right] \\ &= \sum_\omega (\hat{x}_\omega \cos\omega t + \hat{y}_\omega \sin\omega t), \end{aligned} \tag{9.137a}$$

and similarly,

$$\hat{Y} = \sum_\omega(-\hat{x}_\omega \sin\omega t + \hat{y}_\omega \cos\omega t). \tag{9.137b}$$

Clearly $\hat{x}_\omega$ and $\hat{y}_\omega$ are not themselves the Fourier amplitudes, $\hat{X}_\omega$ and $\hat{Y}_\omega$, of $\hat{X}$ and $\hat{Y}$. In fact, the four operators $\hat{x}_\omega$, $\hat{y}_\omega$, $\hat{x}_{-\omega}$, and $\hat{y}_{-\omega}$ all contribute to the Fourier amplitudes $\hat{X}_\omega$ and $\hat{Y}_\omega$. The relationship is illustrated in Fig. 9.5. In the mathematics, we simply make the substitution $\omega \to -\omega$ in appropriate places in (9.137a) and (9.137b), to obtain

$$\hat{X} = \sum_\omega \hat{X}_\omega e^{-i\omega t}, \qquad \hat{Y} = \sum_\omega \hat{Y}_\omega e^{-i\omega t}, \tag{9.138}$$

with

$$\begin{aligned}\hat{X}_\omega &= \tfrac{1}{2}[(\hat{x}_\omega + i\hat{y}_\omega) + (\hat{x}_{-\omega} - i\hat{y}_{-\omega})] \\ &= \tfrac{1}{2}\left[\hat{\alpha}_\omega e^{-i\frac{1}{2}(\psi-\pi/2)} + \hat{\alpha}^\dagger_{-\omega} e^{i\frac{1}{2}(\psi-\pi/2)}\right], \end{aligned} \tag{9.139a}$$

and

$$\begin{aligned}\hat{Y}_\omega &= -i\tfrac{1}{2}[(\hat{x}_\omega + i\hat{y}_\omega) - (\hat{x}_{-\omega} - i\hat{y}_{-\omega})] \\ &= \tfrac{1}{2}\left[\hat{\alpha}_\omega e^{-i\frac{1}{2}(\psi+\pi/2)} + \hat{\alpha}^\dagger_{-\omega} e^{i\frac{1}{2}(\psi+\pi/2)}\right]. \end{aligned} \tag{9.139b}$$

We are finally in a position to decompose the fluctuations in quadrature phase amplitudes into a spectrum. Using (9.139a) and (9.139b), together with the explicit solutions for the degenerate parametric oscillator (9.134a) and (9.134b), the Fourier components of the quadrature phase amplitudes are given for the X quadrature by

$$\begin{aligned}\hat{X}_\omega &= \tfrac{1}{2}\left[\hat{\alpha}_\omega e^{-i\frac{1}{2}(\psi-\pi/2)} + \hat{\alpha}^\dagger_{-\omega} e^{i\frac{1}{2}(\psi-\pi/2)}\right] \\ &= -\sqrt{\frac{c}{2L'}}\sqrt{2\kappa}\frac{\kappa(1+\lambda) - i\omega}{(\kappa - i\omega)^2 - \kappa^2\lambda^2}\tfrac{1}{2}\left[\hat{f}_\omega e^{-i\frac{1}{2}(\psi-\pi/2)} + \hat{f}^\dagger_{-\omega} e^{i\frac{1}{2}(\psi-\pi/2)}\right] \\ &= -\sqrt{\frac{c}{2L'}}\sqrt{2\kappa}\frac{1}{2}\frac{\hat{f}_\omega e^{-i\frac{1}{2}(\psi-\pi/2)} + \hat{f}^\dagger_{-\omega} e^{i\frac{1}{2}(\psi-\pi/2)}}{\kappa(1-\lambda) - i\omega}, \end{aligned} \tag{9.140a}$$

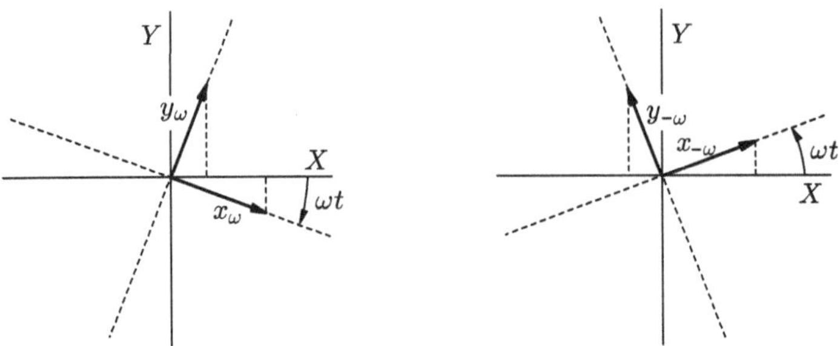

Fig. 9.5. Phase-space pictures showing the relationship between the Fourier amplitudes X_ω and Y_ω, defined by (9.139a) and (9.139b), and the single-mode quadrature phase amplitudes $\hat{x}_\omega$, $\hat{y}_\omega$, $\hat{x}_{-\omega}$, and $\hat{y}_{-\omega}$ defined by (9.136a) and (9.136b)

and in a similar fashion, for the Y quadrature by

$$\hat{Y}_\omega = -\sqrt{\frac{c}{2L'}}\sqrt{2\kappa}\frac{1}{2}\frac{\hat{f}_\omega e^{-i\frac{1}{2}(\psi+\pi/2)} + \hat{f}^\dagger_{-\omega} e^{i\frac{1}{2}(\psi+\pi/2)}}{\kappa(1+\lambda) - i\omega}. \tag{9.140b}$$

Using the free-field variances and covariances (9.133), and introducing the density of states $L'/\pi c$, the *spectra of intracavity field fluctuations* are then given by

$$(\Delta X)^2_<(\omega) = (L'/\pi c)\langle \hat{X}^\dagger_\omega \hat{X}_\omega\rangle = \frac{1}{4}\frac{2\bar{n}+1}{1-\lambda}\frac{\kappa(1-\lambda)/\pi}{[\kappa(1-\lambda)]^2+\omega^2}, \tag{9.141a}$$

$$(\Delta Y)^2_<(\omega) = (L'/\pi c)\langle \hat{Y}^\dagger_\omega \hat{Y}_\omega\rangle = \frac{1}{4}\frac{2\bar{n}+1}{1+\lambda}\frac{\kappa(1+\lambda)/\pi}{[\kappa(1+\lambda)]^2+\omega^2}, \tag{9.141b}$$

where the subscript $<$ indicates that the expressions hold for the degenerate parametric oscillator *below* threshold.

We aim to derive a spectrum of squeezing; are the spectra (9.141a) and (9.141b) what we might expect for X- and Y-quadrature spectra of squeezing? Perhaps, at this point, we should look ahead to the results (10.55a) and (10.55b), which tell us that for the degenerate parametric oscillator below threshold, the quadrature phase variances, without Fourier decomposition, are given by

$$(\Delta X)^2_< = \frac{1}{4}\frac{2\bar{n}+1}{1-\lambda} \quad \text{and} \quad (\Delta Y)^2_< = \frac{1}{4}\frac{2\bar{n}+1}{1+\lambda}.$$

Squeezing is evidenced by the fact that $(\Delta Y)^2_< < 1/4$ for $\lambda \neq 0$ and $\bar{n} = 0$. We note then that when they are integrated over all frequencies, (9.141a) and (9.141b) reproduce these results; so in this respect, at least, (9.141a) and (9.141b) are satisfactory spectral decompositions of the quadrature phase amplitude fluctuations. To reinforce the point, using (9.140a), we can show that $\langle \hat{X}^\dagger_{\omega''} \hat{X}_{\omega'} \rangle \propto \delta_{\omega'',\omega'}$, and hence that

$$\begin{aligned}
(\Delta X)^2_<(\omega) &= (L'/\pi c)\langle \hat{X}^\dagger_\omega \hat{X}_\omega \rangle \\
&= \frac{1}{2\pi}\int_{-\infty}^{\infty} d\tau \sum_{\omega'} \langle \hat{X}^\dagger_{\omega'} \hat{X}_{\omega'} \rangle e^{i(\omega-\omega')\tau} \\
&= \frac{1}{2\pi}\int_{-\infty}^{\infty} d\tau e^{i\omega\tau} \left\langle \left(\sum_{\omega''} \hat{X}^\dagger_{\omega''} e^{i\omega'' t} \right) \left(\sum_{\omega'} \hat{X}_{\omega'} e^{-i\omega'(t+\tau)} \right) \right\rangle \\
&= \frac{1}{2\pi}\int_{-\infty}^{\infty} d\tau e^{i\omega\tau} \langle \hat{X}(t)\hat{X}(t+\tau)\rangle,
\end{aligned} \tag{9.142a}$$

and similarly,

$$\begin{aligned}
(\Delta Y)^2_<(\omega) &= (L'/\pi c)\langle \hat{Y}^\dagger_\omega \hat{Y}_\omega \rangle \\
&= \frac{1}{2\pi}\int_{-\infty}^{\infty} d\tau e^{i\omega\tau} \langle \hat{Y}(t)\hat{Y}(t+\tau)\rangle.
\end{aligned} \tag{9.142b}$$

Thus, $(\Delta X)^2_<(\omega)$ and $(\Delta Y)^2_<(\omega)$ are Fourier transforms of the autocorrelation functions for quadrature phase amplitudes—just what we would expect of a spectral decomposition.

In spite of all this, though, the spectra (9.141a) and (9.141b) are not what we want for spectra of squeezing. To appreciate why, we might look at Fig. 9.6. Squeezing is concerned with the reduction of fluctuations below their level in the vacuum state. Figure 9.6 compares $(\Delta X)^2_<(\omega)$ and $(\Delta Y)^2_<(\omega)$, for $\bar{n} = 0$, with the spectrum

$$(\Delta X)^2(\omega) = (\Delta Y)^2(\omega) = \frac{1}{4}\frac{\kappa/\pi}{\kappa^2 + \omega^2} \tag{9.143}$$

calculated for the vacuum state—i.e., with $\bar{n} = 0$ and $\lambda = 0$. We see that there is, indeed, a frequency-dependent amplification and deamplification of the vacuum fluctuations across the cavity line, such that fluctuations of the Y-quadrature phase amplitude fall below their value in the vacuum state. We are hardly looking at an extension, to many frequencies, of the phenomenon described by (9.36), however; certainly, the horizontal line in the figure, at $\kappa(\Delta X)^2(\omega) = \kappa(\Delta Y)^2(\omega) = 1/4$, cannot be the appropriate vacumm level. In fact, we have a clear indication that we are on the wrong track from the vacuum state Lorentzians without considering parametric amplification at all. Remember that ω denotes a displacement away from the cavity resonance frequency ω_C. It is then apparent that it is possible in principle—even with $\lambda = 0$—to reduce the fluctuations at a chosen frequency $\omega_0 = \omega_C + \omega$ to whatever level we desire by simply detuning ω_C sufficiently far from ω_0. Moreover, in doing this, we reduce the fluctuations in *both* quadrature phase amplitudes at once.

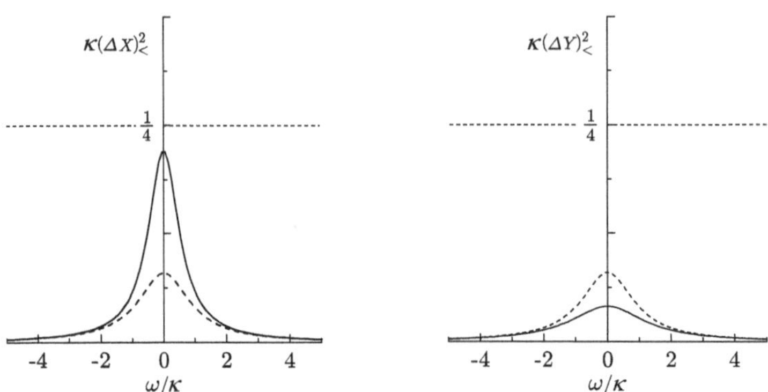

Fig. 9.6. Frequency decomposition of the fluctuations in intracavity quadrature phase amplitudes for the degenerate parametric oscillator below threshold. Spectra (9.141a) and (9.141b) are plotted for $\bar{n} = 0$ and $\lambda = 0.4$. The *dashed curves* show the same spectra calculated with the subharmonic mode in the vacuum state ($\lambda = 0$)

Of course, all of this has nothing to do with squeezing. What we are seeing here is the action of the cavity as a filter for the vacuum field entering through its partially transparent mirrors. Equation 9.143 decomposes the single (quasi-)mode variance $(\Delta X)^2 = (\Delta Y)^2 = \frac{1}{4}$ into a spectrum. Since, by definition, the integral of the spectrum over all frequencies is equal to $\frac{1}{4}$, the noise strength $(\Delta X)^2 \Delta\omega = (\Delta Y)^2 \Delta\omega$ in *any* finite bandwidth $\Delta\omega$ is guaranteed to be less than $\frac{1}{4}$.

Note 9.13. There is no violation of a fundamental uncertainty requirement if $(\Delta X)^2(\omega)$ and $(\Delta Y)^2(\omega)$ are both zero simultaneously, since $\hat{X}_\omega$ and $\hat{Y}_\omega$ are not conjugate variables obeying a canonical commutation relation (they are also not Hermitian). In fact, from their explicit definition (Eqs. 9.140), we find $[\hat{X}_\omega, \hat{Y}_\omega] = 0$. Only after $\hat{X}_\omega$ and $\hat{Y}_\omega$ have been integrated over all frequencies do we recover the commutator $[\hat{X}, \hat{Y}] = \frac{1}{2}i$, and hence the uncertainty relation $\Delta X \Delta Y \geq \frac{1}{4}$. It should be noted that although the filtering effect revealed by $\hat{X}_\omega$ and $\hat{Y}_\omega$ is not related to squeezing, it does have interesting physical consequences; for example, it is responsible for the phenomenon of cavity-inhibited spontaneous emission (see Note 13.4).

9.3.2 Definition of the Spectrum of Squeezing

We have seen that the spectrum of squeezing is not just a Fourier decomposition of the fluctuations in the quadrature phase amplitudes of the intracavity field, as might be suggested by simple analogy with the optical spectrum. Clearly we need to think a little harder about what a spectrum of squeezing, a *measured* spectrum of squeezing, might be. In Sect. 9.3.3 we define it in an operational manner, basing the definition on a specific measurement scheme.

Before beginning the analysis, it might be helpful to see what that definition is, and how it differs from (9.142).

We write $S_X(\omega)$ and $S_Y(\omega)$ for the *source-field spectra of squeezing detected with unit efficiency.* Then the spectra obeying an uncertainty relation analogous to (9.35) are

$$1+\eta\zeta S_X(\omega)=1+\eta\zeta(2\kappa)4\int_{-\infty}^{\infty}d\tau e^{i\omega\tau}\langle:\hat{X}(t)\hat{X}(t+\tau):\rangle, \tag{9.144a}$$

$$1+\eta\zeta S_Y(\omega)=1+\eta\zeta(2\kappa)4\int_{-\infty}^{\infty}d\tau e^{i\omega\tau}\langle:\hat{Y}(t)\hat{Y}(t+\tau):\rangle, \tag{9.144b}$$

where $\eta\le 1$ is a detection efficiency, $\zeta<1$ is a collection efficiency, and $\langle::\rangle$ denotes averages that are normal- and time-ordered; thus, $\sqrt{1+\eta\zeta S_X(\omega)}$ and $\sqrt{1+\eta\zeta S_Y(\omega)}$ are the frequency-resolved generalizations of ΔX and ΔY in Sect. 9.2.2. Actually, they are scaled to correspond to $2\Delta X$ and $2\Delta Y$, so that the uncertainty relation is $\sqrt{1+\eta\zeta S_X(\omega)}\sqrt{1+\eta\zeta S_Y(\omega)}\ge 1$. There is squeezing at frequency ω in the X- or Y-quadrature phase amplitude if $1+\eta\zeta S_{X,Y}(\omega)<1$, or equivalently $S_{X,Y}(\omega)<0$.

On comparing (9.144a) and (9.144b) with (9.142a) and (9.142b), we see two differences over and above the unimportant scaling by a factor of four. First, there is a further scaling by $(2\kappa)\times 2\pi$. This is a consequence of the different normalizations implied by the different definitions of the two types of spectra. The spectra (9.142a) and (9.142b) specify a noise level *per mode* as a function of the mode frequency; there is a background noise level of $1+\eta\zeta S_{X,Y}(\omega)=1$ per mode (the vacuum noise level), which is modified when the spectrum of squeezing differs from zero. The spectra (9.142a) and (9.142b), on the other hand, decompose the fluctuations of a *single quasi-mode* into a spectrum. These spectra produce the noise level of the single quasi-mode when integrated over all frequencies. For the vacuum state, the normalization required is $4\int_{-\infty}^{\infty}d\omega(\Delta X)^2_<(\omega)=4\int_{-\infty}^{\infty}d\omega(\Delta Y)^2_<(\omega)=1$. The scale factor $(2\kappa)\times 2\pi$ is therefore also not a fundamental thing.

The second difference is the more important one. This is the specification of normal ordering and time ordering in (9.143a) and (9.143b)—absent from (9.142a) and (9.142b)—together with the $+1$ outside the integrals. This difference arises from the answer to the question we now pose concerning the measurement scheme that defines the spectrum of squeezing. It is, of course, precisely the ambiguity of operator ordering that advises us not to rely on naive generalizations of classical formulae to account for measurements made on a quantum field.

Note 9.14. In the limit of large thermal photon number, $\bar{n}\gg 1$, the spectra $(\Delta X)^2_<(\omega)$ and $(\Delta Y)^2_<(\omega)$ (Eqs. 9.141), and $S^<_X(\omega)$ and $S^<_Y(\omega)$ (Eqs. 10.61) differ by the scale factor $(2\kappa)\times 4\times 2\pi$ only. In this classical limit, the differences in operator ordering are unimportant—the source-field spectrum of squeezing *is*, aside from the scaling, just the Fourier decomposition of the amplitude fluctuations inside the cavity.

Exercise 9.8. The solutions for a and $a^\dagger$ in terms of free-field operators (Eqs. 9.129 and 9.134) can be used to derive a number of useful results. As an example, use these solutions to derive the commutators (10.70); thus, provide an alternative derivation to the one based on operator ordering conventions for different phase-space representations given in Chap. 10. From (10.70), one can determine the correlations between the free field and the source field (Eqs. 9.124). Use the solutions for a and $a^\dagger$ to verify (9.124) with $\hat{O} \equiv a$ and $\hat{O} \equiv a^\dagger$.

Exercise 9.9. Use the commutators (10.70) to prove $\langle[\hat{X}(t), \hat{X}(t+\tau)]\rangle = \langle[\hat{Y}(t), \hat{Y}(t+\tau)]\rangle = 0$; hence, show that (9.142a) and (9.142b) give the spectra of quadrature phase amplitude fluctuations obtained by Fourier transforming correlation functions calculated in the Wigner representation. In contrast, (9.144a) and (9.144b) are Fourier transforms of the correlation functions calculated in the P representation.

9.3.3 Homodyne Detection: The Source-Field Spectrum of Squeezing

Let us now analyze the measurement scheme that gives definition to the spectrum of squeezing. We begin by considering the measurement that defines the *source-field* spectrum of squeezing. This spectrum applies in situations where the source field carries all the real photons illuminating the detector (the free-field state is the vacuum state). Having understood the source-field spectrum, we include free-field contributions in Sect. 9.3.5. We follow the treatment given by Carmichael [9.21].

Since squeezing is a phase-dependent phenomenon, clearly any scheme designed to measure it must introduce a phase reference. Homodyne detection is then a natural candidate. In homodyne detection, a strong local oscillator (coherent field) is added to the field to be measured (the source field) and continuous photoelectric detection is performed on the sum. We therefore consider continuous photoelectric detection of the field

$$\hat{\mathcal{E}}(t) \equiv e^{-i\omega_C t}[\hat{\mathcal{E}}_{\mathrm{lo}} + \sqrt{\zeta}\sqrt{2\kappa}\Delta\tilde{a}(t')], \tag{9.145}$$

where $\langle\hat{\mathcal{E}}_{\mathrm{lo}}\rangle = \mathcal{E}_{\mathrm{lo}} = |\mathcal{E}_{\mathrm{lo}}|e^{i\theta}$ is the local oscillator amplitude, t' is a retarded time [see, for example, the definitions below (9.121b) and (9.123b)], ζ is the collection efficiency, and $\sqrt{2\kappa}$ scales the amplitude of the source field so that $2\kappa\Delta\tilde{a}^\dagger\tilde{a}(t)$ has units of photon flux; $\hat{\mathcal{E}}(t)$ and $\hat{\mathcal{E}}_{\mathrm{lo}}$ also have photon flux units. We assume $\langle\tilde{a}\rangle = 0$; if this is not so we may consider the nonzero $\langle\tilde{a}\rangle$ to be absorbed into the definition of $\mathcal{E}_{\mathrm{lo}}$.

The probability of photoelectric emission depends on the intensity $(\hat{\mathcal{E}}^\dagger\hat{\mathcal{E}})(t)$, which is sensitive to the relative phase between the local oscillator and the fluctuating amplitude of the source. More generally, there is a distribution in the number of photoelectric counts recorded in an interval $t - \Delta t$ to t, given by

$$p(n_t, t, \Delta t) = \left\langle : \frac{[\eta(\hat{\mathcal{E}}^\dagger\hat{\mathcal{E}})(t)]^{n_t} \Delta t^{n_t}}{n_t!} \exp[-\eta(\hat{\mathcal{E}}^\dagger\hat{\mathcal{E}})(t)\Delta t] : \right\rangle, \tag{9.146}$$

where n_t denotes the number of counts, η is the detection efficiency, and Δt is assumed to be much less than the correlation time of the fluctuations. This distribution follows from the standard theory of photoelectron counting due to Mandel [9.3], Glauber [9.1], and Kelley and Kleiner [9.2]; note how normal and time ordering enters directly in this theory. Now, while we could define the spectrum of squeezing in terms of such a photoelectron counting distribution, in practice the high photon flux of the local oscillator makes the actual counting of photoelectrons inappropriate. Under these circumstances, we stay closer to the physics defining the spectrum of squeezing in terms of an analog photocurrent. This requires us to say something about how the intensity operator $(\hat{\mathcal{E}}^\dagger\hat{\mathcal{E}})(t)$ is connected to the photocurrent $i(t)$: the measurement signal, a time series of real numbers.

Note 9.15. The counting formula (9.146) gives a Poisson distribution when $(\hat{\mathcal{E}}^\dagger\hat{\mathcal{E}})(t)$ is replaced by the classical intensity $|\mathcal{E}(t)|^2$. Physically, the Poisson distribution results from the random emission of photoelectrons from a photocathode illuminated by a constant classical intensity. In quantum-mechanical language, the prescribed normal ordering ensures the result that an initial coherent state $|\mathcal{E}(0)\rangle$, with $\hat{\mathcal{E}}(t)|\mathcal{E}(0)\rangle = \mathcal{E}(t)|\mathcal{E}(0)\rangle$, acts just like the classical intensity $|\mathcal{E}(t)|^2$.

Let us assume that a single photoelectric detection event produces a current pulse of width τ_d and amplitude Ge/τ_d, where e is the electronic charge and G is the gain, as illustrated in Fig. 9.7a. Figure 9.7b shows how the photocurrent

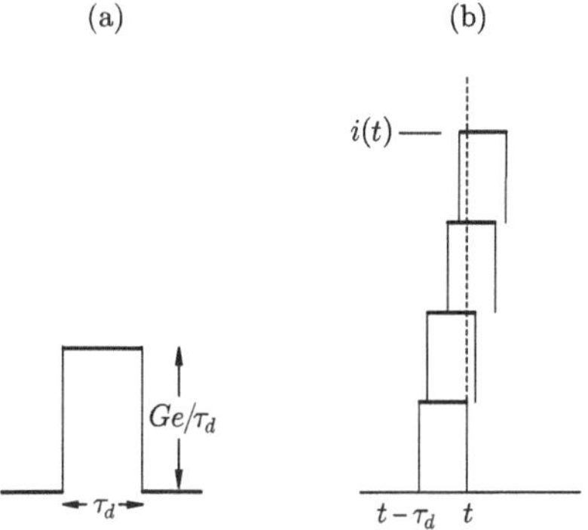

Fig. 9.7. (**a**) Current pulse produced by a single photodetection event. (**b**) Construction of the instantaneous photocurrent $i(t)$ from the overlap of current pulses initiated during the interval $t - \tau_d$ to t

rent

$$i(t) = n_t \frac{Ge}{\tau_d} \tag{9.147}$$

is formed from the overlap of the n_t current pulses initiated during the interval $t - \tau_d$ to t; $i(t)$ is a classical stochastic process and n_t is a random variable. We now use the photoelectron counting formula (9.146) to relate the classical fluctuations of $i(t)$ to the quantum fluctuations of the detected field. In order to derive the spectrum of photocurrent fluctuations, we will need the autocorrelation function $\overline{i(t)i(t+\tau)}$. First, though, just to see how things work, it is easier to calculate the photocurrent variance. We therefore begin with

$$\begin{aligned}
&\overline{i(t)i(t)} - \left(\overline{i(t)}\right)^2 \\
&= \left(\frac{Ge}{\tau_d}\right)^2 \left[\sum_{n_t} n_t^2 p(n_t, t, \tau_d) - \left(\sum_{n_t} n_t p(n_t, t, \tau_d)\right)^2\right] \\
&= \left(\frac{Ge}{\tau_d}\right)^2 \left\{\sum_{n_t} [n_t(n_t - 1) + n_t] \left\langle : \frac{[\eta(\hat{\mathcal{E}}^\dagger\hat{\mathcal{E}})(t)]^{n_t} \tau_d^{n_t}}{n_t!} \exp[-\eta(\hat{\mathcal{E}}^\dagger\hat{\mathcal{E}})(t)\tau_d] : \right\rangle\right. \\
&\quad \left. - \left(\sum_{n_t} n_t \left\langle : \frac{[\eta(\hat{\mathcal{E}}^\dagger\hat{\mathcal{E}})(t)]^{n_t} \tau_d^{n_t}}{n_t!} \exp[-\eta(\hat{\mathcal{E}}^\dagger\hat{\mathcal{E}})(t)\tau_d] : \right\rangle\right)^2\right\} \\
&= \left(\frac{Ge}{\tau_d}\right)^2 \left[\eta^2 \langle \hat{\mathcal{E}}^\dagger(t)\hat{\mathcal{E}}^\dagger(t)\hat{\mathcal{E}}(t)\hat{\mathcal{E}}(t)\rangle \tau_d^2 + \eta \langle \hat{\mathcal{E}}^\dagger(t)\hat{\mathcal{E}}(t)\rangle \tau_d - \eta^2 \langle \hat{\mathcal{E}}^\dagger(t)\hat{\mathcal{E}}(t)\rangle^2 \tau_d^2\right].
\end{aligned} \tag{9.148}$$

After substituting for the field operator from (9.145) and taking the strong local oscillator limit, we find

$$\begin{aligned}
\overline{i(t)i(t)} - \left(\overline{i(t)}\right)^2 &= (Ge)^2 \Big\{\eta^2 |\mathcal{E}_{\mathrm{lo}}|^4 + \eta^2 |\mathcal{E}_{\mathrm{lo}}|^2 \zeta(2\kappa) \big[4\langle \Delta\tilde{a}^\dagger(t)\Delta\tilde{a}(t)\rangle \\
&\quad + e^{-2i\theta} \langle \Delta\tilde{a}(t)\Delta\tilde{a}(t)\rangle + e^{2i\theta} \langle \Delta\tilde{a}^\dagger(t)\Delta\tilde{a}^\dagger(t)\rangle\big] \\
&\quad + \eta |\mathcal{E}_{\mathrm{lo}}|^2 \tau_d^{-1} - \eta^2 \big[|\mathcal{E}_{\mathrm{lo}}|^4 + 2|\mathcal{E}_{\mathrm{lo}}|^2 \zeta(2\kappa) \langle \Delta\tilde{a}^\dagger(t)\Delta\tilde{a}(t)\rangle\big]\Big\} \\
&= (Ge)^2 \eta^2 |\mathcal{E}_{\mathrm{lo}}|^2 \zeta(2\kappa) 4\langle : \Delta\hat{A}_\theta(t)\Delta\hat{A}_\theta(t) : \rangle \\
&\quad + (Ge)^2 \eta |\mathcal{E}_{\mathrm{lo}}|^2 \tau_d^{-1};
\end{aligned} \tag{9.149}$$

$\Delta\hat{A}_\theta \equiv \hat{A}_\theta - \langle \hat{A}_\theta \rangle$, where $\hat{A}_\theta$ is defined in (9.26).

For a given local oscillator phase θ, the noise in the photocurrent depends on the field fluctuations described by the quadrature phase operator $\Delta\hat{A}_\theta$; measurements of different quadrature phase amplitudes are selected by varying the angle θ. How, then, do we set the level of squeezing in a quantitative fashion? The point of reference is set by considering what, in classical language, is a source field without any fluctuations. Assume the subharmonic mode is in a coherent state—it might as well be the vacuum state. Then $\langle : \Delta\hat{A}_\theta(t)\Delta\hat{A}_\theta(t) : \rangle$ vanishes, and the photocurrent fluctuates with variance $(Ge)^2\eta|\mathcal{E}_{\mathrm{lo}}|^2\tau_d^{-1}$. This is the *shot noise* arising from the detection of the local oscillator intensity $|\mathcal{E}_{\mathrm{lo}}|^2$. Squeezed light has $\langle : \Delta\hat{A}_\theta(t)\Delta\hat{A}_\theta(t) : \rangle < 0$; it reduces the photocurrent variance below this shot noise (vacuum state) level. The degree of squeezing is therefore defined by the size of the photocurrent variance relative to the shot noise level. Do we simply take the ratio of the two terms in (9.149), though? Probably not, because this ratio depends on τ_d, and the shot noise always dominates in the limit $\tau_d \to 0$. The reason for this is that the photocurrent variance is the integral, over all frequencies, of the power spectrum

$$P_\theta(\omega) = \frac{1}{\pi}\int_0^\infty d\tau \cos\omega\tau \lim_{t\to\infty}\left[\overline{i(t)i(t+\tau)} - \left(\overline{i(t)}\right)^2\right]. \tag{9.150}$$

Thus, the shot noise term in (9.149) corresponds to a noise level in frequency space of $(Ge)^2\eta|\mathcal{E}_{\mathrm{lo}}|^2/2\pi$ per unit bandwidth, multiplied by a bandwidth $2\pi/\tau_d$. The bandwidth is infinite when $\tau_d \to 0$.

To define the spectrum of squeezing we compare the contributions to the photocurrent fluctuations in frequency space. The details can be found in [9.11]. They result in a fairly obvious generalization of (9.149). In the limit $\tau_d \to 0$, the correlation function needed to calculate $P_\theta(\omega)$ is given by

$$\begin{aligned}\overline{i(t)i(t+\tau)} - \left(\overline{i(t)}\right)^2 &= (Ge)^2\eta^2|\mathcal{E}_{\mathrm{lo}}|^2\zeta(2\kappa)4\langle : \Delta\hat{A}_\theta(t)\Delta\hat{A}_\theta(t+\tau) : \rangle \\ &\quad + (Ge)^2\eta|\mathcal{E}_{\mathrm{lo}}|^2\delta(\tau),\end{aligned} \tag{9.151}$$

where the normal and time ordering in $\langle : \Delta\hat{A}_\theta(t)\Delta\hat{A}_\theta(t+\tau) : \rangle$ are both relevant. After taking the Fourier transform (9.150), we then have

$$P_\theta(\omega) = P_{\mathrm{hom}}^\theta(\omega) + P_{\mathrm{shot}}, \tag{9.152}$$

where

$$P_{\mathrm{hom}}^\theta(\omega) \equiv (Ge)^2\eta^2|\mathcal{E}_{\mathrm{lo}}|^2\zeta(2\kappa)\frac{4}{\pi}\int_0^\infty d\tau \cos\omega\tau \lim_{t\to\infty}\langle : \Delta\hat{A}_\theta(t)\Delta\hat{A}_\theta(t+\tau) : \rangle, \tag{9.153a}$$

and

$$P_{\mathrm{shot}}(\omega) \equiv (Ge)^2\eta|\mathcal{E}_{\mathrm{lo}}|^2/2\pi. \tag{9.153b}$$

The *source-field spectrum of squeezing* is defined by

$$\begin{aligned}\bar{S}_\theta(\omega) \equiv \eta\zeta S_\theta(\omega) &\equiv \frac{P_\theta(\omega) - P_{\mathrm{shot}}}{P_{\mathrm{shot}}} \\ &= \eta\zeta(2\kappa)8\int_0^\infty d\tau \cos\omega\tau \lim_{t\to\infty}\langle : \Delta\hat{A}_\theta(t)\Delta\hat{A}_\theta(t+\tau) : \rangle.\end{aligned} \tag{9.154}$$

The *spectrum of photocurrent fluctuations* is given in terms of the source-field spectrum of squeezing by

$$P_\theta(\omega)/P_{\text{shot}} = 1 + \eta\zeta S_\theta(\omega). \tag{9.155}$$

This is the spectrum given by the expressions (9.144a) and (9.144b).

We see here how the operator order in the definition of the spectrum of squeezing is determined by the need to make a connection between the *quantum field* radiated by the source and the *classical stochastic current* that appears as data when a homodyne measurement of the field is made. Photoelectric emission provides the physical connection between the two. For the moment let us leave things there; we will return to the issue of operator order in Sect. 9.3.6.

Note 9.16. Actual squeezing measurements are made with balanced detectors. We have used a single unbalanced detector to simplify the analysis. Balanced homodyne detection is described in the paper by Yuen and Chan [9.22]. It is the optical analog of a well-known microwave technique. The slightly more elaborate scheme is a little more tedious to analyze, but, ignoring technical noise on the local oscillator, leads to the same result for the spectrum of photocurrent fluctuations. It is treated in Sect. 18.2.2.

9.3.4 The Source-Field Spectrum of Squeezing with Unit Efficiency

We have included two efficiencies in our calculation, which scale the source-field spectrum of squeezing and decrease the observed shot noise reduction: the detection efficiency η and the photon flux collection efficiency ζ. These two efficiencies represent two occurrences of the one problem. The reduction in shot noise results from the detection of correlated photons. In a realistic measurement, however, the homodyne detector will generally not detect every correlated photon emitted by the source. Photons are missed at random unless η and ζ are both unity; this results in a breaking up of correlated photon pairs and a smaller shot noise reduction than is available in principle. So far as photon flux collection efficiency is concerned, setting $\zeta = 1$ assumes that the signal added to the local oscillator carries the total flux lost by the subharmonic mode through dissipation—the flux lost by absorption in the crystal as well as the radiated flux carried by the fields (9.120a) and (9.122a) (the photon flux transmitted through the mirrors). A more realistic choice for ζ is $\zeta(2\kappa) = \gamma_{a1}$, or $\zeta(2\kappa) = \gamma_{a2}$, which corresponds to one or other of the radiated fields being detected. Then the *source-field spectrum of squeezing detected at a single cavity output* is

$$\begin{aligned}\bar{S}_\theta(\omega) &= \eta \frac{\gamma_{a\mu}}{\gamma_{a1}+\gamma_{a2}+\gamma_{a\alpha}} S_\theta(\omega) \\ &= \eta\gamma_{a\mu} 8 \int_0^\infty d\tau \cos\omega\tau \lim_{t\to\infty} \langle : \Delta\hat{A}_\theta(t)\Delta\hat{A}_\theta(t+\tau) : \rangle,\end{aligned} \tag{9.156}$$

where $\mu = 1, 2$.

Of course it is not possible, in practice, to recover photons that are absorbed in the crystal. Nevertheless, we can, at least conceptually, construct a measurement scheme that combines all three cavity outputs and, thus, realizes (within the limitations set by η) the full potential of the source for shot noise reduction. It is interesting to see how this is done, since it demonstrates the sense in which the source-field spectrum with unit detection efficiency, $S_\theta(\omega)$, is well defined. In fact, the scheme we now analyze could be extended to treat the detection efficiency in the same way as we treat the loss in the crystal. For simplicity, however, let us just imagine that $\eta = 1$.

The measurement scheme is illustrated in Fig. 9.8. The cavity output fields are detected separately to produce the photocurrents $i_1(t)$, $i_2(t)$, and $i_\alpha(t)$. These currents are then added to arrive at the current $i(t)$. Let us look at the fluctuations in $i(t)$ as a function of the relative strengths of the three local oscillator fields. We specify these relative strengths by the unit vector

$$\boldsymbol{f} \equiv (f_1, f_2, f_\alpha), \qquad f_1^2 + f_2^2 + f_\alpha^2 = 1, \tag{9.157}$$

writing $(\mu = 1, 2, \alpha)$

$$e^{i\phi_t}\sqrt{t}\mathcal{E}_{\rm lo}^\mu = f_\mu e^{-i\omega_C t}|\mathcal{E}_{\rm lo}|e^{i\theta}, \tag{9.158}$$

where $t \ll 1$ and ϕ_t denote the transmission coefficient and phase shift on transmission of the three mirrors used to combine the local oscillator fields with the cavity outputs (see Fig. 9.8).

The correlation functions for the three individual photocurrents $i_\mu(t)$, $i = 1, 2, \alpha$, are given by (9.151) as $(\eta = 1)$

$$\begin{aligned}\overline{i_\mu(t)i_\mu(t+\tau)} - \left(\overline{i_\mu(t)}\right)^2 &= (Ge)^2(f_\mu|\mathcal{E}_{\rm lo}|)^2\gamma_{a\mu}4\langle : \Delta\hat{A}_\theta(t)\Delta\hat{A}_\theta(t+\tau) : \rangle \\ &\quad + (Ge)^2(f_\mu|\mathcal{E}_{\rm lo}|)^2\delta(\tau).\end{aligned} \tag{9.159}$$

On adding the photocurrents, we cannot simply add the correlation functions to get the result for the total current $i(t)$. We do expect the shot noise terms to add in this way, but we must be careful about the homodyne terms, as correlated photoelectron counts might occur at different detectors. Consider, for example, how the derivation of the photocurrent variance (Eqs. 9.145–9.149) is modified. In place of the averages over $n_t^2 = n_t(n_t - 1) + n_t$ and n_t

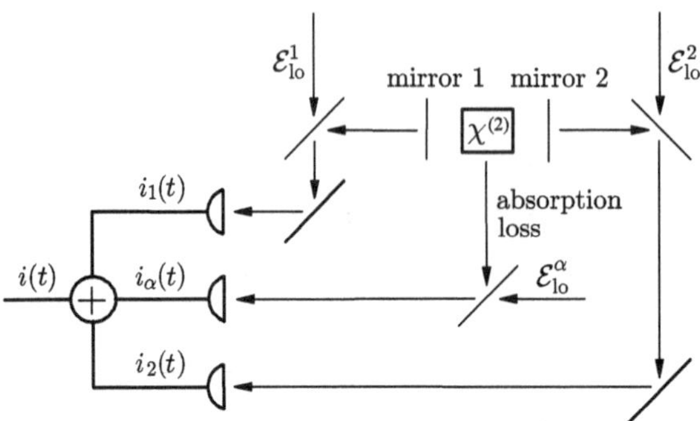

Fig. 9.8. Measurement scheme used to define the source-field spectrum of squeezing with unit efficiency

in (9.148), we now need averages over

$$(n_t + m_t + k_t)^2 = [n_t(n_t - 1) + n_t] + [m_t(m_t - 1) + m_t] + [k_t(k_t - 1) + k_t] \\ + 2n_tm_t + 2m_tk_t + 2k_tn_t$$

and $n_t + m_t + k_t$, where n_t, m_t, and k_t denote the numbers of photoelectric emissions at the three detectors during the interval $t - \tau_d$ to t. The averages are to be taken with respect to the joint photoelectron counting distribution ($\eta = 1$)

$$\begin{aligned} p(n_t, m_t, k_t, t, \tau_d) \\ = \Bigg\langle : \frac{[(\hat{\mathcal{E}}_1^\dagger\hat{\mathcal{E}}_1)(t)]^{n_t}\tau_d^{n_t}}{n_t!} \exp[-(\hat{\mathcal{E}}_1^\dagger\hat{\mathcal{E}}_1)(t)\tau_d] \frac{[(\hat{\mathcal{E}}_2^\dagger\hat{\mathcal{E}}_2)(t)]^{m_t}\tau_d^{m_t}}{m_t!} \\ \times \exp[-(\hat{\mathcal{E}}_2^\dagger\hat{\mathcal{E}}_2)(t)\tau_d] \frac{[(\hat{\mathcal{E}}_\alpha^\dagger\hat{\mathcal{E}}_\alpha)(t)]^{k_t}\tau_d^{k_t}}{k_t!} \exp[-(\hat{\mathcal{E}}_\alpha^\dagger\hat{\mathcal{E}}_\alpha)(t)\tau_d] : \Bigg\rangle, \end{aligned} \quad (9.160)$$

where ($\mu = 1, 2, \alpha$)

$$\hat{\mathcal{E}}_\mu(t) \equiv e^{-i\omega_C t}\big[\mathcal{E}_{\rm lo}^\mu + \sqrt{\gamma_{a\mu}}\Delta\tilde{a}(t')\big]. \quad (9.161)$$

The calculation is performed in much the same way as before. The main new element is the need to compute averages of the products n_tm_t, m_tk_t, and k_tn_t—these averages do not factorize; they account for the correlations between photoelectron counts at different detectors. Omitting the details, the final result for the photocurrent correlation function is ($\eta = 1$)

$$\begin{aligned} \overline{i(t)i(t+\tau)} - \left(\overline{i(t)}\right)^2 = (Ge)^2|\mathcal{E}_{\rm lo}|^2\big(f_1\sqrt{\gamma_{a1}} + f_2\sqrt{\gamma_{a2}} + f_\alpha\sqrt{\gamma_{a\alpha}}\big)^2 \\ \times 4\langle : \Delta\hat{A}_\theta(t)\Delta\hat{A}_\theta(t+\tau) : \rangle + (Ge)^2|\mathcal{E}_{\rm lo}|^2\delta(\tau). \end{aligned} \quad (9.162)$$

From this result, the source-field spectrum of squeezing for the three detector arrangement, with $\eta = 1$, is

$$\bar{S}_\theta(\omega) = (\boldsymbol{f}\cdot\boldsymbol{g})^2(2\kappa)8\int_0^\infty d\tau \cos\omega\tau \lim_{t\to\infty}\langle : \Delta\hat{A}_\theta(t)\Delta\hat{A}_\theta(t+\tau) : \rangle, \tag{9.163}$$

where $\boldsymbol{g} \equiv \big(\sqrt{\gamma_{a1}/2\kappa}, \sqrt{\gamma_{a2}/2\kappa}, \sqrt{\gamma_{a\alpha}/2\kappa}\big)$.

On setting $\boldsymbol{f} = (1,0,0)$ or $\boldsymbol{f} = (0,1,0)$ we recover the definition (9.156) of the source-field spectrum of squeezing at a single cavity output. The *source-field spectrum of squeezing with unit efficiency* is recovered when the relative strengths of the local oscillators are chosen to maximize $|\bar{S}_\theta(\omega)|$. This is achieved with the choice

$$\boldsymbol{f} = \boldsymbol{f}_{\rm unit} \equiv \boldsymbol{g} \equiv \Big(\sqrt{\gamma_{a1}/2\kappa}, \sqrt{\gamma_{a2}/2\kappa}, \sqrt{\gamma_{a\alpha}/2\kappa}\Big). \tag{9.164}$$

In this case $\boldsymbol{f}\cdot\boldsymbol{g} = 1$ and (9.163) corresponds to (9.154) taken with $\eta = \zeta = 1$.

9.3.5 Free-Field Contributions

We are not finished yet with variations on the spectrum of squeezing. In general the source-field spectrum of squeezing, including collection and detection efficiency loss, is not the spectrum observed in an experiment. It is apparent, from (9.120a) and (9.122a) for example, that we have been a little glib in the way we added the local oscillator and source fields when writing (9.145) and (9.161). For the detection of a single cavity output, we should really be writing ($\mu = 1,2$)

$$\hat{\mathcal{E}}_\mu(t) \equiv e^{-i\omega_C t}\big[\hat{\mathcal{E}}_{\rm lo} + \sqrt{c/2L'}\tilde{r}_{a\mu f}(t') + \sqrt{\gamma_{a\mu}}\Delta\tilde{a}(t')\big], \tag{9.165}$$

with the free-field term included; thus, we still have to account for free-field contributions. Formally, the free-field term is needed to preserve the commutation relations. We have omitted it for simplicity because often—leaving the formal niceties aside—the free-field contributes nothing to the spectrum of squeezing. This is not always the case, though, and now we must understand when the free field contributes and when it does not. We must also determine what the free field contributes in those cases where it cannot be ignored.

Now, including the free-field term, the full *spectrum of squeezing* is defined by ($\mu = 1,2$)

$$\begin{aligned}\bar{S}_\theta(\omega) \equiv \eta 8\int_0^\infty d\tau \cos\omega\tau \lim_{t\to\infty}\Big\langle : \big[\sqrt{c/2L'}\hat{R}^\theta_{a\mu f}(t) + \sqrt{\gamma_{a\mu}}\Delta\hat{A}_\theta(t)\big] \\ \times \big[\sqrt{c/2L'}\hat{R}^\theta_{a\mu f}(t+\tau) + \sqrt{\gamma_{a\mu}}\Delta\hat{A}_\theta(t+\tau)\big] : \Big\rangle,\end{aligned} \tag{9.166}$$

where

$$\hat{R}^{\theta}_{a\mu f} \equiv \tfrac{1}{2}(\tilde{r}_{a\mu f}e^{-i\theta} + \tilde{r}^{\dagger}_{a\mu f}e^{i\theta}). \tag{9.167}$$

It is assumed here that the free field has zero mean, or if there is a nonzero mean, that it has been absorbed by adding it into the local oscillator amplitude. Twelve additional terms now appear on the right-hand side of (9.156). They introduce correlation functions, written in normal-ordered, time-ordered (for $\tau \geq 0$) form,

$$\left.\begin{aligned}
&\langle \tilde{r}^{\dagger}_{a\mu f}(t)\tilde{r}_{a\mu f}(t+\tau)\rangle,\ \langle \tilde{r}^{\dagger}_{a\mu f}(t+\tau)\tilde{r}_{a\mu f}(t)\rangle,\\
&\langle \tilde{r}^{\dagger}_{a\mu f}(t)\tilde{r}^{\dagger}_{a\mu f}(t+\tau)\rangle,\ \langle \tilde{r}_{a\mu f}(t+\tau)\tilde{r}_{a\mu f}(t)\rangle,\\
&\langle \tilde{r}^{\dagger}_{a\mu f}(t)\Delta\tilde{a}(t+\tau)\rangle,\quad \langle \Delta\tilde{a}^{\dagger}(t+\tau)\tilde{r}_{a\mu f}(t)\rangle,\\
&\langle \tilde{r}^{\dagger}_{a\mu f}(t)\Delta\tilde{a}^{\dagger}(t+\tau)\rangle,\ \langle \Delta\tilde{a}(t+\tau)\tilde{r}_{a\mu f}(t)\rangle,\\
&\langle \Delta\tilde{a}^{\dagger}(t)\tilde{r}_{a\mu f}(t+\tau)\rangle,\ \langle \tilde{r}^{\dagger}_{a\mu f}(t+\tau)\Delta\tilde{a}(t)\rangle,\\
&\langle \Delta\tilde{a}^{\dagger}(t)\tilde{r}^{\dagger}_{a\mu f}(t+\tau)\rangle,\quad \langle \tilde{r}_{a\mu f}(t+\tau)\Delta\tilde{a}(t)\rangle.
\end{aligned}\right\} \tag{9.168}$$

If the reservoir that carries the free field is in the vacuum state, all of these terms vanish and the spectrum of squeezing, $\bar{\mathcal{S}}_{\theta}(\omega)$, reduces to the source-field spectrum of squeezing $\bar{S}_{\theta}(\omega)$. The vanishing of the first ten correlation functions is clearly guaranteed by the normal ordering and time ordering; this ordering of the operators results in expressions with $\tilde{r}_{a\mu f}(t)$ and $\tilde{r}^{\dagger}_{a\mu f}(t)$ acting to the right and left, respectively, on the vacuum state. The last two correlation functions are seen to vanish by invoking (9.124b). In fact, these correlation functions are zero, for $\tau \geq 0$, even if the reservoir state is not the vacuum state. The result follows because $\Delta\tilde{a}(t)$ and $\Delta\tilde{a}^{\dagger}(t)$ cannot depend on free-field operators evaluated at *later* times.

In summary, then, in our model, free-field contributions vanish if $\bar{n}_{a\mu} = 0$. They do not vanish when $\bar{n}_{a\mu} \neq 0$. In the latter case they can be calculated using the method of Sect. 7.3.3. The explicit calculation is left as an exercise.

Exercise 9.10. When $\bar{n}_{a\mu} \neq 0$, correlations between the free field and source field can be expressed in terms of source-field correlation functions using (9.124a) and (9.124b). Use the results of Sect. 10.2.2 to derive the required source-field correlation functions; hence show that the *spectra of squeezing for the degenerate parametric oscillator below threshold* are given by ($\mu = 1, 2$)

$$\begin{aligned}
\bar{\mathcal{S}}_X(\omega) = \eta\Bigg[&\frac{\omega_C+\omega}{\omega_C}\bar{n}(\omega_C+\omega, T_\mu) + \frac{\omega_C-\omega}{\omega_C}\bar{n}(\omega_C-\omega, T_\mu)\\
&+\left(\frac{2\bar{n}+\lambda}{1-\lambda} - 2\bar{n}_{a\mu}\right)\frac{2\gamma_{a\mu}\kappa(1-\lambda)}{[\kappa(1-\lambda)]^2+\omega^2}\Bigg],
\end{aligned} \tag{9.169a}$$

and

$$\bar{S}_Y(\omega) = \eta\left[\frac{\omega_C+\omega}{\omega_C}\bar{n}(\omega_C+\omega,T_\mu) + \frac{\omega_C-\omega}{\omega_C}\bar{n}(\omega_C-\omega,T_\mu)\right.$$
$$\left. + \left(\frac{2\bar{n}-\lambda}{1+\lambda} - 2\bar{n}_{a\mu}\right)\frac{2\gamma_{a\mu}\kappa(1+\lambda)}{[\kappa(1+\lambda)]^2+\omega^2}\right], \tag{9.169b}$$

where $\bar{n}(\omega_C+\omega,T_\mu)$ and $\bar{n}(\omega_C-\omega,T_\mu)$ are mean photon numbers evaluated in the thermal states (9.92a). Note that we recover the spectra of squeezing quoted in Chap. 10 (Eqs. 10.61a and 10.61b) by setting $\eta=1$, $\gamma_{a\mu}=2\kappa$, and $\bar{n}_{a\mu}=\bar{n}(\omega_C+\omega,T_\mu)=\bar{n}(\omega_C-\omega,T_\mu)=0$. Under these conditions it is still possible to keep a nonzero value of $\bar{n}=\bar{n}_{a\alpha}$; though in doing this, strictly we should read $\gamma_{a\mu}=2\kappa$ as $\gamma_{a\mu}\approx 2\kappa$, since if $\bar{n}_{a\mu}\neq 0$, it is permitted to have $\gamma_{a\alpha}\ll\gamma_{a\mu}$ but not $\gamma_{a\alpha}=0$.

9.3.6 Vacuum Fluctuations

The definition we have given for the spectrum of squeezing is based upon homodyne detection of the squeezed light. By taking this operational approach we gain a clear picture of what the spectrum of squeezing means in terms of photocurrent fluctuations. As the photocurrent is a classical quantity, we can always conjure up a picture of its fluctuations. It is tempting to extend the picture to the field, to regard the fluctuating current to be a direct "mapping" (measurement) of fluctuations in the quantized electromagnetic field. At this point we must exercise some caution. Certainly we can visualize the field if it carries classical (thermal, for example) fluctuations; we have already observed that the spectrum of squeezing has a natural interpretation in terms of field fluctuations in this case (see Note 9.14). Can we construct a mental picture of the *nonclassical* fluctuations of the electromagnetic field though—one to match our picture of what the photocurrent is doing? If we can, on what basis is the picture to be constructed? What is the mathematical correspondence between the fluctuations of the photocurrent and the "visualized" fluctuations of the field?

We will now discuss these questions using the three detector measurement scheme used to define the spectrum (9.163) (Fig. 9.8), but extending our previous treatement to include the free-field term introduced in (9.166). Combining the two expressions, equations (9.163) and (9.166), we have

$$\bar{S}_\theta(\omega)$$
$$= \eta 8\int_0^\infty d\tau\cos\omega\tau\lim_{t\to\infty}\left\langle :\left[\sqrt{c/2L'}\sum_\mu f_\mu\hat{R}^\theta_{a\mu f}(t) + \sqrt{2\kappa}\boldsymbol{f}\cdot\boldsymbol{g}\Delta\hat{A}_\theta(t)\right]\right.$$
$$\left.\times\left[\sqrt{c/2L'}\sum_\nu f_\nu\hat{R}^\theta_{a\nu f}(t+\tau) + \sqrt{2\kappa}\boldsymbol{f}\cdot\boldsymbol{g}\Delta\hat{A}_\theta(t+\tau)\right]:\right\rangle, \tag{9.170}$$

where $\boldsymbol{f}$ and $\boldsymbol{g}$ are defined by (9.157) and (9.164), respectively, and the f_μ, $\mu=1,2,\alpha$, define the local oscillator strengths $\mathcal{E}^\mu_{\rm lo}$ (Eq. 9.158). We consider an ideal measurement, with $\eta=1$ and $\boldsymbol{f}\cdot\boldsymbol{g}=\boldsymbol{f}_{\rm unit}\cdot\boldsymbol{g}=1$. Then the *spectrum of squeezing with unit efficiency*, including the free-field term, is given by

$$\mathcal{S}_\theta(\omega) = 8\int_0^\infty d\tau \cos\omega\tau \lim_{t\to\infty}\Big\langle :\Big(\sqrt{c/2L'}\hat{F}_\theta(t) + \sqrt{2\kappa}\Delta\hat{A}_\theta(t)\Big) \times \Big(\sqrt{c/2L'}\hat{F}_\theta(t+\tau) + \sqrt{2\kappa}\Delta\hat{A}_\theta(t+\tau)\Big) : \Big\rangle, \tag{9.171}$$

where

$$\hat{F}_\theta \equiv \tfrac{1}{2}\big(\hat{\tilde{f}} e^{-i\theta} + \hat{\tilde{f}}^\dagger e^{i\theta}\big), \tag{9.172}$$

with combined free field

$$\hat{\tilde{f}} \equiv \big(\sqrt{\gamma_{a1}}\tilde{r}_{a1f} + \sqrt{\gamma_{a2}}\tilde{r}_{a2f} + \sqrt{\gamma_{a\alpha}}\,\tilde{r}_{a\alpha f}\big)/\sqrt{2\kappa}; \tag{9.173a}$$

alternatively, using (9.128), the combined free field is expressed in the form

$$\hat{\tilde{f}} = \sum_\omega \sqrt{\frac{\omega_C + \omega}{\omega_C}}\,\hat{f}_\omega e^{-i\omega\tau}, \tag{9.173b}$$

where $\hat{f}_\omega$ is defined in (9.131).

As we observed in Sect. 9.3.5, when the reservoirs—the carriers of the free fields—are in the vacuum state, (9.171) reduces, by virtue of the normal and time ordering, to the source-field spectrum of squeezing. This is the spectrum computed in the work of Walls and coworkers [9.23, 9.24, 9.25]. We are now going to relate this spectrum to one defined without normal and time ordering, the spectrum used widely in the work of others [9.18, 9.22, 9.26, 9.27].

The question posed in the introduction can be answered in the affirmative: yes, we can construct a mental picture, a visualization, of the fluctuations of the quantized electromagnetic field. The formal basis for the construction is provided by the quantum–classical correspondence, by adopting one of the phase-space representations we met in Chaps. 3 and 4. In Chap. 10 we will develop the phase-space analysis of the degenerate parametric oscillator in all its details. For the moment it is enough to anticipate just one or two of the results we meet there.

Equation 9.171 states that the spectrum of squeezing is the Fourier transform of the normal-ordered, time-ordered correlation function of a quadrature phase operator of the quantized cavity output field, the field $\sqrt{c/2L'}\hat{\tilde{f}} + \sqrt{2\kappa}\Delta\tilde{a}$. Since it is the Glauber–Sudarshan P representation that evaluates normal-ordered, time-ordered correlation functions as "classical" integrals, the Fokker–Planck equation in this representation (and its associated stochastic differential equation) should provide the desired visualization of the fluctuations of the field. As shown in Chap. 10, however, there is a problem here. The Fokker–Planck equation in the P representation (Eqs. 10.51 and 10.74 with $\sigma = 1$) does not possess positive semidefinite diffusion. The P distribution must therefore be a generalized function (Sect. 3.1.3 and Exercise 9.1), and as such, it seems that it cannot provide any sort of classical visualization of the field fluctuations. There is, nevertheless, a ready solution to this problem. The Fokker–Planck equations in both the Wigner and Q representations (Eqs. 10.51 and 10.74 with $\sigma = 0$ and $\sigma = -1$) do have positive semidefinite diffusion. Perhaps, then, we can base our visualization of the field fluctuations on one of these representations. If we aim to do this, we must first change the operator ordering in the expression for the spectrum of squeezing so that it corresponds to the ordering appropriate to the chosen representation (Sects. 4.1.1 and 4.1.4). Thus, let us reorder the operators in (9.171) to clarify the connection between the spectrum of squeezing and the Wigner stochastic representation of the electromagnetic field.

Note 9.17. The statements made here might convey a somewhat oversimplified view of the situation. There is, in general, no guarantee that either the Wigner or the Q representation will provide a stochastic visualization of the field fluctuations; indeed, for the degenerate parametric oscillator they do so only in the small-noise limit (Sect. 10.1.2); moreover, as we will see shortly (Chaps. 11 and 12), the P representation can be generalized so that it too provides a stochastic visualization of the field fluctuations. The broader message then is that "classical" visualizations of the fluctuations of a quantum field are not unique. Different representations provide different visualizations, the variety being a function largely of the ingenuity of those inventing the pictures. Of course, there is always the deeper question of just how "classical," or physically well-defined, a particular visualization might be—i.e., how well does it fit into a general theory of measurements?

The normal-ordered, time-ordered averages that appear in the expression for the spectrum of squeezing may be written as as averages without normal and time ordering by introducing appropriate commutators. There are four operator products to consider:

$$\begin{aligned}&\langle{:}\,\hat{F}_\theta(t)\hat{F}_\theta(t+\tau){:}\rangle\\&=\langle\hat{F}_\theta(t)\hat{F}_\theta(t+\tau)\rangle+\tfrac{1}{4}\big\langle\big[\hat{\tilde{f}}(t+\tau)e^{-2i\theta}+\hat{\tilde{f}}^\dagger(t+\tau),\hat{\tilde{f}}(t)\big]\big\rangle,\end{aligned}\tag{9.174a}$$

$$\begin{aligned}&\langle{:}\,\hat{F}_\theta(t)\Delta\hat{A}_\theta(t+\tau){:}\rangle\\&=\langle\hat{F}_\theta(t)\Delta\hat{A}_\theta(t+\tau)\rangle+\tfrac{1}{4}\big\langle\big[\Delta\tilde{a}(t+\tau)e^{-2i\theta}+\Delta\tilde{a}^\dagger(t+\tau),\hat{\tilde{f}}(t)\big]\big\rangle\rangle,\end{aligned}\tag{9.174b}$$

$$\begin{aligned}&\langle{:}\,\Delta\hat{A}_\theta(t)\hat{F}_\theta(t+\tau){:}\rangle\\&=\langle\Delta\hat{A}_\theta(t)\hat{F}_\theta(t+\tau)\rangle+\tfrac{1}{4}\big\langle\big[\hat{\tilde{f}}(t+\tau)e^{-2i\theta}+\hat{\tilde{f}}^\dagger(t+\tau),\Delta\tilde{a}(t)\big]\big\rangle,\end{aligned}\tag{9.174c}$$

$$\begin{aligned}&\langle{:}\,\Delta\hat{A}_\theta(t)\Delta\hat{A}_\theta(t+\tau){:}\rangle\\&=\langle\Delta\hat{A}_\theta(t)\Delta\hat{A}_\theta(t+\tau)\rangle+\tfrac{1}{4}\big\langle\big[\Delta\tilde{a}(t+\tau)e^{-2i\theta}+\Delta\tilde{a}^\dagger(t+\tau),\Delta\tilde{a}(t)\big]\big\rangle.\end{aligned}\tag{9.174d}$$

The free-field commutator that appears on the right-hand side of (9.174a) can be evaluated using (9.173b) and the boson commutation relations for the

Fourier components $\hat{f}_\omega$ (Eq. 9.132); thus, with the help of the quasimonochromatic condition $\omega \ll \omega_C$, we may write

$$\begin{aligned}\big\langle\big[\hat{\tilde{f}}(t+\tau)e^{-2i\theta}+\hat{\tilde{f}}^\dagger(t+\tau),\hat{\tilde{f}}(t)\big]\big\rangle &= \sum_\omega \frac{\omega_C+\omega}{\omega_C}\langle[f_\omega^\dagger, f_\omega]\rangle e^{i\omega\tau}\\ &= -(L'/\pi c)\int_{-\infty}^{\infty} d\omega \frac{\omega_C+\omega}{\omega_C}e^{i\omega\tau}\\ &= -(2L'/c)\delta(\tau). \end{aligned} \tag{9.175}$$

The next two relations, (9.174b) and (9.174c), call for expectations of commutators between source-field operators and free-field operators. They may be expressed in terms of source-field commutator expectations alone using (9.173a) and the correlation functions (9.124a) and (9.124b). Specifically, we have

$$\begin{aligned}&\big\langle\big[\Delta\tilde{a}(t+\tau)e^{-2i\theta}+\Delta\tilde{a}^\dagger(t+\tau),\hat{\tilde{f}}(t)\big]\big\rangle\\ &= \begin{cases} -\sqrt{2\kappa}\sqrt{2L'/c}\Big[\big\langle\big[\Delta\tilde{a}(t+\tau)e^{-2i\theta}+\Delta\tilde{a}^\dagger(t+\tau),\Delta\tilde{a}(t)\big]\big\rangle\Big] & \tau>0,\\ -\frac{1}{2}\sqrt{2\kappa}\sqrt{2L'/c}\Big[\big\langle\big[\Delta\tilde{a}(t+\tau)e^{-2i\theta}+\Delta\tilde{a}^\dagger(t+\tau),\Delta\tilde{a}(t)\big]\big\rangle\Big] & \tau=0,\end{cases}\end{aligned} \tag{9.176a}$$

and

$$\begin{aligned}&\big\langle\big[\hat{\tilde{f}}(t+\tau)e^{-2i\theta}+\hat{\tilde{f}}^\dagger(t+\tau),\Delta\tilde{a}(t)\big]\big\rangle\\ &= \begin{cases} 0 & \tau>0,\\ -\frac{1}{2}\sqrt{2\kappa}\sqrt{2L'/c}\Big[\big\langle\big[\Delta\tilde{a}(t+\tau)e^{-2i\theta}+\Delta\tilde{a}^\dagger(t+\tau),\Delta\tilde{a}(t)\big]\big\rangle\Big] & \tau=0.\end{cases}\end{aligned} \tag{9.176b}$$

We notice now that the commutator appearing on the right-hand sides of (9.176a) and (9.176b) appears also on the right-hand side of (9.174d), but there with the opposite sign. Hence, on substituting (9.174a)–(9.174d) into (9.171) and using (9.175), and (9.176a) and (9.176b), a cancellation of this source-field commutator occurs, which gives

$$\begin{aligned}S_\theta(\omega)+1 = 8\int_0^\infty d\tau\cos\omega\tau \lim_{t\to\infty}\Big\langle&\big[\sqrt{c/2L'}\hat{F}_\theta(t)+\sqrt{2\kappa}\Delta\hat{A}_\theta(t)\big]\\ &\times\big[\sqrt{c/2L'}\hat{F}_\theta(t+\tau)+\sqrt{2\kappa}\Delta\hat{A}_\theta(t+\tau)\big]\Big\rangle.\end{aligned} \tag{9.177}$$

The free-field commutator (9.175) contributes the $+1$ on the left-hand side of this expression, the Fourier transform of $(c/2L')\langle[\hat{\tilde{f}}(t),\hat{\tilde{f}}^\dagger(t+\tau)]\rangle = \delta(\tau)$. Apparently, the $+1$ arises from free-field vacuum fluctuations.

If we now compare (9.177) with (9.155) we see that the vacuum fluctuation contribution in (9.177) stands in for the photocurrent shot noise in (9.155).

Since the latter expression is explicitly normal ordered, vacuum fluctuations contribute nothing to it. It explicitly includes the shot noise, however, which arises from the self-correlation of photopulses making up the photocurrent (Fig. 9.7); correlations *between* photopulses cause $\mathcal{S}_\theta(\omega)+1$ to deviate from the shot noise level [9.21]. Now, on the basis of (9.177), we are permitted an alternative picture, one in which all fluctuations—those expressed as photocurrent shot noise included—are considered to reside in the electromagnetic field prior to its detection.

We can take the picture one step further by noting that the expectation of the total field commutator—source field plus free field—vanishes:

$$\Big\langle\big[\hat{F}_\theta(t)+\sqrt{2\kappa}\sqrt{2L'/c}\Delta\hat{A}_\theta(t),\hat{F}_\theta(t+\tau)+\sqrt{2\kappa}\sqrt{2L'/c}\Delta\hat{A}_\theta(t+\tau)\big]\Big\rangle=0; \tag{9.178}$$

in effect, the total field behaves as a free field (in an altered state). With the aid of this result, we can re-express (9.177) as the Fourier transform

$$\begin{aligned}\mathcal{S}_\theta(\omega)+1=4(\pi c/L')\frac{1}{2\pi}\int_{-\infty}^{\infty}d\tau e^{i\omega\tau}\lim_{t\to\infty}\Big\langle\big[\hat{F}_\theta(t)+\sqrt{2\kappa}\sqrt{2L'/c}\Delta\hat{A}_\theta(t)\big]\\ \times\big[\hat{F}_\theta(t+\tau)+\sqrt{2\kappa}\sqrt{2L'/c}\Delta\hat{A}_\theta(t+\tau)\big]\Big\rangle.\end{aligned} \tag{9.179}$$

Then, since the averages appearing on the right-hand side of (9.179) can be calculated as phase-space averages in the Wigner representation, it follows that

$$\begin{aligned}&S_\theta(\omega)+1\\&=4\times(\pi c/L')\times\begin{pmatrix}\text{the variance of quadrature phase amplitude}\\ \text{fluctuations per unit bandwidth (in photon number units)}\\ \text{in the Wigner stochastic representation of the field.}\end{pmatrix}\end{aligned} \tag{9.180}$$

The factor of 4 scales the single-mode quadrature variance of $1/4$ to unity, while the factor $\pi c/L'$ is the mode spacing in frequency space. The latter may be omitted with the words "per unit bandwidth" replaced by "per mode."

Note 9.18. The Wigner representation actually gives correlation functions in symmetrized time order (Eq. 4.124). Strictly, then, the spectrum of quadrature phase amplitude fluctuations calculated in this representation is the average of (9.179) and the same expression with reverse operator order. This issue of time order is unimportant, however, because of the vanishing commutator expectation (9.178).

Exercise 9.11. Use (9.173b) and the correlation functions (9.124) to derive the commutator expectation (9.178).

Equation 9.180 provides the formal basis for the visualization of squeezed field fluctuations most commonly encountered in the squeezing community. In the Wigner representation, every mode of the field has a stochastic amplitude which even in the vacuum state shows a nonzero level of noise. Squeezing deamplifies—and amplifies—this vacuum noise level. In this picture, squeezed light is not emitted from the source as energy deposited against a background of nothing. Rather, the source sits embedded within a background of vacuum fluctuations, which it takes in, transforms, and reemits, with the net result being a modification of the background vacuum fluctuations. In this *transformation*, rather than emission imagery, the vacuum fluctuations are like a processed substance. One must be careful not to confuse the end of the processing with some point part of the way through, though. Thus it is the transformed fluctuations outside the cavity and not the fluctuations inside the cavity that are of concern in a standard squeezing measurement. Even though, aside from its scaling, the right-hand side of (9.179) has the same form as the Fourier integrals used to compute $(\Delta X)^2(\omega)$ and $(\Delta Y)^2(\omega)$ in (9.142a) and (9.142b), the two types of spectra are completely different. Part of the difference comes from the fact that the cavity acts as a filter which suppresses vacuum fluctuations (and thermal fluctuations) at frequencies outside the cavity linewidth (see the discussion above Note 9.13). This is not the complete story, though. The spectrum of field fluctuations outside the cavity, (9.179), is not simply the sum of a vacuum noise level (the $+1$), a thermal noise level for the free field, and a noise level for real photons emitted through the cavity mirrors. The squeezing comes about because there are correlations between the fluctuations of the free-field modes driving the cavity and the fluctuating intracavity source-field amplitude. The presence of these correlations is clear from the operator expressions for the source-field Fourier components (9.140a) and (9.140b); in these the free-field operators $\hat{f}_\omega$ and $\hat{f}^\dagger_{-\omega}$ explicitly appear. The Wigner representation captures the same correlations (for symmetric operator ordering) in a c-number relationship between the source field and the free field.

9.3.7 Squeezing in the Wigner Representation: A Comment on Interpretation

In order to see exactly what the Wigner representation has to say about the spectrum of squeezing, let us restate the principal results from the last few sections with the field operators replaced by phase-space variables. For simplicity, we take $\gamma_{a1} = \gamma_{a\alpha} = 0$; thus, we consider a single-ended cavity with output field traveling to the right:

$$\hat{\mathcal{E}}_{\rightarrow}(z,t) = \sqrt{c/2L'}r_{a2f}(t') + \sqrt{2\kappa}a(t'), \tag{9.181}$$

$z > \ell + d$ and $2\kappa = \gamma_{a2}$. We also ignore the frequency dependence of the thermal photon number, setting $\bar{n}_{a1}(\omega_C + \omega, T_1) = \bar{n}(\omega_C, T_1) \equiv \bar{n}$.

We now follow the development in Sect. 9.3.1. First, expanding the fields as in (9.128) and (9.129), the operator expression (9.181) is replaced by

$$\mathcal{E}_{\rightarrow}(z,t) = \sum_{\omega}\left(\sqrt{c/2L'}f_{\omega} + \sqrt{2\kappa}\alpha_{\omega}\right)e^{-i(\omega_C+\omega)t'}, \tag{9.182}$$

$ct' \equiv ct-(z-\ell-d)$, where $\mathcal{E}_{\rightarrow}(z,t)$, f_{ω}, and α_{ω} are phase-space variables in the Wigner representation (Sects. 4.1.4 and 4.3.1). Then the Fourier amplitudes of the source field are related to those of the free field through phase space versions of the steady-state equations (9.130a) and (9.130b). We have

$$0 = -(\kappa - i\omega)\alpha_{\omega} - ie^{i\psi}\kappa\lambda\alpha^*_{-\omega} - \sqrt{c/2L'}\sqrt{2\kappa}f_{\omega}, \tag{9.183a}$$
$$0 = -(\kappa - i\omega)\alpha^*_{-\omega} + ie^{-i\psi}\kappa\lambda\alpha_{\omega} - \sqrt{c/2L'}\sqrt{2\kappa}f^*_{-\omega}, \tag{9.183b}$$

with the operator expectations (9.133) replaced by

$$\left(\overline{f_{\omega}f_{\omega'}}\right)_W = \left(\overline{f^*_{\omega}f^*_{\omega'}}\right)_W = 0, \tag{9.184a}$$
$$\left(\overline{f^*_{\omega}f_{\omega'}}\right)_W = (\bar{n} + \tfrac{1}{2})\delta_{\omega,\omega'}, \tag{9.184b}$$

where $(\;)_W$ denotes the phase-space average in the Wigner representation; thus, we have used

$$\left(\overline{f^*_{\omega}f_{\omega}}\right)_W = \tfrac{1}{2}(\langle\hat{f}^{\dagger}_{\omega}\hat{f}_{\omega}\rangle + \langle\hat{f}_{\omega}\hat{f}^{\dagger}_{\omega}\rangle) = \bar{n} + \tfrac{1}{2}. \tag{9.185}$$

Finally, solving (9.183a) and (9.183b), and introducing

$$X^{\omega}_f \equiv \tfrac{1}{2}[f_{\omega}e^{-i\frac{1}{2}(\psi-\pi/2)} + f^*_{-\omega}e^{i\frac{1}{2}(\psi-\pi/2)}], \tag{9.186a}$$
$$Y^{\omega}_f \equiv \tfrac{1}{2}[f_{\omega}e^{-i\frac{1}{2}(\psi+\pi/2)} + f^*_{-\omega}e^{i\frac{1}{2}(\psi+\pi/2)}], \tag{9.186b}$$

gives the intracavity field quadrature phase amplitudes [phase space versions of (9.139a) and (9.139b)]

$$X_{\omega} = -\sqrt{\frac{c}{2L'}}\sqrt{2\kappa}\frac{X^{\omega}_f}{\kappa(1-\lambda) - i\omega}, \tag{9.187a}$$

and

$$Y_{\omega} = -\sqrt{\frac{c}{2L'}}\sqrt{2\kappa}\frac{Y^{\omega}_f}{\kappa(1+\lambda) - i\omega}, \tag{9.187b}$$

which are the phase space versions of (9.140a) and (9.140b).

Equations 9.182–9.187 define the visualization of squeezed vacuum fluctuations within the Wigner representation. Vacuum fluctuations enter through the variance of $\frac{1}{2}$ per free-field mode added to the thermal photon number in (9.184b) and (9.185). These fluctuations, together with the thermal fluctuations, drive the parametric oscillator through the terms f_{ω} and $f^*_{-\omega}$ in (9.183a) and (9.183b). The oscillator response is then added to the free field to produce the total field in (9.182). The net result is a transformed spectrum of quadrature phase amplitude fluctuations. The two spectra are given by

$$\begin{aligned}\frac{2L'}{c}\left(\overline{\left|\sqrt{c/2L'}X_f^\omega+\sqrt{2\kappa}X_\omega\right|^2}\right)_W &= \left(\overline{\left|X_f^\omega+2\kappa\frac{X_f^\omega}{\kappa(1-\lambda)-i\omega}\right|^2}\right)_W\\ &= \left|\frac{\kappa(1+\lambda)+i\omega}{\kappa(1-\lambda)-i\omega}\right|^2\overline{|X_f^\omega|^2}\\ &= \frac{1}{4}(2\bar{n}+1)\frac{[\kappa(1+\lambda)]^2+\omega^2}{[\kappa(1-\lambda)]^2+\omega^2},\end{aligned} \tag{9.188a}$$

and

$$\frac{2L'}{c}\left(\overline{\left|\sqrt{c/2L'}Y_f^\omega+\sqrt{2\kappa}Y_\omega\right|^2}\right)_W = \frac{1}{4}(2\bar{n}+1)\frac{[\kappa(1-\lambda)]^2+\omega^2}{[\kappa(1+\lambda)]^2+\omega^2}, \tag{9.188b}$$

which, apart from an overall scale factor, agree with the spectra of photocurrent fluctuations calculated from (9.154) and (9.155), or (9.155) and (9.179); specifically, (9.188a) and (9.188b) are the spectra given by (9.169a) and (9.169b) for the specified conditions of a single-ended cavity.

This phase-space visualization is very appealing and a useful way to think about squeezed light. Some words of caution and qualification are called for, however. First a word of caution against a possible misconception. Squeezed states are usually discussed with reference to a single mode, with the Wigner distribution (9.47b) plotted to illustrate the squeezing. It is tempting to view the spectra (9.188a) and (9.188b) in single-mode terms—as the frequency-dependent variances of a single-mode Wigner function. Strictly, this is not correct. These spectra describe properties of a two-mode Wigner function (other than at $\omega = 0$), since squeezing arises from the correlated fluctuations of the pair of modes with frequencies $\omega_C + \omega$ and $\omega_C - \omega$. This is clear from a number of the above equations. Fundamentally, it goes back to the issue illustrated in Fig. 9.5 and expressed in definitions (9.139) and (9.186) of the frequency-dependent quadrature phase amplitudes: two modes contribute to the modulation of a quadrature phase amplitude at a given frequency ω, those with the symmetrically displaced frequencies $\omega_C \pm \omega$.

Then there is a qualification which is possibly more important. It concerns the identification of a formal variance of field fluctuations, as in (9.188a) and (9.188b), with the variance of a photocurrent produced by photoelectric detection. Should we accept the literal interpretation that the photocurrent fluctuations are simply a proportional response to the visualized field fluctuations, as they appear within the Wigner representation? What is more, should we do so by *fiat*, without even analyzing the photoelectric detection as a physical processes? The answer, surely, is negative on both counts. Suppose, for example, we consider the optical spectrum, and persist with the literal interpretation that the measured quantity is a proportional response to the field fluctuations in the Wigner representation. In this case we solve (9.183a) in the steady state for

$$\alpha_\omega = -\sqrt{\frac{c}{2L'}}\sqrt{2\kappa}\frac{(\kappa - i\omega)f_\omega - ie^{i\psi}\kappa\lambda f^*_{-\omega}}{(\kappa - i\omega)^2 - \kappa^2\lambda^2}, \tag{9.189}$$

and compute the spectrum

$$\begin{aligned}
&\frac{2L'}{c}\left(\overline{\left|\sqrt{c/2L'}f_\omega + \sqrt{2\kappa}\alpha_\omega\right|^2}\right)_W \\
&= \left|1 - 2\kappa\frac{\kappa - i\omega}{(\kappa - i\omega)^2 - \kappa^2\lambda^2}\right|^2 \left(\overline{f^*_\omega f_\omega}\right)_W + 4\kappa\left|\frac{ie^{i\psi}\kappa\lambda}{(\kappa - i\omega)^2 - \kappa^2\lambda^2}\right|^2 \left(\overline{f^*_{-\omega}f_{-\omega}}\right)_W \\
&= (\bar{n} + \tfrac{1}{2})\frac{1}{2}\left\{\frac{[\kappa(1+\lambda)]^2 + \omega^2}{[\kappa(1-\lambda)]^2 + \omega^2} + \frac{[\kappa(1-\lambda)]^2 + \omega^2}{[\kappa(1+\lambda)]^2 + \omega^2}\right\} \\
&= \bar{n} + \tfrac{1}{2} + (2\bar{n} + 1)\left\{\frac{\kappa^2\lambda}{[\kappa(1-\lambda)]^2 + \omega^2} - \frac{\kappa^2\lambda}{[\kappa(1+\lambda)]^2 + \omega^2}\right\}.
\end{aligned} \tag{9.190}$$

The result is of course incorrect, since the $+\frac{1}{2}$, the vacuum fluctuation variance, should not contribute to the optical spectrum. The correct calculation is made in terms of normal-ordered field operators and yields the optical spectrum

$$\begin{aligned}
&\frac{2L'}{c}\left\langle\left(\sqrt{c/2L'}\hat{f}^\dagger_\omega + \sqrt{2\kappa}\hat{\alpha}^\dagger_\omega\right)\left(\sqrt{c/2L'}\hat{f}_\omega + \sqrt{2\kappa}\hat{\alpha}_\omega\right)\right\rangle \\
&= \left|1 - 2\kappa\frac{\kappa - i\omega}{(\kappa - i\omega)^2 - \kappa^2\lambda^2}\right|^2 \langle\hat{f}^\dagger_\omega\hat{f}_\omega\rangle + 4\kappa\left|\frac{ie^{i\psi}\kappa\lambda}{(\kappa - i\omega)^2 - \kappa^2\lambda^2}\right|^2 \langle\hat{f}_{-\omega}\hat{f}^\dagger_{-\omega}\rangle \\
&= \bar{n}\frac{1}{2}\left\{\frac{[\kappa(1+\lambda)]^2 + \omega^2}{[\kappa(1-\lambda)]^2 + \omega^2} + \frac{[\kappa(1-\lambda)]^2 + \omega^2}{[\kappa(1+\lambda)]^2 + \omega^2}\right\} \\
&\quad + \frac{\kappa^2\lambda}{[\kappa(1-\lambda)]^2 + \omega^2} - \frac{\kappa^2\lambda}{[\kappa(1+\lambda)]^2 + \omega^2} \\
&= \bar{n} + (2\bar{n} + 1)\left\{\frac{\kappa^2\lambda}{[\kappa(1-\lambda)]^2 + \omega^2} - \frac{\kappa^2\lambda}{[\kappa(1+\lambda)]^2 + \omega^2}\right\}.
\end{aligned} \tag{9.191}$$

The difference is that where $\langle\hat{f}^\dagger_\omega\hat{f}_\omega\rangle$ and $\langle\hat{f}_{-\omega}\hat{f}^\dagger_{-\omega}\rangle = \langle\hat{f}^\dagger_{-\omega}\hat{f}_{-\omega}\rangle + 1$ appear in the second line of (9.191), a literal interpretation of the Wigner phase-space variables substitutes $\left(\overline{f^*_\omega f_\omega}\right)_W = \langle\hat{f}^\dagger_\omega\hat{f}_\omega\rangle + \frac{1}{2}$ and $\left(\overline{f^*_{-\omega}f_{-\omega}}\right)_W = \langle\hat{f}^\dagger_{-\omega}\hat{f}_{-\omega}\rangle + \frac{1}{2}$. Field *operators* are needed to keep this sort of book-keeping straight when changing from one measurement scheme to another—in this case from homodyne detection to photoelectron counting.

In summary, the qualification is this. The Wigner visualization of squeezed vacuum fluctuations is appealing because it supports an inclination towards naive realism; it provides a picture of realistic field fluctuations, simply transcribed through measurement as the observed photocurrent fluctuations; it presents a simple view of something happening in the field to cause what is seen to happen in the photocurrent. The appealing realism is a deception, though, since it is precisely the connection between the quantum field—with its operators and Hilbert space—and the classical photocurrent—a time series of real numbers—that raises the difficult questions of interpretation and measurement in quantum mechanics. The constructed visualization of squeezed vacuum fluctuations simply ignores these questions, basing its credibility on the fact that it does, indeed, provide a solution to the technical question of calculating a spectrum of squeezing.

For a broader view of things we might note that there is a version of "quantum" electrodynamics known as stochastic electrodynamics (SED) which claims to do away with field quantization altogether by adding to the classical Maxwell field a stochastic (and realistic) vacuum field [9.28, 9.29, 9.30, 9.31, 9.32, 9.33, 9.34]. Adopted literally, the Wigner visualization of squeezed vacuum fluctuations accepts the claim of SED. It is beyond the scope of this book to weigh the evidence for and against this claim. Suffice it to say that it is not considered valid by proponents of conventional quantum electrodynamics, even though it can often have illuminating things to say about the physics of squeezed light [9.35, 9.36].

As a final point, we might take note of our reliance on the linearity of the equations of motion (9.130) and (9.183) when setting up the Wigner visualization of squeezed vacuum fluctuations. As we will see shortly, phase-space treatments of the parametric oscillator meet with serious difficulties in the nonlinear regime. The topic is addressed initially in Chap. 10 and more explicitly in Chap. 12.

Note 9.19. The comments at the beginning of this chapter assert that the Glauber–Sudarshan P representation is special amongst the phase-space representations when we come to consider "classical" visualizations of the electromagnetic field. How is this distinction to be defended in the face of the Wigner visualization of squeezed vacuum fluctuations? There is, in fact, a clear difference between the physical status of "classical" phase-space descriptions in the P and Wigner representations. When such a description exists in the P representation, this means that the classical field concept can be used throughout the entire analysis of some experimental scenario, even in treating the photoelectric detection involved in the measurement process. Thus, the complete scenario can be understood on the basis of classical fields. When the Wigner representation is used, the measurement stage is first analyzed for a *quantized* field, and only after the correct *operator* expression describing the measurement has been obtained is this expression recast in phase-space language. Of course, an operator expression can be recast in any number of representations. The relevant question is whether the classical field concept can, or cannot, be adopted a priori (fundamentally) for the analysis of an entire experimental scenario, including—most importantly in fact—the measurement stage; it is the Glauber–Sudarshan P representation that provides a test of whether it can or it cannot. From a historical perspective, the Glauber–Sudarshan P representation puts the proposal of Bohr, Kramers, and Slater [9.37] to the test [9.38].

Exercise 9.12. Derive an expression for the spectrum of squeezing $\mathcal{S}_\theta(\omega)$ and the spectrum of photocurrent fluctuations $P_\theta(\omega)$ based on the evaluation of correlation functions in the Q representation.

10

The Degenerate Parametric Oscillator II: Phase-Space Analysis in the Small-Noise Limit

We turn now to an analysis of the degenerate parametric oscillator within the phase-space representations. Overall, the program is similar to the one carried out for the laser in Chap. 8, but with one fundamentally new element encountered here. At the start of the exercise, our premise is that the phase-space approach will lead to an acceptable Fokker–Planck equation. This premise will be seen to be incorrect, and two new topics emerge from the problems encountered. First, a new representation is introduced, the so-called positive P representation, formulated as a generalization of the Glauber–Sudarshan P representation to rescue the overall strategy of the quantum–classical correspondence in the small-noise limit; we make use of the positive P representation here, though its systematic development is postponed until Chap. 11. Second, the issue of quantum fluctuations outside the small-noise limit becomes a concern, since for large noise the positive P representation encounters difficulties of its own. These difficulties are explored in some detail in Chap. 12.

10.1 Phase-Space Formalism for the Degenerate Parametric Oscillator

Analysis of the master equation (9.97) is made difficult by the nonlinear coupling between the subharmonic mode and the pump mode. Without this coupling ($g = 0$) the subharmonic mode is described by the master equation for the damped harmonic oscillator (Eq. 1.73) and the pump mode by the master equation for the driven damped harmonic oscillator (Eq. 9.102). Both of these equations may be solved by a number of methods, but there is no known analytic solution to the full equation with the nonlinear coupling. Our aim is to make progress on the basis of a phase-space formulation, by invoking approximations similar to those used for the laser. First, we convert the master equation into phase-space form. We aim to do this for the P, the Q, and the Wigner representations. Let us begin with the calculation within the P representation.

10.1.1 Phase-Space Equation of Motion in the P Representation

The definitions and results of Sect. 3.2.1 generalize to two modes in an obvious manner. The two-mode characteristic function is denoted $\chi_N(z, z^*, w, w^*)$ and the corresponding distribution as $P(\alpha, \alpha^*, \beta, \beta^*)$. We have previously derived the phase-space equation of motion for the harmonic oscillator, and most of the terms that appear in the master equation for the degenerate parametric oscillator (Eq. 9.97) were encountered there. Thus, by referring to the former equation of motion (Eq. 3.47), we can immediately write

$$\begin{aligned}\frac{\partial P}{\partial t} = &\left[(\kappa + i\omega_C)\frac{\partial}{\partial \alpha}\alpha + (\kappa - i\omega_C)\frac{\partial}{\partial \alpha^*}\alpha^* + 2\kappa\bar{n}\frac{\partial^2}{\partial \alpha \partial \alpha^*}\right.\\ &\left. +(\kappa_p + i2\omega_C)\frac{\partial}{\partial \beta}\beta + (\kappa_p - i2\omega_C)\frac{\partial}{\partial \beta^*}\beta^* + 2\kappa_p\bar{n}_p\frac{\partial^2}{\partial \beta \partial \beta^*}\right]P\\ &+\left(\frac{\partial P}{\partial t}\right)_{ab} + \left(\frac{\partial P}{\partial t}\right)_{\text{drive}}.\end{aligned} \tag{10.1}$$

There are two new terms, the last two terms on the right-hand side of (10.1), which are contributed, respectively, by the mode-coupling and the pump-mode-driving terms in (9.97). To derive their explicit form, we start from the corresponding expressions in the equation of motion for the characteristic function χ_N:

$$\begin{aligned}\left(\frac{\partial \chi_N}{\partial t}\right)_{ab} = &\operatorname{tr}\{[(g/2)(a^{\dagger 2}b\rho - \rho a^{\dagger 2}b) - (g/2)(a^2b^\dagger\rho - \rho a^2 b^\dagger)]\\ &\times e^{iz^*a^\dagger}e^{iza}e^{iw^*b^\dagger}e^{iwb}\},\end{aligned} \tag{10.2}$$

and

$$\begin{aligned}\left(\frac{\partial \chi_N}{\partial t}\right)_{\text{drive}} = &\operatorname{tr}\{[-i\bar{\mathcal{E}}_0 e^{-i2\omega_C t}(b^\dagger\rho - \rho b^\dagger) - i\bar{\mathcal{E}}_0^* e^{i2\omega_C t}(b\rho - \rho b)]\\ &\times e^{iz^*a^\dagger}e^{iza}e^{iw^*b^\dagger}e^{iwb}\}.\end{aligned} \tag{10.3}$$

The right-hand sides of (10.2) and (10.3) must be rewritten as partial derivatives of χ_N. Following the strategy of Sect. 3.2.2, with the help of relations

(3.78), we have

$$\left(\frac{\partial \chi_N}{\partial t}\right)_{ab} = (g/2)\left[\left(\frac{\partial}{\partial(iz^*)} + iz\right)^2 \frac{\partial}{\partial(iw)} - \frac{\partial^2}{\partial(iz^*)^2}\left(\frac{\partial}{\partial(iw)} + iw^*\right)\right.$$
$$\left. + \left(\frac{\partial}{\partial(iz)} + iz^*\right)^2 \frac{\partial}{\partial(iw^*)} - \frac{\partial^2}{\partial(iz)^2}\left(\frac{\partial}{\partial(iw^*)} + iw\right)\right]\chi_N$$
$$= (g/2)\left[2iz\frac{\partial^2}{\partial(iz^*)\partial(iw)} - iw^*\frac{\partial^2}{\partial(iz^*)^2} + (iz)^2\frac{\partial}{\partial(iw)}\right.$$
$$\left. + 2iz^*\frac{\partial^2}{\partial(iz)\partial(iw^*)} - iw\frac{\partial^2}{\partial(iz)^2} + (iz^*)^2\frac{\partial}{\partial(iw^*)}\right]\chi_N, \tag{10.4}$$

and

$$\left(\frac{\partial \chi_N}{\partial t}\right)_{\text{drive}} = \left[-i\bar{\mathcal{E}}_0 e^{-i2\omega_C t}(iw) + i\bar{\mathcal{E}}_0^* e^{i2\omega_C t}(iw^*)\right]\chi_N. \tag{10.5}$$

Then, substituting the Fourier transform of $P(\alpha, \alpha^*, \beta, \beta^*)$ for $\chi_N(z, z^*, w, w^*)$ and following steps in parallel to those leading from (3.83) to (3.85), the two new terms in the phase-space equation of motion are

$$\left(\frac{\partial P}{\partial t}\right)_{ab} = (g/2)\left[-2\frac{\partial}{\partial\alpha}\alpha^*\beta + \frac{\partial}{\partial\beta^*}\alpha^{*2} + \frac{\partial^2}{\partial\alpha^2}\beta - 2\frac{\partial}{\partial\alpha^*}\alpha\beta^*\right.$$
$$\left. + \frac{\partial}{\partial\beta}\alpha^2 + \frac{\partial^2}{\partial\alpha^{*2}}\beta^*\right]P, \tag{10.6}$$

and

$$\left(\frac{\partial P}{\partial t}\right)_{\text{drive}} = i\left(\frac{\partial}{\partial\beta}\bar{\mathcal{E}}_0 e^{-i2\omega_C t} - \frac{\partial}{\partial\beta^*}\bar{\mathcal{E}}_0^* e^{i2\omega_C t}\right)P. \tag{10.7}$$

Equations 10.1, 10.6, and 10.7 give the *phase-space equation of motion for the degenerate parametric oscillator in the P representation*:

$$\frac{\partial P}{\partial t} = \left\{\frac{\partial}{\partial\alpha}\left[(\kappa + i\omega_C)\alpha - g\alpha^*\beta\right] + \frac{\partial}{\partial\alpha^*}\left[(\kappa - i\omega_C)\alpha^* - g\alpha\beta^*\right]\right.$$
$$+ \frac{\partial}{\partial\beta}\left[(\kappa_p + i2\omega_C)\beta + (g/2)\alpha^2 + i\bar{\mathcal{E}}_0 e^{-i2\omega_C t}\right]$$
$$+ \frac{\partial}{\partial\beta^*}\left[(\kappa_p - i2\omega_C)\beta^* + (g/2)\alpha^{*2} - i\bar{\mathcal{E}}_0^* e^{i2\omega_C t}\right]$$
$$\left. + (g/2)\left(\frac{\partial^2}{\partial\alpha^2}\beta + \frac{\partial^2}{\partial\alpha^{*2}}\beta^*\right) + 2\kappa\bar{n}\frac{\partial^2}{\partial\alpha\partial\alpha^*} + 2\kappa_p\bar{n}_p\frac{\partial^2}{\partial\beta\partial\beta^*}\right\}P. \tag{10.8}$$

This equation has the fortunate property that it only involves derivatives up to second order. It therefore seems that we can formulate a classical stochastic model for the degenerate parametric oscillator without having to use the system size expansion to justify a truncation of higher-order derivatives. In this respect, we are in a better position than we were with laser theory. Although we may not be able to solve the nonlinear stochastic differential equations corresponding to (10.8) analytically, numerical simulation of the equations should be feasible. Unfortunately, however, Eq. 10.8 presents us with a new difficulty: this equation is not a true Fokker–Planck equation, because it does not possess positive definite diffusion. To see this, we introduce real and imaginary parts of the phase-space variables,

$$\alpha = x + iy, \qquad \alpha^* = x - iy, \tag{10.9a}$$

$$\beta = u + iv, \qquad \beta^* = u - iv. \tag{10.9b}$$

Then, expressed in terms of the real variables x, y, u, and v, the second-order derivatives in phase-space equation (10.8) are

$$\begin{aligned}
&(g/2)\left(\frac{\partial^2}{\partial\alpha^2}\beta + \frac{\partial^2}{\partial\alpha^{*2}}\beta^*\right) + \kappa\bar{n}\frac{\partial^2}{\partial\alpha\partial\alpha^*} + \kappa_p\bar{n}_p\frac{\partial^2}{\partial\beta\partial\beta^*} \\
&= \frac{1}{2}\frac{\partial^2}{\partial x^2}\big[\kappa\bar{n} + (g/2)u\big] + \frac{1}{2}\frac{\partial^2}{\partial y^2}\big[\kappa\bar{n} - (g/2)u\big] \\
&\quad + (g/2)\frac{\partial^2}{\partial x\partial y}v + \frac{1}{2}\kappa_p\bar{n}_p\left(\frac{\partial^2}{\partial u^2} + \frac{\partial^2}{\partial v^2}\right).
\end{aligned} \tag{10.10}$$

Hence our "Fokker–Planck" equation has diffusion matrix

$$\boldsymbol{D}(x,y,u,v) = \begin{pmatrix} \kappa\bar{n} + (g/2)u & (g/2)v & 0 & 0 \\ (g/2)v & \kappa\bar{n} - (g/2)u & 0 & 0 \\ 0 & 0 & \kappa_p\bar{n}_p & 0 \\ 0 & 0 & 0 & \kappa_p\bar{n}_p \end{pmatrix}. \tag{10.11}$$

Positive definite diffusion requires that the quadratic form

$$\begin{aligned}
&\boldsymbol{z}^T\boldsymbol{D}\boldsymbol{z} \\
&= \big[\kappa\bar{n} + (g/2)u\big]z_1^2 + \big[\kappa\bar{n} - (g/2)u\big]z_2^2 + (gv)z_1z_2 + \kappa_p\bar{n}_p(z_3^2 + z_4^2)
\end{aligned} \tag{10.12}$$

be positive for all choices of the vector $\boldsymbol{z}$. Clearly this is not so for all points throughout the phase space; it is not so for $v = 0$ and $|u| > 2\kappa\bar{n}/g$, for example. More generally, a necessary and sufficient condition for $\boldsymbol{D}$ to be positive definite is that the complex amplitude $\beta = u + iv$ lies within the circle $|\beta|^2 = u^2 + v^2 = 4\bar{n}(\kappa/g)^2$—i.e., for the diffusion matrix to be positive definite, we require

$$|\beta|^2 = u^2 + v^2 < 4\bar{n}(\kappa/g)^2. \tag{10.13}$$

Exercise 10.1. Prove that the quadratic form $az_1^2 + bz_2^2 + 2cz_1z_2$ is positive definite if and only if $a > 0$, $b > 0$, and $ab > c^2$. Inequality (10.13) follows from this result.

Actually, a diffusion matrix is acceptable if it is merely positive semidefinite; it is permissible for $\boldsymbol{z}^T\boldsymbol{D}\boldsymbol{z}$ to vanish. This happens when the vector $\boldsymbol{z}$ points in a direction in which locally the diffusion is zero (when $\boldsymbol{z}$ is an eigenvector of $\boldsymbol{D}$ with zero eigenvalue). Choosing $\bar{n}_p = 0$, the diffusion matrix (10.11) is at best positive semidefinite, since then there is no diffusion in the u and z directions. Thus, the problem with the phase-space equation of motion for the degenerate parametric oscillator in the P representation is that it only possesses positive semidefinite diffusion within the region $|\beta|^2 \leq 4\bar{n}(\kappa/g)^2$. For a parametric oscillator at optical frequencies, the diffusion matrix is effectively not positive semidefinite anywhere in phase space, since at room temperature $\bar{n} \sim 10^{-44}$, while $\kappa/g \sim 10^4$; inequality (10.13) is violated for almost all $|\beta|^2$. (At threshold we might approximate the physically relevant values of $|\beta|^2$ by the undepleted pump photon number $|\beta|^2 = |\bar{\mathcal{E}}_0/\kappa_p|^2 \sim 10^8$.)

Note 10.1. The parameter g/κ plays an important role in determining the size of the quantum fluctuations in the degenerate parametric oscillator. To demonstrate the physical significance of this parameter, we use (9.80) and (9.54) to write

$$\left(\frac{g}{\kappa}\right)^2 = \frac{\hbar 2\omega_C}{\epsilon_0 A\bar{L}}\left(\frac{\mathcal{F}}{\pi}\frac{\omega_C \ell \chi^{(2)}}{n^{3/2}c}\right)^2 \cos^2(\phi_p - 2\phi), \tag{10.14}$$

and (9.81), (9.64), and (9.59) to write

$$\left|\frac{\bar{\mathcal{E}}_0}{\kappa_p}\right|^2 = \lambda^2\frac{\epsilon_0 A\bar{L}}{\hbar 2\omega_C}\left(\frac{\pi}{\mathcal{F}}\frac{n^{3/2}c}{\omega_C \ell \chi^{(2)}}\right)^2 \cos^{-2}(\phi_p - 2\phi). \tag{10.15}$$

The quantity $|\bar{\mathcal{E}}_0/\kappa_p|^2$ is the mean photon number in the pump mode assuming there is no pump depletion (Eq. 9.103). Now, setting $\lambda = 1$ in (10.15), we have

$$\left(\frac{g}{\kappa}\right)^2 = |\bar{\mathcal{E}}_0^{\mathrm{thr}}/\kappa_p|^{-2} = (n_p^{\mathrm{thr}})^{-1}, \tag{10.16}$$

where we introduce the *threshold photon number*

$$n_p^{\mathrm{thr}} = \frac{\epsilon_0 A\bar{L}}{\hbar 2\omega_C}\left(\frac{\pi}{\mathcal{F}}\frac{n^{3/2}c}{\omega_C \ell \chi^{(2)}}\right)^2 \cos^{-2}(\phi_p - 2\phi); \tag{10.17}$$

n_p^{thr} is the undepleted pump-mode photon number required to reach threshold—typically $n_p^{\mathrm{thr}} \sim 10^8$.

We might have anticipated that there would be a problem with the phase-space formulation for the degenerate parametric oscillator in the P representation, since we have seen that degenerate parametric amplification generates a squeezed state, a state for which a positive, nonsingular P distribution does not exist. The Q and Wigner distributions, on the other hand, do exist for a squeezed state as ordinary positive functions (Exercise 8.1). Perhaps, then, we should base our phase-space analysis of the degenerate parametric oscillator on one of these representations. Let us now see what the phase-space equation of motion corresponding to master equation (9.97) looks like in the Q and Wigner representations.

10.1.2 Phase-Space Equations of Motion in the Q and Wigner Representations

Fokker–Planck equations for the damped harmonic oscillator in the Q and Wigner representations were derived in Chap. 4. Drawing upon these results (Eqs. 4.14 and 4.37), as in (10.1) we can immediately write

$$\begin{aligned}
\frac{\partial Q, W}{\partial t}
&= \biggl\{ (\kappa + i\omega_C)\frac{\partial}{\partial \alpha}\alpha + (\kappa - i\omega_C)\frac{\partial}{\partial \alpha^*}\alpha^* + 2\kappa\bigl[\bar{n} + \tfrac{1}{2}(1-\sigma)\bigr]\frac{\partial^2}{\partial\alpha\partial\alpha^*} \\
&\quad + (\kappa_p + i2\omega_C)\frac{\partial}{\partial\beta}\beta + (\kappa_p - i2\omega_C)\frac{\partial}{\partial\beta^*}\beta^* \\
&\quad + 2\kappa_p\bigl[\bar{n}_p + \tfrac{1}{2}(1-\sigma)\bigr]\frac{\partial^2}{\partial\beta\partial\beta^*}\biggr\} Q, W + \left(\frac{\partial Q, W}{\partial t}\right)_{ab} + \left(\frac{\partial Q, W}{\partial t}\right)_{\mathrm{drive}},
\end{aligned} \tag{10.18}$$

where σ takes the values -1 and 0 for Q and W, respectively. The mode-coupling and pump-mode-driving terms may be obtained from those in the P representation (Eqs. 10.4 and 10.5) using relationships (4.9) and (4.34) connecting the different characteristic functions; thus, we obtain

$$
\begin{aligned}
&\left(\frac{\partial\chi_{A},\chi_{S}}{\partial t}\right)_{ab}\\
&= e^{-\frac{1}{2}(1-\sigma)|z|^2} e^{-\frac{1}{2}(1-\sigma)|w|^2}\left(\frac{\partial\chi_N}{\partial t}\right)_{ab}\\
&= (g/2)\Bigg[2iz\left(\frac{\partial}{\partial(iz^*)}-\tfrac{1}{2}(1-\sigma)iz\right)\left(\frac{\partial}{\partial(iw)}-\tfrac{1}{2}(1-\sigma)iw^*\right)\\
&\quad - iw^*\left(\frac{\partial}{\partial(iz^*)}-\tfrac{1}{2}(1-\sigma)iz\right)^2 + (iz)^2\left(\frac{\partial}{\partial(iw)}-\tfrac{1}{2}(1-\sigma)iw^*\right)\\
&\quad + 2iz^*\left(\frac{\partial}{\partial(iz)}-\tfrac{1}{2}(1-\sigma)iz^*\right)\left(\frac{\partial}{\partial(iw^*)}-\tfrac{1}{2}(1-\sigma)iw\right)\\
&\quad - iw\left(\frac{\partial}{\partial(iz)}-\tfrac{1}{2}(1-\sigma)iz^*\right)^2 + (iz^*)^2\left(\frac{\partial}{\partial(iw^*)}-\tfrac{1}{2}(1-\sigma)iw\right)\Bigg]\chi_{A},\chi_{S},
\end{aligned}
\tag{10.19}
$$

and

$$
\begin{aligned}
\left(\frac{\partial\chi_{A},\chi_{S}}{\partial t}\right)_{\text{drive}} &= e^{-\frac{1}{2}(1-\sigma)|z|^2} e^{-\frac{1}{2}(1-\sigma)|w|^2}\left(\frac{\partial\chi_N}{\partial t}\right)_{\text{drive}}\\
&= \left[-i\bar{\mathcal{E}}_0 e^{-i2\omega_C t}(iw) + i\bar{\mathcal{E}}_0^* e^{i2\omega_C t}(iw^*)\right]\chi_{A},\chi_{S}.
\end{aligned}
\tag{10.20}
$$

Then substituting the explicit values of σ into (10.19), in the Q representation we have

$$
\begin{aligned}
\left(\frac{\partial\chi_A}{\partial t}\right)_{ab} = (g/2)\Bigg[&2iz\frac{\partial^2}{\partial(iz^*)\partial(iw)} - iw^*\frac{\partial^2}{\partial(iz^*)^2} - (iz)^2\frac{\partial}{\partial(iw)}\\
&+2iz^*\frac{\partial^2}{\partial(iz)\partial(iw^*)} - iw\frac{\partial^2}{\partial(iz)^2} - (iz^*)^2\frac{\partial}{\partial(iw^*)}\Bigg]\chi_A,
\end{aligned}
\tag{10.21a}
$$

and in the Wigner representation,

$$
\begin{aligned}
\left(\frac{\partial\chi_S}{\partial t}\right)_{ab} = (g/2)\Bigg[&2iz\frac{\partial^2}{\partial(iz^*)\partial(iw)} - iw^*\frac{\partial^2}{\partial(iz^*)^2} - \tfrac{1}{4}(iz)^2(iw^*)\\
&+2iz^*\frac{\partial^2}{\partial(iz)\partial(iw^*)} - iw\frac{\partial^2}{\partial(iz)^2} - \tfrac{1}{4}(iz^*)^2(iw)\Bigg]\chi_S.
\end{aligned}
\tag{10.21b}
$$

Passage from the equations of motion for the characteristic functions, χ_A and χ_S, to equations of motion for the distributions Q and W is made with the substitutions

$$iz \to -\frac{\partial}{\partial\alpha}, \qquad iz^* \to -\frac{\partial}{\partial\alpha^*}, \qquad \frac{\partial}{\partial(iz)} \to \alpha, \qquad \frac{\partial}{\partial(iz^*)} \to \alpha^*,$$
$$iw \to -\frac{\partial}{\partial\beta}, \qquad iw^* \to -\frac{\partial}{\partial\beta^*}, \qquad \frac{\partial}{\partial(iw)} \to \beta, \qquad \frac{\partial}{\partial(iw^*)} \to \beta^*,$$

where derivatives are to be placed to the left in each term. With these substitutions, from (10.20), we obtain

$$\left(\frac{\partial Q, W}{\partial t}\right)_{\text{drive}} = \left(i\bar{\mathcal{E}}_0 e^{-i2\omega_C t}\frac{\partial}{\partial\beta} - i\bar{\mathcal{E}}_0^* e^{i2\omega_C t}\frac{\partial}{\partial\beta^*}\right)Q, W, \tag{10.22}$$

while from (10.21a) and (10.21b), respectively, we obtain

$$\left(\frac{\partial Q}{\partial t}\right)_{ab} = (g/2)\left(-2\frac{\partial}{\partial\alpha}\alpha^*\beta + \frac{\partial}{\partial\beta^*}\alpha^{*2} \right.$$
$$\left. -\frac{\partial^2}{\partial\alpha^2}\beta - 2\frac{\partial}{\partial\alpha^*}\alpha\beta^* + \frac{\partial}{\partial\beta}\alpha^2 - \frac{\partial^2}{\partial\alpha^{*2}}\beta^*\right)Q, \tag{10.23a}$$

and

$$\left(\frac{\partial W}{\partial t}\right)_{ab} = (g/2)\left(-2\frac{\partial}{\partial\alpha}\alpha^*\beta + \frac{\partial}{\partial\beta^*}\alpha^{*2} \right.$$
$$\left. +\frac{1}{4}\frac{\partial^3}{\partial\alpha^2\partial\beta^*} - 2\frac{\partial}{\partial\alpha^*}\alpha\beta^* + \frac{\partial}{\partial\beta}\alpha^2 + \frac{1}{4}\frac{\partial^3}{\partial\alpha^{*2}\partial\beta}\right)W. \tag{10.23b}$$

Finally, substituting (10.22) and (10.23a) into (10.18), the *phase-space equation of motion for the degenerate parametric oscillator in the Q representation* is given by

$$\begin{aligned}\frac{\partial Q}{\partial t} = \Bigg\{ &\frac{\partial}{\partial\alpha}\left[(\kappa + i\omega_C)\alpha - g\alpha^*\beta\right] + \frac{\partial}{\partial\alpha^*}\left[(\kappa - i\omega_C)\alpha^* - g\alpha\beta^*\right] \\ &+ \frac{\partial}{\partial\beta}\left[(\kappa_p + i2\omega_C)\beta + (g/2)\alpha^2 + i\bar{\mathcal{E}}_0 e^{-i2\omega_C t}\right] \\ &+ \frac{\partial}{\partial\beta^*}\left[(\kappa_p - i2\omega_C)\beta^* + (g/2)\alpha^{*2} - i\bar{\mathcal{E}}_0^* e^{i2\omega_C t}\right] \\ &- (g/2)\left(\frac{\partial^2}{\partial\alpha^2}\beta + \frac{\partial^2}{\partial\alpha^{*2}}\beta^*\right) + 2\kappa(\bar{n}+1)\frac{\partial^2}{\partial\alpha\partial\alpha^*} \\ &+ 2\kappa_p(\bar{n}_p + 1)\frac{\partial^2}{\partial\beta\partial\beta^*}\Bigg\}Q. \end{aligned} \tag{10.24}$$

Similarly, from (10.22), (10.23b), and (10.18), the *phase-space equation of motion for the degenerate parametric oscillator in the Wigner representation* is given by

$$\begin{aligned}\frac{\partial W}{\partial t} = \bigg\{ & \frac{\partial}{\partial \alpha}\big[(\kappa + i\omega_C)\alpha - g\alpha^*\beta\big] + \frac{\partial}{\partial \alpha^*}\big[(\kappa - i\omega_C)\alpha^* - g\alpha\beta^*\big] \\ & + \frac{\partial}{\partial \beta}\big[(\kappa_p + i2\omega_C)\beta + (g/2)\alpha^2 + i\bar{\mathcal{E}}_0 e^{-i2\omega_C t}\big] \\ & + \frac{\partial}{\partial \beta^*}\big[(\kappa_p - i2\omega_C)\beta^* + (g/2)\alpha^{*2} - i\bar{\mathcal{E}}_0^* e^{i2\omega_C t}\big] \\ & + 2\kappa\big(\bar{n} + \tfrac{1}{2}\big)\frac{\partial^2}{\partial \alpha \partial \alpha^*} + 2\kappa_p\big(\bar{n}_p + \tfrac{1}{2}\big)\frac{\partial^2}{\partial \beta \partial \beta^*} \\ & + \frac{1}{4}(g/2)\left(\frac{\partial^3}{\partial \alpha^2 \partial \beta^*} + \frac{\partial^3}{\partial \alpha^{*2}\partial \beta}\right)\bigg\} W. \end{aligned} \tag{10.25}$$

Do these equations offer a resolution to our problem with non-positive-semidefinite diffusion? Before we answer this question, the first thing to notice is that the Wigner representation yields an equation of motion with third-order derivatives; the exact truncation at second order obtained in (10.8) [and (10.24)] has been lost in (10.25). Perhaps the third-order derivatives may be dropped, though, on the basis of a system size expansion. In this case the Wigner representation does give a Fokker–Planck equation with positive definite diffusion; the diffusion matrix is the diagonal matrix

$$\boldsymbol{D} = \mathrm{diag}\big[\kappa\big(\bar{n} + \tfrac{1}{2}\big), \kappa\big(\bar{n} + \tfrac{1}{2}\big), \kappa_p\big(\bar{n}_p + \tfrac{1}{2}\big), \kappa_p\big(\bar{n}_p + \tfrac{1}{2}\big)\big]. \tag{10.26}$$

The dropping of third-order derivatives is valid in the limit of small quantum noise (Sects. 10.2.1 and 10.2.2), and the resulting Fokker–Planck equation underlies the visualization of squeezed vacuum fluctuations discussed in Sect. 9.3.7. Thus, in the small-noise limit, the problem of non-positive-semidefinite diffusion can be avoided by using the Wigner representation in place of the P representation. The solution is not a fundamental one, though, because it only applies when the small-noise truncation is valid (see also Note 9.19).

Note 10.2. Stochastic methods for dealing with third-order derivatives do in fact exist. Gardiner [10.1] provides a short discussion of this subject.

What now is the situation with the Q representation? The phase-space equation of motion preserves the exact truncation of derivatives at second order. It also has a modified diffusion matrix which might well be positive semidefinite. It is similar to matrix (10.11) but with larger terms along the diagonal.

Specifically, from (10.24), we obtain

$$\boldsymbol{D}(x,y,u,v) = \begin{pmatrix} \kappa(\bar{n}+1)-(g/2)u & -(g/2)v & 0 & 0 \\ -(g/2)v & \kappa(\bar{n}+1)+(g/2)u & 0 & 0 \\ 0 & 0 & \kappa_p(\bar{n}_p+1) & 0 \\ 0 & 0 & 0 & \kappa_p(\bar{n}_p+1) \end{pmatrix}. \tag{10.27}$$

Thus, to replace inequality (10.13), the requirement for positive semidefinite diffusion is

$$|\beta|^2 = u^2 + v^2 \le 4(\bar{n}+1)(\kappa/g)^2. \tag{10.28}$$

Once again, this requirement cannot be satisfied everywhere in phase space. There is, however, an important difference between (10.28) and the requirement $|\beta|^2 \le 4\bar{n}(\kappa/g)^2$ for positive semidefinite diffusion in the P representation. If we assume that the quantum fluctuations of the pump field are small, we may identify β with the classically determined pump field amplitude

$$\begin{aligned} |\beta|^2 = |\langle b\rangle|^2 &= \frac{\epsilon_0 A\bar{L}}{\hbar 2\omega_C}|\mathcal{E}_p|^2 \\ &= \frac{\epsilon_0 A\bar{L}}{2\hbar\omega_C}\left(\frac{\kappa}{K'}\right)^2 |\bar{\mathcal{E}}_p|^2 \\ &= \frac{\epsilon_0 A\bar{L}}{2\hbar\omega_C}\left(\frac{\pi}{\mathcal{F}}\frac{n^{3/2}c}{\omega_C \ell \chi^{(2)}}\right)^2 \cos^{-2}(\phi_p - 2\phi)|\bar{\mathcal{E}}_p|^2 \\ &= (\kappa/g)^2|\bar{\mathcal{E}}_p|^2, \end{aligned} \tag{10.29}$$

where we have used (9.60b), (9.53), (9.54), and (10.14). The largest steady-state value of $|\bar{\mathcal{E}}_p|$ is $|\bar{\mathcal{E}}_p^{\rm ss}| = 1$. This value is reached at threshold and holds everywhere above threshold (Eqs. 9.75–9.77). We may therefore write, using (10.29), $|\beta_{\rm ss}|^2 \le (\kappa/g)^2 < 4(\bar{n}+1)(\kappa/g)^2$, which shows that inequality (10.28) is satisfied in the vicinity of the steady state; it is satisfied both below and above threshold and for all values of $\bar{n}$. It follows that the problem of non-positive-semidefinite diffusion can be avoided by using the Q representation. As with the Wigner representation, however, small noise must be assumed. Again, the resolution is not genuinely fundamental.

In summary, the results obtained using the Q and Wigner representations are mixed. Neither representation provides the ideal phase-space equation of motion—a nonlinear Fokker–Planck equation which has positive semidefinite diffusion everywhere in phase space. In the Wigner representation, the phase-space equation has derivatives beyond second order, while in the Q representation a Fokker–Planck equation is obtained, but one possessing non-positive-semidefinite diffusion outside the circle $|\beta| = 2\sqrt{\bar{n}+1}\kappa/g$. Both representations do, however, provide a solid foundation for a linearized treatment of quantum fluctuations based upon the system size expansion.

10.2 Squeezing: Quantum Fluctuations in the Small-Noise Limit

10.2.1 System Size Expansion Far from Threshold

Although further analysis with the P representation would appear to be problematic, we carry out the system size expansion for all three representations; all are included by starting from the general phase-space equation of motion

$$\frac{\partial F_\sigma}{\partial t} = L_\sigma\left(\alpha, \alpha^*, \beta, \beta^*, \frac{\partial}{\partial \alpha}, \frac{\partial}{\partial \alpha^*}, \frac{\partial}{\partial \beta}, \frac{\partial}{\partial \beta^*}, t\right) F_\sigma, \tag{10.30}$$

with

$$\begin{aligned} L_\sigma&\left(\alpha, \alpha^*, \beta, \beta^*, \frac{\partial}{\partial \alpha}, \frac{\partial}{\partial \alpha^*}, \frac{\partial}{\partial \beta}, \frac{\partial}{\partial \beta^*}, t\right) \\ \equiv& \frac{\partial}{\partial \alpha}\left[(\kappa + i\omega_C)\alpha - g\alpha^*\beta\right] + \frac{\partial}{\partial \alpha^*}\left[(\kappa - i\omega_C)\alpha^* - g\alpha\beta^*\right] \\ &+ \frac{\partial}{\partial \beta}\left[(\kappa_p + i2\omega_C)\beta + (g/2)\alpha^2 + i\bar{\mathcal{E}}_0 e^{-i2\omega_C t}\right] \\ &+ \frac{\partial}{\partial \beta^*}\left[(\kappa_p - i2\omega_C)\beta^* + (g/2)\alpha^{*2} - i\bar{\mathcal{E}}_0^* e^{i2\omega_C t}\right] \\ &+ 2\kappa\left[\bar{n} + \tfrac{1}{2}(1-\sigma)\right]\frac{\partial^2}{\partial \alpha \partial \alpha^*} + 2\kappa_p\left[\bar{n}_p + \tfrac{1}{2}(1-\sigma)\right]\frac{\partial^2}{\partial \beta \partial \beta^*} \\ &+ \sigma(g/2)\left(\frac{\partial^2}{\partial \alpha^2}\beta + \frac{\partial^2}{\partial \alpha^{*2}}\beta^*\right) + \frac{1}{4}(1-\sigma^2)(g/2)\left(\frac{\partial^3}{\partial \alpha^2 \partial \beta^*} + \frac{\partial^3}{\partial \alpha^{*2}\partial \beta}\right), \end{aligned} \tag{10.31}$$

where σ takes the values $+1$, 0, and -1, with definitions

$$\left.\begin{aligned} F_{+1} &\equiv P \\ F_0 &\equiv W \\ F_{-1} &\equiv Q \end{aligned}\right\}. \tag{10.32}$$

We must first scale the field amplitudes following the prescription (5.39). A natural choice for the system size parameter is the undepleted pump photon number at threshold $n_p^{\rm thr} = (\kappa/g)^2$ (Eq. 10.16). The powers of $n_p^{\rm thr}$ used in the scaling are to be chosen self-consistently in the manner of Sect. 8.1.1. The details need not be repeated here. It seems reasonable to adopt the scaling that worked for the laser away from threshold ($p = q = 1/2$), while for fine-tuning we are guided by (9.60) and (9.80) [and the comparison between (9.1) and (9.82)]. Thus, for the subharmonic mode we write [with ξ and ψ defined in (9.61), (9.62), and (9.81)]

$$\sqrt{\xi/2}e^{-i\frac{1}{2}(\psi-\pi/2)}\alpha = \left(n_p^{\text{thr}}\right)^{1/2}\bar{\alpha}, \quad \sqrt{\xi/2}e^{i\frac{1}{2}(\psi-\pi/2)}\alpha^* = \left(n_p^{\text{thr}}\right)^{1/2}\bar{\alpha}^*, \tag{10.33}$$

with

$$\bar{\alpha} = \langle\bar{a}(t)\rangle + \left(n_p^{\text{thr}}\right)^{-1/2} z, \tag{10.34a}$$

$$\bar{\alpha}^* = \langle\bar{a}^\dagger(t)\rangle + \left(n_p^{\text{thr}}\right)^{-1/2} z^*, \tag{10.34b}$$

where

$$\sqrt{\xi/2}e^{-i\frac{1}{2}(\psi-\pi/2)}a = \left(n_p^{\text{thr}}\right)^{1/2}\bar{a}, \quad \sqrt{\xi/2}e^{i\frac{1}{2}(\psi-\pi/2)}a^\dagger = \left(n_p^{\text{thr}}\right)^{1/2}\bar{a}^\dagger, \tag{10.35}$$

and similarly for the pump mode, we write

$$e^{-i\frac{1}{2}(\psi-\pi/2)}\beta = \left(n_p^{\text{thr}}\right)^{1/2}\bar{\beta}, \qquad e^{i\frac{1}{2}(\psi-\pi/2)}\beta^* = \left(n_p^{\text{thr}}\right)^{1/2}\bar{\beta}^*, \tag{10.36}$$

with

$$\bar{\beta} = \langle\bar{b}(t)\rangle + \left(n_p^{\text{thr}}\right)^{-1/2} w, \tag{10.37a}$$

$$\bar{\beta}^* = \langle\bar{b}^\dagger(t)\rangle + \left(n_p^{\text{thr}}\right)^{-1/2} w^*, \tag{10.37b}$$

where

$$e^{-i\frac{1}{2}(\psi-\pi/2)}b = \left(n_p^{\text{thr}}\right)^{1/2}\bar{b}, \qquad e^{i\frac{1}{2}(\psi-\pi/2)}b^\dagger = \left(n_p^{\text{thr}}\right)^{1/2}\bar{b}^\dagger. \tag{10.38}$$

Then the phase-space distribution in scaled variables is defined by

$$\bar{F}_\sigma(z,z^*,w,w^*,t) \equiv \xi^{-1}F_\sigma\big(\alpha(z,t),\alpha^*(z^*,t),\beta(w,t),\beta^*(w^*,t),t\big), \tag{10.39}$$

and satisfies the equation of motion

$$\begin{aligned}
\frac{\partial\bar{F}_\sigma}{\partial t} &= \xi^{-1}\left(\frac{\partial F_\sigma}{\partial\alpha}\frac{\partial\alpha}{\partial t} + \frac{\partial F_\sigma}{\partial\alpha^*}\frac{\partial\alpha^*}{\partial t} + \frac{\partial F_\sigma}{\partial\beta}\frac{\partial\beta}{\partial t} + \frac{\partial F_\sigma}{\partial\beta^*}\frac{\partial\beta^*}{\partial t} + \frac{\partial F_\sigma}{\partial t}\right) \\
&= \left(n_p^{\text{thr}}\right)^{1/2}\left(\frac{\partial\bar{F}_\sigma}{\partial z}\frac{d\langle\bar{a}(t)\rangle}{dt} + \text{c.c.}\right) + \left(n_p^{\text{thr}}\right)^{1/2}\left(\frac{\partial\bar{F}_\sigma}{\partial w}\frac{d\langle\bar{b}(t)\rangle}{dt} + \text{c.c.}\right) \\
&\quad + \frac{\partial}{\partial t}\left(\xi^{-1}F_\sigma\right) \\
&= \left(n_p^{\text{thr}}\right)^{1/2}\left(\frac{\partial\bar{F}_\sigma}{\partial z}\frac{d\langle\bar{a}(t)\rangle}{dt} + \text{c.c.}\right) + \left(n_p^{\text{thr}}\right)^{1/2}\left(\frac{\partial\bar{F}_\sigma}{\partial w}\frac{d\langle\bar{b}(t)\rangle}{dt} + \text{c.c.}\right) \\
&\quad + \bar{L}_\sigma\left(z,z^*,w,w^*,\frac{\partial}{\partial z},\frac{\partial}{\partial z^*},\frac{\partial}{\partial w},\frac{\partial}{\partial w^*},t\right)\bar{F}_\sigma,
\end{aligned} \tag{10.40}$$

where

$$\bar{L}_\sigma\left(z,z^*,w,w^*,\frac{\partial}{\partial z},\frac{\partial}{\partial z^*},\frac{\partial}{\partial w},\frac{\partial}{\partial w^*},t\right)$$
$$\equiv L_\sigma\bigg(\alpha(z,t),\alpha^*(z^*,t),\beta(w,t),\beta^*(w^*,t),\sqrt{\xi/2}e^{-i\frac{1}{2}(\psi-\pi/2)}\frac{\partial}{\partial z},$$
$$\sqrt{\xi/2}e^{i\frac{1}{2}(\psi-\pi/2)}\frac{\partial}{\partial z^*},e^{-i\frac{1}{2}(\psi-\pi/2)}\frac{\partial}{\partial w},e^{i\frac{1}{2}(\psi-\pi/2)}\frac{\partial}{\partial w^*},t\bigg). \qquad (10.41)$$

Substituting for L_σ from (10.31), our aim is to identify the macroscopic law that governs the mean behavior and the Fokker–Planck equation that describes the fluctuations about the mean. The substitution yields

$$\frac{\partial\bar{F}_\sigma}{\partial t}$$
$$=\left(n_p^{\rm thr}\right)^{1/2}\bigg\{\frac{\partial\bar{F}_\sigma}{\partial z}\left[\frac{d\langle\bar{a}(t)\rangle}{dt}+(\kappa+i\omega_C)\langle\bar{a}(t)\rangle-\kappa\langle\bar{a}^\dagger(t)\rangle\langle\bar{b}(t)\rangle\right]+\text{c.c.}$$
$$+\frac{\partial\bar{F}_\sigma}{\partial w}\left[\frac{d\langle\bar{b}(t)\rangle}{dt}+(\kappa_p+i2\omega_C)\langle\bar{b}(t)\rangle+\kappa_p\big(\langle\bar{a}(t)\rangle^2-\lambda e^{-i2\omega_Ct}\big)\right]+\text{c.c.}\bigg\}$$
$$+\bigg\{\frac{\partial}{\partial z}\Big[(\kappa+i\omega_C)z-\kappa\Big(\langle\bar{a}^\dagger(t)\rangle w+\langle\bar{b}(t)\rangle z^*+\left(n_p^{\rm thr}\right)^{-1/2}z^*w\Big)\Big]+\text{c.c.}$$
$$+\frac{\partial}{\partial w}\Big[(\kappa_p+i2\omega_C)w+\kappa_p\Big(2\langle\bar{a}(t)\rangle z+\left(n_p^{\rm thr}\right)^{-1/2}z^2\Big)\Big]+\text{c.c.}$$
$$+\frac{1}{4}\xi\sigma\kappa\left[\frac{\partial^2}{\partial z^2}\Big(\langle\bar{b}(t)\rangle+\left(n_p^{\rm thr}\right)^{-1/2}w\Big)+\text{c.c.}\right]$$
$$+\xi\kappa\big[\bar{n}+\tfrac{1}{2}(1-\sigma)\big]\frac{\partial^2}{\partial z\partial z^*}+2\kappa_p\big[\bar{n}_p+\tfrac{1}{2}(1-\sigma)\big]\frac{\partial^2}{\partial w\partial w^*}$$
$$+\left(n_p^{\rm thr}\right)^{-1/2}\frac{1}{16}\xi(1-\sigma^2)\kappa\bigg(\frac{\partial^3}{\partial z^2\partial w^*}+\text{c.c.}\bigg)\bigg\}\bar{F}_\sigma. \qquad (10.42)$$

The terms multiplied by $\left(n_p^{\rm thr}\right)^{1/2}$ must vanish, otherwise there is a divergence in the limit $n_p^{\rm thr}\to\infty$; hence we arrive at the macroscopic law, the *degenerate parametric oscillator equations without fluctuations*:

$$\kappa^{-1}\frac{d\langle\tilde{\bar{a}}\rangle}{dt}=-\langle\tilde{\bar{a}}\rangle+\langle\tilde{\bar{a}}^\dagger\rangle\langle\tilde{\bar{b}}\rangle, \qquad (10.43a)$$

$$\kappa^{-1}\frac{d\langle\tilde{\bar{a}}^\dagger\rangle}{dt}=-\langle\tilde{\bar{a}}^\dagger\rangle+\langle\tilde{\bar{a}}\rangle\langle\tilde{\bar{b}}^\dagger\rangle, \qquad (10.43b)$$

$$\kappa_p^{-1}\frac{d\langle\tilde{\bar{b}}\rangle}{dt} = -\langle\tilde{\bar{b}}\rangle - \langle\tilde{\bar{a}}\rangle^2 + \lambda, \tag{10.43c}$$

$$\kappa_p^{-1}\frac{d\langle\tilde{\bar{b}}^\dagger\rangle}{dt} = -\langle\tilde{\bar{b}}^\dagger\rangle - \langle\tilde{\bar{a}}^\dagger\rangle^2 + \lambda, \tag{10.43d}$$

where

$$\begin{aligned} \tilde{\bar{a}} &\equiv e^{i\omega_C t}\bar{a}, & \tilde{\bar{a}}^\dagger &\equiv e^{-i\omega_C t}\bar{a}^\dagger, \\ \tilde{\bar{b}} &\equiv e^{i2\omega_C t}\bar{b}, & \tilde{\bar{b}}^\dagger &\equiv e^{-i2\omega_C t}\bar{b}^\dagger. \end{aligned} \tag{10.44}$$

We then drop terms of order $\left(n_p^{\rm thr}\right)^{-1/2}$, and what remains is the *linearized Fokker–Planck equation for the degenerate parametric oscillator*:

$$\begin{aligned} &\frac{\partial\tilde{\bar{F}}_\sigma}{\partial t} \\ &= \Bigg\{\kappa\frac{\partial}{\partial\tilde{z}}\big[\tilde{z} - \langle\tilde{\bar{a}}^\dagger(t)\rangle\tilde{w} - \langle\tilde{\bar{b}}(t)\rangle\tilde{z}^*\big] + \kappa\frac{\partial}{\partial\tilde{z}^*}\big[\tilde{z}^* - \langle\tilde{\bar{a}}(t)\rangle\tilde{w}^* - \langle\tilde{\bar{b}}^\dagger(t)\rangle\tilde{z}\big] \\ &\quad + \kappa_p\frac{\partial}{\partial\tilde{w}}\big[\tilde{w} + 2\langle\tilde{\bar{a}}(t)\rangle\tilde{z}\big] + \kappa_p\frac{\partial}{\partial\tilde{w}^*}\big[\tilde{w}^* + 2\langle\tilde{\bar{a}}^\dagger(t)\rangle\tilde{z}^*\big] \\ &\quad + \frac{1}{4}\xi\sigma\kappa\left(\frac{\partial^2}{\partial\tilde{z}^2}\langle\tilde{\bar{b}}(t)\rangle + \frac{\partial^2}{\partial\tilde{z}^{*2}}\langle\tilde{\bar{b}}^\dagger(t)\rangle\right) \\ &\quad + \xi\kappa\big[\bar{n} + \tfrac{1}{2}(1-\sigma)\big]\frac{\partial^2}{\partial\tilde{z}\partial\tilde{z}^*} + 2\kappa_p\big[\bar{n}_p + \tfrac{1}{2}(1-\sigma)\big]\frac{\partial^2}{\partial\tilde{w}\partial\tilde{w}^*}\Bigg\}\tilde{\bar{F}}_\sigma, \end{aligned} \tag{10.45}$$

where

$$\tilde{\bar{F}}_\sigma(\tilde{z},\tilde{z}^*,\tilde{w},\tilde{w}^*,t) \equiv \bar{F}_\sigma\big(z(\tilde{z},t), z^*(\tilde{z}^*,t), w(\tilde{w},t), w^*(\tilde{w}^*,t), t\big), \tag{10.46}$$

and we have written

$$\begin{aligned} z &= e^{-i\omega_C t}\tilde{z}, & z^* &= e^{i\omega_C t}\tilde{z}^*, \\ w &= e^{-i2\omega_C t}\tilde{w}, & w^* &= e^{i2\omega_C t}\tilde{w}^*. \end{aligned} \tag{10.47}$$

10.2.2 Quantum Fluctuations Below Threshold

Equations 10.43a–10.43d are equivalent to the equations of motion for the field amplitudes $\bar{\mathcal{E}}$ and $\bar{\mathcal{E}}_p$ in the classical treatment of the degenerate parametric oscillator (Eqs. 9.63a and 9.63b). Steady-state solutions below, at, and above threshold are given by (9.75), (9.76), and (9.77), respectively, with

$$\langle\tilde{\bar{a}}\rangle_{\rm ss} = \bar{\mathcal{E}}_{\rm ss}, \qquad \langle\tilde{\bar{a}}^\dagger\rangle_{\rm ss} = \bar{\mathcal{E}}_{\rm ss}^*, \tag{10.48a}$$

$$\langle\tilde{\bar{b}}\rangle_{\rm ss} = \bar{\mathcal{E}}_p^{\rm ss}, \qquad \langle\tilde{\bar{b}}^\dagger\rangle_{\rm ss} = (\bar{\mathcal{E}}_p^{ss})^*. \tag{10.48b}$$

Our interest here is in the quantum fluctuations about these steady states. We look first at the fluctuations below threshold, where the squeezing experiments of Wu et al. [10.2, 10.3] were carried out.

The steady-state solution to the mean value equations 10.43a–10.43d is $\langle\tilde{a}(t)\rangle = \langle\tilde{a}^\dagger(t)\rangle = 0$, $\langle\tilde{\bar{b}}(t)\rangle = \langle\tilde{\bar{b}}^\dagger(t)\rangle = \lambda$. Substituting this solution into (10.45), the Fokker–Planck equation is found to be separable, with solution

$$\tilde{\bar{F}}_\sigma(\tilde{z},\tilde{z}^*,\tilde{w},\tilde{w}^*,t) = \tilde{\bar{X}}_\sigma(\tilde{z}_1,t)\tilde{\bar{Y}}_\sigma(\tilde{z}_2,t)\tilde{\bar{U}}_\sigma(\tilde{w}_1,t)\tilde{\bar{V}}_\sigma(\tilde{w}_2,t), \tag{10.49}$$

where

$$\begin{aligned} \tilde{z} &= \tilde{z}_1 + i\tilde{z}_2, & \tilde{z}^* &= \tilde{z}_1 - i\tilde{z}_2, \\ \tilde{w} &= \tilde{w}_1 + i\tilde{w}_2, & \tilde{w}^* &= \tilde{w}_1 - i\tilde{w}_2. \end{aligned} \tag{10.50}$$

Fluctuations about the steady state are described by the *Fokker–Planck equation below threshold for quadrature phase amplitudes of the subharmonic mode of the degenerate parametric oscillator*,

$$\kappa^{-1}\frac{\partial\tilde{\bar{X}}_\sigma}{\partial t} = \left\{(1-\lambda)\frac{\partial}{\partial\tilde{z}_1}\tilde{z}_1 + \frac{1}{8}\xi\left[2\bar{n}+1-\sigma(1-\lambda)\right]\frac{\partial^2}{\partial\tilde{z}_1^2}\right\}\tilde{\bar{X}}_\sigma, \tag{10.51a}$$

$$\kappa^{-1}\frac{\partial\tilde{\bar{Y}}_\sigma}{\partial t} = \left\{(1+\lambda)\frac{\partial}{\partial\tilde{z}_2}\tilde{z}_2 + \frac{1}{8}\xi\left[2\bar{n}+1-\sigma(1+\lambda)\right]\frac{\partial^2}{\partial\tilde{z}_2^2}\right\}\tilde{\bar{Y}}_\sigma, \tag{10.51b}$$

and the *Fokker–Planck equation below threshold for quadrature phase amplitudes of the pump mode of the degenerate parametric oscillator*,

$$\kappa_p^{-1}\frac{\partial\tilde{\bar{U}}_\sigma}{\partial t} = \left\{\frac{\partial}{\partial\tilde{w}_1}\tilde{w}_1 + \frac{1}{4}(2\bar{n}_p+1-\sigma)\frac{\partial^2}{\partial\tilde{w}_1^2}\right\}\tilde{\bar{U}}_\sigma, \tag{10.52a}$$

$$\kappa_p^{-1}\frac{\partial\tilde{\bar{V}}_\sigma}{\partial t} = \left\{\frac{\partial}{\partial\tilde{w}_2}\tilde{w}_2 + \frac{1}{4}(2\bar{n}_p+1-\sigma)\frac{\partial^2}{\partial\tilde{w}_2^2}\right\}\tilde{\bar{V}}_\sigma. \tag{10.52b}$$

Clearly, the pump mode simply fluctuates as an oscillator in thermal equilibrium, and we focus our attention on the subharmonic mode fluctuations. Note first that the drift terms in (10.51a) and (10.51b) yield deterministic equations

$$\dot{\tilde{z}}_1 = -\kappa(1-\lambda)\tilde{z}_1, \qquad \dot{\tilde{z}}_2 = -\kappa(1+\lambda)\tilde{z}_2, \tag{10.53}$$

where the terms $+\kappa\lambda\tilde{z}_1$ and $-\kappa\lambda\tilde{z}_2$ on the right-hand sides describe amplification and deamplification of the quadrature phase amplitudes, respectively, as seen for the parametric amplifier in (9.22) and (9.28). The gain is less than the loss below threshold, so the fluctuations do not initiate the growth of a mean amplitude $\tilde{z}_1$. Nevertheless, they experience a phase-dependent decay, which results in some form of squeezing, a phase-dependent distribution of deviations from the steady state.

Let us more specifically consider fluctuations in the quadrature phase operators $\hat{X} \equiv \hat{A}_\theta$ and $\hat{Y} \equiv \hat{A}_{\theta+\pi/2}$, where $\theta = \frac{1}{2}(\psi - \pi/2)$, with ψ defined by

(9.62) and (9.81). Since $\langle\tilde{a}\rangle = \langle\tilde{a}^\dagger\rangle = 0$, we may use (9.26) and the scaling relations (10.33)–(10.35) to write

$$\begin{aligned}(\Delta X)^2 &= \tfrac{1}{4}\big\langle\big[\tilde{a}e^{-i\frac{1}{2}(\psi-\pi/2)} + \tilde{a}^\dagger e^{i\frac{1}{2}(\psi-\pi/2)}\big]^2\big\rangle \\ &= \tfrac{1}{4}(2/\xi)n_p^{\rm thr}\langle(\tilde{\tilde{a}} + \tilde{\tilde{a}}^\dagger)^2\rangle \\ &= \tfrac{1}{4}(2/\xi)n_p^{\rm thr}\langle\tilde{\tilde{a}}^2 + \tilde{\tilde{a}}^{\dagger 2} + \tilde{\tilde{a}}\tilde{\tilde{a}}^\dagger + \tilde{\tilde{a}}^\dagger\tilde{\tilde{a}}\rangle \\ &= (2/\xi)\big(\,\overline{\tilde{z}_1\tilde{z}_1}\,\big)_{\tilde{\tilde{X}}_\sigma} + \tfrac{1}{4}\sigma, \end{aligned} \tag{10.54a}$$

where the last step follows by noting that phase-space averages in the P, Q, and Wigner representations give normal-ordered, antinormal-ordered, and symmetric-ordered operator averages, respectively. From a similar calculation,

$$(\Delta Y)^2 = (2/\xi)\big(\,\overline{\tilde{z}_2\tilde{z}_2}\,\big)_{\tilde{\tilde{Y}}_\sigma} + \tfrac{1}{4}\sigma. \tag{10.54b}$$

It is now trivial to use (5.102a), together with drift and diffusion coefficients taken from (10.51a) and (10.51b), to obtain the *intracavity quadrature phase amplitude fluctuations for the degenerate parametric oscillator below threshold*:

$$(\Delta X)_< \equiv (\Delta X)_{\rm ss}^< = \frac{1}{2}\sqrt{\frac{2\bar{n}+1}{1-\lambda}}, \tag{10.55a}$$

$$(\Delta Y)_< \equiv (\Delta Y)_{\rm ss}^< = \frac{1}{2}\sqrt{\frac{2\bar{n}+1}{1+\lambda}}. \tag{10.55b}$$

We see that fluctuations in the deamplified Y-quadrature phase amplitude are smaller than in the vacuum state (Eq. 9.35) whenever $\lambda > 2\bar{n}$; for $\bar{n} = 0$, the limit at threshold is $\Delta Y = \frac{1}{2}(1/\sqrt{2}) < \frac{1}{2}$. On the other hand, fluctuations in the amplified X-quadrature phase amplitude are larger, and diverge as the threshold is approached. An analogous divergence was seen for the laser (Eqs. 8.52), where we concluded that it arises from a breakdown of the system size expansion. Threshold fluctuations for the degenerate parametric oscillator are treated in Sect. 10.2.4. For the remainder of this section, let us see what our results for the fluctuations below threshold can tell us about non-positive-definite diffusion and squeezed states.

Although the choice of representation enters explicitly into (10.54a) and (10.54b), the results for $(\Delta X)_<$ and $(\Delta Y)_<$ are independent of σ. The $+\frac{1}{4}\sigma$ in (10.54a) and (10.54b) cancels with the σ-dependence in the variances of the phase-space variables. This of course is as it should be; different representations cannot produce different answers for the same operator average. The special significance of the Wigner representation is made clear here. It corresponds to the choice $\sigma = 0$; thus, fluctuations of the Wigner phase-space variables relate directly to the picture of squeezed fluctuations presented in Fig. 9.2 (see also Sect. 9.3.7).

At this point we are drawn to an unexpected conclusion. We wrote down the results for $(\Delta X)_<$ and $(\Delta Y)_<$ from the relationship (5.102a), which gives the steady-state covariance matrix for a linear Fokker–Planck equation in terms of its drift and diffusion matrices. This was done without asking if the diffusion matrix was positive semidefinite or not. With $\sigma = 0$ or -1, the diffusion constant in the Fokker–Planck equation (10.51b) is positive; but with $\sigma = 1$ it is not (whenever $\lambda > 2\bar{n}$). Thus, if we accept the results given by (10.54a) and (10.54b) when $\sigma = 0$ or -1, it appears that the P representation ($\sigma = 1$) also gives the correct results, even though the Fokker–Planck equation in this representation does not have positive semidefinite diffusion. Looking more closely, there is certainly a problem when the diffusion constant is negative, because, with $\sigma = 1$, the steady-state distribution given by (5.80) and (10.51b) is

$$\tilde{\tilde{Y}}_{+1}(\tilde{z}_2) \propto \exp\left(-\tilde{z}_2^2 \frac{4}{\xi} \frac{1+\lambda}{2\bar{n}-\lambda}\right). \tag{10.56}$$

This expression diverges as $|\tilde{z}_2| \to \infty$ if $\lambda > 2\bar{n}$. The distribution violates the normalization requirement (3.16) and cannot possibly give the characteristic function via a Fourier transform as it should (Eq. 3.73). From another perspective, the stochastic differential equations corresponding to Fokker–Planck equations (10.51a) and (10.51b) are

$$d\tilde{z}_1 = -(1-\lambda)\tilde{z}_1(\kappa dt) + \tfrac{1}{2}\sqrt{\xi\kappa\big[2\bar{n}+1-\sigma(1-\lambda)\big]}\,dW_{\tilde{z}_1}, \tag{10.57a}$$

$$d\tilde{z}_2 = -(1+\lambda)\tilde{z}_2(\kappa dt) + \tfrac{1}{2}\sqrt{\xi\kappa\big[2\bar{n}+1-\sigma(1+\lambda)\big]}\,dW_{\tilde{z}_2}. \tag{10.57b}$$

With $\sigma = 1$ and $\lambda > 2\bar{n}$, the fluctuation term driving the *real* variable $\tilde{z}_2$ is pure imaginary! Clearly there are problems—at least some sort of inconsistency in the formalism.

For the moment we set the formal questions aside, being satisfied with the observation that the right answers are obtained. We will soon tidy up the formalism by modifying the P representation, introducing the so-called positive P representation (Chap. 11). After this is done, we find the following somewhat surprising statement is true, in line with what is suggested by the above example:

Results for quantum averages derived using the formulae of Sects. 5.2.2–5.2.5 are correct even when the formulae are applied to a "Fokker–Planck" equation whose diffusion matrix is not positive semidefinite.

Now what can be said about the quantum state of the subharmonic field? It is clear from (10.55a) and (10.55b) that it is not a minimum uncertainty state, not a squeezed state; the uncertainly product is $(\Delta X)_<(\Delta Y)_< = \frac{1}{4}(2\bar{n}+1)/\sqrt{1-\lambda^2}$, and always greater than $\frac{1}{4}$. When Milburn and Walls [10.4] first analyzed this model, the minimum value $\Delta Y = \frac{1}{2}(1/\sqrt{2})$ obtained from (10.55b) was a bit of a disappointment. The fluctuation is reduced by only a factor $1/\sqrt{2}$ from its vacuum state value. Perhaps, however, this is not so different from what one might expect. If we take the interaction time for the fields inside the cavity to be the photon lifetime, then $\lambda = 1$ corresponds to a squeeze parameter of $r = \lambda/2 = 0.5$ (see Note 9.7). From this we would predict $\Delta Y = e^{-r}\frac{1}{2} = e^{-0.5}\frac{1}{2} \approx 0.30 \approx \frac{1}{2}(1/\sqrt{2})$, essentially just what the more sophisticated calculation carried out above has obtained. Although λ is larger above threshold, we should not hope for increased squeezing there. Once the parametric oscillator turns on, the pump field depletes to its threshold value; therefore, even above threshold, the squeeze parameter would never exceed $r = 0.5$.

Fortunately this pessimistic outlook is the result of an oversimplification. It is important to recognize that the cavity mode that carries the subharmonic field is actually a quasimode—i.e., it has a linewidth; also that it is the field escaping through the cavity mirrors and not the intracavity field that is measured. The complete picture is revealed by the fluctuations in the Fourier components of the subharmonic field escaping the cavity (see Sect. 9.3). In place of (10.55a) and (10.55b), we must look at the fluctuation amplitudes $\frac{1}{2}\sqrt{1 + S_\theta(\omega)}$, with $\theta = \frac{1}{2}(\psi \mp \pi/2)$, where

$$S_\theta(\omega) = (2\kappa)8\int_0^\infty d\tau \cos\omega\tau \lim_{t\to\infty} \langle :\Delta\hat{A}_\theta(t)\Delta\hat{A}_\theta(t+\tau): \rangle \tag{10.58}$$

is the *source-field spectrum of squeezing (with unit efficiency)* (Sect. 9.3.4); in this expression $\langle :: \rangle$ denotes the normal-ordered, time-ordered average and $\Delta\hat{A}_\theta \equiv \hat{A}_\theta - \langle \hat{A}_\theta \rangle$, with $\hat{A}_\theta$ defined by (9.26). Setting $\theta = \frac{1}{2}(\psi - \pi/2)$, and with $S_X(\omega) \equiv S_{\frac{1}{2}(\psi-\pi/2)}(\omega)$, we have

$$\begin{aligned} S_X(\omega) &= (2\kappa)8\int_0^\infty d\tau \cos\omega\tau \lim_{t\to\infty} \Big\langle :\tfrac{1}{2}\big[\tilde{a}(t)e^{-i\frac{1}{2}(\psi-\pi/2)} + \tilde{a}^\dagger(t)e^{i\frac{1}{2}(\psi-\pi/2)}\big] \\ &\quad \times \tfrac{1}{2}\big[\tilde{a}(t+\tau)e^{-i(\frac{1}{2}(\psi-\pi/2)} + \tilde{a}^\dagger(t+\tau)e^{i\frac{1}{2}(\psi-\pi/2)}\big] : \Big\rangle \\ &= (2\kappa)8\int_0^\infty d\tau \cos\omega\tau(2/\xi)n_p^{\rm thr} \lim_{t\to\infty} \Big\langle :\tfrac{1}{2}\big[\tilde{\bar{a}}(t) + \tilde{\bar{a}}^\dagger(t)\big] \\ &\quad \times \tfrac{1}{2}\big[\tilde{\bar{a}}(t+\tau) + \tilde{\bar{a}}^\dagger(t+\tau)\big] : \Big\rangle \\ &= (2\kappa)8\int_0^\infty d\tau \cos\omega\tau(2/\xi) \lim_{t\to\infty} \big(\overline{\tilde{z}_1(t)\tilde{z}_1(t+\tau)}\big)_{\tilde{\bar{X}}_{+1}}. \end{aligned} \tag{10.59a}$$

In a similar manner, with $\theta = \frac{1}{2}(\psi + \pi/2)$ and $S_Y(\omega) \equiv S_{\frac{1}{2}(\psi+\pi/2)}(\omega)$, we find

$$S_Y(\omega) = (2\kappa)8\int_0^\infty d\tau \cos\omega\tau(2/\xi) \lim_{t\to\infty} \big(\overline{\tilde{z}_2(t)\tilde{z}_2(t+\tau)}\big)_{\tilde{\bar{Y}}_{+1}}. \tag{10.59b}$$

The P representation ($\sigma = +1$) is used in these expressions to give normal-ordered, time-ordered correlation functions.

With the help of (5.93), (5.102a), and the Fokker–Planck equations (10.51a) and (10.51b), we obtain the correlation functions

$$\lim_{t\to\infty}\left(\overline{\tilde{z}_1(t)\tilde{z}_1(t+\tau)}\right)_{\tilde{\tilde{X}}_{+1}} = e^{-\kappa(1-\lambda)|\tau|}(\xi/2)\frac{1}{4}\frac{2\bar{n}+\lambda}{1-\lambda}, \tag{10.60a}$$

$$\lim_{t\to\infty}\left(\overline{\tilde{z}_2(t)\tilde{z}_2(t+\tau)}\right)_{\tilde{\tilde{Y}}_{+1}} = e^{-\kappa(1+\lambda)|\tau|}(\xi/2)\frac{1}{4}\frac{2\bar{n}-\lambda}{1+\lambda}; \tag{10.60b}$$

hence,

$$S_X^<(\omega) = \frac{2\bar{n}+\lambda}{1-\lambda}\frac{4\kappa^2(1-\lambda)}{[\kappa(1-\lambda)]^2+\omega^2}, \tag{10.61a}$$

$$S_Y^<(\omega) = \frac{2\bar{n}-\lambda}{1+\lambda}\frac{4\kappa^2(1+\lambda)}{[\kappa(1+\lambda)]^2+\omega^2}. \tag{10.61b}$$

Thus, we obtain the *quadrature phase amplitude fluctuations of the cavity output for the degenerate parametric oscillator below threshold*:

$$\frac{1}{2}\sqrt{1+S_X^<(\omega)} = \frac{1}{2}\sqrt{\frac{8\kappa^2\bar{n}+[\kappa(1+\lambda)]^2+\omega^2}{[\kappa(1-\lambda)]^2+\omega^2}}, \tag{10.62a}$$

$$\frac{1}{2}\sqrt{1+S_Y^<(\omega)} = \frac{1}{2}\sqrt{\frac{8\kappa^2\bar{n}+[\kappa(1-\lambda)]^2+\omega^2}{[\kappa(1+\lambda)]^2+\omega^2}}. \tag{10.62b}$$

In these expressions we find a close connection with minimum uncertainty squeezed states. If $\bar{n}=0$, the minimum uncertainty condition $\frac{1}{2}\sqrt{1+S_X^<(\omega)}$ $\times\frac{1}{2}\sqrt{1+S_Y^<(\omega)}=\frac{1}{4}$ is satisfied at each frequency; furthermore, the squeezing at line center becomes perfect as threshold is approached: $\frac{1}{2}\sqrt{1+S_Y^<(0)}\to 0$ and $\frac{1}{2}\sqrt{1+S_X^<(0)}\to\infty$ as $\lambda\to 1$.

Note 10.3. These results computed from (10.58) assume that both optical outputs are detected with unit efficiency (Sect. 9.3.4) and that $\gamma_{a\alpha}\ll\gamma_{1\mu}$ $(\mu=1,2)$. Also, since only the source-field spectrum of squeezing is considered, if $\bar{n}$ is taken to be nonzero, the nonzero value must be associated with losses in the crystal—i.e., $\bar{n}=\bar{n}_{a\alpha}$ $(\bar{n}_{a1}=\bar{n}_{a2}=0)$. Spectra of squeezing for more general conditions are given in (9.169a) and (9.169b).

The squeezing of one quadrature phase amplitude of the subharmonic field shows up in its optical spectrum. Well below threshold, we might expect this spectrum to be a Lorentzian with the cavity linewidth, just as it is for the laser below threshold (Eqs. 8.69 and 8.70). In fact this is not so. We calculate the optical spectrum from the Fourier transform of the normalized first-order correlation function

$$\begin{aligned} g_<^{(1)}(\tau) &= \left(\langle a^\dagger a\rangle_<\right)^{-1}\left[\lim_{t\to\infty}\langle a^\dagger(t)a(t+\tau)\rangle\right] \\ &= \left(\langle a^\dagger a\rangle_<\right)^{-1}e^{-i\omega_C\tau}(2/\xi)\Big[\lim_{t\to\infty}\left(\overline{\tilde{z}_1(t)\tilde{z}_1(t+\tau)}\right)_{\tilde{\tilde{X}}_{+1}} \\ &\quad + \lim_{t\to\infty}\left(\overline{\tilde{z}_2(t)\tilde{z}_2(t+\tau)}\right)_{\tilde{\tilde{Y}}_{+1}}\Big], \end{aligned} \tag{10.63}$$

where, using (9.29), (9.31), and (10.55a) and (10.55b), *the mean photon number in the subharmonic mode* is given by

$$\begin{aligned}\langle a^\dagger a\rangle_< \equiv \langle a^\dagger a\rangle_{ss}^< &= \langle(\hat{X} - i\hat{Y})(\hat{X} + i\hat{Y})\rangle_{ss}^< \\ &= \langle\hat{X}^2\rangle_{ss}^< + \langle\hat{Y}^2\rangle_{ss}^< + i\langle[\hat{X}, \hat{Y}]\rangle_{ss}^< \\ &= (\Delta X)_<^2 + (\Delta Y)_<^2 - \tfrac{1}{2} \\ &= \frac{1}{2}\frac{2\bar{n} + \lambda^2}{1 - \lambda^2}. \end{aligned} \tag{10.64}$$

Then, since correlation functions (10.60a) and (10.60b) are real and symmetric in τ, by comparing the expressions (10.59a) and (10.59b) with (10.63), we are able to write the *optical spectrum for the degenerate parametric oscillator below threshold* as

$$\begin{aligned} T_<(\omega) &= \frac{1}{\pi}\int_0^\infty d\tau \cos[(\omega - \omega_C)\tau]|g_<^{(1)}(\tau)| \\ &= (\langle a^\dagger a\rangle_<)^{-1}\frac{S_X^<(\omega - \omega_C) + S_Y^<(\omega - \omega_C)}{16\kappa\pi} \\ &= \frac{1}{2}\frac{1 - \lambda^2}{2\bar{n} + \lambda^2}\left\{\frac{2\bar{n} + \lambda}{1 - \lambda}\frac{\kappa(1 - \lambda)/\pi}{[\kappa(1 - \lambda)]^2 + (\omega - \omega_C)^2}\right. \\ &\qquad \left. + \frac{2\bar{n} - \lambda}{1 + \lambda}\frac{\kappa(1 + \lambda)/\pi}{[\kappa(1 + \lambda)]^2 + (\omega - \omega_C)^2}\right\}. \end{aligned} \tag{10.65}$$

Specifically, well below threshold ($\lambda \ll 1$), we have

$$T_<(\omega) = \frac{1}{2\bar{n} + \lambda}\left[2\bar{n}\frac{\kappa/\pi}{\kappa^2 + (\omega - \omega_C)^2} + \lambda^2\frac{2\kappa^3/\pi}{[\kappa^2 + (\omega - \omega_C)^2]^2}\right]. \tag{10.66}$$

The thermal noise contribution to this spectrum is, indeed, a Lorentzian with the cavity linewidth; but the term associated with the quantum fluctuations is a Lorentzian *squared.* The square arises because the quantum contributions to $S_X^<(\omega - \omega_C)$ and $S_Y^<(\omega - \omega_C)$ *subtract* in (10.65); quantum fluctuations in the Y-quadrature phase amplitude enter (10.65) with negative weight because they are squeezed. This nonclassical effect is discussed in detail by Rice and Carmichael [10.5] who show that it explains a feature in the spectrum of resonance fluorescence noted by Mollow some thirty years ago [10.6].

Note 10.4. It is not always possible to decompose the optical spectrum into the sum of two spectra of squeezing as in (10.65). It is clear that the decomposition will not always work, because the spectrum of squeezing is symmetric about $\omega = 0$, by definition, while the optical spectrum need not be symmetric about ω_C. The decomposition is possible here because fluctuations in the quadrature phase variables $\tilde{z}_1$ and $\tilde{z}_2$ are uncorrelated; thus, in (10.63) the correlation function $e^{i\omega_C\tau}g^{(1)}(\tau)$ is real and given by a sum of correlation functions describing the independent fluctuations in $\tilde{z}_1$ and $\tilde{z}_2$.

At the conclusion of this section we are in a position to illustrate a point raised some time ago about the evaluation of two-time averages within the phase-space representations. Two-time averages calculated "classically" in different representations correspond to quantum averages with different operator orderings. When two times are involved, we generally do not know the commutation relations needed to reorder the operators so that the average can be evaluated in the representation of our choosing [see the discussion below (4.128)]. The spectrum of squeezing (10.58) and the optical spectrum (10.65) are defined in terms of normal-ordered, time-ordered operator averages; we therefore calculated these spectra using diffusion constants taken from Fokker–Planck equations (10.51a) and (10.51b) with the choice $\sigma = +1$ (using the P representation). Without knowing the mean commutators $\lim_{t\to\infty}\langle[a(t+\tau), a^\dagger(t)]\rangle$ and $\lim_{t\to\infty}\langle[a(t+\tau), a(t)]\rangle$, we cannot reorder the operator averages to calculate $S_X^{\lessgtr}(\omega)$, $S_Y^{\lessgtr}(\omega)$, and $T_<(\omega)$ in the Q and Wigner representations. In fact, things work exactly the other way around; two-time averages calculated in the Q and Wigner representations tell us the mean commutators for operators evaluated at unequal times:

Exercise 10.2. Working from (4.100), (4.113), and (4.124), respectively, show that the P representation gives

$$\lim_{t\to\infty}\langle\tilde{a}^\dagger(t)\tilde{a}(t+\tau)\rangle = \frac{1}{4}\left[e^{-\kappa(1-\lambda)|\tau|}\frac{2\bar{n}+\lambda}{1-\lambda} + e^{-\kappa(1+\lambda)|\tau|}\frac{2\bar{n}-\lambda}{1+\lambda}\right], \quad (10.67a)$$

$$\lim_{t\to\infty}\langle\tilde{a}(t+|\tau|)\tilde{a}(t)\rangle = e^{i(\psi-\frac{\pi}{2})}\frac{1}{4}\left[e^{-\kappa(1-\lambda)|\tau|}\frac{2\bar{n}+\lambda}{1-\lambda} - e^{-\kappa(1+\lambda)|\tau|}\frac{2\bar{n}-\lambda}{1+\lambda}\right], \quad (10.67b)$$

that the Q representation gives

$$\lim_{t\to\infty}\langle\tilde{a}(t+\tau)\tilde{a}^\dagger(t)\rangle = \frac{1}{4}\left[e^{-\kappa(1-\lambda)|\tau|}\frac{2\bar{n}+2-\lambda}{1-\lambda} + e^{-\kappa(1+\lambda)|\tau|}\frac{2\bar{n}+2+\lambda}{1+\lambda}\right], \quad (10.68a)$$

$$\lim_{t\to\infty}\langle\tilde{a}(t)\tilde{a}(t+|\tau|)\rangle = e^{i(\psi-\frac{\pi}{2})}\frac{1}{4}\left[e^{-\kappa(1-\lambda)|\tau|}\frac{2\bar{n}+2-\lambda}{1-\lambda} - e^{-\kappa(1+\lambda)|\tau|}\frac{2\bar{n}+2+\lambda}{1+\lambda}\right], \quad (10.68b)$$

and that the Wigner representation gives

$$\lim_{t\to\infty} \tfrac{1}{2}\left[\langle \tilde{a}^\dagger(t)\tilde{a}(t+\tau)\rangle + \langle \tilde{a}(t+\tau)\tilde{a}^\dagger(t)\rangle\right]$$
$$= \frac{1}{4}\left[e^{-\kappa(1-\lambda)|\tau|}\frac{2\bar{n}+1}{1-\lambda} + e^{-\kappa(1+\lambda)|\tau|}\frac{2\bar{n}+1}{1+\lambda}\right], \tag{10.69a}$$

$$\lim_{t\to\infty} \tfrac{1}{2}\left[\langle \tilde{a}(t)\tilde{a}(t+\tau)\rangle + \langle \tilde{a}(t+\tau)\tilde{a}(t)\rangle\right]$$
$$= e^{i(\psi-\frac{\pi}{2})}\frac{1}{4}\left[e^{-\kappa(1-\lambda)|\tau|}\frac{2\bar{n}+1}{1-\lambda} - e^{-\kappa(1+\lambda)|\tau|}\frac{2\bar{n}+1}{1+\lambda}\right]. \tag{10.69b}$$

Hence obtain the mean commutators

$$\lim_{t\to\infty} \langle [\tilde{a}(t+\tau), \tilde{a}^\dagger(t)]\rangle = \tfrac{1}{2}\left[e^{-\kappa(1-\lambda)|\tau|} + e^{-\kappa(1+\lambda)|\tau|}\right], \tag{10.70a}$$

and

$$\lim_{t\to\infty} \langle [\tilde{a}(t+|\tau|), \tilde{a}(t)]\rangle = -e^{i(\psi-\frac{\pi}{2})}\tfrac{1}{2}\left[e^{-\kappa(1-\lambda)|\tau|} - e^{-\kappa(1+\lambda)|\tau|}\right]. \tag{10.70b}$$

Note that (10.70a) and (10.70b) give the expected results $\langle [\tilde{a}, \tilde{a}^\dagger]\rangle = 1$ and $\langle [\tilde{a}, \tilde{a}]\rangle = 0$ at $\tau = 0$. Note also that (10.67a)–(10.69b) are self-consistent: the two-time averages given by the Wigner representation are averages of those given by the P and Q representations.

Note 10.5. It would not be correct to conclude that the Q and Wigner representations cannot be used to calculate normal-ordered, time-ordered averages. What we have just noted concerns calculations carried out entirely within the phase-space formalism—the calculation of a two-time operator average as a two-time phase-space average. Alternatively, one can invoke the quantum regression formula—in the form (1.107)–(1.109)—and solve equations of motion for the desired correlation functions. The initial conditions for these equations require us to calculate certain one-time averages. These can be evaluated in the P, Q, or Wigner representations, using known commutation relations to switch between the different operator orderings. In the above example, the mean quadrature phase amplitudes obey the equations of motion

$$\langle \dot{\hat{X}}\rangle = -\kappa(1-\lambda)\langle \hat{X}\rangle, \qquad \langle \dot{\hat{Y}}\rangle = -\kappa(1+\lambda)\langle \hat{Y}\rangle. \tag{10.71}$$

These follow as averages of the stochastic differential equations (10.57a) and (10.57b). According to the quantum regression formula, (10.67a)–(10.70b) can then be derived by solving equations of motion for two-time averages in the same form. All that is needed to get started are the initial conditions, which may be derived from (10.54a) and (10.54b) using whichever representation one chooses; the $+\frac{1}{4}\sigma$ accounts for the necessary commutators.

10.2.3 Quantum Fluctuations Above Threshold

The treatment of the laser above threshold called for a rather complicated application of the system size expansion based on amplitude and phase variables. Treatment of the degenerate parametric oscillator is simpler, since the subharmonic field above threshold has a well-defined phase (Eqs. 9.77). It is then possible to use the same system size expansion as below threshold. The only change is that the fluctuations are now to be computed for the new steady state solutions

$$\langle \tilde{\bar{a}}(t) \rangle = \langle \tilde{\bar{a}}^\dagger(t) \rangle = \pm\sqrt{\lambda - 1}, \tag{10.72a}$$

$$\langle \tilde{\bar{b}}(t) \rangle = \langle \tilde{\bar{b}}^\dagger(t) \rangle = 1. \tag{10.72b}$$

Substituting these into Fokker–Planck equation (10.45), the resulting equation is again separable, but now with a solution in the form

$$\tilde{\bar{F}}_\sigma(\tilde{z}, \tilde{z}^*, \tilde{w}, \tilde{w}^*, t) = \tilde{\bar{\mathcal{X}}}_\sigma(\tilde{z}_1, \tilde{w}_1, t)\tilde{\bar{\mathcal{Y}}}_\sigma(\tilde{z}_2, \tilde{w}_2, t). \tag{10.73}$$

Above threshold, fluctuations in the X-quadrature phase amplitudes of the subharmonic and pump fields are coupled, and described by the *Fokker–Planck equation above threshold for the X-quadrature phase amplitudes of the degenerate parametric oscillator,*

$$\begin{aligned}\frac{\partial \tilde{\bar{\mathcal{X}}}_\sigma}{\partial t} = &\left\{ \kappa \frac{\partial}{\partial \tilde{z}_1}\left(\mp \sqrt{\lambda - 1}\tilde{w}_1\right) + \kappa_p \frac{\partial}{\partial \tilde{w}_1}\left(\tilde{w}_1 \pm 2\sqrt{\lambda-1}\tilde{z}_1\right)\right.\\ &\left. + \frac{1}{8}\xi\kappa(2\bar{n}+1)\frac{\partial^2}{\partial \tilde{z}_1^2} + \frac{1}{4}\kappa_p(2\bar{n}_p + 1 - \sigma)\frac{\partial^2}{\partial \tilde{w}_1^2}\right\}\tilde{\bar{\mathcal{X}}}_\sigma,\end{aligned} \tag{10.74a}$$

while fluctuations in the Y-quadrature phase amplitudes are also coupled, and described by the *Fokker-Planck equation above threshold for the Y-quadrature phase amplitudes of the degenerate parametric oscillator,*

$$\begin{aligned}\frac{\partial \tilde{\bar{\mathcal{Y}}}_\sigma}{\partial t} = &\left\{ \kappa \frac{\partial}{\partial \tilde{z}_2}\left(2\tilde{z}_2 \mp \sqrt{\lambda - 1}\tilde{w}_2\right) + \kappa_p \frac{\partial}{\partial \tilde{w}_2}\left(\tilde{w}_2 \pm 2\sqrt{\lambda-1}\tilde{z}_2\right)\right.\\ &\left. + \frac{1}{8}\xi\kappa(2\bar{n}+1-2\sigma)\frac{\partial^2}{\partial \tilde{z}_2^2} + \frac{1}{4}\kappa_p(2\bar{n}_p + 1 - \sigma)\frac{\partial^2}{\partial \tilde{w}_2^2}\right\}\tilde{\bar{\mathcal{Y}}}_\sigma.\end{aligned} \tag{10.74b}$$

Since equations in two dimensions must be considered here, the mechanics of the calculation are a little more involved. There is no fundamental difference, however, from the analysis of fluctuations below threshold. The details are left as an exercise.

Exercise 10.3. With $\hat{X} \equiv \hat{A}_\theta$, $\hat{Y} \equiv \hat{A}_{\theta+\pi/2}$, $\hat{X}_p \equiv \hat{B}_\theta$, and $\hat{Y}_p \equiv \hat{B}_{\theta+\pi/2}$, where $\hat{A}_\theta$ is defined by (9.26) (with $\hat{B}_\theta$ similarly defined) and $\theta = \frac{1}{2}(\psi - \frac{\pi}{2})$, use (5.102a), and (10.74a) and (10.74b), to show that the *covariance matrix in quadrature phase amplitudes for the degenerate parametric oscillator above* *threshold* is given by

$$(\Delta X)^2_> \equiv \langle(\hat{X} - \langle\hat{X}\rangle)^2\rangle^>_{\mathrm{ss}} = (2/\xi)\big[\big(\overline{\tilde{z}_1\tilde{z}_1}\big)_{\tilde{\mathcal{X}}_\sigma} + \tfrac{1}{4}\sigma\big]$$

$$= \frac{1}{4}\left[\frac{1}{2}\frac{2\bar{n}+1}{\lambda-1} + \frac{\kappa(2\bar{n}+1)+\kappa_p(2\bar{n}_p+1)}{\kappa_p}\right], \tag{10.75a}$$

$$(\Delta Y)^2_> \equiv \langle(\hat{Y} - \langle\hat{Y}\rangle)^2\rangle^>_{\mathrm{ss}} = (2/\xi)\big[\big(\overline{\tilde{z}_2\tilde{z}_2}\big)_{\tilde{\mathcal{Y}}_\sigma} + \tfrac{1}{4}\sigma\big]$$

$$= \frac{1}{4}\frac{\frac{1}{2}(2\kappa\lambda+\kappa_p)(2\bar{n}+1)+\kappa_p(\lambda-1)(2\bar{n}_p+1)}{(2\kappa+\kappa_p)\lambda}, \tag{10.75b}$$

and

$$(\Delta X_p)^2_> \equiv \langle(\hat{X}_p - \langle\hat{X}_p\rangle)^2\rangle^>_{\mathrm{ss}} = \big(\overline{\tilde{w}_1\tilde{w}_1}\big)_{\tilde{\mathcal{X}}_\sigma} + \tfrac{1}{4}\sigma$$

$$= \frac{1}{4}\frac{\kappa(2\bar{n}+1)+\kappa_p(2\bar{n}_p+1)}{\kappa_p}, \tag{10.76a}$$

$$(\Delta Y_p)^2_> \equiv \langle(\hat{Y}_p - \langle\hat{Y}_p\rangle)^2\rangle^>_{\mathrm{ss}} = \big(\overline{\tilde{w}_2\tilde{w}_2}\big)_{\tilde{\mathcal{Y}}_\sigma} + \tfrac{1}{4}\sigma$$

$$= \frac{1}{4}\frac{\kappa(\lambda-1)(2\bar{n}+1)+(2\kappa+\kappa_p\lambda)(2\bar{n}_p+1)}{(2\kappa+\kappa_p)\lambda}, \tag{10.76b}$$

and

$$\langle(\hat{X} - \langle\hat{X}\rangle)(\hat{X}_p - \langle\hat{X}_p\rangle)\rangle^>_{\mathrm{ss}} = \sqrt{2/\xi}\big(\overline{\tilde{z}_1\tilde{w}_1}\big)_{\tilde{\mathcal{X}}_\sigma}$$

$$= \mp\frac{1}{4}\sqrt{\frac{\kappa}{2\kappa_p}}\frac{2\bar{n}+1}{\sqrt{\lambda-1}}, \tag{10.77a}$$

$$\langle(\hat{Y} - \langle\hat{Y}\rangle)(\hat{Y}_p - \langle\hat{Y}_p\rangle)\rangle^>_{\mathrm{ss}} = \sqrt{2/\xi}\big(\overline{\tilde{z}_2\tilde{w}_2}\big)_{\tilde{\mathcal{Y}}_\sigma}$$

$$= \mp\frac{1}{4}\sqrt{\frac{2\kappa_p}{\kappa}}\frac{\kappa\sqrt{\lambda-1}\big[\frac{1}{2}(2\bar{n}+1)-(2\bar{n}_p+1)\big]}{(2\kappa+\kappa_p)\lambda}. \tag{10.77b}$$

Note that all results are independent of σ.

We see from (10.75b) and (10.76b) that the subharmonic and pump fields are both squeezed above threshold: setting the thermal photon numbers to zero, $(\Delta Y)_>$ varies monotonically from $\frac{1}{2}(1/\sqrt{2})$ at $\lambda = 1$, to $\frac{1}{2}\sqrt{(\kappa+\kappa_p)/(2\kappa+\kappa_p)}$ in the limit $\lambda \to \infty$; $(\Delta Y_p)_>$ varies monotonically from $\frac{1}{2}$ at $\lambda = 1$ to

$\frac{1}{2}\sqrt{(\kappa+\kappa_p)/(2\kappa+\kappa_p)}$ for $\lambda \to \infty$. These are the intracavity results. To obtain spectra of squeezing like (10.61a) and (10.61b), we must work with the correlation matrices

$$\boldsymbol{C}_{\tilde{\bar{\mathcal{X}}}_\sigma}(\tau) = \lim_{t\to\infty}\begin{pmatrix} \big(\overline{\tilde{z}_1(t)\tilde{z}_1(t+\tau)}\big)_{\tilde{\bar{\mathcal{X}}}_\sigma} & \big(\overline{\tilde{z}_1(t)\tilde{w}_1(t+\tau)}\big)_{\tilde{\bar{\mathcal{X}}}_\sigma} \\ \big(\overline{\tilde{w}_1(t)\tilde{z}_1(t+\tau)}\big)_{\tilde{\bar{\mathcal{X}}}_\sigma} & \big(\overline{\tilde{w}_1(t)\tilde{w}_1(t+\tau)}\big)_{\tilde{\bar{\mathcal{X}}}_\sigma} \end{pmatrix}, \tag{10.78a}$$

and

$$\boldsymbol{C}_{\tilde{\bar{\mathcal{Y}}}_\sigma}(\tau) = \lim_{t\to\infty}\begin{pmatrix} \big(\overline{\tilde{z}_2(t)\tilde{z}_2(t+\tau)}\big)_{\tilde{\bar{\mathcal{Y}}}_\sigma} & \big(\overline{\tilde{z}_2(t)\tilde{w}_2(t+\tau)}\big)_{\tilde{\bar{\mathcal{Y}}}_\sigma} \\ \big(\overline{\tilde{w}_2(t)\tilde{z}_2(t+\tau)}\big)_{\tilde{\bar{\mathcal{Y}}}_\sigma} & \big(\overline{\tilde{w}_2(t)\tilde{w}_2(t+\tau)}\big)_{\tilde{\bar{\mathcal{Y}}}_\sigma} \end{pmatrix}. \tag{10.78b}$$

These are evaluated using (Eq. 5.93b)

$$\boldsymbol{C}_{\tilde{\bar{\mathcal{X}}}_{+1}}(\tau) = \boldsymbol{C}_{\tilde{\bar{\mathcal{X}}}_{+1}}(0)\exp(\boldsymbol{A}_{\mathcal{X}}^T\tau), \tag{10.79a}$$

$$\boldsymbol{C}_{\tilde{\bar{\mathcal{Y}}}_{+1}}(\tau) = \boldsymbol{C}_{\tilde{\bar{\mathcal{Y}}}_{+1}}(0)\exp(\boldsymbol{A}_{\mathcal{Y}}^T\tau), \tag{10.79b}$$

where $\boldsymbol{C}_{\tilde{\bar{\mathcal{X}}}_{+1}}(0)$ and $\boldsymbol{C}_{\tilde{\bar{\mathcal{Y}}}_{+1}}(0)$ are to be computed from the right-hand sides of (10.75a)–(10.77b) (with $\sigma = +1$) and the drift matrices are

$$\boldsymbol{A}_{\mathcal{X}} = \begin{pmatrix} 0 & \pm\kappa\sqrt{\lambda-1} \\ \mp 2\kappa_p\sqrt{\lambda-1} & -\kappa_p \end{pmatrix}, \tag{10.80a}$$

$$\boldsymbol{A}_{\mathcal{Y}} = \begin{pmatrix} -2\kappa & \pm\kappa\sqrt{\lambda-1} \\ \mp 2\kappa_p\sqrt{\lambda-1} & -\kappa_p \end{pmatrix}. \tag{10.80b}$$

From the definition of the source-field spectrum of squeezing (with unit efficiency) (Eq. 10.58) we then have

$$\begin{aligned} S_X(\omega) &= (2\kappa)8\int_0^\infty d\tau\cos\omega\tau(2/\xi)\Big[\boldsymbol{C}_{\tilde{\bar{\mathcal{X}}}_{+1}}(\tau)\Big]_{11} \\ &= (2\kappa)8\mathrm{Re}\int_0^\infty d\tau e^{i\omega\tau}(2/\xi)\Big[\boldsymbol{C}_{\tilde{\bar{\mathcal{X}}}_{+1}}(0)\exp(\boldsymbol{A}_{\mathcal{X}}^T\tau)\Big]_{11} \\ &= (2\kappa)8\mathrm{Re}\int_0^\infty d\tau(2/\xi)\Big[\boldsymbol{C}_{\tilde{\bar{\mathcal{X}}}_{+1}}(0)\exp\big[(\boldsymbol{A}_{\mathcal{X}}^T + i\omega\boldsymbol{I}_2)\tau\big]\Big]_{11} \\ &= (2\kappa)8\mathrm{Re}\Big[(2\xi)\boldsymbol{C}_{\tilde{\bar{\mathcal{X}}}_{+1}}(0)(\boldsymbol{A}_{\mathcal{X}}^T + i\omega\boldsymbol{I}_2)^{-1}\Big]_{11}, \end{aligned} \tag{10.81a}$$

and in a similar manner,

$$S_Y(\omega) = (2\kappa)8\mathrm{Re}\Big[(2\xi)\boldsymbol{C}_{\tilde{\bar{\mathcal{Y}}}_{+1}}(0)(\boldsymbol{A}_{\mathcal{Y}}^T + i\omega\boldsymbol{I}_2)^{-1}\Big]_{11}. \tag{10.81b}$$

Similarly, for the pump mode we have the expressions

$$S_{X_p}(\omega) = (2\kappa_p)8\mathrm{Re}\Big[(2\xi)\boldsymbol{C}_{\tilde{\tilde{\mathcal{X}}}_{+1}}(0)(\boldsymbol{A}_{\mathcal{X}}^T + i\omega\boldsymbol{I}_2)^{-1}\Big]_{22}, \tag{10.82a}$$

and

$$S_{Y_p}(\omega) = (2\kappa_p)8\mathrm{Re}\Big[(2\xi)\boldsymbol{C}_{\tilde{\tilde{\mathcal{Y}}}_{+1}}(0)(\boldsymbol{A}_{\mathcal{Y}}^T + i\omega\boldsymbol{I}_2)^{-1}\Big]_{22}. \tag{10.82b}$$

Although these spectra can be developed explicitly in analytic form, the results are sufficiently complicated that there is little to be gained from them. It is more efficient to compute spectra numerically from the formal expressions.

Exercise 10.4. Write a computer program to compute the spectra of squeezing $S_X^>(\omega)$, $S_Y^>(\omega)$, $S_{X_p}^>(\omega)$, and $S_{Y_p}^>(\omega)$, and use it to study the parameter dependence of squeezing in the degenerate parametric oscillator above threshold. Also compute the optical spectra $T_>(\omega)$ and $T_p^>(\omega)$, obtaining them from the spectra of squeezing in the manner used to derive (10.65).

10.2.4 Quantum Fluctuations at Threshold

We saw in Sec. 8.2 that the system size expansion used to treat the laser below threshold breaks down as threshold is approached. The breakdown is evidenced by a divergence of fluctuations, where the source of the divergence traces back to the fact that at least one eigenvalue of the deterministic equations of motion (two in the laser case) vanishes at threshold. We can expect similar behavior in the degenerate parametric oscillator. Indeed, the results derived so far do show divergences as threshold is approached: below threshold divergences occur in (10.55a) and (10.64), and above threshold the expressions (10.75a) and (10.77a) diverge. The eigenvalue responsible for these divergences is $\Lambda_1 = -\kappa(1-\lambda)$, the eigenvalue governing the decay of the real part of the subharmonic field [after removal of the phase $\frac{1}{2}(\psi - \frac{\pi}{2})$]; Λ_1 is negative below threshold, vanishes at threshold, and becomes positive above threshold, where it determines the growth rate of the amplified fluctuations that switch the degenerate parametric oscillator from the unstable vacuum state to a stable state of sustained oscillation.

Because Λ_1 vanishes, we need a new version of the system size expansion to treat threshold fluctuations. First, let us determine where the expansion based on the scaling (10.33)–(10.38) breaks down. To this end, we return to the phase-space equation of motion (10.42) and substitute the solution $\langle\tilde{\bar{a}}(t)\rangle = \langle\tilde{\bar{a}}^\dagger(t)\rangle = 0$, $\langle\tilde{\bar{b}}(t)\rangle = \langle\tilde{\bar{b}}^\dagger(t)\rangle = \lambda$, which describes the macroscopic steady state both below and at threshold. Then in a rotating frame, defined by (10.47), the phase-space equation of motion is

$$\begin{aligned}\frac{\partial \tilde{\tilde{F}}_\sigma}{\partial t} = \bigg\{ & \kappa \frac{\partial}{\partial \tilde{z}} \Big[\tilde{z} - \lambda \tilde{z}^* - \big(n_p^{\rm thr}\big)^{-1/2} \tilde{z}^* \tilde{w} \Big] + \text{c.c.} \\ & + \kappa_p \frac{\partial}{\partial \tilde{w}} \Big[\tilde{w} + \big(n_p^{\rm thr}\big)^{-1/2} \tilde{z}^2 \Big] + \text{c.c.} \\ & + \frac{1}{4} \xi \sigma \kappa \Big[\frac{\partial^2}{\partial \tilde{z}^2} \Big(\lambda + \big(n_p^{\rm thr}\big)^{-1/2} \tilde{w} \Big) + \text{c.c.} \Big] \\ & + \xi \kappa \big[\bar{n} + \tfrac{1}{2}(1-\sigma) \big] \frac{\partial^2}{\partial \tilde{z} \partial \tilde{z}^*} + 2\kappa_p \big[\bar{n}_p + \tfrac{1}{2}(1-\sigma) \big] \frac{\partial^2}{\partial \tilde{w} \partial \tilde{w}^*} \\ & + \big(n_p^{\rm thr}\big)^{-1/2} \frac{1}{16} \xi (1-\sigma^2) \kappa \Big(\frac{\partial^3}{\partial \tilde{z}^2 \partial \tilde{w}^*} + \text{c.c.} \Big) \bigg\} \tilde{\tilde{F}}_\sigma . \end{aligned} \tag{10.83}$$

Written as it is, in terms of complex field amplitudes, this equation does not appear to meet with any special problems at threshold. When rewritten in terms of real variables, though, it becomes

$$\begin{aligned} & \frac{\partial \tilde{\tilde{F}}_\sigma}{\partial t} \\ & = \bigg\{ \kappa \frac{\partial}{\partial \tilde{z}_1} \Big[(1-\lambda) \tilde{z}_1 - \big(n_p^{\rm thr}\big)^{-1/2} (\tilde{z}_1 \tilde{w}_1 + \tilde{z}_2 \tilde{w}_2) \Big] \\ & + \kappa \frac{\partial}{\partial \tilde{z}_2} \Big[(1+\lambda) \tilde{z}_2 + \big(n_p^{\rm thr}\big)^{-1/2} (\tilde{z}_1 \tilde{w}_2 - \tilde{w}_1 \tilde{z}_2) \Big] \\ & + \kappa_p \frac{\partial}{\partial \tilde{w}_1} \Big[\tilde{w}_1 + \big(n_p^{\rm thr}\big)^{-1/2} (\tilde{z}_1^2 - \tilde{z}_2^2) \Big] \\ & + \kappa_p \frac{\partial}{\partial \tilde{w}_2} \Big[\tilde{w}_2 + \big(n_p^{\rm thr}\big)^{-1/2} 2 \tilde{z}_1 \tilde{z}_2 \Big] \\ & + \frac{1}{8} \xi \sigma \kappa \bigg[\Big(\frac{\partial^2}{\partial \tilde{z}_1^2} - \frac{\partial^2}{\partial \tilde{z}_2^2} \Big) \Big(\lambda + \big(n_p^{\rm thr}\big)^{-1/2} \tilde{w}_1 \Big) + 2 \big(n_p^{\rm thr}\big)^{-1/2} \frac{\partial^2}{\partial \tilde{z}_1 \partial \tilde{z}_2} \tilde{w}_2 \bigg] \\ & + \frac{1}{8} \xi \kappa (2\bar{n} + 1 - \sigma) \Big(\frac{\partial^2}{\partial \tilde{z}_1^2} + \frac{\partial^2}{\partial \tilde{z}_2^2} \Big) + \frac{1}{4} \kappa_p (2\bar{n}_p + 1 - \sigma) \Big(\frac{\partial^2}{\partial \tilde{w}_1^2} + \frac{\partial^2}{\partial \tilde{w}_2^2} \Big) \\ & + \big(n_p^{\rm thr}\big)^{-1/2} \frac{1}{64} \xi (1-\sigma^2) \kappa \bigg[\Big(\frac{\partial^2}{\partial \tilde{z}_1^2} - \frac{\partial^2}{\partial \tilde{z}_2^2} \Big) \frac{\partial}{\partial \tilde{w}_1} + 2 \frac{\partial^2}{\partial \tilde{z}_1 \partial \tilde{z}_2} \frac{\partial}{\partial \tilde{w}_2} \bigg] \bigg\} \tilde{\tilde{F}}_\sigma . \end{aligned} \tag{10.84}$$

In this form it is clear that at threshold, when $\lambda = 1$, or more generally when $1 - \lambda \sim \big(n_p^{\rm thr}\big)^{-1/2} \tilde{w}_1$, we cannot justify dropping the nonlinear drift in the $\tilde{z}_1$ direction. In this region the nonlinearity is needed to damp the fluctuations in $\tilde{z}_1$, which otherwise grow without bound. Our task then is to find a scaling of the variables that gives a systematic treatment of fluctuations when the nonlinear terms in (10.84) are retained to lowest order. Since $\tilde{z}_1$ becomes a *"slow" variable* near threshold, damped by the gradient of an ever shallower potential as the threshold is approached, the required scaling cannot be finalized until all *"fast" variables* have been adiabatically eliminated; thus, as in the laser case (Sect. 8.2.1), the system size expansion at threshold introduces a separation of time scales for the fluctuations, in addition to the usual separation of the fluctuations from the macroscopic law.

There is no reason to anticipate a change in the scaling of variables $\tilde{z}_2$, $\tilde{w}_1$, and $\tilde{w}_2$, since the linear drift vanishes in the $\tilde{z}_1$ direction only. We therefore make only one change to the scaling (10.33)–(10.38); in place of (10.34), we write

$$e^{i\omega_C t}\bar{\alpha} = \left(n_p^{\rm thr}\right)^{-q}\tilde{z}_1 + i\left(n_p^{\rm thr}\right)^{-1/2}\tilde{z}_2, \tag{10.85a}$$

$$e^{-i\omega_C t}\bar{\alpha}^* = \left(n_p^{\rm thr}\right)^{-q}\tilde{z}_1 - i\left(n_p^{\rm thr}\right)^{-1/2}\tilde{z}_2, \tag{10.85b}$$

where q is to be chosen self-consistently in the manner outlined in Sects. 5.1.3 and 5.1.4. The change calls for the replacement

$$\tilde{z}_1 \to \left(n_p^{\rm thr}\right)^{-q+1/2}\tilde{z}_1 \tag{10.86}$$

in (10.84). The phase-space equation of motion then reads, setting $\lambda = 1$,

$$\begin{aligned}
\frac{\partial\bar{\tilde{F}}_\sigma}{\partial t} = \Bigg\{ & \kappa\frac{\partial}{\partial\tilde{z}_1}\left[-\left(n_p^{\rm thr}\right)^{-1/2}\tilde{z}_1\tilde{w}_1 - \left(n_p^{\rm thr}\right)^{q-1}\tilde{z}_2\tilde{w}_2\right] \\
& + \kappa\frac{\partial}{\partial\tilde{z}_2}\left[2\tilde{z}_2 + \left(n_p^{\rm thr}\right)^{-q}\tilde{z}_1\tilde{w}_2 - \left(n_p^{\rm thr}\right)^{-1/2}\tilde{w}_1\tilde{z}_2\right] \\
& + \kappa_p\frac{\partial}{\partial\tilde{w}_1}\left[\tilde{w}_1 + \left(n_p^{\rm thr}\right)^{-2q+1/2}\tilde{z}_1^2 - \left(n_p^{\rm thr}\right)^{-1/2}\tilde{z}_2^2\right] \\
& + \kappa_p\frac{\partial}{\partial\tilde{w}_2}\left[\tilde{w}_2 + \left(n_p^{\rm thr}\right)^{-q}2\tilde{z}_1\tilde{z}_2\right] \\
& + \frac{1}{8}\xi\sigma\kappa\Bigg[\left(\left(n_p^{\rm thr}\right)^{2q-1}\frac{\partial^2}{\partial\tilde{z}_1^2} - \frac{\partial^2}{\partial\tilde{z}_2^2}\right)\left(1 + \left(n_p^{\rm thr}\right)^{-1/2}\tilde{w}_1\right) \\
& +2\left(n_p^{\rm thr}\right)^{q-1}\frac{\partial^2}{\partial\tilde{z}_1\partial\tilde{z}_2}\tilde{w}_2\Bigg] \\
& + \frac{1}{8}\xi\kappa(2\bar{n}+1-\sigma)\left(\left(n_p^{\rm thr}\right)^{2q-1}\frac{\partial^2}{\partial\tilde{z}_1^2} + \frac{\partial^2}{\partial\tilde{z}_2^2}\right) \\
& + \frac{1}{4}\kappa_p(2\bar{n}_p+1-\sigma)\left(\frac{\partial^2}{\partial\tilde{w}_1^2} + \frac{\partial^2}{\partial\tilde{w}_2^2}\right) \\
& + \left(n_p^{\rm thr}\right)^{-1/2}\frac{1}{64}\xi(1-\sigma^2)\kappa\Bigg[\left(\left(n_p^{\rm thr}\right)^{2q-1}\frac{\partial^2}{\partial\tilde{z}_1^2} - \frac{\partial^2}{\partial\tilde{z}_2^2}\right)\frac{\partial}{\partial\tilde{w}_1} \\
& +2\left(n_p^{\rm thr}\right)^{q-1/2}\frac{\partial^2}{\partial\tilde{z}_1\partial\tilde{z}_2}\frac{\partial}{\partial\tilde{w}_2}\Bigg]\Bigg\}\bar{\tilde{F}}_\sigma,
\end{aligned} \tag{10.87}$$

where we have defined

$$\tilde{\tilde{F}}_\sigma(\tilde{z}_1, \tilde{z}_2, \tilde{w}_1, \tilde{w}_2, t)$$
$$\equiv (2/\xi)\left(n_p^{\text{thr}}\right)^{-q+1/2} F_\sigma\big(\alpha(\tilde{z}_1, \tilde{z}_2, t), \alpha^*(\tilde{z}_1, \tilde{z}_2, t), \beta(\tilde{w}_1, \tilde{w}_2, t), \beta^*(\tilde{w}_1, \tilde{w}_2, t), t\big). \tag{10.88}$$

Now it is certain that q is smaller than a half, since $q = 1/2$ was used by the scaling away from threshold and the threshold fluctuations are certainly increased in size. On this basis, the third-order derivatives in (10.87) can be dropped as they were away from threshold, along with many of the nonlinear terms. The result is a separable Fokker–Planck equation, with

$$\tilde{\tilde{F}}_\sigma(\tilde{z}_1, \tilde{z}_2, \tilde{w}_1, \tilde{w}_2, t) = \tilde{\tilde{\mathcal{X}}}_\sigma(\tilde{z}_1, \tilde{w}_1, t)\tilde{\tilde{Y}}_\sigma(\tilde{z}_2, t)\tilde{\tilde{V}}_\sigma(\tilde{w}_2, t). \tag{10.89}$$

The Y-quadrature phase amplitudes of the subharmonic and pump fields fluctuate independently and obey the same equations as below threshold (with $\lambda = 1$); i.e., we have *Fokker–Planck equations at threshold for the Y-quadrature phase amplitudes of the degenerate parametric oscillator*

$$\kappa^{-1}\frac{\partial \tilde{\tilde{Y}}_\sigma}{\partial t} = \left[2\frac{\partial}{\partial \tilde{z}_2}\tilde{z}_2 + \frac{1}{8}\xi(2\bar{n} + 1 - 2\sigma)\frac{\partial^2}{\partial \tilde{z}_2^2}\right]\tilde{\tilde{Y}}_\sigma, \tag{10.90a}$$

$$\kappa_p^{-1}\frac{\partial \tilde{\tilde{V}}_\sigma}{\partial t} = \left[\frac{\partial}{\partial \tilde{w}_2}\tilde{w}_2 + \frac{1}{4}(2\bar{n}_p + 1 - \sigma)\frac{\partial^2}{\partial \tilde{w}_2^2}\right]\tilde{\tilde{V}}_\sigma. \tag{10.90b}$$

Fluctuations in the X-quadrature phase amplitudes are coupled. They are described by the Fokker–Planck equation

$$\begin{aligned}\frac{\partial \tilde{\tilde{\mathcal{X}}}_\sigma}{\partial t} = \bigg\{&-\kappa\frac{\partial}{\partial \tilde{z}_1}\left(n_p^{\text{thr}}\right)^{-1/2}\tilde{z}_1\tilde{w}_1 + \kappa_p\frac{\partial}{\partial \tilde{w}_1}\left[\tilde{w}_1 + \left(n_p^{\text{thr}}\right)^{-2q+1/2}\tilde{z}_1^2\right]\\ &+\left(n_p^{\text{thr}}\right)^{2q-1}\frac{1}{8}\xi\kappa(2\bar{n}+1)\frac{\partial^2}{\partial \tilde{z}_1^2} + \frac{1}{4}\kappa_p(2\bar{n}_p + 1 - \sigma)\frac{\partial^2}{\partial \tilde{w}_1^2}\bigg\}\tilde{\tilde{\mathcal{X}}}_\sigma.\end{aligned} \tag{10.90c}$$

From this point we need only consider (10.90c).

It would of course be an error to drop all terms involving the variable $\tilde{z}_1$ in this equation simply because they appear multiplied by negative powers of n_p^{thr}. We know from the divergences observed in the linearized results that fluctuations in $\tilde{z}_1$ are important. In fact most powers of n_p^{thr} appearing in (10.90c) will be absorbed into the scaling of time used to separate the slow fluctuations in $\tilde{z}_1$ from the fast fluctuations in $\tilde{w}_1$. To uncover the separation of timescales, hence fix the value of q, we consider the stochastic differential equations equivalent to (10.90c). These are, following the prescription given in Sect. 5.3.5,

$$d\tilde{z}_1 = \kappa\left(n_p^{\text{thr}}\right)^{-1/2}\tilde{z}_1\tilde{w}_1 dt + \left(n_p^{\text{thr}}\right)^{q-1/2}\tfrac{1}{2}\sqrt{\xi\kappa(2\bar{n}+1)}dW_{\tilde{z}_1}, \tag{10.91a}$$

and

$$d\tilde{w}_1 = -\kappa_p\Big[\tilde{w}_1 + \big(n_p^{\rm thr}\big)^{-2q+1/2}\tilde{z}_1^2\Big]dt + \sqrt{\tfrac{1}{2}\kappa_p(2\bar{n}_p+1-\sigma)}dW_{\tilde{w}_1}. \quad (10.91b)$$

From these equations we see that the evolution of the *"slow" variable*, $\tilde{z}_1$, occurs on a timescale $\sim \big(n_p^{\rm thr}\big)^{1/2}\kappa^{-1}$ that is much longer (as $n_p^{\rm thr} \to \infty$) than that for the *"fast" variable* $\tilde{w}_1$, which is $\sim \kappa_p^{-1}$. We may therefore adiabatically eliminate $\tilde{w}_1$ from (10.91a); thus, setting the left-hand side of (10.91b) to zero, we write

$$\tilde{w}_1 dt = -\big(n_p^{\rm thr}\big)^{-2q+1/2}\tilde{z}_1^2 dt + \kappa_p^{-1}\sqrt{\tfrac{1}{2}\kappa_p(2\bar{n}_p+1-\sigma)}dW_{\tilde{w}_1}, \quad (10.92)$$

and substituting this result into (10.91a), and dropping the noise term of order $\big(n_p^{\rm thr}\big)^{-1/2}$ compared to the term of order $\big(n_p^{\rm thr}\big)^{q-1/2}$, we arrive at the stochastic differential equation

$$d\tilde{z}_1 = -\kappa\big(n_p^{\rm thr}\big)^{-2q}\tilde{z}_1^3 dt + \big(n_p^{\rm thr}\big)^{q-1/2}\tfrac{1}{2}\sqrt{\xi\kappa(2\bar{n}+1)}dW_{\tilde{z}_1}, \quad (10.93a)$$

with equivalent Fokker–Planck equation

$$\kappa^{-1}\frac{\partial \bar{\tilde{X}}_\sigma}{\partial t} = \left[\big(n_p^{\rm thr}\big)^{-2q}\frac{\partial}{\partial \tilde{z}_1}\tilde{z}_1^3 + \big(n_p^{\rm thr}\big)^{2q-1}\frac{1}{8}\xi(2\bar{n}+1)\frac{\partial^2}{\partial \tilde{z}_1^2}\right]\bar{\tilde{X}}_\sigma. \quad (10.93b)$$

Now q may be chosen so that the two terms on the right-hand side of (10.93b) are of the same order. This requires

$$q = \tfrac{1}{4}, \quad (10.94)$$

which is the scaling used for the laser at threshold (Eq. 8.85).

We are now in a position to separate Fokker–Planck equation (10.90c) into a pair of equations, one describing slow fluctuations of the subharmonic field and the other describing fast fluctuations of the pump. Note, though, that while this separation based on timescales may be made, fluctuations in the two modes are not entirely independent. While the subharmonic field fluctuates independently of the pump, the converse is not true; pump fluctuations are conditioned on the value of the X-quadrature phase amplitude of the subharmonic. We therefore write the joint distribution as

$$\bar{\tilde{\mathcal{X}}}_\sigma(\tilde{z}_1,\tilde{w}_1,t) = \bar{\tilde{X}}_\sigma(\tilde{z}_1,t)\bar{\tilde{U}}_\sigma(\tilde{w}_1,t|\tilde{z}_1), \quad (10.95)$$

where, from (10.93b), the *Fokker–Planck equation for the X-quadrature phase amplitude of the subharmonic mode of the degenerate parametric oscillator at threshold* is

$$\left[\big(n_p^{\rm thr}\big)^{-1/2}\kappa\right]^{-1}\frac{\partial \bar{\tilde{X}}_\sigma}{\partial t} = \left[\frac{\partial}{\partial \tilde{z}_1}\tilde{z}_1^3 + \frac{1}{8}\xi(2\bar{n}+1)\frac{\partial^2}{\partial \tilde{z}_1^2}\right]\bar{\tilde{X}}_\sigma, \quad (10.96a)$$

and, from (10.91b), the *Fokker–Planck equation for the X-quadrature phase amplitude of the pump mode of the degenerate parametric oscillator at threshold* is

$$\kappa_p^{-1}\frac{\partial\tilde{\tilde{U}}_\sigma}{\partial t} = \left[\frac{\partial}{\partial\tilde{w}_1}(\tilde{w}_1+\tilde{z}_1^2)+\frac{1}{4}(2\bar{n}_p+1-\sigma)\frac{\partial^2}{\partial\tilde{w}_1^2}\right]\tilde{\tilde{U}}_\sigma. \tag{10.96b}$$

Note the coupling of the two equations through the term $+\tilde{z}_1^2$ in (10.96b).

Steady-state solutions to Fokker–Planck equations (10.90a), (10.90b), (10.96a), and (10.96b) are readily obtained using results from Chap. 5. First, with the help of the potential solution (5.30), for the solution to (10.96a) we obtain

$$\tilde{\tilde{X}}_\sigma(\tilde{z}_1) = \frac{2}{\Gamma(\frac{1}{4})}\left(\frac{2\kappa_p}{\kappa}\frac{1}{2\bar{n}+1}\right)^{\frac{1}{4}}\exp\left(-\frac{2\kappa_p}{\kappa}\frac{\tilde{z}_1^4}{2\bar{n}+1}\right). \tag{10.97a}$$

Solutions to the remaining three equations then follow from the asymptotic form of the conditional distribution (5.18): for (10.96b) we obtain

$$\tilde{\tilde{U}}_\sigma(\tilde{w}_1|\tilde{z}_1) = \sqrt{\frac{2}{\pi}}\frac{1}{\sqrt{2\bar{n}_p+1-\sigma}}\exp\left(-\frac{1}{2}\frac{4(\tilde{w}_1+\tilde{z}_1^2)^2}{2\bar{n}_p+1-\sigma}\right), \tag{10.97b}$$

while the steady-state solutions to (10.90a) and (10.90b) are

$$\tilde{\tilde{Y}}_\sigma(\tilde{z}_2) = \sqrt{\frac{2}{\pi}}\sqrt{\frac{4\kappa_p}{\kappa}}\frac{1}{\sqrt{2\bar{n}+1-2\sigma}}\exp\left(-\frac{1}{2}\frac{4\kappa_p}{\kappa}\frac{4\tilde{z}_2^2}{2\bar{n}+1-2\sigma}\right), \tag{10.97c}$$

and

$$\tilde{\tilde{V}}_\sigma(\tilde{w}_2) = \sqrt{\frac{2}{\pi}}\frac{1}{\sqrt{2\bar{n}+1-\sigma}}\exp\left(-\frac{1}{2}\frac{4\tilde{w}_2^2}{2\bar{n}_p+1-\sigma}\right). \tag{10.97d}$$

From these solutions the covariances are obtained. We first consider the *covariance matrix for the Y quadrature phase amplitudes of the degenerate parametric oscillator at threshold*:

$$\begin{aligned}(\Delta Y)^2_{\rm thr} \equiv \langle(\hat{Y}-\langle\hat{Y}\rangle)^2\rangle_{\rm ss}^{\rm thr} &= (2/\xi)\left[\left(\overline{\tilde{z}_2\tilde{z}_2}\right)_{\tilde{\tilde{Y}}_\sigma}+\tfrac{1}{4}\sigma\right]\\ &= \tfrac{1}{8}(2\bar{n}+1),\end{aligned} \tag{10.98a}$$

$$\begin{aligned}(\Delta Y_p)^2_{\rm thr} \equiv \langle(\hat{Y}_p-\langle\hat{Y}_p\rangle)^2\rangle_{\rm ss}^{\rm thr} &= \left(\overline{\tilde{w}_2\tilde{w}_2}\right)_{\tilde{\tilde{V}}_\sigma}+\tfrac{1}{4}\sigma\\ &= \tfrac{1}{4}(2\bar{n}_p+1),\end{aligned} \tag{10.98b}$$

and

$$\langle(\hat{Y}-\langle\hat{Y}\rangle)(\hat{Y}_p-\langle\hat{Y}_p\rangle)\rangle_{\rm ss}^{\rm thr} = \sqrt{2/\xi}\left(\overline{\tilde{z}_2\tilde{w}_2}\right)_{\tilde{\tilde{Y}}_\sigma\tilde{\tilde{V}}_\sigma} = 0. \tag{10.99}$$

There is nothing of great interest here. These results are recovered by setting $\lambda = 1$ in the corresponding below- and above-threshold expressions

(Eq. 10.55b and Eqs. 10.75b, 10.76b, and 10.77b). In particular, as there is no change of stability for the Y-quadrature phase amplitudes, there is no threshold enhancement of their fluctuations.

Fluctuations in the X-quadrature phase amplitudes are of more interest. The X-quadrature amplitude of the subharmonic does change stability at threshold. Moreover, as we have just noted, its fluctuations are correlated with those of the pump. The correlation is illustrated by Fig. 10.1, where the distribution $\tilde{\tilde{\mathcal{X}}}_\sigma(\tilde{z}_1, \tilde{w}_1)$ is plotted. Here the threshold fluctuations anticipate the development of two peaks above threshold—two peaks to identify two allowed phases for the subharmonic field (Fig. 9.3). In this, the fluctuations of $\tilde{z}_1$ and $\tilde{w}_1$ are correlated. The correlation provides a hint of the developing pump depletion above threshold (Eq. 9.77b). Derivation of the covariance matrix that shows these results is left as an exercise.

Exercise 10.5. Use (10.97a) and (10.97b) to derive the *covariance matrix for the X-quadrature phase amplitudes of the degenerate parametric oscillator at*

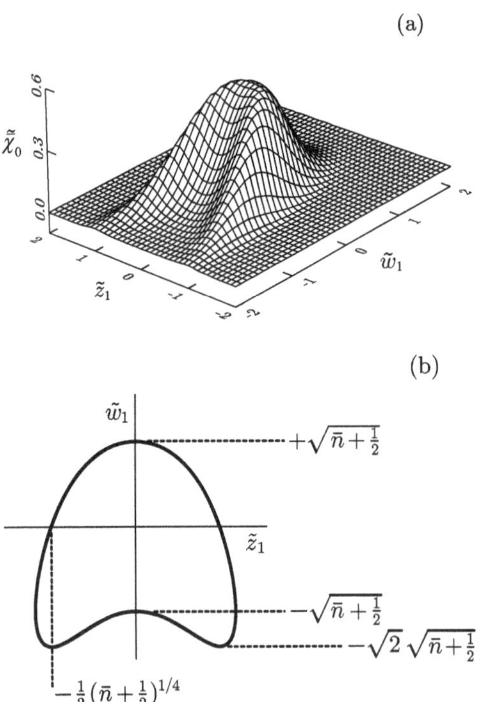

Fig. 10.1. (**a**) The distribution $\tilde{\tilde{\mathcal{X}}}_\sigma(\tilde{z}_1, \tilde{w}_1) = \tilde{\tilde{X}}_\sigma(\tilde{z}_1)\tilde{\tilde{U}}_\sigma(\tilde{w}_1|\tilde{z}_1)$, for $\bar{n} = \bar{n}_p = 0$, $\kappa = \kappa_p$, and $\sigma = 0$. (**b**) The contour $\tilde{\tilde{\mathcal{X}}}_\sigma(\tilde{z}_1, \tilde{w}_1) = e^{-1}\tilde{\tilde{\mathcal{X}}}_\sigma(0,0)$, for $\bar{n} = \bar{n}_p$, $\kappa = \kappa_p$, and $\sigma = 0$

threshold. First, show that

$$\begin{aligned}(\Delta X)^2_{\rm thr} \equiv \langle(\hat{X}-\langle\hat{X}\rangle)^2\rangle^{\rm thr}_{\rm ss} &= (2/\xi)\big[\big(\,\overline{\tilde{z}_1\tilde{z}_1}\,\big)_{\tilde{\bar{X}}_\sigma} + \tfrac{1}{4}\sigma\big] \\ &= \big(n_p^{\rm thr}\big)^{1/2}\frac{\Gamma(\frac{3}{4})}{\Gamma(\frac{1}{4})}\left(\frac{2\kappa_p}{\kappa}\frac{1}{2\bar{n}+1}\right)^{\frac{1}{2}}(2\bar{n}+1),\end{aligned} \tag{10.100a}$$

and

$$\begin{aligned}(\Delta X_p)^2_{\rm thr} \equiv \langle(\hat{X}_p-\langle\hat{X}_p\rangle)^2\rangle^{\rm thr}_{\rm ss} &= \big(\,\overline{\tilde{w}_1\tilde{w}_1}\,\big)_{\tilde{U}_\sigma\tilde{\bar{X}}_\sigma} + \tfrac{1}{4}\sigma \\ &= \tfrac{1}{4}(2\bar{n}_p+1) + (\kappa/2\kappa_p)\tfrac{1}{8}(2\bar{n}+1).\end{aligned} \tag{10.100b}$$

Note the enhancement of $(\Delta X)^2_{\rm thr}$, which is of order unity away from threshold and of order $\big(n_p^{\rm thr}\big)^{1/2}$ at threshold. With regard to the correlations between $\tilde{z}_1$ and $\tilde{w}_1$, it is clear that $\langle(\hat{X}-\langle\hat{X}\rangle)(\hat{X}_p-\langle\hat{X}_p\rangle)\rangle^{\rm thr}_{\rm ss} = 0$; the result follows by symmetry. It is instructive, however, to express the zero as the sum of positive and negative terms (averages over $\tilde{z}_1 < 0$ and $\tilde{z}_1 > 0$), each of which represents the extrapolation to threshold of the nonvanishing correlations in one above-threshold steady state (Eq. 10.77a). Thus, second, show that

$$\begin{aligned}&\langle(\hat{X}-\langle\hat{X}\rangle)(\hat{X}_p-\langle\hat{X}_p\rangle)\rangle^{\rm thr}_{\rm ss} \\ &\quad= \sqrt{2/\xi}\big(\,\overline{\tilde{z}_1\tilde{w}_1}\,\big)_{\tilde{\bar{X}}_\sigma\tilde{U}_\sigma} \\ &\quad= -\big(n_p^{\rm thr}\big)^{1/4}\frac{2}{\Gamma(\frac{1}{4})}\left(\frac{2\kappa_p}{\kappa}\frac{1}{2\bar{n}+1}\right)^{\frac{1}{4}}\frac{1}{4}\sqrt{\frac{\kappa}{2\kappa_p}}(2\bar{n}+1) \\ &\qquad+ \big(n_p^{\rm thr}\big)^{1/4}\frac{2}{\Gamma(\frac{1}{4})}\left(\frac{2\kappa_p}{\kappa}\frac{1}{2\bar{n}+1}\right)^{\frac{1}{4}}\frac{1}{4}\sqrt{\frac{\kappa}{2\kappa_p}}(2\bar{n}+1).\end{aligned} \tag{10.101}$$

From (10.100a) we obtain the *average photon number in the subharmonic mode at threshold*:

$$\langle a^\dagger a\rangle_{\rm thr} \equiv \langle a^\dagger a\rangle^{\rm thr}_{\rm ss} = \frac{\Gamma(\frac{3}{4})}{\Gamma(\frac{1}{4})}\sqrt{\frac{2\kappa_p}{\kappa}}\sqrt{n_p^{\rm thr}(2\bar{n}+1)}, \tag{10.102}$$

and (10.98b) and (10.100b) give the correction due to fluctuations to the *average photon number in the pump mode at threshold*:

$$\begin{aligned}\langle b^\dagger b\rangle_{\rm thr} - n_p^{\rm thr} \equiv \langle b^\dagger b\rangle^{\rm thr}_{\rm ss} - n_p^{\rm thr} &= (\Delta X_p)^2_{\rm thr} + (\Delta Y_p)^2_{\rm thr} - \tfrac{1}{2} \\ &= \bar{n}_p + (\kappa/2\kappa_p)\tfrac{1}{8}(2\bar{n}+1).\end{aligned} \tag{10.103}$$

In conclusion, we might summarize what we have learned about fluctuations of the quadrature phase amplitudes for the degenerate parametric oscillator, combining the results (10.98)–(10.101) with those derived in Sects. 10.2.2

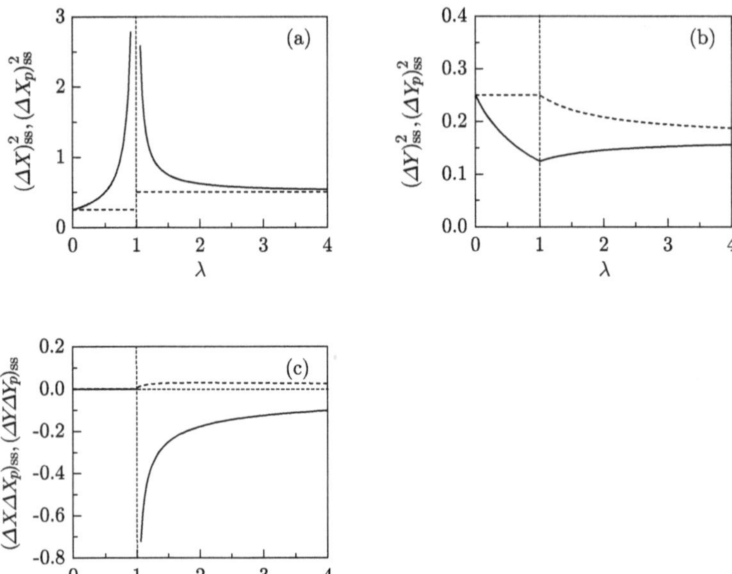

Fig. 10.2. Steady-state fluctuations as a function of pump parameter for the degenerate parametric oscillator, with $\kappa = \kappa_p$ and $\bar{n} = \bar{n}_p = 0$: (**a**) X-quadrature variances, (**b**) Y-quadrature variances, (**c**) cross-correlation between subharmonic and pump modes. In (**a**) and (**b**), *solid lines* are for the subharmonic mode and *dashed lines* for the pump mode. In (**c**) the *solid* and *dashed lines* are for the X- and Y-quadrature phase amplitudes, respectively

and 10.2.3. The resulting picture of the threshold transition is illustrated by Fig. 10.2. Note how the curves change discontinuously at threshold in the limit $n_p^{\text{thr}} \to \infty$.

Note 10.6. The distribution $\tilde{\tilde{X}}_\sigma(\tilde{z}_1)$ is independent of σ. This suggests that the P, Q and Wigner distributions describing fluctuations in the X-quadrature phase amplitude of the subharmonic field all have the same form, which should not be the case, since their moments correspond to operator averages in different orders. The reason for the apparent error is that the system size expansion only provides an approximate answer for $\tilde{\tilde{X}}_\sigma(\tilde{z}_1)$. The scaling defined by (10.33) and (10.85) (with $q = 1/4$) gives α to an accuracy of order $\sqrt{2/\xi}\left(n_p^{\text{thr}}\right)^{1/4}$ when we treat the fluctuations in $\tilde{z}_1$ to an accuracy of order unity. The results obtained are therefore not accurate enough to distinguish between quantum averages differing at the one-quantum level ($\alpha \sim 1$). To achieve one-quantum accuracy, we would need to include first-order corrections to $\tilde{z}_1$, writing $\tilde{z}_1 = \tilde{z}_1^0 + \left(n_p^{\text{thr}}\right)^{-1/4}\tilde{z}_1^1$.

11

The Positive P Representation

The positive P representation generalizes the Glauber–Sudarshan P representation. It was introduced by Drummond and Gardiner [11.1] specifically to deal with the problem of phase-space equations of motion in the Fokker–Planck form that do not possess positive semidefinite diffusion. The degenerate parametric oscillator provides one of the simplest examples. Although this representation has been used extensively by its originators and their colleagues—especially by Drummond and collaborators to treat squeezing in optical fibers [11.2, 11.3, 11.4, 11.5, 11.6, 11.7, 11.8] and more recently quantum gases [11.9, 11.10, 11.11, 11.12]—it is probably fair to say that it is poorly understood outside this circle of the initiated. One reason for this, certainly, is that problems with non-positive-semidefinite diffusion can often be avoided using either the Wigner or the Q representation, as we have seen in Sect. 10.1.2; why would one become entangled in the mysteries of negative diffusion if it can be avoided? Beyond this, any inclination against the representation is reinforced by the appearance that there is something unreasonable, or at least unnecessarily complicated, about it: unreasonable because the positive P representation allows real quantities to be driven by imaginary noise; unnecessarily complicated because it accomplishes this feat by representing each field mode by a pair of complex phase-space variables instead of just one—the positive P distribution is defined within a phase space that has double the usual number of dimensions.

As it turns out, certain applications of the positive P representation do face fundamental difficulties, so there is justification for caution; we look in some detail at the difficulties in Chap. 12. The difficulties are restricted, however, to the highly nonlinear regime where the system size expansion breaks down, and in applications where the expansion is justified there is absolutely no reason for reservations. Here the positive P representation holds no more mystery than the Glauber–Sudarshan representation; moreover, it is just as easy to calculate with, despite its doubled dimensions. This chapter deals with these standard applications. We learn what the positive P representation is and how it is used in the linearized treatment of fluctuations—for example, to justify overlooking the negative diffusion in Fokker–Planck equation (10.51b) (for $\bar{n} = 0$ and $\sigma = 1$).

What should we say, however, about the reasonable objection that since a stochastic description of the degenerate parametric oscillator can be formulated within the Wigner or Q representation, why should we be concerned with the positive P representation at all? There are three principal reasons. First, as we noted in the introduction to Chap. 9, the Glauber–Sudarshan P representation classifies optical fields: there are *classical fields*, which admit a stochastic description within the Glauber–Sudarshan representation and whose statistical properties can therefore be understood using classical statistical optics and the semiclassical theory of photoelectric detection; and there are *nonclassical fields*, which do not admit such a description and understanding. Thus, the close relationship between the positive P representation and the Glauber–Sudarshan representation makes it the natural one to use to highlight the differences between classical and nonclassical fields.

The second reason is simply one of completeness and interest. What we saw in the previous chapter surely raises our curiosity; how did we manage to obtain correct results from a Fokker–Planck equation with a negative diffusion constant? The answer must teach us something about the meaning of "diffusion" in the quantum–classical correspondence.

The third, and perhaps the most important reason of all, is that the success of the Wigner and Q representations is a little misleading. Both representations meet with similar difficulties outside the small-noise limit (see the summary at the end of Sect. 10.1.2). In fact, the strategy adopted by the positive P representation to deal with non-positive-definite diffusion can be applied to these representations as well; though, if it is to be applied outside the small-noise limit, the issues discussed in Chap. 12 must be considered.

11.1 The Positive P Representation

The ideas behind the positive P representation follow quite naturally from the way in which the Glauber–Sudarshan P representation breaks down for the degenerate parametric oscillator. To see how we might *invent* the positive P representation, let us refresh our memories about the difficulty with the stochastic differential equations (10.57). Setting $\bar{n} = 0$ and $\sigma = +1$, these equations read

$$d\tilde{z}_1 = -(1-\lambda)\tilde{z}_1(\kappa dt) + \tfrac{1}{2}\sqrt{\xi\kappa\lambda}\,dW_{\tilde{z}_1}, \tag{11.1a}$$

$$d\tilde{z}_2 = -(1+\lambda)\tilde{z}_2(\kappa dt) + i\tfrac{1}{2}\sqrt{\xi\kappa\lambda}\,dW_{\tilde{z}_2}. \tag{11.1b}$$

In the second, a *pure imaginary* noise term acts as a source for the *real* phase-space variable $\tilde{z}_2$; as the diffusion constant in Fokker–Planck equation (10.51b) is negative, its square root is pure imaginary, hence the noise in the stochastic differential equation is imaginary too. Alternatively, writing the stochastic differential equations in terms of complex variables, $\tilde{z} = \tilde{z}_1 + i\tilde{z}_2$ and $\tilde{z}^* = \tilde{z}_1 - i\tilde{z}_2$, we have

$$d\tilde{z} = -(\tilde{z} - \lambda\tilde{z}^*)(\kappa dt) + \sqrt{(\xi/2)\kappa\lambda}dW_{\tilde{z}}, \tag{11.2a}$$

$$d\tilde{z}^* = -(\tilde{z}^* - \lambda\tilde{z})(\kappa dt) + \sqrt{(\xi/2)\kappa\lambda}dW_{\tilde{z}^*}, \tag{11.2b}$$

with new statistically independent Wiener increments

$$dW_{\tilde{z}} \equiv \tfrac{1}{\sqrt{2}}\big(dW_{\tilde{z}_1} - dW_{\tilde{z}_2}\big), \qquad dW_{\tilde{z}^*} \equiv \tfrac{1}{\sqrt{2}}\big(dW_{\tilde{z}_1} + dW_{\tilde{z}_2}\big). \tag{11.3}$$

Here the inconsistency appears in different form: since $dW_{\tilde{z}}$ and $dW_{\tilde{z}^*}$ are independent [since $dW_{\tilde{z}^*} \neq (dW_{\tilde{z}})^*$], because of their noise terms the stochastic differential equations do not preserve the conjugacy of $\tilde{z}$ and $\tilde{z}^*$.

Definitely, then, something is wrong. On the other hand, one can calculate formally with these equations and get sensible results; for example, from (11.1a) and (11.1b), using (5.144b), we find

$$(2\kappa)^{-1}\frac{d\langle\tilde{z}_1^2\rangle}{dt} = -(1-\lambda)\langle\tilde{z}_1^2\rangle + (\xi/2)\tfrac{1}{4}\lambda, \tag{11.4a}$$

$$(2\kappa)^{-1}\frac{d\langle\tilde{z}_2^2\rangle}{dt} = -(1+\lambda)\langle\tilde{z}_2^2\rangle - (\xi/2)\tfrac{1}{4}\lambda, \tag{11.4b}$$

with steady-state solutions

$$\langle\tilde{z}_1^2\rangle_{\rm ss} = (\xi/2)\frac{1}{4}\frac{\lambda}{1-\lambda} = (\xi/2)\big[(\Delta X)^2_< - \tfrac{1}{4}\big], \tag{11.5a}$$

$$\langle\tilde{z}_2^2\rangle_{\rm ss} = -(\xi/2)\frac{1}{4}\frac{\lambda}{1+\lambda} = (\xi/2)\big[(\Delta Y)^2_< - \tfrac{1}{4}\big], \tag{11.5b}$$

where $(\Delta X)^2_<$ and $(\Delta Y)^2_<$ are the quadrature phase amplitude variances (10.55a) and (10.55b) (with $\bar{n} = 0$). The results $\langle\tilde{z}_1^2\rangle_{\rm ss}$ and $\langle\tilde{z}_2^2\rangle_{\rm ss}$ are correct for operator averages in normal order. Thus, although this success is slim evidence indeed, possibly it is telling us that the main problem with the stochastic differential equations (11.1a)–(11.2b) is one of notation. Perhaps we should accept that $\tilde{z}_2$ is complex, and $\tilde{z}$ and $\tilde{z}^*$ are not complex conjugate. Perhaps we should write (11.2a) and (11.2b) in the similar, though fundamentally different form

$$d\tilde{z} = -(\tilde{z} - \lambda\tilde{z}_*)(\kappa dt) + \sqrt{(\xi/2)\kappa\lambda}dW_{\tilde{z}}, \tag{11.6a}$$

$$d\tilde{z}_* = -(\tilde{z}_* - \lambda\tilde{z})(\kappa dt) + \sqrt{(\xi/2)\kappa\lambda}\,dW_{\tilde{z}_*}, \tag{11.6b}$$

where $\tilde{z}$ and $\tilde{z}_*$ are two *independent* complex variables, two phase-space variables (four real dimensions) associated with the one field mode. The subscript $*$ is used to remind us that we still require $\tilde{z}$ and $\tilde{z}_*$ to be complex conjugate in the mean—i.e., with $\langle\tilde{z}_*\rangle = \langle\tilde{z}\rangle^*$: a relationship that is indeed preserved by (11.6a) and (11.6b).

Here then is the hint to set us on the path to inventing the positive P representation. If the change in notation is permissible, it must be possible to justify it. Moreover, if we aim to find the justification, we must set up the formalism so that there are independent variables α and α_* ($\alpha_* \neq \alpha^*$) associated with each field mode from the beginning. In essence, all we have to do is make the change of notation when formulating the standard Glauber–Sudarshan P representation and see where this leads.

11.1.1 The Characteristic Function and Associated Distribution

Our ultimate aim is to treat the two-mode degenerate parametric oscillator model—master equation (9.97). To simplify the notation, let us first define the positive P representation for a single mode. We replace complex conjugate variables by independent complex variables in the definition of the normal-ordered characteristic function (3.70). Thus, we define the *characteristic function in the positive P representation*

$$\chi_N(z, z_*) \equiv \mathrm{tr}\big(\rho e^{iz_* a^\dagger} e^{iza}\big), \tag{11.7}$$

from which normal-ordered averages are calculated by taking derivatives in the usual way:

$$\begin{aligned} \langle a^{\dagger p} a^q \rangle &= \mathrm{tr}\big(\rho a^{\dagger p} a^q\big) \\ &= \frac{\partial^{p+q}}{\partial(iz_*)^p \partial(iz)^q} \chi_N(z, z_*)\bigg|_{z=z_*=0}. \end{aligned} \tag{11.8}$$

The important change is that the characteristic function is now analytic in the variables z and z_*. The analyticity should be noted, since it is this property that eventually allows us to construct a Fokker–Planck equation with positive semidefinite diffusion.

Note 11.1. The variables z and z_* in (11.7) and (11.8) should not be confused with the $\tilde{z}$ and $\tilde{z}_*$ in (11.6a) and (11.6b). The latter are phase-space variables introduced when making the system size expansion (Eqs. 10.33–10.38); they are arguments of the phase-space distribution, not the characteristic function.

Now the Glauber–Sudarshan P distribution is the two-dimensional Fourier transform of the normal-ordered characteristic function (Eq. 3.72). We might expect, as the natural generalization, to define the positive P distribution as the four-dimensional Fourier transform of (11.7). In fact this is not what we want. We must be guided by the use made of the relationship between P and χ_N. The essential side of the relationship is the integral expression (3.73), which gives χ_N in terms of P, not its inverse. It is this expression, together with the formula (3.71), that allows operator averages to be calculated as phase-space integrals. We therefore generalize this expression. We associate a distribution $P(\alpha,\alpha_*)$ with ρ by assuming that there exists an expansion for $\chi_N(z,z_*)$ in the form

$$\chi_N(z,z_*) = \int d^2\alpha \int d^2\alpha_*\, P(\alpha,\alpha_*)e^{iz_*\alpha_*}e^{iz\alpha}. \tag{11.9}$$

If such an expansion exists, the phase-space result for normal-ordered averages remains intact; from (11.8) and (11.9), we may write

$$\begin{aligned}\langle a^{\dagger p}a^q\rangle &= \frac{\partial^{p+q}}{\partial(iz_*)^p\partial(iz)^q}\int d^2\alpha \int d^2\alpha_*\, P(\alpha,\alpha_*)e^{iz_*\alpha_*}e^{iz\alpha}\bigg|_{z=z_*=0}\\ &= \left(\overline{\alpha_*^p\alpha^q}\right)_P,\end{aligned} \tag{11.10a}$$

with

$$\left(\overline{\alpha_*^p\alpha^q}\right)_P \equiv \int d^2\alpha \int d^2\alpha_*\, \alpha_*^p\alpha^q P(\alpha,\alpha_*). \tag{11.10b}$$

It is important now to note that although (11.9) may look like a Fourier transform, it is not. In order for it to be a Fourier transform, two additional exponentials, $e^{i(z_*)^*(\alpha_*)^*}$ and $e^{iz^*\alpha^*}$, would need to appear in the integrand. In their absence, the relationship between χ_N and P goes in one direction only; (11.9) cannot be inverted to write P in terms of χ_N.

As with the Glauber–Sudarshan representation, the characteristic function is not the only route to a definition. Drummond and Gardiner [11.1] introduce the positive P representation in a different manner. They propose a generalization of the diagonal expansion for ρ presented in Sect. 3.1.2 (Eq. 3.15), writing

$$\rho = \int d^2\alpha \int d^2\alpha_*\, \frac{|\alpha\rangle\langle(\alpha_*)^*|}{\langle(\alpha_*)^*|\alpha\rangle}P(\alpha,\alpha_*). \tag{11.11}$$

It is easy to show that the existence of such an expansion for ρ implies that an expansion for χ_N exists in the form (11.9). The converse is also true, since χ_N specifies ρ uniquely.

From (11.11) we see that there is a close relationship between the expansion proposed by Drummond and Gardiner and Glauber's R representation. The R representation (Eq. 3.13) is recovered from (11.11) by choosing

$$\begin{aligned}P(\alpha,\alpha_*) &= \frac{1}{\pi^2}\langle(\alpha_*)^*|\alpha\rangle e^{-\frac{1}{2}|\alpha|^2}e^{-\frac{1}{2}|\alpha_*|^2}R\big(\alpha^*,(\alpha_*)^*\big)\\ &= \frac{1}{\pi^2}\langle(\alpha_*)^*|\alpha\rangle\langle\alpha|\rho|(\alpha_*)^*\rangle,\end{aligned} \tag{11.12}$$

where (Eq. 3.14) $R\big(\alpha^*,(\alpha_*)^*\big) \equiv e^{\frac{1}{2}|\alpha|^2}e^{\frac{1}{2}|\alpha_*|^2}\langle\alpha|\rho|(\alpha_*)^*\rangle$. Alternatively, the choice (11.12) for $P(\alpha,\alpha_*)$ follows directly from (11.7) and (11.9). We simply insert the expansion of the unit operator (3.9) into (11.7) twice; thus, the characteristic function is written as

$$\begin{aligned}\chi_N(z,z_*) &= \mathrm{tr}\left[\rho\left(\frac{1}{\pi}\int d^2\alpha_* \,|(\alpha_*)^*\rangle\langle(\alpha_*)^*|\right)e^{iz_*a^\dagger}e^{iza}\left(\frac{1}{\pi}\int d^2\alpha\,|\alpha\rangle\langle\alpha|\right)\right]\\ &= \int d^2\alpha\int d^2\alpha_*\left[\frac{1}{\pi^2}\langle(\alpha_*)^*|\alpha\rangle\langle\alpha|\rho|(\alpha_*)^*\rangle\right]e^{iz_*\alpha_*}e^{iz\alpha}.\end{aligned} \tag{11.13}$$

Comparing (11.9) yields the expression (11.12) for $P(\alpha,\alpha_*)$.

The fact that $P(\alpha,\alpha_*)$ and $\chi_N(z,z_*)$ are not Fourier transforms of one another is of central importance. It follows from this that (11.7) and (11.9) do not *define* the positive P distribution in the way the similar relationships from Chaps. 3 and 4 define the Glauber–Sudarshan P, the Q, and the Wigner distributions. Most importantly, $P(\alpha,\alpha_*)$ is not unique. For any density operator ρ there are many functions that allow $\chi_N(z,z_*)$ to be expanded in the form (11.9). Even the positive character of the distribution is not yet defined. The latter is apparent from (11.12). It is clear that $(1/\pi^2)\langle(\alpha_*)^*|\alpha\rangle\langle\alpha|\rho|(\alpha_*)^*\rangle$ need not be positive or even real: consider the vacuum state, $\rho=|0\rangle\langle0|$, for which $(1/\pi^2)\langle(\alpha_*)^*|\alpha\rangle\langle\alpha|\rho|(\alpha_*)^*\rangle=(1/\pi^2)e^{-|\alpha|^2}e^{-|\alpha_*|^2}e^{\alpha\alpha_*}$; this function takes on values that are both negative and complex for appropriate choices of $\alpha\alpha_*$.

At this stage, then, the definition of the positive P representation is not complete. What we have defined so far is a *generalized P representation.* The positivity of the distribution must be imposed as an additional constraint. We must also set out an explicit procedure for constructing such a positive distribution. Thus, in the most general terms, *a positive P representation* for ρ is provided by any $P(\alpha,\alpha_*)$, as specified above, that is both real and positive. Since (11.9) and (11.7) guarantee that

$$\int d^2\alpha\int d^2\alpha_*\,P(\alpha,\alpha_*)=\chi_N(0,0)=1, \tag{11.14}$$

such a distribution satisfies all the properties expected of a probability distribution. Referring to a *positive* as distinct from a generalized P representation is appropriate because, as it turns out, it is always possible to find a real, positive function for $P(\alpha,\alpha_*)$. Let us now prove this by explicit construction. Our proof varies only slightly from the one given by Drummond and Gardiner [11.1].

Using the Glauber–Sudarshan representation, we first expand the density operator in diagonal form,

$$\rho=\frac{1}{\pi}\int d^2\lambda|\lambda\rangle\langle\lambda|P_{GS}(\lambda,\lambda^*), \tag{11.15}$$

where we temporarily adopt the notation P_{GS} to distinguish the Glauber–Sudarshan distribution from the positive P distribution. This diagonal expansion can always be made, although $P_{GS}(\lambda, \lambda^*)$ may need to be a generalized function [see the discussion below (3.32)]. Now substituting (11.15) into (11.7), the characteristic function is given by

$$\chi_N(z, z_*) = \mathrm{tr}\left[\left(\int d^2\lambda |\lambda\rangle\langle\lambda| P_{GS}(\lambda, \lambda^*)\right) e^{iz_* a^\dagger} e^{iza}\right]$$
$$= \int d^2\lambda P_{GS}(\lambda, \lambda^*) e^{iz_*\lambda^*} e^{iz\lambda}. \tag{11.16}$$

Then write (see Note 11.2 for the proof)

$$e^{iz_*\lambda^*} = \frac{1}{2\pi}\int d^2\alpha_*\, e^{iz_*\alpha_*} e^{-\frac{1}{2}|\alpha_* - \lambda^*|^2}, \tag{11.17a}$$
$$e^{iz\lambda} = \frac{1}{2\pi}\int d^2\alpha\, e^{iz\alpha} e^{-\frac{1}{2}|\alpha - \lambda|^2}, \tag{11.17b}$$

and we obtain

$$\chi_N(z, z_*)$$
$$= \int d^2\alpha \int d^2\alpha_* \left[\frac{1}{4\pi^2}\int d^2\lambda P_{GS}(\lambda, \lambda^*) e^{-\frac{1}{2}|\alpha_* - \lambda^*|^2} e^{-\frac{1}{2}|\alpha - \lambda|^2}\right]$$
$$\times e^{iz_*\alpha_*} e^{iz\alpha}$$
$$= \int d^2\alpha \int d^2\alpha_* \left[\frac{1}{4\pi} e^{-\frac{1}{4}|\alpha - (\alpha_*)^*|^2} \frac{1}{\pi}\int d^2\lambda P_{GS}(\lambda, \lambda^*) e^{-|\lambda - \frac{1}{2}(\alpha + (\alpha_*)^*)|^2}\right]$$
$$\times e^{iz_*\alpha_*} e^{iz\alpha}$$
$$= \int d^2\alpha \int d^2\alpha_* \left[\frac{1}{4\pi}|\langle(\alpha_*)^*/2|\alpha/2\rangle|^2 Q\big(\tfrac{1}{2}(\alpha + (\alpha_*)^*), \tfrac{1}{2}(\alpha^* + \alpha_*)\big)\right]$$
$$\times e^{iz_*\alpha_*} e^{iz\alpha}, \tag{11.18}$$

where the last step follows from (3.8) and (4.7); hence, we achieve an expansion of the characteristic function in the form (11.9), from which, by definition, the density operator possesses the generalized P representation

$$P(\alpha, \alpha_*) = \frac{1}{4\pi}|\langle(\alpha_*)^*/2|\alpha/2\rangle|^2 Q\big(\tfrac{1}{2}(\alpha + (\alpha_*)^*), \tfrac{1}{2}(\alpha^* + \alpha_*)\big). \tag{11.19}$$

The representation is, however, a *positive* representation, since the Q distribution, given by the diagonal coherent state matrix elements of ρ (Eq. 4.6), is explicitly a real and positive function.

Note 11.2. The proof of (11.17a) and (11.17b) proceeds as follows. Consider in particular (11.17b). Making the transformation $\alpha - \lambda \to \alpha = re^{i\phi}$, we may write

$$
\begin{aligned}
\text{R.H.S. of (11.17b)} &= e^{iz\lambda}\frac{1}{2\pi}\int d^2\alpha\, e^{iz\alpha}e^{-\frac{1}{2}|\alpha|^2} \\
&= e^{iz\lambda}\frac{1}{2\pi}\int_0^\infty dr\int_0^{2\pi} d\phi\sum_{m=0}^{\infty}\frac{(iz)^m}{m!}r^{m+1}e^{im\phi}e^{-\frac{1}{2}r^2} \\
&= e^{iz\lambda}\int_0^\infty drre^{-\frac{1}{2}r^2} \\
&= e^{iz\lambda}.
\end{aligned}
$$

Even with $P(\alpha,\alpha_*)$ restricted to be a real and positive function, the representation remains nonunique. Our construction, Eq. 11.19, defines *a* positive P distribution, not *the* positive P distribution. To illustrate this point, we might note that any density operator that possesses a positive, nonsingular Glauber–Sudarshan P distribution, also admits the positive P representation

$$P(\alpha,\alpha_*) = P_{GS}(\alpha,\alpha^*)\delta^{(2)}\big(\alpha-(\alpha_*)^*\big). \tag{11.20}$$

For example, for the thermal state (3.26), using (11.20) and the Glauber–Sudarshan distribution (3.27), we obtain

$$P(\alpha,\alpha_*) = \frac{1}{\pi\langle\hat{n}\rangle}\exp\left(-\frac{|\alpha|^2}{\langle\hat{n}\rangle}\right)\delta^{(2)}\big(\alpha-(\alpha_*)^*\big), \tag{11.21a}$$

quite a different positive P representation to the one obtained from (11.19) [and (4.21)]—i.e.,

$$P(\alpha,\alpha_*) = \frac{1}{4\pi^2(1+\langle\hat{n}\rangle)}e^{-\frac{1}{4}|\alpha-(\alpha_*)^*|^2}\exp\left(-\frac{1}{4}\frac{|\alpha+(\alpha_*)^*|^2}{1+\langle\hat{n}\rangle}\right). \tag{11.21b}$$

Exercise 11.1. Show by explicit calculation that the positive P representations (11.21a) and (11.21b) give the same results for normal-ordered operator averages.

At first sight the nonuniqueness of the positive P representation might appear to be a weakness; but along with nonuniqueness comes flexibility, and it is this flexibility that suggests we might use the positive P representation to construct acceptable Fokker–Planck equations in situations where the Glauber–Sudarshan representation fails. It is just possible that amongst all possible positive P distributions, $P(\alpha,\alpha_*,t)$, representing (nonuniquely) the density operator $\rho(t)$, one may be found at each time t such that the family of distributions, for different times, satisfies a Fokker–Planck equation. Of course, there is no reason for the distributions that accomplish this trick to be those defined by (11.19).

Exercise 11.2. Use (11.19) and (4.10) to obtain a positive P distribution for the Fock state $|l\rangle$. Show that this distribution gives the correct normal-ordered operator averages

$$\langle l|a^{\dagger p}a^q|l\rangle = \begin{cases} \delta_{p,q} l(l-1)\cdots(l-q+1) & q \le l, \\ 0 & q > l. \end{cases} \tag{11.22}$$

Show that the same results are obtained from the complex-valued distribution (11.12).

11.1.2 Fokker–Planck Equation for the Degenerate Parametric Oscillator

We are now ready to derive the Fokker–Planck equation for the degenerate parametric oscillator in the positive P representation. Our starting point is the master equation (9.97) and the characteristic function (11.7), generalized to the two-mode situation in the obvious way. Proceeding in the standard fashion [Sects. 3.2.2 and 10.1.1], we first derive the equation of motion for the characteristic function

$$\frac{\partial \chi_N}{\partial t} = D\left(z, z_*, w, w_*, \frac{\partial}{\partial z}, \frac{\partial}{\partial z_*}, \frac{\partial}{\partial w}, \frac{\partial}{\partial w_*}\right)\chi_N, \tag{11.23}$$

with

$$\begin{aligned} D&\left(z, z_*, w, w_*, \frac{\partial}{\partial z}, \frac{\partial}{\partial z_*}, \frac{\partial}{\partial w}, \frac{\partial}{\partial w_*}\right) \\ &\equiv -z\left[(\kappa + i\omega_C)\frac{\partial}{\partial z} + ig\frac{\partial^2}{\partial z_*\partial w}\right] - z_*\left[(\kappa - i\omega_C)\frac{\partial}{\partial z_*} + ig\frac{\partial^2}{\partial z \partial w_*}\right] \\ &\quad - w\left[(\kappa_p + i2\omega_C)\frac{\partial}{\partial w} - i(g/2)\frac{\partial^2}{\partial z^2} - \bar{\mathcal{E}}_0 e^{-i2\omega_C t}\right] \\ &\quad - w_*\left[(\kappa_p - i2\omega_C)\frac{\partial}{\partial w_*} - i(g/2)\frac{\partial^2}{\partial z_*^2} - \bar{\mathcal{E}}_0^* e^{i2\omega_C t}\right] \\ &\quad + i(g/2)\left(z^2\frac{\partial}{\partial w} + z_*^2\frac{\partial}{\partial w_*}\right) - 2\kappa\bar{n}zz_* - 2\kappa_p\bar{n}_p ww_*. \end{aligned} \tag{11.24}$$

Here the algebraic manipulations carry through as they did before, the only change being one of notation, $z^* \to z_*$, $w^* \to w_*$; up to this point we meet nothing new. Substituting the expansion (11.9) for χ_N into (11.23), we then have

$$\begin{aligned} \int d^2\alpha \int d^2\alpha_* \int d^2\beta \int d^2\beta_* \, &\frac{\partial P(\alpha, \alpha_*, \beta, \beta_*, t)}{\partial t} e^{iz_*\alpha_*} e^{iz\alpha} e^{iw_*\beta_*} e^{iw\beta} \\ &= \int d^2\alpha \int d^2\alpha_* \int d^2\beta \int d^2\beta_* \, P(\alpha, \alpha_* \beta, \beta_*, t) \\ &\quad \times D\left(z, z_*, w, w_*, \frac{\partial}{\partial z}, \frac{\partial}{\partial z_*}, \frac{\partial}{\partial w}, \frac{\partial}{\partial w_*}\right) e^{iz_*\alpha_*} e^{iz\alpha} e^{iw_*\beta_*} e^{iw\beta}. \end{aligned} \tag{11.25}$$

Acting with the derivatives to the right on the product of exponentials, we effect the replacements

$$\frac{\partial}{\partial z} \to i\alpha, \qquad \frac{\partial}{\partial z_*} \to i\alpha_*, \qquad \frac{\partial}{\partial w} \to i\beta, \qquad \frac{\partial}{\partial w_*} \to i\beta_*.$$

It is from here that the derivation branches in a new direction.

If we were to follow strictly the derivation from Sect. 3.2., we would now make the replacements

$$z \to -i\frac{\partial}{\partial \alpha}, \qquad z_* \to -i\frac{\partial}{\partial \alpha_*}, \qquad w \to -i\frac{\partial}{\partial \beta}, \qquad w_* \to -i\frac{\partial}{\partial \beta_*}.$$

Here we do much the same thing but with a twist. Recall that the characteristic function in the positive P representation is analytic. The analyticity is passed through to (11.25) in the product of exponentials. This product is an analytic function of α, α_*, β, and β_*, which it is not if $\alpha_* \to \alpha^*$, $\beta_* \to \beta^*$. Given the analyticity, we may interpret derivatives taken with respect to the phase-space variables as arbitrary linear combinations of derivatives taken with respect their real and imaginary parts. Thus, we make the proposed replacements in the form

$$z \to -i\partial_\alpha, \qquad z_* \to -i\partial_{\alpha_*}, \qquad w \to -i\partial_\beta, \qquad w_* \to -i\partial_{\beta_*},$$

with definitions

$$\partial_\alpha \equiv \left(\mu\frac{\partial}{\partial x} - i\nu\frac{\partial}{\partial y}\right), \qquad \partial_{\alpha_*} \equiv \left(\mu\frac{\partial}{\partial \mathrm{x}} - i\nu\frac{\partial}{\partial \mathrm{y}}\right), \tag{11.26a}$$

$$\partial_\beta \equiv \left(\mu\frac{\partial}{\partial u} - i\nu\frac{\partial}{\partial v}\right), \qquad \partial_{\beta_*} \equiv \left(\mu\frac{\partial}{\partial \mathrm{u}} - i\nu\frac{\partial}{\partial \mathrm{v}}\right), \tag{11.26b}$$

where μ and ν are real numbers satisfying $\mu + \nu = 1$, and

$$\alpha = x + iy, \qquad \alpha_* = \mathrm{x} + i\mathrm{y}, \tag{11.27a}$$

$$\beta = u + iv, \qquad \beta_* = \mathrm{u} + i\mathrm{v}. \tag{11.27b}$$

In this way we introduce a degree of flexibility into the equation of motion for $P(\alpha, \alpha_*, \beta, \beta_*, t)$. Ultimately, we plan to make use of this flexibility by choosing the parameters μ and ν, term by term, so as to force the equation of motion for $P(\alpha, \alpha_*, \beta, \beta_*, t)$ to have positive semidefinite diffusion. By this strategy the positivity of the distribution is guaranteed and the path to a unique distribution is defined.

Having made all replacements, $D\left(z, z_*, w, w_*, \frac{\partial}{\partial z}, \frac{\partial}{\partial z_*}, \frac{\partial}{\partial w}, \frac{\partial}{\partial w_*}\right)$ is replaced by the differential operator in phase-space variables

$$
\begin{aligned}
&L^{+}(\alpha,\alpha_*,\beta,\beta_*,\partial_\alpha,\partial_{\alpha_*},\partial_\beta,\partial_{\beta_*})\\
&\qquad\equiv -\big[(\kappa+i\omega_C)\alpha-g\alpha_*\beta\big]\partial_\alpha-\big[(\kappa-i\omega_C)\alpha_*-g\alpha\beta^*\big]\partial_{\alpha_*}\\
&\qquad\quad-\big[(\kappa_p+i2\omega_C)\beta+(g/2)\alpha^2+i\bar{\mathcal{E}}_0e^{-i2\omega_Ct}\big]\partial_\beta\\
&\qquad\quad-\big[(\kappa_p-i2\omega_C)\beta_*+(g/2)\alpha_*^2-i\bar{\mathcal{E}}_0^*e^{i2\omega_Ct}\big]\partial_{\beta_*}\\
&\qquad\quad+(g/2)(\beta\partial_\alpha^2+\beta_*\partial_{\alpha_*}^2)+2\kappa\bar{n}\partial_\alpha\partial_{\alpha_*}+2\kappa_p\bar{n}_p\partial_\beta\partial_{\beta_*}. \qquad (11.28)
\end{aligned}
$$

Then integrating by parts on the right-hand side of (11.25) and assuming that boundary terms at infinity vanish, we arrive at a generalization of (3.85),

$$
\begin{aligned}
&\int d^2\alpha\int d^2\alpha_*\int d^2\beta\int d^2\beta_*\,e^{iz_*\alpha_*}e^{iz\alpha}e^{iw_*\beta_*}e^{iw\beta}\frac{\partial P}{\partial t}\\
&\qquad=\int d^2\alpha\int d^2\alpha_*\int d^2\beta\int d^2\beta_*\,e^{iz_*\alpha_*}e^{iz\alpha}e^{iw_*\beta_*}e^{iw\beta}\\
&\qquad\quad\times L(\alpha,\alpha_*,\beta,\beta_*,\partial_\alpha,\partial_{\alpha_*},\partial_\beta,\partial_{\beta_*})P, \qquad (11.29)
\end{aligned}
$$

where

$$
\begin{aligned}
&L(\alpha,\alpha_*,\beta,\beta_*,\partial_\alpha,\partial_{\alpha_*},\partial_\beta,\partial_{\beta_*})\\
&\qquad\equiv \partial_\alpha\big[(\kappa+i\omega_C)\alpha-g\alpha_*\beta\big]+\partial_{\alpha_*}\big[(\kappa-i\omega_C)\alpha_*-g\alpha\beta_*\big]\\
&\qquad\quad+\partial_\beta\big[(\kappa_p+i2\omega_C)\beta+(g/2)\alpha^2+i\bar{\mathcal{E}}_0e^{-i2\omega_Ct}\big]\\
&\qquad\quad+\partial_{\beta_*}\big[(\kappa_p-i2\omega_C)\beta_*+(g/2)\alpha_*^2-i\bar{\mathcal{E}}_0^*e^{i2\omega_Ct}\big]\\
&\qquad\quad+(g/2)(\partial_\alpha^2\beta+\partial_{\alpha_*}^2\beta_*)+2\kappa\bar{n}\partial_\alpha\partial_{\alpha_*}+2\kappa_p\bar{n}_p\partial_\beta\partial_{\beta_*}. \qquad (11.30)
\end{aligned}
$$

A sufficient condition for (11.28) to be satisfied is that the distribution satisfy the *phase-space equation of motion for the degenerate parametric oscillator within the generalized P representation*

$$
\frac{\partial P}{\partial t}=L(\alpha,\alpha_*,\beta,\beta_*,\partial_\alpha,\partial_{\alpha_*},\partial_\beta,\partial_{\beta_*})P. \qquad (11.31)
$$

This equation incorporates the flexibility introduced through the definitions (11.26a) and (11.26b); it is not yet an equation of motion defining a *positive P* representation. It involves derivatives up to second order and may be viewed as a *generalized Fokker–Planck equation.*

Note 11.3. Since the integration in (11.29) is not a Fourier transform, we cannot pass from it to (11.31) via the inversion of a Fourier transform as we did, for example, at the end of Sect. 3.2.2. There is no cause to worry that (11.31) is only a sufficient condition for (11.29) to hold, however. Our interest is in finding *a* (not *the*) $P(\alpha,\alpha_*,\beta,\beta_*,t)$ that represents $\rho(t)$; we already know from the discussion in Sect. 11.1.1 that the distribution is not unique. We should, on the other hand, be a little worried about the assumption that boundary terms vanish in the integration by parts connecting (11.25) and (11.29). At least one example where this is demonstrably not the case is known [11.13], and for it the solution to the resulting Fokker–Planck equation gives physically incorrect results. We return to this issue at the end of Chap. 12.

On comparing the differential operator (11.30) with the right-hand side of (10.8), we find that our generalized phase-space equation of motion has the same form as the phase-space equation of motion in the Glauber–Sudarshan P representation, but with the advertised difference of notation: α_* and β_* are not assumed complex conjugates of α and β, and the partial derivatives have the arbitrary form defined in (11.26a) and (11.26b). In substance, one might say, we have still done nothing new; it is always possible to write down such a generalized equation of motion knowing the equation in the Glauber–Sudarshan representation. Only when the form of each derivative is specifically set will the Fokker–Planck equation in the positive P representation be defined. The setting of the derivatives must be such that the solution to (11.31) is a real, positive function, a function possessing all the properties of a classical probability density. Then the Fokker–Planck equation will define a classical stochastic process whose moments reproduce all normal-ordered operator averages (Eq. 11.10).

Note 11.4. In addition to the real, positive solution sought, the generalized phase-space equation of motion possesses solutions as complex valued functions. These do not, of course, provide a description in terms of a classical stochastic process, but they do provide an alternative representation of the quantum mechanics. A representation of ρ in terms of a complex valued function is known as a *complex P representation.* The representation is also due to Drummond and Gardiner [11.1]. Within it, normal-ordered operator averages are given by a formula like (11.10) with the integration taken around a contour in the complex plane. Examples of its use can be found in the work of Drummond et al. [11.14] and Walls et al. [11.15].

Our task is to replace the differential operator L by an operator L_{pos} in which each of the derivatives is taken with a specific choice of μ and ν (Eqs. 11.26a and 11.26b). We first write the operator L in compact form; from (11.30), we may write

$$L(\boldsymbol{X}, \boldsymbol{X}') \equiv -\partial_i A_i(\boldsymbol{X}) + \tfrac{1}{2}\partial_i\partial_j D_{ij}(\boldsymbol{X}), \tag{11.32}$$

where $\partial_i \equiv \partial_{X_i}$, $\boldsymbol{X}$ and $\boldsymbol{X}'$ are column vectors of phase-space variables and derivatives with respect to these variables,

$$\boldsymbol{X} \equiv \begin{pmatrix} \alpha \\ \alpha_* \\ \beta \\ \beta_* \end{pmatrix}, \qquad \boldsymbol{X}' \equiv \begin{pmatrix} \partial_\alpha \\ \partial_{\alpha_*} \\ \partial_\beta \\ \partial_{\beta_*} \end{pmatrix}, \tag{11.33}$$

and the functions $A_i(\boldsymbol{X})$ and $D_{ij}(\boldsymbol{X})$ define the generalized drift vector

$$\boldsymbol{A}(\boldsymbol{X}) \equiv \begin{pmatrix} -(\kappa + i\omega_C)\alpha + g\alpha_*\beta \\ -(\kappa - i\omega_C)\alpha_* + g\alpha\beta_* \\ -(\kappa_p + i2\omega_C)\beta - (g/2)\alpha^2 - i\bar{\mathcal{E}}_0 e^{-i2\omega_C t} \\ -(\kappa_p - i2\omega_C)\beta_* - (g/2)\alpha_*^2 + i\bar{\mathcal{E}}_0^* e^{i2\omega_C t} \end{pmatrix}, \tag{11.34a}$$

and diffusion matrix

$$\boldsymbol{D}(\boldsymbol{X}) \equiv \begin{pmatrix} g\beta & 2\kappa\bar{n} & 0 & 0 \\ 2\kappa\bar{n} & g\beta_* & 0 & 0 \\ 0 & 0 & 0 & 2\kappa_p\bar{n}_p \\ 0 & 0 & 2\kappa_p\bar{n}_p & 0 \end{pmatrix}. \tag{11.34b}$$

In (11.32) and throughout this section we adopt the convention that repeated indices imply summation.

Now, it is always possible to decompose a symmetric matrix like the generalized diffusion matrix $\boldsymbol{D}(\boldsymbol{X})$ as a product

$$\boldsymbol{D}(\boldsymbol{X}) = \boldsymbol{B}(\boldsymbol{X})\boldsymbol{B}(\boldsymbol{X})^T, \tag{11.35}$$

where $\boldsymbol{B}(\boldsymbol{X})$ is a complex matrix. The decomposition is not unique since we can insert the product $\boldsymbol{M}\boldsymbol{M}^T = \boldsymbol{I}$ between $\boldsymbol{B}(\boldsymbol{X})$ and $\boldsymbol{B}(\boldsymbol{X})^T$, where $\boldsymbol{M}$ is any orthogonal matrix. For the $\boldsymbol{D}(\boldsymbol{X})$ defined in (11.34b), the decomposition may be made with

$$\boldsymbol{B}(\boldsymbol{X}) = \begin{pmatrix} \boldsymbol{S}(\beta, \beta_*) & 0 \\ 0 & \boldsymbol{P} \end{pmatrix}, \tag{11.36}$$

where

$$\boldsymbol{S}(\beta, \beta_*) \equiv \begin{pmatrix} \sqrt{g\beta/2 + \sqrt{\beta/\beta_*}\kappa\bar{n}} & \sqrt{g\beta/2 - \sqrt{\beta/\beta_*}\kappa\bar{n}} \\ \sqrt{g\beta_*/2 + \sqrt{\beta_*/\beta}\kappa\bar{n}} & -\sqrt{g\beta_*/2 - \sqrt{\beta_*/\beta}\kappa\bar{n}} \end{pmatrix}, \tag{11.37a}$$

and

$$\boldsymbol{P} \equiv \begin{pmatrix} \sqrt{\kappa_p\bar{n}_p} & i\sqrt{\kappa_p\bar{n}_p} \\ \sqrt{\kappa_p\bar{n}_p} & -i\sqrt{\kappa_p\bar{n}_p} \end{pmatrix}. \tag{11.37b}$$

Exercise 11.3. One strategy for making the decomposition (11.35) is to begin with the quadratic form $\boldsymbol{z}^T\boldsymbol{D}(\boldsymbol{X})\boldsymbol{z}$. The decomposition is made by writing the quadratic form as a sum of squares, $\boldsymbol{w}^T\boldsymbol{w}$, for some $\boldsymbol{w} = \boldsymbol{N}\boldsymbol{z}$; the transpose of $\boldsymbol{N}$ is the matrix $\boldsymbol{B}(\boldsymbol{X})$. Use this approach to arrive at the decomposition defined by (11.36)–(11.37b). Starting again with the quadratic form, use the method of completing the square to find an alternative decomposition of $\boldsymbol{D}(\boldsymbol{X})$. Find the orthogonal matrix that connects the two decompositions.

Equations 11.32–11.36 are defined within a phase space of four independent complex variables. Alternatively, we may view the phase space as a space of eight real variables. The Fokker–Planck equation for the positive P distribution is defined within this eight-dimensional space. To construct it explicitly, we first separate $\boldsymbol{A}(\boldsymbol{X})$ and $\boldsymbol{B}(\boldsymbol{X})$ into real and imaginary parts, writing

$$\boldsymbol{A}(\boldsymbol{X}) = \boldsymbol{A}_R(\boldsymbol{x}) + i\boldsymbol{A}_I(\boldsymbol{x}), \tag{11.38a}$$

$$\boldsymbol{B}(\boldsymbol{X}) = \boldsymbol{B}_R(\boldsymbol{x}) + i\boldsymbol{B}_I(\boldsymbol{x}), \tag{11.38b}$$

where $\boldsymbol{x}$ is a real eight-component vector of phase-space variables:

$$\boldsymbol{x} \equiv \begin{pmatrix} \boldsymbol{X}_R \\ \boldsymbol{X}_I \end{pmatrix}, \qquad \boldsymbol{X}_R \equiv \begin{pmatrix} x \\ \mathrm{x} \\ u \\ \mathrm{u} \end{pmatrix}, \qquad \boldsymbol{X}_I \equiv \begin{pmatrix} y \\ \mathrm{y} \\ v \\ \mathrm{v} \end{pmatrix}. \tag{11.39}$$

We also define the vector of derivatives with respect to these variables

$$\boldsymbol{x}' \equiv \begin{pmatrix} \boldsymbol{X}'_R \\ \boldsymbol{X}'_I \end{pmatrix}, \qquad \boldsymbol{X}'_R \equiv \begin{pmatrix} \partial/\partial x \\ \partial/\partial \mathrm{x} \\ \partial/\partial u \\ \partial/\partial \mathrm{u} \end{pmatrix}, \qquad \boldsymbol{X}'_I \equiv \begin{pmatrix} \partial/\partial y \\ \partial/\partial \mathrm{y} \\ \partial/\partial v \\ \partial/\partial \mathrm{v} \end{pmatrix}. \tag{11.40}$$

The differential operator $L_{\mathrm{pos}}(\boldsymbol{x}, \boldsymbol{x}')$ is then defined by fixing the form of each derivative in (11.32) (choosing μ and ν) so that

$$L(\boldsymbol{X}, \boldsymbol{X}') \to L_{\mathrm{pos}}(\boldsymbol{x}, \boldsymbol{x}') \equiv -\frac{\partial}{\partial x_i} A_i^{(2)}(\boldsymbol{x}) + \frac{1}{2}\frac{\partial^2}{\partial x_i \partial x_j} D_{ij}^{(2)}(\boldsymbol{x}), \tag{11.41}$$

with the $A_i^{(2)}(\boldsymbol{x})$ components of an eight-dimensional *real* drift vector $\boldsymbol{A}^{(2)}(\boldsymbol{x})$, and the $D_{ij}^{(2)}(\boldsymbol{x})$ elements of an eight-dimensional (real) diffusion matrix $\boldsymbol{D}^{(2)}(\boldsymbol{x})$ that is explicitly positive semidefinite. This is achieved by taking either $\mu, \nu = 1, 0$ or $\mu, \nu = 0, 1$ in (11.26a) and (11.26b), to obtain

$$-\partial_i A_i(\boldsymbol{X}) \to -\frac{\partial}{\partial X_i^R} A_R^i(\boldsymbol{x}) - \frac{\partial}{\partial X_i^I} A_I^i(\boldsymbol{x}), \tag{11.42a}$$

and

$$\begin{aligned} \tfrac{1}{2}\partial_i \partial_j D_{ij}(\boldsymbol{X}) \to & \frac{1}{2}\frac{\partial^2}{\partial X_i^R \partial X_j^R} B_R^{ik}(\boldsymbol{x}) B_R^{jk}(\boldsymbol{x}) \\ & + \frac{1}{2}\frac{\partial^2}{\partial X_i^R \partial X_j^I} B_R^{ik}(\boldsymbol{x}) B_I^{jk}(\boldsymbol{x}) \\ & + \frac{1}{2}\frac{\partial^2}{\partial X_i^I \partial X_j^R} B_I^{ik}(\boldsymbol{x}) B_R^{jk}(\boldsymbol{x}) \\ & + \frac{1}{2}\frac{\partial^2}{\partial X_i^I \partial X_j^I} B_I^{ik}(\boldsymbol{x}) B_I^{jk}(\boldsymbol{x}). \end{aligned} \tag{11.42b}$$

The choices for μ and ν are guided first by the requirement that $L_{\mathrm{pos}}(\boldsymbol{x},\boldsymbol{x}')$ be real. Thus we choose $\mu,\nu=1,0$ when ∂_i acts on the real part of $A_i(\boldsymbol{X})$ and $\mu,\nu=0,1$ when ∂_i acts on the imaginary part of $A_i(\boldsymbol{X})$; we obtain (11.42a), or

$$\boldsymbol{A}^{(2)}(\boldsymbol{x})\equiv\begin{pmatrix}\boldsymbol{A}_R(\boldsymbol{x})\\ \boldsymbol{A}_I(\boldsymbol{x})\end{pmatrix}. \tag{11.43}$$

The choices for μ and ν when $\partial_i\partial_j$ acts on $D_{ij}(\boldsymbol{X})$ are guided also by the requirement that $\boldsymbol{D}^{(2)}(\boldsymbol{x})$ be positive semidefinite. With a little trial and error we arrive at (11.42b), or

$$\boldsymbol{D}^{(2)}(\boldsymbol{x})\equiv\boldsymbol{B}^{(2)}(\boldsymbol{x})\boldsymbol{B}^{(2)}(\boldsymbol{x})^T, \tag{11.44a}$$

with

$$\boldsymbol{B}^{(2)}(\boldsymbol{x})\equiv\begin{pmatrix}\boldsymbol{B}_R(\boldsymbol{x}) & 0\\ \boldsymbol{B}_I(\boldsymbol{x}) & 0\end{pmatrix}. \tag{11.44b}$$

To verify that (11.44a) and (11.44b) do indeed define a positive semidefinite matrix, consider the quadratic form

$$\begin{pmatrix}\boldsymbol{z}\\ \boldsymbol{w}\end{pmatrix}^T\boldsymbol{D}^{(2)}(\boldsymbol{x})\begin{pmatrix}\boldsymbol{z}\\ \boldsymbol{w}\end{pmatrix}=\left(\boldsymbol{B}_R^T\boldsymbol{z}+\boldsymbol{B}_I^T\boldsymbol{w}\right)^T\left(\boldsymbol{B}_R^T\boldsymbol{z}+\boldsymbol{B}_I^T\boldsymbol{w}\right)+0^2. \tag{11.45}$$

This is either positive or zero for all choices of $\boldsymbol{z}$ and $\boldsymbol{w}$; thus, the diffusion matrix in eight dimensions, $\boldsymbol{D}^{(2)}(\boldsymbol{x})$, is positive semidefinite by construction, no matter the explicit form of $\boldsymbol{B}^{(2)}(\boldsymbol{x})$.

Collecting together our results, from (11.31) and (11.41), we arrive at the *Fokker–Planck equation for the degenerate parametric oscillator in the positive P representation*:

$$\frac{\partial P}{\partial t}=L_{\mathrm{pos}}(\boldsymbol{x},\boldsymbol{x}')P, \tag{11.46}$$

where $L_{\mathrm{pos}}(\boldsymbol{x},\boldsymbol{x}')$ is defined by the drift vector $\boldsymbol{A}^{(2)}(\boldsymbol{x})$ and diffusion matrix $\boldsymbol{D}^{(2)}(\boldsymbol{x})$ given in (11.43) and (11.44), and (11.34)–(11.39). This result is expressed more compactly by the equivalent Ito stochastic differential equations. From (5.149), the *stochastic differential equations for the degenerate parametric oscillator in the positive P representation* are

$$d\boldsymbol{x}=\boldsymbol{A}^{(2)}(\boldsymbol{x})dt+\boldsymbol{B}^{(2)}(\boldsymbol{x})d\boldsymbol{W}^{(2)}, \tag{11.47}$$

$d\boldsymbol{W}^{(2)}\equiv(d\boldsymbol{W}_R,d\boldsymbol{W}_I)^T$, or equivalently, from (11.39), (11.43), and (11.44b),

$$d\boldsymbol{X}_R=\boldsymbol{A}_R(\boldsymbol{X}_R,\boldsymbol{X}_I)dt+\boldsymbol{B}_R(\boldsymbol{X}_R,\boldsymbol{X}_I)d\boldsymbol{W}, \tag{11.48a}$$

$$d\boldsymbol{X}_I=\boldsymbol{A}_I(\boldsymbol{X}_R,\boldsymbol{X}_I)dt+\boldsymbol{B}_I(\boldsymbol{X}_R,\boldsymbol{X}_I)d\boldsymbol{W}, \tag{11.48b}$$

where $d\boldsymbol{W}=d\boldsymbol{W}_R$; the Wiener increments $d\boldsymbol{W}_I$ do not enter the equations because the second column of $\boldsymbol{B}^{(2)}(\boldsymbol{x})$ is zero (Eq. 11.44b). For a still more compact form, (11.48a) and (11.48b) may be written in terms of the four complex phase-space variables α, α_*, β, and β_*,

$$d\boldsymbol{X} = \boldsymbol{A}(\boldsymbol{X})dt + \boldsymbol{B}(\boldsymbol{X})d\boldsymbol{W}. \tag{11.49}$$

Thus we revert to the notation of (11.33)–(11.34b).

Admittedly, all of this seems quite complicated. The complication is largely an illusion, though, since the calculation just performed merely gives formal justification to a simple result. In the introduction to this section we suggested (below Eqs. 11.5) that perhaps there was nothing substantially wrong with the stochastic differential equations obtained in the Glauber–Sudarshan representation, in spite of their complex noise. Certainly, the notation was inconsistent, because variables introduced as complex conjugates would not remain so in the presence of the complex noise. We proposed, nevertheless, that the stochastic differential equations might be accepted and the notation changed ($\alpha^* \to \alpha_*$, $\beta^* \to \beta_*$) to include the extra dimensions that the stochastic trajectories apparently wanted to explore. What we have done in deriving (11.49) is demonstrate that the suggestion was correct. In order to see more explicitly that this is so, let us continue by deriving a set of stochastic differential equations in the Glauber–Sudarshan P representation, by simply overlooking the fact that the diffusion matrix obtained is not positive semidefinite.

Within the Glauber–Sudarshan representation, the phase-space equation of motion for the degenerate parametric oscillator has the form (11.31) with the differential operator $L(\alpha, \alpha_*, \beta, \beta_*, \partial_\alpha, \partial_{\alpha_*}, \partial_\beta, \partial_{\beta_*})$ replaced by

$$L(\boldsymbol{Y}, \boldsymbol{Y}') \equiv -\frac{\partial}{\partial Y_i} A_i(\boldsymbol{Y}) + \frac{1}{2}\frac{\partial^2}{\partial Y_i \partial Y_j} D_{ij}(\boldsymbol{Y}), \tag{11.50}$$

where

$$\boldsymbol{Y} \equiv \begin{pmatrix} \alpha \\ \alpha^* \\ \beta \\ \beta^* \end{pmatrix}, \qquad \boldsymbol{Y}' \equiv \begin{pmatrix} \partial/\partial\alpha \\ \partial/\partial\alpha^* \\ \partial/\partial\beta \\ \partial/\partial\beta^* \end{pmatrix}. \tag{11.51}$$

In terms of the real and imaginary parts of α and β, we write this as

$$L(\boldsymbol{Y}, \boldsymbol{Y}') \equiv L(\boldsymbol{y}, \boldsymbol{y}') = -\frac{\partial}{\partial y_i} A_i^{(1)}(\boldsymbol{y}) + \frac{1}{2}\frac{\partial^2}{\partial y_i \partial y_j} D_{ij}^{(1)}(\boldsymbol{y}), \tag{11.52}$$

where

$$\boldsymbol{y} \equiv \begin{pmatrix} x \\ y \\ u \\ v \end{pmatrix}, \qquad \boldsymbol{y}' \equiv \begin{pmatrix} \partial/\partial x \\ \partial/\partial y \\ \partial/\partial u \\ \partial/\partial v \end{pmatrix}, \tag{11.53}$$

and from the transformation between the complex variables and their real and imaginary parts,

$$\boldsymbol{Y} = \begin{pmatrix} 1 & i & 0 & 0 \\ 1 & -i & 0 & 0 \\ 0 & 0 & 1 & i \\ 0 & 0 & i & -i \end{pmatrix} \boldsymbol{y}, \qquad \boldsymbol{Y}' = \frac{1}{2} \begin{pmatrix} 1 & -i & 0 & 0 \\ 1 & i & 0 & 0 \\ 0 & 0 & 1 & -i \\ 0 & 0 & 1 & i \end{pmatrix} \boldsymbol{y}', \tag{11.54}$$

we have the drift vector

$$\boldsymbol{A}^{(1)}(\boldsymbol{y}) = \frac{1}{2} \begin{pmatrix} 1 & 1 & 0 & 0 \\ -i & i & 0 & 0 \\ 0 & 0 & 1 & 1 \\ 0 & 0 & -i & i \end{pmatrix} \boldsymbol{A}[\boldsymbol{Y}(\boldsymbol{y})], \tag{11.55}$$

and diffusion matrix

$$\boldsymbol{D}^{(1)}(\boldsymbol{y}) = \boldsymbol{B}^{(1)}(\boldsymbol{y})\boldsymbol{B}^{(1)}(\boldsymbol{y})^T, \tag{11.56a}$$

with

$$\boldsymbol{B}^{(1)}(\boldsymbol{y}) = \frac{1}{2} \begin{pmatrix} 1 & 1 & 0 & 0 \\ -i & i & 0 & 0 \\ 0 & 0 & 1 & 1 \\ 0 & 0 & -i & i \end{pmatrix} \boldsymbol{B}[\boldsymbol{Y}(\boldsymbol{y})]. \tag{11.56b}$$

If the diffusion matrix is not positive semidefinite, the matrix $\boldsymbol{B}^{(1)}(\boldsymbol{y})$ will not be real. If we are prepared, however, to overlook this apparent problem, the "Fokker–Planck" operator (11.51) defines the Ito stochastic differential equations

$$d\boldsymbol{y} = \frac{1}{2} \begin{pmatrix} 1 & 1 & 0 & 0 \\ -i & i & 0 & 0 \\ 0 & 0 & 1 & 1 \\ 0 & 0 & -i & i \end{pmatrix} \{\boldsymbol{A}[\boldsymbol{Y}(\boldsymbol{y})]dt + \boldsymbol{B}[\boldsymbol{Y}(\boldsymbol{y})]d\boldsymbol{W}\}, \tag{11.57}$$

or for the complex variables $\boldsymbol{Y}$, using (11.54),

$$d\boldsymbol{Y} = \boldsymbol{A}(\boldsymbol{Y})dt + \boldsymbol{B}(\boldsymbol{Y})d\boldsymbol{W}. \tag{11.58}$$

Equation 11.58 is the same as (11.49), except that in the former, according to the original construction, $\boldsymbol{Y}$ contains two pairs of complex conjugate variables, while the four complex variables of the latter are independent. Note, though, that if $\boldsymbol{B}^{(1)}(\boldsymbol{y})$ is not real, the two pairs of initially conjugate variables are not preserved as complex conjugates by (11.58). Thus, while we can calculate with this equation and get correct results, there is an inconsistency between its derivation and the behavior it describes. The more complicated derivation of (11.49) removes the inconsistency; (11.49) is the same equation as (11.58), but it is derived with nonconjugate variables in mind from the outset. In summary we may state the following:

The stochastic differential equations in the positive P representation are obtained by replacing pairs of complex conjugate variables by pairs of independent complex variables in the stochastic differential equations in the Glauber–Sudarshan P representation; the latter are derived in the standard fashion, relaxing the requirement for positive semidefinite diffusion.

Of course the stochastic differential equations must eventually be solved, and the task again appears to be a complicated one in the positive P representation; the equations are defined in a space of eight dimensions rather than the four needed in the Q and Wigner representations. Nevertheless, we will find that because of the intimate relationship between (11.49) and (11.58) calculations in the positive P representation are often no more arduous than those in the Glauber–Sudarshan P representation.

11.1.3 Linear Theory of Quantum Fluctuations

Now that we have formulated the phase-space description of the degenerate parametric oscillator within the positive P representation, our first task is to treat the small-noise limit. There is little point in rederiving the results of Sect. 10.2. In this section our interest is in more formal aspects of the linearized treatment of fluctuations; specifically, in the relationship between the treatment in the positive P and Glauber–Sudarshan representations. Our main objective is to prove, and hopefully illuminate, the statement made below (10.57b); i.e., that results from Sects. 5.2.2–5.2.5 can be applied to a "Fokker–Planck" equation derived within the Glauber–Sudarshan representation, even when the "diffusion matrix" is not positive semidefinite.

To begin we introduce scaled variables corresponding to those of (10.33)–(10.38); we replace the variables $\boldsymbol{X}$ by the scaled variables $\bar{\boldsymbol{X}}$, where

$$\begin{pmatrix} \sqrt{\xi/2}e^{-i\frac{1}{2}(\psi-\pi/2)}\alpha \\ \sqrt{\xi/2}e^{i\frac{1}{2}(\psi-\pi/2)}\alpha_* \\ e^{-i\frac{1}{2}(\psi-\pi/2)}\beta \\ e^{i\frac{1}{2}(\psi-\pi/2)}\beta_* \end{pmatrix} = \left(n_p^{\rm thr}\right)^{1/2} \begin{pmatrix} \bar{\alpha} \\ \bar{\alpha}_* \\ \bar{\beta} \\ \bar{\beta}_* \end{pmatrix} \equiv \left(n_p^{\rm thr}\right)^{1/2} \bar{\boldsymbol{X}}; \tag{11.59}$$

ξ and ψ are defined in (9.61) and (9.62). Fluctuations about the steady state are described by variables $\boldsymbol{Z}$, of order unity, where we write

$$\bar{\boldsymbol{X}} = \begin{pmatrix} e^{-i\omega_C t}\langle\tilde{\bar{a}}\rangle_{\rm ss} \\ e^{i\omega_C t}\langle\tilde{\bar{a}}^\dagger\rangle_{\rm ss} \\ e^{-i2\omega_C t}\langle\tilde{\bar{b}}\rangle_{\rm ss} \\ e^{i2\omega_C t}\langle\tilde{\bar{b}}^\dagger\rangle_{\rm ss} \end{pmatrix} + \left(n_p^{\rm thr}\right)^{-1/2}\boldsymbol{Z}, \qquad \boldsymbol{Z} \equiv \begin{pmatrix} z \\ z_* \\ w \\ w_* \end{pmatrix}. \tag{11.60}$$

There are now two ways in which we might proceed. The more direct of the two approaches works from the stochastic differential equations (11.49), taking advantage of the complex notation to automatically select the phase-space averages that correspond to normal-ordered, time-ordered averages of the quantum-mechanical operators. The alternative approach uses the Fokker–Planck equation (11.46), written in terms of the real and imaginary parts of the phase-space variables. The notation is less convenient, but the approach has the advantage that it makes the role of the doubling of dimensions particularly clear. Specifically, it shows why, within the linearized treatment, we never actually need to do any calculations in twice as many dimensions. Both have their uses, so let us look at the two approaches in turn.

Consider first the stochastic differential equations (11.49), with $\boldsymbol{A}(\boldsymbol{X})$ defined by (11.34a), and $\boldsymbol{B}(\boldsymbol{X})$ by (11.36) and (11.37). After making a change of variables, the *linearized stochastic differential equations for the degenerate parametric oscillator in the positive P representation* are given by

$$d\tilde{\boldsymbol{Z}} = \bar{\boldsymbol{J}}_{\rm ss}\tilde{\boldsymbol{Z}}dt + \bar{\boldsymbol{B}}_{\rm ss}d\boldsymbol{W}, \qquad \tilde{\boldsymbol{Z}} \equiv \begin{pmatrix} e^{i\omega_C t}z \\ e^{-i\omega_C t}z_* \\ e^{i2\omega_C t}w \\ e^{-i2\omega_C t}w_* \end{pmatrix}, \tag{11.61}$$

where the Jacobian matrix in the steady state is (Eqs. 11.106 and 11.34a)

$$\bar{\boldsymbol{J}}_{\rm ss} \equiv \begin{pmatrix} -\kappa & \kappa\langle\bar{\tilde{b}}\rangle_{\rm ss} & \kappa\langle\bar{\tilde{a}}^\dagger\rangle_{\rm ss} & 0 \\ \kappa\langle\bar{\tilde{b}}^\dagger\rangle_{\rm ss} & -\kappa & 0 & \kappa\langle\bar{\tilde{a}}\rangle_{\rm ss} \\ -2\kappa_p\langle\bar{\tilde{a}}\rangle_{\rm ss} & 0 & -\kappa_p & 0 \\ 0 & -2\kappa_p\langle\bar{\tilde{a}}^\dagger\rangle_{\rm ss} & 0 & -\kappa_p \end{pmatrix}, \tag{11.62}$$

and, using (11.36) and (11.37), we have

$$\bar{\boldsymbol{B}}_{\rm ss} = \begin{pmatrix} \bar{\boldsymbol{S}}_{\rm ss} & 0 \\ 0 & \bar{\boldsymbol{P}}_{\rm ss} \end{pmatrix}, \tag{11.63}$$

with

$$\bar{\boldsymbol{S}}_{\rm ss} \equiv \sqrt{\frac{1}{2}\xi\kappa}\begin{pmatrix} \sqrt{\langle\bar{\tilde{b}}\rangle_{\rm ss}/2 + e^{i\phi_{\rm ss}}\bar{n}} & \sqrt{\langle\bar{\tilde{b}}\rangle_{\rm ss}/2 - e^{i\phi_{\rm ss}}\bar{n}} \\ \sqrt{\langle\bar{\tilde{b}}^\dagger\rangle_{\rm ss}/2 + e^{-i\phi_{\rm ss}}\bar{n}} & -\sqrt{\langle\bar{\tilde{b}}^\dagger\rangle_{\rm ss}/2 - e^{-i\phi_{\rm ss}}\bar{n}} \end{pmatrix}, \tag{11.64a}$$

and

$$\bar{\boldsymbol{P}}_{\rm ss} \equiv \begin{pmatrix} \sqrt{\kappa_p\bar{n}_p} & i\sqrt{\kappa_p\bar{n}_p} \\ \sqrt{\kappa_p\bar{n}_p} & -i\sqrt{\kappa_p\bar{n}_p} \end{pmatrix}. \tag{11.64b}$$

Note that the phase $\phi_{\rm ss} \equiv \arg\big(\langle\bar{\tilde{b}}\rangle_{\rm ss}\big)$ is zero for the steady states (10.43a)–(10.43d). From (11.35) and (11.63), we obtain the steady-state "diffusion matrix"

$$\bar{\boldsymbol{B}}_{\rm ss}\bar{\boldsymbol{B}}_{\rm ss}^T = \bar{\boldsymbol{D}}_{\rm ss} = \begin{pmatrix} \frac{1}{2}\xi\kappa\langle\bar{\tilde{b}}\rangle_{\rm ss} & \xi\kappa\bar{n} & 0 & 0 \\ \xi\kappa\bar{n} & \frac{1}{2}\xi\kappa\langle\bar{\tilde{b}}^\dagger\rangle_{\rm ss} & 0 & 0 \\ 0 & 0 & 0 & 2\kappa_p\bar{n}_p \\ 0 & 0 & 2\kappa_p\bar{n}_p & 0 \end{pmatrix}. \tag{11.65}$$

Recall that this "diffusion matrix" is not positive semidefinite when it is interpreted as the diffusion matrix of a Fokker–Planck equation in the four-dimensional phase space of the Glauber–Sudarshan representation. Within the positive P representation, on the other hand, it defines a positive semidefinite diffusion matrix $\bar{\boldsymbol{D}}_{\mathrm{ss}}^{(2)}$ in eight dimensions (Eqs. 11.44). We meet with the explicit relationship between $\bar{\boldsymbol{D}}_{\mathrm{ss}}$ and $\bar{\boldsymbol{D}}_{\mathrm{ss}}^{(2)}$ shortly.

Now, the central results of the linear theory of fluctuations are the equations of motion for the vector of mean values $(\overline{\tilde{\boldsymbol{Z}}})_{\tilde{P}}$ and for the autocorrelation matrix

$$\boldsymbol{C}(t',t) \equiv (\overline{\tilde{\boldsymbol{Z}}(t')\tilde{\boldsymbol{Z}}(t)^T})_{\tilde{P}}. \tag{11.66}$$

According to the quantum–classical correspondence set up in Sect. 11.1.1, together with its generalization to two-time averages, it is these quantities that correspond to the normal-ordered, or normal-ordered, time-ordered operator averages. We have seen how to derive equations of motion for such quantities in Chap. 5 (Sect. 5.3.6). There, the stochastic variables were considered real, but the derivations of Sect. 5.3.6 remain unchanged if the variables are complex. Hence, we may take over (5.154) and (5.156) directly, writing the equation of motion for mean values

$$\frac{d}{dt}(\overline{\tilde{\boldsymbol{Z}}})_{\tilde{P}} = \bar{\boldsymbol{J}}_{\mathrm{ss}}(\overline{\tilde{\boldsymbol{Z}}})_{\tilde{P}}, \tag{11.67}$$

and for the autocorrelation matrix,

$$\frac{d}{dt'}\boldsymbol{C}(t',t) = \begin{cases} \bar{\boldsymbol{J}}_{\mathrm{ss}}\boldsymbol{C}(t',t) & t' > t \\ \boldsymbol{C}(t',t)\bar{\boldsymbol{J}}_{\mathrm{ss}}^T & t' < t \end{cases}, \tag{11.68}$$

while, from (5.157), the covariance matrix $\boldsymbol{C}(t,t)$ satisfies the equation of motion

$$\frac{d}{dt}\boldsymbol{C}(t,t) = \bar{\boldsymbol{J}}_{\mathrm{ss}}\boldsymbol{C}(t,t) + \boldsymbol{C}(t,t)\bar{\boldsymbol{J}}_{\mathrm{ss}}^T + \bar{\boldsymbol{D}}_{\mathrm{ss}}. \tag{11.69}$$

The important thing to note about these equations is that they involve 4×4, not 8×8, matrices. In fact, $\bar{\boldsymbol{J}}_{\mathrm{ss}}$ and $\bar{\boldsymbol{D}}_{\mathrm{ss}}$ are precisely the matrices obtained within the Glauber–Sudarshan P representation by overlooking the fact that, in the four-dimensional phase-space, $\bar{\boldsymbol{D}}_{\mathrm{ss}}$ is not positive semidefinite. Thus, we quickly prove what we set out to prove. Indeed, the calculations involved in solving (11.67)–(11.69) are simply those performed in Sects. 10.2.2 and 10.2.3; within the linearized theory, there is no need to calculate anything in eight dimensions.

Note 11.5. Extending the quantum–classical correspondence of Sect. 11.1.1 to the calculation of two-time averages requires a little rethinking of the calculation in the Glauber–Sudarshan representation (Sect. 4.3.3). We must recognize that the positive P distribution is not unique; in particular, while we have the relationship (11.9) expressing χ_N in terms of P, we do not have a corresponding inverse relationship, and a direct transcription of the development of Sect. 4.3.3 requires an inverse to arrive at the equivalent of (4.97). Something has to change at this point. Essentially, the difference is that now we prove the results for two-time averages in the reverse direction. First, we assume there exists a family of positive P distributions (parameterized by t) that satisfy a Fokker–Planck equation and represent the density operator at every time. These provide *a* representation of $\rho(t)$, though the representation is not unique. We then show that two-time averages evaluated within the phase-space formalism from the joint distribution $P(\alpha, \alpha_*, t+\tau; \alpha^0, \alpha_*^0, t)$ are equal to the normal-ordered, time-ordered operator averages given by the master equation and the quantum regression formula. This may be done by following the argument of Sect. 4.3.3, except that with regard to the use of (4.97), instead of arguing that $P(\alpha, \alpha_*, t)\alpha_*^p \alpha^q$ is *the* positive P function representing the operator $a^q \rho(t) a^{\dagger p}$, we can only argue that it provides *a* representation.

Exercise 11.4. Obtain drift and diffusion matrices for the degenerate parametric oscillator below and above threshold in the Glauber–Sudarshan P representation from Fokker–Planck equations (10.51), (10.52), and (10.74). Change to the complex variables (10.50) and show that the matrices obtained in this way are those given by (11.62) and (11.65).

Since 4×4 matrices are all that is needed for calculations, it seems that the four additional dimensions introduced by the positive P representation have somehow disappeared. The reason for this is that the representation actually *embeds* a four-dimensional quantum-mechanical problem in an eight-dimensional classical stochastic process. Only a subset of the moments available in the eight dimensions contain the solution to the quantum-mechanical problem; a lot of additional baggage is carried along. The baggage is necessary if the quantum dynamics is to be represented by a classical evolution of the Fokker–Planck type, but it has no direct connection to quantities of physical interest. Within the linearized treatment, an analytical solution is obtained that allows the embedded physical information to be extracted in a simple way, without explicit reference to the additional embedding dimensions. Mathematically, the stochastic differential equation approach avoids the full eight dimensions because we can work with (11.61) alone, without reference to the complex conjugate equation; all physically relevant moments are computed from the variables $\tilde{\boldsymbol{Z}}$, which do not couple dynamically to the conjugate variables. The trick is so slick that it is easy to miss what is really going on. It helps to repeat the calculation leading to (11.67)–(11.69), starting from the Fokker–Planck equation this time.

To explicitly write out the Fokker–Planck equation in the positive P representation we must use the eight-dimensional space of real variables. Using (11.46) and (11.41), the *linearized Fokker–Planck equation for the degenerate parametric oscillator in the positive P representation* is given by

$$\frac{\partial \tilde{\bar{\boldsymbol{P}}}}{\partial t} = \big(- \tilde{\boldsymbol{z}}'^T \bar{\boldsymbol{J}}_{\rm ss}^{(2)} \tilde{\boldsymbol{z}} + \tfrac{1}{2} \tilde{\boldsymbol{z}}'^T \bar{\boldsymbol{D}}_{\rm ss}^{(2)} \tilde{\boldsymbol{z}}' \big) \tilde{\bar{\boldsymbol{P}}}, \tag{11.70}$$

where $\tilde{\boldsymbol{z}}$ is the eight-dimensional vector of phase-space variables,

$$\tilde{\boldsymbol{z}} \equiv \begin{pmatrix} \tilde{\boldsymbol{Z}}_R \\ \tilde{\boldsymbol{Z}}_I \end{pmatrix}, \qquad \tilde{\boldsymbol{Z}}_R \equiv \begin{pmatrix} \tilde{z}_1 \\ \tilde{z}_{*1} \\ \tilde{w}_1 \\ \tilde{w}_{*1} \end{pmatrix}, \qquad \tilde{\boldsymbol{Z}}_I \equiv \begin{pmatrix} \tilde{z}_2 \\ \tilde{z}_{*2} \\ \tilde{w}_2 \\ \tilde{w}_{*2} \end{pmatrix}, \tag{11.71}$$

and $\tilde{\boldsymbol{z}}'$ is the vector of derivatives with respect to these variables,

$$\tilde{\boldsymbol{z}}' \equiv \begin{pmatrix} \tilde{\boldsymbol{Z}}'_R \\ \tilde{\boldsymbol{Z}}'_I \end{pmatrix}, \qquad \tilde{\boldsymbol{Z}}'_R \equiv \begin{pmatrix} \partial/\partial\tilde{z}_1 \\ \partial/\partial\tilde{z}_{*1} \\ \partial/\partial\tilde{w}_1 \\ \partial/\partial\tilde{w}_{*1} \end{pmatrix}, \qquad \tilde{\boldsymbol{Z}}'_I \equiv \begin{pmatrix} \partial/\partial\tilde{z}_2 \\ \partial/\partial\tilde{z}_{*2} \\ \partial/\partial\tilde{w}_2 \\ \partial/\partial\tilde{w}_{*2} \end{pmatrix}. \tag{11.72}$$

The drift and diffusion matrices are defined in terms of the real and imaginary parts of matrices (11.62) and (11.63). For the drift matrix in eight dimensions, we have

$$\bar{\boldsymbol{J}}^{(2)}_{\text{ss}} = \begin{pmatrix} \bar{\boldsymbol{J}}^R_{\text{ss}} & -\bar{\boldsymbol{J}}^I_{\text{ss}} \\ \bar{\boldsymbol{J}}^I_{\text{ss}} & \bar{\boldsymbol{J}}^R_{\text{ss}} \end{pmatrix}, \tag{11.73}$$

and for the diffusion matrix,

$$\bar{\boldsymbol{D}}^{(2)}_{\text{ss}} = \begin{pmatrix} \bar{\boldsymbol{B}}^R_{\text{ss}} & 0 \\ \bar{\boldsymbol{B}}^I_{\text{ss}} & 0 \end{pmatrix} \begin{pmatrix} (\bar{\boldsymbol{B}}^R_{\text{ss}})^T & (\bar{\boldsymbol{B}}^I_{\text{ss}})^T \\ 0 & 0 \end{pmatrix}. \tag{11.74}$$

We apply the results of Chap. 5 (Sect. 5.2) directly to (11.70).

All together, in the eight-dimensional phase space, there are eight mean values and sixty-four correlation functions. From (5.90) and (5.93), these satisfy the equations of motion

$$\frac{d}{dt}\big(\,\overline{\tilde{\boldsymbol{z}}}\,\big)_{\tilde{\bar{P}}} = \bar{\boldsymbol{J}}^{(2)}_{\text{ss}}\big(\,\overline{\tilde{\boldsymbol{z}}}\,\big)_{\tilde{\bar{P}}}, \tag{11.75}$$

and

$$\frac{d}{dt'}\boldsymbol{C}^{(2)}(t',t) = \begin{cases} \bar{\boldsymbol{J}}^{(2)}_{\text{ss}}\boldsymbol{C}^{(2)}(t',t) & t' > t \\ \boldsymbol{C}^{(2)}(t',t)\big(\bar{\boldsymbol{J}}^{(2)}_{\text{ss}}\big)^T & t' < t \end{cases}, \tag{11.76}$$

where $\boldsymbol{C}^{(2)}(t',t)$ is the autocorrelation matrix in eight dimensions,

$$\boldsymbol{C}^{(2)}(t',t) \equiv \big(\,\overline{\tilde{\boldsymbol{z}}(t')\tilde{\boldsymbol{z}}(t)^T}\,\big)_{\tilde{\bar{P}}}, \tag{11.77}$$

and from (5.99), the covariance matrix $\boldsymbol{C}^{(2)}(t,t)$ satisfies the equation of motion

$$\frac{d}{dt}\boldsymbol{C}^{(2)}(t,t) = \bar{\boldsymbol{J}}^{(2)}_{\text{ss}}\boldsymbol{C}^{(2)}(t,t) + \boldsymbol{C}^{(2)}(t,t)\big(\bar{\boldsymbol{J}}^{(2)}_{\text{ss}}\big)^T + \bar{\boldsymbol{D}}^{(2)}_{\text{ss}}. \tag{11.78}$$

The central point now is the observation that it is not necessary to solve these equations in eight dimensions. The solutions in eight dimensions contain a large amount of unnecessary information, information that is not directly related to the physical problem of interest. The relevant moments are all averages of the variables

$$\tilde{\boldsymbol{Z}} = (\boldsymbol{I}_4 \; i\boldsymbol{I}_4)\tilde{\boldsymbol{z}}, \tag{11.79}$$

where $\boldsymbol{I}_4$ is the 4×4 identity matrix. The physically relevant correlation functions are therefore particular combinations of the correlation functions contained in $\boldsymbol{C}^{(2)}(t', t)$; they are contained in the contracted autocorrelation matrix

$$\boldsymbol{C}(t', t) = (\boldsymbol{I}_4 \; i\boldsymbol{I}_4)\boldsymbol{C}^{(2)}(t', t)\begin{pmatrix} \boldsymbol{I}_4 \\ i\boldsymbol{I}_4 \end{pmatrix}. \tag{11.80}$$

It is now straightforward to use (11.79) and (11.80) to contract the equations of motion (11.75), (11.76), and (11.78) into a set of equations in four dimensions. To do this we note that the drift and diffusion matrices may be contracted in a similar way to $\tilde{\boldsymbol{z}}$ and $\boldsymbol{C}^{(2)}(t', t)$:

$$\begin{aligned} (\boldsymbol{I}_4 \; i\boldsymbol{I}_4)\bar{\boldsymbol{J}}^{(2)}_{\mathrm{ss}} &= (\boldsymbol{I}_4 \; i\boldsymbol{I}_4)\begin{pmatrix} \bar{\boldsymbol{J}}^R_{\mathrm{ss}} & -\bar{\boldsymbol{J}}^I_{\mathrm{ss}} \\ \bar{\boldsymbol{J}}^I_{\mathrm{ss}} & \bar{\boldsymbol{J}}^R_{\mathrm{ss}} \end{pmatrix} \\ &= (\boldsymbol{I}_4 \; i\boldsymbol{I}_4)\bar{\boldsymbol{J}}_{\mathrm{ss}}, \end{aligned} \tag{11.81}$$

and

$$\begin{aligned} (\boldsymbol{I}_4 \; i\boldsymbol{I}_4)\bar{\boldsymbol{D}}^{(2)}_{\mathrm{ss}}\begin{pmatrix} \boldsymbol{I}_4 \\ i\boldsymbol{I}_4 \end{pmatrix} &= (\boldsymbol{I}_4 \; i\boldsymbol{I}_4)\begin{pmatrix} \bar{\boldsymbol{B}}^R_{\mathrm{ss}} & 0 \\ \bar{\boldsymbol{B}}^I_{\mathrm{ss}} & 0 \end{pmatrix}\begin{pmatrix} (\bar{\boldsymbol{B}}^R_{\mathrm{ss}})^T & (\bar{\boldsymbol{B}}^I_{\mathrm{ss}})^T \\ 0 & 0 \end{pmatrix}\begin{pmatrix} \boldsymbol{I}_4 \\ i\boldsymbol{I}_4 \end{pmatrix} \\ &= (\bar{\boldsymbol{B}}_{\mathrm{ss}} 0)\begin{pmatrix} \bar{\boldsymbol{B}}^T_{\mathrm{ss}} \\ 0 \end{pmatrix} \\ &= \bar{\boldsymbol{D}}_{\mathrm{ss}}. \end{aligned} \tag{11.82}$$

Using (11.79)–(11.82), we readily recover the equations of motion (11.67)–(11.69).

11.2 Miscellaneous Topics

11.2.1 Alternative Approaches to the Linear Theory of Quantum Fluctuations

Before leaving the subject of linear fluctuation theory, we should consider the question of alternatives to the positive P representation. There are in fact quite a number. We met one in Chap. 10, where in Sect. 10.1 we saw that the Wigner and Q representations yield positive semidefinite diffusion for the degenerate parametric oscillator when the Glauber–Sudarshan P representation does not [see the discussion below (10.11), (10.26), and (10.27)]; thus, choosing a different phase-space representation can eliminate the need to deal with non-positive-semidefinite diffusion. A second alternative is to work directly with the equations of motion for operator averages and correlation functions. Generally these quantities obey infinite hierarchies of coupled equations. In a linearized treatment, though, such a hierarchy may be truncated at second-order, on the basis of the system size expansion, and the resulting equations solved to find relationships equivalent to those of Chap. 5. Since operator averages and correlations functions are what we eventually want in any case—the diffusion process only being a path to this end—we might argue for this more direct approach independently of any problem with non-positive-semidefinite diffusion. Popular versions are the so-called input–output theories, which deal directly with the multimode fields carried by the reservoirs [11.16, 11.17, 11.18, 11.19], fields like those defined in (9.120a)–(9.123b). We will not develop the operator-based approach in detail. The main ideas should be clear after working through the following example.

Exercise 11.5. Show that the master equation for the degenerate parametric oscillator (Eq. 9.97) gives the following equations of motion for first-order averages:

$$\kappa^{-1}\frac{d\langle\tilde{\tilde{a}}\rangle}{dt} = -\langle\tilde{\tilde{a}}\rangle + \langle\tilde{\tilde{a}}^\dagger\tilde{\tilde{b}}\rangle, \tag{11.83a}$$

$$\kappa^{-1}\frac{d\langle\tilde{\tilde{a}}^\dagger\rangle}{dt} = -\langle\tilde{\tilde{a}}^\dagger\rangle + \langle\tilde{\tilde{a}}\tilde{\tilde{b}}^\dagger\rangle, \tag{11.83b}$$

$$\kappa_p^{-1}\frac{d\langle\tilde{\tilde{b}}\rangle}{dt} = -\langle\tilde{\tilde{b}}\rangle - \langle\tilde{\tilde{a}}^2\rangle + \lambda, \tag{11.83c}$$

$$\kappa_p^{-1}\frac{d\langle\tilde{\tilde{b}}^\dagger\rangle}{dt} = -\langle\tilde{\tilde{b}}^\dagger\rangle - \langle\tilde{\tilde{a}}^{\dagger 2}\rangle + \lambda. \tag{11.83d}$$

Unlike the corresponding equations in the linearized treatment of fluctuations (Eqs. 10.43a–10.43d), these couple to the equations of motion for second-order operator averages. Show that the second-order averages couple to those of third-order through the equations

$$\left.\begin{aligned}
(2\kappa)^{-1}\frac{d\langle\tilde{\tilde{a}}^\dagger\tilde{\tilde{a}}\rangle}{dt} &= -\langle\tilde{\tilde{a}}^\dagger\tilde{\tilde{a}}\rangle + \tfrac{1}{2}\big(\langle\tilde{\tilde{a}}^{\dagger 2}\tilde{\tilde{b}}\rangle + \text{c.c.}\big) + (n_p^{\text{thr}})^{-1}(\xi/2)\bar{n},\\
(2\kappa)^{-1}\frac{d\langle\tilde{\tilde{a}}^2\rangle}{dt} &= -\langle\tilde{\tilde{a}}^2\rangle + \langle\tilde{\tilde{a}}^\dagger\tilde{\tilde{a}}\tilde{\tilde{b}}\rangle + (n_p^{\text{thr}})^{-1}(\xi/4)\langle\tilde{\tilde{b}}\rangle,\\
(2\kappa)^{-1}\frac{d\langle\tilde{\tilde{a}}^{\dagger 2}\rangle}{dt} &= -\langle\tilde{\tilde{a}}^{\dagger 2}\rangle + \langle\tilde{\tilde{a}}^\dagger\tilde{\tilde{a}}\tilde{\tilde{b}}^\dagger\rangle + (n_p^{\text{thr}})^{-1}(\xi/4)\langle\tilde{\tilde{b}}^\dagger\rangle,
\end{aligned}\right\} \tag{11.84a}$$

$$\left.\begin{aligned}
(2\kappa_p)^{-1}\frac{d\langle\tilde{\bar{b}}^\dagger\tilde{\bar{b}}\rangle}{dt} &= -\langle\tilde{\bar{b}}^\dagger\tilde{\bar{b}}\rangle - \tfrac{1}{2}\left[\langle(\tilde{\bar{a}}^2-\lambda)\tilde{\bar{b}}^\dagger\rangle + \text{c.c.}\right] + (n_p^{\text{thr}})^{-1}\bar{n}_p,\\
(2\kappa_p)^{-1}\frac{d\langle\tilde{\bar{b}}^2\rangle}{dt} &= -\langle\tilde{\bar{b}}^2\rangle - \langle(\tilde{\bar{a}}^2-\lambda)\tilde{\bar{b}}\rangle,\\
(2\kappa_p)^{-1}\frac{d\langle\tilde{\bar{b}}^{\dagger 2}\rangle}{dt} &= -\langle\tilde{\bar{b}}^{\dagger 2}\rangle - \langle(\tilde{\bar{a}}^{\dagger 2}-\lambda)\tilde{\bar{b}}^\dagger\rangle,
\end{aligned}\right\}\qquad(11.84\text{b})$$

and

$$\left.\begin{aligned}
(\kappa+\kappa_p)^{-1}\frac{d\langle\tilde{\bar{a}}^\dagger\tilde{\bar{b}}\rangle}{dt} &= -\langle\tilde{\bar{a}}^\dagger\tilde{\bar{b}}\rangle + (1+\xi)^{-1}\left[\xi\langle\tilde{\bar{a}}\tilde{\bar{b}}^\dagger\tilde{\bar{b}}\rangle - \langle\tilde{\bar{a}}^\dagger(\tilde{\bar{a}}^2-\lambda)\rangle\right],\\
(\kappa+\kappa_p)^{-1}\frac{d\langle\tilde{\bar{a}}\tilde{\bar{b}}^\dagger\rangle}{dt} &= -\langle\tilde{\bar{a}}\tilde{\bar{b}}^\dagger\rangle + (1+\xi)^{-1}\left[\xi\langle\tilde{\bar{a}}^\dagger\tilde{\bar{b}}^\dagger\tilde{\bar{b}}\rangle - \langle(\tilde{\bar{a}}^{\dagger 2}-\lambda)\tilde{\bar{a}}\rangle\right],\\
(\kappa+\kappa_p)^{-1}\frac{d\langle\tilde{\bar{a}}\tilde{\bar{b}}\rangle}{dt} &= -\langle\tilde{\bar{a}}\tilde{\bar{b}}\rangle + (1+\xi)^{-1}\left[\xi\langle\tilde{\bar{a}}^\dagger\tilde{\bar{b}}^2\rangle - \langle(\tilde{\bar{a}}^2-\lambda)\tilde{\bar{a}}\rangle\right],\\
(\kappa+\kappa_p)^{-1}\tfrac{d\langle\tilde{\bar{a}}^\dagger\tilde{\bar{b}}^\dagger\rangle}{dt} &= -\langle\tilde{\bar{a}}^\dagger\tilde{\bar{b}}^\dagger\rangle + (1+\xi)^{-1}\left[\xi\langle\tilde{\bar{a}}\tilde{\bar{b}}^{\dagger 2}\rangle - \langle\tilde{\bar{a}}^\dagger(\tilde{\bar{a}}^{\dagger 2}-\lambda)\rangle\right].
\end{aligned}\right\}\qquad(11.84\text{c})$$

Now introduce the scaling from the phase-space implementation of the system size expansion (Eqs. 10.33–10.38) in operator form, writing

$$\bar{a} = \langle\bar{a}\rangle + (n_p^{\text{thr}})^{-1/2}\hat{z},\qquad(11.85\text{a})$$

$$\bar{a}^\dagger = \langle\bar{a}^\dagger\rangle + (n_p^{\text{thr}})^{-1/2}\hat{z}^\dagger,\qquad(11.85\text{b})$$

and

$$\bar{b} = \langle\bar{b}\rangle + (n_p^{\text{thr}})^{-1/2}\hat{w},\qquad(11.86\text{a})$$

$$\bar{b}^\dagger = \langle\bar{b}^\dagger\rangle + (n_p^{\text{thr}})^{-1/2}\hat{w}^\dagger,\qquad(11.86\text{b})$$

where $\hat{z}$, $\hat{z}^\dagger$, $\hat{w}$, and $\hat{w}^\dagger$ are displaced annihilation and creation operators describing fluctuations about the mean. Assuming $(n_p^{\text{thr}})^{-1/2} \ll 1$, show that to dominant order the equations of motion for first-order averages reduce to (10.43a)–(10.43d), and that second-order averages of the fluctuation operators satisfy

$$\left.\begin{aligned}
(2\kappa)^{-1}\frac{d\langle\tilde{\hat{z}}^\dagger\tilde{\hat{z}}\rangle}{dt} &= -\langle\tilde{\hat{z}}^\dagger\tilde{\hat{z}}\rangle + \tfrac{1}{2}\left(\langle\tilde{\bar{a}}^\dagger\rangle\langle\tilde{\hat{z}}^\dagger\tilde{\hat{w}}\rangle + \langle\tilde{\bar{b}}\rangle\langle\tilde{\hat{z}}^{\dagger 2}\rangle + \text{c.c.}\right) + (\xi/2)\bar{n},\\
(2\kappa)^{-1}\frac{d\langle\tilde{\hat{z}}^2\rangle}{dt} &= -\langle\tilde{\hat{z}}^2\rangle + \langle\tilde{\bar{a}}^\dagger\rangle\langle\tilde{\hat{z}}\tilde{\hat{w}}\rangle + \langle\tilde{\bar{b}}\rangle\langle\tilde{\hat{z}}^\dagger\tilde{\hat{z}}\rangle + (\xi/4)\langle\tilde{\bar{b}}\rangle,\\
(2\kappa)^{-1}\frac{d\langle\tilde{\hat{z}}^{\dagger 2}\rangle}{dt} &= -\langle\tilde{\hat{z}}^{\dagger 2}\rangle + \langle\tilde{\bar{a}}\rangle\langle\tilde{\hat{z}}^\dagger\tilde{\hat{w}}^\dagger\rangle + \langle\tilde{\bar{b}}^\dagger\rangle\langle\tilde{\hat{z}}^\dagger\tilde{\hat{z}}\rangle + (\xi/4)\langle\tilde{\bar{b}}^\dagger\rangle,
\end{aligned}\right\}\qquad(11.87\text{a})$$

$$\left.\begin{aligned}
(2\kappa_p)^{-1}\frac{d\langle\tilde{\tilde{w}}^\dagger\tilde{\tilde{w}}\rangle}{dt} &= -\langle\tilde{\tilde{w}}^\dagger\tilde{\tilde{w}}\rangle - \big(\langle\tilde{\bar{a}}\rangle\langle\tilde{\tilde{z}}\tilde{\tilde{w}}^\dagger\rangle + \text{c.c.}\big) + \bar{n}_p,\\
(2\kappa_p)^{-1}\frac{d\langle\tilde{\tilde{w}}^2\rangle}{dt} &= -\langle\tilde{\tilde{w}}^2\rangle - 2\langle\tilde{\bar{a}}\rangle\langle\tilde{\tilde{z}}\tilde{\tilde{w}}\rangle,\\
(2\kappa_p)^{-1}\frac{d\langle\tilde{\tilde{w}}^{\dagger 2}\rangle}{dt} &= -\langle\tilde{\tilde{w}}^{\dagger 2}\rangle - 2\langle\tilde{\bar{a}}^\dagger\rangle\langle\tilde{\tilde{z}}^\dagger\tilde{\tilde{w}}^\dagger\rangle,
\end{aligned}\right\} \tag{11.87b}$$

and

$$\left.\begin{aligned}
(\kappa+\kappa_p)^{-1}\frac{d\langle\tilde{\tilde{z}}^\dagger\tilde{\tilde{w}}\rangle}{dt} &= -\langle\tilde{\tilde{z}}^\dagger\tilde{\tilde{w}}\rangle + (1+\xi)^{-1}\big[\xi\big(\langle\tilde{\bar{a}}\rangle\langle\tilde{\tilde{w}}^2\rangle + \langle\tilde{\bar{b}}^\dagger\rangle\langle\tilde{\tilde{z}}\tilde{\tilde{w}}\rangle\big)\\
&\qquad -2\langle\tilde{\bar{a}}\rangle\langle\tilde{\tilde{z}}^\dagger\tilde{\tilde{z}}\rangle\big],\\
(\kappa+\kappa_p)^{-1}\frac{d\langle\tilde{\tilde{z}}\tilde{\tilde{w}}^\dagger\rangle}{dt} &= -\langle\tilde{\tilde{z}}\tilde{\tilde{w}}^\dagger\rangle + (1+\xi)^{-1}\big[\xi\big(\langle\tilde{\bar{a}}^\dagger\rangle\langle\tilde{\tilde{w}}^{\dagger 2}\rangle + \langle\tilde{\bar{b}}\rangle\langle\tilde{\tilde{z}}^\dagger\tilde{\tilde{w}}^\dagger\rangle\big)\\
&\qquad -2\langle\tilde{\bar{a}}^\dagger\rangle\langle\tilde{\tilde{z}}^\dagger\tilde{\tilde{z}}\rangle\big],\\
(\kappa+\kappa_p)^{-1}\frac{d\langle\tilde{\tilde{z}}\tilde{\tilde{w}}\rangle}{dt} &= -\langle\tilde{\tilde{z}}\tilde{\tilde{w}}\rangle + (1+\xi)^{-1}\big[\xi\big(\langle\tilde{\bar{a}}^\dagger\rangle\langle\tilde{\tilde{w}}^2\rangle + 2\langle\tilde{\bar{b}}\rangle\langle\tilde{\tilde{z}}^\dagger\tilde{\tilde{w}}\rangle\big)\\
&\qquad -2\langle\tilde{\bar{a}}\rangle\langle\tilde{\tilde{z}}^2\rangle\big],\\
(\kappa+\kappa_p)^{-1}\frac{d\langle\tilde{\tilde{z}}^\dagger\tilde{\tilde{w}}^\dagger\rangle}{dt} &= -\langle\tilde{\tilde{z}}^\dagger\tilde{\tilde{w}}^\dagger\rangle + (1+\xi)^{-1}\big[\xi\big(\langle\tilde{\bar{a}}\rangle\langle\tilde{\tilde{w}}^{\dagger 2}\rangle + 2\langle\tilde{\bar{b}}^\dagger\rangle\langle\tilde{\tilde{z}}\tilde{\tilde{w}}^\dagger\rangle\big)\\
&\qquad -2\langle\tilde{\bar{a}}^\dagger\rangle\langle\tilde{\tilde{z}}^{\dagger 2}\rangle\big].
\end{aligned}\right\} \tag{11.87c}$$

Note that the equations of motion for second-order averages depend on the solutions for first-order averages, but no longer couple to third-order averages. Verify that (11.87a)–(11.87c) are equivalent to the equation of motion for the covariance matrix in the positive P representation (Eq. 11.69). (Operators referring to the same field mode have been written in normal order to obtain this equivalence.)

Carrying out the linearization at the level of the operator equations underlines the fact that the positive P representation is merely a bookkeeping device for operator averages. In this approach there is no call for the diffusion matrix $\bar{\boldsymbol{D}}_{\text{ss}}$ that appears in (11.69) [implicitly in (11.87a)–(11.87c)] to be positive semidefinite; it is the attempt to recover the operator averages from a classical stochastic process that leads to the requirement for positive semidefinite diffusion. There, of course, should be no surprise here. The quantum–classical correspondence was initially set up as a formal device for computing operator averages; we noted, for example, in Sect. 3.1.3 that something as formal as a generalized function might be needed to make the correspondence work. This raises the question: can we work with non-positive-definite diffusion in the Glauber–Sudarshan P representation and do the bookkeeping using a generalized function? The answer is affirmative—indeed we can. In this we find yet another—a third—version of linear fluctuation theory.

As an illustration of the generalized function approach, let us consider the one-dimensional Fokker–Planck equation

$$\frac{\partial P}{\partial t} = \left(-A\frac{\partial}{\partial x}x + \frac{1}{2}D\frac{\partial^2}{\partial x^2}\right)P. \tag{11.88}$$

When the diffusion constant D is a positive number (and A is negative) this equation has a Gaussian steady-state solution, as described in Sects. 5.1.1 and 5.1.2. If D is negative, the Gaussian diverges as $x^2 \to \infty$ and is no longer acceptable as a normalizable probability distribution. On the other hand, if generalized functions are permitted, the usual Gaussian solution of (11.88) is not unique; we may also find a solution in the form

$$P_{\text{ss}}(x) = \sum_{n=0}^{\infty} c_n \delta^{(n)}(x), \tag{11.89}$$

where $\delta^{(n)}(x)$ is the nth derivative of the δ-function [see the discussion below (3.32b)]. To determine the coefficients c_n, we substitute (11.89) into (11.88) with the time derivative set to zero. This yields

$$\sum_{n=0}^{\infty} c_n\big[-A\big(\delta^{(n)}(x) + x\delta^{(n+1)}\big) + \tfrac{1}{2}D\delta^{(n+2)}(x)\big] = 0. \tag{11.90}$$

Multiplying throughout by x^m and integrating over x gives

$$\begin{aligned}
&\int_{-\infty}^{\infty} dx x^m \sum_{n=0}^{\infty} c_n\big[-A\big(\delta^{(n)}(x) + x\delta^{(n+1)}\big) + \tfrac{1}{2}D\delta^{(n+2)}(x)\big] = 0 \\
&\Rightarrow -A\big[(-1)^m m! c_m + (-1)^{m+1}(m+1)! c_m\big] + \tfrac{1}{2}D(-1)^m m! c_{m-2} = 0 \\
&\Rightarrow Amc_m + \tfrac{1}{2}Dc_{m-2} = 0.
\end{aligned} \tag{11.91}$$

Thus we arrive at the steady-state distribution

$$\begin{aligned}
P_{\text{ss}}(x) &= \sum_{k=0}^{\infty}\left(-\frac{D}{2A}\right)^k \frac{1}{k!}\frac{1}{2^k}\delta^{(2k)}(x) \\
&= \exp\left[-\tfrac{1}{2}(D/2A)\frac{d^2}{dx^2}\right]\delta(x).
\end{aligned} \tag{11.92}$$

In fact, all we have done is demonstrate the equivalence—previously proved in Note 4.2—between a Gaussian distribution and an infinite sum of derivatives of the δ-function.

The distribution (11.92) has one advantage over the usual Gaussian; it is a normalized distribution for both positive and negative values of D. It readily yields the result that all odd-order moments of x are zero, while the even-order moments are given by

$$\langle x^{2k}\rangle = \left(-\frac{D}{2A}\right)^k \frac{(2k)!}{k!2^k}. \tag{11.93}$$

For D positive and A negative, these are the moments of the usual normalizable Gaussian distribution. The generalized function tells us that the same moments hold, formally, when D is negative—some of the moments, e.g., the variance, simply become negative.

Of course, a negative variance makes no sense if x takes on the observable values of a physical variable, in the sense of a classical ensemble, since it must then be a real number. The phase-space of the quantum–classical correspondence is not, however, a space of observable values. At least it makes no fundamental claim to this effect. The quantum–classical correspondence provides a formal connection between phase-space moments and ordered operator averages. If it is to be interpreted as anything more, the interpretation must be established through a theory of measurement.

The sense in which a negative variance is *formally* acceptable is demonstrated by the example of the degenerate parametric oscillator. Below threshold, in the Glauber–Sudarshan P representation, Fokker–Planck equations (10.51a) and (10.51b) require that the steady-state P distribution satisfy ($\bar{n}=0$)

$$\left[(1-\lambda)\frac{\partial}{\partial\tilde{z}_1}\tilde{z}_1 + \frac{1}{8}\xi\lambda\frac{\partial^2}{\partial\tilde{z}_1^2}\right]\bar{\tilde{X}}_{+1} = 0, \tag{11.94a}$$

and

$$\left[(1+\lambda)\frac{\partial}{\partial\tilde{z}_2}\tilde{z}_2 - \frac{1}{8}\xi\lambda\frac{\partial^2}{\partial\tilde{z}_2^2}\right]\bar{\tilde{Y}}_{+1} = 0. \tag{11.94b}$$

Equation 11.94b has non-positive-semidefinite diffusion. Nonetheless, it is solved by a generalized function in the form (11.92). The solution yields the phase-space moments

$$\left(\overline{\tilde{z}_1^2}\right)_{\bar{\tilde{X}}_{+1}} = (\xi/2)\frac{1}{4}\frac{\lambda}{1-\lambda}, \qquad \left(\overline{\tilde{z}_2^2}\right)_{\bar{\tilde{Y}}_{+1}} = -(\xi/2)\frac{1}{4}\frac{\lambda}{1+\lambda}. \tag{11.95}$$

The negative variance $\left(\overline{\tilde{z}_2^2}\right)_{\bar{\tilde{Y}}_{+1}}$ is physically acceptable because it corresponds to the *normal-ordered* operator average

$$\langle : [(-i/2)(\tilde{a}-\tilde{a}^\dagger)]^2 : \rangle = \tfrac{1}{4}\left(2\langle\tilde{a}^\dagger\tilde{a}\rangle - \langle\tilde{a}^2\rangle - \langle\tilde{a}^{\dagger 2}\rangle\right);$$

of course this normal-ordered average is permitted to be negative. The negative variance will only contribute *a part* of any physical quantity that is necessarily positive, the remaining parts offsetting its negative value. This is so, for example, in the expression for the mean photon number:

$$
\begin{aligned}
\langle a^\dagger a\rangle_< &= (2/\xi)\Big[\big(\overline{\tilde{z}_1^2}\big)_{\tilde{\bar{X}}_{+1}} + \big(\overline{\tilde{z}_2^2}\big)_{\tilde{\bar{Y}}_{+1}}\Big] \\
&= \frac{1}{4}\frac{\lambda}{1-\lambda} - \frac{1}{4}\frac{\lambda}{1+\lambda} \\
&= \frac{1}{2}\frac{\lambda^2}{1-\lambda^2},
\end{aligned} \tag{11.96}
$$

which reproduces (10.64).

Time dependence can also be handled using generalized functions. A generalized function can be found to replace the Gaussian Green function solution to a linear Fokker–Planck equation with non-positive-semidefinite diffusion.

Exercise 11.6. Show that (11.88) has the Green function solution

$$
P(x,t|x_0,0) = \exp\left[\tfrac{1}{2}(D/2A)(e^{2At}-1)\frac{d^2}{dx^2}\right]\delta(x - x_0e^{At}). \tag{11.97}
$$

Note 11.6. By referring to (5.80)–(5.81b), it is not difficult to guess the multidimensional generalization of the steady-state distribution (11.92). It is given by

$$
P_{\mathrm{ss}}(\boldsymbol{x}) = \exp\big(\tfrac{1}{2}\boldsymbol{x}'^T\boldsymbol{Q}_{\mathrm{ss}}\boldsymbol{x}'\big)\delta(\boldsymbol{x}), \tag{11.98}
$$

with

$$
\boldsymbol{A}\boldsymbol{Q}_{\mathrm{ss}} + \boldsymbol{Q}_{\mathrm{ss}}\boldsymbol{A}^T = -\boldsymbol{D}, \tag{11.99}
$$

where $\boldsymbol{D}$ need not be positive semidefinite. The formal generalization of (5.102a) is made by calculating the steady-state covariance matrix

$$
\begin{aligned}
\boldsymbol{C}_{\mathrm{ss}} &= \int_{-\infty}^{\infty} dx_1\cdots dx_n \boldsymbol{x}\boldsymbol{x}^T \exp\big(\tfrac{1}{2}\boldsymbol{x}'^T\boldsymbol{Q}_{\mathrm{ss}}\boldsymbol{x}'\big)\delta(\boldsymbol{x}) \\
&= \int_{-\infty}^{\infty} dx_1\cdots dx_n \boldsymbol{x}\boldsymbol{x}^T \big(\tfrac{1}{2}\boldsymbol{x}'^T\boldsymbol{Q}_{\mathrm{ss}}\boldsymbol{x}'\big)\delta(\boldsymbol{x}) \\
&= \int_{-\infty}^{\infty} dx_1\cdots dx_n \boldsymbol{x}\boldsymbol{x}^T \frac{1}{2}\sum_{i,j=1}^{n}(\boldsymbol{Q}_{\mathrm{ss}})_{ij}\frac{\partial^2}{\partial x_i\partial x_j}\delta(\boldsymbol{x}) \\
&= \boldsymbol{Q}_{\mathrm{ss}}.
\end{aligned} \tag{11.100}
$$

The second line follows because only second-order derivatives contribute.

Other variations on the theme of linear fluctuation theory exist. In one final example, Yuen and Tombesi [11.20] note that equations like (11.69) may be absorbed, formally, into classical Langevin theory when $\bar{\boldsymbol{D}}_{\mathrm{ss}}$ is not positive semidefinite. Their approach is essentially that of the positive P representation, but it avoids explicit reference to the $i = \sqrt{-1}$ used to connect a negative operator average to a positive phase-space variance in an equation like (11.1b).

11.2.2 Dynamical Stability of the Classical Phase Space

The positive P representation uses double the number of phase-space dimensions. The extra dimensions are introduced to accommodate the quantum noise. In the presence of non-positive-semidefinite diffusion, some noise sources effectively "ask" for new dimensions; in (11.1b), for example, the noise term $i\frac{1}{2}\sqrt{\xi\kappa\lambda}dW_{\tilde{z}_2}$ "asks" that the real variable $\tilde{z}_2$ be given an imaginary part. We will designate the standard phase space the *classical phase space* and its extension the *nonclassical phase space.* When diffusion in the Glauber–Sudarshan P representation is not positive semidefinite, fluctuations in the positive P representation extend into the nonclassical phase space. It is then natural to ask whether the classical phase space is stable (under the deterministic dynamics) to perturbations into the nonclassical dimensions. If not, doubling the dimensions will have a more profound effect on the dynamics than we might expect; small excursions into the nonclassical phase space will grow exponentially, possibly leading to unforeseen nonlinear behavior involving the nonclassical degrees of freedom.

We investigate this question by linearizing the deterministic equations about an arbitrary point in the classical phase space and looking at the eigenvalues of the linearized dynamics. First, we transform the deterministic equations of motion derived in Sect. 11.1.2 into a rotating frame, thus removing their explicit time dependence. To replace (11.33) and (11.34a), we define the vector of phase-space variables

$$\tilde{\boldsymbol{X}} \equiv \begin{pmatrix} \tilde{\alpha} \\ \tilde{\alpha}_* \\ \tilde{\beta} \\ \tilde{\beta}_* \end{pmatrix} \equiv \begin{pmatrix} e^{i\omega_C t}\alpha \\ e^{-i\omega_C t}\alpha_* \\ e^{i2\omega_C t}\beta \\ e^{-i2\omega_C t}\beta_* \end{pmatrix}, \tag{11.101}$$

and the drift vector

$$\tilde{\boldsymbol{A}}(\tilde{\boldsymbol{X}}) \equiv \begin{pmatrix} -\kappa\tilde{\alpha} + g\tilde{\alpha}_*\tilde{\beta} \\ -\kappa\tilde{\alpha}_* + g\tilde{\alpha}\tilde{\beta}_* \\ -\kappa_p\tilde{\beta} - (g/2)\tilde{\alpha}^2 - i\bar{\mathcal{E}}_0 \\ -\kappa_p\tilde{\beta}_* - (g/2)\tilde{\alpha}_*^2 + i\bar{\mathcal{E}}_0^* \end{pmatrix}. \tag{11.102}$$

The *deterministic equations of motion for the degenerate parametric oscillator in the positive P representation* are then given by

$$\frac{d\tilde{\boldsymbol{X}}}{dt} = \tilde{\boldsymbol{A}}(\tilde{\boldsymbol{X}}), \tag{11.103a}$$

$$\frac{d\tilde{\boldsymbol{X}}^*}{dt} = \left(\tilde{\boldsymbol{A}}(\tilde{\boldsymbol{X}})\right)^*. \tag{11.103b}$$

We linearize the equations about an arbitrary point $\tilde{\boldsymbol{X}}_0 = (\tilde{\alpha}, \tilde{\alpha}^*, \tilde{\beta}, \tilde{\beta}^*)^T$ located within the classical phase space. Writing $\tilde{\boldsymbol{X}}$ as the sum of the fixed vector $\tilde{\boldsymbol{X}}_0$ and a perturbation $\delta\tilde{\boldsymbol{X}}$,

$$\tilde{\boldsymbol{X}} = \tilde{\boldsymbol{X}}_0 + \delta\tilde{\boldsymbol{X}}, \tag{11.104}$$

the linearized equations of motion are

$$\frac{d\delta\tilde{\boldsymbol{X}}}{dt} = \boldsymbol{J}(\tilde{\boldsymbol{X}}_0)\delta\tilde{\boldsymbol{X}} + \tilde{\boldsymbol{A}}(\tilde{\boldsymbol{X}}_0), \tag{11.105a}$$

$$\frac{d(\delta\tilde{\boldsymbol{X}})^*}{dt} = \big(\boldsymbol{J}(\tilde{\boldsymbol{X}}_0)\big)^*(\delta\tilde{\boldsymbol{X}})^* + \big(\tilde{\boldsymbol{A}}(\tilde{\boldsymbol{X}}_0)\big)^*, \tag{11.105b}$$

where $\boldsymbol{J}(\tilde{\boldsymbol{X}})$ is the Jacobian matrix

$$\boldsymbol{J}(\tilde{\boldsymbol{X}}) \equiv \begin{pmatrix} \partial_{\tilde{\alpha}}\tilde{A}_{\tilde{\alpha}}(\tilde{\boldsymbol{X}}) & \partial_{\tilde{\alpha}_*}\tilde{A}_{\tilde{\alpha}}(\tilde{\boldsymbol{X}}) & \partial_{\tilde{\beta}}\tilde{A}_{\tilde{\alpha}}(\tilde{\boldsymbol{X}}) & \partial_{\tilde{\beta}_*}\tilde{A}_{\tilde{\alpha}}(\tilde{\boldsymbol{X}}) \\ \partial_{\tilde{\alpha}}\tilde{A}_{\tilde{\alpha}_*}(\tilde{\boldsymbol{X}}) & \partial_{\tilde{\alpha}_*}\tilde{A}_{\tilde{\alpha}_*}(\tilde{\boldsymbol{X}}) & \partial_{\tilde{\beta}}\tilde{A}_{\tilde{\alpha}_*}(\tilde{\boldsymbol{X}}) & \partial_{\tilde{\beta}_*}\tilde{A}_{\tilde{\alpha}_*}(\tilde{\boldsymbol{X}}) \\ \partial_{\tilde{\alpha}}\tilde{A}_{\tilde{\beta}}(\tilde{\boldsymbol{X}}) & \partial_{\tilde{\alpha}_*}\tilde{A}_{\tilde{\beta}}(\tilde{\boldsymbol{X}}) & \partial_{\tilde{\beta}}\tilde{A}_{\tilde{\beta}}(\tilde{\boldsymbol{X}}) & \partial_{\tilde{\beta}_*}\tilde{A}_{\tilde{\beta}}(\tilde{\boldsymbol{X}}) \\ \partial_{\tilde{\alpha}}\tilde{A}_{\tilde{\beta}_*}(\tilde{\boldsymbol{X}}) & \partial_{\tilde{\alpha}_*}\tilde{A}_{\tilde{\beta}_*}(\tilde{\boldsymbol{X}}) & \partial_{\tilde{\beta}}\tilde{A}_{\tilde{\beta}_*}(\tilde{\boldsymbol{X}}) & \partial_{\tilde{\beta}_*}\tilde{A}_{\tilde{\beta}_*}(\tilde{\boldsymbol{X}}) \end{pmatrix}. \tag{11.106}$$

Our interest is in the eigenvalues of $\boldsymbol{J}(\tilde{\boldsymbol{X}}_0)$, which determine the stability of the point $\tilde{\boldsymbol{X}}_0$.

We are interested specifically in the growth or decay of perturbations out of the classical phase space. In order to isolate this behavior, it is helpful to make a change of variables. We introduce a set of *classical phase-space variables*,

$$\tilde{\boldsymbol{X}}_{\mathrm{C}} \equiv \frac{1}{2}\begin{pmatrix} \tilde{\alpha} + (\tilde{\alpha}_*)^* \\ \tilde{\alpha}^* + \tilde{\alpha}_* \\ \tilde{\beta} + (\tilde{\beta}_*)^* \\ \tilde{\beta}^* + \tilde{\beta}_* \end{pmatrix} = \frac{1}{2}\left[\tilde{\boldsymbol{X}} + \begin{pmatrix} 0 & 1 & 0 & 0 \\ 1 & 0 & 0 & 0 \\ 0 & 0 & 0 & 1 \\ 0 & 0 & 1 & 0 \end{pmatrix}\tilde{\boldsymbol{X}}^*\right], \tag{11.107a}$$

and a set of *nonclassical phase-space variables*

$$\tilde{\boldsymbol{X}}_{\mathrm{NC}} \equiv \frac{1}{2}\begin{pmatrix} \tilde{\alpha} - (\tilde{\alpha}_*)^* \\ -\tilde{\alpha}^* + \tilde{\alpha}_* \\ \tilde{\beta} - (\tilde{\beta}_*)^* \\ -\tilde{\beta}^* + \tilde{\beta}_* \end{pmatrix} = \frac{1}{2}\left[\tilde{\boldsymbol{X}} - \begin{pmatrix} 0 & 1 & 0 & 0 \\ 1 & 0 & 0 & 0 \\ 0 & 0 & 0 & 1 \\ 0 & 0 & 1 & 0 \end{pmatrix}\tilde{\boldsymbol{X}}^*\right]. \tag{11.107b}$$

The *classical phase space* is the subspace in which $\tilde{\boldsymbol{X}}_{\mathrm{NC}} = 0$. Within this subspace the variables $\tilde{\boldsymbol{X}}_{\mathrm{C}}$ satisfy the same conjugacy relations as in the Glauber–Sudarshan P representation. We now transform (11.105a) and (11.105b) into equations of motion for the perturbations $\delta\tilde{\boldsymbol{X}}_{\mathrm{C}}$ and $\delta\tilde{\boldsymbol{X}}_{\mathrm{NC}}$. Inverting (11.107a) and (11.107b) yields

$$\tilde{\boldsymbol{X}} = \tilde{\boldsymbol{X}}_{\mathrm{C}} + \tilde{\boldsymbol{X}}_{\mathrm{NC}}, \tag{11.108a}$$

and

$$\tilde{\boldsymbol{X}}^* = \begin{pmatrix} 0 & 1 & 0 & 0 \\ 1 & 0 & 0 & 0 \\ 0 & 0 & 0 & 1 \\ 0 & 0 & 1 & 0 \end{pmatrix}\big(\tilde{\boldsymbol{X}}_{\mathrm{C}} - \tilde{\boldsymbol{X}}_{\mathrm{NC}}\big). \tag{11.108b}$$

Then from (11.105a) and (11.105b), and (11.107a)–(11.108b), the equations of motion for perturbations within the classical phase space, $\delta\tilde{\boldsymbol{X}}_{\mathrm{C}}$, and out of the classical phase space, $\delta\tilde{\boldsymbol{X}}_{\mathrm{NC}}$, are

$$\frac{d}{dt}\begin{pmatrix}\delta\tilde{\boldsymbol{X}}_{\mathrm{C}}\\ \delta\tilde{\boldsymbol{X}}_{\mathrm{NC}}\end{pmatrix} = \frac{1}{2}\begin{pmatrix}\boldsymbol{J}(\tilde{\boldsymbol{X}}_0)+\boldsymbol{J}'(\tilde{\boldsymbol{X}}_0) & \boldsymbol{J}(\tilde{\boldsymbol{X}}_0)-\boldsymbol{J}'(\tilde{\boldsymbol{X}}_0)\\ \boldsymbol{J}(\tilde{\boldsymbol{X}}_0)-\boldsymbol{J}'(\tilde{\boldsymbol{X}}_0) & \boldsymbol{J}(\tilde{\boldsymbol{X}}_0)+\boldsymbol{J}'(\tilde{\boldsymbol{X}}_0)\end{pmatrix}\begin{pmatrix}\delta\tilde{\boldsymbol{X}}_{\mathrm{C}}\\ \delta\tilde{\boldsymbol{X}}_{\mathrm{NC}}\end{pmatrix} + \begin{pmatrix}\boldsymbol{C}_{+}(\tilde{\boldsymbol{X}}_0)\\ \boldsymbol{C}_{-}(\tilde{\boldsymbol{X}}_0)\end{pmatrix}, \tag{11.109}$$

where $\boldsymbol{C}_{+}(\tilde{\boldsymbol{X}}_0)$ and $\boldsymbol{C}_{-}(\tilde{\boldsymbol{X}}_0)$ are constant vectors,

$$\begin{aligned}\boldsymbol{C}_{\pm}(\tilde{\boldsymbol{X}}_0) &\equiv \frac{1}{2}\left[\tilde{\boldsymbol{A}}(\tilde{\boldsymbol{X}}_0) \pm \begin{pmatrix}0&1&0&0\\1&0&0&0\\0&0&0&1\\0&0&1&0\end{pmatrix}\left(\tilde{\boldsymbol{A}}(\tilde{\boldsymbol{X}}_0)\right)^*\right]\\ &= \frac{1}{2}\left[\begin{pmatrix}\tilde{A}_{\tilde{\alpha}}(\tilde{\boldsymbol{X}}_0)\\ \tilde{A}_{\tilde{\alpha}_*}(\tilde{\boldsymbol{X}}_0)\\ \tilde{A}_{\tilde{\beta}}(\tilde{\boldsymbol{X}}_0)\\ \tilde{A}_{\tilde{\beta}_*}(\tilde{\boldsymbol{X}}_0)\end{pmatrix} \pm \begin{pmatrix}\left(\tilde{A}_{\tilde{\alpha}_*}(\tilde{\boldsymbol{X}}_0)\right)^*\\ \left(\tilde{A}_{\tilde{\alpha}}(\tilde{\boldsymbol{X}}_0)\right)^*\\ \left(\tilde{A}_{\tilde{\beta}_*}(\tilde{\boldsymbol{X}}_0)\right)^*\\ \left(\tilde{A}_{\tilde{\beta}}(\tilde{\boldsymbol{X}}_0)\right)^*\end{pmatrix}\right],\end{aligned} \tag{11.110}$$

and we have introduced the transformed Jacobian matrix

$$\begin{aligned}\boldsymbol{J}'(\tilde{\boldsymbol{X}}) &\equiv \begin{pmatrix}0&1&0&0\\1&0&0&0\\0&0&0&1\\0&0&1&0\end{pmatrix}\left(\boldsymbol{J}(\tilde{\boldsymbol{X}})\right)^*\begin{pmatrix}0&1&0&0\\1&0&0&0\\0&0&0&1\\0&0&1&0\end{pmatrix}\\ &= \begin{pmatrix}\left(\partial_{\tilde{\alpha}_*}\tilde{A}_{\tilde{\alpha}_*}(\tilde{\boldsymbol{X}})\right)^* & \left(\partial_{\tilde{\alpha}}\tilde{A}_{\tilde{\alpha}_*}(\tilde{\boldsymbol{X}})\right)^* & \left(\partial_{\tilde{\beta}_*}\tilde{A}_{\tilde{\alpha}_*}(\tilde{\boldsymbol{X}})\right)^* & \left(\partial_{\tilde{\beta}}\tilde{A}_{\tilde{\alpha}_*}(\tilde{\boldsymbol{X}})\right)^*\\ \left(\partial_{\tilde{\alpha}_*}\tilde{A}_{\tilde{\alpha}}(\tilde{\boldsymbol{X}})\right)^* & \left(\partial_{\tilde{\alpha}}\tilde{A}_{\tilde{\alpha}}(\tilde{\boldsymbol{X}})\right)^* & \left(\partial_{\tilde{\beta}_*}\tilde{A}_{\tilde{\alpha}}(\tilde{\boldsymbol{X}})\right)^* & \left(\partial_{\tilde{\beta}}\tilde{A}_{\tilde{\alpha}}(\tilde{\boldsymbol{X}})\right)^*\\ \left(\partial_{\tilde{\alpha}_*}\tilde{A}_{\tilde{\beta}_*}(\tilde{\boldsymbol{X}})\right)^* & \left(\partial_{\tilde{\alpha}}\tilde{A}_{\tilde{\beta}_*}(\tilde{\boldsymbol{X}})\right)^* & \left(\partial_{\tilde{\beta}_*}\tilde{A}_{\tilde{\beta}_*}(\tilde{\boldsymbol{X}})\right)^* & \left(\partial_{\tilde{\beta}}\tilde{A}_{\tilde{\beta}_*}(\tilde{\boldsymbol{X}})\right)^*\\ \left(\partial_{\tilde{\alpha}_*}\tilde{A}_{\tilde{\beta}}(\tilde{\boldsymbol{X}})\right)^* & \left(\partial_{\tilde{\alpha}}\tilde{A}_{\tilde{\beta}}(\tilde{\boldsymbol{X}})\right)^* & \left(\partial_{\tilde{\beta}_*}\tilde{A}_{\tilde{\beta}}(\tilde{\boldsymbol{X}})\right)^* & \left(\partial_{\tilde{\beta}}\tilde{A}_{\tilde{\beta}}(\tilde{\boldsymbol{X}})\right)^*\end{pmatrix}.\end{aligned} \tag{11.111}$$

Note that if $\tilde{\boldsymbol{X}}_0$ happens to be a steady state, the constants $\boldsymbol{C}_{+}(\tilde{\boldsymbol{X}}_0)$ and $\boldsymbol{C}_{-}(\tilde{\boldsymbol{X}}_0)$ vanish, since in this case $\tilde{\boldsymbol{A}}(\tilde{\boldsymbol{X}}_0) = 0$.

Now we assume the point $\tilde{\boldsymbol{X}}_0$ to lie within the classical phase space, and may therefore read the definitions of $\boldsymbol{J}(\tilde{\boldsymbol{X}}_0)$, $\boldsymbol{J}'(\tilde{\boldsymbol{X}}_0)$, and $\boldsymbol{C}_{\pm}(\tilde{\boldsymbol{X}}_0)$ with $\tilde{\alpha}_*$ replaced by $\tilde{\alpha}^*$ and $\tilde{\beta}_*$ replaced by $\tilde{\beta}^*$. It follows that $\boldsymbol{J}'(\tilde{\boldsymbol{X}}_0) = \boldsymbol{J}(\tilde{\boldsymbol{X}}_0)$,

$\boldsymbol{C}_{+}(\tilde{\boldsymbol{X}}_0) = \tilde{\boldsymbol{A}}(\tilde{\boldsymbol{X}}_0)$, and $\boldsymbol{C}_{-}(\tilde{\boldsymbol{X}}_0) = 0$. Then, from (11.109), the perturbations $\delta\tilde{\boldsymbol{X}}_{\mathrm{C}}$ and $\delta\tilde{\boldsymbol{X}}_{\mathrm{NC}}$ undergo a decoupled evolution: the *deterministic equations of motion within the positive P representation linearized about a point located in the classical phase space* are

$$\frac{d}{dt}\begin{pmatrix}\delta\tilde{\boldsymbol{X}}_{\mathrm{C}}\\ \delta\tilde{\boldsymbol{X}}_{\mathrm{NC}}\end{pmatrix} = \begin{pmatrix}\boldsymbol{J}(\tilde{\boldsymbol{X}}_0) & 0\\ 0 & \boldsymbol{J}(\tilde{\boldsymbol{X}}_0)\end{pmatrix}\begin{pmatrix}\delta\tilde{\boldsymbol{X}}_{\mathrm{C}}\\ \delta\tilde{\boldsymbol{X}}_{\mathrm{NC}}\end{pmatrix} + \begin{pmatrix}\tilde{\boldsymbol{A}}(\tilde{\boldsymbol{X}}_0)\\ 0\end{pmatrix}. \tag{11.112}$$

According to this equation, the classical and nonclassical perturbations are not only decoupled; their dynamical evolution is governed by the same Jacobian matrix, $\boldsymbol{J}(\tilde{\boldsymbol{X}}_0)$. Thus, for each eigenvalue governing motion within the classical phase space, there is an equal eigenvalue governing motion out of the classical phase space and into the nonclassical dimensions. Since $\boldsymbol{J}(\tilde{\boldsymbol{X}}_0)$ is also the Jacobian matrix from the semiclassical theory of the degenerate parametric oscillator (Eqs. 9.63 or 10.43), we conclude that the stability of the classical phase space is tied directly to the linearized semiclassical theory. One result of this is that the classical phase space is stable in the vicinity of stable steady states; for example, those shown in Fig. 9.3. This is reassuring, as we aim to use the positive P representation to describe quantum fluctuations about such states; one would hardly expect the representation to convert a stable steady state into an unstable steady state.

We also find, however, that the classical phase space is unstable wherever there is instability in the semiclassical theory. With reference to Fig. 9.3, for example, above threshold the classical phase space is unstable in the vicinity of the vacuum state. The implications of this are not yet clear. There is no obvious reason to conclude that such instability will lead to a problem. We should keep in mind, though, that whenever there are unstable phase-space regions in the semiclassical theory, the *global* nonlinear dynamics within the doubled dimensions of the positive P representation may well be important, because in the presence of noise, it cannot be guaranteed that the stochastic trajectory will remain close to any trajectory of the semiclassical equations of motion. Indeed, we will discover later (Sect. 12.2) that when it does not, the positive P representation can run into serious problems.

11.2.3 Preservation of Conjugacy for Stochastic Averages

The formal derivation of the stochastic differential equations in the positive P representation (Eq. 11.49) should be sufficient to satisfy us that the equations are valid in the sense that they yield normal-ordered operator averages as ensemble averages over trajectories. It is helpful, nevertheless, to explicitly see that they do not lead to obvious problems. One problem that might be foreseen concerns the question of conjugacy: since α and α_* are no longer complex conjugates, can we be sure that their averages over an ensemble of trajectories are going to be complex conjugates? More generally, can we be sure that trajectory averages preserve the conjugacy requirement

$$\langle a^{\dagger p} a^q b^{\dagger p'} b^{q'} \rangle = \langle a^{\dagger q} a^p b^{\dagger q'} b^{p'} \rangle^* ? \tag{11.113}$$

Within the positive P representation the requirement is

$$\begin{aligned} &\int d^2\alpha \int d^2\alpha_* \int d^2\beta \int d^2\beta_* \, \alpha_*^p \alpha^q \beta_*^{p'} \beta^{q'} P(\alpha, \alpha_*, \beta, \beta_*) \\ &= \left(\int d^2\alpha \int d^2\alpha_* \int d^2\beta \int d^2\beta_* \, \alpha_*^q \alpha^p \beta_*^{q'} \beta^{p'} P(\alpha, \alpha_*, \beta, \beta_*) \right)^* , \end{aligned} \tag{11.114}$$

which holds if the P distribution satisfies the symmetry relation

$$P(\alpha, \alpha_*, \beta, \beta_*) = P\big((\alpha_*)^*, \alpha^*, (\beta_*)^*, \beta^*\big), \tag{11.115}$$

where the latter follows by making the change of variables $\alpha \to (\alpha_*)^*$, $\alpha_* \to \alpha^*$, $\beta \to (\beta_*)^*$, $\beta_* \to \beta^*$ in (11.114), and noting that $P(\alpha, \alpha_*, \beta, \beta_*)$ is real. Note that (11.115) does not require that states satisfying $\alpha_* = \alpha^*$ and $\beta_* = \beta^*$ alone occur with nonzero probability; this is a sufficient but not a necessary condition to satisfy (11.115) and gives back the Glauber–Sudarshan representation. More generally, (11.115) assumes that pairs of states, $\alpha, \alpha_*, \beta, \beta_*$ and $(\alpha_*)^*, \alpha^*, (\beta_*)^*, \beta^*$, occur with equal probability (equal probability density). The positive P distribution constructed from the Q representation (Eq. 11.19), as one example, exhibits the symmetry (11.115).

The question is not whether or not distributions satisfying (11.115) exist, but whether or not the Fokker–Planck equation in the positive P representation (Eq. 11.46) preserves this symmetry if it holds initially. Equivalently, we can ask the same question of the stochastic differential equations (11.49), which are easier to work with than the Fokker–Planck equation since they have a convenient form in complex notation. In this section we show that if the distribution from which the initial conditions for trajectories are chosen obeys (11.115), then the stochastic differential equations in the positive P representation do, indeed, preserve the conjugacy requirement over time. For the stochastic differential equations, the requirement takes the form

$$\Big(\overline{(\alpha_*^p \alpha^q \beta_*^{p'} \beta^{q'})(t)} \Big)_P = \Big(\overline{(\alpha_*^q \alpha^p \beta_*^{q'} \beta^{p'})(t)} \Big)_P^*, \tag{11.116}$$

where the average is taken, ideally, over an infinite ensemble of independent realizations or trajectories.

In order to make the demonstration clear, it is helpful to define certain special trajectories associated with a given realization of the stochastic differential equations. In the first place, a particular trajectory is defined by an initial condition $\boldsymbol{X}_0$ and a vector of Wiener paths $\boldsymbol{W}$. We denote such a trajectory by

$$T(\boldsymbol{X}_0, \boldsymbol{W}) \equiv \big(\alpha(t), \alpha_*(t), \beta(t), \beta_*(t)\big); \tag{11.117}$$

using (11.49), and (11.33), (11.34a), and (11.36), the explicit form of the trajectory is generated as a realization of the *stochastic differential equations for the degenerate parametric oscillator in the positive P representation*:

$$\begin{aligned} d\alpha = &\big[-(\kappa + i\omega_C)\alpha + g\alpha_*\beta\big]dt \\ &+ \left(\sqrt{g\beta/2 + \sqrt{\beta/\beta_*}\kappa\bar{n}}dW_\alpha + \sqrt{g\beta/2 - \sqrt{\beta/\beta_*}\kappa\bar{n}}dW_{\alpha_*}\right), \end{aligned} \tag{11.118a}$$

$$\begin{aligned} d\alpha_* = &\big[-(\kappa - i\omega_C)\alpha_* + g\alpha\beta_*\big]dt \\ &+ \left(\sqrt{g\beta_*/2 + \sqrt{\beta_*/\beta}\kappa\bar{n}}dW_\alpha - \sqrt{g\beta_*/2 - \sqrt{\beta_*/\beta}\kappa\bar{n}}dW_{\alpha_*}\right), \end{aligned} \tag{11.118b}$$

$$\begin{aligned} d\beta = &\big[-(\kappa_p + i2\omega_C)\beta - (g/2)\alpha^2 - i\bar{\mathcal{E}}_0 e^{-i2\omega_C t}\big]dt \\ &+ \sqrt{\kappa_p\bar{n}_p}(dW_\beta + idW_{\beta_*}), \end{aligned} \tag{11.118c}$$

$$\begin{aligned} d\beta_* = &\big[-(\kappa_p - i2\omega_C)\beta_* - (g/2)\alpha_*^2 + i\bar{\mathcal{E}}_0^* e^{i2\omega_C t}\big]dt \\ &+ \sqrt{\kappa_p\bar{n}_p}(dW_\beta - idW_{\beta_*}), \end{aligned} \tag{11.118d}$$

where the initial condition $\boldsymbol{X}_0$ and the Wiener paths $\boldsymbol{W}$ are

$$\boldsymbol{X}_0 \equiv \begin{pmatrix} \alpha^0 \\ \alpha_*^0 \\ \beta^0 \\ \beta_*^0 \end{pmatrix}, \qquad \boldsymbol{W} \equiv \begin{pmatrix} W_\alpha \\ W_{\alpha_*} \\ W_\beta \\ W_{\beta_*} \end{pmatrix}. \tag{11.119}$$

Given a particular realization of (11.118a)–(11.118d), a particular trajectory (11.117), we define the two related trajectories

$$\begin{aligned} &T_I(\boldsymbol{X}_0, \boldsymbol{W}) \\ &\equiv \big(\alpha^I(t), \alpha_*^I(t), \beta^I(t), \beta_*^I(t)\big) \equiv \Big(\big(\alpha_*(t)\big)^*, \big(\alpha(t)\big)^*, \big(\beta_*(t)\big)^*, \big(\beta(t)\big)^*\Big), \end{aligned} \tag{11.120a}$$

and

$$T_A(\boldsymbol{X}_0, \boldsymbol{W}) \equiv \big(\alpha^A(t), \alpha_*^A(t), \beta^A(t), \beta_*^A(t)\big) \equiv T(\boldsymbol{X}_0^I, \boldsymbol{W}), \tag{11.120b}$$

where

$$\boldsymbol{X}_0^I \equiv \begin{pmatrix} (\alpha_*^0)^* \\ (\alpha^0)^* \\ (\beta_*^0)^* \\ (\beta^0)^* \end{pmatrix}. \tag{11.121}$$

The initial state $\boldsymbol{X}_0^I$ is referred to as the *image* of the state $\boldsymbol{X}_0$; for states satisfying $\alpha_* = \alpha^*$ and $\beta_* = \beta^*$, the state and its image are the same. The trajectory $T_I(\boldsymbol{X}_0, \boldsymbol{W})$ is the *image trajectory*, the trajectory obtained by taking the image of every state visited by $T(\boldsymbol{X}_0, \boldsymbol{W})$, and $T_A(\boldsymbol{X}_0, \boldsymbol{W})$ is the *associated trajectory*, the trajectory evolved from the initial image $\boldsymbol{X}_0^I$ by the same realization of Wiener paths as the one generating $T(\boldsymbol{X}_0, \boldsymbol{W})$.

Now, to prove (11.116), we must show that

$$T_I(\boldsymbol{X}_0, \boldsymbol{W}) = T_A(\boldsymbol{X}_0, \boldsymbol{W}') \equiv T(\boldsymbol{X}_0^I, \boldsymbol{W}'), \tag{11.122}$$

where

$$\boldsymbol{W}' \equiv \begin{pmatrix} W_\alpha \\ -W_{\alpha_*} \\ W_\beta \\ W_{\beta_*} \end{pmatrix}. \tag{11.123}$$

With this result, since $\boldsymbol{X}_0^I$ and $\boldsymbol{X}_0$ occur with equal probability to $\boldsymbol{X}_0$ (Eq. 11.115 for the initial distribution), and since the Wiener path $\boldsymbol{W}'$ is realized with equal probability to the path $\boldsymbol{W}$, then in an infinite ensemble of trajectories, $T(\boldsymbol{X}_0, \boldsymbol{W})$ and $T_I(\boldsymbol{X}_0, \boldsymbol{W})$ must be present with equal weight. We may therefore divide the ensemble average into two sums, one over trajectories $T(\boldsymbol{X}_0, \boldsymbol{W})$ and the other over their images, to write

$$\begin{aligned}
&\Big(\overline{(\alpha_*^p \alpha^q \beta_*^{p'} \beta^{q'})(t)}\Big)_P \\
&= \lim_{N\to\infty} \frac{1}{N} \sum_{j=1}^{N/2} \Big\{ \Big[(\alpha_*(t))^p (\alpha(t))^q (\beta_*(t))^{p'} (\beta(t))^{q'} \Big]_j \\
&\quad + \Big[(\alpha(t))^{*p} (\alpha_*(t))^{*q} (\beta(t))^{*p'} (\beta_*(t))^{*q'} \Big]_j \Big\} \\
&= \lim_{N\to\infty} \frac{1}{N} \sum_{j=1}^{N/2} \Big\{ \big[(\alpha_*^p \alpha^q \beta_*^{p'} \beta^{q'})(t) \big]_j + \big[(\alpha_*^q \alpha^p \beta_*^{q'} \beta^{p'})(t) \big]_j^* \Big\},
\end{aligned} \tag{11.124a}$$

where j labels the individual trajectories and N is the size of the ensemble. In a similar way

$$\begin{aligned}
&\Big(\overline{(\alpha_*^q \alpha^p \beta_*^{q'} \beta^{p'})(t)}\Big)_P \\
&= \lim_{N\to\infty} \frac{1}{N} \sum_{j=1}^{N/2} \Big\{ \big[(\alpha_*^q \alpha^p \beta_*^{q'} \beta^{p'})(t) \big]_j + \big[(\alpha_*^p \alpha^q \beta_*^{p'} \beta^{q'})(t) \big]_j^* \Big\}.
\end{aligned} \tag{11.124b}$$

The conjugacy requirement, in the form (11.116), follows from (11.124a) and (11.124b).

It remains to prove (11.122). To do this we note that the image trajectory defined by (11.120b) satisfies the stochastic differential equations

$$d\alpha^I = \left[-(\kappa + i\omega_C)\alpha^I + g\alpha_*^I\beta^I\right]dt + \left(\sqrt{g\beta^I/2 + \sqrt{\beta^I/\beta_*^I}\kappa\bar{n}}dW_\alpha - \sqrt{g\beta^I/2 - \sqrt{\beta^I/\beta_*^I}\kappa\bar{n}}dW_{\alpha_*}\right), \tag{11.125a}$$

$$d\alpha_*^I = \left[-(\kappa - i\omega_C)\alpha_*^I + g\alpha^I\beta_*^I\right]dt + \left(\sqrt{g\beta_*^I/2 + \sqrt{\beta_*^I/\beta^I}\kappa\bar{n}}dW_\alpha + \sqrt{g\beta_*^I/2 - \sqrt{\beta_*^I/\beta^I}\kappa\bar{n}}dW_{\alpha_*}\right), \tag{11.125b}$$

$$d\beta^I = \left[-(\kappa_p + i2\omega_C)\beta^I - (g/2)(\alpha^I)^2 - i\bar{\mathcal{E}}_0 e^{-i2\omega_C t}\right]dt + \sqrt{\kappa_p\bar{n}_p}(dW_\beta + idW_{\beta_*}), \tag{11.125c}$$

$$d\beta_*^I = \left[-(\kappa_p - i2\omega_C)\beta_*^I - (g/2)(\alpha_*^I)^2 + i\bar{\mathcal{E}}_0^* e^{i2\omega_C t}\right]dt + \sqrt{\kappa_p\bar{n}_p}(dW_\beta - idW_{\beta_*}). \tag{11.125d}$$

These equations are obtained by taking complex conjugates of (11.118a)–(11.118d) and interchanging the first with the second equation and the third with the fourth equation. Observe now that the image trajectory $T_I(\boldsymbol{X}_0, \boldsymbol{W})$ and the associated trajectory $T_A(\boldsymbol{X}_0, \boldsymbol{W})$ evolve from the same initial state—the state $\boldsymbol{X}_0^I$—but the image trajectory satisfies (11.125a)–(11.125d) while the associated trajectory satisfies (11.118a)–(11.118d). The two sets of equations are the same except for the appearance of $-dW_{\alpha_*}$ and $+dW_{\alpha_*}$ in (11.125a) and (11.125b) where $+dW_{\alpha_*}$ and $-dW_{\alpha_*}$ appear in (11.118a) and (11.118b). Offsetting this difference with the introduction of $\boldsymbol{W}'$, we conclude that $T_I(\boldsymbol{X}_0, \boldsymbol{W}) = T_A(\boldsymbol{X}_0, \boldsymbol{W}')$. This completes the proof of (11.122).

Note 11.7. The square roots appearing in front of the Wiener increments in (11.118a) and (11.118b) must be interpreted using a consistent convention by choosing the same root in each term; for example, taking an argument $z \equiv re^{i\phi}$ under the square root, with $-\pi < \phi \le \pi$, choose the root $\sqrt{r}e^{i\phi/2}$. Keeping this convention in mind, we then trace the nonclassical character of the noise to the minus sign in front of the second square root in (11.118b). Consider, for example, the case with β and β_* as complex conjugates. Then for $-\pi < \phi_\pm < \pi$ (but not for $\phi_\pm = \pi$), the noise terms in (11.118a) and (11.118b) are, respectively,

$$\sqrt{z_+}dW_\alpha + \sqrt{z_-}dW_{\alpha_*} = \sqrt{r_+}e^{i\phi_+/2}dW_\alpha + \sqrt{r_-}e^{i\phi_-/2}dW_{\alpha_*}, \tag{11.126a}$$

$$\sqrt{z_+^*}dW_\alpha - \sqrt{z_-^*}dW_{\alpha_*} = \sqrt{r_+}e^{-i\phi_+/2}dW_\alpha - \sqrt{r_-}e^{-i\phi_-/2}dW_{\alpha_*}. \tag{11.126b}$$

These are not complex conjugates, due to the minus sign; thus, if initially there is conjugacy, $\alpha = (\alpha_*)^*$ and $\beta = (\beta_*)^*$, it is not preserved at later times. Note that (11.126a) and (11.126b) do not hold on the boundary of the range of phase, for $\phi_+ = \pi$ or $\phi_- = \pi$. These cases are special because we write $(e^{i\pi})^* = e^{i\pi}$, not $(e^{i\pi})^* = e^{-i\pi}$ ($-\pi$ lies outside the chosen range of phase). The special cases apply when the effect of the pump on the noise is removed by setting $\beta = \beta_* = 0$, $\sqrt{\beta/\beta_*} = \sqrt{\beta_*/\beta} = 1$. Then we replace (11.126a) and (11.126b) by

$$\sqrt{\kappa\bar{n}}dW_\alpha + \sqrt{-\kappa\bar{n}}dW_{\alpha_*} = \sqrt{\kappa\bar{n}}(dW_\alpha + idW_{\alpha_*}), \tag{11.127a}$$

$$\sqrt{\kappa\bar{n}}dW_\alpha - \sqrt{-\kappa\bar{n}}dW_{\alpha_*} = \sqrt{\kappa\bar{n}}(dW_\alpha - idW_{\alpha_*}). \tag{11.127b}$$

These are complex conjugates, thermal noise terms like those in (11.118c) and (11.118d). Figure 11.1 illustrates how the phase factors multiplying dW_{α_*} in (11.118a) and (11.118b) vary as ϕ_- is varied continuously from zero to 2π (through intervals $[0, \pi]$ and $(-\pi, 0]$). There is a discontinuity at $\phi_- = \pi$, which, although it appears strange, amounts to a change of sign only; such a sign change is irrelevant, since the Wiener increment dW_{α_*} is positive or negative with equal probabilities at each step of the integration. It is only for $\phi_- = \pi$ that the phase factors depicted in Fig. 11.1 are conjugates of one another.

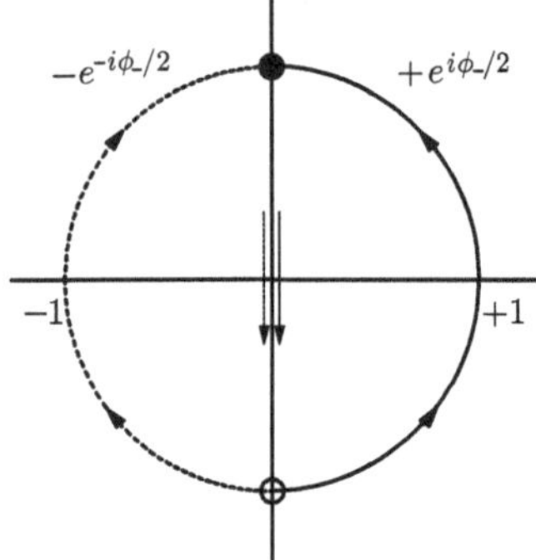

Fig. 11.1. Behavior of the phase factors multiplying dW_{α_*} in (11.126a) and (11.126b) as ϕ_- changes from zero to 2π. For $\phi_- = \pi$ the phase factors are written as $+e^{i\phi_-/2} \to +i$ and $-e^{-i\phi_-/2} \to -i$. As ϕ_- moves incrementally away from π in either direction, one or other of the phase factors crosses the discontinuity

12

The Degenerate Parametric Oscillator III: Phase-Space Analysis Outside the Small-Noise Limit

The small-noise limit has received a great deal of attention in quantum optics, and not without good reason. Until quite recently, in cases of experimental interest, the system size expansion has always been well justified. In the case of the laser, for example, its validity is governed by the saturation photon number $n_{\rm sat} = \gamma_h(\gamma_\uparrow + \gamma_\downarrow)/8g^2$; for the degenerate parametric oscillator by the threshold pump photon number $n_p^{\rm thr} = (\kappa/g)^2$. Each of these numbers is determined by the ratio of a dissipation rate and a fundamental coupling constant. When the dissipation rate dominates, the ratio is large; $n_{\rm sat}$ and $n_p^{\rm thr}$ are typically on the order of 10^8 (see the final paragraph of Sect. 7.1.4 and Eq. 10.17).

The experimental situation has changed in recent years. With the use of Rydberg atoms and superconducting microwave cavities, dipole coupling strengths that are much larger than dissipation rates can now be realized. Also, at optical frequencies, the development of high-finesse cavities allows a similar regime of strong dipole coupling to be entered. These strong coupling conditions define the field of cavity QED. Cavity QED systems have small system size, hence large quantum noise. Treatment of their quantum fluctuations must be taken beyond the small-noise limit. The remaining chapters of the book explore various aspects of this topic. The subject of cavity QED itself is taken up in Chaps. 13–16. In the present chapter, we continue with the positive P representation and the degenerate parametric oscillator, to see where the quantum–classical correspondence leads when it is taken beyond the small-noise limit.

The quantum–classical correspondence provides a visualization in phase space. In the small-noise limit, the dynamical picture is that of a "fuzz ball" of noise carried along classical trajectories, trajectories obtained from classical nonlinear dynamics; the "fuzz ball" accounts for the quantum fluctuations. The picture is not qualitatively changed in the presence of weak nonlinearities—the threshold regions of the laser and degenerate parametric oscillator, for example; there the analysis is based on the system size expansion also (Sects. 8.2 and 10.2.4) and is still limited by the small-noise assumption. Interesting new territory is reached, however, when the photon number needed to reach threshold is on the order of unity or less. Then the small-noise assumption fails. The quantum fluctuations are large and intertwined with the classical nonlinearity in a fundamental way [see the discussion below (7.78)]. Phase-space trajectories no longer have any relevance here. A manifestly quantum-mechanical picture of the fluctuations must take their place. In this chapter we track this transition for the degenerate parametric oscillator example.

12.1 The Degenerate Parametric Oscillator with Adiabatic Elimination of the Pump

There is probably little progress to be made by searching for an analytical solution to the Fokker–Planck equation (11.46). Certainly, we could start with numerical simulations of the equivalent stochastic differential equations (Eqs. 11.118a–11.118d), but even this may be too large a step to take if we wish to gain some understanding—to achieve something more than a blind numerical calculation of operator averages. The stochastic differential equations describe a nonlinear evolution in eight dimensions. Could a reduction in the number of dimensions be achieved?

Apparently, the minimum number of dimensions in the positive P representation is four, the four dimensions demanded by the two complex variables, α and α_*, that represent a single mode of the field. If the field amplitude in the classical dimensions is real, however, and if, in addition, the fluctuations are confined to one nonclassical dimension only, then a two-dimensional phase space is a possibility. As it turns out, this simplest case can be realized for the degenerate parametric oscillator. However, before we can see how to achieve the reduction from four dimensions to two, we must first perform a reduction from eight dimensions to four. This is done by adiabatically eliminating the pump to obtain a set of stochastic differential equations for the subharmonic mode alone.

There are a number of ways to carry out the adiabatic elimination. We may work, for example, entirely within the phase-space representation, as we did when adiabatically eliminating the atomic variables in our treatment of the laser (Sects. 8.1.4, 8.2.1, and 8.3.2). In the present situation this method provides a quick route to the desired stochastic differential equations. Thus, we first derive the equations in this way. Then, for the sake of learning something new, we see how the adiabatic elimination can be carried through at the level of the master equation itself.

12.1.1 Adiabatic Elimination in the Stochastic Differential Equations

Since we are primarily interested in the nonclassical noise, let us for simplicity set $\bar{n} = 0$. The stochastic differential equations for the coupled pump and subharmonic modes are then given by Eqs. 11.118a–11.118d:

$$d\tilde{\alpha} = [-\kappa\tilde{\alpha} + g\tilde{\alpha}_*\tilde{\beta}]dt + \sqrt{g\tilde{\beta}}dW_\alpha, \tag{12.1a}$$

$$d\tilde{\alpha}_* = [-\kappa\tilde{\alpha}_* + g\tilde{\alpha}\tilde{\beta}_*]dt + \sqrt{g\tilde{\beta}_*}dW_{\alpha_*}, \tag{12.1b}$$

$$d\tilde{\beta} = [-\kappa_p\tilde{\beta} - (g/2)\tilde{\alpha}^2 - i\bar{\mathcal{E}}_0]dt, \tag{12.1c}$$

$$d\tilde{\beta}_* = [-\kappa_p\tilde{\beta}_* - (g/2)\tilde{\alpha}_*^2 + i\bar{\mathcal{E}}_0^*]dt, \tag{12.1d}$$

where we have adopted a rotating frame (Eq. 11.101) and made the substitutions $(dW_\alpha + dW_{\alpha_*})/\sqrt{2} \to dW_\alpha$ and $(dW_\alpha - dW_{\alpha_*})/\sqrt{2} \to dW_{\alpha_*}$; these substitutions are permitted because $(dW_\alpha + W_{\alpha_*})/\sqrt{2}$ and $(dW_\alpha - dW_{\alpha_*})/\sqrt{2}$ are independent Wiener increments. Assuming now that the pump mode decays much more quickly than the subharmonic mode ($\xi = \kappa/\kappa_p \ll 1$), we solve (12.1c) and (12.1d) for $\tilde{\beta}$ and $\tilde{\beta}_*$ by setting the increments on the left-hand side to zero; thus, introducing scaled variables $\bar{\tilde{\alpha}}$ and $\bar{\tilde{\alpha}}_*$, defined as in (10.33), with the help of (10.14)–(10.16) we obtain

$$g\tilde{\beta} = \kappa e^{i(\psi-\pi/2)}(\lambda - \bar{\tilde{\alpha}}^2), \qquad g\tilde{\beta}_* = \kappa e^{-i(\psi-\pi/2)}(\lambda - \bar{\tilde{\alpha}}_*^2). \tag{12.2}$$

Substituting these results into (12.1a) and (12.1b), we arrive at the *stochastic differential equations for the degenerate parametric oscillator in the positive P representation with adiabatic elimination of the pump (and $\bar{n} = 0$)*:

$$d\bar{\tilde{\alpha}} = [-\bar{\tilde{\alpha}} + \bar{\tilde{\alpha}}_*(\lambda - \bar{\tilde{\alpha}}^2)]d\bar{t} + \left(2\xi^{-1}n_p^{\text{thr}}\right)^{-1/2}\sqrt{\lambda - \bar{\tilde{\alpha}}^2}d\bar{W}_\alpha, \tag{12.3a}$$

$$d\bar{\tilde{\alpha}}_* = [-\bar{\tilde{\alpha}}_* + \bar{\tilde{\alpha}}(\lambda - \bar{\tilde{\alpha}}_*^2)]d\bar{t} + \left(2\xi^{-1}n_p^{\text{thr}}\right)^{-1/2}\sqrt{\lambda - \bar{\tilde{\alpha}}_*^2}d\bar{W}_{\alpha_*}, \tag{12.3b}$$

where

$$\bar{t} \equiv \kappa t. \tag{12.4}$$

The corresponding phase-space equation of motion is

$$\begin{aligned}\frac{\partial\bar{\tilde{P}}}{\partial\bar{t}} = \Bigg\{ &\frac{\partial}{\partial\bar{\tilde{\alpha}}}[\bar{\tilde{\alpha}} - \bar{\tilde{\alpha}}_*(\lambda - \bar{\tilde{\alpha}}^2)] + \frac{\partial}{\partial\bar{\tilde{\alpha}}_*}[\bar{\tilde{\alpha}}_* - \bar{\tilde{\alpha}}(\lambda - \bar{\tilde{\alpha}}_*^2)] \\ &+ \frac{1}{2}\left(2\xi^{-1}n_p^{\text{thr}}\right)^{-1}\left[\frac{\partial^2}{\partial\bar{\tilde{\alpha}}^2}(\lambda - \bar{\tilde{\alpha}}^2) + \frac{\partial^2}{\partial\bar{\tilde{\alpha}}_*^2}(\lambda - \bar{\tilde{\alpha}}_*^2)\right]\Bigg\}\bar{\tilde{P}},\end{aligned} \tag{12.5}$$

where, as explained in Sect. 11.1.2, a specific interpretation of derivatives is assumed so that (12.5) is a Fokker–Planck equation with positive semidefinite diffusion—now in four dimensions instead of eight. The derivation of the explicit Fokker–Planck equation is left as an exercise.

Exercise 12.1. Show that the *Fokker–Planck equation for the degenerate parametric oscillator in the positive P representation with adiabatic elimination of the pump (and $\bar{n} = 0$)* is given by (11.46) and (11.41), where

$$\boldsymbol{x} \equiv \begin{pmatrix} \tilde{\bar{x}} \\ \tilde{\bar{\mathrm{x}}} \\ \tilde{\bar{y}} \\ \tilde{\bar{\mathrm{y}}} \end{pmatrix}, \qquad \boldsymbol{x}' \equiv \begin{pmatrix} \partial/\partial\tilde{\bar{x}} \\ \partial/\partial\tilde{\bar{\mathrm{x}}} \\ \partial/\partial\tilde{\bar{y}} \\ \partial/\partial\tilde{\bar{\mathrm{y}}} \end{pmatrix}, \tag{12.6}$$

with drift matrix

$$\boldsymbol{A}^{(2)}(\boldsymbol{x}) = \begin{pmatrix} -\tilde{\bar{x}} + \tilde{\bar{\mathrm{x}}}(\lambda - \tilde{\bar{x}}^2 + \tilde{\bar{y}}^2) + 2\tilde{\bar{\mathrm{y}}}\tilde{\bar{x}}\tilde{\bar{y}} \\ -\tilde{\bar{\mathrm{x}}} + \tilde{\bar{x}}(\lambda - \tilde{\bar{\mathrm{x}}}^2 + \tilde{\bar{\mathrm{y}}}^2) + 2\tilde{\bar{y}}\tilde{\bar{\mathrm{x}}}\tilde{\bar{\mathrm{y}}} \\ -\tilde{\bar{y}} + \tilde{\bar{\mathrm{y}}}(\lambda - \tilde{\bar{y}}^2 + \tilde{\bar{x}}^2) + 2\tilde{\bar{\mathrm{x}}}\tilde{\bar{y}}\tilde{\bar{x}} \\ -\tilde{\bar{\mathrm{y}}} + \tilde{\bar{y}}(\lambda - \tilde{\bar{\mathrm{y}}}^2 + \tilde{\bar{\mathrm{x}}}^2) + 2\tilde{\bar{x}}\tilde{\bar{\mathrm{y}}}\tilde{\bar{\mathrm{x}}} \end{pmatrix}, \tag{12.7}$$

and diffusion matrix

$$\boldsymbol{D}^{(2)}(\boldsymbol{x}) = \left(2\xi^{-1}n_p^{\mathrm{thr}}\right)^{-1} \begin{pmatrix} R^2 & 0 & RI & 0 \\ 0 & R_*^2 & 0 & R_*I_* \\ RI & 0 & I^2 & 0 \\ 0 & R_*I_* & 0 & I_*^2 \end{pmatrix}, \tag{12.8}$$

where

$$R \equiv \mathrm{Re}\left(\sqrt{\lambda - (\tilde{\bar{x}} + i\tilde{\bar{y}})^2}\right), \qquad I \equiv \mathrm{Im}\left(\sqrt{\lambda - (\tilde{\bar{x}} + i\tilde{\bar{y}})^2}\right), \tag{12.9a}$$

and

$$R_* \equiv \mathrm{Re}\left(\sqrt{\lambda - (\tilde{\bar{\mathrm{x}}} + i\tilde{\bar{\mathrm{y}}})^2}\right), \qquad I_* \equiv \mathrm{Im}\left(\sqrt{\lambda - (\tilde{\bar{\mathrm{x}}} + i\tilde{\bar{\mathrm{y}}})^2}\right). \tag{12.9b}$$

Show explicitly that the diffusion matrix is positive semidefinite.

It is possible to work backwards from the phase-space equation of motion (12.5) to deduce the single-mode master equation to which it corresponds. Thus, the *master equation for the degenerate parametric oscillator with adiabatic elimination of the pump (and $\bar{n} = 0$)* is given by

$$\begin{aligned} \dot{\rho} = {} & -i\omega_C[a^\dagger a, \rho] - i\frac{g}{2\kappa_p}[\bar{\mathcal{E}}_0 e^{-i2\omega_C t}a^{\dagger 2} + \bar{\mathcal{E}}_0^* e^{i2\omega_C t}a^2, \rho] \\ & + \kappa(2a\rho a^\dagger - a^\dagger a\rho - \rho a^\dagger a) \\ & + \frac{g^2}{4\kappa_p}(2a^2\rho a^{\dagger 2} - a^{\dagger 2}a^2\rho - \rho a^{\dagger 2}a^2). \end{aligned} \tag{12.10}$$

We can discern the physics of the adiabatic elimination more clearly here than in the stochastic differential equations. The elimination has two distinct effects, revealed by the second and last terms on the right-hand side of (12.10).

The former describes the driving of the subharmonic mode by the pump. It replaces the term proportional to $(g/2)$ in the full master equation (Eq. 9.97); for the operator b in that equation it substitutes the coherent state amplitude $\langle b(t)\rangle_{\rm ss} = -i(\bar{\mathcal{E}}_0/\kappa_p)e^{-i2\omega_C t}$, the amplitude that would be established in an empty cavity in steady state (Eq. 9.103). This seems sensible; at least the appearance of the steady-state amplitude makes sense. However, the substitution $b \to -i(\bar{\mathcal{E}}_0/\kappa_p)e^{-i2\omega_C t}$ cannot be the whole story, since it allows the pump mode amplitude to continue to increase with $|\bar{\mathcal{E}}_0|$ above threshold where we know that pump depletion must set in (Eqs. 9.76b and 9.77b). The nonlinear physics needed to account for pump depletion is contained in the last term on the right-hand side of (12.10).

To see how the depletion term arises, we must first identify the process that reduces the pump mode energy when the subharmonic mode becomes excited. Contrary to what we might expect, the process is one that creates, rather than annihilates, a photon in the pump mode—specifically, it is the process that converts two subharmonic photons into a pump photon through the term $-i\hbar a^2 b^\dagger$ in Hamiltonian (9.79). This photon creation process can account for a *loss* of pump mode energy because it interferes, destructively, with the direct creation of pump photons by the term $\hbar\bar{\mathcal{E}}_0 e^{-i2\omega_C t}$ in (9.79). The interference is seen explicitly in the equation of motion (9.63a). From here, the physical origin of the last term on the right-hand side of (12.10) can be traced as follows. Under the adiabatic elimination, a pump photon created by the annihilation of two subharmonic photons is immediately lost from the cavity before its conversion back into subharmonic photons can occur. The net effect is a two-photon loss from the subharmonic mode, the last term on the right-hand side of (12.10). The loss is nonlinear, having a rate that increases with the subharmonic mode energy. It therefore forestalls the energy explosion that would occur above threshold in the absence of pump depletion.

Exercise 12.2. Show that within the positive P representation, master equation (12.10) does, indeed, yield the phase-space equation of motion given in (12.5).

A procedure that takes us from master equation (9.97) to master equation (12.10) via the phase-space equation of motion (10.8), the stochastic differential equations (12.1a–12.1d), the stochastic differential equations (12.3a) and (12.3b), and the phase-space equation of motion (12.5), seems like a roundabout way to carry out an adiabatic elimination. Of course, all of this effort has not been expended in order to arrive at (12.10). Our main interest is in the stochastic differential equations (12.3a) and (12.3b) themselves. Before doing anything with these equations, though, it would be nice to see that it is possible to pass directly from master equation (9.97) to master equation (12.10). To help with the calculation, we first make a short diversion to familiarize ourselves with some of the intricacies of the superoperator notation.

12.1.2 A Note About Superoperators

We met the basic idea of a superoperator in Chap. 1, where in Sect. 1.5.1 the notation $e^{\mathcal{L}t}$, where $\mathcal{L}$ is a superoperator, provided a compact way of writing the formal solution to the master equation and related equations of motion. We have not actually used the properties of superoperators in any calculations, though. It is now our intention to do so, and it is time to meet some elementary properties of superoperators and to see what they allow us to do.

Superoperators act on operators to produce new operators, just as operators act on vectors to produce new vectors. The difference to keep in mind is that superoperators "embrace" their arguments, as indicated in the following examples:

$$(a^{\dagger 2}b\,\cdot)\hat{O} \equiv a^{\dagger 2}b\hat{O}, \qquad (a\cdot a^{\dagger})\hat{O} \equiv a\hat{O}a^{\dagger}, \qquad (\cdot\, b^{\dagger}b)\hat{O} \equiv \hat{O}b^{\dagger}b. \tag{12.11}$$

$\hat{O}$ denotes the operator on which the superoperator—the parentheses and its contents—acts. We adopt the convention that the action is always on the operator, or operator expression, to the immediate right of the parentheses; although action to the left is sometimes defined too [12.1]. The dot is an essential part of the superoperator definition. It indicates where, within the parentheses, the argument is to be placed. In cases like the first example of (12.11), the notation appears redundant, since the effect of the superoperator is the same as ordinary operator multiplication. The advantage of the notation comes from situations like those illustrated in the other examples, where it allows the *nesting of operators* to be expressed as a superoperator product. Without such a device, writing out the nested operators explicitly can become very tedious.

Superoperators are added in an obvious way—for example, to obtain the right-hand side of the typical master equation. The meaning of a superoperator product is similarly straightforward. To illustrate what is involved, let us factorize the examples given; we may write

$$(a^{\dagger 2}b\,\cdot) = (a^{\dagger 2}\,\cdot)(b\,\cdot), \qquad (a\cdot a^{\dagger}) = (a\,\cdot)(\cdot\, a^{\dagger}), \qquad (\cdot\, b^{\dagger}b) = (\cdot\, b)(\cdot\, b^{\dagger}). \tag{12.12}$$

In each case, the equality is verified by imagining an operator argument placed to the right on each side of the expression and working from right to left, substituting the appropriate operator, or parenthesized product of operators, for the dots. Of course the superoperators in a product do not generally commute. They do, however, if every operator appearing in the definition of one of the superoperators commutes with every operator appearing in the definition of the other. A trivial example is given by the product

$$\begin{aligned}(a\cdot a^{\dagger})(b\cdot b^{\dagger}) &= (ab\cdot b^{\dagger}a^{\dagger})\\ &= (ba\cdot a^{\dagger}b^{\dagger})\\ &= (b\cdot b^{\dagger})(a\cdot a^{\dagger}).\end{aligned} \tag{12.13}$$

Given a superoperator $\mathcal{S}$, we associate with it a *conjugate superoperator* $\mathcal{S}^\dagger$, with

$$(\mathcal{S}\hat{O})^\dagger \equiv \mathcal{S}^\dagger \hat{O}^\dagger, \tag{12.14}$$

where $\hat{O}$ is again an arbitrary operator argument. As an example, we have

$$\begin{aligned}[(a^{\dagger 2}b\,\cdot)\hat{O}]^\dagger &= (a^{\dagger 2}b\hat{O})^\dagger \\ &= (\hat{O}^\dagger b^\dagger a^2) \\ &= (\cdot\, b^\dagger a^2)\hat{O},\end{aligned} \tag{12.15}$$

and thus

$$(a^{\dagger 2}b\,\cdot)^\dagger = (\cdot\, b^\dagger a^2). \tag{12.16}$$

The rule, therefore, is that the operators on either side of the dot are conjugated and the positions of the operators relative to the dot are exchanged. It follows that for two superoperators $\mathcal{S}_1$ and $\mathcal{S}_2$,

$$(\mathcal{S}_1\mathcal{S}_2)^\dagger = \mathcal{S}_1^\dagger\mathcal{S}_2^\dagger, \tag{12.17}$$

as in

$$\begin{aligned}[(a\cdot a^\dagger)(\cdot\, a^\dagger a)]^\dagger &= (a\cdot a^\dagger a a^\dagger)^\dagger \\ &= (a a^\dagger a\cdot a^\dagger) \\ &= (a\cdot a^\dagger)(a^\dagger a\cdot) \\ &= (a\cdot a^\dagger)^\dagger(\cdot\, a^\dagger a)^\dagger.\end{aligned} \tag{12.18}$$

Note that contrary to what one might expect, the order of the superoperators on the right-hand side of (12.17) is not reversed.

Note 12.1. The conjugate superoperator defined by (12.14) is not the usual Hermitian conjugate or adjoint obtained from $\mathrm{tr}(\hat{O}_1\mathcal{S}\hat{O}_2) \equiv \mathrm{tr}(\hat{O}_2^\dagger\mathcal{S}^\dagger\hat{O}_1^\dagger)$—or equivalently, $\mathcal{S}\hat{O} \equiv (\hat{O}^\dagger\mathcal{S}^\dagger)$, with an appropriate definition of superoperator action to the left. See [12.1] for a more complete discussion of the conjugation of superoperators. In that reference, the superoperator we denote by $\mathcal{S}^\dagger$ is called the "associated" superoperator. It is this type of conjugation that is of use to us in the next section.

Knowing how to construct products of superoperators allows their commutators to be derived. Three particular examples, which we will find useful, are:

$$[(b\cdot b^\dagger),(b\,\cdot)] = (b\cdot b^\dagger)(b\,\cdot) - (b\,\cdot)(b\cdot b^\dagger) = 0, \tag{12.19a}$$

a trivial example; followed by a slightly less trivial one,

$$\begin{aligned}[(b^\dagger b\,\cdot),(b\,\cdot)] &= (b^\dagger b\,\cdot)(b\,\cdot) - (b\,\cdot)(b^\dagger b\,\cdot) \\ &= (b^\dagger b^2\,\cdot) - (b b^\dagger b\,\cdot) \\ &= -(b\,\cdot);\end{aligned} \tag{12.19b}$$

and thirdly,

$$\begin{aligned}[(\cdot\, b^\dagger b), (b\,\cdot)] &= (\cdot\, b^\dagger b)(b\,\cdot) - (b\,\cdot)(\cdot\, b^\dagger b) \\ &= 0. \end{aligned} \tag{12.19c}$$

Taking the conjugate of each, using (12.14) and (12.17), yields

$$[(b\cdot b^\dagger), (\cdot\, b^\dagger)] = 0, \qquad [(b^\dagger b\,\cdot), (\cdot\, b^\dagger)] = 0, \qquad [(\cdot\, b^\dagger b), (\cdot\, b^\dagger)] = -(\cdot\, b^\dagger). \tag{12.19d}$$

The derivation of other useful commutators is left as an exercise.

Exercise 12.3. Derive the commutation relations

$$[(b\cdot b^\dagger), (b^\dagger\,\cdot)] = (\cdot\, b^\dagger), \qquad [(b^\dagger b\,\cdot), (b^\dagger\,\cdot)] = (b^\dagger\,\cdot), \qquad [(\cdot\, b^\dagger b), (b^\dagger\,\cdot)] = 0, \tag{12.19e}$$

and

$$[(b\cdot b^\dagger), (\cdot\, b)] = (b\,\cdot), \qquad [(b^\dagger b\,\cdot), (\cdot\, b)] = 0, \qquad [(\cdot\, b^\dagger b), (\cdot\, b)] = (\cdot\, b), \tag{12.19f}$$

where the latter are obtained as conjugates of the former.

From their commutation relations we may write equations of motion for superoperators that are quite analogous to the more familiar Heisenberg equations of motion for ordinary operators. Thus, the superoperator formally defined by the expression

$$\mathcal{S}'(t) \equiv e^{-\mathcal{L}t}\mathcal{S}e^{\mathcal{L}t} \tag{12.20}$$

obeys the equation of motion

$$\frac{d\mathcal{S}'}{dt} = [\mathcal{S}', \mathcal{L}]. \tag{12.21}$$

This equation of motion is solved in much the same way as a Heisenberg equation of motion. As an example, let us take $\mathcal{L} = \mathcal{L}_p \equiv \kappa_p(2b\cdot b^\dagger - b^\dagger b\,\cdot - \cdot\, b^\dagger b)$ and $\mathcal{S}'(0) = (b\,\cdot)$. Then

$$\begin{aligned}\frac{d(b\,\cdot)'}{dt} &= \kappa_p[(b\,\cdot)', 2(b\cdot b^\dagger)' - (b^\dagger b\,\cdot)' - (\cdot\, b^\dagger b)'] \\ &= -\kappa_p(b\,\cdot)', \end{aligned} \tag{12.22}$$

and the solution is clearly

$$e^{-\mathcal{L}_p t}(b\,\cdot)e^{\mathcal{L}_p t} = e^{-\kappa_p t}(b\,\cdot). \tag{12.23}$$

The example is particularly simple. It is more usual to find that the equation of motion for one superoperator couples it to others, in which case further equations of motion must be added until a closed set is obtained. In the following section we need the solution to one example of this type.

Exercise 12.4. Take $\mathcal{L} = \mathcal{L}_p \equiv \kappa_p(2b\cdot b^\dagger - b^\dagger b\,\cdot - \cdot\, b^\dagger b)$ and $\mathcal{S}'(0) = (b^\dagger\,\cdot)$, and solve the equation of motion for $(b^\dagger\,\cdot)'(t)$. Hence show that

$$e^{-\mathcal{L}_p t}(b^\dagger\,\cdot)e^{\mathcal{L}_p t} = e^{\kappa_p t}(b^\dagger\,\cdot) + (e^{-\kappa_p t} - e^{\kappa_p t})(\cdot\, b^\dagger). \tag{12.24}$$

12.1.3 Adiabatic Elimination in the Master Equation

The discussion at the end of Sect. 12.1.1 shows that the physics of the adiabatic elimination may be summarized in the following two observations: first, for the purpose of exciting the subharmonic, the pump mode appears to be in a coherent state of amplitude $-i\bar{\mathcal{E}}_0/2\kappa_p$; second, the pump in addition mediates a two-photon loss from the subharmonic mode. Considering these observations, and in light of the underlying assumption of the adiabatic elimination—that the pump field has far larger bandwidth than the subharmonic—we see a parallel emerging with the system plus reservoir approach to damping discussed in Sect. 1.3. In the case of the adiabatic elimination, the system S is the subharmonic mode, while the broad bandwidth of the pump field justifies its treatment as a reservoir R. The reservoir is not, however, in a thermal state; it is externally driven and excited to a monochromatic coherent state. Nonetheless, the apparent connection with the system plus reservoir approach to damping suggests a method of adiabatic elimination at the level of the master equation.

The idea is to follow the formal development of Sects. 1.3.1 and 1.3.2 to obtain an equation of motion for the density operator

$$\tilde{\sigma}(t) \equiv \mathrm{tr}_p[\tilde{\rho}(t)] = \mathrm{tr}_p[\tilde{\rho}_D(t)], \tag{12.25}$$

where

$$\tilde{\rho}_D(t) \equiv D_p(i\bar{\mathcal{E}}_0/\kappa_p)\tilde{\rho}(t)D_p^\dagger(i\bar{\mathcal{E}}_0/\kappa_p). \tag{12.26}$$

Equation 12.25 is the analog of (1.17), with $\tilde{\rho}_D(t)$ in place of the density operator for $S \oplus R$, the trace over R in (1.17) replaced by a trace over the pump mode, and the reduced density operator for the subharmonic mode, $\tilde{\sigma}(t)$, replacing the density operator for S. The displacement (12.26) has been introduced so that the pump mode may be treated as a reservoir in the vacuum state (see Note 9.9). We also adopt the interaction picture, $\rho(t) = e^{-i[(\omega_C a^\dagger a + 2\omega_C b^\dagger b)t]}\tilde{\rho}(t)e^{i[(\omega_C a^\dagger a + 2\omega_C b^\dagger b)t]}$, which removes the explicit time-dependence of the pump-mode driving field. [In this section we use different notations, ρ and σ, to distinguish the two-mode and the one-mode density operators. The density operator denoted by $\rho(t)$ in (12.10) corresponds to $\sigma(t) = e^{-i\omega_C a^\dagger a t}\tilde{\sigma}(t)e^{i\omega_C a^\dagger a\, t}$ in the present notation.]

The principal change from the calculation of Sects. 1.3.1 and 1.3.2 is that the starting equation of motion is the master equation (9.97) rather than the Liouville equation (1.19). Written in the interaction picture, and after making the displacement (12.26), this two-mode master equation is

$$\dot{\tilde{\rho}}_D = (\mathcal{L}_s + \mathcal{L}_p + \mathcal{L}_{sp})\tilde{\rho}_D, \tag{12.27}$$

where

$$\mathcal{L}_s \equiv -i\frac{g}{2\kappa_p}[\bar{\mathcal{E}}_0 a^{\dagger 2} + \bar{\mathcal{E}}_0^* a^2, \cdot\,] + \kappa(2a \cdot a^\dagger - a^\dagger a \cdot - \cdot a^\dagger a), \tag{12.28a}$$

$$\mathcal{L}_p \equiv \kappa_p(2b \cdot b^\dagger - b^\dagger b \cdot - \cdot b^\dagger b), \tag{12.28b}$$

$$\mathcal{L}_{sp} \equiv (g/2)[a^{\dagger 2} b - a^2 b^\dagger, \cdot\,]. \tag{12.28c}$$

The goal is to derive an equation of motion for the one-mode density operator $\tilde{\sigma}(t)$ under the assumption

$$\tilde{\rho}_D(t) \approx \tilde{\sigma}(t)\big(|0\rangle\langle 0|\big)_p. \tag{12.29}$$

It is of course not sufficient to simply substitute the *ansatz* (12.29) into (12.27) and take the trace, since this yields only $\dot{\tilde{\sigma}} = \mathcal{L}_s\tilde{\sigma}$, omitting the two-photon loss required to account for pump depletion. Rather, as in Sect. 1.3.1, we must introduce the *ansatz* at a higher order of perturbation theory.

The first step is to isolate the interaction between the subharmonic and pump modes. In the earlier calculation, the interaction between S and R was isolated by the transformation (1.20). Here we make the similar transformation

$$\bar{\rho}_D(t) \equiv e^{-(\mathcal{L}_s+\mathcal{L}_p)t}\tilde{\rho}_D(t), \tag{12.30}$$

to obtain

$$\dot{\bar{\rho}}_D = \bar{\mathcal{L}}_{sp}(t)\bar{\rho}_D, \tag{12.31}$$

with the superoperator

$$\bar{\mathcal{L}}_{sp}(t) \equiv e^{-(\mathcal{L}_s+\mathcal{L}_p)t}\mathcal{L}_{sp}e^{(\mathcal{L}_s+\mathcal{L}_p)t} \tag{12.32}$$

taking the role of the interaction picture Liouvillian $(1/i\hbar)[\tilde{H}_{SR}(t),\cdot\,]$. To parallel (1.26), we have

$$\mathrm{tr}_p[\bar{\rho}_D(t)] = e^{-\mathcal{L}_s t}\tilde{\sigma}\_a(t) \equiv \bar{\sigma}(t). \tag{12.33}$$

Then, after integrating (12.31) once formally, substituting the integral form for $\bar{\rho}_D(t)$ on the right-hand side of the equation, and taking the trace over the pump mode, we arrive at the equation of motion

$$\dot{\bar{\sigma}} = \mathrm{tr}_p[\bar{\mathcal{L}}_{sp}(t)\rho_D(0)] + \int_0^t dt'\mathrm{tr}_p[\bar{\mathcal{L}}_{sp}(t)\bar{\mathcal{L}}_{sp}(t')\bar{\rho}_D(t')]. \tag{12.34}$$

This equation is the analog of (1.27).

The purpose of the displacement (12.26) now becomes apparent. To a good approximation the displaced pump mode is in the vacuum state. Therefore, the first term on the right-hand side of (12.34) is zero; it takes the expectation of the pump field in the vacuum state $\big(|0\rangle\langle 0|\big)_p$. Although a proof of the result is hardly needed, it does give some practice in the manipulation of superoperators.

Exercise 12.5. Show that

$$\begin{aligned}\mathrm{tr}_p[\bar{\mathcal{L}}_{sp}(t)\rho_D(0)] &= (g/2)\big[e^{-\mathcal{L}_s t}{}_p\langle 0|(a^{\dagger 2}b - a^2b^\dagger)|0\rangle_p e^{\mathcal{L}_s t}\sigma(0) \\ &\quad - \sigma(0)_p\langle 0|(a^{\dagger 2}b - a^2b^\dagger)|0\rangle_p\big] \\ &= 0. \end{aligned} \tag{12.35}$$

It is now safe to introduce the *ansatz* (12.29). Doing so brings us to an equation of motion in the Born approximation corresponding to master equation (1.30):

$$\dot{\bar{\sigma}} = \int_0^t dt' \mathrm{tr}_p\big[\bar{\mathcal{L}}_{sp}(t)\bar{\mathcal{L}}_{sp}(t')\bar{\sigma}(t')\big(|0\rangle\langle 0|\big)_p\big]. \tag{12.36}$$

To complete the adiabatic elimination we have only to find the explicit form of $\bar{\mathcal{L}}_{sp}(t)$ and to carry out the integration over time.

Recall that any two superoperators commute when the sets of operators entering their definitions commute (Eq. 12.13). Using this property and some straightforward factorizations (Eq. 12.12), we can pass from (12.28c) and (12.32) to

$$\begin{aligned}\bar{\mathcal{L}}_{sp}(t) &= (g/2)e^{-(\mathcal{L}_s+\mathcal{L}_p)t}\big[(a^{\dagger 2}b\,\cdot) - (a^2b^\dagger\,\cdot) - (\cdot\, a^{\dagger 2}b) + (\cdot\, a^2b^\dagger)\big]e^{(\mathcal{L}_s+\mathcal{L}_p)t} \\ &= (g/2)e^{-(\mathcal{L}_s+\mathcal{L}_p)t}\big[(a^{\dagger 2}\cdot)(b\,\cdot) - (a^2\cdot)(b^\dagger\,\cdot) - (a^2\cdot)^\dagger(b^\dagger\,\cdot)^\dagger \\ &\quad + (a^{\dagger 2}\cdot)^\dagger(b\,\cdot)^\dagger\big]e^{(\mathcal{L}_s+\mathcal{L}_p)t} \\ &= (g/2)[\bar{\mathcal{S}}_1(t)\bar{\mathcal{P}}_1(t) - \bar{\mathcal{S}}_2(t)\bar{\mathcal{P}}_2(t) + \bar{\mathcal{S}}_1^\dagger(t)\bar{\mathcal{P}}_1^\dagger(t) - \bar{\mathcal{S}}_2^\dagger(t)\bar{\mathcal{P}}_2^\dagger(t)], \end{aligned} \tag{12.37}$$

where

$$\begin{aligned}\bar{\mathcal{S}}_1(t) &\equiv e^{-\mathcal{L}_s t}(a^{\dagger 2}\cdot)e^{\mathcal{L}_s t}, & \bar{\mathcal{S}}_2(t) &\equiv e^{-\mathcal{L}_s t}(a^2\cdot)e^{\mathcal{L}_s t}, \\ \bar{\mathcal{P}}_1(t) &\equiv e^{-\mathcal{L}_p t}(b\,\cdot)e^{\mathcal{L}_p t}, & \bar{\mathcal{P}}_2(t) &\equiv e^{-\mathcal{L}_p t}(b^\dagger\,\cdot)e^{\mathcal{L}_p t}. \end{aligned} \tag{12.38a}$$

The explicit time dependences of $\bar{\mathcal{P}}_1(t)$ and $\bar{\mathcal{P}}_2(t)$ are given by (12.23) and (12.24), and the conjugates are obtained from (12.14); thus, we have

$$\bar{\mathcal{P}}_1(t) = e^{-\kappa_p t}(b\,\cdot), \tag{12.39a}$$

$$\bar{\mathcal{P}}_1^\dagger(t) = e^{-\kappa_p t}(\cdot\, b^\dagger), \tag{12.39b}$$

$$\bar{\mathcal{P}}_2(t) = e^{\kappa_p t}(b^\dagger\,\cdot) + (e^{-\kappa_p t} - e^{\kappa_p t})(\cdot\, b^\dagger), \tag{12.39c}$$

$$\bar{\mathcal{P}}_2^\dagger(t) = e^{\kappa_p t}(\cdot\, b) + (e^{-\kappa_p t} - e^{\kappa_p t})(b\,\cdot). \tag{12.39d}$$

All that remains is some tedious algebra.

On substituting the explicit form of $\bar{\mathcal{L}}_{sp}(t)\bar{\mathcal{L}}_{sp}(t')$ in (12.36), a total of thirty-six superoperator products are produced. Many terms are zero, though, due to the assumption that the displaced pump mode is in the vacuum state. When making the substitution, our eyes should be on the superoperators $(b\,\cdot)$,

$(b\,\cdot)$, $(b^\dagger\,\cdot)$, and $(\cdot\, b)$, and how they act upon the state $\big(|0\rangle\langle 0|\big)_p$. For the first two, the result of their action on the vacuum state is zero; this eliminates four of the six terms that enter (12.36) at the first order in the interaction, so we obtain

$$\begin{aligned}&\bar{\mathcal{L}}_{sp}(t')\bar{\sigma}(t')\big(|0\rangle\langle 0|\big)_p\\ &\qquad = -(g/2)e^{\kappa_p t'}\big[\bar{\mathcal{S}}_2(t')(b^\dagger\,\cdot) + \bar{\mathcal{S}}_2^\dagger(t')(\cdot\, b)\big]\bar{\sigma}(t')\big(|0\rangle\langle 0|\big)_p. \end{aligned} \tag{12.40}$$

There are then twelve rather than thirty-six terms at the second order. Four of these vanish for the same reason, and the result at second order is

$$\begin{aligned}&\bar{\mathcal{L}}_{sp}(t)\bar{\mathcal{L}}_{sp}(t')\bar{\sigma}(t')\big(|0\rangle\langle 0|\big)_p\\
&\quad = -(g/2)^2 e^{\kappa_p t'}\Big\{\big[e^{-\kappa_p t}\bar{\mathcal{S}}_1(t)(b\,\cdot) - e^{\kappa_p t}\bar{\mathcal{S}}_2(t)(b^\dagger\,\cdot)\\
&\quad - e^{\kappa_p t}\bar{\mathcal{S}}_2^\dagger(t)(\cdot\, b) - (e^{-\kappa_p t} - e^{-\kappa_p t})\bar{\mathcal{S}}_2^\dagger(t)(b\,\cdot)\big]\bar{\mathcal{S}}_2(t')(b^\dagger\,\cdot)\\
&\quad + \big[-e^{\kappa_p t}\bar{\mathcal{S}}_2(t)(b^\dagger\,\cdot) - (e^{-\kappa_p t} - e^{\kappa_p t})\bar{\mathcal{S}}_2(t)(\cdot\, b^\dagger)\\
&\quad + e^{-\kappa_p t}\bar{\mathcal{S}}_1^\dagger(t)(\cdot\, b^\dagger) - e^{\kappa_p t}\bar{\mathcal{S}}_2^\dagger(t)(\cdot\, b)\big]\bar{\mathcal{S}}_2^\dagger(t')(\cdot\, b)\Big\}\bar{\sigma}(t')\big(|0\rangle\langle 0|\big)_p\\
&\quad = -(g/2)^2\Big\{e^{-\kappa_p(t-t')}\big[\big(\bar{\mathcal{S}}_1(t)\bar{\mathcal{S}}_2(t') - \bar{\mathcal{S}}_2^\dagger(t)\bar{\mathcal{S}}_2(t')\big)(bb^\dagger\,\cdot) + \text{s.c.}\big]\\
&\quad - e^{\kappa_p(t+t')}\big[\bar{\mathcal{S}}_2(t)\bar{\mathcal{S}}_2(t')(b^{\dagger 2}\cdot)\\
&\quad + \bar{\mathcal{S}}_2^\dagger(t)\bar{\mathcal{S}}_2(t')\big((b^\dagger\,\cdot\, b) - (bb^\dagger\,\cdot)\big) + \text{s.c.}\big]\Big\}\bar{\sigma}(t')\big(|0\rangle\langle 0|\big)_p,\end{aligned} \tag{12.41}$$

where s.c. denotes the superoperator conjugate defined in (12.14).

The surviving terms in (12.41) are of two types, those proportional to $e^{-\kappa_p(t-t')}$ and those proportional to $e^{\kappa_p(t+t')}$. The latter appear to diverge, but vanish when the trace in (12.36) is taken. The time integral over the former is performed in the adiabatic limit, where we take

$$e^{-\kappa_p(t-t')}\bar{\mathcal{S}}_2(t')\bar{\sigma}(t') \to 2\kappa_p^{-1}\delta(t-t')\bar{\mathcal{S}}_2(t')\bar{\sigma}(t'), \tag{12.42}$$

which is the analog of the Markov approximation (1.36). Thus, from (12.36), (12.41), (12.42), and (12.38a), we arrive at the equation of motion

$$\dot{\bar{\sigma}} = \frac{g^2}{4\kappa_p}e^{-\mathcal{L}_s t}\big[2(a^2\cdot a^{\dagger 2}) - (a^{\dagger 2}a^2\cdot) - (\cdot\, a^{\dagger 2}a^2)\big]e^{\mathcal{L}_s t}\bar{\sigma}. \tag{12.43}$$

Inverting the transformations (12.33) and into the interaction picture, writing $\sigma(t) = e^{-i\omega_C a^\dagger a t}\tilde{\sigma}(t)e^{i\omega_C a^\dagger a t} = e^{-i\omega_C a^\dagger a t}[e^{\mathcal{L}_s t}\bar{\sigma}(t)e^{-\mathcal{L}_s t}]e^{i\omega_C a^\dagger a t}$, we recover the master equation with adiabatic elimination of the pump (12.10).

12.1.4 Numerical Simulation of the Stochastic Differential Equations

We now return to our main agenda, the numerical simulation of the stochastic differential equations (12.3a) and (12.3b). Of course these equations do not possess a solution as a continuous function of time like a set of ordinary differential equations do. The procedure is to generate an ensemble of trajectories, $T(\boldsymbol{X}_0, \boldsymbol{W}) \equiv \big(\alpha(t), \alpha_*(t)\big)$,

$$\boldsymbol{X}_0 \equiv \begin{pmatrix} \alpha^0 \\ \alpha^0_* \end{pmatrix}, \qquad \boldsymbol{W} \equiv \begin{pmatrix} W_\alpha \\ W_{\alpha_*} \end{pmatrix}, \tag{12.44}$$

with each trajectory obtained by the Ito stochastic integration of (12.3a) and (12.3b) for the same initial condition $\boldsymbol{X}_0$ and a particular realization of the two-dimensional Wiener path $\boldsymbol{W}$. So long as the ensemble is sufficiently large, the average of $\big(\alpha_*(t)\big)^p\big(\alpha(t)\big)^q$ over the ensemble will be a good approximation to the moment $\big(\overline{(\alpha_*^p\alpha^q)(t)}\,\big)_P = \langle(a^{\dagger p}a^q)(t)\rangle$.

The simplest integration algorithm is provided by the one-step Euler method, where, for a time step of $\Delta\bar{t}$, the solution is advanced by writing

$$\tilde{\bar{\alpha}}(t + \Delta t) = \tilde{\bar{\alpha}}(t) + \Delta\tilde{\bar{\alpha}}(t), \tag{12.45a}$$

$$\tilde{\bar{\alpha}}_*(t + \Delta t) = \tilde{\bar{\alpha}}_*(t) + \Delta\tilde{\bar{\alpha}}_*(t), \tag{12.45b}$$

with $\Delta\tilde{\bar{\alpha}}(t)$ and $\Delta\tilde{\bar{\alpha}}_*(t)$ computed from (12.3a) and (12.3b), taking $d\bar{t} \to \Delta\bar{t}$, $d\bar{W}_\alpha \to \Delta\bar{W}_\alpha$, and $d\bar{W}_{\alpha_*} \to \Delta\bar{W}_{\alpha_*}$; the Wiener increments $\Delta\bar{W}_\alpha$ and $\Delta\bar{W}_{\alpha_*}$ are statistically independent Gaussian-distributed random numbers of zero mean and variance $\Delta\bar{t}$ (see Sect. 5.3.3). More sophisticated algorithms exist that are superior to the Euler algorithm in either their accuracy or stability [12.2, 12.3, 12.4, 12.5, 12.6, 12.7, 12.8]. Needless to say, when compared with the situation for ordinary differential equations, the numerical analysis is more difficult, and there certainly is not an equivalent range of packaged, general purpose integrators available. So far as accuracy is concerned, we should remember that the computed moments are going to be compromised by the finite size of the ensemble (sampling error) as well as by the limited accuracy of the stochastic integration. In many cases, this sampling error will be the limiting factor. Stability of the integration is another matter. The one-step Euler method is particularly sensitive to instability, and may require a very small time step in order to be stable, a demand that can easily increase the time needed for an accurate simulation to the point that the computation is effectively impossible. We will discover in fact that stability, or rather instability, is potentially a real worry for the integration of (12.3a) and (12.3b). Nevertheless, let us forge ahead using the Euler method. For an initial attempt at the integration, its tendency towards instability can actually be an advantage, in so far as it serves as a diagnostic; perhaps it is just as well to encourage any potential numerical instability, given that the physics has very little to say about what to expect from stochastic differential equations in the positive P representation.

We are interested in behavior outside the small-noise limit. Up to now we have used the threshold photon number in the pump mode, n_p^{thr}, as the system size parameter. Having taken the adiabatic limit, the threshold photon number in the subharmonic mode is the more natural parameter to use. This is apparent from the noise terms in (12.3a) and (12.3b), both of which have (aside from the constant factor $\Gamma(\frac{1}{4})/\Gamma(\frac{3}{4})$) the coefficient $\langle a^\dagger a\rangle_{\mathrm{thr}}^{-1}$ (Eq. 10.102). In the following we therefore compare results for different values of $\langle a^\dagger a\rangle_{\mathrm{thr}}$.

Figures 12.1 and 12.2 illustrate the behavior in two different operating regimes. The first displays results at threshold, where the quantities plotted are computed as ensemble averages from the expressions

$$\langle a^\dagger a\rangle_{\mathrm{thr}}^{-1}\langle (a^\dagger a)(t)\rangle = \left[\frac{\Gamma(\frac{1}{4})}{\Gamma(\frac{3}{4})}\right]^2 \langle a^\dagger a\rangle_{\mathrm{thr}}\big(\overline{(\tilde{\bar{\alpha}}_*\tilde{\bar{\alpha}})(t)}\,\big)_{P}, \tag{12.46a}$$

$$\left(n_p^{\mathrm{thr}}\right)^{-1/2} e^{-i\frac{1}{2}(\psi-\pi/2)}\langle\tilde{b}(t)\rangle = \lambda - \big(\,\overline{\tilde{\bar{\alpha}}^2(t)}\,\big)_{P}, \tag{12.46b}$$

$$(\Delta Y)^2(t) - \tfrac{1}{4} = -\tfrac{1}{4}\Big[\big(\,\overline{(\tilde{\bar{\alpha}}-\tilde{\bar{\alpha}}_*)^2(t)}\,\big)_P - \big(\,\overline{(\tilde{\bar{\alpha}}-\tilde{\bar{\alpha}}_*)(t)}\,\big)_P^2\Big]. \tag{12.46c}$$

Since each of the averages should be real, only the real parts are shown. Of course the imaginary parts are not precisely zero when computed from an ensemble of finite size. They are seen to decrease, though, as the number of trajectories is increased, and in practice deviations from the conjugacy requirement (11.113)—proved for an infinite ensemble in Sect. 11.2.3—serve as a test to decide when the ensemble is sufficiently large.

The mean photon number at threshold in the subharmonic mode was calculated using the small-noise assumption in Sect. 10.2.4. The first thing to note from the simulations is that the small-noise result is not reached by any of the curves in Fig. 12.1a; in the most extreme case—for $\langle a^\dagger a\rangle_{\mathrm{thr}} = 0.25$—the steady-state photon number is only one half that predicted by the small-noise approximation (system size expansion to lowest nonvanishing order). This large deviation decreases rather quickly with increasing $\langle a^\dagger a\rangle_{\mathrm{thr}}$, to become something less than 5% for $\langle a^\dagger a\rangle_{\mathrm{thr}} = 10$. These changes in photon number are explained by Fig. 12.1b, where the mean amplitude of the pump field is plotted. For small values of the system size parameter, the amplitude of the pump is significantly depleted, falling below the $\langle\tilde{\bar{b}}\rangle_{\mathrm{ss}} = \lambda = 1$ obtained by neglecting the fluctuations (Eqs. 10.43). A further consequence of this depletion appears in Fig. 12.1c, where the steady-state squeezing (integrated over frequency) is smaller than the predicted $(\Delta Y)^2_{\mathrm{thr}} = 1/8$ (Eq. 10.98a). Finally, note the scaling of time in Figs. 12.1a and 12.1b; the timescale reflects the critical slowing down of fluctuations in the dimension that has become unstable—the amplified X-quadrature phase amplitude of the subharmonic field (Eq. 10.96a).

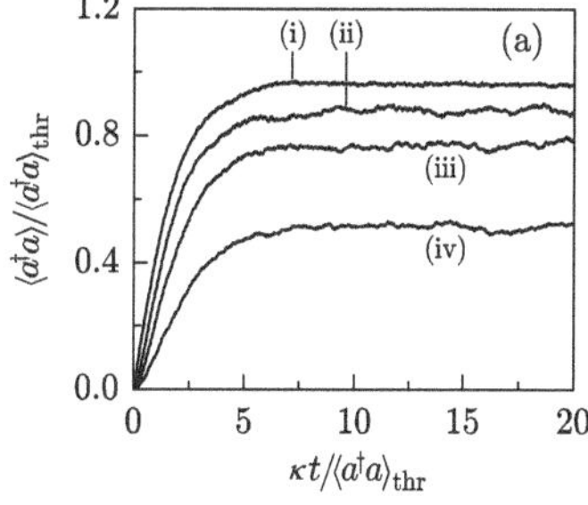

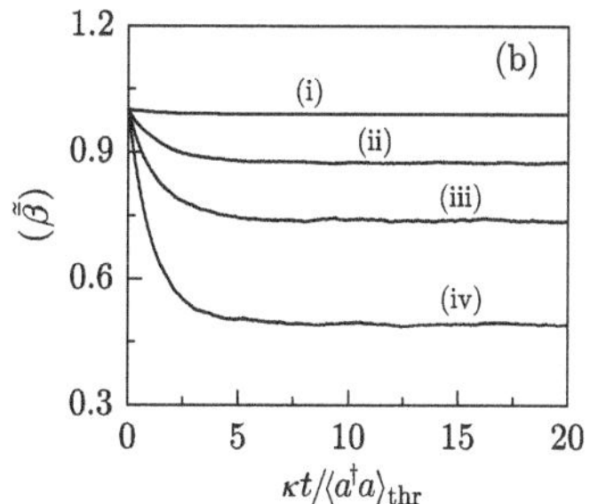

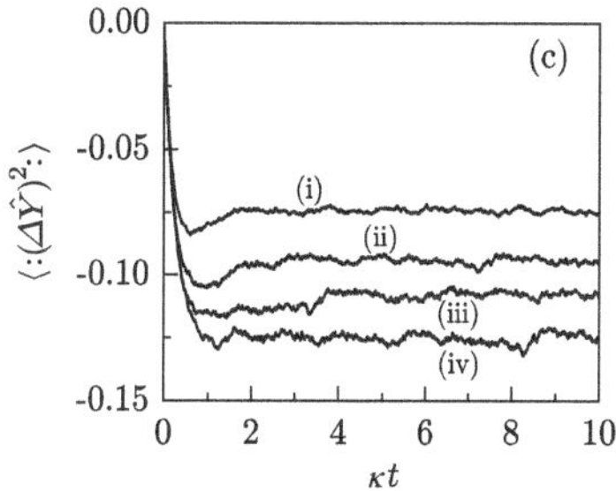

Fig. 12.1. Quantum fluctuations for the degenerate parametric oscillator at threshold as a function of threshold photon number. Each curve averages over an ensemble of 10,000 trajectories generated by simulating (12.3a) and (12.3b) for $\lambda = 1$ and $\tilde{\alpha}(0) = \tilde{\alpha}_*(0) = 0$, with time step $\Delta\bar{t} = 10^{-4}$: (**a**) mean photon number in the subharmonic mode (Eq. 12.46a), (**b**) mean amplitude of the pump mode (Eq. 12.46b), and (**c**) normal-ordered variance of the squeezed quadrature phase amplitude of the subharmonic mode (Eq. 12.46c). *Curves* (i), (ii), (iii), and (iv) are for threshold photon numbers $\langle a^\dagger a\rangle_{\rm thr} = 10$, 1, 0.5, and 0.25, respectively

Depletion of the pump amplitude occurs because of its dependence on the squared amplitude of the subharmonic field. Specifically, away from the small-noise limit, the average $\left(\overline{\tilde{\alpha}^2}\right)_P$ in (12.2) is not negligible compared to $\lambda = 1$. In fact, this depletion of the pump is also included in the small-noise analysis of Sect. 10.2.4. It enters through the conditional distribution (10.97b), whose mean, $\tilde{w}_1 = -\tilde{z}_1^2$, depends on the squared amplitude of the subharmonic mode, itself distributed according to (10.97a). The point is illustrated by the plot of the conditional distribution in Fig. 10.1, which shows an average fluctuation of the pump amplitude that is negative, rather than zero. The small-noise analysis is limited, nevertheless, because it fails to feed the depletion back into the calculation of fluctuations for the subharmonic. The approximation is acceptable when the system size parameter is large, but not more generally, as Fig. 12.1 shows.

No doubt the results of Fig. 12.1 could be reproduced by systematically extending the system size expansion beyond the lowest nonvanishing order. Let us not follow up on this idea right now (see Sect. 12.1.9), however, since we are about to discover, after a little more probing with numerical simulations, that the steady-state solution for the positive P distribution can be given to all orders in a closed form expression.

Note 12.2. The methods of many-body quantum theory have been used to treat system size effects in the degenerate parametric oscillator [12.9, 12.10, 12.11]. It is reasonably expected that results obtained using these methods would also be reproduced by a systematic extension of the system size expansion to higher order.

For a second demonstration of what can be done with stochastic differential equations, Fig. 12.2 displays some results for above-threshold operation. The quantities plotted are (12.46b) and (12.46c), and also

$$\left(2\xi^{-1}n_p^{\mathrm{thr}}\right)^{-1}\langle(a^\dagger a)(t)\rangle = \left(\overline{(\tilde{\bar{\alpha}}_*\tilde{\bar{\alpha}})(t)}\right)_P, \tag{12.47a}$$

and

$$\begin{aligned}\left(2\xi^{-1}n_p^{\mathrm{thr}}\right)^{-2}&[\langle(a^\dagger a)^2(t)\rangle - \langle(a^\dagger a)(t)\rangle^2] \\ &= \left(\overline{(\tilde{\bar{\alpha}}_*\tilde{\bar{\alpha}})^2(t)}\right)_P - \left(\overline{(\tilde{\bar{\alpha}}_*\tilde{\bar{\alpha}})(t)}\right)_P^2 + \left(2\xi^{-1}n_p^{\mathrm{thr}}\right)^{-1}\left(\overline{(\tilde{\bar{\alpha}}_*\tilde{\bar{\alpha}})(t)}\right)_P.\end{aligned} \tag{12.47b}$$

For the plots of Figs. 12.2a and 12.2b, $\langle a^\dagger a\rangle_{\mathrm{thr}}$ is large, comparable to what would be typical in an experiment (see Note 10.1). The purpose here is to illustrate the "turn on" of the degenerate parametric oscillator, where a macroscopic number of photons grows in the subharmonic mode, initiated by quantum fluctuations about the unstable vacuum state. The decay of an unstable state amplifies fluctuations up to the macroscopic level (Sect. 5.1.4), and it is only in the early stages of the amplification that the linearization of Sects. 10.2.1 and 10.2.2 holds. Analytical methods for treating the amplification process exist. They allow one to calculate such things as the *mean first passage time* for the system to reach a specified locally stable state [12.12]. A good discussion of the difficulties of the problem is given by van Kampen [12.13], who illustrates the issues with an exact solution for a simplified model of a symmetric bistable potential.

We may leave aside the sophisticated analytical methods. If we are not too worried about quantitative agreement, a crude approximation does a reasonable job of getting at least the orders of magnitude right. After the "turn on" of the degenerate parametric oscillator, the variable $\frac{1}{2}(\tilde{\bar{\alpha}} + \tilde{\bar{\alpha}}_*)$ has acquired one of two possible macroscopic values—either $+\sqrt{\lambda-1}$ or $-\sqrt{\lambda-1}$. Before the "turn on," this variable is zero; but it is driven away from zero by a noise source with variance $\left(2\xi^{-1}n_p^{\mathrm{thr}}\right)^{-1}(\lambda/2)\bar{t} = [\Gamma(\frac{3}{4})/\Gamma(\frac{1}{4})]^2\langle a^\dagger a\rangle_{\mathrm{thr}}^{-2}(\lambda/2)\bar{t}$ (in the linear regime), and the fluctuations are amplified with a gain coefficient $\lambda-1$. The argument based on these observations is that the "turn on" really gets under way after the time, $\bar{t}_{\mathrm{on}}$, at which the exponential growth from an

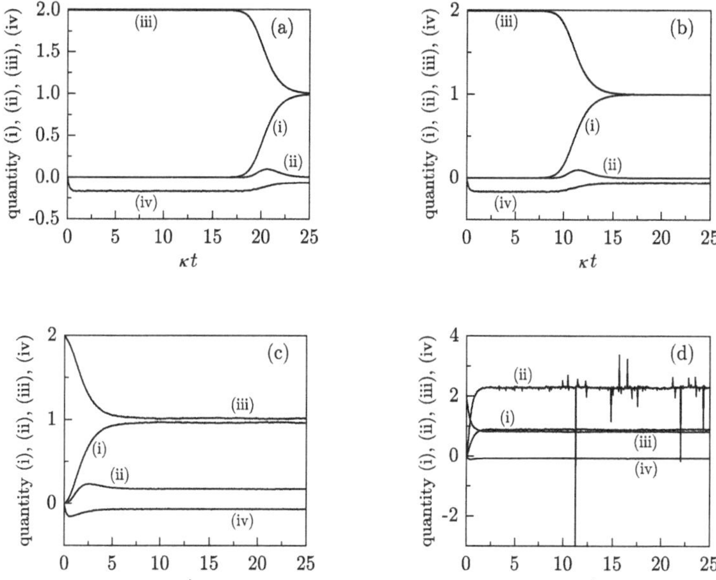

Fig. 12.2. Quantum fluctuations for the degenerate parametric oscillator above threshold as a function of threshold photon number. Each curve averages over an ensemble of 10,000 trajectories generated by simulating (12.3a) and (12.3b) for $\lambda = 2$ and $\tilde{\bar{\alpha}}(0) = \tilde{\bar{\alpha}}_*(0) = 0$, with time step $\Delta\bar{t} = 10^{-4}$: $\langle a^\dagger a\rangle_{\rm thr} = 10^8$ (**a**), 10^4 (**b**), 1 (**c**), and 0.25 (**d**). *Curve* (i) is the mean photon number in the subharmonic mode (Eq. 12.47a); *curve* (ii) is the photon number variance in the subharmonic mode (Eq. 12.47b); *curve* (iii) is the mean amplitude of the pump (Eq. 12.46b); and *curve* (iv) is the normal-ordered variance of the squeezed quadrature phase amplitude of the subharmonic mode (Eq. 12.46c)

initial condition of the typical size provided by pure diffusion (up to time $\bar{t}_{\rm on}$) would give $\frac{1}{2}[\tilde{\bar{\alpha}}(\bar{t}_{\rm on}) + \tilde{\bar{\alpha}}_*(\bar{t}_{\rm on})] \sim 1$. Thus, $\bar{t}_{\rm on}$ is the solution to

$$\left[\frac{\Gamma(3/4)}{\Gamma(1/4)}\right]^2 \langle a^\dagger a\rangle_{\rm thr}^{-2}(\lambda/2)\bar{t}_{\rm on}e^{2(\lambda-1)\bar{t}_{\rm on}} = 1. \tag{12.48}$$

Iteration with a calculator quickly gives $\bar{t}_{\rm on} \sim 18.1$ for the parameters of Fig. 12.2a and $\bar{t}_{\rm on} \sim 9.2$ for the parameters of Fig. 12.2b. These estimates are not at all bad for the time at which the mean photon number in the subharmonic mode starts to grow [curve (i)], simultaneously depleting the mean amplitude of the pump [curve (ii)]. Note how the squeezing is reduced as pump depletion sets in [curve (iv)]. Before the pump is depleted, formula (10.55b)—derived from the linearized analysis below threshold—continues to hold [12.14]. Of course the decay of the unstable state is actually stochastic, and there is a distribution of "turn on" times. Evidence of this appears as a peak in the photon number variance—curve (iii) of Fig. 12.2.

Turning now to Figs. 12.2c and 12.2d, the system size parameter is small enough that we would expect to see deviations from the small-noise analysis, similar to those seen in Fig. 12.1. Indeed, the subharmonic mode photon number falls below the predicted $\left(2\xi^{-1}n_p^{\text{thr}}\right)^{-1}\langle a^\dagger a\rangle_{\text{ss}} = \langle\tilde{\bar{a}}^\dagger\rangle_{\text{ss}}\langle\tilde{\bar{a}}\rangle_{\text{ss}} = \lambda - 1 = 1$ (Eq. 10.72a) and the mean pump amplitude is no longer depleted to $\left(n_p^{\text{thr}}\right)^{-1/2}e^{-i\frac{1}{2}(\psi-\pi/2)}\langle\tilde{b}\rangle_{\text{ss}} = \langle\tilde{\bar{b}}\rangle_{\text{ss}} = 1$ (Eq. 10.72b). The largest change from the predictions of the small-noise analysis is the absence of a distinct peak in the plot of photon number variance; instead, the variance grows and settles to a nonzero value, indicating that full-scale intensity fluctuations persist in the steady state.

It would appear from these results that the positive P representation can, indeed, take us into the regime of large quantum noise. In Fig. 12.2d, in particular, the system size parameter is sufficiently small that the macroscopic equations (10.43a–10.43d) no longer provide a meaningful starting point for an understanding of the physics. From curves (i) and (ii), for example, we deduce that $\langle a^\dagger a\rangle_{\text{ss}} < \langle\tilde{a}\rangle_{\text{ss}}^2$, with the quantity on the right-hand side of the inequality real—a relationship that cannot possibly hold for classical random variables. Of course, in itself this relationship is not particularly special; it is what accounts for the existence of squeezing and holds, therefore, in Figs. 12.2a–c as well. What *is* special about Fig. 12.2d, though, is the scale of the nonclassical fluctuations. Here, the fluctuations have the same scale as the nonlinear physics predicted by the macroscopic law. Thus, the picture of classical nonlinear dynamics plus "fuzz" (even squeezed "fuzz") must surely beak down.

Before we become too enthusiastic, however, we need to take note of one other feature of Fig. 12.2d, one which raises some concern. What should we make of the "spikes" that occur along curve (iv)? These "spikes" are the most dramatic features of the whole figure, and are quite anomalous compared with the regular behavior seen elsewhere in Figs. 12.1 and 12.2. Anomalous behavior of this kind was first seen by Carmichael et al. [12.15] in their treatment of absorptive bistability outside the small-noise limit. The authors traced the "spikes" to a few trajectories within an ensemble that make wide excursions into the nonclassical phase-space. They suggested that the excursions might be nonphysical manifestations of a modified nonlinear dynamics introduced by doubling the phase-space dimensions. As we have seen (Sect. 11.2.2), the classical phase space can indeed be unstable to perturbations in the nonclassical phase-space variables. Of course, it could be that our numerical integration is at fault; we should be particularly cautious now with the Euler method. Even granting this qualification, though, the consensus reached from work to date is that the "spikes" are a genuine feature of the nonlinear dynamical equations [12.16, 12.17, 12.18]. Our task now is to understand their origin and consequences in detail. We start with an analysis of the deterministic nonlinear dynamics in the extended phase space.

12.1.5 Deterministic Dynamics in the Extended Phase Space

With the noise terms neglected, (12.3a) and (12.3b) are coupled nonlinear ordinary differential equations. It is possible to solve these equations and, from the solution, gain some understanding of the origin of the "spikes." Written in terms of the new complex variables

$$q \equiv \tilde{\bar{\alpha}}/\tilde{\bar{\alpha}}_*, \qquad p \equiv 1/\tilde{\bar{\alpha}}\tilde{\bar{\alpha}}_*, \tag{12.49}$$

the equations to be solved are

$$\frac{dq}{d\bar{t}} = \lambda(1-q^2), \tag{12.50a}$$

$$\frac{dp}{d\bar{t}} = -p[\lambda(q+1/q)-2]+2. \tag{12.50b}$$

From (12.50a), we have

$$dq\left(\frac{1}{1+q}+\frac{1}{1-q}\right) = 2\lambda d\bar{t}, \tag{12.51}$$

whose direct integration gives the solution for q,

$$q(\bar{t}) = \frac{M(\bar{t})}{P(\bar{t})}, \tag{12.52}$$

with

$$M(\bar{t}) \equiv [1+q(0)]e^{\lambda\bar{t}} - [1-q(0)]e^{-\lambda\bar{t}}, \tag{12.53a}$$

$$P(\bar{t}) \equiv [1+q(0)]e^{\lambda\bar{t}} + [1-q(0)]e^{-\lambda\bar{t}}. \tag{12.53b}$$

A formal integration of (12.50b) yields

$$p(\bar{t}) = p(0)e^{-Q(\bar{t})} + 2e^{-Q(\bar{t})}\int_0^{\bar{t}} d\bar{t}' e^{Q(\bar{t}')}, \tag{12.54}$$

with

$$Q(\bar{t}) \equiv \int_0^{\bar{t}} d\bar{t}'\{\lambda[q(\bar{t}')+1/q(\bar{t}')]-2\}. \tag{12.55}$$

We can obtain an explicit expression for $Q(\bar{t})$ using (12.52), (12.53a), and (12.53b); thus, we write

$$q(\bar{t}) + \frac{1}{q(\bar{t})} = \frac{1}{\lambda}\frac{d}{d\bar{t}}\ln[M(\bar{t})P(\bar{t})], \tag{12.56}$$

to obtain

$$e^{Q(\bar{t})} = e^{-2\bar{t}}\frac{M(\bar{t})P(\bar{t})}{4q(0)}. \tag{12.57}$$

Substituting this result into (12.54) gives the solution for p,

$$p(\bar{t}) = e^{2\bar{t}} \frac{4q(0)}{M(\bar{t})P(\bar{t})} \left\{ p(0) + \frac{1}{4q(0)} \times \left[[1+q(0)]^2 \frac{1-e^{-2(1-\lambda)\bar{t}}}{1-\lambda} - [1-q(0)]^2 \frac{1-e^{-2(1+\lambda)\bar{t}}}{1+\lambda} \right] \right\}. \tag{12.58}$$

Finally, inverting the transformation (12.49) yields the *solution to the deterministic equations of motion within the positive P representation for the degenerate parametric oscillator with adiabatic elimination of the pump*:

$$\tilde{\bar{\alpha}}(\bar{t}) = \frac{[\tilde{\bar{\alpha}}(0) + \tilde{\bar{\alpha}}_*(0)]e^{-(1-\lambda)\bar{t}} + [\tilde{\bar{\alpha}}(0) - \tilde{\bar{\alpha}}_*(0)]e^{-(1+\lambda)\bar{t}}}{\sqrt{4 + [\tilde{\bar{\alpha}}(0) + \tilde{\bar{\alpha}}_*(0)]^2 \Lambda_-(\bar{t}) - [\tilde{\bar{\alpha}}(0) - \tilde{\bar{\alpha}}_*(0)]^2 \Lambda_+(\bar{t})}}, \tag{12.59a}$$

and

$$\tilde{\bar{\alpha}}_*(\bar{t}) = \frac{[\tilde{\bar{\alpha}}(0) + \tilde{\bar{\alpha}}_*(0)]e^{-(1-\lambda)\bar{t}} - [\tilde{\bar{\alpha}}(0) - \tilde{\bar{\alpha}}_*(0)]e^{-(1+\lambda)\bar{t}}}{\sqrt{4 + [\tilde{\bar{\alpha}}(0) + \tilde{\bar{\alpha}}_*(0)]^2 \Lambda_-(\bar{t}) - [\tilde{\bar{\alpha}}(0) - \tilde{\bar{\alpha}}_*(0)]^2 \Lambda_+(\bar{t})}}, \tag{12.59b}$$

where

$$\Lambda_-(\bar{t}) \equiv \frac{1-e^{-2(1-\lambda)\bar{t}}}{1-\lambda}, \qquad \Lambda_+(\bar{t}) \equiv \frac{1-e^{-2(1+\lambda)\bar{t}}}{1+\lambda}. \tag{12.60}$$

Our aim is to use this solution to construct a picture of the deterministic flow, particularly within the nonclassical phase space.

It is difficult to deduce very much by looking at the general expressions. We therefore simplify the situation by considering particular initial conditions. There are four simple cases for which the phase space trajectories are straight lines. The first two give trajectories confined to the classical phase space: we take initial condition $\tilde{\bar{\alpha}}(0) = \tilde{\bar{\alpha}}_*(0) = x$, for which the phase space trajectory is given by

$$\tilde{\bar{\alpha}}(\bar{t}) = \tilde{\bar{\alpha}}_*(\bar{t}) = \frac{xe^{-(1-\lambda)\bar{t}}}{\sqrt{1 + x^2[1-e^{-2(1-\lambda)\bar{t}}]/(1-\lambda)}}, \tag{12.61a}$$

and initial condition $\tilde{\bar{\alpha}}(0) = -\tilde{\bar{\alpha}}_*(0) = iy$, which gives the trajectory

$$\tilde{\bar{\alpha}}(\bar{t}) = -\tilde{\bar{\alpha}}_*(\bar{t}) = i\frac{ye^{-(1+\lambda)\bar{t}}}{\sqrt{1 + y^2[1-e^{-2(1+\lambda)\bar{t}}]/(1+\lambda)}}. \tag{12.61b}$$

With a change of sign the trajectories are confined to the nonclassical phase space (at least initially, see below): for $\tilde{\bar{\alpha}}(0) = -\tilde{\bar{\alpha}}_*(0) = x$ we obtain the trajectory

$$\tilde{\bar{\alpha}}(\bar{t}) = -\tilde{\bar{\alpha}}_*(\bar{t}) = \frac{xe^{-(1+\lambda)\bar{t}}}{\sqrt{1 - x^2[1-e^{-2(1+\lambda)\bar{t}}]/(1+\lambda)}}, \tag{12.62a}$$

and for $\tilde{\bar{\alpha}}(0) = \tilde{\bar{\alpha}}_*(0) = iy$, the trajectory

$$\tilde{\bar{\alpha}}(\bar{t}) = \tilde{\bar{\alpha}}_*(\bar{t}) = i\frac{ye^{-(1-\lambda)\bar{t}}}{\sqrt{1 - y^2[1 - e^{-2(1-\lambda)\bar{t}}]/(1-\lambda)}}. \tag{12.62b}$$

There is an important difference between the first and second pair of trajectories. In (12.61a) and (12.61b) the expression under the square root is always positive, and it is easily seen that the trajectories asymptotically approach the steady states $\tilde{\bar{\alpha}}_{\rm ss} = \tilde{\bar{\alpha}}_*^{\rm ss} = 0$, for $\lambda < 1$, and $\tilde{\bar{\alpha}}_{\rm ss} = \tilde{\bar{\alpha}}_*^{\rm ss} = (x/|x|)\sqrt{\lambda - 1}$, for $\lambda > 1$. The expressions under the square root in (12.62a) and (12.62b), on the other hand, are *not* always positive. These expressions can vanish: the square root in (12.62a) vanishes at time

$$\bar{t} = -\frac{1}{2(1+\lambda)}\ln\left[1 - \frac{1+\lambda}{x^2}\right],$$

if $|x| > \sqrt{1+\lambda}$, and in (12.62b) it vanishes at

$$\bar{t} = -\frac{1}{2(1-\lambda)}\ln\left[1 - \frac{1-\lambda}{y^2}\right],$$

if $|y| > \sqrt{1-\lambda}$, for $\lambda < 1$, and $y \neq 0$, for $\lambda > 1$. When the square root vanishes, the phase space trajectory passes to infinity, and it does so in a finite time. If we apply expressions (12.62a) and (12.62b) for later times, the divergent trajectory returns from infinity, now within the classical phase space. In the case of (12.62a), the trajectory returns to the origin along the line $\tilde{\bar{\alpha}}(\bar{t}) = -\tilde{\bar{\alpha}}_*(\bar{t})$ with both variables pure imaginary; for (12.62b), the returning trajectory follows the line $\tilde{\bar{\alpha}}(\bar{t}) = \tilde{\bar{\alpha}}_*(\bar{t})$ with both variables real; in the latter case the trajectory returns to the origin for $\lambda < 1$ and to the steady state $(y/|y|)\sqrt{\lambda - 1}$ for $\lambda > 1$.

Even in just four dimensions it is difficult to construct a global picture of how the deterministic flow is organized. There are certainly other initial conditions that produce divergences; for example, with $\tilde{\bar{\alpha}}(0) = -\big(\tilde{\bar{\alpha}}_*(0)\big)^* = x + iy$, for $\lambda < 1$ all initial states outside the ellipse $x^2/(1+\lambda) + y^2/(1-\lambda) = 1$ produce diverging trajectories, while for $\lambda > 1$, all trajectories except those beginning on the line segment $|x| \leq \sqrt{1+\lambda}$, $y = 0$, diverge. For a more complete picture, we can plot the flow in two-dimensional cross-sections of the four-dimensional phase space. The real plane, $\tilde{\bar{y}} = \tilde{\bar{\mathrm{y}}} = 0$, and the imaginary plane, $\tilde{\bar{x}} = \tilde{\bar{\mathrm{x}}} = 0$, $[\tilde{\bar{\alpha}} = \tilde{\bar{x}} + i\tilde{\bar{y}}, \tilde{\bar{\alpha}}_* = \tilde{\bar{\mathrm{x}}} + i\tilde{\bar{\mathrm{y}}}]$ are particularly instructive, because they contain all the steady states and the existence of diverging trajectories is tied to the fact that, in the extended phase space, new unstable steady states appear; thus, below threshold, the one physical steady state (Eq. 9.68) is complemented by four nonphysical steady states located in the nonclassical phases space, while above threshold there are two nonphysical steady states and three physical steady states (Eqs. 9.68 and 9.74). While the stability of the physical steady states continues to follow the prediction of semiclassical theory (see Sect. 11.2.2 and Exercise 9.3), the nonphysical steady states are all unstable.

The complete set of steady states is obtained by setting the deterministic parts of (12.3a) and (12.3b) to zero. This gives

$$\left.\begin{aligned}\tilde{\tilde{\alpha}}_{\mathrm{ss}}(1+\tilde{\tilde{\alpha}}_{\mathrm{ss}}\tilde{\tilde{\alpha}}_*^{\mathrm{ss}}) &= \lambda\tilde{\tilde{\alpha}}_*^{\mathrm{ss}}\\ \tilde{\tilde{\alpha}}_*^{\mathrm{ss}}(1+\tilde{\tilde{\alpha}}_{\mathrm{ss}}\tilde{\tilde{\alpha}}_*^{\mathrm{ss}}) &= \lambda\tilde{\tilde{\alpha}}_{\mathrm{ss}}\end{aligned}\right\}\Rightarrow \tilde{\tilde{\alpha}}_{\mathrm{ss}} = \pm\tilde{\tilde{\alpha}}_*^{\mathrm{ss}}, \tag{12.63}$$

where the implication follows by taking the ratio of the two equations. Then substituting $\tilde{\tilde{\alpha}}_*^{\mathrm{ss}} = \pm\tilde{\tilde{\alpha}}_{\mathrm{ss}}$ back into the steady-state equations, we have

$$\tilde{\tilde{\alpha}}_{\mathrm{ss}}(1-\lambda+\tilde{\tilde{\alpha}}_{\mathrm{ss}}^2) = 0, \qquad \tilde{\tilde{\alpha}}_*^{\mathrm{ss}} = \tilde{\tilde{\alpha}}_{\mathrm{ss}}, \tag{12.64a}$$

$$\tilde{\tilde{\alpha}}_{\mathrm{ss}}(1+\lambda-\tilde{\tilde{\alpha}}_{\mathrm{ss}}^2) = 0, \qquad \tilde{\tilde{\alpha}}_*^{\mathrm{ss}} = -\tilde{\tilde{\alpha}}_{\mathrm{ss}}, \tag{12.64b}$$

with a full complement of five solutions

$$\left.\begin{aligned}&\tilde{\tilde{\alpha}}_{\mathrm{ss}} = \tilde{\tilde{\alpha}}_*^{\mathrm{ss}} = 0 && \mathrm{(I)}\\ &\tilde{\tilde{\alpha}}_{\mathrm{ss}} = \tilde{\tilde{\alpha}}_*^{\mathrm{ss}} = \pm i\sqrt{1-\lambda} = \pm\sqrt{\lambda-1} && \mathrm{(II,III)}\\ &\tilde{\tilde{\alpha}}_{\mathrm{ss}} = -\tilde{\tilde{\alpha}}_*^{\mathrm{ss}} = \pm\sqrt{1+\lambda} && \mathrm{(IV,V)}\end{aligned}\right\}. \tag{12.65}$$

Steady state (I) is physical below threshold, and (I), (II), and (III) are physical above threshold. Steady states (IV) and (V) are always nonphysical, and (II) and (III) are nonphysical below threshold.

Figures 12.3 and 12.4 show how the deterministic flow is organized around these steady states. The fate of trajectories repelled by the nonphysical steady states is the focus of our interest. Below threshold (Fig. 12.3) the situation is quite complex. Consider the trajectories repelled by steady state (V). Some, like T_5, execute wide loops that terminate on the physical steady state (I), remaining within the real plane throughout. Others, like trajectory T_4, diverge after a finite time; they leave the real plane at infinity, to reenter the imaginary plane and follow a track like trajectory $T_{4'}$, before diverging again at a later time, reentering the real plane this time, to be eventually attracted, just like T_5, to steady state (I). There are also trajectories like T_2 that diverge, leave the real plane, and do not return to it. These trajectories are also attracted to steady state (I), but from within the imaginary plane, like trajectory $T_{2'}$. The trajectories labeled T_1, T_3, and T_6 demark boundaries between these different kinds of behavior.

As λ is increased, nonphysical steady states (II) and (III), located in the imaginary plane, move towards the physical steady state (I) (Fig. 12.3b). They collide with the physical steady state at threshold and emerge above threshold (in the real plane) as new stable physical steady states. The deterministic flow is simplified somewhat in the process (Fig. 12.4). Above threshold, all trajectories that diverge and leave the real plane—like T_2 and T_3 in Fig. 12.4a—eventually return to it, where they terminate on steady states (II) and (III).

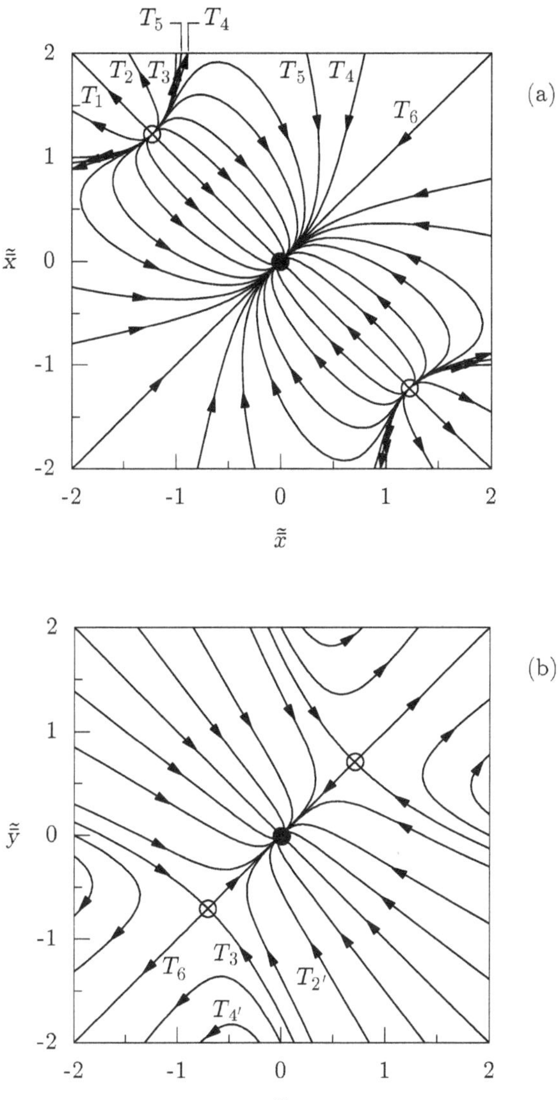

Fig. 12.3. Trajectories generated by the deterministic parts of (12.3a) and (12.3b) plotted in the planes (**a**) $\tilde{y} = \tilde{\tilde{y}} = 0$ and (**b**) $\tilde{x} = \tilde{\tilde{x}} = 0$, for $\lambda = 0.5$. The *filled* and *open circles* indicate physical and nonphysical steady states, respectively

Note 12.3. In order to match trajectories moving from the real to the imaginary plane or vice versa, a consistent convention must be applied to the square roots in (12.59a) and (12.59b). When a trajectory leaves one plane (at infinity) and enters the other, the expression under the square root changes sign, from positive to negative. If a trajectory moves from the real plane to the imaginary plane, we interpret the square root after the move with $\sqrt{-1} = +i$. For a trajectory moving from the imaginary plane to the real plane we take $\sqrt{-1} = -i$. With this convention there is no net change in the sign of the square root for trajectories like T_4 that leave the real plane and return at a later time; the trajectory returns to the quadrant (defined by the diagonals in the plane) from which it left.

Having seen this analysis of the deterministic flow, the "spikes" in Fig. 12.2d are hardly a surprise anymore. The deterministic equations generate numerous diverging trajectories; moreover, the divergences are strong, as they set in over a finite time. Admittedly, the word "numerous" is rather loose, since we have not even determined the dimension of the phase space volume containing all initial conditions that diverge. Nevertheless, the existence of a set of diverging trajectories implies the existence of a neighborhood within which trajectories make large excursions into the nonclassical phase space without actually diverging. Unless the quantum noise cleverly avoids these regions, we can certainly expect "spikes"—and perhaps even "diseases."

All of this raises the possibility that there is something fundamentally wrong with the use we are making of the positive P representation. We will see shortly that this can indeed be the case; for certain examples, simulations within the positive P representation yield the wrong results (Sect. 12.2.1). The degenerate parametric oscillator with adiabatic elimination of the pump is not such an example, however, in so far as the noise terms in (12.3a) and (12.3b) *do* cleverly avoid the dangerous regions of the extended phase space—or at least they should, if only the simulation were performed correctly; the "spikes" in Fig. 12.2d are numerical artifacts. Before we look into the genuine difficulties that arise with other examples, we should spend some time understanding how a correct integration of the stochastic differential equations works out in this case—for arbitrarily large quantum noise, despite the dangers revealed in Figs. 12.3 and 12.4.

12.1.6 Steady-State Solution for the Positive P Distribution

Let us imagine a stochastic trajectory beginning from the phase space origin $\tilde{\alpha}(0) = \tilde{\alpha}_*(0) = 0$. Initially, the noise terms in (12.3a) and (12.3b) are real. They generate a Brownian motion in the vicinity of the origin confined to the real plane $\tilde{y} = \tilde{\bar{y}} = 0$. The deterministic flow is also confined to this plane for excursions not too far away from the origin (Figs. 12.3a and 12.4a). Clearly, then, the stochastic trajectory remains within the real plane, at least during some initial short period of time.

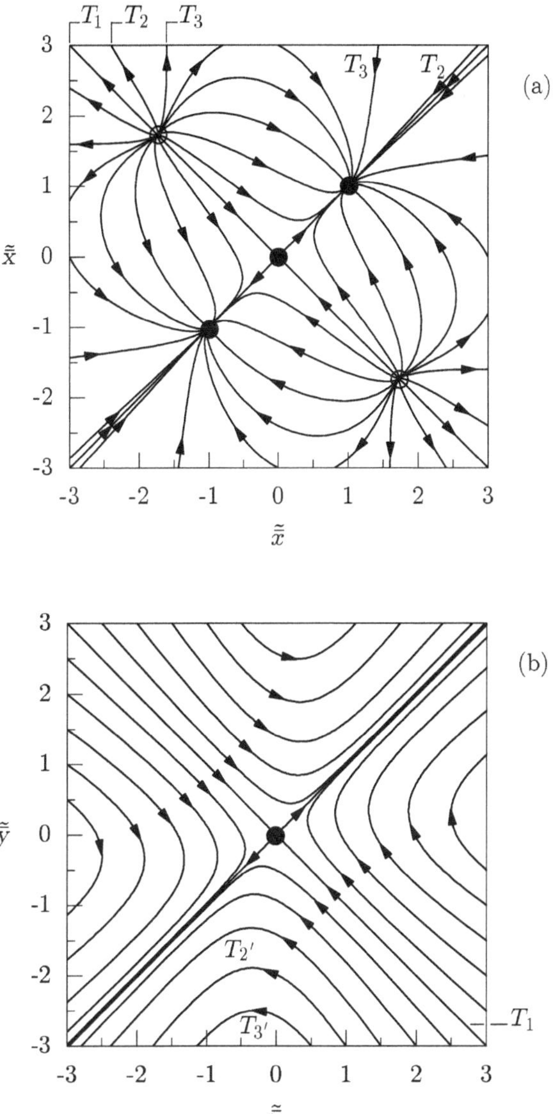

Fig. 12.4. Trajectories generated by the deterministic parts of (12.3a) and (12.3b) plotted in the planes (**a**) $\tilde{\tilde{y}} = \tilde{\bar{y}} = 0$ and (**b**) $\tilde{\tilde{x}} = \tilde{\bar{x}} = 0$, for $\lambda = 2$. The *filled* and *open circles* indicate physical and nonphysical steady states, respectively

There are two paths by which the trajectory might leave: (i) either $\tilde{\tilde{x}}$ or $\tilde{\tilde{\mathrm{x}}}$ becomes larger than $\sqrt{\lambda}$ so that one or other of the noise terms becomes pure imaginary, or (ii) the trajectory diverges, switching between the real and imaginary planes in the manner of T_2 from Figs. 12.3a and 12.4a. In fact, the second path may be eliminated in favor of the first, because all divergent regions of the phase space lie outside the square—$|\tilde{\tilde{x}}| \leq \sqrt{\lambda}$, $|\tilde{\tilde{\mathrm{x}}}| \leq \sqrt{\lambda}$—within which both noise terms are real. But, the first path may then be eliminated as well, since it is not possible to cross the boundary of this square: on it, the quantum noise is directed entirely *along* the boundary, while the deterministic flow is everywhere *inwards*, back towards one of the physical steady states. Thus, a trajectory beginning at the origin is strictly confined to a bounded two-dimensional phase space, the square $|\tilde{\tilde{x}}| \leq \sqrt{\lambda}$, $|\tilde{\tilde{\mathrm{x}}}| \leq \sqrt{\lambda}$, $\tilde{\tilde{y}} = \tilde{\tilde{\mathrm{y}}} = 0$; the same is true for any trajectory beginning within the square. The "spikes" of Fig. 12.2d occur only because the finite time step used for the numerical simulation allows trajectories to "jump" out of the bounded region, as shown by Carmichael and Wolinsky [12.19]. This is verified by adopting a simple strategy to enforce the confinement.

Exercise 12.6. Implement a numerical simulation of (12.3a) and (12.3b) in the real plane $\tilde{\tilde{y}} = \tilde{\tilde{\mathrm{y}}} = 0$ by including the reflecting boundary condition $\tilde{\tilde{x}} \to (\tilde{\tilde{x}}/|\tilde{\tilde{x}}|)[\sqrt{\lambda} - (|\tilde{\tilde{x}}| - \sqrt{\lambda})]$ if $|\tilde{\tilde{x}}| > \sqrt{\lambda}$, $\tilde{\tilde{\mathrm{x}}} \to (\tilde{\tilde{\mathrm{x}}}/|\tilde{\tilde{\mathrm{x}}}|)[\sqrt{\lambda} - (|\tilde{\tilde{\mathrm{x}}}| - \sqrt{\lambda})]$ if $|\tilde{\tilde{\mathrm{x}}}| > \sqrt{\lambda}$; the test for reflection is applied after $\tilde{\tilde{x}}$ and $\tilde{\tilde{\mathrm{x}}}$ are updated at the end of each time step. Use the simulation to reproduce Fig. 12.2d without any "spikes" and show that the numerical integration remains well behaved for even smaller values of $\langle a^\dagger a \rangle_{\mathrm{thr}}$.

Exercise 12.7. Implement a numerical simulation of the stochastic differential equations without adiabatic elimination of the pump (with $\bar{n} = 0$) (Eqs. 12.1a–12.1d). Investigate how the results presented in Figs. 12.1 and 12.2 change for $\xi \sim 0.1$–1.0. Is there a similar confinement of phase space trajectories in the general case? Can very small values of $\langle a^\dagger a \rangle_{\mathrm{thr}}$ be handled without generating "spikes"?

Having recognized that stochastic trajectories are strictly confined to the square in the real plane $|\tilde{\tilde{x}}| \leq \sqrt{\lambda}$, $|\tilde{\tilde{\mathrm{x}}}| \leq \sqrt{\lambda}$, it is convenient to adopt the scaled variables

$$\theta \equiv \tilde{\tilde{x}}/\sqrt{\lambda}, \qquad \vartheta \equiv \tilde{\tilde{\mathrm{x}}}/\sqrt{\lambda}, \tag{12.66}$$

with θ and ϑ confined to the unit square. Figure 12.5 shows the square $|\theta| \leq 1$, $|\vartheta| \leq 1$ in relation to the steady states and deterministic flow. Note that all physical steady states lie within the square and all nonphysical steady states lie outside it. The positive P distribution is

$$\tilde{\tilde{\mathcal{P}}}(\theta, \vartheta, \bar{t}) \equiv \lambda \tilde{\tilde{P}}(\sqrt{\lambda}\theta \sqrt{\lambda}\vartheta, \bar{t}), \tag{12.67}$$

with zero measure outside the unit square. From (12.5), it satisfies the *Fokker–Planck equation for the degenerate parametric oscillator within the positive*

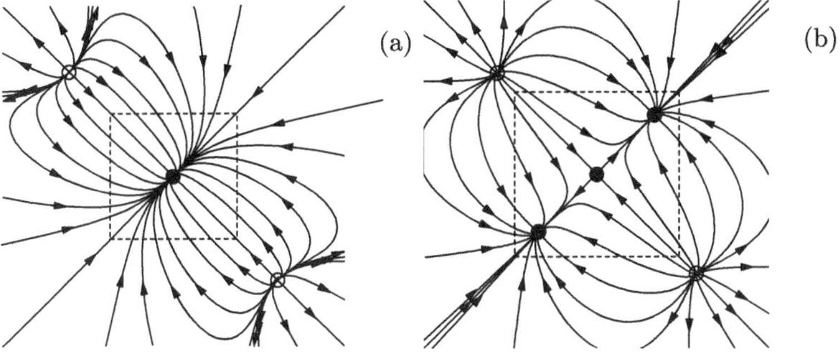

Fig. 12.5. Deterministic flow in the real plane $\tilde{\tilde{y}} = \tilde{\tilde{\bar{y}}} = 0$ in relation to the unit square $|\theta| \leq 1$, $|\vartheta| \leq 1$ (*dashed lines*): (**a**) below threshold ($\lambda = 0.5$), (**b**) above threshold ($\lambda = 2$). All physical steady states (*filled circles*) lie inside the square, while nonphysical steady states (*open circles*) lie outside

P representation with adiabatic elimination of the pump (and $\bar{n} = 0$) and quantum noise confined to the real plane:

$$\begin{aligned}\frac{\partial \tilde{\tilde{\mathcal{P}}}}{\partial \bar{t}} = \bigg\{ & \frac{\partial}{\partial \theta}[\theta - \lambda\vartheta(1-\theta^2)] + \frac{\partial}{\partial \vartheta}[\vartheta - \lambda\theta(1-\vartheta^2)] \\ & + \frac{1}{2}\left(2\xi^{-1}n_p^{\rm thr}\right)^{-1}\left[\frac{\partial^2}{\partial\theta^2}(1-\theta^2) + \frac{\partial^2}{\partial\vartheta^2}(1-\vartheta^2)\right]\bigg\}\tilde{\tilde{\mathcal{P}}}. \end{aligned} \quad (12.68)$$

Conveniently, we now have a Fokker–Planck equation in two, rather than the original four, phase space dimensions.

We aim to solve (12.68) in the steady state. Note first that we may account for the change in the probability density over time by considering it to follow from the flow of a probability current $\boldsymbol{J}_{\tilde{\tilde{\mathcal{P}}}}(\theta, \vartheta, t)$. We rewrite (12.68) as

$$\frac{\partial \tilde{\tilde{\mathcal{P}}}}{\partial \bar{t}} = -\mathrm{div}\boldsymbol{J}_{\tilde{\tilde{\mathcal{P}}}}, \quad (12.69)$$

with

$$\boldsymbol{J}_{\tilde{\tilde{\mathcal{P}}}} \equiv \begin{pmatrix} -\left[\theta - \lambda\vartheta(1-\theta^2) + \frac{1}{2}\left(2\xi^{-1}n_p^{\rm thr}\right)^{-1}\dfrac{\partial}{\partial\theta}(1-\theta^2)\right]\tilde{\tilde{\mathcal{P}}} \\ -\left[\vartheta - \lambda\theta(1-\vartheta^2) + \frac{1}{2}\left(2\xi^{-1}n_p^{\rm thr}\right)^{-1}\dfrac{\partial}{\partial\vartheta}(1-\vartheta^2)\right]\tilde{\tilde{\mathcal{P}}} \end{pmatrix}. \quad (12.70)$$

Then one way for a steady state to be obtained is for the probability current to vanish everywhere within the unit square. We know, in fact, that on the boundary, the perpendicular component of the current does vanish. In addition, a symmetry argument allows us to discount the possibility of a steady nonzero current circulating around the square; for this to happen the symmetry between clockwise and counterclockwise circulation must be broken. Thus, there is every reason to expect a steady state solution satisfying $\boldsymbol{J}_{\tilde{\tilde{\mathcal{P}}}}(\theta, \vartheta) = 0$.

For a general Fokker–Planck equation (Eq. 5.1), the necessary and sufficient conditions for such a solution to exist are given by a relationship connecting the components of the drift vector, $A_i(\boldsymbol{x})$, and diffusion matrix, $D_{ij}(\boldsymbol{x})$. These so-called *potential conditions* can be found, for example, in [12.20]. In the present case, the existence of a solution satisfying $\boldsymbol{J}_{\tilde{\tilde{\mathcal{P}}}}(\theta,\vartheta)=0$ requires

$$\frac{\partial}{\partial\theta}\ln\tilde{\tilde{\mathcal{P}}} = -\left(2\xi^{-1}n_p^{\mathrm{thr}}-1\right)\frac{2\theta}{1-\theta^2}+\lambda\left(2\xi^{-1}n_p^{\mathrm{thr}}\right)2\vartheta, \tag{12.71a}$$

$$\frac{\partial}{\partial\vartheta}\ln\tilde{\tilde{\mathcal{P}}} = -\left(2\xi^{-1}n_p^{\mathrm{thr}}-1\right)\frac{2\vartheta}{1-\vartheta^2}+\lambda\left(2\xi^{-1}n_p^{\mathrm{thr}}\right)2\theta. \tag{12.71b}$$

If the two equations are to be consistent, the partial derivative of the first with respect to ϑ must equal the partial derivative of the second with respect to θ. This rather simple potential condition is satisfied. Thus, the equations may be integrated to obtain

$$\ln\tilde{\tilde{\mathcal{P}}} =\left(2\xi^{-1}n_p^{\mathrm{thr}}-1\right)\ln(1-\theta^2)+\lambda\left(2\xi^{-1}n_p^{\mathrm{thr}}\right)2\vartheta\theta+F(\vartheta), \tag{12.72a}$$

$$\ln\tilde{\tilde{\mathcal{P}}} =\left(2\xi^{-1}n_p^{\mathrm{thr}}-1\right)\ln(1-\vartheta^2)+\lambda\left(2\xi^{-1}n_p^{\mathrm{thr}}\right)2\theta\vartheta+G(\theta), \tag{12.72b}$$

where the functions $F(\vartheta)$ and $G(\theta)$ are chosen to make the two results consistent with one another. We arrive at an analytical expression for the *steady-state positive P distribution for the degenerate parametric oscillator with adiabatic elimination of the pump (and $\bar{n}=0$)*, in the form

$$\tilde{\tilde{\mathcal{P}}}(\theta,\vartheta)=\mathcal{N}[(1-\theta^2)(1-\vartheta^2)]^{(2\xi^{-1}n_p^{\mathrm{thr}}-1)}\exp\left[\lambda\left(2\xi^{-1}n_p^{\mathrm{thr}}\right)2\theta\vartheta\right]; \tag{12.73}$$

$\mathcal{N}$ is a normalization constant such that $\int_{-1}^{1}d\theta\int_{-1}^{1}d\vartheta\tilde{\tilde{\mathcal{P}}}(\theta,\vartheta)=1$.

12.1.7 Quantum Fluctuations and System Size

What does the exact solution for the positive P distribution have to say about quantum fluctuations and the way they change with system size? The first step towards an answer is to recover the results of the small-noise analysis derived in Sects. 10.2.2–10.2.4. This is most easily accomplished using the expression for the logarithm of $\tilde{\tilde{\mathcal{P}}}(\theta,\vartheta)$,

$$\ln\tilde{\tilde{\mathcal{P}}}=(2\xi^{-1}n_p^{\mathrm{thr}}-1)[\ln(1-\theta^2)+\ln(1-\vartheta^2)]+(2\xi^{-1}n_p^{\mathrm{thr}})\lambda2\theta\vartheta+\text{constant}. \tag{12.74}$$

We expand the logarithm in different ways, depending on the region of operation—below, at, or above threshold. The justification for the expansion is that $2\xi^{-1}n_p^{\mathrm{thr}}$ is very large in the small-noise limit. It is therefore necessary to represent the logarithm accurately only near its maximum or maxima above threshold; away from a maximum, once the value of the logarithm falls by an order of magnitude, the distribution falls by the factor e^{-10}—it might just as well be set to zero. Maxima and minima of the logarithm are located—to lowest order in $\left(2\xi^{-1}n_p^{\rm thr}\right)^{-1}$—by the steady states (I)–(V) of (12.65), with maxima and minima corresponding, respectively, to stable and unstable steady states. It is convenient to use classical and nonclassical phase space variables as defined in (11.107a) and (11.107b). We make the change of variables

$$\Theta_{\rm C} \equiv \tfrac{1}{2}(\theta + \vartheta), \qquad \Theta_{\rm NC} \equiv \tfrac{1}{2}(\theta - \vartheta). \tag{12.75}$$

Thus, our goal is to derive expressions below, at, and above threshold, for the steady-state positive P distribution

$$\bar{\tilde{\Pi}}(\Theta_C, \Theta_{NC}) \equiv 2\bar{\tilde{\mathcal{P}}}[\theta(\Theta_C, \Theta_{NC}), \vartheta(\Theta_C, \Theta_{NC})], \tag{12.76}$$

with $\bar{\tilde{\mathcal{P}}}(\theta, \vartheta)$ given by (12.73), under the small-noise assumption $2\xi^{-1}n_p^{\rm thr} = [\Gamma(\frac{1}{4})/\Gamma(\frac{3}{4})]^2 \langle a^\dagger a\rangle_{\rm thr}^2 \gg 1$.

Below threshold ($\lambda < 1$): There is a single stable steady state for $\lambda < 1$, the state (I) of (12.65),

$$\theta_{\rm ss} = \vartheta_{\rm ss} = 0. \tag{12.77}$$

When expanding $\ln \bar{\tilde{\mathcal{P}}}$ about this steady state we expand $\ln(1-\theta^2)$ and $\ln(1-\vartheta^2)$ to lowest order. This yields

$$\ln \bar{\tilde{\mathcal{P}}} = 2\xi^{-1}n_p^{\rm thr}(-\theta^2 - \vartheta^2 + 2\lambda\theta\vartheta) + \text{constant}, \tag{12.78a}$$

or in terms of classical and nonclassical variables,

$$\begin{aligned} \ln \bar{\tilde{\Pi}} &= 4\xi^{-1}n_p^{\rm thr}[-(\Theta_{\rm C}^2 + \Theta_{\rm NC}^2) + \lambda(\Theta_{\rm C}^2 - \Theta_{\rm NC}^2)] + \text{constant} \\ &= 4\xi^{-1}n_p^{\rm thr}[-(1-\lambda)\Theta_{\rm C}^2 - (1+\lambda)\Theta_{\rm NC}^2] + \text{constant}. \end{aligned} \tag{12.78b}$$

Hence, below threshold, the normalized steady-state positive P distribution in the small-noise limit is

$$\bar{\tilde{\Pi}}(\Theta_{\rm C}, \Theta_{\rm NC}) = \frac{1}{2\pi} \frac{\sqrt{1-\lambda^2}}{\frac{1}{4}(2\xi^{-1}n_p^{\rm thr})^{-1}} \exp\left[-\frac{1}{2}\frac{(1-\lambda)\Theta_{\rm C}^2 + (1+\lambda)\Theta_{\rm NC}^2}{\frac{1}{4}(2\xi^{-1}n_p^{\rm thr})^{-1}}\right]. \tag{12.79}$$

Aside from a different scaling of variables, the result is essentially the product of the steady-state solutions to (10.51a) and (10.51b) (with $\bar{n} = 0$ and $\sigma = 1$). There is, however, one important difference: $\Theta_{\rm NC}$ is associated with the variable $i\tilde{z}_2$ in (10.51b), rather than with $\tilde{z}_2$ itself. By releasing α_* from the constraint $\alpha_* = \alpha^*$ in the definition $\Theta_{\rm NC} = \frac{1}{2}(\tilde{\bar{\alpha}} - \tilde{\bar{\alpha}}_*)/\sqrt{\lambda}$ (Eqs. 11.27a, 12.66, and 12.75), the positive P representation works with the scaled variable $\left(n_p^{\rm thr}\right)^{1/2}\Theta_{\rm NC} \leftrightarrow (i\tilde{z}_2)\sqrt{\lambda}$ as a real quantity. Distribution (12.79) is therefore normalized with respect to $\Theta_{\rm NC}$, while (10.51b) has no normalizable steady-state solution in the variable $\tilde{z}_2$.

At threshold ($\lambda = 1$): At threshold, the term in $\Theta_{\rm C}^2$ vanishes in (12.78b). While $\ln\tilde{\tilde{\mathcal{P}}}$ still has its maximum at $\theta = \vartheta = 0$, it is now necessary to expand $\ln(1+\theta^2)$ and $\ln(1+\vartheta^2)$ to higher order to obtain a normalizable distribution. At the next order, we have

$$\ln\tilde{\tilde{\mathcal{P}}} = 2\xi^{-1}n_p^{\rm thr}\big(-\theta^2 - \vartheta^2 + 2\theta\vartheta - \tfrac{1}{2}\theta^4 - \tfrac{1}{2}\vartheta^4\big) + \text{constant}, \tag{12.80a}$$

and

$$\begin{aligned}\ln\tilde{\tilde{\Pi}} &= 2\xi^{-1}n_p^{\rm thr}\big[-4\Theta_{\rm NC}^2 - \tfrac{1}{2}(\Theta_{\rm C}+\Theta_{\rm NC})^4 - \tfrac{1}{2}(\Theta_{\rm C}-\Theta_{\rm NC})^4\big] + \text{constant}\\ &= 2\xi^{-1}n_p^{\rm thr}(-4\Theta_{\rm NC}^2 - \Theta_{\rm C}^4 - \Theta_{\rm NC}^4 - 6\Theta_{\rm C}^2\Theta_{\rm NC}^2) + \text{constant}.\end{aligned} \tag{12.80b}$$

Three new terms appear in place of the one that has vanished: $-(2\xi^{-1}n_p^{\rm thr})\Theta_{\rm C}^4$, $-(2\xi^{-1}n_p^{\rm thr})\Theta_{\rm NC}^4$, and $-(2\xi^{-1}n_p^{\rm thr})6\Theta_{\rm C}^2\Theta_{\rm NC}^2$. Of the three, only those that make a significant contribution near the origin need be kept. Clearly, we have $\Theta_{\rm C} \sim \big(2\xi^{-1}n_p^{\rm thr}\big)^{-1/4}$ and $\Theta_{\rm NC} \sim \big(2\xi^{-1}n_p^{\rm thr}\big)^{-1/2}$. Thus, the orders are $\Theta_{\rm C}^4 \sim \big(2\xi^{-1}n_p^{\rm thr}\big)^{-1}$, $\Theta_{\rm NC}^4 \sim \big(2\xi^{-1}n_p^{\rm thr}\big)^{-2}$, and $\Theta_{\rm C}^2\Theta_{\rm NC}^2 \sim \big(2\xi^{-1}n_p^{\rm thr}\big)^{-3/2}$; the three new terms are of three different orders in system size. It is therefore acceptable to drop the smaller two. Retaining only the term $\Theta_{\rm C}^4$, at threshold, the normalized steady-state positive P distribution in the small-noise limit is

$$\tilde{\tilde{\Pi}}(\Theta_{\rm C},\Theta_{\rm NC}) = \frac{4}{\sqrt{\pi}\Gamma(\frac{1}{4})}\frac{1}{\big(2\xi^{-1}n_p^{\rm thr}\big)^{-3/4}}\exp\left[-\frac{\Theta_{\rm C}^4 + 4\Theta_{\rm NC}^2}{\big(2\xi^{-1}n_p^{\rm thr}\big)^{-1}}\right]. \tag{12.81}$$

The result reproduces the product of (10.97a) and (10.97b) (for $\bar{n} = 0$ and $\sigma = 1$). Similar comments regarding the relationship between the variable $\Theta_{\rm NC}$ in (12.81) and $\tilde{z}_2$ in (10.97b) apply.

Note 12.4. By including the term in $\Theta_{\rm C}^4$ we constrain the fluctuations in the X-quadrature phase amplitude to a finite value, removing the divergence (at $\omega = 0$) from the spectrum in the linearized treatment of fluctuations (Eq. 10.62a). If there is to be no violation of the Heisenberg uncertainty relations, the spectrum of fluctuations in the Y-quadrature phase amplitude cannot vanish at $\omega = 0$, as it does in (10.62b); strictly, the squeezing at zero frequency cannot be perfect for $\lambda = 1$. Apparently, (12.81) does not respect this requirement, since it makes no changes to the distribution of fluctuations in the Y-quadrature phase amplitude (variable $\Theta_{\rm NC}$) compared to what one obtains by setting $\lambda = 1$ in (12.79). As it turns out, it is necessary to keep the term $6\Theta_{\rm C}^2\Theta_{\rm NC}^2$ in (12.80b) in order to take the correction to the degree of squeezing into account (see Sect. 12.1.9).

Above threshold ($\lambda > 1$): There are two stable steady states for $\lambda > 1$, the states (II) and (III) of (12.65),

$$\theta_{\rm ss} = \vartheta_{\rm ss} = \Theta_0 \equiv \pm\sqrt{1 - 1/\lambda}. \tag{12.82}$$

We make the expansion of $\ln\tilde{\tilde{\mathcal{P}}}$ about either steady state as

$$\begin{aligned}\ln\tilde{\tilde{\mathcal{P}}} &= 2\xi^{-1}n_p^{\text{thr}}\{\ln[1-\Theta_0^2-(\theta-\Theta_0)^2+2\Theta_0(\theta-\Theta_0)]\\ &\quad+\ln[1-\Theta_0^2-(\vartheta-\Theta_0)^2+2\Theta_0(\vartheta-\Theta_0)]\\ &\quad+2\lambda[\Theta_0^2+(\theta-\Theta_0)(\vartheta-\Theta_0)-\Theta_0(\vartheta+\theta-2\Theta_0)]\}+\text{constant}\\ &= 2\xi^{-1}n_p^{\text{thr}}\{\ln[1-\lambda(\theta-\Theta_0)^2+2\lambda\Theta_0(\theta-\Theta_0)]\\ &\quad+\ln[1-\lambda(\vartheta-\Theta_0)^2+2\lambda\Theta_0(\vartheta-\Theta_0)]\\ &\quad+2\lambda[(\theta-\Theta_0)(\vartheta-\Theta_0)-\Theta_0(\theta+\vartheta-2\Theta_0)]\}+\text{constant},\end{aligned}\tag{12.83}$$

where in the second line we have substituted $1-\Theta_0^2=1/\lambda$ and moved $4\xi^{-1}n_p^{\text{thr}}(-\ln\lambda+\lambda\Theta_0^2)$ into the constant. Now, on expanding the logarithms, the linear terms cancel and to lowest order

$$\begin{aligned}\ln\tilde{\tilde{\mathcal{P}}} &= 2\xi^{-1}n_p^{\text{thr}}\lambda\{-(2\lambda-1)[(\theta-\Theta_0)^2+(\vartheta-\Theta_0)^2]\\ &\quad+2(\theta-\Theta_0)(\vartheta-\Theta_0)\}+\text{constant},\end{aligned}\tag{12.84a}$$

or in terms of classical and nonclassical variables,

$$\begin{aligned}\ln\tilde{\tilde{\Pi}} &= 4\xi^{-1}n_p^{\text{thr}}\lambda\{-(2\lambda-1)[(\Theta_{\text{C}}-\Theta_0)^2+\Theta_{\text{NC}}^2]\\ &\quad+(\Theta_{\text{C}}-\Theta_0)^2-\Theta_{\text{NC}}^2\}+\text{constant}\\ &= 8\xi^{-1}n_p^{\text{thr}}\lambda[-(\lambda-1)(\Theta_{\text{C}}-\Theta_0)^2-\lambda\Theta_{\text{NC}}^2]+\text{constant}.\end{aligned}\tag{12.84b}$$

There are two symmetrically placed steady states; thus, the positive P distribution has two peaks of equal weight. Normalizing the local distribution, from (12.84b), we write

$$\begin{aligned}&\tilde{\tilde{\Pi}}_{\text{local}}(\Theta_{\text{C}},\Theta_{\text{NC}})\\ &= \frac{1}{2\pi}\frac{\lambda\sqrt{\lambda(\lambda-1)}}{\frac{1}{8}\left(2\xi^{-1}n_p^{\text{thr}}\right)^{-1}}\exp\left[-\frac{1}{2}\frac{\lambda(\lambda-1)(\Theta_{\text{C}}-\Theta_0)^2+\lambda^2\Theta_{\text{NC}}^2}{\frac{1}{8}\left(2\xi^{-1}n_p^{\text{thr}}\right)^{-1}}\right].\end{aligned}\tag{12.85}$$

After taking the different scalings of the variables into account, the variances of Θ_{C} and Θ_{NC} reproduce the previously calculated quadrature phase amplitude variances (10.75a) and (10.75b).

Having recovered our previous results for the limit of large system size, we are now set to see how the steady-state distribution (12.73) changes as the system size is decreased. It is possible to take the mathematical analysis further in this direction. First, however, it is helpful to see the changes through a series of numerical simulations and plots. Figures 12.6–12.8 plot the distribution $\tilde{\tilde{\mathcal{P}}}$ for $\lambda=2$ (above threshold) and five different values of the system size parameter. Each plot is complemented with a typical realization of the corresponding stochastic process, to give some impression of how the fluctuations develop in time.

To provide an initial point of reference, Fig. 12.6 illustrates the situation for a large system size—a correspondingly small quantum noise. The scale of Fig. 12.6a is the entire unit square. At this scale the quantum fluctuations do not show up at all. The trajectory begins with the decay of the unstable steady state, $\theta = \vartheta = 0$, and apparently moves in a deterministic way to the stable steady state $\theta = \vartheta = \sqrt{1 - 1/\lambda}$; the trajectory appears to be confined within the classical phase space all the while—i.e., $\Theta_{\rm NC} = 0$. Of course, there are actually small fluctuations, which, in particular, initiate the decay of the unstable state. The inset—also Fig. 12.6b—illustrates how the fluctuations are distributed once the final steady state is reached. Their size is consistent with the amplitudes

$$\frac{\left(2\xi^{-1}n_p^{\rm thr}\right)^{-1/2}}{2\sqrt{2\lambda(\lambda-1)}} = \frac{[\Gamma(\frac{3}{4})/\Gamma(\frac{1}{4})]\langle a^\dagger a\rangle_{\rm thr}^{-1}}{2\sqrt{2\lambda(\lambda-1)}} \sim 8 \times 10^{-6},$$

for $\Theta_{\rm C}$, and

$$\frac{\left(2\xi^{-1}n_p^{\rm thr}\right)^{-1/2}}{2\sqrt{2}\lambda} = \frac{[\Gamma(\frac{3}{4})/\Gamma(\frac{1}{4})]\langle a^\dagger a\rangle_{\rm thr}^{-1}}{2\sqrt{2}\lambda} \sim 6 \times 10^{-6},$$

for $\Theta_{\rm NC}$, obtained from (12.85).

The basic idea of the system size expansion is that the fluctuations increase with decreasing system size. This, certainly, is what is observed. Yet while the amplitude of the fluctuations increases, the organization of the dynamics continues to reflect the influence of the steady states and, more generally, the deterministic flow within the classical phase space. Surprisingly, even for $\langle a^\dagger a\rangle_{\rm thr} \sim 1$, it is reasonable to take the classical nonlinear dynamics as the starting point for understanding the physics (Fig. 12.7a). If $\langle a^\dagger a\rangle_{\rm thr}$ becomes still smaller, though, we begin to notice the changes illustrated in Fig. 12.7b. Much greater use is made of the nonclassical dimension, to such an extent that the attraction for the classical phase space begins to disappear. Also, while the boundary of the unit square is completely irrelevant in Fig. 12.6—at no stage does $\tilde{\tilde{\mathcal{P}}}(\theta, \vartheta, t)$ have significant weight near the boundary—Fig. 12.7b shows a pronounced effect from the constraint imposed by the boundary. Indeed, eventually the constraint completely controls the behavior, with no reference being made at all to the classical nonlinear dynamics. Thus, in Fig. 12.8, the probability density is attracted to the boundary, away from the classical phase space, where it concentrates at the corners of the unit square. In fact, $\tilde{\tilde{\mathcal{P}}}(\theta, \vartheta)$ diverges on the boundary for $2\xi^{-1}n_p^{\rm thr} < 1 \Rightarrow \langle a^\dagger a\rangle_{\rm thr} < \Gamma(\frac{3}{4})/\Gamma(\frac{1}{4}) \approx 0.33$. This explains why the "spiking" in Fig. 12.2d arises for a system size that is just a little smaller than this value.

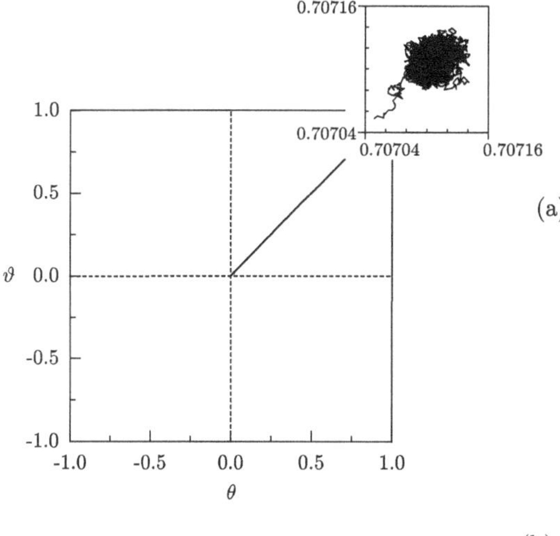

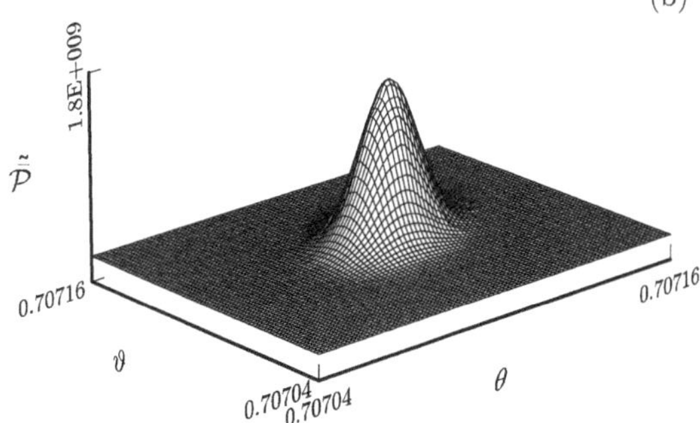

Fig. 12.6. (**a**) One realization of the decay, above threshold, of the unstable steady state $\theta = \vartheta = 0$, for the limit of large system size. (**b**) The positive P distribution (Eq. 12.73) in the vicinity of the stable steady state $\theta = \vartheta = \sqrt{1 - 1/\lambda}$. The trajectory in (**a**) begins from the initial condition $\theta = \vartheta = 0$ and was generated from (12.3a) and (12.3b) using the Euler algorithm with $\Delta t = 10^{-4}$ and $t_{\max} = 100$. The parameters are $\lambda = 2$ and $\langle a^\dagger a\rangle_{\text{thr}} = 10^4$

The message conveyed by Figs. 12.6–12.8 is, first, that when the system size becomes small enough, quantum fluctuations are no longer organized around classical nonlinear dynamics. This might also be true of an extremely noisy classical system, though, if by "classical nonlinear dynamics" we mean the deterministic behavior displayed in the absence of noise. Thus, a second observation is even more important: there is *qualitative* change in the nature of the fluctuations, from the "fuzz ball" notion of diffusion in phase space to something decidedly more quantum mechanical, requiring new tricks for its mathematical description and a new conceptual framework for its understand-

(a)

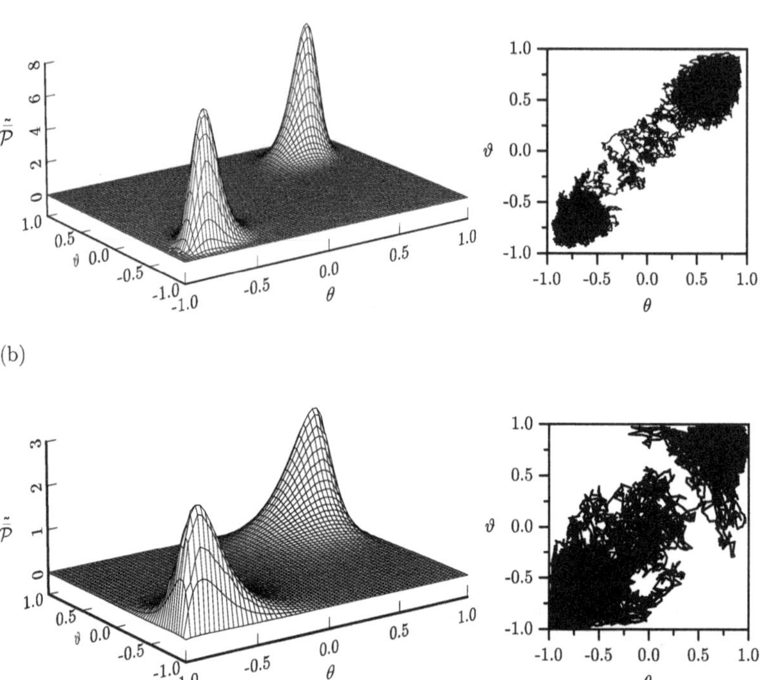

Fig. 12.7. Development of the steady-state positive P distribution with decreasing system size: for $\lambda = 2$ and (**a**) $\langle a^\dagger a \rangle_{\rm thr} = 1.0$, (**b**) $\langle a^\dagger a \rangle_{\rm thr} = 0.5$. Each plot of the distribution (Eq. 12.73) is accompanied by a single realization of the corresponding stochastic process; trajectories begin from the initial condition $\theta = \vartheta = 0$ and were generated from (12.3a) and (12.3b) using the Euler algorithm with (**a**) $\Delta t = 5 \times 10^{-5}$, $t_{\rm max} = 400$ and (**b**) $\Delta t = 5 \times 10^{-5}$, $t_{\rm max} = 100$

ing. To underline the point, we have only a little, if any, trouble interpreting the distribution of Fig. 12.6b; the only unfamiliar feature is the small variance of the nonclassical variable (stochastic trajectories make small excursions into the nonclassical dimension). A satisfactory strategy for accommodating oneself to this is to replace the positive P distribution by a Wigner distribution (see Sect. 10.1.2); excursions into the nonclassical dimension are then replaced by a narrowing of the background distribution of vacuum fluctuations, all within the usual phase space. What, however, are we to make of Fig. 12.8b? Once we understand these fluctuations, it will be clear that in the limit of small system size, the Wigner function too must exhibit quantum features of an entirely new sort—something beyond the paradigm of classical nonlinear dynamics plus "fuzz."

(a)

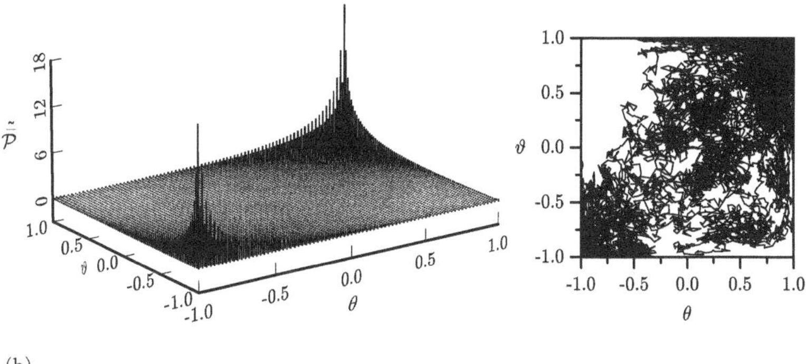

(b)

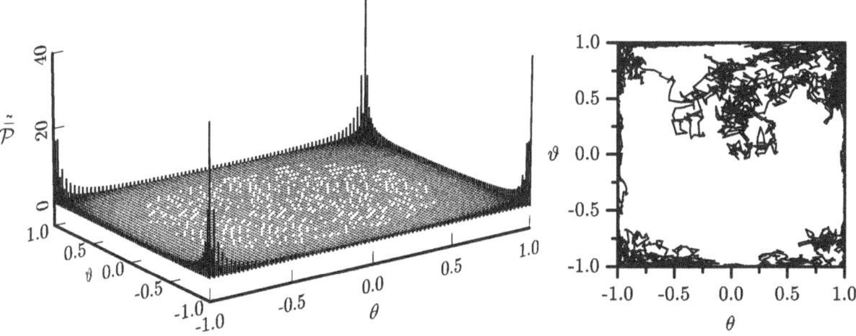

Fig. 12.8. As in Fig. 12.7: for $\lambda = 2$ and (**a**) $\langle a^\dagger a \rangle_{\rm thr} = 0.25$, $\Delta t = 10^{-6}$, $t_{\max} = 20$; (**b**) $\langle a^\dagger a \rangle_{\rm thr} = 0.025$, $\Delta t = 10^{-8}$, $t_{\max} = 0.4$

Note 12.5. When we speak of an increase in the size of the fluctuations, the increase must be measured in appropriately scaled variables. This can be confusing, since from another point of view the size of the quantum fluctuations always remains the same—it is set by the amplitude of a one-quantum field. In Figs. 12.6–12.8, the scale set by the unit square is the scale of the nonlinear physics. What the figures illustrate is the transition from a situation in which many quanta are needed to reach the intracavity energy density that turns on the nonlinearity to a situation in which a single quantum produces the same energy density. In an alternative view, using unscaled variables, the size of the quantum noise remains the same, while the range in phase space characterizing the nonlinearity shrinks down to meet the scale of the quantum noise.

It is largely fortuitous that the positive P representation has given such a graphic illustration of the transition from large to a small system size. As we will see shortly, the representation offers no guarantee of such a pleasing outcome, and it is merely a coincidental matching of the physical example to the mathematical technique that allows things to work out so well in this case. Even with the success, the physics behind the stochastic process on the unit square remains unclear. It is clear enough for a large system size where the classical nonlinear dynamics serves as a guide; but when we consider trajectories like those in Fig. 12.8, there is little to indicate *what* physical process has replaced the classical nonlinear dynamics plus "fuzz."

One clue can be found with little extra effort. To add to the approximate expressions (12.79), (12.81), and (12.85) for the limit of large system size, we can write down an approximate expression for the steady-state positive P distribution in the opposite limit. Let us take $2\xi^{-1}n_p^{\rm thr} \to 0$, $\lambda \to \infty$, with $2\xi^{-1}n_p^{\rm thr}\lambda$ constant, and assume that the singular expression given by (12.73) may be written in terms of δ functions as

$$\tilde{\bar{\mathcal{P}}}(\theta\vartheta) = \frac{1}{2}\left[\frac{\delta(\theta-1)\delta(\vartheta-1)+\delta(\theta+1)\delta(\vartheta+1)}{1+e^{-4|A|^2}} + \frac{\delta(\theta-1)\delta(\vartheta+1)+\delta(\theta+1)\delta(\vartheta-1)}{1+e^{4|A|^2}}\right]. \tag{12.86}$$

Then, to determine the corresponding density operator, we turn to (11.11). From this equation, it appears that the first two products of δ functions in (12.86) represent diagonal terms in a coherent state expansion of the density operator—the others, interferences between these coherent states. Thus, after inverting the various transformations of variables (Eqs. 10.33, 11.27a, 12.66, and 12.75), and with the help of (3.8), the corresponding density operator is

$$\tilde{\rho} = \frac{1}{2}\left[\frac{|A\rangle\langle A| + |-A\rangle\langle -A|}{1+e^{-4|A|^2}} + e^{2|A|^2}\frac{|A\rangle\langle -A| + |-A\rangle\langle A|}{1+e^{4|A|^2}}\right], \tag{12.87}$$

with

$$A \equiv e^{i\frac{1}{2}(\psi-\pi/2)}\sqrt{2\xi^{-1}n_p^{\rm thr}\lambda} = e^{i\frac{1}{2}(\psi-\pi/2)}\sqrt{\lambda}\sqrt{\frac{2\kappa\kappa_p}{g^2}} = \sqrt{\frac{-ig\bar{\mathcal{E}}_i/2\kappa_p}{g^2/4\kappa_p}}, \tag{12.88}$$

where, to help express the coherent state amplitude A in terms of fundamental parameters we have used (9.62), (9.64), (9.80), and (9.81). It may not be obvious that the δ functions are legitimate, but a calculation of normal-ordered averages shows that (12.86) and the density operator deduced from it are, indeed, correct.

Exercise 12.8. Use (11.10) and the expression (12.73) for the steady-state positive P distribution to derive the following result for the general normal-ordered average:

$$\langle a^{\dagger p}a^q\rangle_{\rm ss} = 0, \tag{12.89a}$$

$p+q$ odd, and for $p+q$ even

$$\begin{aligned}
&\langle a^{\dagger p} a^q \rangle_{ss} \\
&= (e^{i\omega_C t}A^*)^p (e^{-i\omega_C t}A)^q \left\{ \sum_{k\geq 0 \text{ even}} \frac{(2|A|^2)^k}{k!} \left[B\left(\frac{k+1}{2}, 2\xi^{-1} n_p^{\text{thr}}\right)\right]^2 \right\}^{-1} \\
&\times \sum_{\substack{k\geq 0 \text{ even} \\ \text{or} \\ k>0 \text{ odd}}} \frac{(2|A|^2)^k}{k!} B\left(\frac{p+k+1}{2}, 2\xi^{-1} n_p^{\text{thr}}\right) B\left(\frac{q+k+1}{2}, 2\xi^{-1} n_p^{\text{thr}}\right),
\end{aligned} \tag{12.89b}$$

where $B(\mu,\nu)$ denotes the Beta function and the second summation is taken over even k for p and q even, and odd k for p and q odd. Verify that in the limit $2\xi^{-1}n_p^{\text{thr}} \to 0$, $\lambda \to \infty$, with $2\xi^{-1}n_p^{\text{thr}}\lambda$ constant, these averages agree with those obtained from (12.86). Write a computer program to evaluate (12.89b), and use it to study the dependence of the average photon number and the fluctuation of quadrature phase amplitudes on system size. (Note that (12.89b) is a special case of Eq. 4.6 in [12.21].)

A physical process that can explain the quantum fluctuations in the limit of small system size is suggested by introducing the even and odd superpositions of coherent states

$$|A_{\text{even}}\rangle \equiv \frac{1}{\sqrt{1+e^{-2|A|^2}}} \frac{1}{\sqrt{2}} (|A\rangle + |-A\rangle), \tag{12.90a}$$

$$|A_{\text{odd}}\rangle \equiv \frac{1}{\sqrt{1-e^{-2|A|^2}}} \frac{1}{\sqrt{2}} (|A\rangle - |-A\rangle), \tag{12.90b}$$

and writing density operator (12.87) as

$$\tilde{\rho} = \frac{1}{2}\left\{\left[1+\frac{1}{\cosh(2|A|^2)}\right]|A_{\text{even}}\rangle\langle A_{\text{even}}| + \left[1-\frac{1}{\cosh(2|A|^2)}\right]|A_{\text{odd}}\rangle\langle A_{\text{odd}}|\right\}. \tag{12.91}$$

This representation in terms of coherent state superpositions is diagonal, and a plausible suggestion for the dynamics is a two-state rate process which produces this density operator in the steady state from a balancing of transitions between $|A_{\text{even}}\rangle$ and $|A_{\text{odd}}\rangle$. We should work this out further; but whatever the underlying dynamic, the distinctly quantum mechanical character of the behavior illustrated in Fig. 12.8b is already apparent from (12.91)—with $2|A|^2 = 2[\Gamma(\frac{1}{4})/\Gamma(\frac{3}{4})]^2 \langle a^\dagger a\rangle_{\text{thr}}^2 \lambda \approx 3.6\times 10^{-3}$, to a first approximation, the steady state of the subharmonic mode is the superposition state $|A_{\text{even}}\rangle$.

12.1.8 Quantum Dynamics Beyond Classical Trajectories plus "Fuzz"

The superposition state corresponding to Fig. 12.8b is not particularly dramatic. It is not a macroscopic superposition of the sort that arises in relation to Schrödinger's famous cat. The mean photon number is much less than unity, and, according to (12.91), cannot be increased without changing the superposition into a mixture. This is effectively what happens in moving from Fig. 12.8b to Fig. 12.8a, with a corresponding increase in $2|A|^2$ by two orders of magnitude. In fact, the status of the proposed superposition state is more marginal still. Consider the mean photon numbers in the coherent state components: for $|A_{\rm even}\rangle$ we have

$$\begin{aligned}\bar{n}_{\rm even} \equiv \langle A_{\rm even}|a^\dagger a|A_{\rm even}\rangle &= \frac{1}{1+e^{-2|A|^2}}\frac{1}{2}(\langle A| + \langle -A|)a^\dagger a(|A\rangle + |-A\rangle)\\ &= |A|^2\frac{1-e^{-2|A|^2}}{1+e^{-2|A|^2}}\\ &= |A|^2\tanh|A|^2, \end{aligned}\tag{12.92a}$$

and similarly, for $|A_{\rm odd}\rangle$,

$$\bar{n}_{\rm odd} \equiv \langle A_{\rm odd}|a^\dagger a|A_{\rm odd}\rangle = |A|^2\coth|A|^2. \tag{12.92b}$$

Now with $|A|^2 \ll 1$, from (12.92a), we obtain $\bar{n}_{\rm even} \sim |A|^4$, a very small number indeed; moreover, $\bar{n}_{\rm odd} \sim |A|^2$. It follows that even in this limit where Fig. 12.8b appears to represent a coherent state superposition, both components of the mixture (12.91) contribute to the photon number expectation at the same order in $|A|^2$. Thus, the positive P distribution of Fig. 12.8b does not actually represent a coherent state superposition at all. Nevertheless, the suggestion made about the dynamics does turn out to be correct, and it is here, in the dynamics, that the quantum nature of the limit of small system size shows itself most clearly.

In this section we intend to develop the suggestion further—i.e., the idea of a two-state rate process for transitions between $|A_{\rm even}\rangle$ and $|A_{\rm odd}\rangle$. To begin, let us add a little more substance to the suggestion by writing

$$\begin{aligned}1+\frac{1}{\cosh(2|A|^2)} &= \frac{2\cosh^2|A|^2}{\cosh(2|A|^2)}\\ &= \tanh(2|A|^2)\coth|A|^2\\ &= 2\frac{\coth|A|^2}{\tanh|A|^2+\coth|A|^2}\\ &= 2\frac{\bar{n}_{\rm even}}{\bar{n}_{\rm even}+\bar{n}_{\rm odd}}, \end{aligned}\tag{12.93a}$$

and similarly,

$$1-\frac{1}{\cosh(2|A|^2)} = 2\frac{\bar{n}_{\rm odd}}{\bar{n}_{\rm even}+\bar{n}_{\rm odd}}. \tag{12.93b}$$

Then (12.91) may be rewritten as

$$\tilde{\rho} = \frac{\bar{n}_{\rm odd}}{\bar{n}_{\rm even} + \bar{n}_{\rm odd}} |A_{\rm even}\rangle\langle A_{\rm even}| + \frac{\bar{n}_{\rm even}}{\bar{n}_{\rm even} + \bar{n}_{\rm odd}} |A_{\rm odd}\rangle\langle A_{\rm odd}|, \tag{12.94}$$

which is the steady-state limit of the time-dependent density operator

$$\tilde{\rho}(t) = p_{\rm even}(t)|A_{\rm even}\rangle\langle A_{\rm even}| + p_{\rm odd}(t)|A_{\rm odd}\rangle\langle A_{\rm odd}|, \tag{12.95}$$

with probabilities $p_{\rm even}(t)$ and $p_{\rm odd}(t)$ governed by the rate equations

$$\frac{dp_{\rm even}}{dt} = -2\kappa\bar{n}_{\rm even}p_{\rm even} + 2\kappa\bar{n}_{\rm odd}p_{\rm odd}, \tag{12.96a}$$

$$\frac{dp_{\rm odd}}{dt} = -2\kappa\bar{n}_{\rm odd}p_{\rm odd} + 2\kappa\bar{n}_{\rm even}p_{\rm even}. \tag{12.96b}$$

In (12.96a) and (12.96b), the transition rates, $2\kappa\bar{n}_{\rm even}$ and $2\kappa\bar{n}_{\rm odd}$, are the rates of emission of subharmonic photons from the cavity, with the subharmonic mode in coherent states $|A_{\rm even}\rangle$ and $|A_{\rm odd}\rangle$, respectively. If we add to this the observation that $a|A_{\rm even}\rangle \propto |A_{\rm odd}\rangle$ and $a|A_{\rm odd}\rangle \propto A_{\rm even}$, a fairly complete picture emerges. The steady state (12.95) is maintained by the two-state rate process

$$\cdots|A_{\rm even}\rangle\underset{2\kappa\bar{n}_{\rm even}}{\longrightarrow}|A_{\rm odd}\rangle\underset{2\kappa\bar{n}_{\rm odd}}{\longrightarrow}|A_{\rm even}\rangle\underset{2\kappa\bar{n}_{\rm even}}{\longrightarrow}|A_{\rm odd}\rangle\cdots,$$

where the transitions occur at random and are accompanied by the emission from the cavity of a single subharmonic photon. This *is* a stochastic process; but it is not the stochastic process offered up to us directly by the positive P representation.

We would like to have a more convincing demonstration that this is what the master equation actually describes. Confirmation can be obtained in the following way. The basic idea is to develop the solution to the master equation, $\dot{\rho} = \mathcal{L}\rho$, in a perturbation series using a superoperator Dyson expansion. The physics enters in dividing the superoperator $\mathcal{L}$ into a part that generates free evolution to an unperturbed steady state, and an interaction part, which "perturbs" this steady state. For the example of the degenerate parametric oscillator with adiabatic elimination of the pump, the superoperator $\mathcal{L}$ is defined by the right-hand side of (12.10), after transforming to the interaction picture ($\rho = e^{-i\omega_C a^\dagger a t}\tilde{\rho}e^{i\omega_C a^\dagger a t}$) to remove the time dependence of the driving field.

Our interest is specifically with the limit of small system size. It is this limit that dictates how the decomposition of $\mathcal{L}$ should be made. Taking the system size parameter as $2\xi^{-1}n_p^{\rm thr} = 2\kappa_p\kappa/g^2 = \frac{1}{2}\kappa(g^2/4\kappa_p)^{-1}$, we see that a large system is one in which one-photon decay dominates two-photon decay [compare the coefficients of the decay terms in (12.10)], while in a small system the relative sizes of one- and two-photon decay are reversed. In the limit of small system size, the "perturbation" is therefore the one-photon decay (the emission of subharmonic photons from the cavity) and the unperturbed steady states are generated by the two-photon decay (the emission of pump photons from the cavity), balanced against the excitation by the external pump field. With this in mind, we decompose (12.10) as

$$\dot{\tilde{\rho}} = (\mathcal{D}_1 + \mathcal{C}_1 + \mathcal{L}_2)\tilde{\rho} \tag{12.97}$$

with

$$\mathcal{D}_1 \equiv 2\kappa(a \cdot a^\dagger), \tag{12.98a}$$

$$\mathcal{C}_1 \equiv -\kappa(a^\dagger a \cdot + \cdot a^\dagger a), \tag{12.98b}$$

$$\begin{aligned}\mathcal{L}_2 \equiv & -i\frac{g}{2\kappa_p}[\bar{\mathcal{E}}_i a^{\dagger 2} + \bar{\mathcal{E}}_i^* a^2, \cdot] \\ & + \frac{g^2}{4\kappa_p}(2a^2 \cdot a^{\dagger 2} - a^{\dagger 2}a^2 \cdot - \cdot a^{\dagger 2}a^2),\end{aligned} \tag{12.98c}$$

where the principal step taken to this point is the identification of $\mathcal{L}_2$ as the generator of the unperturbed steady states. To be convinced that $\mathcal{L}_2$ does, indeed, generate appropriate steady states, we note that $|A_{\rm even}\rangle$ and $|A_{\rm odd}\rangle$ are eigenstates of a^2, with

$$a^2|A_{\rm even}\rangle = A^2|A_{\rm even}\rangle, \qquad a^2|A_{\rm odd}\rangle = A^2|A_{\rm odd}\rangle. \tag{12.99}$$

It follows that $|A_{\rm even}\rangle$ and $|A_{\rm odd}\rangle$ are steady states with respect to $\mathcal{L}_2$:

$$\begin{aligned}&\mathcal{L}_2\big(|A_{\rm even}\rangle\langle A_{\rm even}|\big) \\ &\quad = -i\frac{g}{2\kappa_p}[\bar{\mathcal{E}}_i(a^{\dagger 2}\cdot) - \bar{\mathcal{E}}_i^*(\cdot\, a^2) + \bar{\mathcal{E}}_i^* A^2 - \bar{\mathcal{E}}_i A^{*2}]\big(|A_{\rm even}\rangle\langle A_{\rm even}|\big) \\ &\qquad + \frac{g^2}{4\kappa_p}[2|A|^4 - A^2(a^{\dagger 2}\cdot) - A^{*2}(\cdot\, a^2)]\big(|A_{\rm even}\rangle\langle A_{\rm even}|\big) \\ &\quad = 0,\end{aligned} \tag{12.100a}$$

and similarly

$$\mathcal{L}_2\big(|A_{\rm odd}\rangle\langle A_{\rm odd}|\big) = 0, \tag{12.100b}$$

where the definition (12.88) has been used.

Turning now to the superoperators $\mathcal{D}_1$ and $\mathcal{C}_1$, we note for the first that $\mathcal{D}_1\big(|A_{\rm even}\rangle\langle A_{\rm even}|\big) \propto |A_{\rm odd}\rangle\langle A_{\rm odd}|$ and $\mathcal{D}_1\big(|A_{\rm odd}\rangle\langle A_{\rm odd}|\big) \propto |A_{\rm even}\rangle\langle A_{\rm even}|$. Thus, $\mathcal{D}_1$ describes the one-photon emissions associated with the interaction part of $\mathcal{L}$. There remains the superoperator $\mathcal{C}_1$: where should this piece of $\mathcal{L}$ go, into the free evolution or the interaction? It seems that $\mathcal{L}_2$ and $\mathcal{D}_1$ already account for the two parts of the proposed rate process, and we might consider neglecting $\mathcal{C}_1$ altogether. After all, if added to $\mathcal{L}_2$ it should not alter the unperturbed steady states a great deal, and it does not appear to be involved in the transitions between steady states so there seems to be no reason to add it to

$\mathcal{D}_1$ either. It must be recognized, though, that in speaking of $\mathcal{D}_1$ as a "perturbation" the word is used rather loosely. A perturbation must be "weak," and in the limit of small system size $\mathcal{D}_1$ and $\mathcal{C}_2$ are certainly both small compared to $\mathcal{L}_2$. But "weak" means that the perturbation may be treated to low order in an *evolution over short enough time*—and we intend to allow the evolution to proceed add infinitum, certainly for a time sufficiently long that very many photons are emitted ($2\kappa\langle a^\dagger a\rangle t \gg 1$). In this situation, it would be better, perhaps, to argue for the decomposition of $\mathcal{L}$ on the basis of a separation of time scales; we are separating a process that is fast—evolution to a steady state under $\mathcal{L}_2$—from a much slower process—emission of subharmonic photons under $\mathcal{D}_1$. While slower, the second process will have a large enough effect over a sufficiently long time. We should therefore be cautious about neglecting $\mathcal{C}_1$, which is in fact of the same order as $\mathcal{D}_1$. As the latter is so clearly satisfactory for describing the one-photon emissions, for the moment let us add $\mathcal{C}_1$ to $\mathcal{L}_2$ and see what can be done after the Dyson expansion has been made.

Following through now with the proposed strategy, we define

$$\mathcal{D} \equiv \mathcal{D}_1, \tag{12.101a}$$

$$\mathcal{C} \equiv \mathcal{L}_2 + \mathcal{C}_1 \tag{12.101b}$$

and make a Dyson expansion [12.22] of the density operator in the form

$$\begin{aligned}\tilde{\rho}(t) = \sum_{n=0}^{\infty} \int_0^t dt_n \cdots \int_0^{t_3} dt_2 \int_0^{t_2} dt_1 e^{\mathcal{C}(t-t_n)} \mathcal{D} e^{\mathcal{C}(t_n - t_{n-1})} \\ \cdots \mathcal{D} e^{\mathcal{C}(t_2-t_1)} \mathcal{D} e^{\mathcal{C}t_1} \big(|0\rangle\langle 0|\big).\end{aligned} \tag{12.102}$$

The expansion is exact; at this stage no approximation has been made.

Our task is to recast (12.102) so that it may be seen as a generalized sum over n transitions between states $|A_{\rm even}\rangle$ and $|A_{\rm odd}\rangle$, occuring at the ordered sequence of times $0 \le t_1 \le t_2 \cdots \le t_n$, the dynamic suggested by (12.94)–(12.96b). The formal structure for such an evolution is already there, but we must show that the free propagation, $e^{\mathcal{C}(t_k - t_{k-1})}$, really does form the steady states $|A_{\rm even}\rangle$ and $|A_{\rm odd}\rangle$; that $\mathcal{D}$ generates transitions between these steady states once formed is already clear. Beyond this, we must also demonstrate that the correct probabilities, $p_{\rm even}(t)$ and $p_{\rm odd}(t)$, can emerge from this expression. Note that it is only in accomplishing the first task, in the treatment of the free propagation, that an approximation based on the limit of small system size must be used.

Let us begin with the initial vacuum state and work through the first few steps of the evolution, from right to left inside the integrals, up to time t_2. We start with the assumption that the time t_1 of the first photon emission is much larger than the time required to reach a steady state under the propagator $e^{\mathcal{C}t_1}$. If the effect of $\mathcal{C}_1$ in $\mathcal{C}$ is neglected, the steady state reached is certainly $|A_{\rm even}\rangle\langle A_{\rm even}|$, since the repeated application of $\mathcal{L}_2$ must yield a state with even photon number when beginning from the vacuum state. The superoperator $\mathcal{C}_1$ conserves the photon number and cannot change this property. As noted, though, $\mathcal{C}_1$ might have a nonvanishing effect over long enough times, so to be cautious, we make the conservative approximation

$$e^{\mathcal{C}t_1}\big(|0\rangle\langle 0|\big) = e^{\mathcal{C}t_1}\big(|A_{\rm even}\rangle\langle A_{\rm even}|\big). \tag{12.103}$$

In fact, we can see a clear reason now why $\mathcal{C}_1$ should not be neglected: it brings a qualitative, not merely a quantitative, change to the state; without it the norm is conserved, which is not the case when $\mathcal{C}_1$ is retained. We therefore make the additional assumption that the only effect of $\mathcal{C}_1$ is to change the norm of the state. This leads to the *ansatz*

$$e^{\mathcal{C}t_1}\big(|A_{\rm even}\rangle\langle A_{\rm even}|\big) = \mathcal{N}_{\rm even}(t_1)\big(|A_{\rm even}\rangle\langle A_{\rm even}|\big), \tag{12.104}$$

and hence to the equation of motion for the state norm

$$\begin{aligned}\frac{d\mathcal{N}_{\rm even}}{dt}\big(|A_{\rm even}\rangle\langle A_{\rm even}|\big) &= \mathcal{N}_{\rm even}\mathcal{C}\big(|A_{\rm even}\rangle\langle A_{\rm even}|\big)\\ &= -\kappa\mathcal{N}_{\rm even}(a^\dagger a\cdot+\cdot a^\dagger a)\big(|A_{\rm even}\rangle\langle A_{\rm even}|\big)\\ \Rightarrow\quad \frac{d\mathcal{N}_{\rm even}}{dt} &= -2\kappa\bar{n}_{\rm even}\mathcal{N}_{\rm even}.\end{aligned} \tag{12.105}$$

It follows that

$$e^{\mathcal{C}t_1}\big(|A_{\rm even}\rangle\langle A_{\rm even}|\big) = e^{-2\kappa\bar{n}_{\rm even}t_1}\big(|A_{\rm even}\rangle\langle A_{\rm even}|\big). \tag{12.106}$$

Combining (12.103) and (12.106) takes us up to the first transition in the Dyson expansion (12.102).

At this point all of the approximations we need have been made and the sought after sequence of transitions unfolds with little effort. The interaction superoperator $\mathcal{D}$ is applied for the first time at time t_1, where we have

$$\begin{aligned}\mathcal{D}\big(|A_{\rm even}\rangle\langle A_{\rm even}|\big) &= 2\kappa|A|^2\frac{1-e^{-2|A|^2}}{1+e^{-2|A|^2}}\big(|A_{\rm odd}\rangle\langle A_{\rm odd}|\big)\\ &= 2\kappa\bar{n}_{\rm even}\big(|A_{\rm odd}\rangle\langle A_{\rm odd}|\big).\end{aligned} \tag{12.107}$$

An *ansatz* equivalent to (12.105) takes us to time t_2,

$$\begin{aligned}e^{\mathcal{C}(t_2-t_1)}\big(|A_{\rm even}\rangle\langle A_{\rm even}|\big) &= \mathcal{N}_{\rm odd}(t_2-t_1)\big(|A_{\rm odd}\rangle\langle A_{\rm odd}|\big)\\ &= e^{-2\kappa\bar{n}_{\rm odd}(t_2-t_1)}\big(|A_{\rm odd}\rangle\langle A_{\rm odd}|\big),\end{aligned} \tag{12.108}$$

where a second application of $\mathcal{D}$ gives

$$\begin{aligned}\mathcal{D}\big(|A_{\rm odd}\rangle\langle A_{\rm odd}|\big) &= 2\kappa|A|^2\frac{1+e^{-2|A|^2}}{1-e^{-2|A|^2}}\big(|A_{\rm even}\rangle\langle A_{\rm even}|\big)\\ &= 2\kappa\bar{n}_{\rm odd}\big(|A_{\rm even}\rangle\langle A_{\rm even}|\big).\end{aligned} \tag{12.109}$$

From this point the sequence (12.106)–(12.109) is simply repeated. In this way the Dyson expansion yields a mixed state density operator in the form (12.95) with

$$\begin{aligned}p_{\mathrm{even}}(t)\\=\sum_{n\geq 0\ \mathrm{even}}\int_0^t\cdots\int_0^{t_3}\int_0^{t_2}e^{-2\kappa\bar{n}_{\mathrm{even}}(t-t_n)}(2\kappa\bar{n}_{\mathrm{odd}}dt_n)e^{-2\kappa\bar{n}_{\mathrm{odd}}(t_n-t_{n-1})}\\\cdots(2\kappa\bar{n}_{\mathrm{odd}}dt_2)e^{-2\kappa\bar{n}_{\mathrm{odd}}(t_2-t_1)}(2\kappa\bar{n}_{\mathrm{even}}dt_1)e^{-2\kappa\bar{n}_{\mathrm{even}}t_1},\end{aligned}\tag{12.110a}$$

and

$$\begin{aligned}p_{\mathrm{odd}}(t)\\=\sum_{n>0\ \mathrm{odd}}\int_0^t\cdots\int_0^{t_3}\int_0^{t_2}e^{-2\kappa\bar{n}_{\mathrm{odd}}(t-t_n)}(2\kappa\bar{n}_{\mathrm{even}}dt_n)e^{-2\kappa\bar{n}_{\mathrm{even}}(t_n-t_{n-1})}\\\cdots(2\kappa\bar{n}_{\mathrm{odd}}dt_2)e^{-2\kappa\bar{n}_{\mathrm{odd}}(t_2-t_1)}(2\kappa\bar{n}_{\mathrm{even}}dt_1)e^{-2\kappa\bar{n}_{\mathrm{even}}t_1}.\end{aligned}\tag{12.110b}$$

It is clear that the same probabilities are recovered from the two-state stochastic process described by rate equations (12.96a) and (12.96b). The terms inside the integrals multiply a sequence of probabilities, $e^{2\kappa\bar{n}_{\mathrm{even}}(t_k-t_{k-1})}$ or $e^{2\kappa\bar{n}_{\mathrm{odd}}(t_k-t_{k-1})}$, for no transition to occur between t_{k-1} and t_k, by the probabilities, $2\kappa\bar{n}_{\mathrm{even}}dt_k$ or $2\kappa\bar{n}_{\mathrm{odd}}dt_k$, for a transition to take place between t_k and t_k+dt_k; the transition rates alternate in the appropriate way, in accord with the proposed switching backwards and forwards between $|A_{\mathrm{even}}\rangle$ and $|A_{\mathrm{odd}}\rangle$.

Explicit results for $p_{\mathrm{even}}(t)$ and $p_{\mathrm{odd}}(t)$ may be obtained by taking Laplace transforms to help with the evaluation of the integrals and the execution of the sums. If the integrals are evaluated without carrying out the sums, the photon counting distribution for subharmonic photons is obtained:

Exercise 12.9. Show that (12.110a) and (12.110b) reproduce the results,

$$p_{\mathrm{even}}(t)=1-p_{\mathrm{odd}}(t),\tag{12.111a}$$

$$p_{\mathrm{odd}}(t)=\frac{\bar{n}_{\mathrm{even}}}{\bar{n}_{\mathrm{even}}+\bar{n}_{\mathrm{odd}}}\left[1-e^{-2\kappa(\bar{n}_{\mathrm{even}}+\bar{n}_{\mathrm{odd}})t}\right],\tag{12.111b}$$

obtained by direct solution of the rate equations (12.96a) and (12.96b).

Exercise 12.10. Evaluate the integrals in (12.110a) and (12.110b) but do not carry out the sums. Hence show that the counting distribution for subharmonic photons emitted by the cavity in time T is

$$\begin{aligned}P(2k,T)=\frac{(2\kappa\bar{n}_{\mathrm{even}}T)^k(2\kappa\bar{n}_{\mathrm{odd}}T)^k}{2^k k!}\exp[-\kappa(\bar{n}_{\mathrm{even}}+\bar{n}_{\mathrm{odd}})T]\\\times\sqrt{\frac{\pi}{2}}\frac{I_{k-\frac{1}{2}}[\kappa(\bar{n}_{\mathrm{odd}}-\bar{n}_{\mathrm{even}})T]+I_{k+\frac{1}{2}}[\kappa(\bar{n}_{\mathrm{odd}}-\bar{n}_{\mathrm{even}})T]}{[\kappa(\bar{n}_{\mathrm{odd}}-\bar{n}_{\mathrm{even}})T]^{k-\frac{1}{2}}},\end{aligned}\tag{12.112a}$$

and

$$P(2k+1,T) = \frac{(2\kappa\bar{n}_{\rm even}T)^{k+1}(2\kappa\bar{n}_{\rm odd}T)^k}{2^k k!}\exp[-\kappa(\bar{n}_{\rm even}+\bar{n}_{\rm odd})T]$$
$$\times\sqrt{\frac{\pi}{2}}\frac{I_{k+\frac{1}{2}}[\kappa(\bar{n}_{\rm odd}-\bar{n}_{\rm even})T]}{[\kappa(\bar{n}_{\rm odd}-\bar{n}_{\rm even})T]^{k+\frac{1}{2}}}, \tag{12.112b}$$

for $k = 0,1,2,\cdots$, where $\sqrt{\pi/2z}I_{n+\frac{1}{2}}(z)$ is the Modified Spherical Bessel function of the first kind.

The picture of the quantum dynamics gained from this analysis is very different from the one provided by the diffusion process of Fig. 12.8. It does appear, moreover, to offer a better understanding of the physics. The main point is that we are now able to connect the evolution to physical events—the emission of subharmonic photons from the cavity, one by one. In the limit of small system size, each emission brings a large change to the state of the intracavity field, and it is this large, discrete perturbation that gives so much trouble in the phase space approach; we encountered similar difficulties with the damped two-level atom (Sects. 6.1.3 and 6.1.4). By relying on diffusion, the positive P stochastic process effectively anticipates that the quantum fluctuations will arise from an incremental accumulation of perturbations over time—there is no place within its framework for a notion as fundamental as the single quantum event or quantum jump.

Of course, the emission of a photon from a cavity need not necessarily bring about a large change to the quantum state. If, for example, the intracavity field is in the coherent state $|A\rangle$, action by the superoperator $\mathcal{D}$ describes no physical change at all—$\mathcal{D}(|A\rangle\langle A|) = 2\kappa|A\rangle\langle A|$; this is the case even when the photon number is small. Thus, it is the scale of the nonlinearity that distinguishes small from large system size, not merely the number of photons the cavity contains.

In the present example, the large perturbation is the change from a state of even photon number to one of odd photon number, or vice versa. Both before and after, the subharmonic mode is in a superposition of coherent states, and potentially one of high photon number. Thus, coherent state superpositions of high photon number (Schrödinger cat states) are produced, in principle, by our model. Considerations of practice are another question. Keeping track of the even or odd parity would be essential in any scenario that aimed to observe the coherent state superpositions. A measurement like this would of course be extremely difficult, since it is impossible to detect every last photon, and for large photon numbers (flux) there is the difficulty involved in resolving the photon counts in time. A more realistic place to observe the perturbation due to a photon emission is in the regime of low photon numbers, $|A|^2 \ll 1$. Here, also, a dramatic change occurs in the transition $|A_{\rm even}\rangle \to |A_{\rm odd}\rangle$ or $|A_{\rm odd}\rangle \to |A_{\rm even}\rangle$. It is apparent from the wide disparity in the photon number expectations

$$\bar{n}_{\text{even}} = |A|^4 \ll 1, \tag{12.113a}$$
$$\bar{n}_{\text{odd}} = 1. \tag{12.113b}$$

This disparity accounts for some rather simple physics. Photons are created inside the cavity in pairs; $2\kappa\bar{n}_{\text{even}} = 2\kappa|A|^4$ governs the rate at which we expect to detect a "first" emitted photon, indicating that a pair has been created. In the limit of very weak excitation, this rate is small compared to the inverse cavity lifetime. In contrast, $2\kappa\bar{n}_{\text{odd}} = \kappa$ governs the rate at which we expect to see a "second" photon emitted given that a "first" has just been seen. The "second" photon is surely there, as it is created inside the cavity alone with the "first," and it must certainly be emitted in a time of the order of the cavity lifetime. Thus the wide disparity in $\bar{n}_{\text{even}}$ and $\bar{n}_{\text{odd}}$ accommodates the difference between the average time separating successive pairs of photons and the time—on the order of the cavity lifetime—separating the two photons of a pair.

We have conjured up a picture of the photon stream from parametric down conversion: a stream of *photon pairs* separated by a time of order $(2\kappa|A|^4)^{-1}$, with the photons of each pair separated, on average, by $(2\kappa)^{-1}$. It follows that counting photons for a time that is long compared with the average time between pairs should invariably result in an even number of counts. To verify this, we evaluate (12.112a) and (12.112b) in the limit $|A|^2 \to 0$, $\kappa T \to \infty$, with $2\kappa|A|^4T$ constant; Modified Spherical Bessel functions for large argument are given in [12.23]. The result is

$$P(2k,T) \to \frac{(2\kappa|A|^4T)^k}{k!} e^{-2\kappa|A|^4T}, \tag{12.114a}$$

$$\begin{aligned} P(2k+1,T) \to \frac{(2\kappa|A|^4T)^{k+1}}{k!}\frac{e^{-2\kappa|A|^4T}}{2\kappa T} &= |A|^4 P(2k,T) \\ &\to 0, \end{aligned} \tag{12.114b}$$

which agrees with the suggested physical picture.

Note 12.6. Photon counting reveals a similar pairing phenomenon in the limit of large system size. Again, there is the requirement for low mean photon number, so the pairing is seen below threshold. Photon counting distributions have been derived for this regime using the positive P representation and the methods of linearization [12.24, 12.25]. On the experimental side, an indirect observation of photon pairs was recently made by tomographic reconstruction of the photon number distribution from balanced heterodyne measurements [12.26].

12.1.9 Higher-Order Corrections to the Spectrum of Squeezing at Threshold

Before we leave the topic of quantum fluctuations and system size, one last comment on threshold fluctuations is in order. The steady-state positive P distribution derived in Sec. 9.9.7 (Eq. 12.81) incorporates the lowest-order correction to the linearized treatment needed to remove the divergence of fluctuations in the classical (unsqueezed) phase-space variable. It makes no correction to the linearized treatment of the nonclassical, or squeezed fluctuations, however, not even at lowest order (Note 12.4). To include such a correction, two of the three higher-order terms in (12.80b) must be kept—the term $-(2\xi^{-1}n_p^{\mathrm{thr}})\Theta_{\mathrm{C}}^4$, which gives the previously derived result, and also the term $-(2\xi^{-1}n_p^{\mathrm{thr}})6\Theta_{\mathrm{C}}^2\Theta_{\mathrm{NC}}^2$ that couples fluctuations in the two quadrature phase amplitudes. The extra term introduces an additional exponential, $\exp\left[-6\Theta_{\mathrm{C}}^2\Theta_{\mathrm{NC}}^2/\left(2\xi^{-1}n_p^{\mathrm{thr}}\right)^{-1}\right]$, multiplying expression (12.81) for the positive P distribution. We may expand the exponential to lowest order. Thus, correctly normalized to the new higher order of the treatment, the positive P distribution for the degenerate parametric oscillator at threshold becomes

$$\tilde{\bar{\Pi}}(\Theta_{\mathrm{C}},\Theta_{\mathrm{NC}}) = \frac{4}{\sqrt{\pi}\Gamma(\frac{1}{4})}\frac{1}{\left(2\xi^{-1}n_p^{\mathrm{thr}}\right)^{-3/4}}\exp\left[-\frac{\Theta_{\mathrm{C}}^4+4\Theta_{\mathrm{NC}}^2}{\left(2\xi^{-1}n_p^{\mathrm{thr}}\right)^{-1}}\right] \times\left[1-\frac{6\Theta_{\mathrm{C}}^2\Theta_{\mathrm{NC}}^2}{\left(2\xi^{-1}n_p^{\mathrm{thr}}\right)^{-1}}+\frac{3}{4}\frac{\Gamma(\frac{3}{4})}{\Gamma(\frac{1}{4})}\left(2\xi^{-1}n_p^{\mathrm{thr}}\right)^{-1/2}\right]. \quad (12.115)$$

How, then, are the fluctuations of the squeezed quadrature phase amplitude changed? From (12.115), the variance in the nonclassical variable is

$$\begin{aligned}\left(\overline{\Theta_{\mathrm{NC}}^2}\right)_{\tilde{\bar{\Pi}}} &= \left[1+\frac{3}{4}\frac{\Gamma(\frac{3}{4})}{\Gamma(\frac{1}{4})}\left(2\xi^{-1}n_p^{\mathrm{thr}}\right)^{-1/2}\right]\frac{1}{8}\left(2\xi^{-1}n_p^{\mathrm{thr}}\right)^{-1} \\ &\quad -\frac{6}{\left(2\xi^{-1}n_p^{\mathrm{thr}}\right)^{-1}}\frac{\Gamma(\frac{3}{4})}{\Gamma(\frac{1}{4})}\left(2\xi^{-1}n_p^{\mathrm{thr}}\right)^{-1/2}3\left[\frac{1}{8}\left(2\xi^{-1}n_p^{\mathrm{thr}}\right)^{-1}\right]^2 \\ &= \frac{1}{8}\left(2\xi^{-1}n_p^{\mathrm{thr}}\right)^{-1}\left[1-\frac{3}{2}\frac{\Gamma(\frac{3}{4})}{\Gamma(\frac{1}{4})}\left(2\xi^{-1}n_p^{\mathrm{thr}}\right)^{-1/2}\right],\end{aligned} \quad (12.116)$$

from which, after inverting the transformation of variables (Eqs. 10.33, 11.27a, 12.66, and 12.75) and using definitions (9.26) and (9.29), the *variance of fluctuations in the squeezed quadrature phase amplitude at threshold corrected for system size* is given by

$$(\Delta Y)_{\mathrm{thr}}^2 = \left[(\Delta Y)_{\mathrm{thr}}^2\right]_0\left\{1-\tfrac{3}{2}\left(2\xi^{-1}n_p^{\mathrm{thr}}\right)^{-1}\left[(\Delta X)_{\mathrm{thr}}^2\right]_0\right\}. \quad (12.117)$$

In this formula $\left[(\Delta Y)_{\mathrm{thr}}^2\right]_0$ and $\left[(\Delta X)_{\mathrm{thr}}^2\right]_0$ denote the variances calculated at lower order, (10.98a) and (10.100a) respectively. Since $\left[(\Delta X)_{\mathrm{thr}}^2\right]_0$ is of order $\left(2\xi^{-1}n_p^{\mathrm{thr}}\right)^{1/2}$, the correction to the squeezing is of order $\left(2\xi^{-1}n_p^{\mathrm{thr}}\right)^{-1/2}$.

This calculation based on the state-state positive P distribution does not tell us how the system-size correction enters into the spectrum of squeezing; nor does it provide much physical insight into the origin of the correction. To address these issues we return to the stochastic differential equations. First, let us write (12.3a) and (12.3b) in terms of Θ_{C} and Θ_{NC}, which gives ($\lambda = 1$)

$$\begin{aligned} d\Theta_{\mathrm{C}} &= -\Theta_{\mathrm{C}}(\Theta_{\mathrm{C}}^2 - \Theta_{\mathrm{NC}}^2)d\bar{t} \\ &\quad + \tfrac{1}{\sqrt{2}}\left(2\xi^{-1}n_p^{\mathrm{thr}}\right)^{-1/2}\sqrt{1 - \Theta_{\mathrm{C}}^2 - \Theta_{\mathrm{NC}}^2}\,d\bar{W}_{\Theta_{\mathrm{C}}}, \end{aligned} \tag{12.118a}$$

$$\begin{aligned} d\Theta_{\mathrm{NC}} &= -\Theta_{\mathrm{NC}}(2 + \Theta_{\mathrm{C}}^2 - \Theta_{\mathrm{NC}}^2)d\bar{t} \\ &\quad + \tfrac{1}{\sqrt{2}}\left(2\xi^{-1}n_p^{\mathrm{thr}}\right)^{-1/2}\sqrt{1 - \Theta_{\mathrm{C}}^2 - \Theta_{\mathrm{NC}}^2}\,d\bar{W}_{\Theta_{\mathrm{NC}}}, \end{aligned} \tag{12.118b}$$

where $d\bar{W}_{\Theta_{\mathrm{C}}} = (d\bar{W}_\alpha + d\bar{W}_{\alpha_*})/\sqrt{2}$ and $d\bar{W}_{\Theta_{\mathrm{NC}}} = (d\bar{W}_\alpha - d\bar{W}_{\alpha_*})/\sqrt{2}$ are new independent Wiener increments, with variance $d\bar{t}$. Next, we simplify the stochastic differential equations, keeping only those nonlinear terms that contribute at the order included in (12.117). Certainly the decay term $-\Theta_{\mathrm{C}}^3 d\bar{t}$ in (12.118a) must be kept, since it is needed to prevent the divergence of fluctuations in the unsqueezed quadrature phase amplitude. We must also keep Θ_{C}^2 in (12.118b); this is the term that couples fluctuations in the unsqueezed quadrature phase amplitude into the squeezed quadrature phase amplitude. Keeping these terms and neglecting the rest, the stochastic differential equations are

$$d\Theta_{\mathrm{C}} = -\Theta_{\mathrm{C}}^3 d\bar{t} + \tfrac{1}{\sqrt{2}}\left(2\xi^{-1}n_p^{\mathrm{thr}}\right)^{-1/2} d\bar{W}_{\Theta_{\mathrm{C}}}, \tag{12.119a}$$

$$d\Theta_{\mathrm{NC}} = -\Theta_{\mathrm{NC}}(2 + \Theta_{\mathrm{C}}^2)d\bar{t} + \tfrac{1}{\sqrt{2}}\left(2\xi^{-1}n_p^{\mathrm{thr}}\right)^{-1/2}\sqrt{1 - \Theta_{\mathrm{C}}^2}\,d\bar{W}_{\Theta_{\mathrm{NC}}}. \tag{12.119b}$$

Aside from the different scaling, (12.119a) is the same as (10.93a), and the new physics is contained in (12.119b) [compare (10.91b)] where both the decay rate and the noise strength depend on the fluctuating variable Θ_{C}.

There is one further simplification to make. We noted in Sects. 8.2 and 10.2.4 that threshold fluctuations in a single *"slow" variable* show critical slowing down, and this allows other *"fast" variables* to be adiabatically eliminated. In the present situation, it is the fluctuations in Θ_{C} that are slow, having timescale $\left(2\xi^{-1}n_p^{\mathrm{thr}}\right)^{-1/2}\bar{t}$, compared to those in Θ_{NC}. We may therefore view (12.119b) as if it were an autonomous equation, regarding Θ_{C} as being constant in time. In this way we first calculate a spectrum of squeezing that depends conditionally on Θ_{C}, from which the final spectrum is obtained by taking a steady-state average over Θ_{C}. This procedure accomplishes an adiabatic elimination of, and averaging over, the critical fluctuations of the classical variable Θ_{C}.

The completion of the calculation is now straightforward. The positive P distribution corresponding to stochastic differential equations (12.119a) and (12.119b) is written in the factorized form

$$\tilde{\bar{\Pi}}(\Theta_{\mathrm{C}}, \Theta_{\mathrm{NC}}, \bar{t}) = \tilde{\bar{C}}(\Theta_{\mathrm{C}}, \bar{t})\tilde{\bar{N}}(\Theta_{\mathrm{NC}}, \bar{t}|\Theta_{\mathrm{C}}), \tag{12.120}$$

where the distribution in the classical variable satisfies the Fokker–Planck equation

$$\frac{\partial \tilde{\bar{C}}}{\partial \bar{t}} = \left[\frac{\partial}{\partial \Theta_{\mathrm{C}}}\Theta_{\mathrm{C}}^3 + \frac{1}{4}\left(2\xi^{-1}n_p^{\mathrm{thr}}\right)^{-1}\frac{\partial^2}{\partial \Theta_{\mathrm{C}}^2}\right]\tilde{\bar{C}}, \tag{12.121a}$$

while the conditional distribution in the nonclassical variable satisfies the equation

$$\frac{\partial \tilde{\bar{N}}}{\partial \bar{t}} = \left[(2+\Theta_{\mathrm{C}}^2)\frac{\partial}{\partial \Theta_{\mathrm{NC}}}\Theta_{\mathrm{NC}} + \frac{1}{4}\left(2\xi^{-1}n_p^{\mathrm{thr}}\right)^{-1}(1-\Theta_{\mathrm{C}}^2)\frac{\partial^2}{\partial \Theta_{\mathrm{NC}}^2}\right]\tilde{\bar{N}}. \tag{12.121b}$$

If Θ_{C}, the "slow" variable, is considered to be constant, (12.121b) is a simple one-dimensional linear Fokker–Planck equation. From it, using (5.93) and (5.102a), we obtain the autocorrelation

$$\lim_{t\to\infty}\left(\overline{\Theta_{\mathrm{NC}}(t)\Theta_{\mathrm{NC}}(t+\tau)}\right)_{\tilde{\bar{N}}} = e^{-\kappa(2+\Theta_{\mathrm{C}}^2)|\tau|}\left(2\xi^{-1}n_p^{\mathrm{thr}}\right)^{-1}\frac{1}{8}\left(1-\tfrac{3}{2}\Theta_{\mathrm{C}}^2\right). \tag{12.122}$$

The spectrum of squeezing (Eq. 10.58) in the Y-quadrature phase amplitude, conditioned on the value of the X-quadrature phase amplitude, is then

$$S_Y^{\mathrm{thr}}(\omega|\Theta_{\mathrm{C}}) = -\left(1-\tfrac{3}{2}\Theta_{\mathrm{C}}^2\right)\frac{1}{2}\frac{8\kappa\left(1+\frac{1}{2}\Theta_{\mathrm{C}}^2\right)}{\left[2\kappa\left(1+\frac{1}{2}\Theta_{\mathrm{C}}^2\right)\right]^2+\omega^2}. \tag{12.123}$$

This result may be written down from (10.61b), by comparing (12.122) and (10.60b)(with $\lambda = 1$ and $\bar{n} = 0$). Note that both the linewidth and the maximum squeezing, at $\omega = 0$, vary with Θ_{C}. Finally, the spectrum of squeezing is the average of this result taken with respect to the steady-state solution to (12.121a):

$$S_Y^{\mathrm{thr}}(\omega) = \left(\overline{S_Y^{\mathrm{thr}}(\omega|\Theta_{\mathrm{C}})}\right)_{\tilde{\bar{C}}}. \tag{12.124}$$

When the system-size correction is not taken into account, squeezing is perfect on resonance (Eq. 10.62b with $\lambda = 1$ and $\bar{n} = 0$). From (12.123) and (12.124), on the other hand, we have

$$\frac{1}{2}\sqrt{1+S_Y^{\mathrm{thr}}(0)} = \frac{1}{\sqrt{2}}\frac{\Gamma(\frac{3}{4})}{\Gamma(\frac{1}{4})}\left(2\xi^{-1}n_p^{\mathrm{thr}}\right)^{-1/4}. \tag{12.125}$$

Thus, the amplitude of the squeezed fluctuations is nonzero and of order $\left(2\xi^{-1}n_p^{\mathrm{thr}}\right)^{-1/4}$.

Note 12.7. Due to critical slowing down, the spectrum of fluctuations in the X-quadrature phase amplitude has width $\sim \left(2\xi^{-1}n_p^{\mathrm{thr}}\right)^{-1/2}$. The integral over this spectrum gives a variance $(\Delta X)^2_{\mathrm{thr}} \sim \left(2\xi^{-1}n_p^{\mathrm{thr}}\right)^{1/2}$ (Eq. 8.223a); on resonance, the amplitude of the fluctuations is therefore $\sim \left(2\xi^{-1}n_p^{\mathrm{thr}}\right)^{1/2}$. It follows, by taking the product with (12.125), that the resonant mode of the radiated subharmonic field is not, in fact, in a minimum uncertainty state at threshold; the uncertainty product is of the order $\left(2\xi^{-1}n_p^{\text{thr}}\right)^{1/4}$ rather than $\frac{1}{2}\sqrt{1+S_X^{\text{thr}}(0)} \times \frac{1}{2}\sqrt{1+S_Y^{\text{thr}}(0)} = \frac{1}{4}$.

12.2 Difficulties with the Positive *P* Representation

From what we have seen, particularly from the results of Sect. 12.1.6, the positive P representation has served its advertised purpose. It has taken us beyond the picture of quantum noise as merely a "fuzz" added to classical nonlinear dynamics, and demonstrated the role of the system size in this story. Enthusiasm for the approach must be dampened, however, by what we noticed along the way. The issue of the "spikes" in stochastic simulations (Fig. 12.2d) is still a worry. While it is true that for initial conditions taken within the unit square, $|\theta| \leq 1$, $|\vartheta| \leq 1$, they are a numerical artifact, the implications for initial conditions outside the square is still an open question. Conveniently for the analysis of Sect. 12.1.6, all physical steady states lie within the unit square. Initial conditions need not be so restricted, though, and there is certainly no guarantee that a trajectory beginning outside the square can avoid the unstable regions of the extended phase space (Sect. 12.1.5).

In fact, even the confinement of trajectories having initial conditions inside the unit square is slightly misleading, since this property is a structurally unstable feature of the model; it is not retained, for example, by the full set of stochastic differential equations (12.1a–12.1d)—i.e., without adiabatic elimination of the pump; neither does it hold in the presence of detuning where terms $-i\Delta\tilde{\tilde{\alpha}}d\bar{t}$ and $i\Delta\tilde{\tilde{\alpha}}_*d\bar{t}$, respectively, are added to the right-hand sides of (12.3a) and (12.3b). There remains, then, a possibility that in some cases at least the stochastic differential equations derived on the basis of the positive P representation contain a genuine pathology. It is desirable to have a definite answer on an issue of such importance.

A definite answer in the affirmative can be given by considering the degenerate parametric oscillator with adiabatic elimination of the pump for the limit of zero pump amplitude. In this limit, $\lambda \to 0$, the region of bounded trajectories—$\tilde{\tilde{x}} \leq \sqrt{\lambda}$, $\tilde{\tilde{\mathrm{x}}} \leq \sqrt{\lambda}$—shrinks to the origin. Then *all* nontrivial initial conditions lie outside the bounded domain. So far as master equation (12.10) is concerned, the second term on the right-hand side is removed, and only those terms describing the two damping processes remain: the standard one-photon damping at rate 2κ, and the two-photon damping, at rate $g^2/2\kappa_p$, which arises from the up conversion of two photons of frequency ω_C to a photon of frequency $2\omega_C$—which is then rapidly lost from the cavity before its reconversion to a pair of subharmonic photons can occur. Smith and Gardiner discovered that simulations of the positive P stochastic differential equations for this so-called two-photon damping model yield a result that is demonstrably incorrect on straightforward physical grounds [12.17]. Let us look at the evidence and try to understand why in this case a wrong result is obtained.

12.2.1 Technical Difficulties: Two-Photon Damping

The stochastic differential equations in the positive P representation for two-photon damping follow by setting $\lambda = 0$ in (12.3a) and (12.3b):

$$d\tilde{\bar{\alpha}} = -\tilde{\bar{\alpha}}\left[(1 + \tilde{\bar{\alpha}}_*\tilde{\bar{\alpha}})d\bar{t} - i\left(2\xi^{-1}n_p^{\mathrm{thr}}\right)^{-1/2}d\bar{W}_\alpha\right], \tag{12.126a}$$

$$d\tilde{\bar{\alpha}}_* = -\tilde{\bar{\alpha}}_*\left[(1 + \tilde{\bar{\alpha}}_*\tilde{\bar{\alpha}})d\bar{t} - i\left(2\xi^{-1}n_p^{\mathrm{thr}}\right)^{-1/2}d\bar{W}_{\alpha_*}\right]. \tag{12.126b}$$

To appreciate the technical difficulties that arise with this model it is sufficient to consider the decay of the mean photon number, for which we need only the one equation

$$d(\tilde{\bar{\alpha}}_*\tilde{\bar{\alpha}}) = -2(\tilde{\bar{\alpha}}_*\tilde{\bar{\alpha}})\left\{[1 + (\tilde{\bar{\alpha}}_*\tilde{\bar{\alpha}})]d\bar{t} - i\sqrt{\eta}d\bar{W}\right\}, \tag{12.127}$$

where $d\bar{W} \equiv (d\bar{W}_\alpha + d\bar{W}_{\alpha_*})/\sqrt{2}$ is a Wiener increment, variance $d\bar{t}$, and

$$\eta \equiv \frac{g^2/4\kappa_p}{\kappa} = \tfrac{1}{2}\left(2\xi^{-1}n_p^{\mathrm{thr}}\right)^{-1}. \tag{12.128}$$

Figure 12.9 shows the result of numerical simulations for an initial coherent state with amplitude α_0, initial condition $(\tilde{\bar{\alpha}}_*\tilde{\bar{\alpha}})(0) = |\alpha_0|^2/2\eta$. The mean photon number computed from the simulation is compared with the result of a direct numerical integration of the master equation. Clearly there is a problem with the long-time behavior, since the mean photon number does not approach zero as it should. In this respect the stochastic differential equations in the positive P representation disagree dramatically with the master equation from which they are derived.

One possible explanation is that something has gone wrong with the numerics, perhaps because the Euler method was used for the integration. This explanation can be ruled out, though, because the stochastic differential equation can also be solved analytically [12.18], and the same wrong result is obtained. To make the analytical calculation, we re-express the right-hand side of (12.127) as a sum of two separable terms, by writing

$$d(\tilde{\bar{\alpha}}_*\tilde{\bar{\alpha}}) = -2(\tilde{\bar{\alpha}}_*\tilde{\bar{\alpha}})(d\bar{t} - i\sqrt{\eta}d\bar{W}) - 2(\tilde{\bar{\alpha}}_*\tilde{\bar{\alpha}})^2d\bar{t}, \tag{12.129}$$

and introducing the *ansatz*

$$(\tilde{\bar{\alpha}}_*\tilde{\bar{\alpha}}) = \frac{\mu}{\nu}, \tag{12.130}$$

with $\mu(0) = (\tilde{\bar{\alpha}}_*\tilde{\bar{\alpha}})(0)$ and $\nu(0) = 1$. From (12.130), we obtain

$$d(\tilde{\bar{\alpha}}_*\tilde{\bar{\alpha}}) = \frac{1}{\nu}d\mu - \frac{\mu}{\nu^2}d\nu. \tag{12.131}$$

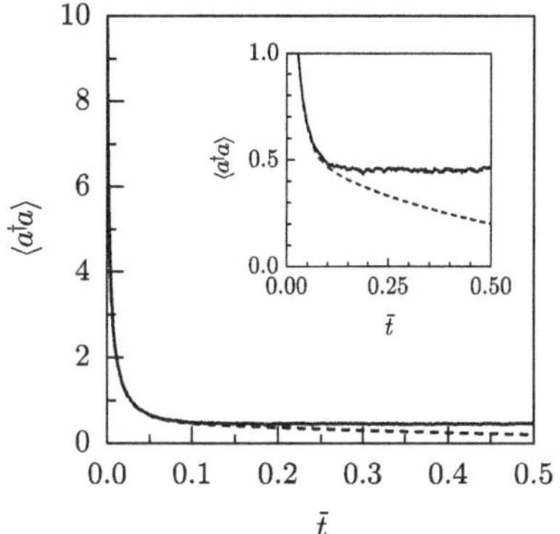

Fig. 12.9. Numerical simulation of the positive P stochastic differential equations for two-photon damping of a coherent state (Eq. 12.127). The *solid curve* is the average over 10,000 stochastic trajectories and the *dashed curve* the result obtained from master equation (12.10) (with $\bar{\mathcal{E}}_0 = 0$): for $|\alpha_0|^2 = 10$ and $\eta = 10$ with time step $\Delta\bar{t} = 10^{-4}$

The two terms on the right-hand side of this equation may be identified with those in (12.129) so long as

$$d\mu = -2\mu(d\bar{t} - i\sqrt{\eta}d\bar{W}), \tag{12.132a}$$

$$d\nu = 2\mu d\bar{t}. \tag{12.132b}$$

Solving (12.127) is reduced in this way to the task of solving the pair of equations (12.132a) and (12.132b). The latter is readily accomplished. To solve for μ, we make a change of variable and consider

$$\begin{aligned} d\ln(\mu) &= \ln(\mu + d\mu) - \ln(\mu) \\ &= \ln(\mu) + d\mu/\mu - \tfrac{1}{2}(d\mu/\mu)^2 - \ln(\mu) \\ &= -2(d\bar{t} - i\sqrt{\eta}d\bar{W}) + 2\eta d\bar{t} \\ &= -2(1-\eta)d\bar{t} + i2\sqrt{\eta}d\bar{W}, \end{aligned}$$

with solution

$$\mu(\bar{t}) = \mu(0)e^{-2(1-\eta)\bar{t}}e^{i2\sqrt{\eta}\bar{W}}. \tag{12.133a}$$

Then, from (12.132b), the solution for ν is

$$\nu(\bar{t}) = 1 + 2\int_0^{\bar{t}} d\bar{t}'\mu(\bar{t}'). \tag{12.133b}$$

Bringing together (12.130), (12.133a), and (12.133b), we arrive at an explicit expression for $(\tilde{\bar{\alpha}}_*\tilde{\bar{\alpha}})(\bar{t})$ in terms of the Wiener process $\bar{W}$; thus, we obtain

$$(\tilde{\bar{\alpha}}_*\tilde{\bar{\alpha}})(\bar{t}) = \frac{(\tilde{\bar{\alpha}}_*\tilde{\bar{\alpha}})(0)e^{-2(1-\eta)\bar{t}}e^{i2\sqrt{\eta}\bar{W}}}{1 + (\tilde{\bar{\alpha}}_*\tilde{\bar{\alpha}})(0)2\int_0^{\bar{t}} d\bar{t}' e^{-2(1-\eta)\bar{t}'}e^{i2\sqrt{\eta}\bar{W}'}}. \tag{12.134}$$

Recall that $\bar{W}$ is a Gaussian-distributed random variable with variance $\bar{t}$ (Sect. 5.3.2); its mean is taken here to be zero.

What we have obtained is perhaps not quite a full analytical solution, since it does not allow a closed-form expression for the mean photon number to be written down. It does, nevertheless, allow a more direct numerical computation to be made. As $\bar{W}$ is Gaussian-distributed with zero mean and variance $\bar{t}$, a sequence of Gaussian-distributed random numbers is all that is needed to compute the mean photon number for any particular time $\bar{t}$. This computation reproduces the wrong result displayed in Fig. 12.9.

Note 12.8. A closed form expression for the mean photon number may in fact be found by expanding (12.134) in a Taylor series. Surprisingly, the expression obtained in this way is correct, in so far as it agrees with a series solution to the infinite hierarchy of coupled equations for normal-ordered operator averages, $\langle(\tilde{\bar{a}}^{\dagger p}\tilde{\bar{a}}^{p})(\bar{t})\rangle$, obtained from the master equation; the hierarchy of moment equations is certainly equivalent to the master equation. Apparently there is a contradiction: it is possible to construct a correct analytical solution, as a Taylor series, when direct computation from (12.134) yields wrong results. The resolution of the apparent contradiction is that the Taylor expansion is not strictly legitimate for arbitrary times. It is valid for times that are sufficiently short, and analytic continuation of the short time result follows the correct physical solution (the dashed curve in Fig. 12.9). As shown below, however, knowing nothing of analytic continuation, the solution followed by the stochastic differential equation encounters a singularity—where the Taylor expansion first fails to converge—which switches it onto a nonphysical path (a consequence of the divergent trajectories of Sect. 12.1.5.).

To provide some insight into what has gone wrong, we can make a simplification by replacing the phase factor $\exp(i2\sqrt{\eta}\bar{W}')$ inside the integral in (12.134) by its value, $\exp(i2\sqrt{\eta}\bar{W})$, at the upper limit of integration. The approximation is perhaps somewhat uncontrolled, and we would not argue that it is good one for producing accurate results. It is nevertheless useful in so far as it defines a simplified model that can be analyzed further analytically. We refer to this model as the "toy" model. It also exhibits pathological behavior, and it helps us uncover the origin of the disagreement in Fig. 12.9 in a clear and convincing way.

After making the replacement $\exp(i2\sqrt{\eta}\bar{W}') \to \exp(i2\sqrt{\eta}\bar{W})$, the integration can be carried out. Thus, to replace the exact solution for photon number (12.134), the "toy" model is defined by

$$(\tilde{\bar{\alpha}}_*\tilde{\bar{\alpha}})_{\text{toy}}(\bar{t}) \equiv \frac{(\tilde{\bar{\alpha}}_*\tilde{\bar{\alpha}})(0)e^{-2(1-\eta)\bar{t}}e^{i2\sqrt{\eta}\bar{W}}}{1+(\tilde{\bar{\alpha}}_*\tilde{\bar{\alpha}})(0)\{\left[1-e^{-2(1-\eta)\bar{t}}\right]/(1-\eta)\}e^{i2\sqrt{\eta}\bar{W}}}. \tag{12.135}$$

Taking the average over the Wiener process, the mean photon number in the "toy" model is

$$\left(\overline{(\alpha_*\alpha)_{\text{toy}}(\bar{t})}\right)_P = \frac{1}{\sqrt{2\pi\bar{t}}}\int_{-\infty}^{\infty} d\bar{W}\frac{A(\bar{t})e^{i2\sqrt{\eta}\bar{W}}}{1+B(\bar{t})e^{i2\sqrt{\eta}\bar{W}}}e^{-\frac{1}{2}\bar{W}^2/\bar{t}}, \tag{12.136}$$

where

$$A(\bar{t}) \equiv |\alpha_0|^2 e^{-2(1-\eta)\bar{t}}, \qquad B(\bar{t}) \equiv 2\eta|\alpha_0|^2 \frac{1-e^{-2(1-\eta)\bar{t}}}{1-\eta}. \tag{12.137}$$

Now the integral has zero imaginary part, as the integrand is odd, while for the real part we find

$$\begin{aligned}
&\big(\overline{(\tilde{\bar{\alpha}}_*\tilde{\bar{\alpha}})_{\rm toy}(\bar{t})}\,\big)_P \\
&= \frac{1}{\sqrt{2\pi\bar{t}}}A(\bar{t})\int_{-\infty}^{\infty} d\bar{W}\frac{\cos(2\sqrt{\eta}\bar{W})+B(\bar{t})}{1+2B(\bar{t})\cos(2\sqrt{\eta}\bar{W})+B(\bar{t})^2}e^{-\frac{1}{2}\bar{W}^2/\bar{t}} \\
&= \frac{1}{\sqrt{2\pi\bar{t}}}\frac{A(\bar{t})}{B(\bar{t})}\int_0^{\infty} d\bar{W}\left[1-\frac{1-B(\bar{t})^2}{1+2B(\bar{t})\cos(2\sqrt{\eta}\bar{W})+B(\bar{t})^2}\right]e^{-\frac{1}{2}\bar{W}^2/\bar{t}} \\
&= \begin{cases} \frac{A(\bar{t})}{B(\bar{t})}\sum_{k=1}^{\infty}(-1)^{k+1}B(\bar{t})^k e^{-k^2 2\eta\bar{t}} & B(\bar{t})^2<1 \\ \frac{A(\bar{t})}{B(\bar{t})}\left[1+\sum_{k=1}^{\infty}(-1)^k B(\bar{t})^{-k}e^{-k^2 2\eta\bar{t}}\right] & B(\bar{t})^2>1 \end{cases},
\end{aligned} \tag{12.138}$$

where the final result is taken from Gradshteyn and Ryzhik [12.27].

The pathological behavior of the "toy" model follows directly from (12.138). Note, in particular, how the this result is conditional on the value of $B(\bar{t})^2$—a quantity that changes with time. From this, the mean photon number evolves *discontinuously* in time; at least, this will be the case whenever $|\alpha_0|^2 > (1-\eta)/2\eta$ [so that $B(\infty)>1$]. The discontinuity occurs at the time

$$\bar{t}_D = \frac{1}{2(\eta-1)}\ln\left[\frac{|\alpha_0|^2+(\eta-1)/2\eta}{|\alpha_0|^2}\right], \tag{12.139}$$

where there is a jump in the value of the mean photon number,

$$D \equiv \big(\overline{(\tilde{\bar{\alpha}}_*\tilde{\bar{\alpha}})_{\rm toy}(\bar{t})}\,\big)_P\Big|_{\bar{t}=\bar{t}_D^+} - \big(\overline{(\tilde{\bar{\alpha}}_*\tilde{\bar{\alpha}})_{\rm toy}(\bar{t})}\,\big)_P\Big|_{\bar{t}=\bar{t}_D^-}, \tag{12.140}$$

given, from (12.138), by the explicit expression

$$\begin{aligned}
D &= A(\bar{t}_D)\left[1+2\sum_{k=1}^{\infty}(-1)^k e^{-k^2\eta\bar{t}_D}\right] \\
&= \big[|\alpha_0|^2+(\eta-1)/2\eta\big]\left\{1+2\sum_{k=1}^{\infty}(-1)^k\left[\frac{|\alpha_0|^2}{|\alpha_0|^2+(\eta-1)/2\eta}\right]^{k^2\eta/(\eta-1)}\right\}.
\end{aligned} \tag{12.141}$$

This discontinuity is the clue to understanding what has gone wrong in the simulation of the full model, Fig. 12.9.

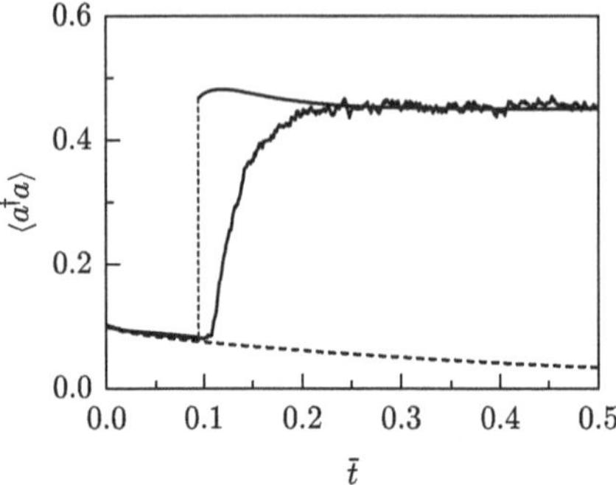

Fig. 12.10. Discontinuity in the "toy" model (Eq. 12.138) compared with a simulation of the positive P stochastic differential equation (12.127). The parameters are $|\alpha_0|^2 = 0.1$ and $\eta = 10$ ($D = 0.385$ and $\bar{t}_D = 0.0947$). The simulation averages 10,000 stochastic trajectories with time step $\Delta\bar{t} = 10^{-4}$. The *dashed curve* shows the correct result obtained from master equation (12.10) (with $\bar{\mathcal{E}}_i = 0$)

For the parameters used in Fig. 12.9, the values of $\bar{t}_D$ and D are $\bar{t}_D = 2.45 \times 10^{-3}$ and $D = 3.48 \times 10^{-15}$. The numbers seem so small as to be of no consequence at all. Changing the initial photon number, though, from $|\alpha_0|^2 = 10$ to $|\alpha_0|^2 = 0.1$, gives $t_D = 9.5 \times 10^{-2}$ and $D = 3.85 \times 0^{-1}$. Then the discontinuity is the dominant feature of the decay, as seen in Fig. 12.10, where the analytical expression (12.138) for the "toy" model is compared with a numerical simulation of the exact solution (12.134). The "toy" model shows a dramatic discontinuity in this case, and although the discontinuity is smoothed out by the exact solution, the pathological behavior remains essentially the same: at time $\bar{t} \sim \bar{t}_D$ both models abruptly switch away from following the expected monotonic decay.

The source of the abrupt change can be traced to the denominators of (12.134) and (12.135). Both produce divergences—similar to those of Sect. 12.1.5—for certain choices of the stochastic path. This does not mean that a simulation necessarily produces many trajectories that race off to infinity. The probability density in the variable $\mathrm{Re}(\tilde{\bar{\alpha}}_*\tilde{\bar{\alpha}})$ can approach zero at infinity, while at the same time there is a finite contribution to the mean from the tail of the distribution—essentially, the approach of the probability density to zero at infinity is too slow. In the case of the positive P representation, a further subtlety must be considered. For it the variable $\mathrm{Re}(\tilde{\bar{\alpha}}_*\tilde{\bar{\alpha}})$ need not be positive, unlike its counterpart, $\tilde{\bar{\alpha}}^*\tilde{\bar{\alpha}}$, in the Glauber–Sudarshan P representation. It is then possible to have an outward probability current at $\mathrm{Re}(\tilde{\bar{\alpha}}_*\tilde{\bar{\alpha}}) = \pm\infty$ balanced by an inward probability current at $\mathrm{Re}(\tilde{\bar{\alpha}}_*\tilde{\bar{\alpha}}) = \mp\infty$. This, in fact, is how the abrupt switching seen in Fig. 12.10 occurs. For the "toy" model, the probability current at infinity is isolated as a δ-function in time, contributing only at time $\bar{t}_D$, while for the full stochastic model it flows during an interval of time around (near) $\bar{t}_D$.

These ideas can be demonstrated in detail for the "toy" model. From (12.135), at any time and for all choices of the random variable $\bar{W}$, the variable $(\tilde{\bar{\alpha}}_* \tilde{\bar{\alpha}})_{\text{toy}}(\bar{t})$ lies on a circle in the complex plane. The particular circle changes over time. Figure 12.11a shows its evolution for the parameters of Fig. 12.10. Initially, so long as $\bar{t}$ is smaller than $\bar{t}_D$, the radius of the circle grows and the circle extends further and further into the negative half plane. At $\bar{t} = \bar{t}_D$, the radius is infinite. For $\bar{t}$ larger than $\bar{t}_D$, the circle shrinks again. The discontinuity results from the behavior connecting $\bar{t}_D^-$ and $\bar{t}_D^+$. At this instant, the circle switches from the negative half plane to the positive half plane. Thus, there is a δ-function-in-time probability current passing from $\text{Re}(\tilde{\bar{\alpha}}_* \tilde{\bar{\alpha}})_{\text{toy}} = -\infty$ to $\text{Re}(\tilde{\bar{\alpha}}_* \tilde{\bar{\alpha}})_{\text{toy}} = +\infty$ at $\bar{t} = \bar{t}_D$. Figure 12.11b plots the probability density in the variable $\text{Re}\big((\tilde{\bar{\alpha}}_* \tilde{\bar{\alpha}})_{\text{toy}}\big)$ for the earliest time in the sequence of Fig. 12.11a. Note that the density has two integrable divergences at the boundaries of its domain. Similar divergences are present at all times. Being integrable, the divergences are of no concern in themselves; they arise from the Jacobian when changing variables from $\bar{W}$ to $\text{Re}\big((\tilde{\bar{\alpha}}_* \tilde{\bar{\alpha}})_{\text{toy}}\big)$. The significant points are: (i) that the divergence at the lower boundary travels in the negative direction to pass from $-\infty$ to $+\infty$ at $\bar{t} = \bar{t}_D$, and (ii) the integrated probability under the lower divergence approaches zero sufficiently slowly, as the divergence moves to $-\infty$, that for $\bar{t} = \bar{t}_D$ the probability current at infinity contributes a finite amount to the mean.

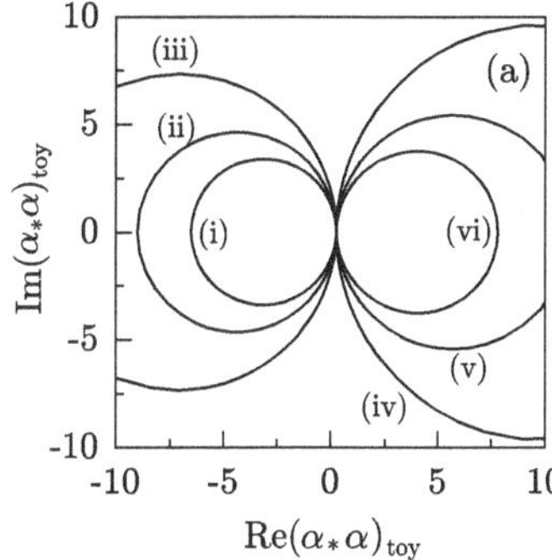

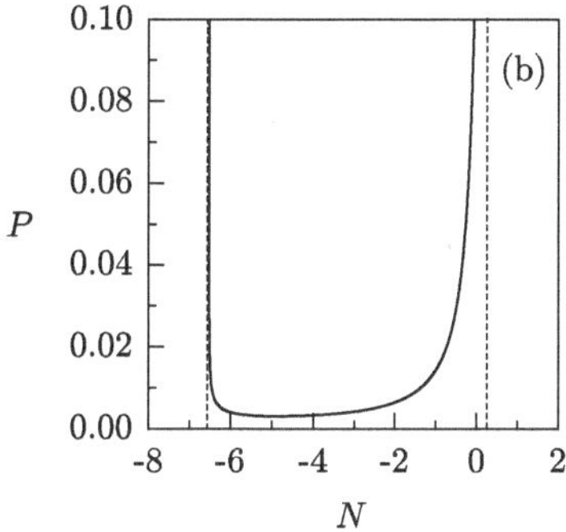

Fig. 12.11. (**a**) Contours $(\alpha_*\alpha)_{\text{toy}}(\theta) = A(\bar{t})/[B(\bar{t}) + e^{-i\theta}]$ (Eq. 12.135) plotted for the parameters of Fig. 12.10 at (i) $\bar{t} = 0.091$, (ii) $\bar{t} = 0.092$, (iii) $\bar{t} = 0.093$, (iv) $\bar{t} = 0.096$, (v) $\bar{t} = 0.097$, and (vi) $\bar{t} = 0.098$. (**b**) The probability distribution $P(N, \bar{t})$ (Eq. 12.142) at $\bar{t} = 0.091$; the distribution diverges at the boundaries $N = A(\bar{t})/[B(\bar{t}) + 1] = 0.2678$ and $N = A(\bar{t})/[B(\bar{t}) - 1] = -6.519$

Exercise 12.11. From (12.135) and the distribution over the random variable $\bar{W}$, show that the probability distribution over $N \equiv \mathrm{Re}\big((\tilde{\bar{\alpha}}_*\tilde{\bar{\alpha}})_{\mathrm{toy}}\big)$ is

$$P(N,\bar{t}) = \frac{A(\bar{t})}{|A(\bar{t}) - 2NB(\bar{t})|}\left|\frac{B(\bar{t})^2 - 1}{[A(\bar{t}) - NB(\bar{t})]^2 - N^2}\right|^{1/2} \frac{1}{\sqrt{2\pi\eta\bar{t}}}$$
$$\times \sum_{k=0}^{\infty} \exp\left[-\frac{1}{2}\frac{\theta(N)^2 + (k2\pi)^2}{4\eta\bar{t}}\right] \cosh\left[k\pi\theta(N)/2\eta\bar{t}\right], \qquad (12.142)$$

with

$$\theta(N) \equiv \cos^{-1}\left\{\frac{N[B(\bar{t})^2 + 1] - A(\bar{t})B(\bar{t})}{A(\bar{t}) - 2NB(\bar{t})}\right\}, \qquad (12.143)$$

where N lies inside the range bounded by $A(\bar{t})/[B(\bar{t})+1]$ and $A(\bar{t})/[B(\bar{t})-1]$.

Returning to our original observation, for the initial condition used in Fig. 12.9, the discontinuity itself is far too small to be noticed. Nonetheless, it is still the heart of the problem. In fact, the behavior in Figs. 12.9 and 12.10 is not as different as it first appears. In both the long-time solution approaches the same wrong steady state value; the only difference is that for the initial condition of Fig. 12.10, a large departure from the correct monotonic decay is needed to reach that state, while a small and gradual deviation of the path suffices for the initial condition of Fig. 12.10.

Of course the ultimate source of the difficulty with the positive P representation must lie somewhere in the derivation of the stochastic differential equation (12.127). From what we have just seen, it is fairly clear where the mistake must occur. We can trace it to the sentence below (11.28): "Integrating by parts on the right-hand side of (11.25) and *assuming that the boundary terms at infinity vanish*, ..." Apparently the stochastic process derived under the assumption that "the boundary terms at infinity vanish" can, in some cases at least, generate boundary terms that do not in fact vanish. Considering the sort of reorganization the deterministic nonlinear dynamics receives—i.e., its extrapolation into the nonclassical phase space (Sect. 12.1.5)—it is likely the boundary term assumption fails very often when the full nonlinear behavior of the stochastic process is put on display for small system size. There is never a problem, on the other hand, in the linear treatment of fluctuations (Sect. 11.1.3); it gives Gaussian distributions, which decay faster than any polynomial at infinity.

As a final word, it is important to say that it is not the positive P representation *per se* that fails when the boundary term assumption is not fulfilled. It remains true that any density operator has a positive P representation—given for example by (11.19). Thus, one can certainly, in principle, construct a family of positive P distributions, parameterized by time, to represent the time evolution of the density operator for two-photon damping. Nevertheless, the utility of the positive P representation (also of the Glauber–Sudarshan, Q, and Wigner representations) relies on more than the existence of a phase-space distribution representing the state at a particular time. The wider aim is to represent the *dynamics*, with the traditional assumption that the representation should take the form of a phase-space diffusion. The latter places a constraint on the family of distributions used to represent a density operator evolving in time: members of the family must be related to one another through a particular type of propagator, through the Fokker–Planck equation.

The positive P representation attempts to satisfy this aim in a clever way. It first abandons the idea that the distribution representing a state must be unique. It then tries to impose the assumption of a phase-space diffusion by selecting amongst the many distributions representing the state at a given time, to construct a unique *family* of distributions, parameterized by time, that satisfy a Fokker–Planck equation. When deriving the Fokker–Planck equation for the degenerate parametric oscillator (Sect. 11.1.2), the selection is made by fixing the interpretation of the partial derivatives, the step between (11.31) and (11.46). Ultimately, then, it is the failure of the dynamical assumption—the attempt to represent the dynamics as a diffusion process—that causes the technical difficulties with the two-photon damping example.

Note 12.9. Technical difficulties with the positive P representation are mitigated to a large extent by the stochastic gauge method introduced by Deuar and Drummond [12.28, 12.29, 12.30]. This approach exploits the nonuniqueness of positive P representations even further. It introduces still more degrees of freedom, setting the stochastic evolution in a still larger phase space. By a judicious choice of the added gauge variables, instabilities and divergences of the sort described can be avoided.

12.2.2 Physical Interpretation: The Anharmonic Oscillator

To add to any technical difficulties, the positive P representation also raises a question of interpretation. As we just noted, one aim of the phase-space approach is to represent quantum dynamics, bringing with it an implicit assumption of a meaningful physical interpretation. For example, stochastic realizations of the Glauber–Sudarshan P representation of a laser field may be considered, in a sense, to be realizations of an actual laser output. The physical interpretation—the "sense"—is grounded in the fact that the multi-time correlation functions of phase-space variables are equal to the normal-ordered, time-ordered correlation functions of field operators. Thus, a connection with photoelectric detection is established: assuming the representation yields a well-behaved stochastic process (positive semidefinite diffusion), individual realizations of that process elicit a response in the semiclassical simulation of photoelectric detection—random photoelectron counts at a rate governed by the stochastic field intensity—that is statistically equivalent to an ensemble of actual detection records.

The same connection might be claimed for the positive P representation, which also generates normal-ordered, time-ordered correlation functions, except that realizations of the positive P stochastic process generally explore the nonclassical dimensions of the extended phase space; generally, $\alpha_*(t) \neq \alpha^*(t)$, so that the stochastic intensity need not be positive definite, or even real, and a straightforward extension of the physical interpretation fails. The consequences of this are not so apparent in the small-noise limit, where the classical nonlinear dynamics provides a framework around which a physical interpretation may be built. As the system size decreases, though, the quantum noise increases and the classical framework dissolves. Then the positive P representation of the dynamics provides little, if any, physical insight, as demonstrated for the example of the degenerate parametric oscillator with adiabatic elimination of the pump in Sects. 12.1.7 and 12.1.8.

It might be said that the discussion of Sects. 12.1.7 and 12.1.8 does not make much of a case against the positive P representation as a source of physical insight. After all, it was pictures like those of Fig. 12.8 that suggested the two-state rate process described in Sect. 12.1.8. Recall, however, the technical difficulties and their resolution, obtained by confining the stochastic trajectories to a bounded region in phase space—the device looks more like a mathematical trick than a consequence of anything physical. Indeed, whenever the positive P representation proves successful outside the small-noise limit, it does appear to manage its success by employing some surprising mathematical trick.

Possibly the best demonstration of its cunning is given by the example considered in this section, the positive P stochastic process for the anharmonic oscillator. Here a rather involved stochastic evolution is used to generate normal-ordered operator averages that are strictly *constant* in time. Furthermore, although the evolution of the quantum state is periodic, the stochastic evolution is a Brownian motion that spreads, monotonically, deeper and deeper into the nonclassical phase space; thus, a quantum state that repeats periodically is represented by a different positive P distribution every time it recurs—the nonuniqueness of the positive P representation is made explicit.

The model Hamiltonian is

$$H = \hbar\omega_C a^\dagger a + \hbar\lambda(a^\dagger a)^2, \tag{12.144}$$

where λ sets the strength of the anharmonicity. One may readily solve for the normal-ordered operator averages as a function of time in either the Schrödinger or Heisenberg pictures. The positive P stochastic process representing the dynamics is also straightforward to obtain. The details of these calculations are unimportant and are left as an exercise.

Exercise 12.12. Consider the model Hamiltonian (12.144). Show that the *normal-ordered operator averages for the anharmonic oscillator prepared in initial coherent state* $|\alpha_0\rangle$ evolve periodically in time, with

$$\langle(\tilde{a}^{\dagger p}\tilde{a}^q)(\bar{t})\rangle = \alpha_0^{*p}\alpha_0^q \exp\{-|\alpha_0|^2\left[1 - e^{i(p-q)\bar{t}}\right]\}e^{i\frac{1}{2}[p(p-1)-q(q-1)]\bar{t}}, \tag{12.145}$$

where

$$a = \tilde{a}e^{-i(\omega_C+\lambda)t}, \qquad a^\dagger = \tilde{a}^\dagger e^{i(\omega_C+\lambda)t}, \tag{12.146}$$

and $\bar{t} \equiv 2\lambda t$.

Exercise 12.13. Derive the *stochastic differential equations in the positive P representation for the anharmonic oscillator*:

$$d\tilde{\alpha} = \tilde{\alpha}[-i(\tilde{\alpha}_*\tilde{\alpha})d\bar{t} + e^{-i\pi/4}d\bar{W}_\alpha], \tag{12.147a}$$

$$d\tilde{\alpha}_* = \tilde{\alpha}_*[i(\tilde{\alpha}_*\tilde{\alpha})d\bar{t} + e^{i\pi/4}d\bar{W}_{\alpha_*}], \tag{12.147b}$$

where $d\bar{W}_\alpha$ and $d\bar{W}_{\alpha_*}$ are independent Wiener increments of zero mean and variance $d\bar{t}$.

The normal-ordered operator averages are periodic with period $\bar{T} = 2\pi$, and, in fact, the anharmonic oscillator provides a particularly simple example of a quantum collapse and revival, with the initial state recurring exactly at times $\bar{t} = \bar{T}, 2\bar{T}, 3\bar{T}, \ldots$ [12.31, 12.32]. Our interest is with the unusual way in which the positive P representation handles these manifestly quantum mechanical dynamics. To this end, it is helpful first to consider the stochastic differential equation for photon number. From (12.147a) and (12.147b), we have

$$d(\tilde{\alpha}_*\tilde{\alpha}) = (\tilde{\alpha}_*\tilde{\alpha})(e^{-i\pi/4}d\bar{W}_\alpha + e^{i\pi/4}d\bar{W}_{\alpha_*}). \tag{12.148}$$

With this equation we can compute all normal-ordered operator averages with $p = q$. If this were our only aim, we might combine the independent noise terms, $e^{-i\pi/4}d\bar{W}_\alpha$ and $e^{i\pi/4}d\bar{W}_\alpha$, in a single complex Wiener increment of variance $e^{-i\pi/2}d\bar{t} + e^{i\pi/2}d\bar{t} = 0$. Thus, (12.148) is reduced to the statement $d(\tilde{\alpha}_*\tilde{\alpha}) = 0$; there is in fact no noise at all on the variable $(\tilde{\alpha}_*\tilde{\alpha})$. The result is hardly a surprise, as the photon number operator, $a^\dagger a$, commutes with Hamiltonian (12.144). It is therefore a constant of the motion, from which it follows that the normal-ordered operator averages with $p = q$ are also all constant in time.

Dropping the noise from (12.148) is not acceptable, however, if we aim to compute operator averages with $p \neq q$. We might wish to compute the mean field amplitude, for example, by simulating (12.147a) and (12.148) as a coupled set of equations [we might equivalently simulate (12.147a) and (12.147b)]. For this the correlation between the noises in the two equations must be kept, since it is easily shown that on setting $(\tilde{\alpha}_*\tilde{\alpha})$ constant in (12.147a), we replace the correct solution for the mean field amplitude,

$$\langle\tilde{a}(\bar{t})\rangle = \alpha_0 \exp[-|\alpha_0|^2(1-\cos\bar{t})]\exp(-i|\alpha_0|^2\sin\bar{t}), \tag{12.149}$$

with its short-time approximation, $\langle\tilde{a}(\bar{t})\rangle = \alpha_0\exp[-i|\alpha_0|^2\bar{t}]$. The collapse and revival has been lost. Clearly, then, we face the following strange and unfamiliar (from classical noise processes) situation:

In the presence of a conserved physical quantity, the positive P stochastic representation of the quantum dynamics generally requires that there be noise on the corresponding phase-space variable, even though every single moment of that variable is constant in time.

To understand in detail how the positive P stochastic process for the anharmonic oscillator works, we consider (12.148) without combining the Wiener increments on the right-hand side. Consider the equation

$$\begin{aligned} d\ln(\tilde{\alpha}_*\tilde{\alpha}) &= \ln[(\tilde{\alpha}_*\tilde{\alpha}) + d(\tilde{\alpha}_*\tilde{\alpha})] - \ln(\tilde{\alpha}_*\tilde{\alpha}) \\ &= \frac{1}{(\tilde{\alpha}_*\tilde{\alpha})}d(\tilde{\alpha}_*\tilde{\alpha}) - \frac{1}{2}\left[\frac{1}{(\tilde{\alpha}_*\tilde{\alpha})}d(\tilde{\alpha}_*\tilde{\alpha})\right]^2 \\ &= e^{-i\pi/4}d\bar{W}_\alpha + e^{i\pi/4}d\bar{W}_{\alpha_*}, \end{aligned} \tag{12.150}$$

whose solution for the initial state $|\alpha_0\rangle$ may be written as

$$\begin{aligned} (\tilde{\alpha}_*\tilde{\alpha})(\bar{t}) &= |\alpha_0|^2 \exp(e^{-i\pi/4}\bar{W}_\alpha + e^{i\pi/4}\bar{W}_{\alpha_*}) \\ &= |\alpha_0|^2 e^{\bar{W}_+} e^{-i\bar{W}_-}, \end{aligned} \tag{12.151}$$

where $\bar{W}_+ = (\bar{W}_\alpha + \bar{W}_{\alpha_*})/\sqrt{2}$ and $\bar{W}_- = (\bar{W}_\alpha - \bar{W}_{\alpha_*})/\sqrt{2}$ are independent Wiener processes that determine, respectively, the amplitude and phase of $(\tilde{\alpha}_*\tilde{\alpha})(\bar{t})$.

The unusual character of the positive P stochastic process is now made clear. By expanding $e^{\bar{W}_+}$ in a power series, it is readily shown that it has the average value $\langle e^{\bar{W}_+}\rangle = e^{\bar{t}/2}$. Thus, the average *amplitude* of $(\tilde{\alpha}_*\tilde{\alpha})(\bar{t})$ grows exponentially, while the average value of the complex number itself remains constant in time. This somewhat surprising combination of results is allowed because $(\tilde{\alpha}_*\tilde{\alpha})$, unlike $|\tilde{\alpha}|^2$ (the photon number phase-space variable in the Glauber–Sudarshan representation), is neither real nor positive definite. While the positive P stochastic process explores more and more of the $(\tilde{\alpha}_*\tilde{\alpha})$-plane over time (the mean amplitude grows exponentially), a cancellation of phases in the ensemble average is nevertheless possible, such that $\left(\overline{(\tilde{\alpha}_*\tilde{\alpha})(\bar{t})}\right)_{\tilde{P}} = |\alpha_0|^2$. The cancellation is put into effect by the phase factor $e^{-i\bar{W}_-}$, whose average value is $\langle e^{-i\bar{W}_-}\rangle = e^{-\bar{t}/2}$; the phase average brings an exponential decay to cancel the exponential growth of the amplitude average.

The principal lesson from the anharmonic oscillator example is that while the stochastic process (12.151) serves, formally, as a generator of operator averages, it does so using a phase-space dynamics that clearly has nothing to do with the physics. Then there is a secondary but important practical lesson: numerical simulations of the stochastic differential equations (12.147a) and (12.147b) are bound to fail for long enough times. This follows from the spreading of the positive P stochastic process in the nonclassical phase space and the phase-averaging in the ensemble average needed to recover physical results. Even before the revival time is reached, $\bar{t} = \bar{T}$, an accurate average in the presence of sampling error calls for ensembles of an impossibly large size.

Exercise 12.14. Show that (12.151) gives normal-ordered, time-ordered operator averages for $p = q$ that are constant in time.

Exercise 12.15. Formally solve the stochastic differential equations for the anharmonic oscillator, (12.147a) and (12.147b); hence verify the result for normal-ordered operator averages (12.145).

13

Cavity QED I: Simple Calculations

Having seen that fluctuations outside the small-noise limit present the phase-space methods with fundamental difficulties, we look in the remaining four chapters at alternative ways of treating quantum fluctuations in this regime. The topic is not entirely new to us, since we handled the problem of resonance fluorescence in Chapter 2 without the need for either phase-space representations or a system size expansion. Of course this example is rather trivial, falling as it does within the class of solvable one-particle problems. Phase-space representations do an excellent job in the opposite limit, where the nonlinear physics builds upon the cooperation of many atoms and many photons, and the fluctuations may be viewed as small perturbations ("fuzz") about classical nonlinear dynamics—the positive P representation even allows the "fuzz" to be squeezed. The real difficulties lie in the intermediate regime, where it may be hard, if not impossible, to solve density matrix equations exactly as we did for resonance fluorescence, and yet a small-noise approximation to a phase-space equation of motion may not be made either. We meet two new techniques in the remaining chapters: a systematic expansion of the density matrix equations for weak excitation (Sects. 16.1 and 16.3.4) and the quantum trajectory method (Chaps. 17–19). There is also more to be done with exact solutions of density matrix equations for systems of small or moderate size, analytically in the former case (Sect. 13.2) and in the latter with the help of a computer (Sect. 16.3.6).

The theme we take up here has been laid out in some detail for the degenerate parametric oscillator in Chap. 12; but this system is not a good candidate for realizing a small system size in the laboratory. A far better one is provided by the near-resonant interaction of atoms with an optical cavity mode. Systems of this type have been studied experimentally for more than two decades. At least one half of that time has focused explicitly on conditions beyond the small-noise limit—this is the regime of so-called cavity quantum electrodynamics (cavity QED). The early studies of atoms in cavities were concerned with optical bistability, and later dynamical instabilities and chaos; we will look at a little of this physics too since it is certainly a part of the overall picture. On the whole, though, it is the quantum fluctuations beyond the small-noise limit that interest us the most, so many things outside this scope we will simply overlook. The classical nonlinear dynamics of atoms in cavities is covered in the books of Gibbs [13.1] and Narducci and Abraham [13.2], while Lugiato [13.3] gives an excellent review of the topic of quantum fluctuations in the small-noise limit.

13.1 System Size and Coupling Strength

To reiterate, when the system size parameter is small, we enter a regime of large fluctuations where the phase space methods based on the Fokker–Planck equation appear, perhaps inevitably, to encounter difficulties. There is a natural scale to define "large"—the scale of the system nonlinearity, or more precisely, the number of elementary excitations (quanta) required to bring the nonlinear physics into play. Thus, the system size parameter is generally a photon number: for the laser, it is the saturation photon number $n_{\mathrm{sat}} = \gamma_h(\gamma_\uparrow + \gamma_\downarrow)/8g^2$ (essentially the square of the threshold photon number); for the degenerate parametric oscillator, it is the threshold photon number, $n_p^{\mathrm{thr}} = (\kappa/g)^2$, for the pump; for the degenerate parametric oscillator with adiabatic elimination of the pump, it is the number $2\xi^{-1}n_p^{\mathrm{thr}} = 2\kappa\kappa_p/g^2$ (essentially the square of the threshold photon number for the subharmonic mode).

In each of these cases the system size parameter is a product of ratios of dissipation rates, or linewidths, and a fundamental coupling constant. Here is a clue to an alternative view of things. In this view, for the words *system size* we substitute *coupling strength* (small system size means large coupling strength) and the reference to *nonlinearity* is understood in terms of *multi-photon transitions*, which take place, of course, within an energy spectrum influenced by the coupling strength. The issue of small versus large becomes one of energy level resolution—a question of the sizes of energy level shifts or splittings (coupling strength) compared with energy level widths. This is a more appropriate language than that of classical nonlinear dynamics plus "fuzz" once we enter the domain of small system size.

We can illustrate these ideas with the degenerate parametric oscillator Hamiltonian

$$H_S \equiv \hbar\omega_C a^\dagger a + \hbar 2\omega_C b^\dagger b + i\hbar(g/2)(a^{\dagger 2}b - a^2 b^\dagger). \tag{13.1}$$

The energy level diagram for the ground state and first two excited states is shown in Fig. 13.1. The eigenstates are

$$|G\rangle = |0\rangle_a|0\rangle_b, \qquad |E_1\rangle = |1\rangle_a|0\rangle_b, \tag{13.2}$$

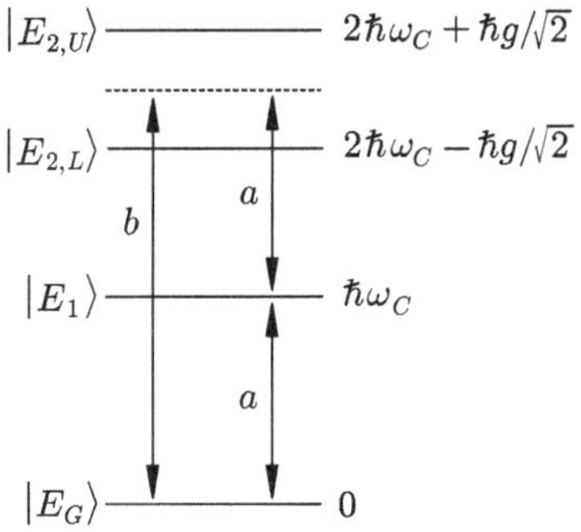

Fig. 13.1. Energy level diagram for the degenerate parametric oscillator showing the ground state, the first excited state, and the second excited state doublet

and

$$|E_{2,U}\rangle = \frac{1}{\sqrt{2}}\big(|2\rangle_a|0\rangle_b - i|0\rangle_a|1\rangle_b\big), \tag{13.3a}$$

$$|E_{2,L}\rangle = \frac{1}{\sqrt{2}}\big(|2\rangle_a|0\rangle_b + i|0\rangle_a|1\rangle_b\big), \tag{13.3b}$$

with corresponding energies,

$$E_G = 0, \qquad E_1 = \hbar\omega_C, \tag{13.4}$$

and a doublet at the second excited state,

$$E_{2,U} = 2\hbar\omega_C + \hbar g/\sqrt{2}, \tag{13.5a}$$

$$E_{2,L} = 2\hbar\omega_C - \hbar g/\sqrt{2}. \tag{13.5b}$$

The system size parameters are $n_p^{\text{thr}} = (\kappa/g)^2$ and $2\xi^{-1}n_p^{\text{thr}} = 2\kappa\kappa_p/g^2$.

Whether the system size parameters are small numbers or large numbers amounts to asking how well the doublet at the second excited state is resolved, with the relevant scale set by the linewidths κ and κ_p. Is the coupling, by this measure, weak or is it strong? We should note that it is the fundamental coupling constant g that is the issue—i.e., the splitting produced by a single pump quantum. Level shifts of much larger size appear at high excitation and might even have dynamical consequences (for example the Rabi oscillation of a two-level atom). It is not, however, the appearance of level shifts per se that interests us in cavity QED. It is the fine scale of the energy spectrum. We ask whether the addition or subtraction of just one quantum, or perhaps a few, is noticeable as observable physics. If the answer is yes, we will say, by definition, that we are dealing with a phenomenon in *cavity QED*—equivalently a phenomenon in the regime of small system size where the treatment of fluctuations can no longer rely on concepts adapted to the small-noise limit.

There are two distinct senses in which the coupling may be strong. The obvious one defines *cavity QED in the nonperturbative limit.* Clearly, if both $g/\kappa > 1$ and $g/\kappa_p > 1$, the doublet at the second excited state is resolved and observable spectroscopically (Sects. 13.3.1 and 13.3.2), and the well-split energy levels create what is essentially a new quantum system. Consider, for example, the subharmonic mode driven at frequency ω_C, so that the transition from the ground state to the first excited state is resonantly excited. Assuming a large enough splitting, both transitions to the second excited state are far from resonance. The lower levels then behave as a two-state system (Sect. 13.3.3), and all of the fluctuation phenomena of Sect. 2.3 should emerge to replace the usual behavior of the degenerate parametric oscillator.

Strong coupling of a slightly different kind underlines the definition of *cavity QED in the perturbative limit.* In this case we take $g/\kappa \gg 1$, $g/\kappa_p \ll 1$, with $g^2/\kappa\kappa_p \sim 1$. These are the appropriate strong coupling conditions when the pump mode is adiabatically eliminated (Sects. 12.1.1 and 12.1.3). They allow that mode to act as a reservoir, opening up a second channel for the loss of subharmonic photons. Thus, master equation (12.10) acquires its two-photon loss term with rate $\sim g^2/\kappa_p$. What we have here, in the language of cavity QED, is an example of cavity-enhanced emission (Sect. 13.2.1). Since the additional loss is nonlinear, its presence is felt in this case even if the coupling is arbitrarily weak; it certainly cannot be neglected for sufficiently high excitation. Once again, though, it is the effect at the level of a few quanta that is of concern in cavity QED. The requirement for an enhanced loss rate at this level is $g^2/\kappa_p \sim \kappa$.

13.2 Cavity QED in the Perturbative Limit

As noted above, the degenerate parametric oscillator is not a suitable system for studying cavity QED in the laboratory. The strong coupling requirements cannot be met. They can for the near-resonant interaction of a cavity mode with an atomic transition. We therefore make our initial survey of cavity QED effects by considering the system illustrated in Fig. 13.2. A single two-level atom is localized at an antinode of a standing-wave cavity that supports one TEM_{00} mode near-resonant with the atom. The coupling to all nonresonant modes is neglected. The cavity has length L and the Gaussian mode waist is w_0. The system S, so defined, is described by the *Jaynes–Cummings Hamiltonian* [13.4, 13.5],

$$\begin{aligned} H_S &= H_A + H_F + H_{AF} \\ &\equiv \tfrac{1}{2}\hbar\omega_A\sigma_z + \hbar\omega_C a^\dagger a + i\hbar g(a^\dagger\sigma_- - a\sigma_+), \end{aligned} \tag{13.6}$$

with dipole coupling constant

$$g \equiv \sqrt{\frac{\omega_C d^2}{2\hbar\epsilon_0 V_Q}}, \qquad V_Q \equiv \pi(w_0/2)^2 L, \tag{13.7}$$

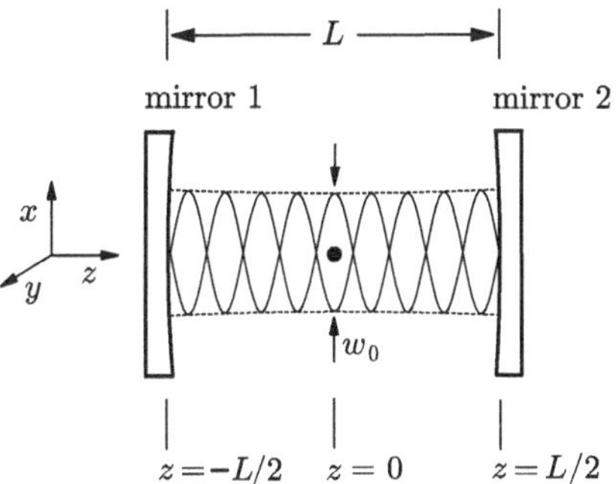

Fig. 13.2. Schematic diagram of single-atom cavity QED. The atom is localized on axis at an antinode of the standing-wave TEM_{00} mode

where V_Q is the mode volume. The cavity field operator is

$$\hat{\boldsymbol{E}}(\boldsymbol{r},t) = i\sqrt{\frac{\hbar\omega_C}{2\epsilon_0 V_Q}}\cos[(\omega_C/c)z+\phi_C]e^{-(x^2+y^2)/w_0^2} \times \left[\boldsymbol{e}_0\tilde{a}(t)e^{-i(\omega_C t-\phi_C')} - \boldsymbol{e}_0^*\tilde{a}^\dagger(t)e^{i(\omega_C t-\phi_C')}\right], \tag{13.8}$$

where $\phi_C' \equiv \arg(-\boldsymbol{e}_0^*\cdot\boldsymbol{d}_{12})$.

Note 13.1. We may alternatively write V_Q as $V_Q'/2$ with $V_Q' \equiv \pi(w_0/\sqrt{2})^2 L$, where the explicit factor of $1/2$ is the spatial average of $\cos^2(k_C z)$. This is what was done in (9.82a); in that expression the factor $A\bar{L}$ corresponds to the volume V_Q'. Note that V_Q' is the volume of a cylinder whose radius is equal to the half-width of the Gaussian transverse profile of the mode *amplitude*; when the $1/2$ is absorbed into the definition of the mode volume, V_Q is the volume of a cylinder whose radius is equal to the half-width of the transverse profile of the mode *intensity*.

Setting out the rest of the model repeats things we have done before. The atom and cavity mode are both coupled to radiative reservoirs. For the cavity mode there is loss through the mirrors, with transmission coefficients T_1 and T_2, and there is spontaneous emission from the atom which exits through the sides of the cavity. The Hamiltonian for the complete reservoir R is

$$H_R = H_R^{a1} + H_R^{a2} + H_R^A, \tag{13.9}$$

with $(\mu = 1,2)$

$$H_R^{a\mu} \equiv \sum_j \hbar\omega_j r_{a\mu j}^\dagger r_{a\mu j}, \tag{13.10a}$$

$$H_R^A \equiv \sum_{\boldsymbol{k},\lambda} \hbar\omega_k r_{\boldsymbol{k},\lambda}^\dagger r_{\boldsymbol{k},\lambda}, \tag{13.10b}$$

and the interaction between S and R is

$$H_{SR} = H_{SR}^{a1} + H_{SR}^{a2} + H_{SR}^{A}, \tag{13.11}$$

with $(\mu = 1, 2)$

$$H_{SR}^{a\mu} \equiv \hbar(a\Gamma_{a\mu}^{\dagger} + a^{\dagger}\Gamma_{a\mu}), \tag{13.12a}$$

$$H_{SR}^{A} \equiv \hbar(\sigma_{-}\Gamma_{A}^{\dagger} + \sigma_{+}\Gamma_{A}), \tag{13.12b}$$

where

$$\Gamma_{a\mu}^{\dagger} \equiv \sum_{j} \kappa_{a\mu j}^{*} r_{a\mu j}^{\dagger}, \qquad \Gamma_{a\mu} \equiv \sum_{j} \kappa_{a\mu j} r_{a\mu j}, \tag{13.13a}$$

$$\Gamma_{A}^{\dagger} \equiv \sum_{\boldsymbol{k},\lambda} \kappa_{\boldsymbol{k},\lambda}^{*} r_{\boldsymbol{k},\lambda}^{\dagger}, \qquad \Gamma_{A} \equiv \sum_{\boldsymbol{k},\lambda} \kappa_{\boldsymbol{k},\lambda} r_{\boldsymbol{k},\lambda}. \tag{13.13b}$$

The master equation may be written down from (1.73) and (2.26), noting the issues concerning internal coupling discussed in Sect. 2.3.2. For simplicity we take $\bar{n} = 0$, which is appropriate for optical frequencies, and then our model is defined by the *master equation for single-atom cavity QED at zero temperature* ($\bar{n} = 0$)

$$\begin{aligned}\dot{\rho} = &-i\tfrac{1}{2}\omega_{A}[\sigma_{z}, \rho] - i\omega_{C}[a^{\dagger}a, \rho] + g[a^{\dagger}\sigma_{-} - a\sigma_{+}, \rho] \\ &+ \frac{\gamma}{2}(2\sigma_{-}\rho\sigma_{+} - \sigma_{+}\sigma_{-}\rho - \rho\sigma_{+}\sigma_{-}) \\ &+ \kappa(2a\rho a^{\dagger} - a^{\dagger}a\rho - \rho a^{\dagger}a),\end{aligned} \tag{13.14}$$

with spontaneous emission rate

$$\gamma = \frac{1}{4\pi\epsilon_{0}} \frac{4\omega_{A}^{3} d_{12}^{2}}{3\hbar c^{3}}, \tag{13.15}$$

the Einstein A coefficient, and cavity mode damping rate

$$\kappa \equiv \tfrac{1}{2}(\gamma_{a1} + \gamma_{a2}), \tag{13.16}$$

with $(\mu = 1, 2)$

$$\gamma_{a\mu} \equiv 2\pi g_{a\mu}(\omega_{C})|\kappa_{a\mu}(\omega_{C})|^{2}. \tag{13.17}$$

Written in terms of the mirror transmission coefficients (see the paragraph below Note 7.10), the cavity linewidth is

$$\kappa = \tfrac{1}{2}(T_{1} + T_{2})c/2L. \tag{13.18}$$

Note 13.2. Strictly speaking γ is not the Einstein A coefficient, since the sum over modes in the expressions for Γ_{A} and $\Gamma_{A}^{\dagger}$ excludes those modes whose k-vectors lie within the solid angle subtended by the cavity. These are accounted for in the interaction Hamiltonian H_{AF} and the coupling of the cavity mode to its own two reservoirs. By taking γ to be the full Einstein A coefficient, we specialize to the case where the solid angle subtended by the cavity mode is very small. Cavities of this type are used quite widely in cavity QED experiments at optical frequencies [13.6, 13.7, 13.8, 13.9].

The output fields, the fields transmitted though the cavity mirrors, are excitations of the reservoir. They may be expanded in a similar fashion to those of Chap. 9 [Sect. 9.2.5, with $AL' \to 2V_Q' \equiv \pi(w_0/2)^2 L'$]. Written in photon flux units (strictly the square root of photon flux), there is the forward-propagating field ($z > L/2$)

$$\hat{\mathcal{E}}_{\rightarrow}(z,t) = \sqrt{c/2L'}r_{a2f}(t') + \sqrt{\gamma_{a2}}a(t'), \tag{13.19}$$

where $ct' \equiv ct - (z - L/2)$, and the backward-propagating field ($z < -L/2$)

$$\hat{\mathcal{E}}_{\leftarrow}(z,t) = \sqrt{c/2L'}r_{a1f}(t') + \sqrt{\gamma_{a1}}a(t'), \tag{13.20}$$

where $ct' \equiv ct + (z + L/2)$. Both are composed of a free field and a source field radiated by the cavity QED system. The free fields are expanded in reservoir modes as

$$r_{a2f}(t') \equiv -ie^{i(\phi_R - \phi_T)}e^{-i[(\omega_C/c)L/2 + \phi_C + \phi_C']} \sum_j \sqrt{\frac{\omega_j}{\omega_C}} r_{a2j}(0)e^{-i\omega_j t'}, \tag{13.21a}$$

$$r_{a1f}(t') \equiv -ie^{i(\phi_R - \phi_T)}e^{-i[(\omega_C/c)L/2 - \phi_C + \phi_C']} \sum_j \sqrt{\frac{\omega_j}{\omega_C}} r_{a1j}(0)e^{-i\omega_j t'}, \tag{13.21b}$$

where ϕ_R and ϕ_T are phase changes upon reflection and transmission at the mirrors, and ϕ_C and ϕ_C' are defined by (13.8).

It might be helpful to recall the phase relationships encountered when setting up cavity modes for the laser and the degenerate parametric oscillator. Firstly, energy conservation at the mirrors requires (Note 7.12)

$$\phi_R - \phi_T = \pi/2. \tag{13.22a}$$

Then there is the resonance condition for the cavity mode (Eq. 9.9),

$$2[\phi_R + (\omega_C/c)L] = N2\pi, \qquad N \text{ an integer}, \tag{13.22b}$$

and finally, from the boundary condition at $L/2$ (Eq. 9.10),

$$\phi_R + [(\omega_C/c)L + 2\phi_C] = M2\pi, \qquad M \text{ an integer}. \tag{13.22c}$$

13.2.1 Cavity-Enhanced Spontaneous Emission

In 1946 Purcell proposed the first scheme for realizing enhanced spontaneous emission [13.10]. No cavity was involved, but the main idea was the same: if a two-level system is coupled to a damped oscillator with high enough coupling strength, and the oscillator has weak enough damping, a new loss channel can be opened up to significantly increase the overall radiative decay rate of the two-level system. Purcell had a radiofrequency transition in mind, and the damped oscillator was a resonant electronic circuit. The idea has since been implemented at both microwave [13.11] and optical frequencies [13.12, 13.13] as true *cavity-enhanced* emission, with the role of the oscillator taken by a mode of an electromagnetic cavity. Purcell's idea also works in the absence of an actual cavity, but near a reflecting surface [13.14, 13.15], where one can think of the enhancement as being due to constructive interference between an outgoing and a reflected radiation reaction field.

Cavity-enhanced spontaneous emission is an effect in the perturbative limit of cavity QED; it is explained within perturbation theory. While it is desirable that the damping of the oscillator should not be too fast, its decay time must nevertheless be short enough for the oscillator to be treated as a reservoir; any photon transferred to the oscillator must be dissipated before being reabsorbed by the two-level system—alternatively, the oscillator quasimode bandwidth must be much larger than the linewidth of the two-level system, thus permitting the application of Fermi's golden rule.

It is often said that the enhanced emission rate is caused by an increase in the density of states; but while the density of states is certainly an important factor in Fermi's golden rule, this way of speaking is not strictly correct. Imagine the cavity of Fig. 13.2, embedded in a one-dimension universe, as in Fig. 9.4. Surely the presence of the cavity hardly changes the density of states. The length of the cavity is negligible ($L/L' \to 0$), and the density of states is essentially $2L'/\pi c$ whether there is a cavity present or not. What changes are the mode functions, hence the mode coupling strengths to the two-level atom. There are some modes, near-resonant with the cavity, that have much greater amplitudes inside than outside the cavity, and there are other nonresonant modes that have much smaller amplitudes inside than out. If the atom is located inside the cavity, it couples more strongly to those modes that have increased amplitude inside the cavity; it still couples to these modes if located outside the cavity but of course at considerably reduced strength. When the more strongly coupled modes are the main ones for determining the atomic decay rate—i.e., the atomic transition is near-resonant with the cavity—the decay rate is increased. It is thus the enhancement of the electromagnetic field amplitude inside a resonant cavity that increases the spontaneous emission rate. Although the context is quantum mechanical when we speak of spontaneous emission, the mechanism increasing the emission rate is largely classical and could have been anticipated from antenna theory.

In this section we look at the perturbative treatment of cavity-enhanced emission for the system depicted in Fig. 13.2. We approach the problem from two different points of view, the first set in the Schrödinger picture using the master equation approach, and the second based upon Heisenberg equations of motion.

In the first calculation, we adiabatically eliminate the cavity mode from master equation (13.4), following the approach of Sect. 12.1.3. We write the master equation in the interaction picture,

$$\rho = e^{-i[(\frac{1}{2}\omega_A\sigma_z+\omega_C a^\dagger a)t]}\tilde{\rho}e^{i[(\frac{1}{2}\omega_A\sigma_z+\omega_C a^\dagger a)t]},$$

and for exact resonance ($\omega_C = \omega_A$) as

$$\dot{\tilde{\rho}} = (\mathcal{L}_A + \mathcal{L}_a + \mathcal{L}_{Aa})\tilde{\rho}, \tag{13.23}$$

with superoperators

$$\mathcal{L}_A \equiv \frac{\gamma}{2}(2\sigma_-\cdot\sigma_+ - \sigma_+\sigma_-\cdot - \cdot\sigma_+\sigma_-), \tag{13.24a}$$

$$\mathcal{L}_a \equiv \kappa(2a\cdot a^\dagger - a^\dagger a\cdot - \cdot a^\dagger a), \tag{13.24b}$$

$$\mathcal{L}_{Aa} \equiv g[a^\dagger\sigma_- - a\sigma_+, \cdot]. \tag{13.24c}$$

We seek an equation of motion for the reduced density operator of the atom alone,

$$\tilde{\rho}_A(t) \equiv \mathrm{tr}_a[\tilde{\rho}(t)], \tag{13.25}$$

under the assumption that in the *bad-cavity limit*, $\kappa \gg \gamma/2, g$, the density operator approximately factorizes as

$$\tilde{\rho}(t) \approx \tilde{\rho}_A(t)\big(|0\rangle\langle 0|\big)_a. \tag{13.26}$$

Thus we introduce an *ansatz* equivalent to (12.29), where the cavity mode plays the role of a reservoir in the vacuum state. Then, following (12.30)–(12.32), we transform (13.23) to obtain

$$\dot{\bar{\rho}} = \bar{\mathcal{L}}_{Aa}(t)\bar{\rho}, \tag{13.27}$$

with

$$\bar{\rho}(t) \equiv e^{-(\mathcal{L}_A+\mathcal{L}_a)t}\tilde{\rho}(t), \tag{13.28}$$

and

$$\bar{\mathcal{L}}_{Aa}(t) \equiv e^{-(\mathcal{L}_A+\mathcal{L}_a)t}\mathcal{L}_{Aa}e^{(\mathcal{L}_A+\mathcal{L}_a)t}. \tag{13.29}$$

Introducing the trace over the cavity mode and following the earlier calculation to get equations equivalent to (12.26) and (12.37), we arrive at

$$\dot{\bar{\rho}}_A = \int_0^t dt'\mathrm{tr}_a[\bar{\mathcal{L}}_{Aa}(t)\bar{\mathcal{L}}_{Aa}(t')\bar{\rho}(t')], \tag{13.30}$$

where

$$\bar{\mathcal{L}}_{Aa}(t) = -g[\bar{\mathcal{A}}_1(t)\bar{\mathrm{a}}_1(t) - \bar{\mathcal{A}}_2(t)\bar{\mathrm{a}}_2(t) + \bar{\mathcal{A}}_1^\dagger(t)\bar{\mathrm{a}}_1^\dagger(t) - \bar{\mathcal{A}}_2^\dagger(t)\bar{\mathrm{a}}_2^\dagger(t)], \quad (13.31)$$

with

$$\begin{aligned} \bar{\mathcal{A}}_1(t) &\equiv e^{-\mathcal{L}_A t}(\sigma_+ \cdot)e^{\mathcal{L}_A t}, & \bar{\mathcal{A}}_2(t) &\equiv e^{-\mathcal{L}_A t}(\sigma_- \cdot)e^{\mathcal{L}_A t}, \\ \bar{\mathrm{a}}_1(t) &\equiv e^{-\mathcal{L}_a t}(a\ \cdot)e^{\mathcal{L}_a t}, & \bar{\mathrm{a}}_2(t) &\equiv e^{-\mathcal{L}_a t}(a^\dagger \cdot)e^{\mathcal{L}_a t}. \end{aligned} \quad (13.32a)$$

It remains for us to find the explicit time dependences of the cavity mode operators, and then to take the trace and evaluate the integrals in the manner described below (12.39). For $\bar{\mathrm{a}}_1(t)$, $\bar{\mathrm{a}}_2(t)$ and their conjugates, using (12.23), (12.24), and (12.14), we obtain

$$\bar{\mathrm{a}}_1(t) = e^{-\kappa t}(a\ \cdot), \quad (13.33a)$$

$$\bar{\mathrm{a}}_1^\dagger(t) = e^{-\kappa t}(\cdot\ a^\dagger), \quad (13.33b)$$

$$\bar{\mathrm{a}}_2(t) = e^{\kappa t}(a^\dagger \cdot) + (e^{-\kappa t} - e^{\kappa t})(\cdot\ a^\dagger), \quad (13.33c)$$

$$\bar{\mathrm{a}}_2^\dagger(t) = e^{\kappa t}(\cdot\ a) + (e^{-\kappa t} - e^{\kappa t})(a\ \cdot). \quad (13.33d)$$

Finally, after retracing the steps from (12.40) to (12.43), we reach the equation of motion

$$\dot{\tilde{\rho}}_A = \frac{g^2}{\kappa}e^{-\mathcal{L}_A t}\big[2(\sigma_- \cdot \sigma_+) - (\sigma_+\sigma_- \cdot) - (\cdot\ \sigma_+\sigma_-)\big]e^{\mathcal{L}_A t}\tilde{\rho}_A; \quad (13.34)$$

thus, inverting the transformation (13.28), the *master equation for cavity-enhanced (on resonance) spontaneous emission at zero temperature* ($\bar{n} = 0$) is

$$\dot{\rho}_A = -i\tfrac{1}{2}\omega_A[\sigma_z, \rho_A] + \frac{\gamma}{2}(1 + 2C_1)(2\sigma_-\rho_A\sigma_+ - \sigma_+\sigma_-\rho_A - \rho_A\sigma_+\sigma_-), \quad (13.35)$$

where

$$2C_1 \equiv 2\frac{g^2}{\gamma\kappa} \quad (13.36)$$

is the *spontaneous emission enhancement factor.*

Perhaps the technical detail in this approach to the adiabatic elimination is a little complex; but when the algebra is all over, the physics that goes into the enhanced spontaneous emission rate is rather trivial: the square of the coupling constant is just what we would expect from a perturbative calculation in the spirit of Fermi's golden rule, while the inverse cavity linewidth comes from the integration of the cavity quasimode correlation function over time—in the density of states way of thinking, it is 1 mode divided by the mode linewidth.

Exercise 13.1. Repeat the adiabatic elimination of the cavity mode for nonzero temperature ($\bar{n} \neq 0$). Show that (13.33a)–(13.33d) are replaced by the expressions

$$\bar{\mathrm{a}}_1(t) = [(1+\bar{n})e^{-\kappa t} - \bar{n}e^{\kappa t}](a\,\cdot) - \bar{n}(e^{-\kappa t} - e^{\kappa t})(\cdot\, a), \tag{13.37a}$$

$$\bar{\mathrm{a}}_1^\dagger(t) = [(1+\bar{n})e^{-\kappa t} - \bar{n}e^{\kappa t}](\cdot\, a^\dagger) - \bar{n}(e^{-\kappa t} - e^{\kappa t})(a^\dagger\cdot), \tag{13.37b}$$

$$\bar{\mathrm{a}}_2(t) = [(1+\bar{n})e^{\kappa t} - \bar{n}e^{-\kappa t}](a^\dagger\cdot) + (1+\bar{n})(e^{-\kappa t} - e^{\kappa t})(\cdot\, a^\dagger), \tag{13.37c}$$

$$\bar{\mathrm{a}}_2^\dagger(t) = [(1+\bar{n})e^{\kappa t} - \bar{n}e^{-\kappa t}](\cdot\, a) + (1+\bar{n})(e^{-\kappa t} - e^{\kappa t})(a\,\cdot). \tag{13.37d}$$

Hence derive the *master equation for cavity-enhanced (on resonance) spontaneous emission at nonzero temperature*:

$$\begin{aligned}\dot{\rho}_A = {} & -i\tfrac{1}{2}\omega_A[\sigma_z, \rho_A] \\ & + \frac{\gamma}{2}(1+2C_1)(1+\bar{n})(2\sigma_-\rho_A\sigma_+ - \sigma_+\sigma_-\rho_A - \rho_A\sigma_+\sigma_-) \\ & + \frac{\gamma}{2}(1+2C_1)\bar{n}(2\sigma_+\rho_A\sigma_- - \sigma_-\sigma_+\rho_A - \rho_A\sigma_-\sigma_+).\end{aligned} \tag{13.38}$$

The treatment of spontaneous emission in Sect. 2.2 produced a frequency shift Δ (Eq. 2.22) to go along with the spontaneous decay rate γ (Eq. 2.21). No frequency shift has been found in the above treatment of cavity-enhanced spontaneous emission. This is not because there is none in principle. Indeed, it is apparent from (2.22) that the shift obtained for ordinary spontaneous emission would be zero if $g(\boldsymbol{k})|\kappa(\boldsymbol{k},\lambda)|^2$ were a symmetric function about the atomic resonance frequency $kc = \omega_A$. We have found no cavity-enhanced frequency shift (see [13.13] for example) for essentially the same reason: we took the cavity mode resonant with the atom, which makes its lineshape symmetric about ω_A. We now rederive the cavity-enhanced emission rate from a Heisenberg picture point of view. Here we include a detuning of the cavity mode from the atom so that the frequency shift will appear.

We make the calculation on the basis of coupled moment equations, working towards modified optical Bloch equations to describe the atomic decay. From master equation (13.14), the equations of motion for the coupled cavity mode amplitude, atomic polarization, and inversion are

$$\frac{d\langle a\rangle}{dt} = -(\kappa + i\omega_C)\langle a\rangle + g\langle\sigma_-\rangle, \tag{13.39a}$$

$$\frac{d\langle a^\dagger\rangle}{dt} = -(\kappa - i\omega_C)\langle a^\dagger\rangle + g\langle\sigma_+\rangle, \tag{13.39b}$$

$$\frac{d\langle\sigma_-\rangle}{dt} = -\left(\frac{\gamma}{2} + i\omega_A\right)\langle\sigma_-\rangle + g\langle\sigma_z a\rangle, \tag{13.39c}$$

$$\frac{d\langle\sigma_+\rangle}{dt} = -\left(\frac{\gamma}{2} - i\omega_A\right)\langle\sigma_+\rangle + g\langle a^\dagger\sigma_z\rangle, \tag{13.39d}$$

$$\frac{d\langle\sigma_z\rangle}{dt} = -\gamma\big(\langle\sigma_z\rangle + 1\big) - 2g\big(\langle\sigma_+ a\rangle + \langle a^\dagger\sigma_-\rangle\big). \tag{13.39e}$$

The Heisenberg equation of motion for the cavity field amplitude [corresponding to (7.130) and (9.126)] is

$$\dot{a} = -(\kappa + i\omega_C)a + g\sigma_- - \sqrt{c/2L'}\sqrt{2\kappa}r_{af}, \tag{13.40}$$

where

$$r_{af} \equiv (1/\sqrt{2\kappa})\big(\sqrt{\gamma_{a1}}r_{a1f} + \sqrt{\gamma_{a2}}r_{a2f}\big), \tag{13.41}$$

with the free fields defined explicitly in (13.21a) and (13.21b).

Our approach is to adiabatically eliminate the cavity field by formally integrating the Heisenberg equation of motion and substituting the result into (13.39c)–(13.39e) thus obtaining a closed set of equations for the atom. The formal integration runs parallel to the integration used in the derivation of the scattered field in resonance fluorescence (Sect. 2.3.1) and the laser output field (Sect. 7.3.1). There, however, the reservoir modes were the focus of the attention; here it is the system operator that couples to the reservoir. The result of the formal integration of (13.40) is

$$\begin{aligned}\tilde{a}(t) = a(0)e^{-\kappa(1+i\Delta_C)t} &+ g\int_0^t dt'\tilde{\sigma}_-(t-t')e^{-\kappa(1+i\Delta_C)t'}\\ &- \sqrt{c/2L'}\sqrt{2\kappa}\int_0^t dt' e^{i\omega_A(t-t')}r_{af}(t-t')e^{-\kappa(1+i\Delta_C)t'},\end{aligned} \tag{13.42}$$

where we introduce the atom-cavity detuning,

$$\Delta_C \equiv \frac{\omega_C - \omega_A}{\kappa}, \tag{13.43}$$

and slowly-varying operators

$$\begin{aligned}\tilde{a} &\equiv a e^{i\omega_A t}, & \tilde{a}^\dagger &\equiv a^\dagger e^{-i\omega_A t},\\ \tilde{\sigma}_- &\equiv \sigma_- e^{i\omega_A t}, & \tilde{\sigma}_+ &\equiv \sigma_+ e^{-i\omega_A t}.\end{aligned} \tag{13.44a}$$

Note that although $\tilde{a}$ began its life as a system operator in master equation (13.14), we should read (13.42) as an expression relating a reservoir field to a source field after the fashion of (2.75) and (7.105) (also see Note 13.5); the second and third terms on the right-hand side of (13.42) are the source-field and free-field contributions, respectively. As noted above, in the perturbative regime of cavity QED, the cavity quasimode acts as a reservoir for the atom.

Our principal concern is with the mean-value equations for the atom. Written in terms slowly-varying operators, from (13.39c)–(13.39e), these are

$$\frac{d\langle\tilde{\sigma}_-\rangle}{dt} = -\frac{\gamma}{2}\langle\tilde{\sigma}_-\rangle + g\langle\sigma_z\tilde{a}\rangle, \tag{13.45a}$$

$$\frac{d\langle\tilde{\sigma}_+\rangle}{dt} = -\frac{\gamma}{2}\langle\tilde{\sigma}_+\rangle + g\langle\tilde{a}^\dagger\sigma_z\rangle, \tag{13.45b}$$

$$\frac{d\langle\sigma_z\rangle}{dt} = -\gamma\big(\langle\sigma_z\rangle + 1\big) - 2g\big(\langle\tilde{\sigma}_+\tilde{a}\rangle + \langle\tilde{a}^\dagger\tilde{\sigma}_-\rangle\big). \tag{13.45c}$$

Although the operator products could be written in either order, there is a definite advantage to adopting normal order as shown. With this ordering of the operator products, it follows immediately that the free-field terms may be dropped when (13.42) is substituted into (13.45a)–(13.45c); in every case, upon evaluating the expectation the free-field operators act upon the vacuum state of the reservoir R. With the operator products written in any other order, we would need to account explicitly for correlations between the free-field operators $r_{af}(t-t')$ and $r_{af}^{\dagger}(t-t')$ and the source operators $\tilde{\sigma}_-(t)$, $\tilde{\sigma}_+(t)$, and $\sigma_z(t)$ (see Sect. 7.3). As things are written, in (13.45a) for example, we have

$$\langle(\sigma_z\tilde{a})(t)\rangle = \langle\sigma_z(t)a(0)\rangle e^{-\kappa(1+i\Delta_C)t} + g\int_0^t dt'\langle\sigma_z(t)\tilde{\sigma}_-(t-t')\rangle e^{-\kappa(1+\Delta_C)t'}. \tag{13.46}$$

In the bad-cavity limit the cavity field correlation time is very short compared to the timescale for the decay of the atom. The first term on the right-hand side decays rapidly to zero, while, inside the integral, the atomic correlation function may be evaluated at $t' = 0$. Thus, in the adiabatic or Markov approximation, we can write

$$\begin{aligned}\langle(\sigma_z\tilde{a})(t)\rangle &= g\langle\sigma_z(t)\tilde{\sigma}_-(t)\rangle \int_0^\infty dt' e^{-\kappa(1+i\Delta_C)t'} \\ &= \frac{g}{\kappa}\frac{1-i\Delta_C}{1+\Delta_C^2}\langle(\sigma_z\tilde{\sigma}_-)(t)\rangle.\end{aligned} \tag{13.47a}$$

In a similar fashion, for the other three operator products, we have

$$\langle(\tilde{a}^\dagger\sigma_z)(t)\rangle = \frac{g}{\kappa}\frac{1+i\Delta_C}{1+\Delta_C^2}\langle(\tilde{\sigma}_+\sigma_z)(t)\rangle, \tag{13.47b}$$

$$\langle(\tilde{\sigma}_+\tilde{a})(t)\rangle = \frac{g}{\kappa}\frac{1-i\Delta_C}{1+\Delta_C^2}\langle(\tilde{\sigma}_+\tilde{\sigma}_-)(t)\rangle, \tag{13.47c}$$

$$\langle(\tilde{a}^\dagger\tilde{\sigma}_-)(t)\rangle = \frac{g}{\kappa}\frac{1+i\Delta_C}{1+\Delta_C^2}\langle(\tilde{\sigma}_+\tilde{\sigma}_-)(t)\rangle. \tag{13.47d}$$

Then, using (2.25a) and (2.132),

$$\langle\sigma_z\tilde{a}\rangle = -\frac{g}{\kappa}\frac{1-i\Delta_C}{1+\Delta_C^2}\langle\tilde{\sigma}_-\rangle, \tag{13.48a}$$

$$\langle\tilde{a}^\dagger\sigma_z\rangle = -\frac{g}{\kappa}\frac{1+i\Delta_C}{1+\Delta_C^2}\langle\tilde{\sigma}_+\rangle, \tag{13.48b}$$

$$\langle\tilde{\sigma}_+\tilde{a}\rangle = \frac{1}{2}\frac{g}{\kappa}\frac{1-i\Delta_C}{1+\Delta_C^2}\big(\langle\sigma_z\rangle+1\big), \tag{13.48c}$$

$$\langle\tilde{a}^\dagger\tilde{\sigma}_-\rangle = \frac{1}{2}\frac{g}{\kappa}\frac{1+i\Delta_C}{1+\Delta_C^2}\big(\langle\tilde{\sigma}_z\rangle+1\big). \tag{13.48d}$$

Finally, after substituting (13.48a)–(13.48d) into the right-hand side of (13.45a)–(13.45c), we arrive at the *optical Bloch equations for cavity-enhanced spontaneous emission at zero temperature* ($\bar{n}=0$):

$$\frac{d\langle\sigma_-\rangle}{dt} = -\left(\frac{\gamma'}{2} + i\omega'_A\right)\langle\sigma_-\rangle, \tag{13.49a}$$

$$\frac{d\langle\sigma_+\rangle}{dt} = -\left(\frac{\gamma'}{2} - i\omega'_A\right)\langle\sigma_+\rangle, \tag{13.49b}$$

$$\frac{d\langle\sigma_z\rangle}{dt} = -\gamma'\big(\langle\sigma_z\rangle + 1\big), \tag{13.49c}$$

where, with the atom-cavity detuning included, the *cavity-enhanced emission rate* is

$$\gamma' = \gamma\left(1 + 2C_1\frac{1}{1+\Delta_C^2}\right), \tag{13.50}$$

and we obtain the advertised *cavity-induced frequency shift*,

$$\omega'_A - \omega_A = -\frac{\gamma}{2}2C_1\frac{\Delta_C}{1+\Delta_C^2}. \tag{13.51}$$

Of course, by including an atom-cavity detuning, the same results are obtained from the adiabatic elimination within the master equation.

Exercise 13.2. Repeat the derivation of master equation (13.35) with the cavity mode detuned from the resonance frequency of the atom; hence derive the cavity-enhanced emission rate (13.50) and the cavity-induced frequency shift (13.51) in the Schrödinger picture approach.

The spontaneous emission enhancement factor can be expressed in terms of the solid angle subtended by the cavity mode. From its definition in (13.36), using (13.7) and (13.15), and noting that $\omega_C \approx \omega_A$, we write

$$\begin{aligned} 2C_1 &\equiv \frac{2g^2}{\gamma\kappa} \\ &= 2\frac{\omega_A(\boldsymbol{e}_0\cdot\boldsymbol{d}_{12})^2}{2\hbar\epsilon_0\pi(w_0/2)^2L}4\pi\epsilon_0\frac{3\hbar c^3}{4\omega_A^3 d_{12}^2}\frac{2L}{(T_1+T_2)c} \\ &= \frac{2}{T_1+T_2}\frac{(\boldsymbol{e}_0\cdot\boldsymbol{d}_{12})^2}{d_{12}^2/3}\left(\frac{\lambda}{\pi w_0}\right)^2. \end{aligned} \tag{13.52}$$

Note then that the TEM_{00} cavity mode has an expanding width as a function of distance from the midpoint along the cavity axis [13.16],

$$w(z) = w_0\sqrt{1+\frac{z^2}{z_0^2}}, \qquad z_0 \equiv \frac{\pi w_0^2}{\lambda}. \tag{13.53}$$

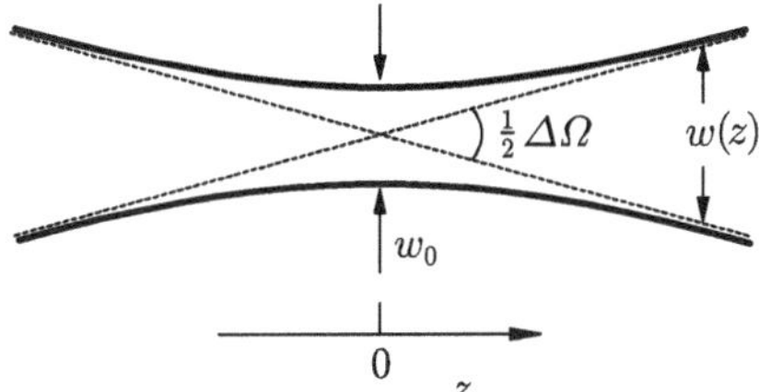

Fig. 13.3. Sketch of the solid angle subtended by an expanding Gaussian beam

The relationship between the expanding width and solid angle is illustrated in Fig. 13.3. For large z/z_0, the width increases as $w(z) \approx z\lambda/\pi w_0$; therefore the mode subtends the solid angle

$$\Delta\Omega = \frac{\pi\left[w(z)/\sqrt{2}\right]^2}{z^2} = \frac{1}{\pi}\left(\frac{\lambda}{w_0}\right)^2, \tag{13.54}$$

where $w(z)/\sqrt{2}$ is the radial distance to the $1/e$ point of the field amplitude. It follows from these results that

$$2C_1 = \frac{8}{T_1 + T_2}\frac{(\boldsymbol{e}_0 \cdot \boldsymbol{d}_{12})^2}{d_{12}^2/3}\frac{\Delta\Omega}{4\pi}. \tag{13.55}$$

Typical ultra-high-finesse optical cavities have $T_1 + T_2 \sim 10^{-5}$ (transmission plus absorption) [13.17], for which an enhancement factor of between one or two orders of magnitude is possible even though $\Delta\Omega$ is extremely small (see Note 13.2).

Note 13.3. It may be difficult to achieve a large enhancement factor while also satisfying the requirements of the bad-cavity limit ($\kappa \gg \gamma/2, g$). Large values of $2C_1$ generally move in the direction of nonperturbative cavity QED. Some examples taken from the literature are:

1. $(g, \kappa, \gamma)/2\pi = (0.22, 11, 0.175)$MHz with $2C_1 = 0.05$ [13.12] (perturbative, optical frequency);
2. $(g, \kappa, \gamma)/2\pi = (120, 40, 5.2)$MHz with $2C_1 = 138$ [13.18] (nonperturbative, optical frequency);
3. $(g, \kappa, \gamma)/2\pi = (39, 223, 0.024)$kHz with $2C_1 = 568$ [13.11] (perturbative, microwave frequency);
4. $(g, \kappa, \gamma)/2\pi = (25, 0.36, 0.005)$kHz with $2C_1 = 6.8 \times 10^5$ [13.19] (nonperturbative, microwave frequency).

Note 13.4. When the solid angle subtended by the cavity mode is not negligible, the spontaneous emission rate can be inhibited as well as enhanced. This possibility is apparent from the expression derived for the emission rate (Eq. 13.50). By detuning the cavity sufficiently far from the atomic resonance, the contribution $2C_1/(1+\Delta^2)$ to γ' is turned off. Then if the rate γ that remains is smaller than the free-space Einstein A coefficient (as it is for large solid angle), we speak of *cavity-inhibited spontaneous emission.* For details see [13.20, 13.21, 13.22, 13.12, 13.13].

13.2.2 Cavity-Enhanced Resonance Fluorescence

Let us make a small extension to our model for cavity-enhanced spontaneous emission. We introduce the classical driving field

$$\boldsymbol{E}_0(\boldsymbol{r},t) = \boldsymbol{e}_0 \tfrac{1}{2}\mathcal{E}_0 e^{-(x^2+y^2)/w_0^2} e^{-i[\omega_0(t-z/c)-\phi_0]} + \text{c.c.}, \tag{13.56}$$

which is mode-matched to the cavity and traveling in the positive z direction—i.e., incident upon the left-hand mirror in Fig. 13.2. Then master equation (13.14) is generalized to the *master equation for single-atom cavity QED with coherent driving of the cavity mode*:

$$\begin{aligned}\dot{\rho} = &-i\tfrac{1}{2}\omega_A[\sigma_z,\rho] - i\omega_C[a^\dagger a,\rho] \\ &+ g[a^\dagger\sigma_- - a\sigma_+,\rho] - i[\bar{\mathcal{E}}_0 e^{-i\omega_0 t}a^\dagger + \bar{\mathcal{E}}_0^* e^{i\omega_0 t}a,\rho] \\ &+ \frac{\gamma}{2}(2\sigma_-\rho\sigma_+ - \sigma_+\sigma_-\rho - \rho\sigma_+\sigma_-) \\ &+ \kappa(2a\rho a^\dagger - a^\dagger a\rho - \rho a^\dagger a),\end{aligned} \tag{13.57}$$

where we introduce the scaled driving field amplitude

$$\bar{\mathcal{E}}_0 \equiv e^{i(\phi_T+\phi_0-\phi_C-\phi_C')}\sqrt{\frac{2\epsilon_0 V_Q}{\hbar\omega_C}}(c/2L)\sqrt{T_1}\mathcal{E}_0. \tag{13.58}$$

The relationship between $\bar{\mathcal{E}}_0$ and $\mathcal{E}_0$ is similar to that in (9.81). Its derivation can be traced in detail through Sects. 14.2.2 and 14.3.

We return to master equation (13.57) repeatedly throughout our discussion of single-atom cavity QED in the remainder of this chapter and in Chap. 16. This equation contains a surprising amount of physics, and poses all the difficulties associated with fluctuations outside the small-noise limit. By quantizing the cavity mode, we have formulated a model that takes us beyond the trivial example of free-space resonance fluorescence. Of course, the Jaynes–Cummings Hamiltonian, if it is taken on its own, can be diagonalized (Exercise 13.10), and certainly its diagonalization explains many things in cavity QED. Nevertheless, so far as quantum fluctuations in the presence of coherent driving and damping are concerned, the problems posed by (13.57) cannot be solved analytically in the general case. In certain parameter regimes, though, an analytical treatment is possible. We look at perhaps the simplest example in this section, what might be termed cavity-enhanced resonance fluorescence.

Continuing with the simplification provided by the perturbative limit but now taking the driving field into account, and specializing to the resonant case ($\omega_0 = \omega_C = \omega_A$), the *master equation for cavity-enhanced resonance fluorescence* is

$$\dot{\rho}_A = -i\tfrac{1}{2}\omega_A[\sigma_z, \rho_A] + i\frac{g}{\kappa}[\bar{\mathcal{E}}_0 e^{-i\omega_A t}\sigma_+ + \bar{\mathcal{E}}_0^* e^{i\omega_A t}\sigma_-, \rho_A] \\ + \frac{\gamma}{2}(1 + 2C_1)(2\sigma_-\rho_A\sigma_+ - \sigma_+\sigma_-\rho_A - \rho_A\sigma_+\sigma_-). \tag{13.59}$$

With the addition of the driving field, the cavity field operator (Eq. 13.42) is given by

$$\tilde{a}(t) = -i\frac{\bar{\mathcal{E}}_0}{\kappa} + \frac{g}{\kappa}\tilde{\sigma}_-(t) + \text{v.f.}, \tag{13.60}$$

where there is no need to explicitly write out the vacuum field contribution, denoted here by v.f.

Note 13.5. Master equation (13.57) provides a description within the Schrödinger picture, while the adiabatically eliminated cavity field operator is given as a Heisenberg picture expression. At first sight, combining the two pictures in this way might seem a little confusing; would it not be preferable to work with the Schrödinger picture at all times? In fact, combining the pictures is helpful, because it enables us to break our problem up into two easily tackled parts. The strategy is not something new; we do this, for example, when using expressions like (2.83) and (7.111) to relate an output field operator to an operator of the source; these expressions are given in the Heisenberg picture and (13.59) [also (13.42)] is another of the same kind. Once we are in possession of such expressions, we may work with a master equation in the Schrödinger picture to analyze the source part of the problem. In this way we account for quantum correlations between a system S (source) and its reservoir R (output field) without explicitly constructing a state vector that entangles the two, as would be required if we were to work entirely within the Schrödinger picture. This division of labor relies on the separation of time scales first encountered in the context of the Markov approximation (Sect. 1.3.3). Here we encounter it in a slightly different form, by making an adiabatic elimination.

Master equation (13.59) is identical to the master equation for free-space resonance fluorescence (Eq. 2.96), except that it has an enhanced spontaneous emission rate, $\gamma \to \gamma(1 + 2C_1)$. Everything discussed in Sects. 2.3.3–2.3.5 is transferable. It is important to remember, however, that the master equation and quantum regression formula are used, ultimately, to calculate the statistical properties of outgoing fields. At this level there is a difference between free-space resonance fluorescence and resonance fluorescence from an atom in a cavity. With the cavity in place, the forward- and backward-scattered light constitute distinct output channels, with statistical properties that are measurably different from those of the light scattered out the side of the cavity—which looks like free-space resonance fluorescence—or of the incident laser light. The spectrum (Sect. 2.3.4) only changes in the ratio of the incoherent and coherent intensities. The second-order correlation function (Sect. 2.3.5 and 2.3.6), on the other hand, is modified significantly. It shows either photon antibunching or photon bunching depending on the size of the spontaneous emission enhancement factor.

Consider first the steady-state averages of atomic operators, which may be taken directly from (2.120a) and (2.120b). These are

$$\langle\tilde{\sigma}_{\mp}\rangle_{\mathrm{ss}} = \pm i\frac{1}{\sqrt{2}}e^{\pm i\arg(\bar{\mathcal{E}}_0)}\frac{Y'}{1+Y'^2}, \tag{13.61a}$$

$$\langle\sigma_z\rangle_{\mathrm{ss}} = -\frac{1}{1+Y'^2}, \tag{13.61b}$$

with

$$Y' \equiv \frac{\sqrt{2}(2g|\bar{\mathcal{E}}_0|/\kappa)}{\gamma'} = \frac{|\bar{\mathcal{E}}_0|/\kappa}{\sqrt{n'_{\mathrm{sat}}}}, \tag{13.62}$$

where

$$n'_{\mathrm{sat}} \equiv \frac{\gamma'^2}{8g^2} \tag{13.63}$$

is the saturation photon number calculated from the cavity-enhanced spontaneous emission rate. Then, from (13.60) and (13.61a), the mean amplitude of the intracavity field is

$$\langle\tilde{a}\rangle_{\mathrm{ss}} = -i\frac{\bar{\mathcal{E}}_0}{\kappa}\left(1-\frac{2C_1}{1+2C_1}\frac{1}{1+Y'^2}\right). \tag{13.64}$$

Note that it is not $\langle\tilde{a}\rangle_{\mathrm{ss}}$, but rather $-i\bar{\mathcal{E}}_0/\kappa$, that appears as the amplitude of the field driving the atom in the second term on the right-hand side of (13.59). The effect of radiation reaction—the term $(g/\kappa)\tilde{\sigma}_-(t)$ in (13.60)—is not seen as a *mean* field by the atom; it appears through the cavity enhancement of spontaneous emission.

Before saying something about the fluctuations, let us look at the conservation of photon flux. The sum of all output fluxes should equal the photon flux of the driving field. There are three output channels to consider: forwards scattering, backwards scattering, and off-axis scattering from the sides of the cavity. From (13.19), we write the forward-scattered field, in units of the square root of photon flux, as

$$\hat{\mathcal{E}}_{\rightarrow}(z,t) = \sqrt{\gamma_{a2}}\tilde{a}(t')e^{-i\omega_C t'} + \mathrm{v.f.}, \tag{13.65}$$

where $ct' \equiv ct-(z-L/2)$. The backwards field takes a similar form, but in this case the reservoir carries a coherent excitation, which is needed to account for the direct reflection of the incident field from the input mirror. Derivation of the explicit expression is left as an exercise.

Exercise 13.3. From (13.20), show that the backward-scattered field, expressed in units of the square root of photon flux, is given by

$$\hat{\mathcal{E}}_{\leftarrow}(z,t) = \sqrt{\gamma_{a1}}\left[i\frac{\bar{\mathcal{E}}_0}{\gamma_{a1}} + \tilde{a}(t')\right]e^{-i\omega_C t'} + \text{v.f.}, \tag{13.66}$$

where $ct' \equiv ct + (z + L/2)$ and $\bar{\mathcal{E}}_0$ is defined in (13.58). Note the phase shift on the term $i\bar{\mathcal{E}}_0/\gamma_{a1}$, so that in an empty cavity with symmetric loss ($\gamma_{a1} = \gamma_{a2} = \kappa$) the mean reflected field amplitude is zero.

Now working from (13.60), (13.65), and (13.66), we obtain for the forwards photon flux

$$\begin{aligned}
&\langle \hat{\mathcal{E}}^{\dagger}_{\rightarrow}\hat{\mathcal{E}}_{\rightarrow}\rangle_{\text{ss}} \\
&= \gamma_{a2}\left\{\left(\frac{|\bar{\mathcal{E}}_0|}{\kappa}\right)^2 + \frac{|\bar{\mathcal{E}}_0|}{\kappa}\frac{g}{\kappa}\left[ie^{-i\arg(\bar{\mathcal{E}}_0)}\langle\tilde{\sigma}_-\rangle_{\text{ss}} + \text{c.c.}\right] + \left(\frac{g}{\kappa}\right)^2\langle\tilde{\sigma}_+\tilde{\sigma}_-\rangle_{\text{ss}}\right\},
\end{aligned} \tag{13.67}$$

and for the backwards photon flux

$$\begin{aligned}
&\langle \hat{\mathcal{E}}^{\dagger}_{\leftarrow}\hat{\mathcal{E}}_{\leftarrow}\rangle_{\text{ss}} \\
&= \gamma_{a1}\left\{\left(\frac{\kappa}{\gamma_{a1}} - 1\right)^2\left(\frac{|\bar{\mathcal{E}}_0|}{\kappa}\right)^2 - \left(\frac{\kappa}{\gamma_{a1}} - 1\right)\frac{|\bar{\mathcal{E}}_0|}{\kappa}\frac{g}{\kappa}\left[ie^{-i\arg(\bar{\mathcal{E}}_0)}\langle\tilde{\sigma}_-\rangle_{\text{ss}} + \text{c.c.}\right]\right. \\
&\quad \left. + \left(\frac{g}{\kappa}\right)^2\langle\tilde{\sigma}_+\tilde{\sigma}_-\rangle_{\text{ss}}\right\}.
\end{aligned} \tag{13.68}$$

Photon flux conservation can be demonstrated from these results.

Exercise 13.4. From (13.67) and (13.68), and the flux of photons scattered by the atom out the sides of the cavity, show that photon flux is conserved overall—i.e., that the sum of the three outgoing photon fluxes equals the photon flux, $|\bar{\mathcal{E}}_0|^2/\gamma_{a1}$, of the incident field. The latter is the square of the first term in the bracket on the right-hand side of (13.66).

Let us now turn to the fluctuations. The analysis can be made along the lines of Sects. 2.3.4–2.3.6. Nothing changes for the fluorescence out the sides of the cavity, other than the different scaling of Y' in (13.62) compared with the scaling of Y in (2.112)—i.e., γ is replaced by γ'. As mentioned, though, there are changes for the forward- and backward-scattered fields. For them, we introduce the operators

$$\begin{aligned}
\Delta\tilde{a} &\equiv \tilde{a} - \langle\tilde{a}\rangle_{\text{ss}} \\
&= \frac{g}{\kappa}(\tilde{\sigma}_- - \langle\tilde{\sigma}_-\rangle_{\text{ss}}) + \text{v.f.},
\end{aligned} \tag{13.69a}$$

and

$$\begin{aligned}\Delta\tilde{a}^\dagger &\equiv \tilde{a}^\dagger - \langle\tilde{a}^\dagger\rangle_{\rm ss} \\ &= \frac{g}{\kappa}(\tilde{\sigma}_+ - \langle\tilde{\sigma}_+\rangle_{\rm ss}) + \text{v.f.},\end{aligned} \tag{13.69b}$$

and, as in (2.122)–(2.126), divide the photon flux into coherent and incoherent components. From (13.64) and (13.65), the coherent part of the forwards scattering is

$$\begin{aligned}F_{\rightarrow}^{\rm coh} &= \gamma_{a2}\langle\tilde{a}^\dagger\rangle_{\rm ss}\langle\tilde{a}\rangle_{\rm ss} \\ &= \gamma_{a2}\left(\frac{|\bar{\mathcal{E}}_0|}{\kappa}\right)^2\left(1 - \frac{2C_1}{1+2C_1}\frac{1}{1+Y'^2}\right)^2,\end{aligned} \tag{13.70}$$

and from (13.65a) and (13.65b), and (13.61a) and (13.61b), the incoherent part is

$$\begin{aligned}F_{\rightarrow}^{\rm inc} &= \gamma_{a2}\langle\Delta\tilde{a}^\dagger\Delta\tilde{a}\rangle_{\rm ss} \\ &= \gamma_{a2}\left(\frac{g}{\kappa}\right)^2\left(\langle\tilde{\sigma}_+\tilde{\sigma}_-\rangle_{\rm ss} - \langle\tilde{\sigma}_+\rangle_{\rm ss}\langle\tilde{\sigma}_-\rangle_{\rm ss}\right) \\ &= \frac{\gamma_{a2}}{\gamma_{a1}+\gamma_{a2}}\gamma 2C_1\frac{1}{2}\frac{Y'^4}{(1+Y'^2)^2}.\end{aligned} \tag{13.71}$$

The relative magnitudes of the two pieces tell us whether the quantum fluctuations are large or small. For the purposes of a comparison with free-space resonance fluorescence, it is useful to rewrite (13.70) and (13.71) as

$$\begin{aligned}F_{\rightarrow}^{\rm coh} &= \gamma_{a2}n'_{\rm sat}Y'^2\left(1 - \frac{2C_1}{1+2C_1}\frac{1}{1+Y'^2}\right)^2 \\ &= \gamma_{a2}n'_{\rm sat}\left[\frac{1+Y'^2(1+2C_1)}{1+2C_1}\right]^2\frac{Y'^2}{(1+Y'^2)^2},\end{aligned} \tag{13.72}$$

and

$$\begin{aligned}F_{\rightarrow}^{\rm inc} &= \gamma_{a2}\frac{\gamma}{2\kappa}2C_1\frac{1}{2}\frac{Y'^4}{(1+Y'^2)^2} \\ &= \gamma_{a2}n'_{\rm sat}\left(\frac{2C_1}{1+2C_1}\right)^2\frac{Y'^4}{(1+Y'^2)^2},\end{aligned} \tag{13.73}$$

where we have used the relationship $\gamma/2\kappa = 4C_1 n'_{\rm sat}/(1+2C_1)^2$, which follows from the definition of the saturation photon number (Eq. 13.63) and the enhanced spontaneous emission rate (Eq. 13.50). The sum of the two expressions gives the total forwards photon flux,

$$
\begin{aligned}
\langle\hat{\mathcal{E}}_{\rightarrow}\hat{\mathcal{E}}_{\rightarrow}\rangle_{\rm ss} &= F_{\rightarrow}^{\rm coh} + F_{\rightarrow}^{\rm inc} \\
&= \gamma_{a2}n'_{\rm sat}\frac{Y'^2}{(1+Y'^2)^2}\frac{[1+Y'^2(1+2C_1)]^2+4C_1^2Y'^2}{(1+2C_1)^2} \\
&= \gamma_{a2}n'_{\rm sat}\frac{Y'^2}{(1+Y'^2)^2}\frac{1+Y'^2[1+(1+2C_1)^2]+Y'^4(1+2C_1)^2}{(1+2C_1)^2} \\
&= \gamma_{a2}n'_{\rm sat}Y'^2\frac{(1+2C_1)^{-2}+Y'^2}{1+Y'^2}.
\end{aligned}
\tag{13.74}
$$

Note the factors on the far right of (13.72) and (13.73); these are the free-space results (2.125) and (2.126), evaluated with $\gamma \to \gamma'$. The behavior of the three photon fluxes, (13.72), (13.73), and (13.74), as a function of parameters is illustrated in Fig. 13.4.

When $2C_1$ is small, $F_{\rightarrow}^{\rm coh}$ is approximately linear in Y'^2 and $F_{\rightarrow}^{\rm inc}$ is negligible. In this case, the forwards flux is dominated by the straight-through transmission of the incident light. For large $2C_1$, the incoherent flux is generally a significant contribution. Indeed, for nonsaturating intensities it dominates; now it is the coherent part of the forwards scattering that is negligible, due to *cavity-enhanced absorption*: the straight-through transmission of the incident light interferes destructively with the cavity-enhanced reradiation from the atom. Note also that for large $2C_1$, $F_{\rightarrow}^{\rm inc}/F_{\rightarrow}^{\rm coh} \to 1/Y'^2$. This is exactly the

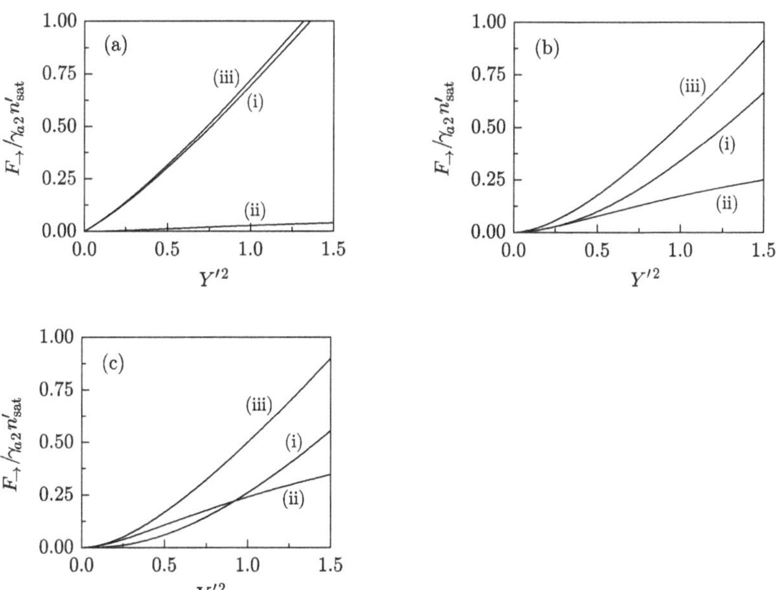

Fig. 13.4. Variation of (i) the coherent, (ii) the incoherent, and (iii) the total photon flux in the forwards direction as a function of the incident intensity Y'^2: for (**a**) $2C_1 = 0.5$, (**b**) $2C_1 = 5.0$, (**c**) $2C_1 = 50.0$

inverse of the ratio for the sideways fluorescence and resonance fluorescence in free space.

Exercise 13.5. When $2C_1$ is small the forwards photon flux is essentially the transmitted flux, $\gamma_{a2}(|\bar{\mathcal{E}}_0|/\kappa)^2$, of the empty cavity. When it is large, there is significant absorption at weak and moderate levels of excitation. Show that at high excitation, so long as the conditions for the adiabatic elimination hold, the deviation from $\gamma_{a2}(|\bar{\mathcal{E}}_0|/\kappa)^2$ is necessarily much less than $\gamma_{a2} \times 1$; thus, the presence of the atom changes the intracavity photon number by much less than one photon. The atom still develops a Rabi oscillation, exchanging one photon backwards and forwards with the driving field. Nevertheless, the photon number *inside* the cavity does not oscillate by a whole photon—i.e., it is not reduced, in the time average, by one half of a photon. This is because the cavity bandwidth must be much larger than the Rabi frequency for the adiabatic elimination to hold.

The most significant difference from ordinary resonance fluorescence is seen in the photon statistics of the forward-scattered light, where dramatic modifications of the usual photon antibunching are seen. In the following section we derive the second-order correlation function of the forward-scattered light in the limit of a weak driving field. Here, for completeness, we quote the result for arbitrary excitation strengths: from Rice and Carmichael [13.23], the *second-order correlation function of forwards photon scattering in cavity-enhanced resonance fluorescence* is given by

$$\begin{aligned} g^{(2)}_{\rightarrow}(\tau) - 1 \\ &= -\frac{8C_1^2}{[1+Y'^2(1+2C_1)^2]^2} e^{-(3\gamma'/4)\tau}\{\left[1-2C_1^2-Y'^2(1+2C_1)^2\right]\cosh\delta'\tau \\ &\quad + \frac{\gamma'/4}{\delta'}\left[1+2C_1^2-Y'^2(1+2C_1)(5+2C_1)\right]\sinh\delta'\tau\}, \end{aligned} \tag{13.75}$$

where

$$\delta' \equiv \frac{\gamma'}{4}\sqrt{1-8Y'^2}. \tag{13.76}$$

This is to compared with (2.152). The result differs from resonance fluorescence in free space because of the interference between the straight-through transmission and the forwards scattering from the atom (Eq. 13.60).

Note 13.6. The cavity might be driven off-axis, in which case the driving field does not interfere with the atomic scattering along the cavity axis. For such non-mode-matched driving, the light emitted through the cavity mirrors has the $g^{(2)}(\tau)$ of resonance fluorescence in free space.

13.2.3 Forwards Photon Scattering in the Weak-Excitation Limit

New features arise in the second-order statistics of the forward-scattered light because of an interference between the straight-through transmission of the coherent driving field and the forward-directed fluorescence from the atom. The results of this interference were first noted by Rice and Carmichael [13.23] for single-atom cavity QED in the perturbative limit. Subsequently, an extension to many atoms and nonperturbative cavity QED was made [13.24, 13.25]; in the latter regime a number of experiments have been performed [13.26, 13.27, 13.9]. Here we analyze the original proposal, specializing, in addition, to weak excitation, which produces the most dramatic effects. We return to this topic in Sects. 15.2.4 and 16.1.4, where the limited treatment of the present section is extended.

We have observed (Note 2.8) that the second-order correlation function factorizes as the product of a probability for a first photodetection, at $\tau = 0$, and the probability for a delayed photodetection, which evolves under the propagator $e^{\mathcal{L}\tau}$. The factorization is a formal consequence of the quantum regression formula, specifically of the formula (1.102), and in the present case, for the forward-scattered field, we use it to write ($\tau \geq 0$)

$$\begin{aligned} g^{(2)}_{\rightarrow}(\tau) &= \frac{\langle \tilde{a}^\dagger(0)\tilde{a}^\dagger(\tau)\tilde{a}(\tau)\tilde{a}(0)\rangle_{\rm ss}}{\langle \tilde{a}^\dagger\tilde{a}\rangle_{\rm ss}^2} \\ &= \frac{{\rm tr}\{\tilde{a}^\dagger(0)\tilde{a}(0)e^{\tilde{\mathcal{L}}\tau}[\tilde{a}(0)\tilde{\rho}_{\rm ss}\tilde{a}^\dagger(0)]\}}{\langle \tilde{a}^\dagger\tilde{a}\rangle_{\rm ss}^2} \\ &= \frac{\langle(\tilde{a}^\dagger\tilde{a})(\tau)\rangle_{\tilde{\rho}(0)=\tilde{\rho}'_{\rm ss}}}{\langle \tilde{a}^\dagger\tilde{a}\rangle_{\rm ss}}, \end{aligned} \tag{13.77}$$

where the initial state $\tilde{\rho}(0) = \tilde{\rho}'_{\rm ss}$ is given by

$$\tilde{\rho}'_{\rm ss} \equiv \frac{\tilde{a}\tilde{\rho}_{\rm ss}\tilde{a}^\dagger}{{\rm tr}(\tilde{a}\tilde{\rho}_{\rm ss}\tilde{a}^\dagger)}; \tag{13.78}$$

this initial state is the *reduced state* prepared under the condition that a photodetection occurs at time $\tau = 0$. We are working in the interaction picture, $\rho = e^{-i[\omega_0(\frac{1}{2}\sigma_z + a^\dagger a)t]}\tilde{\rho}e^{i[\omega_0(\frac{1}{2}\sigma_z + a^\dagger a)t]}$.

Under the adiabatic elimination, (13.60) expresses the forward-scattered field as a sum of the straight-through transmission and the atomic fluorescence. Our calculation is therefore carried out within the Hilbert space of the atom, where, using (13.61a) and (13.61b), the steady-state density matrix is

$$\begin{aligned} (\tilde{\rho}_A)_{\rm ss} &= \frac{1}{2}\frac{2+Y'^2}{1+Y'^2}|1\rangle\langle 1| + \frac{1}{2}\frac{Y'^2}{1+Y'^2}|2\rangle\langle 2| \\ &\quad - i\frac{1}{\sqrt{2}}\frac{Y'}{1+Y'^2}\big[e^{-i\arg(\bar{\mathcal{E}}_0)}|1\rangle\langle 2| - e^{i\arg(\bar{\mathcal{E}}_0)}|2\rangle\langle 1|\big]. \end{aligned} \tag{13.79}$$

The calculation is simplified considerably by specializing to weak excitation, for which this density matrix is approximated as

$$(\tilde{\rho}_A)_{\rm ss} \approx |1\rangle\langle 1| + \tfrac{1}{2}Y'^2|2\rangle\langle 2| \\ - i\frac{1}{\sqrt{2}}Y'\big[e^{-i\arg(\bar{\mathcal{E}}_0)}|1\rangle\langle 2| - e^{i\arg(\bar{\mathcal{E}}_0)}|2\rangle\langle 1|\big]. \tag{13.80}$$

Thus the steady state is approximately a pure state,

$$(\tilde{\rho}_A)_{\rm ss} \approx |\tilde{A}_{\rm ss}\rangle\langle\tilde{A}_{\rm ss}|, \tag{13.81}$$

with

$$|\tilde{A}_{\rm ss}\rangle = |1\rangle + i\frac{1}{\sqrt{2}}Y'e^{i\arg(\bar{\mathcal{E}}_0)}|2\rangle. \tag{13.82}$$

The reduced state (13.78) is then also approximately pure, admitting the factorization

$$(\tilde{\rho}_A)'_{\rm ss} \approx |\tilde{A}'_{\rm ss}\rangle\langle\tilde{A}'_{\rm ss}| \tag{13.83}$$

with

$$|\tilde{A}'_{\rm ss}\rangle \equiv \frac{[-i\bar{\mathcal{E}}_0/\kappa + (g/\kappa)\tilde{\sigma}_-]|\tilde{A}_{\rm ss}\rangle}{\sqrt{\langle\tilde{A}_{\rm ss}|[i\bar{\mathcal{E}}_0^*/\kappa + (g/\kappa)\tilde{\sigma}_+][-i\bar{\mathcal{E}}_0/\kappa + (g/\kappa)\tilde{\sigma}_-]|\tilde{A}_{\rm ss}\rangle}}, \tag{13.84}$$

where we make use of (13.60) once again.

We now develop the explicit form of the reduced state (13.84) to dominant order in the excitation strength. Consider first the unnormalized state, where, using (13.82) and definition (13.62) of Y', we may write

$$\begin{aligned}
&\left(-i\frac{\bar{\mathcal{E}}_0}{\kappa} + \frac{g}{\kappa}\tilde{\sigma}_-\right)|\tilde{A}_{\rm ss}\rangle \\
&\quad = \left(-i\frac{\bar{\mathcal{E}}_0}{\kappa} + \frac{g}{\kappa}\tilde{\sigma}_-\right)\left[|1\rangle + i\frac{1}{\sqrt{2}}Y'e^{i\arg(\bar{\mathcal{E}}_0)}|2\rangle\right] \\
&\quad = -i\left(\frac{\bar{\mathcal{E}}_0}{\kappa} - \frac{g}{\kappa}\frac{1}{\sqrt{2}}Y'e^{i\arg(\bar{\mathcal{E}}_0)}\right)|1\rangle + \frac{\bar{\mathcal{E}}_0}{\kappa}\frac{1}{\sqrt{2}}Y'e^{i\arg(\bar{\mathcal{E}}_0)}|2\rangle \\
&\quad = \left(-i\frac{\bar{\mathcal{E}}_0}{\kappa}\right)\left[\left(1 - \frac{2C_1}{1+2C_1}\right)|1\rangle + i\frac{1}{\sqrt{2}}Y'e^{i\arg(\bar{\mathcal{E}}_0)}|2\rangle\right].
\end{aligned} \tag{13.85}$$

Then to dominant order in the driving field amplitude, the state norm, the square of the denominator in (13.84), is

$$\begin{aligned}
\mathrm{tr}(\tilde{a}\tilde{\rho}_{\rm ss}\tilde{a}^\dagger) &\approx \langle\tilde{A}_{\rm ss}|\left(i\frac{\bar{\mathcal{E}}_0^*}{\kappa} + \frac{g}{\kappa}\tilde{\sigma}_+\right)\left(-i\frac{\bar{\mathcal{E}}_0}{\kappa} + \frac{g}{\kappa}\tilde{\sigma}_-\right)|\tilde{A}_{ss}\rangle \\
&\approx \left(\frac{|\bar{\mathcal{E}}_0|/\kappa}{1+2C_1}\right)^2.
\end{aligned} \tag{13.86}$$

From (13.84)–(13.86), the reduced state is

$$|\tilde{A}'_{\rm ss}\rangle = |1\rangle + i\frac{1}{\sqrt{2}}Ye^{i\arg(\bar{\mathcal{E}}_0)}|2\rangle, \tag{13.87}$$

where

$$Y \equiv \sqrt{2}\frac{2g|\bar{\mathcal{E}}_0|/\kappa}{\gamma} = \frac{|\bar{\mathcal{E}}_0|/\kappa}{\sqrt{n_{\rm sat}}}. \tag{13.88}$$

On comparing (13.87) with the steady state (13.82), we see that the change brought about by conditioning on a photodetection is remarkably simple: one is to simply replace Y' by Y—the cavity enhancement is turned off.

In effect, the enhancement of the spontaneous emission rate which enters the normalization of Y' is turned off by conditioning the atomic state on the detection of a forward-scattered photon; given the photodetection, the polarization of the atom is determined by its decay rate $\gamma/2$ in free space, rather than by the cavity-enhanced rate $\gamma'/2$. The results of this can be dramatic if the cavity enhancement factor is large. Given the changed decay rate, the conditional polarization amplitude is increased from its steady-state value; the cavity field amplitude is decreased, as the polarization is out of phase with the cavity field. For large cavity enhancement, the increase in the polarization is so large that the amplitude of the cavity field changes sign; thus, the cavity field acquires a π phase shift relative to its phase in steady state.

The fact that the conditional cavity field amplitude *decreases* indicates a nonclassical field. This follows because, by classical logic, conditioning on a photodetection should select a subensemble of field amplitudes biased towards larger rather than smaller values; photodetection is more likely when the intensity fluctuates above, not below, its mean. Conditional homodyne detection [13.28, 13.29] may be used to observe the nonclassical field-amplitude fluctuation directly. It appears only indirectly in the photon statistics, through nonclassical features of $g^{(2)}(\tau)$. These nonclassical features appear in two distinct ways, depending on the ratio of Y to Y'. Before describing them, let us complete our calculation by solving for the relaxation back to the steady state.

The regression of the fluctuation—the relaxation of conditional state $|\tilde{A}'_{\rm ss}\rangle$ to the steady state $|\tilde{A}_{\rm ss}\rangle$—is governed by the propagator $e^{\tilde{\mathcal{L}}_A\tau}$, where $\tilde{\mathcal{L}}_A$ is defined by the right-hand side of (13.59). Under the assumption of weak excitation, we make the approximation

$$\begin{aligned}\tilde{\mathcal{L}}_A &\equiv i\frac{g}{\kappa}[\bar{\mathcal{E}}_0\sigma_+ + \bar{\mathcal{E}}_0^*\sigma_-,\cdot] + \frac{\gamma'}{2}(2\sigma_-\cdot\sigma_+ - \sigma_+\sigma_-\cdot - \cdot\sigma_+\sigma_-)\\ &\approx i\frac{g}{\kappa}\bar{\mathcal{E}}_0[\sigma_+,\cdot] - \frac{\gamma'}{2}[\sigma_+\sigma_-,\cdot]_+.\end{aligned} \tag{13.89}$$

This simplification can be verified through an expansion in powers of $\bar{\mathcal{E}}_0$ (see Sect. 16.1.1). Physically, it is justified so long as there is negligible probability for the atom to emit a photon during its evolution back to the steady state.

Thus, for weak excitation, the purity of the state is preserved over time and from (13.77), (13.86), and (13.89) we obtain

$$g_{\rightarrow}^{(2)}(\tau)$$
$$\approx \left(\frac{|\bar{\mathcal{E}}_0|/\kappa}{1+2C_1}\right)^{-2} \langle\tilde{A}(\tau)|\left(i\frac{\bar{\mathcal{E}}_0^*}{\kappa}+\frac{g}{\kappa}\tilde{\sigma}_+\right)\left(-i\frac{\bar{\mathcal{E}}_0}{\kappa}+\frac{g}{\kappa}\tilde{\sigma}_-\right)|\tilde{A}(\tau)\rangle\Bigg|_{|\tilde{A}(0)\rangle=|\tilde{A}'_{\rm ss}\rangle}, \tag{13.90}$$

where the state obeys a Schrödinger equation with non-Hermitian Hamiltonian:

$$\frac{d|\tilde{A}(\tau)\rangle}{d\tau} = \left(i\frac{g}{\kappa}\bar{\mathcal{E}}_0\sigma_+ - \frac{\gamma'}{2}\sigma_+\sigma_-\right)|\tilde{A}(\tau)\rangle. \tag{13.91}$$

Expanding the state as

$$|\tilde{A}(\tau)\rangle = \alpha(\tau)|1\rangle + \beta(\tau)|2\rangle, \tag{13.92}$$

the Schrödinger equation reduces to the coupled equations

$$\dot{\alpha} = 0, \qquad \dot{\beta} = -\frac{\gamma'}{2}\beta + i\frac{g}{\kappa}\bar{\mathcal{E}}_0\alpha, \tag{13.93}$$

which are to be solved—matching conditional state (13.87)—for initial conditions $\alpha(0)=1$, $\beta(0)=i(Y/\sqrt{2})e^{i\arg(\bar{\mathcal{E}}_0)}$. The solution is

$$|\tilde{A}(\tau)\rangle = |1\rangle + i\frac{1}{\sqrt{2}}e^{i\arg(\bar{\mathcal{E}}_0)}\{Ye^{-(\gamma'/2)\tau} + Y'[1-e^{-(\gamma'/2)\tau}]\}|2\rangle. \tag{13.94}$$

Then, to lowest order in the strength of the driving field,

$$\begin{aligned}\left(-i\frac{\bar{\mathcal{E}}_0}{\kappa}+\frac{g}{\kappa}\tilde{\sigma}_-\right)&|\tilde{A}(\tau)\rangle \\ &= -i\frac{\bar{\mathcal{E}}_0}{\kappa}\left\{1-2C_1e^{-(\gamma'/2)\tau} - \frac{2C_1}{1+2C_1}[1-e^{-(\gamma'/2)\tau}]\right\}|1\rangle \\ &= -i\frac{\bar{\mathcal{E}}_0/\kappa}{1+2C_1}[1-4C_1^2e^{-(\gamma'/2)\tau}]|1\rangle,\end{aligned} \tag{13.95}$$

and, from (13.90) and (13.95), we arrive at a simple expression for the *second-order correlation function of forwards photon scattering in cavity-enhanced resonance fluorescence in the weak-excitation limit*:

$$g_{\rightarrow}^{(2)}(\tau) \approx [1-4C_1^2e^{-(\gamma'/2)\tau}]^2. \tag{13.96}$$

Note that this expression is recovered for $Y' \to 0$ from (13.75).

In contrast to free-space resonance fluorescence, the correlation function (13.96) does not generally vanish at zero delay. Returning to the earlier comment, it is, however, the square of a field amplitude that is *decreased* relative

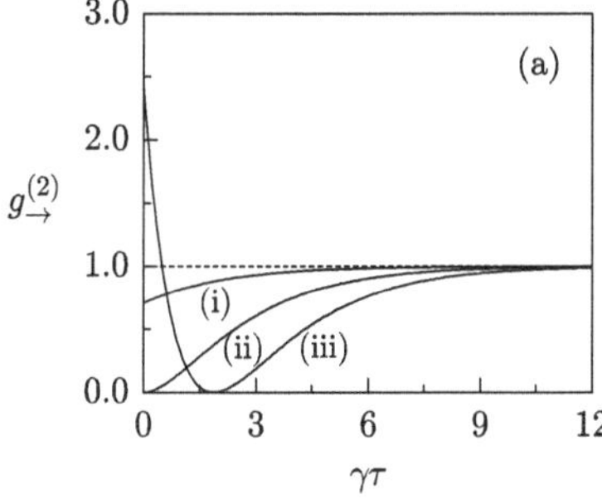

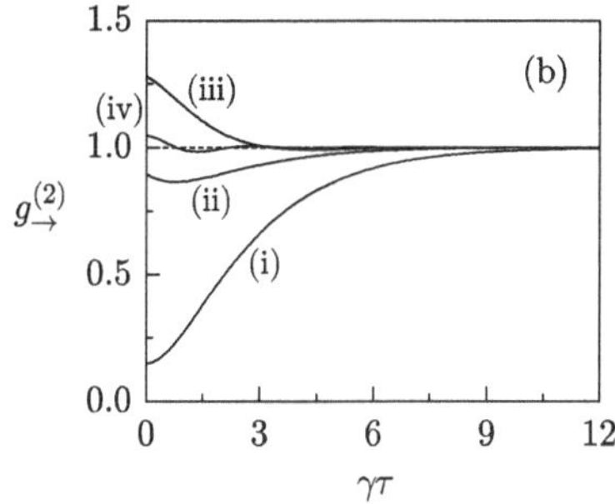

Fig. 13.5. Second-order intensity correlation function of the forward-scattered light in cavity-enhanced resonance fluorescence (Eq. 13.75): (**a**) in the weak-excitation limit ($Y'^2 = 10^{-4}$) with (i) $2C_1 = 0.4$, (ii) $2C_1 = 1.0$, (iii) $2C_1 = 1.6$, and (**b**) with $2C_1 = 1.0$ and (i) $Y'^2 = 10^{-2}$, (ii) $Y'^2 = 10^{-1}$, (iii) $Y'^2 = 1$, and (iv) $Y'^2 = 10$

to its steady-state value. We may write $g^{(2)}(0) = (1+2C_1)^2(1-2C_1)^2$, where the factor $(1+2C_1)^2$ may be identified as arising from the normalization of the correlation function with respect to the mean photon number (Eq. 13.86). The factor $(1-2C_1)$ scales the conditional field amplitude relative to its magnitude in the steady state. For $2C_1 \leq 1$, the phase of the conditional field does not change, though its amplitude is reduced. In this case, $g^{(2)}(\tau)$ exhibits conventional photon antibunching, the behavior illustrated by curves (i) and (ii) of Fig. 13.5a. For $2C_1 > 1$, there is a π phase shift of the conditional field. In such a case the field amplitude passes through zero during its return to the steady state and we observe a zero in $g^{(2)}(\tau)$ at a finite delay—curve (iii) of Fig. 13.5a. Thus, while the conditioning of the field amplitude develops continuously as the spontaneous emission enhancement factor is increased, its nonclassical character appears in two distinct ways in $g^{(2)}(\tau)$. Figure 13.5b illustrates how stimulated emission gradually eliminates these nonclassical features at larger excitation strengths.

Note 13.7. The weak-excitation limit calls for extremely low intracavity photon numbers. The constraint is that the probability of photon emission during relaxation back to the steady state should be very small. From (13.86)–(13.88), this translates into the requirement

$$\langle \tilde{a}^\dagger \tilde{a} \rangle_{ss} \ll \frac{\gamma'}{2\kappa} \frac{1}{2C_1(1+2C_1)^3}. \tag{13.97}$$

The ratio $\gamma'/2\kappa$ is necessarily less then unity for the adiabatic elimination (perturbative treatment) to be valid. Consider then that the spontaneous emission rate is enhanced by an order of magnitude: the intracavity photon number is required to be much smaller than 10^{-4}. The extreme condition arises because, for a large spontaneous emission enhancement factor, $2C_1$, the forwards photon scattering is highly bunched.

13.2.4 A One-Atom "Laser"

The treatment of quantum noise in the laser is an important and instructive example of the use of phase-space methods in the small-noise limit. Outside this limit, the laser provides an instructive example of cavity QED ideas. Much work in this area is concerned with the micromaser [13.30, 13.31, 13.32], a very different device to the one described in Chaps. 7 and 8. More closely related are proposals for practical light sources that operate with reduced power consumption, laser sources designed using the principles of cavity QED. The idea is to eliminate loss due to spontaneous emission into modes other than the laser mode [13.33, 13.34, 13.35]. The figure of merit for this is the so-called *laser β-factor*, the ratio of the spontaneous emission rate into the laser mode to the total spontaneous emission rate. For the model of Sect. 7.1.4 the β-factor may be written as

$$\beta = \frac{2\bar{C}_1}{1+2\bar{C}_1}, \qquad 2\bar{C}_1 = 2\frac{g^2}{\gamma_\downarrow(\gamma_h/2)}. \tag{13.98}$$

$2\bar{C}_1$ is an enhancement factor analogous to (13.36), except that in the case of the laser the inhomogeneous linewidth, $\gamma_h/2$, replaces the cavity linewidth as the dominant reservoir bandwidth. Note that when $\gamma_\uparrow \gg \gamma_\downarrow$ (no population in the lower level of the lasing transition), $2\bar{C}_1 = 1/2n_{\mathrm{sat}}$ (Eq. 7.69c); once again we find a connection between strong coupling, $\beta \to 1$, and a small system size parameter, $n_{\mathrm{sat}} \ll 1$.

There is a great deal that could be said about laser theory in the cavity QED regime, especially on the topic of quantum noise. Cavity QED lasers do not show the generic threshold behavior that follows from the "thermodynamic" limit, $n_{\mathrm{sat}} \to \infty$. Indeed, they have been proposed to operate precisely in the opposite regime, $n_{\mathrm{sat}} \to 0$ ($\beta \to 1$), where they become thresholdless devices [13.33, 13.34, 13.36]. While for small but nonzero n_{sat} a smeared out threshold region might remain, ultimately the noise characteristics of cavity QED lasers depend on the particular system considered. We do not aim to explore these things in detail. Here, we simply observe, for a simplest case, how the quantum noise is changed when the small-noise approximation no longer holds. We show, in particular, that the emitted light becomes nonclassical, exhibiting photon antibunching and sub-Poissonian statistics, as observed in the experiment of McKeever and coworkers [13.37]. We also take the opportunity to say a little more about waiting-time distributions (Sect. 2.3.6).

As the starting point for our model, we adopt the one-atom limit of the laser master equation from Chap. 7 (Eq. 7.93). Setting $\bar{n} = 0$, for simplicity, the one-atom master equation is

$$\begin{aligned}\dot{\rho} = &-i\tfrac{1}{2}\omega_A[\sigma_z,\rho] - i\omega_A[a^\dagger a,\rho] + g[a^\dagger\sigma_- - a\sigma_+,\rho] \\ &+ \kappa(2a\rho a^\dagger - a^\dagger a\rho - \rho a^\dagger a) \\ &+ \frac{\gamma_\downarrow}{2}(2\sigma_-\rho\sigma_+ - \sigma_+\sigma_-\rho - \rho\sigma_+\sigma_-) \\ &+ \frac{\gamma_\uparrow}{2}(2\sigma_+\rho\sigma_- - \sigma_-\sigma_+\rho - \rho\sigma_-\sigma_+).\end{aligned} \tag{13.99}$$

Normally the laser gain is provided by the weak coupling of many atoms to the laser mode. One strongly coupled atom can possibly do a similar job; although for a single atom, the approximation (Sect. 6.2.4) that justifies the truncation of the phase-space equation of motion in atomic variables must certainly be set aside. In fact, to make something close to a conventional laser—a light source dominated by stimulated emission—with a single atom, a good cavity, $\kappa < 2g^2/\gamma_h$, is required, so that a significant intracavity photon number builds up. In the good-cavity limit, the solution of the master equation must be carried out numerically [13.38,13.39]. For something simpler, something that allows for an analytical treatment, we stay with the adiabatic elimination of the cavity mode. From (13.99), this gives

$$\begin{aligned}\dot{\rho}_A = &-i\tfrac{1}{2}\omega_A[\sigma_z,\rho_A] + \frac{\gamma_\downarrow'}{2}(2\sigma_-\rho_A\sigma_+ - \sigma_+\sigma_-\rho_A - \rho_A\sigma_+\sigma_-) \\ &+ \frac{\gamma_\uparrow}{2}(2\sigma_+\rho_A\sigma_- - \sigma_-\sigma_+\rho - \rho_A\sigma_-\sigma_+),\end{aligned} \tag{13.100}$$

where

$$\gamma_\downarrow' \equiv \gamma_\downarrow + \frac{2g^2}{\kappa}, \tag{13.101}$$

and from (13.60) and (13.65), setting $\bar{\mathcal{E}}_0 = 0$, the output field in the forwards direction is

$$\hat{\mathcal{E}}_{\rightarrow}(z,t) = \sqrt{\gamma_{a2}}\frac{g}{\kappa}\sigma_-(t') + \text{v.f.}, \tag{13.102}$$

where $ct' = ct - (z - L/2)$. Of course our model is hardly the model of a genuine laser. More accurately, the cavity is simply used to collect and direct incoherently driven fluorescence from the atom. We are considering an incoherently driven version of cavity-enhanced resonance fluorescence, from which it is already clear that the output field will be nonclassical. Nevertheless, the model explores an instructive limit of laser theory.

Consider first the optical Bloch equations derived from (13.100). These are the closed set of three coupled equations

$$\frac{d\langle\sigma_-\rangle}{dt} = -\left(\frac{\gamma_\uparrow + \gamma_\downarrow'}{2} + i\omega_A\right)\langle\sigma_-\rangle, \tag{13.103a}$$

$$\frac{d\langle\sigma_+\rangle}{dt} = -\left(\frac{\gamma_\uparrow + \gamma_\downarrow'}{2} - i\omega_A\right)\langle\sigma_+\rangle, \tag{13.103b}$$

$$\frac{d\langle\sigma_z\rangle}{dt} = -(\gamma_\uparrow + \gamma_\downarrow')(\langle\sigma_z\rangle - \wp), \tag{13.103c}$$

where we introduce the pump parameter

$$\wp \equiv \frac{\gamma_\uparrow - \gamma'_\downarrow}{\gamma_\uparrow + \gamma'_\downarrow}. \tag{13.104}$$

The steady-state solution to these equations is

$$\langle \sigma_\mp \rangle_{\rm ss} = 0, \qquad \langle \sigma_z \rangle_{\rm ss} = \wp, \tag{13.105}$$

from which, using (13.102), we calculate the forwards-emitted mean photon flux

$$\begin{aligned}\langle \hat{\mathcal{E}}_\rightarrow^\dagger \hat{\mathcal{E}}_\rightarrow \rangle_{\rm ss} &= \gamma_{a2}\left(\frac{g}{\kappa}\right)^2 \langle \sigma_+\sigma_- \rangle_{\rm ss} \\ &= \gamma_{a2}\left(\frac{g}{\kappa}\right)^2 \tfrac{1}{2}(1+\wp) \\ &= \frac{\gamma_{a2}}{2\kappa}\left(\frac{2g^2}{\kappa}\right)\frac{\gamma_\uparrow}{\gamma_\uparrow + \gamma'_\downarrow}. \end{aligned} \tag{13.106}$$

Note now that the pump parameter (13.104) is not simply the one-atom version of the pump parameter from standard laser theory (Eqs. 7.72 and 7.73). In fact, it is less than unity for all $\gamma_\uparrow$, and therefore, consistent with the comment above—"our model is hardly the model of a genuine laser"—this "laser" remains forever below threshold. Thus, (13.106) should be compared with a conventional laser operating below threshold. For the latter, using (7.41), (7.72), and (7.73), we obtain the photon flux

$$\begin{aligned}\langle \hat{\mathcal{E}}_\rightarrow \hat{\mathcal{E}}_\rightarrow \rangle_{\rm ss} &= \gamma_{a2}\frac{C+\frac{1}{2}\wp}{1-\wp} \\ &= \gamma_{a2}\frac{1}{1-\wp}\left(\frac{2Ng^2}{\gamma_h\kappa}\right)\frac{\gamma_\uparrow}{\gamma_\uparrow+\gamma_\downarrow} \\ &= \frac{\gamma_{a2}}{2\kappa}2\kappa\frac{1}{2\kappa(1-\wp)}\left(\frac{2g^2}{\gamma_h/2}\right)\frac{N\gamma_\uparrow}{\gamma_\uparrow+\gamma_\downarrow}. \end{aligned} \tag{13.107}$$

A comparison of (13.107) with (13.106) is instructive.

The common prefactor $\gamma_{a2}/2\kappa$ is the fraction of cavity mode loss in the forwards direction. Consider then the terms on the far right of each expression. Equation 13.107 has an extra factor of N from the number of atoms in the lasing medium. Beyond this trivial difference, where $\gamma'_\downarrow$ appears in (13.106), $\gamma_\downarrow$ appears in (13.107). The two terms have the same function. Both are spontaneous emission rates; but in (13.106) the spontaneous emission rate is enhanced. Here there is a substantial difference. For the one-atom model, the spontaneous emission enhancement *is* the entire effect of the interaction between the atom and cavity mode. In a true laser, on the other hand, the interaction enters in a different way. The cavity does not simply act as a reservoir to modify emission rates. It introduces a second, independent, step to the emission process, by providing a temporary storage place for photons. Thus, to understand (13.107), in comparison with (13.106), we must see the emission of a photon as a two-step sequence: photons are first emitted into the cavity mode, and then, in an independent step, transmitted through the output mirror. In this way, a significant photon number can be built up inside the cavity so that stimulated emission dominates over spontaneous emission. The important factor $1 - \wp$ therefore enters (13.107); as $\wp \to 1$, it accounts for the onset of stimulated emission.

Note that it would be incorrect to replace $\gamma_\downarrow$ and γ_h in (13.107) by $\gamma'_\downarrow$ and γ'_h, as a sort of higher-order correction to standard laser theory. This double-counts the effect of the interaction of the atoms with the cavity mode. Only in the perturbative (bad-cavity) limit is the interaction accounted for in this way.

Note 13.8. The pump rate for this one-atom "laser" is controlled entirely by $\gamma_\uparrow$. The option available to a conventional laser of changing the number of atoms [see the discussion below (7.75)] has been given up. One consequence of this is that increasing the pumping rate in the one-atom "laser" eventually causes the inversion to saturate. When the inversion is plotted as a function of pump rate, the behavior looks a little like a smoothed-out inversion (gain) clamping (Eq. 7.22), but it is actually straightforward saturation.

Our principal aim is to explore the nonclassicality of the one-atom "laser" emission. Rather than calculate $g^{(2)}(\tau)$ again, let us do something a little different and calculate the *waiting-time distribution of photon emissions in the forwards direction* (see Sects. 2.3.6 and 17.3.5)

$$
\begin{aligned}
w_{a2}(\tau) &= \gamma_{a2} \frac{\mathrm{tr}[a^\dagger a e^{\bar{\mathcal{L}}\tau}(a\rho_{\mathrm{ss}}a^\dagger)]}{\langle a^\dagger a\rangle_{\mathrm{ss}}} \\
&= \gamma_{a2}\left(\frac{g}{\kappa}\right)^2 \frac{\mathrm{tr}\{\sigma_+\sigma_- e^{\bar{\mathcal{L}}_A\tau}[\sigma_-(\rho_A)_{\mathrm{ss}}\sigma_+]\}}{\langle\sigma_+\sigma_-\rangle_{\mathrm{ss}}} \\
&= \frac{\gamma_{a2}}{2\kappa}\left(\frac{2g^2}{\kappa}\right)\mathrm{tr}[\sigma_+\sigma_-\bar{\rho}_A(\tau)],
\end{aligned}
\tag{13.108}
$$

where we have defined

$$
\bar{\rho}_A(\tau) \equiv e^{\bar{\mathcal{L}}_A\tau}(\rho_A)'_{\mathrm{ss}},
\tag{13.109}
$$

with reduced state

$$
(\rho_A)'_{\mathrm{ss}} \equiv \frac{\sigma_-(\rho_A)_{\mathrm{ss}}\sigma_+}{\mathrm{tr}[\sigma_-(\rho_A)_{\mathrm{ss}}\sigma_+]} = |1\rangle\langle 1|,
\tag{13.110}
$$

and superoperator

$$\bar{\mathcal{L}}_A \equiv \mathcal{L}_A - \frac{\gamma_{a2}}{2\kappa}\left(\frac{2g^2}{\kappa}\right)\sigma_- \cdot \sigma_+, \tag{13.111a}$$

$$\mathcal{L}_A \equiv -i\tfrac{1}{2}\omega_A[\sigma_z, \cdot] + \frac{\gamma'_\downarrow}{2}(\sigma_- \cdot \sigma_+ - \sigma_+\sigma_- \cdot - \cdot \sigma_+\sigma_-)$$
$$+ \frac{\gamma_\uparrow}{2}(\sigma_+ \cdot \sigma_- - \sigma_-\sigma_+ \cdot - \cdot \sigma_-\sigma_+). \tag{13.111b}$$

Aside from an overall normalization, the difference between the second-order correlation function (Eq. 13.77) and the waiting-time distribution (Eq. 13.108) is that the time evolution of the former is governed by $e^{\mathcal{L}_A\tau}$, while that of the latter is governed by $e^{\bar{\mathcal{L}}_A\tau}$; the change from superoperator $\mathcal{L}_A$ to $\bar{\mathcal{L}}_A$ imposes the condition that no photons are emitted in the forwards direction during the interval τ. The definition given in (13.108) places no constraint on the number of photon emissions in the backwards direction or the fluorescence from the sides of the cavity.

While (13.108)–(13.111) set up a somewhat imposing formalism, the explicit dynamic is that of a simple rate equation or quantum jump process. For the full atomic density operator, we can write

$$\rho_A(t) = p_1(t)|1\rangle\langle 1| + p_2(t)|2\rangle\langle 2|, \tag{13.112}$$

and master equation (13.100) is then conveniently recast as the equivalent rate equations

$$\frac{d\boldsymbol{P}}{dt} = \boldsymbol{M}\boldsymbol{P}, \tag{13.113}$$

where the state of the atom is represented by the vector of probabilities

$$\boldsymbol{P}(t) \equiv \begin{pmatrix} p_1(t) \\ p_2(t) \end{pmatrix}, \tag{13.114}$$

and

$$\boldsymbol{M} \equiv \begin{pmatrix} -\gamma_\uparrow & \gamma'_\downarrow \\ \gamma_\uparrow & -\gamma'_\downarrow \end{pmatrix}. \tag{13.115}$$

The evolution required by (13.109) may be recast in a similar way, with the change from superoperator $\mathcal{L}_A$ to $\bar{\mathcal{L}}_A$ expressed through a different dynamical matrix $\bar{\boldsymbol{M}}$. Thus, we write

$$\bar{\rho}_A(\tau) = \bar{p}_1(\tau)|1\rangle\langle 1| + \bar{p}_2(\tau)|2\rangle\langle 2|, \tag{13.116}$$

and from (13.109) obtain

$$\frac{d\bar{\boldsymbol{P}}}{d\tau} = \bar{\boldsymbol{M}}\bar{\boldsymbol{P}}, \tag{13.117}$$

with

$$\bar{\boldsymbol{P}}(\tau) \equiv \begin{pmatrix} \bar{p}_1(\tau) \\ \bar{p}_2(\tau) \end{pmatrix}, \tag{13.118}$$

and

$$\bar{\boldsymbol{M}} \equiv \begin{pmatrix} -\gamma_\uparrow & \bar{\gamma_\downarrow}' \\ \gamma_\uparrow & -\gamma_\downarrow' \end{pmatrix}. \tag{13.119}$$

The only change is to the element in the upper right corner of the matrix $\boldsymbol{M}$, which is replaced by

$$\bar{\gamma_\downarrow}' \equiv \gamma_\downarrow' - \frac{\gamma_{a2}}{2\kappa}\frac{2g^2}{\kappa} \equiv \gamma_\downarrow + \frac{\gamma_{a1}}{2\kappa}\frac{2g^2}{\kappa}. \tag{13.120}$$

In the new and more transparent notation, the waiting-time distribution of photon emissions in the forwards direction is given by the solution to (13.117), with initial condition (13.110):

$$w_{a2}(\tau) = \frac{\gamma_{a2}}{2\kappa}\left(\frac{2g^2}{\kappa}\right)\bar{p}_2(\tau)|_{\bar{\boldsymbol{P}}(0)=\binom{1}{0}}. \tag{13.121}$$

Note 13.9. Although the notation suggests it, $\bar{\rho}_A(\tau)$ is not a normalized density matrix; the sum $\bar{p}_1(\tau) + \bar{p}_2(\tau)$ is not unity. This sum is in fact the *null-measurement, or no-jump probability*, the probability that no photon is emitted in the forwards direction up to the time τ, given there was a forwards photon emission at $\tau = 0$. Clearly this probability must approach zero asymptotically, a property that is easily verified by noting that, unlike $\boldsymbol{M}$, which has one zero eigenvalue, the eigenvalues of $\bar{\boldsymbol{M}}$ are both nonzero and negative.

Note 13.10. Equations 13.108 and 13.121 are both rather formal. They are useful for calculations but obscure the simple principles at play. The basic ideas apply also to a classical jump process and have nothing, specifically, to do with quantum mechanics. To understand them we might label all γ_{a2}-type jumps, leaving the other jumps unlabeled, and display a single realization of the time sequence of jumps diagrammatically as

$$\begin{array}{llllll} & & \gamma_{a2} & & \gamma_{a2} & \\ |1\rangle \to |2\rangle \cdots\cdots & |2\rangle \to & |1\rangle \cdots\cdots\cdots & |2\rangle \to & |1\rangle \cdots\cdots \\ 0 & t_1 & t_1 + dt_1 & t_2 & t_2 + dt_2 & \end{array}$$

The times $t_1, t_2, \ldots$ are the times of successive γ_{a2}-type jumps and the times of the other jumps are not noted. Now, the rate equations (13.113) are formally solved by

$$\boldsymbol{P}(t) = e^{\boldsymbol{M}t}\boldsymbol{P}(0). \tag{13.122}$$

The solution tells us the unconditional state occupation probabilities. How then do the conditional probabilities $\bar{\boldsymbol{P}}(t)$—the state occupation probabilities given the state was $|1\rangle$ at $t=0$ *and* no γ_{a2}-type jump has occurred up to time t—evolve? Let us answer the question by writing

$$\bar{\boldsymbol{P}}(t) = e^{\bar{\boldsymbol{M}}t}\begin{pmatrix}1\\0\end{pmatrix}. \tag{13.123}$$

What, then, is the matrix $\bar{\boldsymbol{M}}$? Clearly, from the proposed expression (13.123), the probability for the $(k+1)$-th $(a2)$-type jump to occur at time t_{k+1}, given that the k-th occurred at t_k, is

$$\begin{aligned}\text{Prob}&\left(\begin{array}{c|c}k+1\text{th }(a2)\text{–jump} & k\text{th }(a2)\text{–jump}\\ \text{at }t_{k+1} & \text{at }t_k\end{array}\right)\\ &= dt_{k+1}\left(\frac{\gamma_{a2}}{2\kappa}\frac{2g^2}{\kappa}\right)\left[(01)e^{\bar{\boldsymbol{M}}(t_{k+1}-t_k)}\begin{pmatrix}1\\0\end{pmatrix}\right].\end{aligned} \tag{13.124}$$

Hence, taking the product of these probabilities, and summing over the number and times of possible $(a2)$-type jumps up to time t, we obtain an expression for the unconditional probabilities in the form

$$\begin{aligned}\boldsymbol{P}(t) = \sum_{n=0}^{\infty}\int_0^t dt_n\cdots\int_0^{t_3}dt_2\int_0^{t_2}dt_1 e^{\bar{\boldsymbol{M}}(t-t_n)}\boldsymbol{J}_{a2}e^{\bar{\boldsymbol{M}}(t_n-t_{n-1})}\\ \cdots\boldsymbol{J}_{a2}e^{\bar{\boldsymbol{M}}(t_2-t_1)}\boldsymbol{J}_{a2}e^{\bar{\boldsymbol{M}}t_1}\begin{pmatrix}1\\0\end{pmatrix},\end{aligned} \tag{13.125}$$

where

$$\begin{aligned}\boldsymbol{J}_{a2} &\equiv \frac{\gamma_{a2}}{2\kappa}\frac{2g^2}{\kappa}\begin{pmatrix}1\\0\end{pmatrix}(0\;\;0)\\ &= \frac{\gamma_{a2}}{2\kappa}\frac{2g^2}{\kappa}\begin{pmatrix}0&1\\0&0\end{pmatrix}\end{aligned} \tag{13.126}$$

acts as a jump operator (matrix). The generalized sum over jumps in (13.125) is a Dyson expansion, which when reversed gives

$$\boldsymbol{P}(t) = e^{(\bar{\boldsymbol{M}}+\boldsymbol{J}_{a2})t}\begin{pmatrix}1\\0\end{pmatrix}. \tag{13.127}$$

From (13.122), (13.123), and (13.127), we deduce the relationship

$$\bar{\boldsymbol{M}} = \boldsymbol{M} - \boldsymbol{J}_{a2}. \tag{13.128}$$

This is precisely the matrix recovered from the formal expression (13.108) and defined in (13.119).

Returning now to the evaluation of the waiting-time distribution, it is straightforward to solve the pair of coupled equations for the conditional probabilities $\bar{p}_1(\tau)$ and $\bar{p}_2(\tau)$. From (13.117)–(13.119), we find for the excited state probability,

$$\bar{p}_2(\tau)|_{\bar{\boldsymbol{P}}(0)=\binom{1}{0}} = C_+ e^{\lambda_+ \tau} + C_- e^{\lambda_- \tau}, \tag{13.129}$$

with eigenvalues of $\bar{\boldsymbol{M}}$

$$\lambda_\pm = -\frac{\gamma_\uparrow + \gamma_\downarrow'}{2} \pm \bar{\delta}, \tag{13.130}$$

where

$$\bar{\delta} \equiv \sqrt{\left(\frac{\gamma_\uparrow - \gamma_\downarrow'}{2}\right)^2 + \gamma_\uparrow \bar{\gamma}_\downarrow{}'}. \tag{13.131}$$

The initial conditions are

$$C_+ + C_- = 0, \qquad \lambda_+ C_+ + \lambda_- C_- = \gamma_\uparrow. \tag{13.132}$$

Then, using (13.129), (13.130), and (13.132), from (13.121) we obtain the *waiting-time distribution of photon emissions in the forwards direction for incoherently driven cavity-enhanced fluorescence*:

$$w_{a2}(\tau) = \frac{\gamma_{a2}}{2\kappa}\left(\frac{2g^2}{\kappa}\right)\frac{\gamma_\uparrow}{\bar{\delta}} \exp\left(-\frac{\gamma_\uparrow + \gamma_\downarrow'}{2}\tau\right)\sinh \bar{\delta}\tau. \tag{13.133}$$

Note 13.11. This result holds as a *photoelectron* waiting-time distribution—a distribution of waiting times between photoelectric detections—assuming unit quantum efficiency. More generally, for a quantum efficiency $\eta < 1$, one must replace $\gamma_{a_2}/2\kappa$ by $\eta\gamma_{a2}/2\kappa$ (in the prefactor and the definition of $\bar{\gamma}_\downarrow'$). The change follows from the obvious modification of (13.124), which is now a probability of photoelectric detection. See also the work of Carmichael and coworkers on waiting times in resonance fluorescence [13.40].

Our result is similar to the waiting-time distribution for free-space resonance fluorescence (Eq. 2.158). In particular, it vanishes at $\tau = 0$, from which we know that $g^{(2)}(0) = 0$ and the photon emissions are antibunched; of course there is no effect of the kind discussed in Sect. 13.2.3 from interference with the driving field. More can be deduced from the waiting time distribution, however, than from $g^{(2)}(\tau)$. In some instances, for example, although $g^{(2)}(0) = w_{a2}(0) = 0$, in a very real sense the light is nonetheless nearly classical. Let us explore this assertion a little.

We make a comparison between two operating regimes of the one-atom "laser," both of which produce $g^{(2)}(0) = w_{a2}(0) = 0$, but one corresponding to light that is nearly classical and the other to manifestly nonclassical light. First, rewrite the expression for $\bar{\delta}$ (Eq. 13.131) as

$$\begin{aligned}\bar{\delta} \equiv \sqrt{\left(\frac{\gamma_\uparrow - \gamma_\downarrow'}{2}\right)^2 + \gamma_\uparrow \bar{\gamma}_\downarrow{}'} &= \sqrt{\left(\frac{\gamma_\uparrow + \gamma_\downarrow'}{2}\right)^2 + \gamma_\uparrow (\bar{\gamma}_\downarrow{}' - \gamma_\downarrow')} \\ &= \frac{\gamma_\uparrow + \gamma_\downarrow'}{2}\sqrt{1 - 2\frac{\langle \hat{\mathcal{E}}_\rightarrow^\dagger \hat{\mathcal{E}}_\rightarrow \rangle_{\rm ss}}{(\gamma_\uparrow + \gamma_\downarrow')/2}},\end{aligned} \tag{13.134}$$

where we have introduced the photon flux from (13.106) and made use of (13.120). Consider now case (i): a *correlation time much less than the mean time between photon emissions*—i.e., $(\gamma_\uparrow+\gamma_\downarrow')^{-1} \ll \langle\hat{\mathcal{E}}_\rightarrow^\dagger\hat{\mathcal{E}}_\rightarrow\rangle_{\rm ss}^{-1}$. From (13.130) and (13.134), the decay rates are

$$\lambda_+ \approx -\langle\hat{\mathcal{E}}_\rightarrow^\dagger\hat{\mathcal{E}}_\rightarrow\rangle_{\rm ss}, \qquad \lambda_- \approx -(\gamma_\uparrow+\gamma_\downarrow'), \tag{13.135}$$

and from (13.121) and (13.129) the waiting-time distribution is

$$w_{a2}(\tau) \approx \langle\hat{\mathcal{E}}_\rightarrow^\dagger\hat{\mathcal{E}}_\rightarrow\rangle_{\rm ss}\big[e^{-\langle\hat{\mathcal{E}}_\rightarrow^\dagger\hat{\mathcal{E}}_\rightarrow\rangle_{\rm ss}\tau} - e^{-(\gamma_\uparrow+\gamma_\downarrow')\tau}\big]. \tag{13.136}$$

This, to a good approximation, is the waiting-time distribution of a Poisson process (classical light; Note 2.9). It shows only one small deviation from the exponential form for a Poisson process, this to account for the antibunching effect, which is confined to one or two correlation times only. As the ratio of the correlation time to the mean time between photon emissions approaches zero, the area subtracted from a pure exponential decay also approaches zero, as illustrated by curve (i) in Fig. 13.6. In this sense the limit of classical light is approached.

Contrast this behavior with case (ii): *optimal photon antibunching for incoherently driven cavity-enhanced fluorescence*, which occurs for $\bar{\delta}\to 0$. Considering the term under the square root in (13.134), note that

$$2\frac{\langle\hat{\mathcal{E}}_\rightarrow^\dagger\hat{\mathcal{E}}_\rightarrow\rangle_{\rm ss}}{(\gamma_\uparrow+\gamma_\downarrow')/2} = \frac{\gamma_{a2}}{2\kappa}\frac{4(2g^2/\kappa)\gamma_\uparrow}{\gamma_\uparrow+\gamma_\downarrow+2g^2/\kappa}, \tag{13.137}$$

whose maximum value of unity is reached with

$$\gamma_{a2} \gg \gamma_{a1}, \qquad \frac{2g^2}{\kappa} = \gamma_\uparrow \gg \gamma_\downarrow.$$

Under these conditions, the correlation time and the mean time between photon emissions have the same order of magnitude, and the waiting-time distribution is

$$w_{a2}(\tau) = \gamma_\uparrow^2\tau e^{-\gamma_\uparrow\tau}. \tag{13.138}$$

The distribution is illustrated by curve (ii) in Fig. 13.6. It shows a distinct tendency towards perfect photon antibunching—a sequence of photons equally separated in time.

Note 13.12. A measurement of $g^{(2)}(\tau)$ can focus in on the few events where two photons are emitted close together. Such a measurement is therefore able to detect the nonclassical dip in Fig. 13.5a. It provides an example of a conditional measurement. A nonconditional measurement, of the Mandel Q parameter, for example, looks at the photon stream without regard to when a previous photon might have been emitted. The nonclassical dip does leave a mark to be seen in a nonconditional measurement—in a sub-Poissonian photon counting distribution. In the case of curve (i) of Fig. 13.6, however, the

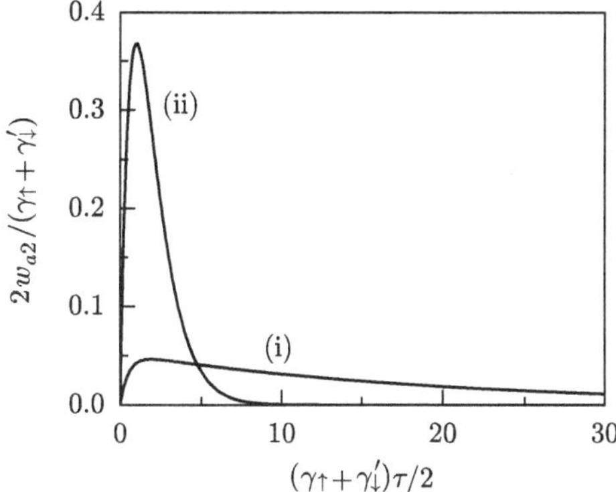

Fig. 13.6. Waiting-time distribution for cavity-enhanced fluorescence (Eq. 13.133): for (i) a photon flux much less than the inverse correlation time, $2\langle\hat{\mathcal{E}}_\rightarrow^\dagger\hat{\mathcal{E}}_\rightarrow\rangle_{\rm ss}/\frac{1}{2}(\gamma_\uparrow+\gamma_\downarrow')=0.1$, and (ii) optimal photon antibunching, $2\langle\hat{\mathcal{E}}_\rightarrow^\dagger\hat{\mathcal{E}}_\rightarrow\rangle_{\rm ss}/\frac{1}{2}(\gamma_\uparrow+\gamma_\downarrow')\to 1$

sub-Poissonian effect is very small. It is large only for conditions like those depicted in the second curve of Fig. 13.6. A practical difficulty can arise even in this case, since a low detection efficiency will turn the waiting-time distribution of curve (ii) into one looking more like curve (i); it reduces the detected flux so that the mean time between photon counts is again much larger than the correlation time.

Exercise 13.6. To complement the waiting-time distribution, one might also be interested in the related conditional distribution, the so-called null-measurement, or no-jump probability (Note 13.9). Show that

$$\mathrm{Prob}\left(\begin{array}{c}\text{no }(a2)\text{–jump}\\ \text{at }\tau=0\end{array}\middle|\begin{array}{c}(a2)\text{–jump}\\ \text{up to }\tau\end{array}\right)$$

$$=\exp\left(-\frac{\gamma_\uparrow+\gamma_\downarrow'}{2}\tau\right)\left(\cosh\bar{\delta}\tau+\frac{\gamma_\uparrow+\gamma_\downarrow'}{2\bar{\delta}}\sinh\bar{\delta}\tau\right). \tag{13.139}$$

13.3 Nonperturbative Cavity QED

Cavity QED in the perturbative limit focuses on the engineering of spontaneous emission—its enhancement or inhibition and angular redistribution. Changes are made at the level of incoherent quantum dynamics: decay rates in a master equation are changed with the possible introduction of perturbative level shifts. The nonperturbative regime extends the range of phenomena to modifications of the coherent quantum dynamics. Here the entire energy level structure is redesigned, such that new single- and multiphoton resonances appear. We begin exploring the possibilities by calculating the spontaneous emission spectrum for one atom in a cavity (Fig. 13.2) without the earlier restriction on dipole coupling strength (the bad-cavity limit of Sect. 13.2.1). Physically, for strong enough dipole coupling, a photon emitted into the cavity mode can be reabsorbed before it is lost by transmission through the cavity mirrors. Clearly, if we are to allow for this possibility, an adiabatic elimination of the cavity mode must not be made.

13.3.1 Spontaneous Emission from a Coupled Atom and Cavity

It is convenient to perform the calculation in a way that anticipates the quantum trajectory theory of Chaps. 17 and 18. To this end, we separate the master equation into two parts, the first containing all terms that act within the subspace of one energy quantum, and the second, those terms that generate transitions to the ground state with the emission of a photon. The decomposition is similar to the one introduced in (12.97) and (12.98), which helped us follow the photon emission sequences of the degenerate parametric oscillator. Making the development in parallel fashion, master equation (13.14) is written as

$$\dot{\rho} = (\mathcal{C} + \mathcal{D})\rho, \tag{13.140}$$

with

$$\mathcal{C} = \mathcal{C}_A + \mathcal{C}_a + \mathcal{C}_{Aa}, \tag{13.141a}$$

$$\mathcal{D} = \mathcal{D}_A + \mathcal{D}_a, \tag{13.141b}$$

where the individual superoperators are

$$\mathcal{C}_A \equiv -i\tfrac{1}{2}\omega_A[\sigma_z, \cdot\,] - \frac{\gamma}{2}[\sigma_+\sigma_-, \cdot\,]_+, \tag{13.142a}$$

$$\mathcal{C}_a \equiv -i\omega_C[a^\dagger a, \cdot\,] - \kappa[a^\dagger a, \cdot\,]_+, \tag{13.142b}$$

$$\mathcal{C}_{Aa} \equiv g[a^\dagger\sigma_- - a\sigma_+, \cdot\,], \tag{13.142c}$$

where $[\,\cdot, \cdot\,]_+$ denotes the anticommutator, and

$$\mathcal{D}_A \equiv \gamma(\sigma_- \cdot \sigma_+), \tag{13.143a}$$

$$\mathcal{D}_a \equiv 2\kappa(a \cdot a^\dagger). \tag{13.143b}$$

Now, for the spontaneous emission problem, the density operator has an initial condition confined within the one-quantum subspace

$$\rho(0) = \big(|2\rangle\langle 2|\big)_A\big(|0\rangle\langle 0|\big)_a. \tag{13.144}$$

We can express the solution, $\rho(t)$, in terms of a pure state $|\bar{\psi}(t)\rangle$ expanded in that subspace and a probability, $P_{\rm spon}(t)$, that a photon has been emitted via one of the two output channels; thus, we can decompose the density operator as

$$\rho(t) = \rho_0(t) + \rho_1(t), \tag{13.145}$$

where

$$\rho_0(t) = P_{\text{spon}}(t)|G\rangle\langle G|, \tag{13.146a}$$

$$\rho_1(t) = |\bar{\psi}(t)\rangle\langle\bar{\psi}(t)|, \tag{13.146b}$$

with ground state

$$|G\rangle \equiv |1\rangle_A|0\rangle_a \tag{13.147a}$$

and expansion within the one-quantum subspace

$$|\bar{\psi}(t)\rangle = \alpha(t)|1\rangle_A|1\rangle_a + \beta(t)|2\rangle_A|0\rangle_a, \tag{13.147b}$$

where $\alpha(t)$ and $\beta(t)$ are probability amplitudes for the excitation of the cavity mode and atom, respectively.

With the decomposition (13.145)–(13.147), the master equation separates into two easily solvable pieces. Substituting this *ansatz* into (13.140), it follows that

$$\dot{\rho}_1 = \mathcal{C}\rho_1, \tag{13.148a}$$

$$\dot{\rho}_0 = \mathcal{D}\rho_1. \tag{13.148b}$$

Then, using (13.141b), (13.143), (13.146), and (13.147), the second of these equations yields an equation of motion for P_{spon},

$$\begin{aligned}\frac{dP_{\text{spon}}}{dt} &= \gamma\langle G|\sigma_-|\bar{\psi}(t)\rangle\langle\bar{\psi}(t)|\sigma_+|G\rangle + 2\kappa\langle G|a|\bar{\psi}(t)\rangle\langle\bar{\psi}(t)|a^\dagger|G\rangle \\ &= \gamma|\beta(t)|^2 + 2\kappa|\alpha(t)|^2,\end{aligned} \tag{13.149}$$

and given the form of $\mathcal{C}$—as a sum of commutators and anticommutators—the first factorizes to give a Schrödinger equation for the one-quantum state consistent with (13.146b):

$$\frac{d|\bar{\psi}\rangle}{dt} = \frac{1}{i\hbar}H_C|\bar{\psi}\rangle, \tag{13.150}$$

with non-Hermitian Hamiltonian

$$\begin{aligned}H_C \equiv{}& \tfrac{1}{2}\hbar(\omega_A - i\gamma)\sigma_+\sigma_- - \tfrac{1}{2}\hbar\omega_A\sigma_-\sigma_+ + \hbar(\omega_C - i\kappa)a^\dagger a \\ &+ i\hbar g(a^\dagger\sigma_- - a\sigma_+).\end{aligned} \tag{13.151}$$

From (13.150) and (13.151), we obtain the equations of motion for the one-quantum amplitudes,

$$\dot{\alpha} = -(\kappa + i\omega_C)\alpha + g\beta, \tag{13.152a}$$

$$\dot{\beta} = -(\gamma/2 + i\omega_A)\beta - g\alpha. \tag{13.152b}$$

Thus, the solution of master equation (13.140) with initial state (13.144) has been reduced to the solution of (13.152a), (13.152b), and (13.149), with initial condition $\alpha(0) = 0$ and $|\beta(0)| = 1$.

In effect, the spontaneous emission problem has been reduced to the equations of motion for damped, coupled harmonic oscillators (in the rotating-wave approximation). Strictly speaking, the oscillator amplitudes are probability amplitudes in the present quantum mechanical context; nevertheless, so far as phenomenology is concerned, anything we can learn from (13.152a) and (13.152b) will be familiar from the classical evolution of a Lorentz oscillator of amplitude β, radiatively coupled to an electromagnetic field oscillator of amplitude α, the two oscillators damped at rates $\gamma/2$ and κ, respectively. Of course, the initial state (13.144) gives a *mean* amplitude to neither oscillator, since these means are the products of β and α with the probability amplitude of state $|G\rangle$, which is zero in the initial state. Nevertheless, this simply requires that the coupled oscillator equations be understood in a statistical sense, where, as we shall see, it is the correlation functions of the amplitudes that have nontrivial initial values. Indeed, recall that according to the quantum regression formula, the correlation functions obey the same equations of motion as the oscillator amplitudes themselves; thus, even with the statistical interpretation, we eventually solve (13.152a) and (13.152b).

We are running a little ahead of ourselves, however. To carry out our program we must solve the coupled oscillator equations twice, the first time simply to find the density operator $\rho(t)$. The exercise is straightforward:

Exercise 13.7. Solve (13.152a) and (13.152b) for initial conditions $\alpha(0) = 0$ and $|\beta(0)| = 1$; i.e., for slowly varying amplitudes

$$\tilde{\alpha} \equiv e^{i\frac{1}{2}(\omega_A + \omega_C)t} e^{-i\arg[\beta(0)]}\alpha, \tag{13.153a}$$

$$\tilde{\beta} \equiv e^{i\frac{1}{2}(\omega_A + \omega_C)t} e^{-i\arg[\beta(0)]}\beta, \tag{13.153b}$$

show that

$$\tilde{\alpha}(t) = e^{-\frac{1}{2}(\kappa + \gamma/2)t} \frac{g}{\delta_1} \sinh \delta_1 t, \tag{13.154a}$$

and

$$\tilde{\beta}(t) = e^{-\frac{1}{2}(\kappa + \gamma/2)t} \left[\cosh \delta_1 t + \frac{\kappa + \gamma/2 + i\kappa\Delta_C}{2\delta_1} \sinh \delta_1 t\right], \tag{13.154b}$$

with

$$\delta_1 \equiv \sqrt{\tfrac{1}{4}(\kappa - \gamma/2 + i\kappa\Delta_C)^2 - g^2}, \tag{13.155}$$

where Δ_C is the detuning (13.43). Hence, obtain the one-quantum part of the density operator (Eqs. 13.146b and 13.147b).

Our goal is to calculate the spontaneous emission spectrum, which, as we will see, follows directly from (13.154a) and (13.154b). Before turning to the spectrum, though, let us briefly consider the ground state part of the density operator, which is fully specified by the probability of emission $P_{\rm spon}(t)$. We integrate (13.149) for the special case of equal damping rates, $\kappa = \gamma/2$. The one-quantum probability amplitudes in this case are

$$\tilde{\alpha}(t) = e^{-(\gamma/2)t} \frac{g}{\sqrt{(\kappa\Delta_C/2)^2 + g^2}} \sin\left[\sqrt{(\kappa\Delta_C/2)^2 + g^2}t\right], \tag{13.156a}$$

and

$$\tilde{\beta}(t) = e^{-(\gamma/2)t} \left\{ \cos\left[\sqrt{(\kappa\Delta_C/2)^2 + g^2}t\right] + i\frac{\kappa\Delta_C/2}{\sqrt{(\kappa\Delta_C/2)^2 + g^2}} \sin\left[\sqrt{(\kappa\Delta_C/2)^2 + g^2}t\right] \right\}. \tag{13.156b}$$

Then the probabilities for the cavity mode and atom to be excited are, respectively,

$$|\alpha(t)|^2 = e^{-\gamma t} \frac{g^2}{(\kappa\Delta_C/2)^2 + g^2} \sin^2\left[\sqrt{(\kappa\Delta_C/2)^2 + g^2}t\right], \tag{13.157a}$$

and

$$|\beta(t)|^2 = e^{-\gamma t} \left\{ \cos^2\left[\sqrt{(\kappa\Delta_C/2)^2 + g^2}t\right] + \frac{(\kappa\Delta_C/2)^2}{(\kappa\Delta_C/2)^2 + g^2} \sin^2\left[\sqrt{(\kappa\Delta_C/2)^2 + g^2}t\right] \right\}. \tag{13.157b}$$

It follows that the probability of a spontaneous emission up to time t is precisely the same as it would be if there were no interaction with the cavity mode; we have

$$|\beta(t)|^2 + |\alpha(t)|^2 = e^{-\gamma t}, \tag{13.158}$$

with probability

$$\begin{aligned} P_{\rm spon}(t) &= \gamma \int_0^t dt' [|\beta(t')|^2 + |\alpha(t')|^2] \\ &= 1 - e^{-\gamma t}. \end{aligned} \tag{13.159}$$

Of course there is a cavity interaction, and because of it the emission is distributed between two output channels—a photon is emitted either from the side of the cavity or along the cavity axis (through the cavity mirrors). Generally the spectra in the two directions are different, and they, like the individual probabilities $|\beta(t)|^2$ and $|\alpha(t)|^2$, depend explicitly on the coupled oscillator dynamics within the one-quantum subspace.

Spontaneous emission is nonstationary process, so the emission spectrum is given by a double integration, after the fashion of (2.49) (see also Sect. 19.3). The spectrum from the sides of the cavity is determined from the autocorrelation of the atomic dipole amplitude,

$$T_{\text{side}}(\omega) = \frac{\gamma}{2\pi}\int_0^\infty dt \int_0^\infty dt' e^{-i\omega(t-t')}\langle\sigma_+(t)\sigma_-(t')\rangle, \tag{13.160a}$$

while the spectrum along the cavity axis follows from the autocorrelation of the cavity field,

$$T_{\text{axis}}(\omega) = \frac{\kappa}{\pi}\int_0^\infty dt \int_0^\infty dt' e^{-i\omega(t-t')}\langle a^\dagger(t)a(t')\rangle. \tag{13.160b}$$

In these expressions the normalization has been chosen so that the integrated spectra reflect the branching ratio into the output channels, with

$$\int_{-\infty}^{\infty} d\omega T_{\text{side}}(\omega) = \gamma\int_0^\infty dt|\beta(t)|^2 = P_{\text{side}}(\infty), \tag{13.161a}$$

$$\int_{-\infty}^{\infty} d\omega T_{\text{axis}}(\omega) = 2\kappa\int_0^\infty dt|\alpha(t)|^2 = P_{\text{axis}}(\infty), \tag{13.161b}$$

where $P_{\text{side}}(\infty)$ and $P_{\text{axis}}(\infty)$ are the probabilities for the eventual emission of a photon from the sides of the cavity and along the cavity axis, respectively, and of course

$$P_{\text{side}}(\infty) + P_{\text{axis}}(\infty) = 1. \tag{13.162}$$

Let us carry out the calculation of the autocorrelation function for the sideways-emitted light. The axial result follows by extension. We must determine the correlation function $\langle\sigma_+(t)\sigma_-(t')\rangle$. The decomposition (13.145) of $\rho(t)$ provides an elegant alternative to the derivation in Sect. 2.2.3. From the quantum regression formula, in the form (1.97), we have

$$\langle\sigma_+(t)\sigma_-(t')\rangle = \text{tr}\{\sigma_- e^{(\mathcal{C}+\mathcal{D})(t'-t)}[\rho(t)\sigma_+]\}, \tag{13.163}$$

which holds for $t' \geq t$, while for $t' < t$

$$\langle\sigma_+(t)\sigma_-(t')\rangle = \langle\sigma_+(t')\sigma_-(t)\rangle^*. \tag{13.164}$$

Now, with $\rho(t)$ expressed through (13.145)–(13.147), we find

$$\rho(t)\sigma_+ = \beta^*(t)|\bar{\psi}(t)\rangle\langle G|, \tag{13.165}$$

and using (13.141)–(13.143),

$$\mathcal{C}\big[|\bar{\psi}(t)\rangle\langle G|\big] = \left[\frac{1}{i\hbar}H_C|\bar{\psi}(t)\rangle\right]\langle G|, \tag{13.166a}$$

$$\mathcal{D}\big[|\bar{\psi}(t)\rangle\langle G|\big] = 0, \tag{13.166b}$$

where H_C is the non-Hermitian Hamiltonian (13.151). It follows then that $(t' \geq t)$

$$
\begin{aligned}
e^{(\mathcal{C}+\mathcal{D})(t'-t)}|\bar{\psi}(t)\rangle\langle G| &= \{e^{(1/i\hbar)H_C(t'-t)}|\bar{\psi}(t)\rangle\}\langle G| \\
&= |\bar{\psi}(t')\rangle\langle G|,
\end{aligned}
\tag{13.167}
$$

where the exponential inside the curly bracket propagates $|\bar{\psi}(t)\rangle$ forwards in time under Schrödinger equation (13.150). Hence, from (13.163), (13.165), (13.167), and (13.147b), we arrive at

$$
\begin{aligned}
\langle \sigma_+(t)\sigma_-(t')\rangle &= \beta^*(t)\mathrm{tr}\big[\sigma_-|\bar{\psi}(t')\rangle\langle G|\big] \\
&= \beta^*(t)\beta(t'),
\end{aligned}
\tag{13.168}
$$

where, although the result is derived for $t' \geq t$, from (13.164) it holds for $t' < t$ as well. Thus, from (13.160a), (13.160b), and (13.168), we are led to a simple expression for the *spontaneous emission spectrum in single-atom cavity QED*:

$$
T_{\mathrm{side}}(\omega) = \frac{\gamma}{2\pi}\left|\int_0^\infty dt e^{i\omega t}\beta(t)\right|^2, \tag{13.169a}
$$

and in a parallel derivation,

$$
T_{\mathrm{axis}}(\omega) = \frac{\kappa}{\pi}\left|\int_0^\infty dt e^{i\omega t}\alpha(t)\right|^2, \tag{13.169b}
$$

where $\alpha(t)$ and $\beta(t)$ obey the coupled oscillator equations of motion (13.152a) and (13.152b).

The result shows that, although the means of the cavity mode and atomic dipole amplitudes are zero, their autocorrelations are not; furthermore, the autocorrelations are the products $\alpha^*(t)\alpha(t')$ and $\beta^*(t)\beta(t')$, respectively. Formally, $\alpha(t)$ and $\beta(t)$ behave as the amplitudes of damped, coupled harmonic oscillators; it is the random phase of β in the initial state (vacuum fluctuations) that leads to the vanishing of the ensemble means.

Exercise 13.8. Write a computer program to calculate the spontaneous emission spectrum by taking fast Fourier transforms of $\alpha(t)$ and $\beta(t)$. Explore the dependence of the spectrum on parameters g/κ, $\gamma/2\kappa$, and Δ_C.

Results showing the cavity enhancement of the spontaneous emission rate are recovered from (13.168a) and (13.168b) in the bad-cavity limit. Assuming $\kappa \gg \gamma/2, g$, we expand δ_1 (Eq. 13.155) in powers of g/κ to obtain

$$
\begin{aligned}
\delta_1 &= \tfrac{1}{2}(\kappa - \gamma/2 + i\kappa\Delta_C)\sqrt{1 + \frac{4g^2}{(\kappa - \gamma/2 + i\kappa\Delta_C)^2}} \\
&\approx \tfrac{1}{2}(\kappa + \gamma/2 + i\kappa\Delta_C) - \frac{g^2}{\kappa}\frac{1 - i\Delta_C}{1 + \Delta_C^2}.
\end{aligned}
\tag{13.170}
$$

Then the coupled dynamics of the probability amplitudes $\tilde{\alpha}(t)$ and $\tilde{\beta}(t)$ are governed by eigenvalues

$$\begin{aligned}\tilde{\lambda}_+ &= -\tfrac{1}{2}(\kappa + \gamma/2) + \delta_1 \\ &\approx -\left(\frac{\gamma}{2} + \frac{g^2}{\kappa}\frac{1}{1+\Delta_C^2}\right) + i\left(\frac{1}{2}\kappa\Delta_C + \frac{g^2}{\kappa}\frac{\Delta_C}{1+\Delta_C^2}\right),\end{aligned} \tag{13.171a}$$

and

$$\tilde{\lambda}_- = -\tfrac{1}{2}(\kappa + \gamma/2) - \delta_1 \approx -\kappa(1 + i\tfrac{1}{2}\Delta_C), \tag{13.171b}$$

or for the evolution of $\alpha(t)$ and $\beta(t)$ (Eqs. 13.153a and 13.153b), the eigenvalues

$$\lambda_+ = \tilde{\lambda}_+ - i\tfrac{1}{2}(\omega_A + \omega_C) \approx \left(\frac{\gamma'}{2} + i\omega_A'\right), \tag{13.172a}$$

$$\lambda_- = \tilde{\lambda}_- - i\tfrac{1}{2}(\omega_A + \omega_C) \approx -(\kappa + i\omega_C), \tag{13.172b}$$

where γ' is the cavity-enhanced spontaneous emission rate (13.50) and ω_A' is the shifted resonance frequency (13.51). In the bad-cavity limit we retain the shift of the atomic resonance frequency, ω_A, because it might be significant compared to the linewidth $\gamma'/2$; the shift of the cavity resonance frequency, ω_C, is assumed to be negligible compared to the linewidth κ. Now the *spectrum of cavity-enhanced spontaneous emission* follows from the solutions for α and β. Using (13.172a) and (13.172b),

$$\alpha(t) \approx \frac{g}{\kappa(1+i\Delta_C)}\left[e^{-(\gamma'/2+i\omega_A')t} - e^{-(\kappa+i\omega_C)t}\right], \tag{13.173a}$$

and

$$\beta(t) \approx e^{-(\gamma'/2+i\omega_A')t}. \tag{13.173b}$$

For the sideways spectrum, the Fourier transform of (13.173b) gives a shifted Lorentzian with cavity-enhanced linewidth. Equation 13.173a yields the same result for the axial spectrum in the vicinity of the atomic resonance, where the first term inside the bracket dominates for $\kappa \gg \gamma'/2$. The branching ratio recovered from (13.161a) and (13.161b) is $P_{\text{side}}(\infty) = \gamma/\gamma'$ and $P_{\text{axis}}(\infty) = 1 - \gamma/\gamma'$.

Note 13.13. There is a symmetric limit to the bad-cavity limit—the good-cavity limit, $\gamma/2 \gg \kappa, g$. Whereas in the bad-cavity limit the cavity enhances the spontaneous emission from the atom, in the good-cavity limit the opposite occurs—the atom enhances cavity mode loss and at the same time shifts the cavity resonance. This particular phenomenon is commonplace, though. It simply amounts to a spoiling of the cavity Q, $\omega_C/2\kappa$, in the presence of an intracavity absorber, and an associated refractive index effect. To recover these effects from the general analysis, it is convenient to introduce the alternative detuning

$$\Delta_A \equiv \frac{\omega_A - \omega_C}{\gamma/2} = -\frac{\kappa}{\gamma/2}\Delta_C. \tag{13.174}$$

Then for $\gamma/2 \gg \kappa, g$, we expand δ_1 (Eq. 13.155) to obtain the approximation

$$\begin{aligned}\delta_1 &= \tfrac{1}{2}[\kappa - \gamma/2 - i(\gamma/2)\Delta_A]\sqrt{1 - \frac{4g^2}{[\kappa - \gamma/2 - i(\gamma/2)\Delta_A]^2}} \\ &\approx \tfrac{1}{2}[\kappa - \gamma/2 - i(\gamma/2)\Delta_A] + \frac{2g^2}{\gamma}\frac{1 - i\Delta_A}{1 + \Delta_A^2}.\end{aligned} \tag{13.175}$$

In place of (13.172a) and (13.172b), the eigenvalues are

$$\lambda_+ = \tilde{\lambda}_+ - i\tfrac{1}{2}(\omega_A + \omega_C) \approx \left(\frac{\gamma}{2} + i\omega_A\right), \tag{13.176a}$$

$$\lambda_- = \tilde{\lambda}_- - i\tfrac{1}{2}(\omega_A + \omega_C) \approx -(\kappa' + i\omega_C'). \tag{13.176b}$$

Thus, we obtain the enhanced cavity decay rate (spoiled cavity Q) due to atomic absorption,

$$\kappa' = \kappa\left(1 + 2C_1\frac{1}{1 + \Delta_A^2}\right), \tag{13.177}$$

together with the shifted cavity resonance frequency,

$$\omega_C' = \omega_C - \kappa 2C_1\frac{\Delta_A}{1 + \Delta_A^2}, \tag{13.178}$$

due to atomic dispersion. The latter may be associated with a single-atom refractive index.

Exercise 13.9. Calculate the spontaneous emission spectrum for the case where the atom is initially unexcited but there is one quantum of excitation in the cavity mode. Compare the spectrum for an initially excited atom.

13.3.2 Vacuum Rabi Splitting

Continuing now with our derivation, from (13.169a), (13.169b), and (13.153)–(13.155), and assuming exact resonance between the cavity mode and atom ($\Delta_C = 0$), we arrive at compact expressions for the *spontaneous emission spectrum in nonperturbative cavity QED*:

$$T_{\mathrm{side}}(\omega)$$
$$= \frac{\gamma}{2\pi}\frac{1}{4|g'|^2}\left|\frac{\frac{1}{2}(\kappa - \gamma/2) + ig'}{\frac{1}{2}(\kappa + \gamma/2) - i(\omega - \omega_A + g')} - \frac{\frac{1}{2}(\kappa - \gamma/2) - ig'}{\frac{1}{2}(\kappa + \gamma/2) - i(\omega - \omega_A - g')}\right|^2, \tag{13.179a}$$

and

$$T_{\text{axis}}(\omega)$$
$$= \frac{\kappa}{\pi}\frac{1}{4|g'|^2}\left|\frac{g}{\frac{1}{2}(\kappa+\gamma/2)-i(\omega-\omega_A+g')} - \frac{g}{\frac{1}{2}(\kappa+\gamma/2)-i(\omega-\omega_A-g')}\right|^2, \tag{13.179b}$$

where

$$g' \equiv \sqrt{g^2 - \tfrac{1}{4}(\kappa-\gamma/2)^2}. \tag{13.180}$$

Generally, the two components have different shapes due to the cross-terms in the square modulus, a consequence of the different phases of oscillation of probability amplitudes $\alpha(t)$ and $\beta(t)$. The cross-terms are negligible, however, in the strong-coupling limit, $g \gg \kappa, \gamma/2$, in which case we find

$$\frac{T_{\text{side}}(\omega)}{\gamma} \approx \frac{T_{\text{axis}}(\omega)}{2\kappa} \approx \frac{T(\omega)}{\gamma+2\kappa}. \tag{13.181}$$

In this limit only the emission probabilities differ; both spectral components approach the so-called *vacuum Rabi spectrum*, or *vacuum Rabi doublet*,

$$T(\omega) = \frac{1}{2}\left[\frac{\frac{1}{2}(\kappa+\gamma/2)/\pi}{\frac{1}{4}(\kappa+\gamma/2)^2+(\omega-\omega_A+g)^2} + \frac{\frac{1}{2}(\kappa+\gamma/2)/\pi}{\frac{1}{4}(\kappa+\gamma/2)^2+(\omega-\omega_A-g)^2}\right]. \tag{13.182}$$

Figure 13.7 shows the development of the vacuum Rabi doublet as the dipole coupling strength is increased.

The vacuum Rabi spectrum may be interpreted as the emission from the first excited state of the Jaynes–Cummings Hamiltonian (Eq. 13.6), where the degenerate first excited state is split by the strong dipole coupling. In the resonant case, $\omega_C = \omega_A$, the energy eigenstates are the one-quantum dressed states (Exercise 13.10)

$$|1,U\rangle = \frac{1}{\sqrt{2}}\big(|2\rangle_A|0\rangle_a + i|1\rangle_A|1\rangle_a\big), \tag{13.183a}$$
$$|1,L\rangle = \frac{1}{\sqrt{2}}\big(|2\rangle_A|0\rangle_a - i|1\rangle_A|1\rangle_a\big), \tag{13.183b}$$

where, from (13.6),

$$(H_A + H_F + H_{AF})|1,U\rangle = (\tfrac{1}{2}\hbar\omega_A + \hbar g)|1,U\rangle, \tag{13.184a}$$
$$(H_A + H_F + H_{AF})|1,L\rangle = (\tfrac{1}{2}\hbar\omega_A - \hbar g)|1,L\rangle, \tag{13.184b}$$

which yield transition frequencies to the ground state $\omega_A + g$ and $\omega_A - g$. The initial state $|2\rangle_A$ is a superposition of the energy eigenstates, so the emission

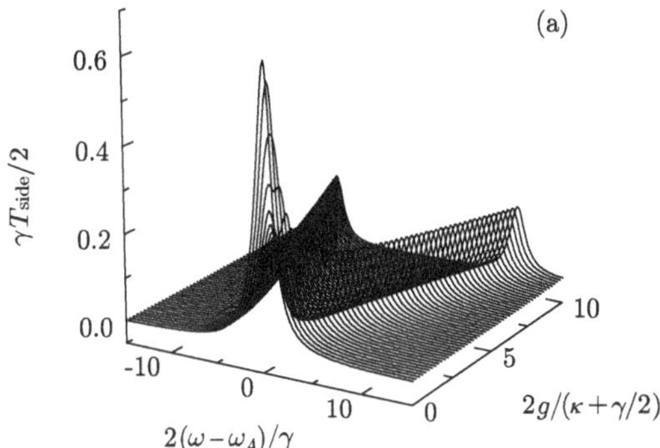

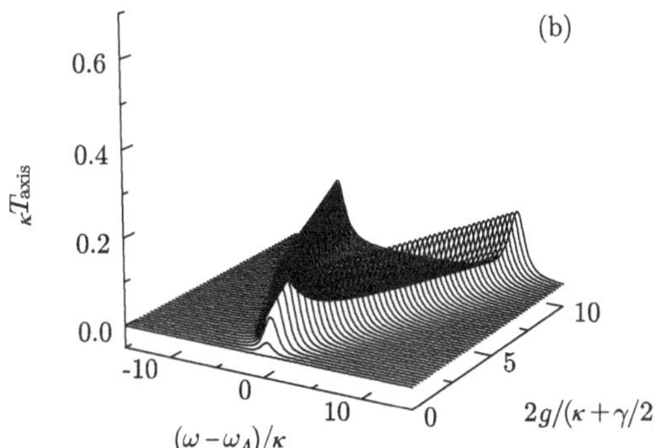

Fig. 13.7. Development of the vacuum Rabi doublet with increasing dipole coupling strength: (**a**) spectrum of spontaneous emission out the sides of the cavity (Eq. 13.179a), and (**b**) spectrum of spontaneous emission along the cavity axis (Eq. 13.179b). The spectra are plotted for $\gamma/2\kappa = 1$

spectrum shows peaks at both transition frequencies. Of course, for a doublet to be observed, the peak widths must be smaller than the level separation. The widths are obtained by noting that each energy eigenstate decays via both output channels—via the transition operator (sideways channel)

$$\sigma_- = |G\rangle\left[\frac{1}{\sqrt{2}}\left(\langle 1,U| + \langle 1,L|\right)\right] + \cdots, \tag{13.185a}$$

and the transition operator (axial channel)

$$a = |G\rangle\left[-i\frac{1}{\sqrt{2}}\left(\langle 1,U| - \langle 1,L|\right)\right] + \cdots. \tag{13.185b}$$

Thus, the vacuum Rabi resonances show the average linewidth $\frac{1}{2}(\kappa + \gamma/2)$. Alternatively, a time-domain argument can be given, where the *linewidth averaging* occurs as the system energy oscillates between the atom (decay rate γ)

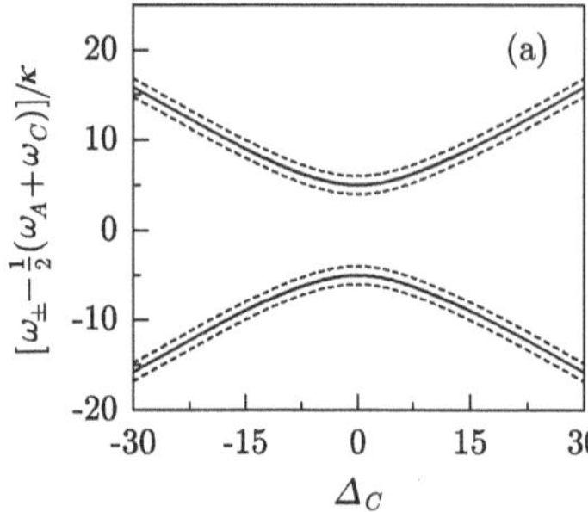

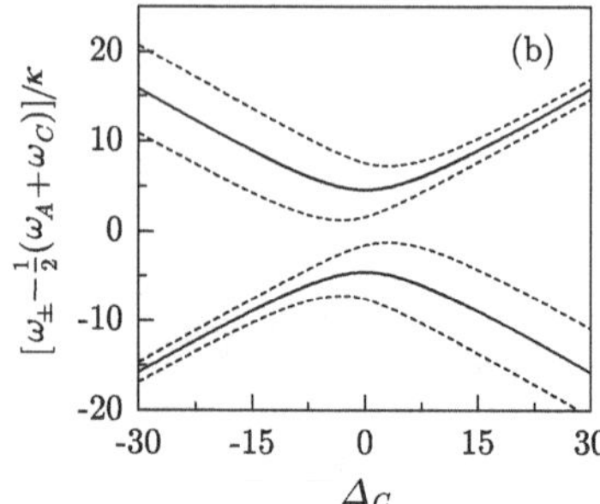

Fig. 13.8. Frequencies (*solid curves*) and widths (*dashed curves*) of the vacuum Rabi resonances plotted as a function of the detuning of the cavity mode from the atomic resonance: for $g/\kappa = 5$ and (**a**) $\gamma/2\kappa = 1$, (**b**) $\gamma/2\kappa = 5$

and the cavity mode (decay rate 2κ). It is interesting to note that the average linewidth is subnatural whenever $\kappa < \gamma/2$. A subnatural linewidth has been observed [13.41] through the measurement of weak-probe transmission spectra in many-atom cavity QED (Sect. 14.4.1).

When viewed as a function of the detuning between the cavity mode and the atom, the vacuum Rabi resonances provide an example of an avoided crossing. The resonance frequencies and widths are given by

$$\omega_\pm \equiv \tfrac{1}{2}(\omega_A + \omega_C) \pm \mathrm{Im}\delta_1, \qquad \gamma_\pm \equiv \tfrac{1}{2}(\kappa + \gamma/2) \mp \mathrm{Re}\delta_1, \tag{13.186}$$

with δ_1 defined in (13.155). These are plotted to illustrate the avoided crossing in Fig. 13.8.

Of course, the terminology "vacuum Rabi resonances" (also "vacuum Rabi splitting"), taken from Sanchez-Mondragon and coworkers [13.42], is something of a misnomer. It is clearly the first excited state of the Jaynes–Cummings Hamiltonian that explains the level splitting, so "one-quantum Rabi splitting" might be better. Other names are also used and more accurately describe the physics—*normal-mode resonances* or *normal-mode splitting* [13.43, 13.6], and, for semiconductor systems, *cavity polaritons* [13.44, 13.45].

Note 13.14. When κ is set to zero $T_{\mathrm{side}}(\omega)$ (Eq. 13.179a) vanishes identically on line center ($\omega = \omega_A$) for all values of g and $\gamma/2$. This is sometimes explained [13.41] in terms of an interference of probability amplitudes for emission from the dressed states $|1, U\rangle$ and $|1, L\rangle$; these states are prepared so as to give destructive interference by the initial condition (13.144). An alternative time-domain explanation follows from the coupled equations (13.152a) and (13.152b). These may be written as ($\omega_C = \omega_A$, $\kappa = 0$)

$$\dot{\tilde{\alpha}} = g\tilde{\beta}, \qquad \dot{\tilde{\beta}} = -(\gamma/2)\tilde{\beta} - g\tilde{\alpha}. \tag{13.187}$$

Note then that the expression (13.169a) for $T_{\mathrm{side}}(\omega)$ writes the transition amplitude at line center as the integral over time of the dipole amplitude

$\tilde{\beta}(t)$; hence, we obtain

$$T_{\text{side}}(\omega_A) = \frac{1}{2\pi}\left|\int_0^\infty dt'\tilde{\beta}(t')\right|^2 = \frac{1}{2\pi}\frac{1}{g^2}|\tilde{\alpha}(\infty) - \tilde{\alpha}(0)|^2 = 0. \qquad (13.188)$$

The integral is guaranteed to vanish because $\tilde{\alpha}(\infty)$ is definitely zero—the photon is definitely emitted—and the initial condition is set with $\tilde{\alpha}(0) = 0$. Thus, the initial condition sets up a particular phased oscillation of the dipole amplitude, which leads to the zero in the spectrum. Seen in this light, the phenomenon is not uniquely quantum-mechanical. There is a counterpart in the decay of coupled classical oscillator amplitudes.

13.3.3 Vacuum Rabi Resonances in the Two-State Approximation

We will see in Sect. 14.4.1 that vacuum Rabi resonances also arise in many-atom cavity QED. It is not that the spontaneous emission problems for one and many atoms are the same; but when excited by a weak probe, the many-atom system shows the same two-peaked spectrum. The connection is made through the coupled oscillator equations (13.152a) and (13.152b). For a weakly excited two-level medium the Maxwell–Bloch equations take the same coupled oscillator form. In fact, the oscillator equations are generic; they describe the coupling of one electromagnetic field mode to a polarized medium in the linear response regime. Details that distinguish one medium from another change parameters, but not the form of the equations; hence the connection between polaritons [13.46] and vacuum Rabi resonances, for example.

Once the nonlinear regime is entered, however, differences appear in the behavior of one versus many atoms. In quantum mechanical terms, they arise through multiphoton transitions. Such transitions depend on the excited state spectrum above the one-quantum level, which is of course different for one compared with many atoms, and for a semiconductor medium. Fluctuations provide a sensitive probe of multiphoton effects, and measurements of $g^{(2)}(\tau)$ (Sects. 15.2.7, 16.1.3, and 16.2) or the incoherent spectrum (Sect. 15.2.6) can detect the influence of nonlinearity—even in a many-atom system—at the two-quanta level. The mean system response is less sensitive, and in a many-atom system remains approximately linear until the cavity contains enough photons to saturate the entire medium [13.47]; thus, the vacuum Rabi resonances persist to high excitation, looking much like the normal modes of coupled harmonic oscillators (Exercise 13.11).

The situation is different for single-atom cavity QED, where the energy spectrum differs from that of coupled harmonic oscillators even at the second excited state. For this case, Tian and Carmichael [13.48] have shown that, for sufficiently strong dipole coupling, each vacuum Rabi resonance behaves like an approximate two-state system, from which one recovers the Mollow spectrum (Sect. 2.3.4) and photon antibunching of resonance fluorescence

(Sects. 2.3.5 and 2.3.6). The predicted photon antibunching has been observed [13.49]. In this section we formulate the two-state approximation of Tian and Carmichael without reproducing their exact numerical results.

The derivation of the full Jaynes–Cummings spectrum is left as an exercise (Exercise 13.10). In the resonant case, there is the ground state (13.147a), and an infinite ladder of excited state doublets ($n = 1, 2, \ldots$)

$$|n, U\rangle = \frac{1}{\sqrt{2}}\big(|2\rangle_A|n-1\rangle_a + i|1\rangle_A|n\rangle_a\big), \tag{13.189a}$$

$$|n, L\rangle = \frac{1}{\sqrt{2}}\big(|2\rangle_A|n-1\rangle_a - i|1\rangle_A|n\rangle_a\big), \tag{13.189b}$$

generalizing (13.183a) and (13.183b). The energies are

$$E_G = -\tfrac{1}{2}\hbar\omega_A, \tag{13.190}$$

and

$$E_{n,U} = (n - \tfrac{1}{2})\hbar\omega_A + \sqrt{n}\hbar g, \tag{13.191a}$$

$$E_{n,L} = (n - \tfrac{1}{2})\hbar\omega_A - \sqrt{n}\hbar g. \tag{13.191b}$$

Consider now that the cavity is coherently driven, as in Sects. 13.2.2 and 13.2.3, but at the frequency of one of the vacuum Rabi resonances rather than the cavity resonance frequency ω_C. For sufficiently large g, and assuming that the strength of the driving field is not too large, the detuning of the transition to the second excited state suggests that we might make a two-state approximation. Choosing excitation of the lower vacuum Rabi resonance (Fig. 13.9), the two-state system comprises the ground state $|G\rangle$ and the excited state $|1, L\rangle$. Formally, we define

$$\hat{\Sigma}_- \equiv |G\rangle\langle 1, L|, \qquad \hat{\Sigma}_+ \equiv |1, L\rangle\langle G|, \tag{13.192}$$

and make the approximation

$$ia \rightarrow \frac{1}{\sqrt{2}}\hat{\Sigma}_-, \qquad -ia^\dagger \rightarrow \frac{1}{\sqrt{2}}\hat{\Sigma}_+,$$

$$\sigma_- \rightarrow \frac{1}{\sqrt{2}}\hat{\Sigma}_-, \qquad \sigma_+ \rightarrow \frac{1}{\sqrt{2}}\hat{\Sigma}_+,$$

Master equation (13.57) becomes ($\omega_0 \rightarrow \omega_A - g$)

$$\begin{aligned}\dot{\rho} = &-i\big[(\tfrac{1}{2}\omega_A - g)\hat{\Sigma}_+\hat{\Sigma}_- - \tfrac{1}{2}\omega_A\hat{\Sigma}_-\hat{\Sigma}_+, \rho\big] \\ &+ \frac{1}{\sqrt{2}}\big[\bar{\mathcal{E}}_0 e^{-i(\omega_A - g)t}\hat{\Sigma}_+ - \bar{\mathcal{E}}_0^* e^{i(\omega_A - g)t}\hat{\Sigma}_-, \rho\big] \\ &+ \tfrac{1}{2}(\kappa + \gamma/2)(2\hat{\Sigma}_-\rho\hat{\Sigma}_+ - \hat{\Sigma}_+\hat{\Sigma}_-\rho - \rho\hat{\Sigma}_+\hat{\Sigma}_-),\end{aligned} \tag{13.193}$$

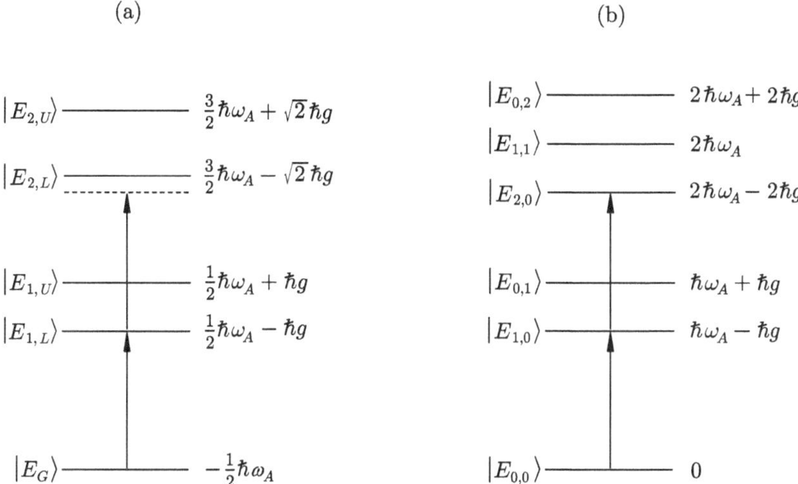

Fig. 13.9. (**a**) Spectrum of the Jaynes–Cummings Hamiltonian showing the ground state and first and second excited-state doublets; the driving field is tuned to the lower vacuum Rabi resonance so the second excited state transition is detuned. (**b**) Corresponding spectrum of coupled harmonic oscillators; the second excited-state transition is tuned

which is the master equation of resonance fluorescence (Eq. 2.96). Thus, everything familiar from resonance fluorescence is reproduced, approximately at least—the Mollow triplet spectrum (Sect. 2.3.4), and the photon antibunching and squeezing (Sects. 2.3.5 and 2.3.6). Since in the two-state approximation the fields scattered into both output channels are proportional to $\hat{\Sigma}_-$, we can expect to see these effects for the axially as well as the sideways-scattered light. The generation of a Mollow triplet spectrum in transmission is a particularly interesting novelty, arising from a splitting of the $|G\rangle$ and $|1, L\rangle$ levels, a *dressing of the dressed states.*

Exercise 13.10. Show that in the presence of detuning the eigenstates and eigenenergies of the Jaynes–Cummings Hamiltonian (Eq. 13.6), the so-called *dressed states* and *dressed energies*, are the ground state $|G\rangle$ (Eq. 13.147a), with energy $E_G = -\frac{1}{2}\hbar\omega_A$, and the excited-state doublets ($n = 1, 2, \ldots$)

$$|n, U\rangle = \frac{1}{\sqrt{2}}\big(C|2\rangle_A|n-1\rangle_a + iD|1\rangle_A|n\rangle_a\big), \tag{13.194a}$$

$$|n, L\rangle = \frac{1}{\sqrt{2}}\big(D|2\rangle_A|n-1\rangle_a - iC|1\rangle_A|n\rangle_a\big), \tag{13.194b}$$

where

$$C \equiv \left[1 + \frac{\frac{1}{2}(\omega_A - \omega_C)}{\sqrt{\frac{1}{4}(\omega_A - \omega_C)^2 + g^2 n}}\right]^{1/2}, \tag{13.195}$$

and

$$D \equiv \left[1 - \frac{\frac{1}{2}(\omega_A - \omega_C)}{\sqrt{\frac{1}{4}(\omega_A - \omega_C)^2 + g^2 n}}\right]^{1/2}, \tag{13.196}$$

with energies

$$E_{n,U} = (n - \tfrac{1}{2})\hbar\omega_C + \hbar\sqrt{\tfrac{1}{4}(\omega_A - \omega_C)^2 + g^2 n}, \tag{13.197a}$$

$$E_{n,L} = (n - \tfrac{1}{2})\hbar\omega_C - \hbar\sqrt{\tfrac{1}{4}(\omega_A - \omega_C)^2 + g^2 n}. \tag{13.197b}$$

Exercise 13.11. Consider the Hamiltonian of coupled harmonic oscillators, with the interaction on resonance,

$$\begin{aligned} H_S &= H_a + H_b + H_{ab} \\ &= \hbar\omega_0 a^\dagger a + \hbar\omega_0 b^\dagger b + i\hbar g(a^\dagger b - a b^\dagger). \end{aligned} \tag{13.198}$$

Show that the eigenenergies and eigenstates are ($N = 1, 2, \ldots$; $M = 1, 2, \ldots$)

$$E_{N,M} = N\hbar(\omega_0 - g) + M\hbar(\omega_0 + g), \tag{13.199}$$

and

$$|E_{N,M}\rangle = \frac{(a^\dagger - ib^\dagger)^N (a^\dagger + ib^\dagger)^M}{\sqrt{2^{N+M} N! M!}} |0\rangle_a |0\rangle_b. \tag{13.200}$$

Draw a sketch to compare the energy spectrum (13.199) with the Jaynes–Cummings spectrum. Why is it that for coherent excitation—interaction Hamiltonian $i\hbar\mathcal{E}(e^{-i\omega t}a^\dagger - e^{i\omega t}a)$, for example—resonances are observed at the normal-mode frequencies $\omega_0 - g$ and $\omega_0 + g$ only?

14

Many Atoms in a Cavity: Macroscopic Theory

We will be familiar with much of the background to many-atom cavity QED from our treatment of laser theory in Chaps. 7 and 8. Laser theory is closely related to the theory of optical bistability, which is also built around the quasi-resonant interaction of atoms and light inside an optical cavity, but considers a passive rather than active medium. In this chapter we develop the semiclassical theory of optical bistability and uncover its most obvious connections with cavity QED—in particular, the many-atom version of the vacuum Rabi doublet. We meet the Maxwell–Bloch equations for a two-level medium interacting with a single mode of a Fabry–Perot, and discuss the relationship between these macroscopic equations and a microscopic description in terms of Hamiltonians and master equations.

To start with, in Sect. 14.1 we review the physical principles that underlie the phenomena of absorptive and dispersive optical bistability, avoiding the detailed description of any particular nonlinear medium. An excellent review of the application of these principles is given in the book by Gibbs [14.1] where a number of specific systems are considered, ranging from the two-level atoms considered in Sects. 14.2–14.4 to optical fibers and excitonic nonlinearities in semiconductors.

14.1 Optical Bistability: Steady-State Transmission of a Nonlinear Fabry–Perot

Consider a Fabry–Perot cavity partially filled with a nonlinear medium, as depicted in Fig. 14.1. A coherent TEM_{00} traveling-wave field, frequency ω_0, is mode-matched into the cavity traveling in the positive z direction. The input field is expanded in the form

$$\boldsymbol{E}_0(\boldsymbol{r},t) = \hat{e}_0 \tfrac{1}{2}\mathcal{E}_0 e^{-(x^2+y^2)/w_0^2} e^{-i[\omega_0(t-z/c)-\phi_0]} + \mathrm{c.c.}, \tag{14.1}$$

$z \leq -L+\ell+d$, where $\hat{e}_0$ is a unit polarization vector, $\mathcal{E}_0$ is the amplitude of the field (a real number), w_0 is the mode waist, and ϕ_0 is an arbitrary phase.

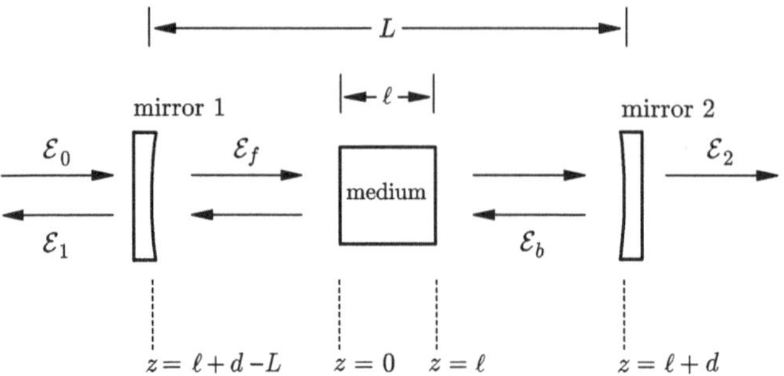

Fig. 14.1. Schematic of the nonlinear Fabry–Perot interferometer. A medium of two-level atoms extends between $z = 0$ and $z = \ell$

There are two output fields, one propagating to the left of the cavity,

$$\boldsymbol{E}_1(\boldsymbol{r},t) = \hat{e}_0 \tfrac{1}{2}\mathcal{E}_1(t)e^{-(x^2+y^2)/w_0^2}e^{-i[\omega_0(t+z/c)-\phi_1(t)]} + \text{c.c.}, \tag{14.2}$$

$z \leq -L + \ell + d$, and the other to the right,

$$\boldsymbol{E}_2(\boldsymbol{r},t) = \hat{e}_0 \tfrac{1}{2}\mathcal{E}_2(t)e^{-(x^2+y^2)/w_0^2}e^{-i[\omega_0(t-z/c)-\phi_2(t)]} + \text{c.c.}, \tag{14.3}$$

$z \geq \ell + d$, where $\mathcal{E}_1(t)e^{i\phi_1(t)}$ and $\mathcal{E}_2(t)e^{i\phi_2(t)}$ are complex amplitudes, assumed to be slowly varying functions of time. Inside the cavity the field is a sum of forwards and backwards traveling waves. We expand these as

$$\boldsymbol{E}_f(\boldsymbol{r},t) = \hat{e}_0 \tfrac{1}{2}\mathcal{E}_f(z,t)e^{-(x^2+y^2)/w_0^2}e^{-i[\omega_0(t+z/c)-\phi_f(z,t)]} + \text{c.c.}, \tag{14.4a}$$

and

$$\boldsymbol{E}_b(\boldsymbol{r},t) = \hat{e}_0 \tfrac{1}{2}\mathcal{E}_b(z,t)e^{-(x^2+y^2)/w_0^2}e^{-i[\omega_0(t-z/c)-\phi_b(z,t)]} + \text{c.c.}, \tag{14.4b}$$

$\ell + d - L \leq z \leq \ell + d$, where $\mathcal{E}_f(z,t)e^{i\phi_f(z,t)}$ and $\mathcal{E}_b(z,t)e^{i\phi_b(z,t)}$ are complex amplitudes, assumed to be slowly varying in space and time. Like $\mathcal{E}_0$ the functions $\mathcal{E}_1(t)$, $\mathcal{E}_2(t)$, $\mathcal{E}_f(z,t)$, and $\mathcal{E}_b(z,t)$ are all real.

Our goal is to determine the output fields, $\boldsymbol{E}_1(\boldsymbol{r},t)$ and $\boldsymbol{E}_2(\boldsymbol{r},t)$, in terms of the input field $\boldsymbol{E}_0(\boldsymbol{r},t)$. Here and in the following section we address this question for the steady state only. Time evolution is considered in Sect. 14.2.2. We start from the equations that relate the steady-state output fields to the input and the steady-state fields inside the cavity. Using (14.1)–(14.4b), and dropping the time argument on the field amplitudes, we have

$$\mathcal{E}_1 e^{i\phi_1} = \sqrt{R_1}\mathcal{E}_0 e^{i[\phi_0+\phi_R-2(\omega_0/c)(L-\ell-d)]} + \sqrt{T_1}\mathcal{E}_b(0)e^{i[\phi_b(0)+\phi_T]}, \tag{14.5a}$$

and

$$\mathcal{E}_2 e^{i\phi_2} = \sqrt{T_2}\mathcal{E}_f(\ell)e^{i[\phi_f(\ell)+\phi_T]}, \tag{14.5b}$$

where ϕ_T is the phase change on transmission at the mirrors. It remains to relate $\mathcal{E}_b(0)e^{i\phi_b(0)}$ and $\mathcal{E}_f(\ell)e^{i\phi_f(\ell)}$ to $\mathcal{E}_0 e^{i\phi_0}$. This is accomplished in two steps.

Consider first the boundary conditions relating the forwards and backwards field amplitudes at either end of the medium. At $z = 0$, the forwards field may be written as a sum of the input field and the backwards field. Taking into account free propagation and transmission and reflection at mirror 1, the boundary condition is

$$\mathcal{E}_f(0)e^{i\phi_f(0)} = \sqrt{T_1}\mathcal{E}_0 e^{i(\phi_0+\phi_T)} + \sqrt{R_1}\mathcal{E}_b(0)e^{i[\phi_b(0)+\phi_R+2(\omega_0/c)(L-\ell-d)]}, \tag{14.6a}$$

where ϕ_R is the phase change on reflection, and $2(\omega_0/c)(L-\ell-d)$ is the phase accumulated through free propagation of the backwards field. At $z = \ell$, the backwards field is related to the forwards field through its free propagation and reflection at mirror 2:

$$\mathcal{E}_b(\ell)e^{i\phi_b(\ell)} = \sqrt{R_2}\mathcal{E}_f(\ell)e^{i[\phi_f(\ell)+\phi_R+2(\omega_0/c)(\ell+d)]}, \tag{14.6b}$$

where $(\omega_0/c)(\ell + d)$ is the phase accumulated through free propagation.

The second step requires us to relate the amplitudes of the forwards and backwards fields at either end of the medium. This calls for the solution of a nonlinear propagation problem. A diversion in this direction would lead us into unnecessary detail at this stage and require us to specialize to a particular model, so we set aside the nonlinear propagation for the time being and simply write

$$\mathcal{E}_f(\ell)e^{i\phi_f(\ell)} = K^f_{\mathrm{abs}}\mathcal{E}_f(0)e^{i[\phi_f(0)+\theta^f_{\mathrm{disp}}]}, \tag{14.7a}$$

$$\mathcal{E}_b(0)e^{i\phi_b(0)} = K^b_{\mathrm{abs}}\mathcal{E}_b(\ell)e^{i[\phi_b(\ell)+\theta^b_{\mathrm{disp}}]}, \tag{14.7b}$$

which holds without loss of generality so long as K^f_{abs}, K^b_{abs}, θ^f_{disp}, and θ^b_{disp} are viewed as functions of the field amplitudes. The amplitude changing factors K^f_{abs} and K^b_{abs} account for absorption in the medium, while θ^f_{disp} and θ^b_{disp} account for dispersion. Fundamentally, each of these is a function of two amplitudes, because in a nonlinear medium the forwards and backwards waves do not propagate independently of one another (see Note 14.1). The dependence on one amplitude is eliminated through the boundary conditions, though, which leaves a dependence on just one of the four field amplitudes $\mathcal{E}_f(0)$, $\mathcal{E}_b(\ell)$, $\mathcal{E}_b(0)$, or $\mathcal{E}_b(\ell)$; we consider the dependence to be on $\mathcal{E}_f(0)$.

We are now in a position to derive the *steady-state relationship between input and output fields for a standing-wave cavity (Fabry-Perot)*. From (14.6a), and using (14.7b), (14.6b), and (14.7a), we obtain

$$\begin{aligned}\mathcal{E}_f(0)e^{\phi_f(0)} &= \sqrt{T_1}\mathcal{E}_0 e^{i(\phi_0+\phi_T)} \\ &\quad + \sqrt{R_1R_2}K^f_{\mathrm{abs}}K^b_{\mathrm{abs}}\mathcal{E}_f(0)e^{i[\phi_f(0)+\theta_0+\theta^f_{\mathrm{disp}}+\theta^b_{\mathrm{disp}}]},\end{aligned} \tag{14.8}$$

where

$$\theta_0 \equiv \operatorname{mod} 2\pi[2(\omega_0/c)L + 2\phi_R] \tag{14.9}$$

is the round-trip phase change due to free propagation. Collecting terms in $\mathcal{E}_f(0)e^{\phi_f(0)}$, we have

$$\mathcal{E}_f(0)e^{i\phi_f(0)} = \frac{\sqrt{T_1}\mathcal{E}_0 e^{i(\phi_0+\phi_T)}}{1-\sqrt{R_1R_2}K^f_{\text{abs}}K^b_{\text{abs}}e^{i(\theta_0+\theta^f_{\text{disp}}+\theta^b_{\text{disp}})}}. \tag{14.10}$$

Implicitly, this is a nonlinear equation to be solved by $\mathcal{E}_f(0)$, through the functional dependence in K^f_{abs}, K^b_{abs}, θ^f_{disp}, and θ^b_{disp}. Given a solution for $\mathcal{E}_f(0)$, the amplitudes of the output fields are

$$\begin{aligned}\mathcal{E}_1 e^{i\phi_1} &= \sqrt{R_1}\mathcal{E}_0 e^{i[\phi_0+\phi_R-2(\omega_0/c)(L-\ell-d)]}\\ &\quad+\sqrt{R_1T_1}K^f_{\text{abs}}K^b_{\text{abs}}\left[\mathcal{E}_f(0)e^{i\phi_f(0)}\right]e^{i[\phi_R+\phi_T+2(\omega_0/c)(\ell+d)+\theta^f_{\text{disp}}+\theta^b_{\text{disp}}]},\end{aligned} \tag{14.11}$$

where we have used (14.5a), (14.7b), (14.6b), and (14.7a), and

$$\mathcal{E}_2 e^{i\phi_2} = \sqrt{T_2}K^f_{\text{abs}}\left[\mathcal{E}_f(0)e^{i\phi_f(0)}\right]e^{i(\phi_T+\theta^f_{\text{disp}})}, \tag{14.12}$$

which follows from (14.5b) and (14.7a). Note that it can be helpful to read (14.10) in the reverse direction: assuming that the functional dependencies of K^f_{abs}, K^b_{abs}, θ^f_{disp}, and θ^b_{disp} on $\mathcal{E}_f(0)$ are given, then (14.10) determines $\mathcal{E}_0 e^{i\phi_0}$ as an explicit function of $\mathcal{E}_f(0)e^{i\phi_f(0)}$; stated another way, the input field and one output field are given explicitly as a function of the second output field.

Note 14.1. Equation 14.10 is equivalent to what one obtains by summing over successive round trips in the cavity. Thus, we have the alternative derivation

$$\begin{aligned}\mathcal{E}_f(0)e^{i\phi_f(0)} &= \sqrt{T_1}\mathcal{E}_0 e^{i(\phi_0+\phi_T)}\Big[1+\sqrt{R_1R_2}K^f_{\text{abs}}K^b_{\text{abs}}e^{i(\theta_0+\theta^f_{\text{disp}}+\theta^b_{\text{disp}})}\\ &\qquad+\left(\sqrt{R_1R_2}K^f_{\text{abs}}K^b_{\text{abs}}\right)^2 e^{2i(\theta_0+\theta^f_{\text{disp}}+\theta^b_{\text{disp}})}+\cdots\Big]\\ &= \frac{\sqrt{T_1}\mathcal{E}_0 e^{i(\phi_0+\phi_T)}}{1-\sqrt{R_1R_2}K^f_{\text{abs}}K^b_{\text{abs}}e^{i(\theta_0+\theta^f_{\text{disp}}+\theta^b_{\text{disp}})}}.\end{aligned} \tag{14.13}$$

Strictly, however, the notion of following successive round trips in this way is a little problematic for a nonlinear medium, since the forwards and backwards waves scatter into one another, through the standing-wave grating written into the medium by the standing-wave field (see Eqs. 14.26a and 14.26b).

Equation 14.10 is the so-called *optical bistability state equation*, at this stage written in rather rudimentary form. Essentially, it is the standard result for the steady-state field inside a Fabry–Perot cavity, generalized to include losses and phase shifts that depend themselves on the amplitude of the intracavity field; the equation amounts to a self-consistency requirement to be satisfied by the intracavity field. Various explicit versions of it appear in the literature, each with a different choice of the functions $K_{\rm abs}^f$, $K_{\rm abs}^b$, $\theta_{\rm disp}^f$, and $\theta_{\rm disp}^b$. Different versions hold for different nonlinear media or different cavity geometries—the equation also applies to a ring cavity, for example, where one simply considers the backwards wave to propagate freely and sets $K_{\rm abs}^b e^{i\theta_{\rm disp}^b} = 1$. A ring cavity is often assumed for theoretical work, since the solution to the nonlinear propagation problem is simpler in this case. Explicit state equations for a two-level medium appear in Sect. 14.2 (Eqs. 14.53–14.56).

Most generally, absorption and dispersion are present at the same time, but for an initial understanding of the physics it is helpful to consider them separately. Consider first a purely absorptive medium, $\theta_{\rm disp}^f \theta_{\rm disp}^b = 0$. We assume the absorption exhibits a saturable nonlinearity, so that as a function of the intracavity intensity,

$$I_f(0) \equiv [\mathcal{E}_f(0)]^2, \tag{14.14}$$

$K_{\rm abs}^f K_{\rm abs}^b$ behaves as shown in Fig. 14.2a. The implicit expression for $I_f(0)$ may then be found by solving (14.10) graphically for fixed $I_0 \equiv \mathcal{E}_0^2$. Thus, introducing the two functions [left- and right-hand sides of (14.10)]

$$g[I_f(0)] \equiv \frac{I_f(0)}{T_1 I_0}, \tag{14.15}$$

and

$$h[I_f(0)] \equiv \frac{1}{\{1 - \sqrt{R_1 R_2} K_{\rm abs}^f [I_f(0)] K_{\rm abs}^b [I_b(0)]\}^2}, \tag{14.16}$$

the solutions for $I_f(0)$ occur at curve crossings, where

$$g[I_f(0)] = h[I_f(0)]. \tag{14.17}$$

Figures 14.2b and 14.2c illustrate what is found at a series of I_0 values for two different choices of the absorption strength.

In Fig. 14.2c there is only one solution for $I_f(0)$ at each value of input intensity I_0; outputs are uniquely determined by the input, as in a linear interferometer. On the other hand, when the unsaturated absorption is sufficiently strong [Fig. 14.2b] there can be three solutions for $I_f(0)$ at certain input intensities I_0. Two of the three solutions are stable (Sect. 14.2.3). There is one stable solution of low intracavity intensity, for which the absorption is unsaturated, hence able, self-consistently, to maintain the low intensity; and there is a second stable solution of sufficiently high intensity to saturate the absorption and maintain the high intensity, self-consistently. This is the phenomenon of *absorptive optical bistability*, which was proposed in a patent application by H. Seidel in 1969 [14.2] and independently by Szoke et al. [14.3].

We turn now to a purely dispersive medium, setting $K_{\rm abs}^f K_{\rm abs}^b = 1$. Equation 14.15 carries over, but in place of (14.16) we now introduce the function

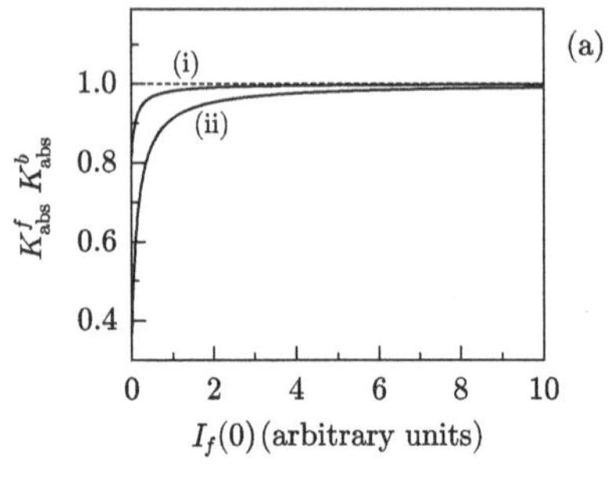

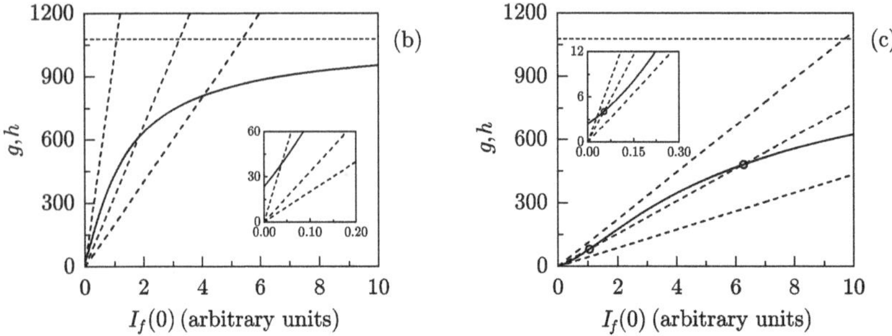

Fig. 14.2. The mechanism of absorptive optical bistability in a saturable medium: (**a**) the saturable function $K^f_{\mathrm{abs}}[I_f(0)]K^b_{\mathrm{abs}}[I_f(0)]$ is plotted for two values of weak-field absorption, smaller in case (i) than case (ii); (**b**) for low absorption [case (i)] the functions $g[I_f(0)]$ (*dashed lines*) and $h[I_f(0)]$ (*solid curve*) have only one intersection for all input intensities and no bistability exists; (**c**) for larger absorption [case (ii)] the functions $g[I_f(0)]$ and $h[I_f(0)]$ have three intersections (*open circles*) for intermediate input intensities and absorptive bistability exists. The decreasing slope of $g[I_f(0)]$ corresponds to increasing intensity I_0; $h[I_f(0)]$ is plotted from (14.16) for $R_1R_2 = 0.94$

[right-hand side of (14.10)]

$$h[I_f(0)] \equiv \frac{1}{\left(1-\sqrt{R_1R_2}\right)^2 + 4\sqrt{R_1R_2}\sin^2\frac{1}{2}\{\theta_0 + \theta^f_{\mathrm{disp}}[I_f(0)] + \theta^b_{\mathrm{disp}}[I_f(0)]\}}, \tag{14.18}$$

where the nonlinearity enters as a phase shift which can tune the cavity through resonance. Figure 14.3a shows results for a saturable phase shift, but in fact a simple Kerr effect is all that is required. Whether or not *dispersive optical bistability* occurs depends on θ_0 and the range of tuning provided by the nonlinear phase shift. Figure 14.3b illustrates the case where no bistability occurs since the nonlinear phase shift is too small. In contrast, Fig. 14.3c shows how a larger nonlinear phase shift results in the tuning of the interferometer through resonance and a range of input intensity yielding three solutions for

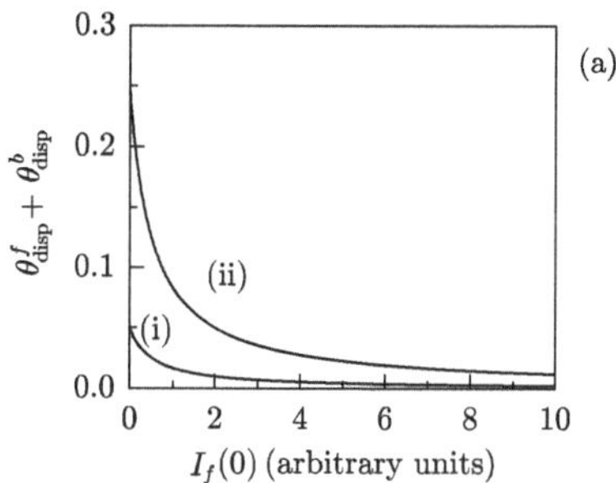

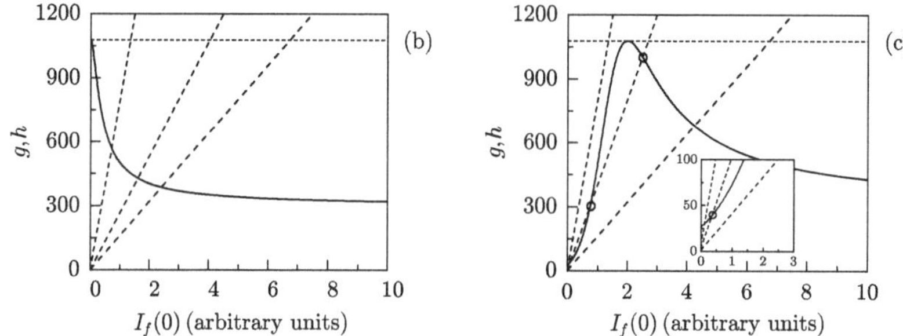

Fig. 14.3. The mechanism of dispersive optical bistability: (**a**) the nonlinear phase shift $\theta^f_{\rm disp}[I_f(0)] + \theta^b_{\rm disp}[I_f(0)]$ (for a saturable medium) is plotted for two values of weak-field dispersion, smaller in case (i) than case (ii); (**b**) for weak dispersion [case (i)] the functions $g[I_f(0)]$ (*dashed lines*) and $h[I_f(0)]$ (*solid curve*) have only one intersection for all input intensities and no bistability exists; (**c**) for stronger dispersion [case (ii)] the nonlinear phase shift tunes the cavity through resonance so that $g[I_f(0)]$ and $h[I_f(0)]$ have three intersections (*open circles*) for intermediate input intensities and dispersive bistability exists. The decreasing slope of $g[I_f(0)]$ corresponds to increasing intensity I_0; $h[I_f(0)]$ is plotted from (14.18) for $R_1R_2 =$ 0.94

$I_f(0)$. The first explanation of this dispersive mechanism appears in the work of McCall, Gibbs, and Venkatessan [14.4, 14.5].

14.2 The Mean-Field Limit for a Homogeneously Broadened Two-Level Medium

Our ultimate interest is in cavity QED, specifically in the quantum fluctuations of cavity QED systems. With this goal in mind, it is useful to make some specializations and simplifications. We specialize to a dilute medium of homogeneously broadened two-level atoms interacting with one mode of a high finesse standing-wave cavity. There are four important pieces to this prescription: a dilute medium, two-level atoms, one mode, and a high finesse cavity. Taken together they greatly simplify the analysis of the model. The pieces are interdependent and stand together as a package. One mode, for example, is generally insufficient. This is readily appreciated, since in an optically dense medium the steady-state amplitudes $\mathcal{E}_f(z)e^{i\phi_f(z)}$ and $\mathcal{E}_b(z)e^{i\phi_b(z)}$ change significantly over the medium length. Clearly, many cavity modes are needed to expand the spatial dependence. Thus, the simplification to one mode requires also that the medium be dilute. Then there is the time domain to consider. In a low finesse cavity the mode frequencies are closely spaced, increasing the possibility that the bandwidth of the fluctuations overlaps adjacent modes; the simplification to one mode also calls for a high cavity finesse. Finally, to close the circle, a high cavity finesse increases the number of round trips per cavity lifetime, providing for a good optical depth even when the medium is dilute.

In this section, we first derive the explicit form of the optical bistability state equation for a dilute medium of homogeneously-broadened two-level atoms. Then we develop a set of Maxwell–Bloch equations to describe the time dependence under single-mode conditions. The review by Lugiato [14.6] is a good place to start for a treatment that goes beyond the simplifications used here.

14.2.1 Steady State

The scenario that ensures that the forwards and backwards wave amplitudes remain essentially constant throughout the medium is referred to as the *mean-field limit.* We introduce the mean-field limit into state equation (14.10) by requiring

$$\left.\begin{aligned} 1-R_1=T_1\ll 1, &\qquad \theta_0\ll 1,\\ 1-R_2=T_2\ll 1, &\qquad \theta^f_{\rm disp}+\theta^b_{\rm disp}\ll 1,\\ 1-K^f_{\rm abs}K^b_{\rm abs}\ll 1. & \end{aligned}\right\} \tag{14.19}$$

Keeping first-order terms in each of the small quantities, the equation is rewritten as

$$\begin{aligned} \sqrt{T_1}\mathcal{E}_0e^{i(\phi_0+\phi_T)} &= \mathcal{E}_f(0)e^{i\phi_f(0)}\left[1-\sqrt{R_1R_2}K^f_{\rm abs}K^b_{\rm abs}e^{i(\theta_0+\theta^f_{\rm disp}+\theta^b_{\rm disp})}\right]\\ &= \mathcal{E}_f(0)e^{i\phi_f(0)}\Big\{1-(1-\tfrac{1}{2}T_1)(1-\tfrac{1}{2}T_2)\\ &\quad\times\left[1-(1-K^f_{\rm abs}K^b_{\rm abs})\right]\left[1+i(\theta_0+\theta^f_{\rm disp}+\theta^b_{\rm disp})\right]\\ &\quad+\text{higher orders}\Big\}\\ &= \mathcal{E}_f(0)e^{i\phi_f(0)}\left[\tfrac{1}{2}(T_1+T_2)\right.\\ &\quad+(1-K^f_{\rm abs}K^b_{\rm abs})-i(\theta_0+\theta^f_{\rm disp}+\theta^b_{\rm disp})\\ &\quad\left.+\text{higher orders}\right]. \end{aligned} \tag{14.20}$$

Now, dividing throughout by $\frac{1}{2}(T_1+T_2)$ and assuming that the resulting ratios of small quantities are all of similar order, we have

$$\frac{\sqrt{T_1}\mathcal{E}_0}{\frac{1}{2}(T_1+T_2)}e^{i(\phi_0+\phi_T)} = \mathcal{E}e^{i\phi_C}\left[1+\frac{1-K^f_{\rm abs}K^b_{\rm abs}}{\frac{1}{2}(T_1+T_2)} - i\frac{\theta_0+\theta^f_{\rm disp}+\theta^b_{\rm disp}}{\frac{1}{2}(T_1+T_2)}\right], \tag{14.21}$$

where we have set (see Eqs. 14.68b, 14.69b, and 14.91a)

$$\mathcal{E}_f(0)e^{i\phi_f(0)} = \mathcal{E}e^{i\phi_C}, \qquad \phi_C = \tfrac{1}{2}(\phi_f-\phi_b). \tag{14.22}$$

From (14.11) and (14.12), the output fields are

$$\mathcal{E}_1e^{i\phi_1} = \sqrt{R_1}\mathcal{E}_0e^{i[\phi_0+\phi_R-2(\omega_0/c)(L-\ell-d)]} + \sqrt{T_1}\mathcal{E}e^{i[\phi_R+\phi_T+\phi_C+2(\omega_0/c)(\ell+d)]}, \tag{14.23}$$

and

$$\mathcal{E}_2e^{i\phi_2} = \sqrt{T_2}\mathcal{E}e^{i(\phi_T+\phi_C)}, \tag{14.24}$$

where again we keep only dominant terms.

The remaining task is to derive an explicit expression for the absorption loss factor, $1-K^f_{\rm abs}K^b_{\rm abs}$, and the phase shift $\theta^f_{\rm disp}+\theta^b_{\rm disp}$. We must solve the nonlinear propagation problem for a dilute, homogeneously broadened two-level medium. To this end, the medium polarization is expanded in forwards and backwards waves in the manner of (14.4a) and (14.4b); we introduce expansions

$$\boldsymbol{P}_f(\boldsymbol{r},t) = \hat{e}_0\tfrac{1}{2}\mathcal{P}_f(\boldsymbol{r},t)e^{-(x^2+y^2)/w_0^2}e^{-i[\omega_0(t+z/c)-\psi_f(z,t)]} + \text{c.c.}, \tag{14.25a}$$

and

$$\boldsymbol{P}_b(\boldsymbol{r},t) = \hat{e}_0\tfrac{1}{2}\mathcal{P}_b(\boldsymbol{r},t)e^{-(x^2+y^2)/w_0^2}e^{-i[\omega_0(t-z/c)-\psi_b(z,t)]} + \text{c.c.}, \tag{14.25b}$$

$0\le z\le\ell$, where $\psi_f(z,t)$ and $\psi_b(z,t)$ are slowly-varying functions of space and time, and $\mathcal{P}_f(\boldsymbol{r},t)$ and $\mathcal{P}_b(\boldsymbol{r},t)$ are slowly varying in time but not in space—the latter in anticipation of a dependence on the standing-wave modulation of the light intensity; note also that in contrast to $\mathcal{E}_f(z,t)$ and $\mathcal{E}_b(z,t)$ in (14.4a) and (14.1b), $\mathcal{P}_f(\boldsymbol{r},t)$ and $\mathcal{P}_b(\boldsymbol{r},t)$ are functions of both longitudinal and transverse spatial coordinates.

We work now from Maxwell's equation for the field (9.14) (with $n=1$), where the polarization $\boldsymbol{P}_f(\boldsymbol{r},t)+\boldsymbol{P}_b(\boldsymbol{r},t)$ replaces the source term on the right-hand side. Since $\mathcal{E}_f(z,t)e^{i\phi_f(z,t)}$ and $\mathcal{E}_b(z,t)e^{i\phi_b(z,t)}$ are slowly varying in z and t, we may use the slowly-varying-amplitude approximation. The resulting equations are similar to (9.16a) and (9.16b), but with the source terms integrated against TEM_{00} traveling-wave mode functions. Setting the time derivatives to zero, the steady-state field amplitudes then satisfy the pair of coupled equations

$$\frac{d\mathcal{E}_f e^{i\phi_f}}{dz} = i\frac{\omega_0}{2\epsilon_0 c}\frac{2}{\pi w_0^2 \ell}\int d^3r e^{-2(x^2+y^2)/w_0^2}\left\{\mathcal{P}_f(\boldsymbol{r})e^{i\psi_f(\boldsymbol{r})} + \mathcal{P}_b(\boldsymbol{r})e^{-i[2(\omega_0/c)z-\psi_b(\boldsymbol{r})]}\right\}, \tag{14.26a}$$

$$\frac{d\mathcal{E}_b e^{i\phi_b}}{dz} = -i\frac{\omega_0}{2\epsilon_0 c}\frac{2}{\pi w_0^2 \ell}\int d^3r e^{-2(x^2+y^2)/w_0^2}\left\{\mathcal{P}_b(\boldsymbol{r})e^{i\psi_b(\boldsymbol{r})} + \mathcal{P}_f(\boldsymbol{r})e^{i[2(\omega_0/c)z+\psi_f(\boldsymbol{r})]}\right\}. \tag{14.26b}$$

Note that the integral over the second term in the curly bracket does not vanish because $\mathcal{P}_f(\boldsymbol{r})e^{i\psi_f(\boldsymbol{r})}$ and $\mathcal{P}_b(\boldsymbol{r})e^{i\psi_b(\boldsymbol{r})}$ are modulated by the standing-wave light intensity—the so-called *population grating* effect.

For a homogeneously broadened two-level medium, explicit expressions for the forwards and backwards polarization amplitudes follow from

$$\boldsymbol{P}_f(\boldsymbol{r},t) + \boldsymbol{P}_b(\boldsymbol{r},t) = \hat{e}_0 Dd\langle\sigma_{j-}(t)\rangle|_{\boldsymbol{r}_j=\boldsymbol{r}} + \text{c.c.}, \tag{14.27}$$

where D is the atomic density, and the interaction of atom j with the intra-cavity field is described by Hamiltonian

$$H_S \equiv \tfrac{1}{2}\hbar\omega_A\sigma_{jz} - d\tfrac{1}{2}e^{-(x_j^2+y_j^2)/w_0^2}\{[\mathcal{E}_f(z_j)e^{i\phi_f(z_j)} + \mathcal{E}_b(z_j)e^{i\phi_b(z_j)}] \times e^{-i\omega_0 t}\sigma_{j+} + \text{c.c.}\}, \tag{14.28}$$

where $\boldsymbol{r}_j \equiv (x_j, y_j, z_j)$ is the location of the atom. Radiative damping and dephasing are described by master equation (2.66) ($\bar{n} = 0$) and are independent of the atomic location. Thus, every atom obeys a similar set of optical Bloch equations (Sect. 2.3.3), the only variation being the dependence of (14.28) on $\boldsymbol{r}_j$. The polarization amplitudes are given by the steady-state solution to these equations with the replacement $\boldsymbol{r}_j \to \boldsymbol{r}$:

Exercise 14.1. Derive the optical Bloch equations for an atom located at $\boldsymbol{r}_j \to \boldsymbol{r}$ from Hamiltonian (14.28) and master equation (2.66); hence show that the forwards and backwards steady-state polarization amplitudes are given by

$$\mathcal{P}_{f,b}(\boldsymbol{r})e^{i\psi_{f,b}(\boldsymbol{r})} = iD\frac{2d^2}{\hbar\gamma_h}\mathcal{E}_{f,b}(z)e^{i\phi_{f,b}(z)}\frac{1+i\Delta}{1+\Delta^2+I(\boldsymbol{r})/I_{\text{sat}}}, \tag{14.29}$$

with detuning from atomic resonance

$$\Delta \equiv \frac{\omega_0 - \omega_A}{\gamma_h/2}, \tag{14.30}$$

saturation intensity

$$I_{\text{sat}} \equiv \frac{\hbar^2 \gamma \gamma_h}{8d^2}, \tag{14.31}$$

and

$$I(\boldsymbol{r}) \equiv e^{-2(x^2+y^2)/w_0^2} \tfrac{1}{4} |\mathcal{E}_f e^{i[(\omega_0/c)z+\phi_f]} + \mathcal{E}_b e^{-i[(\omega_0/c)z-\phi_b]}|^2. \tag{14.32}$$

The population grating comes from the interference term in (14.32). The homogeneous width (half-width at half-maximum) is

$$\gamma_h/2 = (\gamma + \gamma_p)/2, \tag{14.33}$$

which includes nonradiative dephasing at the rate γ_p.

We return now to the coupled equations (14.26a) and (14.26b) with the polarization amplitudes given by (14.29). Since we are considering a dilute medium, we may take $\mathcal{E}_f(z)e^{i\phi_f(z)}$ and $\mathcal{E}_b(z)^{i\phi_b(z)}$ to be independent of z in the prefactor on the right-hand sides of (14.29), but the spatial dependence of $I(\boldsymbol{r})$ must be retained. Then, after integrating over the length of the medium, we obtain

$$\begin{aligned} \mathcal{E}_f(\ell)e^{i\phi_f(\ell)} &= \mathcal{E}_f(0)e^{i\phi_f(0)} \Bigg[1 - \frac{D\omega_0 d^2}{\hbar\gamma_h \epsilon_0 c}(1+i\Delta) \\ &\quad \times \frac{2}{\pi w_0^2}\int_{-\infty}^{\infty} dx \int_{-\infty}^{\infty} dy e^{-2(x^2+y^2)/w_0^2} \int_0^\ell dz \Xi(\boldsymbol{r})\Bigg], \end{aligned} \tag{14.34a}$$

$$\begin{aligned} \mathcal{E}_b(0)e^{i\phi_b(0)} &= \mathcal{E}_b(\ell)e^{i\phi_b(\ell)} \Bigg[1 - \frac{D\omega_0 d^2}{\hbar\gamma_h \epsilon_0 c}(1+i\Delta) \\ &\quad \times \frac{2}{\pi w_0^2}\int_{-\infty}^{\infty} dx \int_{-\infty}^{\infty} dy e^{-2(x^2+y^2)/w_0^2} \int_0^\ell dz \Xi^*(\boldsymbol{r})\Bigg], \end{aligned} \tag{14.34b}$$

with

$$\Xi(\boldsymbol{r}) \equiv \frac{1 + e^{-2i[(\omega_0/c)z + \frac{1}{2}(\phi_f - \phi_b)]}}{1 + \Delta^2 + (|\mathcal{E}|^2/I_{\text{sat}})e^{-2(x^2+y^2)/w_0^2}\cos^2[(\omega_0/c)z + \frac{1}{2}(\phi_f - \phi_b)]}. \tag{14.35}$$

Note the different physical significance of the two terms in the numerator of (14.35). For the first term, which is constant, the integral over z takes an average of the population grating; this accounts for field propagation without coupling between the forwards and backwards waves. The second term oscillates with a period equal to half the optical wavelength, and originates from the polarization $\mathcal{P}_b(\boldsymbol{r})$ in (14.26a) and $\mathcal{P}_f(\boldsymbol{r})$ in (14.26b); it accounts for back-scattering from the population grating.

First, let us consider the integral with respect to z. It is convenient to make a change of variable, with

$$\theta \equiv (\omega_0/c)z + \tfrac{1}{2}(\phi_f - \phi_b), \qquad \theta' \equiv 2\theta, \tag{14.36}$$

and to neglect the difference between the actual length of the medium and the nearest integer multiple of half-wavelengths. Then we can write

$$\begin{aligned}
\int_0^\ell dz \Xi(\boldsymbol{r}) &= \frac{\ell}{\lambda_0/2}\int_{-(\omega_0/c)^{-1}\frac{1}{2}(\phi_f-\phi_b)}^{-(\omega_0/c)^{-1}\frac{1}{2}(\phi_f-\phi_b)+\lambda_0/2} dz \Xi(\boldsymbol{r}) \\
&= \ell\frac{1}{\pi}\int_0^\pi d\theta \frac{1+e^{-2i\theta}}{1+\Delta^2+(|\mathcal{E}|^2/I_{\mathrm{sat}})e^{-2(x^2+y^2)/w_0^2}\cos^2\theta} \\
&= \ell\frac{1}{2\pi}\int_0^{2\pi} d\theta' \frac{1+\cos\theta'}{1+\Delta^2+(|\mathcal{E}|^2/I_{\mathrm{sat}})e^{-2(x^2+y^2)/w_0^2}\frac{1}{2}(1+\cos\theta')},
\end{aligned} \tag{14.37}$$

where the imaginary part of the integral vanishes because the integrand is odd. For the transverse integration we make a second change of variable, introducing ($r^2 = x^2 + y^2$)

$$\bar{\eta} \equiv \frac{|\mathcal{E}|^2/I_{\mathrm{sat}}}{1+\Delta^2}e^{-2r^2/w_0^2}. \tag{14.38}$$

We now have

$$\begin{aligned}
&\frac{2}{\pi w_0^2}\int_{-\infty}^{\infty} dx \int_{-\infty}^{\infty} dy e^{-2(x^2+y^2)/w_0^2}\int_0^\ell dz \Xi(\boldsymbol{r}) \\
&\quad = \ell\frac{1}{|\mathcal{E}|^2/I_{\mathrm{sat}}}\int_0^{\frac{|\mathcal{E}|^2/I_{\mathrm{sat}}}{1+\Delta^2}} d\bar{\eta}\frac{1}{2\pi}\int_0^{2\pi} d\theta' \frac{1+\cos\theta'}{1+\bar{\eta}\frac{1}{2}(1+\cos\theta')} \\
&\quad = \ell\frac{2}{|\mathcal{E}|^2/I_{\mathrm{sat}}}\int_0^{\frac{|\mathcal{E}|^2/I_{\mathrm{sat}}}{1+\Delta^2}} d\bar{\eta}\frac{1}{\bar{\eta}}\frac{1}{2\pi}\int_0^{2\pi} d\theta' \left[1 - \frac{1}{1+\bar{\eta}\frac{1}{2}(1+\cos\theta')}\right] \\
&\quad = \ell\frac{2}{|\mathcal{E}|^2/I_{\mathrm{sat}}}\int_0^{\frac{|\mathcal{E}|^2/I_{\mathrm{sat}}}{1+\Delta^2}} d\bar{\eta}\frac{1}{\bar{\eta}}\left(1 - \frac{1}{\sqrt{1+\bar{\eta}}}\right).
\end{aligned} \tag{14.39}$$

The integral over θ' can be taken from Gradshteyn and Ryzhik [14.7]. Making a further change of variable, with

$$\eta' \equiv \sqrt{1+\bar{\eta}}, \tag{14.40}$$

the integral on the right-hand side of (14.34a) is finally evaluated as

$$
\begin{aligned}
\frac{2}{\pi w_0^2}\int_{-\infty}^{\infty} dx \int_{-\infty}^{\infty} dy e^{-2(x^2+y^2)/w_0^2}\int_0^{\ell} dz \Xi(\boldsymbol{r}) \\
&= \ell \frac{4}{|\mathcal{E}|^2/I_{\mathrm{sat}}}\int_1^{\sqrt{1+\frac{|\mathcal{E}|^2/I_{\mathrm{sat}}}{1+\Delta^2}}} d\eta' \frac{1}{1+\eta'} \\
&= \ell \frac{4}{|\mathcal{E}|^2/I_{\mathrm{sat}}}\ln\left[\frac{1}{2}\left(1+\sqrt{1+\frac{|\mathcal{E}|^2/I_{\mathrm{sat}}}{1+\Delta^2}}\right)\right]. \qquad (14.41)
\end{aligned}
$$

Thus, substituting (14.41) into (14.34a) and (14.34b), we arrive at the solution to the propagation problem relating the forwards and backwards wave amplitudes at either end of the nonlinear medium:

$$
\begin{aligned}
\mathcal{E}_f(\ell)e^{i\phi_f(\ell)} &= \mathcal{E}_f(0)e^{i\phi_f(0)}\left\{1-\frac{D\omega_0 d^2}{\hbar\gamma\epsilon_0 c}\ell \right. \\
&\left. \times(1+i\Delta)\frac{4}{|\mathcal{E}|^2/I_{\mathrm{sat}}}\ln\left[\frac{1}{2}\left(1+\sqrt{1+\frac{|\mathcal{E}|^2/I_{\mathrm{sat}}}{1+\Delta^2}}\right)\right]\right\}, \qquad (14.42a)
\end{aligned}
$$

$$
\begin{aligned}
\mathcal{E}_b(0)e^{i\phi_b(0)} &= \mathcal{E}_b(\ell)e^{i\phi_b(\ell)}\left\{1-\frac{D\omega_0 d^2}{\hbar\gamma\epsilon_0 c}\ell \right. \\
&\left. \times(1+i\Delta)\frac{4}{|\mathcal{E}|^2/I_{\mathrm{sat}}}\ln\left[\frac{1}{2}\left(1+\sqrt{1+\frac{|\mathcal{E}|^2/I_{\mathrm{sat}}}{1+\Delta^2}}\right)\right]\right\}. \qquad (14.42b)
\end{aligned}
$$

Comparing (14.7a) and (14.7b), absorption in the medium is accounted for by the factor

$$
1-K_{\mathrm{abs}}^f K_{\mathrm{abs}}^b = \alpha\ell \frac{4}{|\mathcal{E}|^2/I_{\mathrm{sat}}}\ln\left[\frac{1}{2}\left(1+\sqrt{1+\frac{|\mathcal{E}|^2/I_{\mathrm{sat}}}{1+\Delta^2}}\right)\right], \qquad (14.43a)
$$

and the dispersive phase shift by

$$
-(\theta_{\mathrm{disp}}^f+\theta_{\mathrm{disp}}^b) = \alpha\ell \frac{4\Delta}{|\mathcal{E}|^2/I_{\mathrm{sat}}}\ln\left[\frac{1}{2}\left(1+\sqrt{1+\frac{|\mathcal{E}|^2/I_{\mathrm{sat}}}{1+\Delta^2}}\right)\right]. \qquad (14.43b)
$$

The assumption of a dilute medium requires $\alpha\ell \ll 1$, with *on-resonance weak-field absorption coefficient*

$$
\alpha \equiv \frac{2D\omega_0 d^2}{\hbar\gamma_h\epsilon_0 c}. \qquad (14.44)
$$

Note 14.2. The expression for the absorption coefficient may also be derived from an understanding of the radiative losses caused by spontaneous emission (see Note 2.6). For simplicity, consider a wave propagating in the forwards direction with amplitude $\frac{1}{2}\mathcal{E}_f$. The change in intensity over a propagation distance δz through the medium is given in terms of α by

$$2\epsilon_0 c\delta[(\tfrac{1}{2}\mathcal{E}_f)^2] = 2\epsilon_0 c(\alpha\delta z)(\tfrac{1}{2}\mathcal{E}_f)^2. \tag{14.45}$$

The energy loss rate for a volume $A\delta z$ (cross-section A) is then

$$\delta P = 2\epsilon_0 c(\alpha\delta z)(\tfrac{1}{2}\mathcal{E}_f)^2 A. \tag{14.46}$$

This is the difference in power carried into and out of the elemental volume by the forward-propagating wave. The energy lost is radiated as spontaneous emission from $DA\delta z$ atoms. Since each atom has a probability $(\frac{1}{2}\mathcal{E}_f)^2/2I_{\text{sat}}$ of being in the excited state and each photon carries away an energy $\hbar\omega_A \approx \hbar\omega_0$, alternatively

$$\delta P = \hbar\omega_0(DA\delta z)\gamma\frac{(\frac{1}{2}\mathcal{E}_f)^2}{2I_{\text{sat}}}. \tag{14.47}$$

Equating (14.46) and (14.47) yields

$$\alpha = \frac{1}{2\epsilon_0 c}(\hbar\omega_0 D\gamma)\frac{4d^2}{\hbar^2\gamma\gamma_h} = \frac{2D\omega_0 d^2}{\hbar\gamma_h\epsilon_0 c}. \tag{14.48}$$

We can now write down the explicit relationship between the input and output fields for a homogeneously broadened two-level medium and a TEM_{00} standing-wave cavity mode. From (14.21), (14.43a), and (14.43b), we have

$$\begin{aligned}
&\frac{\sqrt{T_1}\mathcal{E}_0}{\frac{1}{2}(T_1+T_2)}e^{i(\phi_0+\phi_T)} \\
&= \mathcal{E}e^{i\phi_C}\left\{1 + \frac{\alpha\ell}{\frac{1}{2}(T_1+T_2)}\frac{4}{|\mathcal{E}|^2/I_{\text{sat}}}\ln\left[\frac{1}{2}\left(1+\sqrt{1+\frac{|\mathcal{E}|^2/I_{\text{sat}}}{1+\Delta^2}}\right)\right]\right. \\
&\quad\left. - i\frac{\theta_0}{\frac{1}{2}(T_1+T_2)} + i\frac{\alpha\ell}{\frac{1}{2}(T_1+T_2)}\frac{4\Delta}{|\mathcal{E}|^2/I_{\text{sat}}}\ln\left[\frac{1}{2}\left(1+\sqrt{1+\frac{|\mathcal{E}|^2/I_{\text{sat}}}{1+\Delta^2}}\right)\right]\right\}.
\end{aligned} \tag{14.49}$$

Then, to simplify the relation we introduce the *cooperativity parameter*

$$2C \equiv \frac{\alpha\ell}{\frac{1}{2}(T_1+T_2)}, \tag{14.50}$$

cavity detuning

$$\Phi \equiv \frac{\theta_0}{\frac{1}{2}(T_1+T_2)} = \frac{\omega_0-\omega_C}{\kappa}, \tag{14.51}$$

and the scaled field amplitudes

$$Y \equiv \frac{\sqrt{T_1}\mathcal{E}_0}{[\frac{1}{2}(T_1+T_2)]\sqrt{I_{\text{sat}}}}, \qquad X \equiv \frac{|\mathcal{E}|}{\sqrt{I_{\text{sat}}}}. \tag{14.52}$$

Taking the square modulus of (14.49) brings us finally to the *optical bistability state equation for a homogeneously broadened two-level medium in a TEM$_{00}$ mode (Gaussian-mode) standing-wave cavity*:

$$\begin{aligned} Y^2 = X^2 \Bigg\{ & \left(1 + 2C\frac{4}{X^2}\ln\left[\frac{1}{2}\left(1+\sqrt{1+\frac{X^2}{1+\Delta^2}}\right)\right]\right)^2 \\ & + \left(\Phi - 2C\Delta\frac{4}{X^2}\ln\left[\frac{1}{2}\left(1+\sqrt{1+\frac{X^2}{1+\Delta^2}}\right)\right]\right)^2 \Bigg\}. \end{aligned} \tag{14.53}$$

There are two output fields, one reflected and one transmitted, whose amplitudes, using (14.23), (14.24), and (14.52), are given by

$$\begin{aligned} \mathcal{E}_1 e^{i\phi_1} = \sqrt{T_1 I_{\text{sat}}} \Bigg\{ & \frac{\frac{1}{2}(T_1+T_2)}{T_1} Y e^{i[\phi_0+\phi_R-2(\omega_0/c)(L-\ell-d)]} \\ & + X e^{i\arg(\mathcal{E})} e^{i[\phi_R+\phi_T+\phi_C+2(\omega_0/c)(\ell+d)]} \Bigg\}, \end{aligned} \tag{14.54}$$

and

$$\mathcal{E}_2 e^{i\phi_2} = \sqrt{T_2 I_{\text{sat}}} X e^{i\arg(\mathcal{E})} e^{i(\phi_T+\phi_C)}. \tag{14.55}$$

The phase of the intracavity field is

$$\arg(\mathcal{E}) = \phi_0 + \phi_T - \phi_C + \phi_X, \tag{14.56}$$

with

$$\sin(\phi_X) = \frac{X}{Y}\left\{\Phi - 2C\Delta\frac{4}{X^2}\ln\left[\frac{1}{2}\left(1+\sqrt{1+\frac{X^2}{1+\Delta^2}}\right)\right]\right\}, \tag{14.57a}$$

$$\cos(\phi_X) = \frac{X}{Y}\left\{1 + 2C\frac{4}{X^2}\ln\left[\frac{1}{2}\left(1+\sqrt{1+\frac{X^2}{1+\Delta^2}}\right)\right]\right\}, \tag{14.57b}$$

which follows by equating phase factors on the left- and right-hand sides of (14.49).

Exercise 14.2. Other optical bistability state equations appear in the literature, for plane rather than TEM$_{00}$ waves, and ring rather then standing-wave cavities. Derive the following: (i) the *optical bistability state equation for a homogeneously broadened two-level medium in a plane and standing-wave* *cavity,*

$$Y^2 = X^2\left\{\left[1 + 2C\frac{1}{X^2}\left(1 - \sqrt{\frac{1+\Delta^2}{1+\Delta^2+X^2}}\right)\right]^2 + \left[\Phi - 2C\Delta\frac{1}{X^2}\left(1 - \sqrt{\frac{1+\Delta^2}{1+\Delta^2+X^2}}\right)\right]^2\right\}; \tag{14.58}$$

(ii) the *optical bistability state equation for a homogeneously broadened two-level medium in a TEM$_{00}$ mode (Gaussian-mode) ring cavity,*

$$Y^2 = X^2\left\{\left[1 + 2C\frac{1}{X^2}\ln\left(1 + \frac{X^2}{1+\Delta^2}\right)\right]^2 + \left[\Phi - 2C\Delta\frac{1}{X^2}\ln\left(1 + \frac{X^2}{1+\Delta^2}\right)\right]^2\right\}; \tag{14.59}$$

and (iii) the *optical bistability state equation for a homogeneously broadened two-level medium in a plane-wave ring cavity,*

$$Y^2 = X^2\left\{\left(1 + 2C\frac{1}{1+\Delta^2+X^2}\right)^2 + \left(\Phi - 2C\Delta\frac{1}{1+\Delta^2+X^2}\right)^2\right\}. \tag{14.60}$$

Note that the state equations for a ring cavity are for a medium of length 2ℓ and assume $\mathcal{E}_b \to 0$, $\frac{1}{2}\mathcal{E}_f \to |\mathcal{E}|$.

In the mean-field limit the intracavity field amplitude is almost constant throughout the medium; there is only a small change, of order $\alpha\ell \ll 1$, from one end of the medium to the other. Nevertheless, the small change can have a dramatic effect on the cavity transmission if it is accumulated over many round trips—i.e., when the cavity finesse is high, for $\frac{1}{2}(T_1 + T_2) \ll 1$. This balance between absorption loss and cavity finesse is expressed through the cooperativity parameter $2C$. To make a connection with Figs. 14.2 and 14.3, the state equation yields three solutions for X^2 at fixed Y^2 when $2C$ exceeds a threshold value. The threshold is different for the different state equations. To determine it, we first look for the turning points of the input–output curve, $Y^2(X^2)$, given as solutions to the equation

$$dY^2/d(X^2) = 0. \tag{14.61}$$

There are either no solutions or two solutions to this equation depending on whether $2C$ is above or below its threshold value; of course, physically acceptable solutions are real and positive. At threshold, (14.61) has a single degenerate solution—a solution to both (14.61) and

$$d^2Y^2/d(X^2)^2 = 0. \tag{14.62}$$

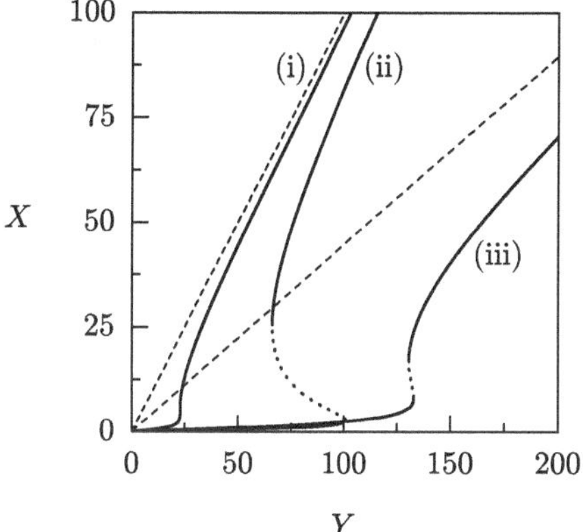

Fig. 14.4. Typical optical bistability input–output curves plotted from state equation (14.53): (i) $2C = 20$, $\varDelta = 0$, $\varPhi = 0$; (ii) $2C = 100$, $\varDelta = 0$, $\varPhi = 0$; (iii) $2C = 120$, $\varDelta = 2$, $\varPhi = -2$. *Dotted regions* are unstable (Sect. 14.2.3) and *dashed lines* mark the asymptotes $X = Y/\sqrt{1+\varPhi^2}$

To illustrate the possibilities, the input–output relation (14.53) is plotted for a selection of parameter values in Fig. 14.4.

Exercise 14.3. Show that the threshold condition for pure absorptive optical bistability ($\varDelta = \varPhi = 0$) is $2C_{\rm thr} = 8$ for state equation (14.60) and $2C_{\rm thr} = 20.1$ for state equation (14.53). The calculation must be done numerically in the latter case.

14.2.2 Maxwell–Bloch Equations

The mean-field limit allows us to make a one-mode expansion of the intracavity field in the steady state. We now turn our attention to the time dependence of the intracavity field. The topic introduces a new consideration, since, if the medium bandwidth is large enough, polarization fluctuations can excite additional cavity modes. Should these modes become unstable, amplification of the fluctuations will replace the stationary steady state by a steady oscillation—several excited modes beating against one another. This possibility is excluded in the *mean-field single-mode limit*, defined by the mean-field limit (14.19) plus

$$\frac{\gamma_h/2}{\varDelta\omega_{\rm fsr}} \ll 1, \tag{14.63}$$

where $\varDelta\omega_{\rm fsr} = \pi c/L$ is the *free spectral range*. The new inequality ensures that the fluctuation bandwidth is much less than the longitudinal mode spacing. Of course, more generally, transverse modes should be considered with the obvious modification of (14.63). Assuming, then, that both (14.19) and (14.63) hold, in this section we retain the single-mode expansion of the intracavity field and derive an equation of motion for the mode amplitude, coupled to optical Bloch equations to account for the time dependence of the two-level medium.

We begin with the full Maxwell equations for the slowly-varying forwards and backwards wave amplitudes. With time derivatives included and the source terms taken from (14.26a) and (14.26b), we write

$$\left(\frac{\partial}{\partial z}+\frac{1}{c}\frac{\partial}{\partial t}\right)\mathcal{E}_f e^{\phi_f}=i\frac{\omega_0}{2\epsilon_0 c\ell}\int_0^1 d\eta\int_0^\ell dz\left\{\mathcal{P}_f e^{i\psi_f}+\mathcal{P}_b e^{-i[2(\omega_0/c)z-\psi_b]}\right\}, \tag{14.64a}$$

$$\left(\frac{\partial}{\partial z}-\frac{1}{c}\frac{\partial}{\partial t}\right)\mathcal{E}_b e^{\phi_b}=-i\frac{\omega_0}{2\epsilon_0 c\ell}\int_0^1 d\eta\int_0^\ell dz\left\{\mathcal{P}_b e^{i\psi_b}+\mathcal{P}_f e^{i[2(\omega_0/c)z+\psi_f]}\right\}, \tag{14.64b}$$

where

$$\eta\equiv e^{-2r^2/w_0^2}\equiv e^{-2(x^2+y^2)/w_0^2}. \tag{14.65}$$

The polarization amplitudes are functions of z, t, and η, while the field amplitude are functions of z and t alone. The medium polarization (Eq. 14.27) and inversion,

$$\mathcal{M}(\boldsymbol{r},t)=D\langle\sigma_{jz}(t)\rangle|_{\boldsymbol{r}_j=\boldsymbol{r}}, \tag{14.66}$$

satisfy the optical Bloch equations

$$\begin{aligned}\left[\frac{\partial}{\partial t}+\frac{\gamma_h}{2}(1-i\Delta)\right]&\left\{\mathcal{P}_f e^{i[(\omega_0/c)z+\psi_f]}+\mathcal{P}_b e^{-i[(\omega_0/c)z-\psi_b]}\right\},\\ &=-i\frac{d^2}{\hbar}\mathcal{M}\left\{\mathcal{E}_f e^{i[(\omega_0/c)z+\phi_f]}+\mathcal{E}_b e^{-i[(\omega_0/c)z-\phi_b]}\right\},\end{aligned} \tag{14.67a}$$

$$\begin{aligned}\frac{\partial\mathcal{M}}{\partial t}+\gamma(\mathcal{M}+D)=i\frac{1}{2\hbar}\eta&\left\{\mathcal{P}_f e^{-i[(\omega_0/c)z+\psi_f]}+\mathcal{P}_b e^{i[(\omega_0/c)z-\psi_b]}\right\}\\ &\times\left\{\mathcal{E}_f e^{i[(\omega_0/c)z+\phi_f]}+\mathcal{E}_b e^{-i[(\omega_0/c)z-\phi_b]}\right\}+\text{c.c.}.\end{aligned} \tag{14.67b}$$

We aim to construct a course-grained time derivative by following the propagation of the field for one cavity round trip (see also Eqs. 9.8 and 9.51). Note, first, that in the mean-field limit, forwards and backwards wave amplitudes are nearly equal and uniform throughout the medium. Thus, on the right-hand sides of the Maxwell equations and in the optical Bloch equations, we may take

$$\mathcal{P}_f(\boldsymbol{r},t)=\mathcal{P}_b(\boldsymbol{r},t)\equiv|\mathcal{P}[\eta,\theta(z),t]|, \tag{14.68a}$$

$$\mathcal{E}_f(z,t)=\mathcal{E}_b(z,t)\equiv|\mathcal{E}(t)|, \tag{14.68b}$$

and introduce

$$\arg\{\mathcal{P}[\eta,\theta(z),t]\} \equiv \tfrac{1}{2}[\psi_f(\boldsymbol{r},t)+\psi_b(\boldsymbol{r},t)], \tag{14.69a}$$

$$\arg[\mathcal{E}(t)] \equiv \tfrac{1}{2}[\phi_f(z,t)+\phi_b(z,t)], \tag{14.69b}$$

where $\theta(z)$ is defined by (14.36). The phase differences $\psi_f - \psi_b$ and $\phi_f - \phi_b$ simply determine the location of the nodes and antinodes of the standing wave, with $\phi_f - \phi_b$ fixed by the boundary conditions (Eq. 14.78) and the same phase written onto the medium polarization; we therefore take

$$\psi_f - \psi_b = \phi_f - \phi_b, \tag{14.70}$$

where both phase differences are independent of z and t.

Clearly we cannot use (14.68a)–(14.69b) on the left-hand sides of the Maxwell equations or in boundary conditions (14.6a) and (14.6b), since our aim is to compute the change—though very small—in the field amplitude over a cavity round trip. To this end, we introduce the retarded times

$$t_- \equiv t - z/c, \qquad t_+ \equiv t + z/c, \tag{14.71}$$

and with

$$\mathcal{E}_f^{\mathrm{ret}}(z,t_-) \equiv \mathcal{E}_f(z,t), \tag{14.72a}$$

$$\phi_f^{\mathrm{ret}}(z,t_-) \equiv \phi_f(z,t), \tag{14.72b}$$

and

$$\mathcal{E}_b^{\mathrm{ret}}(z,t_+) \equiv \mathcal{E}_b(z,t), \tag{14.73a}$$

$$\phi_b^{\mathrm{ret}}(z,t_+) \equiv \phi_b(z,t), \tag{14.73b}$$

write the Maxwell equations, (14.64a) and (14.64b), as

$$\frac{\partial \mathcal{E}_f^{\mathrm{ret}} e^{i\phi_f^{\mathrm{ret}}}}{\partial z} = i\frac{\omega_0}{2\epsilon_0 c} e^{\frac{1}{2}i(\phi_f-\phi_b)} \int_0^1 d\eta \frac{1}{\pi} \int_0^\pi d\theta \mathcal{P}(\eta,\theta)(1+e^{-2i\theta}), \tag{14.74a}$$

$$\frac{\partial \mathcal{E}_b^{\mathrm{ret}} e^{i\phi_b^{\mathrm{ret}}}}{\partial z} = -i\frac{\omega_0}{2\epsilon_0 c} e^{-\frac{1}{2}i(\phi_f-\phi_b)} \int_0^1 d\eta \frac{1}{\pi} \int_0^\pi d\theta \mathcal{P}(\eta,\theta)(1+e^{2i\theta}), \tag{14.74b}$$

where the previous transformation from integration with respect to z to θ has been made (Eqs. 14.36 and 14.37). From (14.67)–(14.70), the *optical Bloch equations* are

$$\frac{\partial \mathcal{P}}{\partial t} = -\frac{\gamma_h}{2}(1-i\Delta)\mathcal{P} - i\frac{d^2}{\hbar}\mathcal{M}\mathcal{E}, \tag{14.75a}$$

$$\frac{\partial \mathcal{M}}{\partial t} = -\gamma(\mathcal{M}+D) + \left(i\frac{2}{\hbar}\eta\cos^2\theta\mathcal{P}^*\mathcal{E} + \mathrm{c.c.}\right), \tag{14.75b}$$

where the spatial dependence of the TEM_{00} mode function is contained in $\eta(x,y)$ (Eq. 14.65) and $\theta(z)$ (Eq. 14.36).

We now follow the cavity round trip starting with the backwards wave as it exits the medium at $z = 0$. The full round trip is comprised of four pieces: (i) free propagation with one reflection back to where the forwards wave enters the medium; (ii) propagation through the medium in the forwards direction; (iii) free propagation with a second reflection to where the backwards wave enters the medium; and (iv) propagation through the medium in the backwards direction. There are four associated propagation delays, as depicted in Fig. 14.5:

$$\begin{aligned} \tau_1 &= 2(L-\ell-d)/c, & \tau_2 &= \tau_1 + \ell/c, \\ \tau_3 &= \tau_2 + 2d/c, & \tau_4 &= \tau_3 + \ell/c. \end{aligned} \tag{14.76}$$

We now write boundary conditions (14.6a) and (14.6b) with the propagation delays included, where for the boundary condition at the cavity input, we have

$$\begin{aligned} \mathcal{E}_f(0,t) = \sqrt{T_1}\mathcal{E}_0 e^{i(\phi_0+\phi_T)} &+ \sqrt{R_1}\mathcal{E}_b(0,t-\tau_1) \\ &\times e^{i[\phi_b(0,t-\tau_1)+2(\omega_0/c)(L-\ell-d)+\phi_R]}, \end{aligned} \tag{14.77a}$$

and at the cavity output,

$$\begin{aligned} \mathcal{E}_b(\ell,t-\tau_2)e^{-i[(\omega_0/c)\ell-\phi_b(\ell,t-\tau_2)]} &= \sqrt{R_2}\mathcal{E}_f(\ell,t-\tau_3) \\ &\quad\times e^{i[(\omega_0/c)\ell+\phi_f(\ell,t-\tau_3)+2(\omega_0/c)d+\phi_R]}. \end{aligned} \tag{14.77b}$$

It follows from (14.77b) that the standing-wave phase is

$$\tfrac{1}{2}(\phi_f - \phi_b) = -[(\omega_0/c)(\ell+d) + \tfrac{1}{2}\phi_R], \tag{14.78}$$

where we neglect the small time difference in the arguments of ϕ_f and ϕ_b. We must also relate the field amplitudes at either end of the medium. This is done by integrating the retarded equations (14.74a) and (14.74b) through the medium. Thus, we also have

$$\mathcal{E}_f(\ell,t-\tau_3)e^{i\phi_f(\ell,t-\tau_3)} = \mathcal{E}_f(0,t-\tau_4)e^{i\phi_f(0,t-\tau_4)} + e^{i\frac{1}{2}(\phi_f-\phi_b)}\Pi, \tag{14.79a}$$

$$\mathcal{E}_b(0,t-\tau_1)e^{i\phi_b(0,t-\tau_1)} = \mathcal{E}_b(\ell,t-\tau_2)e^{i\phi_b(\ell,t-\tau_2)} + e^{-i\frac{1}{2}(\phi_f-\phi_b)}\Pi, \tag{14.79b}$$

where

$$\Pi \equiv i\frac{\omega_0\ell}{\epsilon_0 c}\int_0^1 d\eta \frac{1}{\pi}\int_0^\pi d\theta \cos^2\theta \mathcal{P}. \tag{14.80}$$

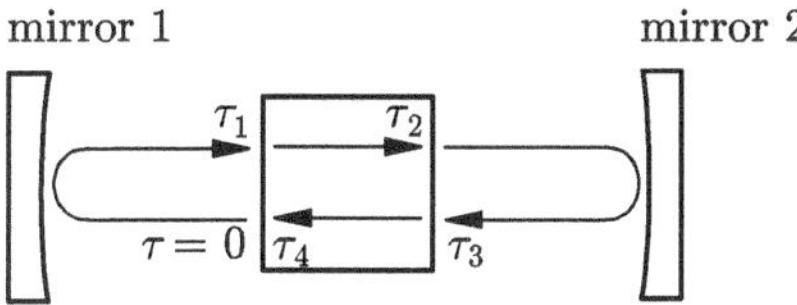

Fig. 14.5. Illustration of the four propagation intervals making up a cavity round trip

Equations 14.77a–14.79b provide everything needed to construct the change in the field amplitude over a cavity round trip. The first three steps are taken beginning with (14.77a) and then using (14.79b) and (14.77b). From these equations we obtain

$$\begin{aligned}
&\mathcal{E}_f(0,t)e^{i\phi_f(0,t)} \\
&= \sqrt{T_1}\mathcal{E}_0 e^{i(\phi_0+\phi_T)} + \sqrt{R_1}e^{i[2(\omega_0/c)(L-\ell-d)+\phi_R]}\big[\mathcal{E}_b(\ell,t-\tau_2)e^{i\phi_b(\ell,t-\tau_2)} \\
&\quad + e^{-i\frac{1}{2}(\phi_f-\phi_b)}\Pi\big] \\
&= \sqrt{T_1}\mathcal{E}_0 e^{i(\phi_0+\phi_T)} + \sqrt{R_1}e^{i[2(\omega_0/c)(L-\ell-d)+\phi_R]} \\
&\quad \times \Big\{\sqrt{R_2}\mathcal{E}_f(\ell,t-\tau_3)e^{i\phi_f(\ell,t-\tau_3)}e^{i[2(\omega_0/c)(\ell+d)+\phi_R]} + e^{-i\frac{1}{2}(\phi_f-\phi_b)}\Pi\Big\},
\end{aligned} \tag{14.81}$$

which, using (14.9) and (14.78), may be written as

$$\begin{aligned}
\mathcal{E}_f(0,t)e^{i\phi_f(0,t)} &= \sqrt{T_1}\mathcal{E}_0 e^{i(\phi_0+\phi_T)} + \sqrt{R_1}e^{i\theta_0}\big[\sqrt{R_2}\mathcal{E}_f(\ell,t-\tau_3)e^{i\phi_f(\ell,t-\tau_3)} \\
&\quad + e^{i\frac{1}{2}(\phi_f-\phi_b)}\Pi\big].
\end{aligned} \tag{14.82}$$

The fourth step is taken using (14.79a). It brings us to a relationship between the forwards field amplitude at the entrance to the medium at times t and $t-\tau_4$, where $\tau_4 = 2L/c$ is the cavity round-trip time:

$$\begin{aligned}
\mathcal{E}_f(0,t)e^{i\phi_f(0,t)} &= \sqrt{T_1}\mathcal{E}_0 e^{i(\phi_0+\phi_T)} + \sqrt{R_1}e^{i\theta_0}\big[\sqrt{R_2}\mathcal{E}_f(0,t-\tau_4)e^{i\phi_f(0,t-\tau_4)} \\
&\quad + e^{i\frac{1}{2}(\phi_f-\phi_b)}(\sqrt{R_2}+1)\Pi\big].
\end{aligned} \tag{14.83}$$

Setting $\phi_f = \frac{1}{2}(\phi_f+\phi_b)+\frac{1}{2}(\phi_f-\phi_b)$ on the left-hand side and using (14.68b) and (14.69b), the round-trip change in field amplitude is

$$\begin{aligned}
\mathcal{E}(t)-\mathcal{E}(t-2L/c) &= \sqrt{T_1}\mathcal{E}_0 e^{i[\phi_0+\phi_T-\frac{1}{2}(\phi_f-\phi_b)]} + \big(\sqrt{R_1R_2}e^{i\theta_0}-1\big)\mathcal{E}(t) \\
&\quad + \sqrt{R_1}\big(\sqrt{R_2}+1\big)e^{i\theta_0}\Pi\big].
\end{aligned} \tag{14.84}$$

Division by $2L/c$ gives the course-grained derivative in time. Thus, keeping only lowest-order terms in the expansion of $\sqrt{R_1} = \sqrt{1-T_1}$, $\sqrt{R_2} = \sqrt{1-T_2}$, and $e^{i\theta_0}$ (Eq. 14.19), we obtain the *single-mode Maxwell equation*

$$\frac{d\mathcal{E}}{dt} = -\kappa(1 - i\Phi)\mathcal{E} + i\frac{\omega_0 \ell}{\epsilon_0 L}\int_0^1 d\eta \frac{1}{\pi}\int_0^\pi d\theta \cos^2\theta \mathcal{P}$$
$$+ (c/2L)\sqrt{T_1}\mathcal{E}_0 e^{i[\phi_0 + \phi_T - \frac{1}{2}(\phi_f - \phi_b)]}, \tag{14.85}$$

where we have used (13.18) and (14.51). Equations 14.85, 14.75a, and 14.75b are the *Maxwell–Bloch equations for a homogeneously broadened two-level medium in a TEM$_{00}$ (Gaussian-mode) standing-wave cavity.*

14.2.3 Stability of the Steady State

The steady-state solution to the Bloch equations (14.75a) and (14.75b) recovers the medium response (14.29); the solution to the Maxwell equation (14.85) reproduces the state equation (14.53). Steady states are not necessarily stable, however. In this section we look briefly at the stability of the steady state.

There is a huge literature on instabilities in passive nonlinear interferometers [14.6, 14.8] and an equal, or possibly even larger body of work on instabilities in laser systems [14.8, 14.9, 14.10]. The instabilities come with some variety: there are *single-mode instabilities*, with the steady state yielding to sustained oscillation of the medium variables coupled to the amplitude of one cavity mode—behavior described by the single-mode Maxwell–Bloch equations; and there are *multimode instabilities*, where the beating of a few, or even many, cavity modes occurs. The entire zoology known from nonlinear dynamics can be found. In the case of multimode instabilities, the work of Ikeda [14.11, 14.12] was particularly influential, since it was through his work that the field of optical bistability first encountered period doubling and chaos [14.13,14.14,14.15]. Ikeda treated the evolution of the cavity field at the level of the individual round trips with the medium adiabatically eliminated. As a result, although the dynamical behavior is complex, Ikeda's model is very simple; one simply iterates a two-dimensional map to construct the amplitude of the intracavity field on successive round trips. For a two-level medium, the Ikeda map can be obtained from the extreme multimode limit of the coupled Maxwell–Bloch equations. In fact, the multimode stability analyses for these equations possess a symmetry that guarantees the existence of so-called Ikeda instabilities whenever optical bistability itself exists [14.16,14.15,14.18]: optical bistability occurs with the cavity tuned near resonance, while the corresponding multimode, or Ikeda, instability occurs when the driving field is tuned approximately midway between two cavity resonances.

A thorough treatment of instabilities would take us too far afield. We should look, though, at the simplest case—the instability that comes as a package with the existence of a bistable transmission. As we have seen, bistability is associated with three curve intersections in Figs. 14.2c and 14.3c, or with the corresponding S-shaped regions of the input–output characteristics of Fig. 14.4. Thus, in a bistable system there are three, not two, coexisting steady states, one of which is unstable. The instability of the third solution is what we now aim to prove.

The dashed central branch of Fig. 14.4 identifies the unstable solution, or equivalently the middle of the three intersections in Figs. 14.2c and 14.3c. A plausible argument for its instability can be given in a few words. Consider the middle intersection in Fig. 2c, for example, where the nonlinearity is provided by saturable absorption. Moving the intracavity intensity to a slightly higher value decreases the strength of the absorption, which produces a further increase in intensity; thus, positive feedback is present to move the intracavity intensity towards the stable high-intensity steady state. Similarly, a slight decrease in the intracavity intensity increases the absorption, providing the positive feedback to move the solution towards the low-intensity steady state. Perhaps the words are not entirely convincing, though, as they make no reference to the actual time-dependence of the cavity field or medium. Certainly, questions about stability should generally make such a reference. Let us address them, more carefully, then, by carrying through a linear stability analysis of the Maxwell–Bloch equations.

For the sake of simplicity, we confine our attention to the single-mode Maxwell–Bloch equations, (14.75a), (14.75b), and (14.85). Steady-state solutions satisfy the five equations

$$\begin{aligned}0 &= \kappa\sqrt{T_1 I_{\text{sat}}}\,Y e^{i[\phi_0+\phi_T-\frac{1}{2}(\phi_f-\phi_b)]}\\&\quad-\kappa(1-i\Phi)\mathcal{E}_{\text{ss}}+i\frac{\omega_0\ell}{2\epsilon_0 L}\int_0^1 d\eta\frac{1}{\pi}\int_0^\pi d\theta\cos^2\theta\mathcal{P}_{\text{ss}},\end{aligned}\tag{14.86a}$$

$$\begin{aligned}0 &= \kappa\sqrt{T_1 I_{\text{sat}}}\,Y e^{-i[\phi_0+\phi_T-\frac{1}{2}(\phi_f-\phi_b)]}\\&\quad-\kappa(1+i\Phi)\mathcal{E}^*_{\text{ss}}-i\frac{\omega_0\ell}{2\epsilon_0 L}\int_0^1 d\eta\frac{1}{\pi}\int_0^\pi d\theta\cos^2\theta\mathcal{P}^*_{\text{ss}},\end{aligned}\tag{14.86b}$$

$$0=-\frac{\gamma_h}{2}(1-i\Delta)\mathcal{P}_{\text{ss}}-i\frac{d^2}{\hbar}\mathcal{M}_{\text{ss}}\mathcal{E}_{\text{ss}},\tag{14.86c}$$

$$0=-\frac{\gamma_h}{2}(1+i\Delta)\mathcal{P}^*_{\text{ss}}+i\frac{d^2}{\hbar}\mathcal{M}_{\text{ss}}\mathcal{E}^*_{\text{ss}},\tag{14.86d}$$

$$0=-\gamma(\mathcal{M}_{\text{ss}}+D)+\left(i\frac{2}{\hbar}\eta^2\cos^2\theta\mathcal{P}^*_{\text{ss}}\mathcal{E}_{\text{ss}}+\text{c.c.}\right).\tag{14.86e}$$

Stability is assessed by considering small perturbations away from the steady state. Generically, these decay or grow exponentially, faster in some phase-space directions than in others. For an arbitrary perturbation, the full time-dependent solution is written as a sum of exponential components; these are the *normal modes*—those specially configured perturbations whose decay (growth) is governed by a single exponential. Thus, we look for solutions to the Maxwell–Bloch equations in the form

$$\mathcal{E}(t) = \mathcal{E}_{\rm ss} + e^{\lambda t}\Delta\mathcal{E}, \tag{14.87a}$$

$$\mathcal{E}^*(t) = \mathcal{E}^*_{\rm ss} + e^{\lambda t}\Delta\mathcal{E}^*, \tag{14.87b}$$

$$\mathcal{P}(\eta,\theta,t) = \mathcal{P}_{\rm ss}(\eta,\theta) + e^{\lambda t}\Delta\mathcal{P}(\eta,\theta), \tag{14.87c}$$

$$\mathcal{P}^*(\eta,\theta,t) = \mathcal{P}^*_{\rm ss}(\eta,\theta) + e^{\lambda t}\Delta\mathcal{P}^*(\eta,\theta), \tag{14.87d}$$

$$\mathcal{M}(\eta,\theta,t) = \mathcal{M}_{\rm ss}(\eta,\theta) + e^{\lambda t}\Delta\mathcal{M}(\eta,\theta). \tag{14.87e}$$

Substituting (14.87a)–(14.87e) into (14.86a)–(14.86e), and keeping up to first-order terms in the perturbations $\Delta\mathcal{E}$, $\Delta\mathcal{E}^*$, $\Delta\mathcal{P}$, $\Delta\mathcal{P}^*$, and $\Delta\mathcal{M}$, they and the decay (growth) rate λ must satisfy the system of linear equations

$$0 = -[\lambda + \kappa(1 - i\Phi)]\Delta\mathcal{E} + i\frac{\omega_0\ell}{2\epsilon_0 L}\int_0^1 d\eta\frac{1}{\pi}\int_0^\pi d\theta\cos^2\theta\Delta\mathcal{P}, \tag{14.88a}$$

$$0 = -[\lambda + \kappa(1 + i\Phi)]\Delta\mathcal{E}^* - i\frac{\omega_0\ell}{2\epsilon_0 L}\int_0^1 d\eta\frac{1}{\pi}\int_0^\pi d\theta\cos^2\theta\Delta\mathcal{P}^*, \tag{14.88b}$$

$$0 = -\left[\lambda + \frac{\gamma_h}{2}(1 - i\Delta)\right]\Delta\mathcal{P} - i\frac{d^2}{\hbar}(\mathcal{E}_{\rm ss}\Delta\mathcal{M} + \mathcal{M}_{\rm ss}\Delta\mathcal{E}), \tag{14.88c}$$

$$0 = -\left[\lambda + \frac{\gamma_h}{2}(1 + i\Delta)\right]\Delta\mathcal{P}^* + i\frac{d^2}{\hbar}(\mathcal{E}^*_{\rm ss}\Delta\mathcal{M} + \mathcal{M}_{\rm ss}\Delta\mathcal{E}^*), \tag{14.88d}$$

$$0 = (\lambda + \gamma)\Delta\mathcal{M} + \left[i\frac{2}{\hbar}\eta^2\cos^2\theta(\mathcal{E}_{\rm ss}\Delta\mathcal{P}^* + \mathcal{P}^*_{\rm ss}\Delta\mathcal{E}) + \text{c.c.}\right]. \tag{14.88e}$$

In general, for arbitrary λ, only the trivial solution holds. The normal modes are the nontrivial solutions. Five such solutions exist, one for each eigenvalue λ_j, $j = 1, 2, \ldots, 5$, of the system of equations (14.88a)–(14.88e).

There is no need to explore the solution to the eigenvalue problem in all its generality. We recall now that a particular steady-state of the Maxwell–Bloch equations is uniquely defined by the intracavity intensity X^2 through the state equation (14.53). This allows us to differentiate each of (14.86a)–(14.86e) with respect to X to obtain a second system of linear equations for the derivatives of $\mathcal{E}_{\rm ss}$, $\mathcal{E}^*_{\rm ss}$, $\mathcal{P}_{\rm ss}$, $\mathcal{P}^*_{\rm ss}$, and $\mathcal{M}_{\rm ss}$, and dY/dX, where Y^2 is the output intensity. Thus, we obtain

$$\begin{aligned} 0 = {} & \kappa\sqrt{T_1 I_{\rm sat}}\frac{dY}{dX}e^{i[\phi_0+\phi_T-\frac{1}{2}(\phi_f-\phi_b)]} \\ & - \kappa(1 - i\Phi)\frac{d\mathcal{E}_{\rm ss}}{dX} + i\frac{\omega_0\ell}{2\epsilon_0 L}\int_0^1 d\eta\frac{1}{\pi}\int_0^\pi d\theta\cos^2\theta\frac{d\mathcal{P}_{\rm ss}}{dX}, \end{aligned} \tag{14.89a}$$

$$\begin{aligned} 0 = {} & \kappa\sqrt{T_1 I_{\rm sat}}\frac{dY}{dX}e^{-i[\phi_0+\phi_T-\frac{1}{2}(\phi_f-\phi_b)]} \\ & - \kappa(1 + i\Phi)\frac{d\mathcal{E}^*_{\rm ss}}{dX} - i\frac{\omega_0\ell}{2\epsilon_0 L}\int_0^1 d\eta\frac{1}{\pi}\int_0^\pi d\theta\cos^2\theta\frac{d\mathcal{P}^*_{\rm ss}}{dX}, \end{aligned} \tag{14.89b}$$

$$0 = -\frac{\gamma_h}{2}(1 - i\Delta)\frac{d\mathcal{P}_{\text{ss}}}{dX} - i\frac{d^2}{\hbar}\left(\mathcal{E}_{\text{ss}}\frac{d\mathcal{M}_{\text{ss}}}{dX} + \mathcal{M}_{\text{ss}}\frac{d\mathcal{E}_{\text{ss}}}{dX}\right), \tag{14.89c}$$

$$0 = -\frac{\gamma_h}{2}(1 + i\Delta)\mathcal{P}^*_{\text{ss}} + i\frac{d^2}{\hbar}\left(\mathcal{E}^*_{\text{ss}}\frac{d\mathcal{M}_{\text{ss}}}{dX} + \mathcal{M}_{\text{ss}}\frac{d\mathcal{E}^*_{\text{ss}}}{dX}\right), \tag{14.89d}$$

$$0 = -\gamma\frac{d\mathcal{M}_{\text{ss}}}{dX} + \left[i\frac{2}{\hbar}\eta^2\cos^2\theta\left(\mathcal{E}_{\text{ss}}\frac{d\mathcal{P}^*_{\text{ss}}}{dX} + \mathcal{P}^*_{\text{ss}}\frac{d\mathcal{E}_{\text{ss}}}{dX}\right) + \text{c.c.}\right]. \tag{14.89e}$$

When $dY/dX = 0$, these equations are the same as (14.88a)–(14.88e) with eigenvalue λ set to zero. It follows that at the turning points of the S-shaped input–output curves of Fig. 14.4, there exists a nontrivial solution to (14.88a)–(14.88e) with eigenvalue $\lambda = 0$ and eigenvector

$$\begin{pmatrix} \Delta\mathcal{E} \\ \Delta\mathcal{E}^* \\ \Delta\mathcal{P} \\ \Delta\mathcal{P}^* \\ \Delta\mathcal{M} \end{pmatrix} = \begin{pmatrix} d\mathcal{E}_{\text{ss}}/dX \\ d\mathcal{E}^*_{\text{ss}}/dX \\ d\mathcal{P}_{\text{ss}}/dX \\ d\mathcal{P}^*_{\text{ss}}/dX \\ d\mathcal{M}_{\text{ss}}/dX \end{pmatrix}. \tag{14.90}$$

The vanishing eigenvalue indicates a change of stability. A perturbative calculation shows that λ changes from a negative value—a stable steady state—for dY/dX small and positive, to a positive value—an unstable steady state—for dY/dX small and negative. Thus, we show that negative slope regions of the input–output curves are unstable. It should be apparent that the proof is quite general; it holds no matter what the equations of motion for the medium might be.

14.3 Relationship Between Macroscopic and Microscopic Variables

It is helpful to see how the variables used in the macroscopic Maxwell–Bloch equations relate to the variables of a microscopic description like that offered by master equation (13.57). We develop the microscopic description in the following two chapters. This section provides a bridge to the Maxwell–Bloch equations of Sect. 14.2.2.

To begin, let us make the identification between phase factors previously introduced

$$\tfrac{1}{2}(\phi_f - \phi_b) \to \phi_C, \tag{14.91a}$$

where ϕ_f and ϕ_b are introduced in (14.4a) and (14.4b), and ϕ_C is introduced in (13.8). We also make the identifications

$$\mathcal{E} \to i\sqrt{\frac{\hbar\omega_C}{2\epsilon_0 V_Q}}\tilde{\alpha}e^{i\phi'_C}, \tag{14.91b}$$

$$\eta \cos\theta \mathcal{P} \to (Dd)\tilde{v}e^{i\phi'_C}, \tag{14.91c}$$

$$\mathcal{M} \to Dm, \tag{14.91d}$$

where $\tilde{\alpha}$, $\tilde{v}$, and m are new field, polarization, and inversion variables, and the phase ϕ'_C is also introduced in (13.8). Written in terms of the new variables, the single-mode Maxwell equation (Eq. 14.85) now reads

$$\frac{d\tilde{\alpha}}{dt} = -\kappa(1 - i\Phi)\tilde{\alpha} + \frac{2Dd\omega_0\ell}{\epsilon_0 L}\sqrt{\frac{2\epsilon_0 V_Q}{\hbar\omega_C}} \int_0^1 d\eta \frac{1}{\pi} \int_0^\pi d\theta \cos\theta \tilde{v} - i\bar{\mathcal{E}}_0, \tag{14.92}$$

where we use the scaled driving field amplitude (13.58). The new form of the optical Bloch equations, (14.75a) and (14.75b), is

$$\frac{d\tilde{v}}{dt} = -\frac{\gamma_h}{2}(1 - i\Delta)\tilde{v} + \sqrt{\frac{\omega_C d^2}{2\hbar\epsilon_0 V_Q}}\eta \cos\theta m\tilde{\alpha}, \tag{14.93a}$$

$$\frac{dm}{dt} = -\gamma(m+1) - 2\sqrt{\frac{\omega_C d}{2\hbar\epsilon_0 V_Q}}\eta \cos\theta(\tilde{v}^*\tilde{\alpha} + \text{c.c.}). \tag{14.93b}$$

Now if we write g_{max}, instead of g, for the dipole coupling constant of an optimally located atom (Eq. 13.7), then the spatially dependent coupling for the TEM_{00} standing-wave mode is

$$g(\boldsymbol{r}) \equiv g_{\text{max}} e^{-(x^2+y^2)/w_0^2} \cos[(\omega_0/c)z + \phi_C], \tag{14.94a}$$

or with the changes of variable (14.36) and (14.65),

$$g[\boldsymbol{r}(\eta,\theta)] = \sqrt{\frac{\omega_C d^2}{2\hbar\epsilon_0 V_Q}}\eta \cos\theta. \tag{14.94b}$$

Using (14.94a) and (14.94b), the optical Bloch equations may be written as

$$\frac{d\tilde{v}}{dt} = -\frac{\gamma_h}{2}(1 - i\Delta)\tilde{v} + g(\boldsymbol{r})m\tilde{\alpha}, \tag{14.95a}$$

$$\frac{dm}{dt} = -\gamma(m+1) - 2g(\boldsymbol{r})(\tilde{v}^*\tilde{\alpha} + \text{c.c.}), \tag{14.95b}$$

and, if we overlook the difference between ω_0 and ω_C, the Maxwell equation takes the form

$$\frac{d\tilde{\alpha}}{dt} = -\kappa(1 - i\Phi)\tilde{\alpha} + \bar{N}_{\text{eff}}\frac{1}{V_Q}\int_{-\infty}^{\infty} dx \int_{-\infty}^{\infty} dy \int_0^\ell dz g(\boldsymbol{r})\tilde{v}(\boldsymbol{r}) - i\bar{\mathcal{E}}_0, \tag{14.96}$$

where we have introduced an *effective number of atoms*, defined as the mean number of atoms in the *effective interaction volume* $\ell V_Q/L$:

$$\bar{N}_{\text{eff}} \equiv D\left(\frac{\ell}{L}V_Q\right). \tag{14.97}$$

Note that the spatial integration to the right of $\bar{N}_{\text{eff}}$ in (14.96) gives the mean—spatially averaged—polarization entering as a source into Maxwell's equation *per atom*.

Equations 14.96, 14.95a, and 14.95b are the *Maxwell–Bloch equations for a spatially continuous distribution of atoms with number density D*. Of course, the two-level medium is actually a collection of localized atoms, not a continuous fluid. Looking on the microscopic level, we should therefore use the density

$$D_{\text{loc}}(\boldsymbol{r}) = \sum_j \delta^{(3)}(\boldsymbol{r} - \boldsymbol{r}_j), \tag{14.98}$$

where $\boldsymbol{r}_j$ denotes the atomic positions. In passing from (14.92) to (14.96), the nonuniform density must be kept inside the spatial integral, which for the density (14.98) becomes a sum over atoms. Thus, we arrive at the *Maxwell–Bloch equations for a collection of homogeneously broadened two-level atoms and one coherently driven mode of the electromagnetic field*:

$$\frac{d\tilde{\alpha}}{dt} = -\kappa(1 - i\Phi)\tilde{\alpha} + \sum_j g(\boldsymbol{r}_j)\tilde{v}_j - i\bar{\mathcal{E}}_0, \tag{14.99a}$$

$$\frac{d\tilde{\alpha}^*}{dt} = -\kappa(1 + i\Phi)\tilde{\alpha}^* + \sum_j g(\boldsymbol{r}_j)\tilde{v}_j^* + i\bar{\mathcal{E}}_0^*, \tag{14.99b}$$

$$\frac{d\tilde{v}_j}{dt} = -\frac{\gamma_h}{2}(1 - i\Delta)\tilde{v}_j + g(\boldsymbol{r}_j)m_j\tilde{\alpha}, \tag{14.99c}$$

$$\frac{d\tilde{v}_j^*}{dt} = -\frac{\gamma_h}{2}(1 + i\Delta)\tilde{v}_j^* + g(\boldsymbol{r}_j)m_j\tilde{\alpha}^*, \tag{14.99d}$$

$$\frac{dm_j}{dt} = -\gamma(m_j + 1) - 2g(\boldsymbol{r}_j)(\tilde{v}_j^*\tilde{\alpha} + \text{c.c.}). \tag{14.99e}$$

One last step completes the connection with master equation (13.57). Introducing the further identifications

$$\tilde{\alpha} \to \langle\tilde{a}\rangle, \qquad \tilde{v}_j \to \langle\tilde{\sigma}_{-}a_{j-}\rangle, \qquad m_j \to \langle\sigma_{jz}\rangle, \tag{14.100}$$

with

$$\tilde{a} \equiv ae^{i\omega_0 t}, \qquad \tilde{a}^\dagger \equiv a^\dagger e^{-i\omega_0 t}, \tag{14.101a}$$

$$\tilde{\sigma}_{j-} \equiv \sigma_{j-}e^{i\omega_0 t}, \qquad \tilde{\sigma}_{j+} \equiv \sigma_{j+}e^{-i\omega_0 t}, \tag{14.101b}$$

we see that (14.99a)–(14.99e) are the factorized mean-value equations derived from master equation (13.57), generalized to the many-atom Hamiltonian

$$\begin{aligned} H_S \equiv{}& \tfrac{1}{2}\hbar\omega_A \sum_j \sigma_{jz} + \hbar\omega_C a^\dagger a + i\hbar \sum_j g(\boldsymbol{r}_j)(a^\dagger\sigma_{j-} - a\sigma_{j+}) \\ &+ \hbar(\bar{\mathcal{E}}_0 e^{-i\omega_0 t}a^\dagger + \bar{\mathcal{E}}_0^* e^{i\omega_0 t}a), \end{aligned} \tag{14.102}$$

and including the dephasing term $(\gamma_p/2)\sum_j(\sigma_{jz}\rho\sigma_{jz} - \rho)$ (Sect. 2.2.4).

Two important relationships determine the importance of quantum fluctuations—i.e., whether a small-noise analysis holds or strong coupling conditions apply. The first involves the so-called cooperativity parameter $2C$. This is an intensive parameter, which may take the same value in a large system, where the small-noise approximation holds, as it does in a small system (strong coupling) where it does not. To see this, we write $2C$ in terms of the spontaneous emission enhancement factor, a measure of the absorption (interaction strength) *per atom*. Thus, from (14.50), (14.44), (13.18), and (14.97), we write

$$\begin{aligned} 2C &\equiv \frac{\alpha \ell}{\frac{1}{2}(T_1 + T_2)} \\ &= 2\frac{D\omega_0 d^2 \ell}{\hbar \gamma_h \epsilon_0 c} \frac{1}{\frac{1}{2}(T_1 + T_2)} \\ &= \frac{\gamma}{\gamma_h} D \left(\frac{\ell}{L} V_Q \right) 2 \left(\frac{\omega_0 d^2}{2\hbar \epsilon_0 V_Q} \right) \frac{1}{\gamma \kappa} \\ &= \frac{\gamma}{\gamma_h} \bar{N}_{\text{eff}} 2C_1^{\max}, \end{aligned} \tag{14.103}$$

where, once again, the difference between ω_0 and ω_C is neglected. The factor $2C_1^{\max}$ is the spontaneous emission enhancement factor (Eq. 13.36) for an optimally coupled atom (Eq. 13.7). For radiatively damped atoms ($\gamma_h = \gamma$), the cooperativity parameter is just this enhancement factor multiplied by the effective number of interacting atoms. It follows that the same value of $2C$ may be realized with large $\bar{N}_{\text{eff}}$ and small $2C_1$—corresponding to a large system size, small noise, and weak coupling—or with small $\bar{N}_{\text{eff}}$ and large $2C_1$—corresponding to small system size, large noise, and strong coupling. As macroscopic equations, the Maxwell–Bloch equations make no distinction between the two cases; they depend on the intensive parameter $2C$ alone. The difference is of primary importance, however, when quantum fluctuations are considered.

Note 14.3. For a TEM_{00} standing-wave mode, $\bar{N}_{\text{eff}}$ is the average number of atoms in a cylinder with radius equal to the half-width, $w_0/2$, of the transverse intensity profile and length ℓ of the medium (Eqs. 14.97 and 13.7; also see Note 13.1). When the density is low and the limit of a single atom interacting with the cavity mode is approached, it is tempting to interpret this number as the actual number of interacting atoms. The number cannot be interpreted so strictly, however. In fact, it is not possible to assign any definite number of interacting atoms—even a mean number—because the interaction volume is unbounded in the direction perpendicular to the cavity axis. Of course, one should not take the number of interacting atoms to be infinite either, since certainly atoms at different locations couple to the cavity mode at different strengths (Eq. 14.94a), and atoms sufficiently far from the cavity axis *are*, for all practical purposes, uncoupled, even though no distinct boundary marks where the coupling turns off. A reasonable estimate of the number of atoms contributing significantly to the interaction is the best one can do. Carmichael and Sanders [14.19] argue that the number is close to an order of magnitude larger than $\bar{N}_{\text{eff}}$.

To complement the effective number of atoms, there is a similar measure for the number of photons. Here, the issue of nonlinearity enters explicitly, as noted before. The relevant measure is the number of photons required to turn on the nonlinearity—in the case of the two-level medium, to saturate the atoms (Sect. 13.1; also see the discussion of Eq. 7.76). As with $\bar{N}_{\text{eff}}$, the number itself is not relevant in the Maxwell–Bloch equations. For them only the intensive parameter, the saturation intensity, matters. The saturation photon number is the relevant quantity when fluctuations are considered, however. The issue may be approached in the same way as with the parameter $2C$. The same saturation intensity might be realized in a large system or a small system—where fluctuations are either a small or a large perturbation. The saturation photon number provides a measure of what is large or small, with the measure tied to the energy density required to turn on the nonlinearity. We convert the energy density $2\epsilon_0 I_{\text{sat}}$ to a *saturation photon number* through the definition

$$\begin{aligned} n_{\text{sat}} &\equiv \frac{2\epsilon_0 I_{\text{sat}} V_Q}{\hbar\omega_C} \\ &= \frac{\gamma\gamma_h}{8}\left(\frac{2\epsilon_0\hbar V_Q}{\omega_C d^2}\right) \\ &= \frac{\gamma\gamma_h}{8g_{\max}^2}, \end{aligned} \tag{14.104}$$

where we have used (14.31) and (13.7). Of course, the degree of saturation experienced by a particular atom depends on its location; only an optimally coupled atom experiences saturation for n_{sat} photons in the cavity.

14.4 Cavity QED with Many Atoms

Cavity QED phenomena are not restricted to single atoms. When strong coupling conditions are meet, there are physical consequences for a system of many atoms, just as there are for one atom. The many-atom case does meet with a new dimension, however, since there is the possibility that the coupling is strong in a collective, many-atom sense, but not for each atom considered alone. Thus, in place of the definitions of strong coupling in terms of the ratios $2g/\gamma$ and g/κ, we consider the ratios $2\sqrt{N}g/\gamma$ and $\sqrt{N}g/\kappa$, where N is the number of atoms. In the perturbative limit, for example, $2Ng^2/\kappa \sim \gamma$ may hold (even $2Ng^2/\kappa \gg \gamma$) while $2g^2/\kappa \ll \gamma$. This may be viewed as a version of strong coupling because $\sqrt{N}g$ does emerge as an effective coupling strength for certain collective radiative effects. In some instances these collective effects have a long history unrelated to cavity QED. Superradiance and superfluorescence are cases in point [14.20,14.21]—or perhaps, more appropriately, cavity-assisted superradiance and superfluorescence [14.22,14.23,14.24]. Of course, from (14.103) and (13.36), with $\gamma_h = \gamma$ and $\bar{N}_{\rm eff} \to N$, the inequality $2Ng^2/\kappa > \gamma$ may be read as $2C > 1$; strong coupling for many atoms is therefore a requirement of absorptive optical bistability (Exercise 14.3). It is achieved in the *perturbative limit* with $\sqrt{N}g/\kappa < 1$ and $\gamma/2\kappa \ll 1$; thus, we encounter collective, or superradiant, cavity-enhanced emission rates in optical bistability in the bad-cavity limit [14.25,14.26,14.27].

The history of *nonperturbative* strong coupling for many atoms is shorter, but this case is encountered in optical bistability as well. It is here that we met with the many-atom version of vacuum Rabi splitting [14.28]. Vacuum Rabi oscillation was first noticed in the context of optical bistability by Carmichael, in his treatment of the linearized theory of quantum fluctuations about stable states [14.29]. The novelty of this work was its avoidance of the usual adiabatic elimination, either of the field (bad-cavity limit) or of the atoms (good-cavity limit). There is a more direct path to the many-atom vacuum Rabi oscillation than the one followed by Carmichael, though. Its frequency-domain counterpart—the vacuum Rabi doublet—sits waiting to be uncovered in the optical bistability state equation. We begin this section by seeing how this is so.

14.4.1 Weak-Probe Transmission Spectra

We use the optical bistability state equation to calculate the transmission spectrum of an optical cavity containing a homogeneously broadened two-level medium in the weak-excitation limit. The atoms and cavity mode are on resonance, with (Eqs. 14.30 and 14.51)

$$\Phi \equiv \frac{\omega_0 - \omega_C}{\kappa} = \frac{\omega_0 - \omega_A}{\kappa} = \frac{\gamma_h}{2\kappa}\Delta. \tag{14.105}$$

The amplitude Y (Eq. 14.52) of the incident field is held constant while its frequency ω_0 is scanned. Specialized to the weak-excitation limit, the state equation (Eq. 14.53, 14.58, 14.59, or 14.60) is

$$Y^2 = X^2\left[\left(1 + \frac{2C}{1+\Delta^2}\right)^2 + \Delta^2\left(\frac{\gamma_h}{2\kappa} - \frac{2C}{1+\Delta^2}\right)^2\right]. \tag{14.106}$$

It is a very simple relationship, accounting for the interplay of linear absorption, $2C/(1+\Delta^2)$, linear dispersion, $2C\Delta/(1+\Delta^2)$, and detuning, $(\Delta\gamma_h/2\kappa)^2$. Despite its simplicity, the interplay produces a remarkable result if the parameter $2C$ is sufficiently large.

For $2C \to 0$ (14.106) yields the Lorentzian line $X^2 = Y^2/(1+\Delta^2)$ of the empty cavity. To display the dependence on $2C$ in the most transparent way, we first expand the right-hand side in powers of $1 + \Delta^2$:

$$
\begin{aligned}
Y^2 &= X^2 \frac{1}{(1+\Delta^2)^2\kappa^2}\Big\{\kappa^2(1+\Delta^2+2C)^2 + \big[(1+\Delta^2)-1\big] \\
&\quad \times \big[(\gamma_h/2)(1+\Delta^2) - 2C\kappa\big]^2\Big\} \\
&= X^2 \frac{1}{(1+\Delta^2)\kappa^2}\Big\{(\kappa+\gamma_h/2)\big[(\kappa-\gamma_h/2)(1+\Delta^2)+4C\kappa\big] \\
&\quad + \big[(\gamma_h/2)(1+\Delta^2) - 2C\kappa\big]^2\Big\} \\
&= X^2 \frac{1}{(1+\Delta^2)\kappa^2}\Big\{(\gamma/2)^2(1+\Delta^2)^2 + \big[\kappa^2-(\gamma_h/2)^2-2C\kappa\gamma_h\big](1+\Delta)^2 \\
&\quad + (\kappa+\gamma_h/2)4C\kappa + 4C^2\kappa^2\Big\}.
\end{aligned} \tag{14.107}
$$

The result is a ratio of polynomials in the detuning

$$
\Delta\omega = \omega_0 - \omega_C = \omega_0 - \omega_A = (\gamma_h/2)\Delta; \tag{14.108}
$$

from (14.107), we have

$$
Y^2 = X^2 \frac{\Delta\omega^4 + \big[\kappa^2 + (\gamma_h/2)^2 - 2C\kappa\gamma_h\big]\Delta\omega^2 + \kappa^2(\gamma_h/2)^2(1+2C)^2}{\kappa^2[(\gamma_h/2)^2+\Delta\omega^2]}. \tag{14.109}
$$

Then, writing the numerator as the product of quadratics

$$
|\Lambda_+ + i\Delta\omega|^2|\Lambda_- + i\Delta\omega|^2 = |\Lambda_+\Lambda_- - \Delta\omega^2 + i\Delta\omega(\Lambda_+ + \Lambda_-)|^2, \tag{14.110}
$$

with

$$
\Lambda_\pm \equiv -\tfrac{1}{2}(\kappa+\gamma_h/2) \pm \sqrt{\tfrac{1}{4}(\kappa-\gamma_h/2)^2 - C\kappa\gamma_h}, \tag{14.111}
$$

we obtain the *weak-probe transmission spectrum of a Lorentzian cavity resonance with linear intracavity absorption and dispersion*:

$$
X^2(\Delta\omega) = Y^2 \frac{\kappa^2[(\gamma_h/2)^2+\Delta\omega^2]}{|\Lambda_+ + i\Delta\omega|^2|\Lambda_- + i\Delta\omega|^2}. \tag{14.112}
$$

As we now see, the many-atom vacuum Rabi doublet emerges in a straightforward way from this expression.

The connection is immediately apparent from (14.111). Returning to the problem of spontaneous emission for an atom in a cavity, we recall the solutions, (13.154a) and (13.154b), to coupled equations (13.152a) and (13.152b), which depend upon the eigenvalues (with $\Delta_C = 0$)

$$
\tilde{\lambda}_\pm = -\tfrac{1}{2}(\kappa+\gamma/2) \pm \sqrt{\tfrac{1}{4}(\kappa-\gamma/2)^2 - g^2}. \tag{14.113}
$$

For comparison, using (14.103) and (13.36), we have

$$
\Lambda_\pm = -\tfrac{1}{2}(\kappa+\gamma_h/2) \pm \sqrt{\tfrac{1}{4}(\kappa-\gamma_h/2)^2 - \bar{N}_{\rm eff}g_{\rm max}^2}. \tag{14.114}
$$

Thus, the same eigenvalues govern intracavity spontaneous emission and the weak-probe transmission spectrum (14.112), but with a collective dipole coupling constant, $\sqrt{\bar{N}_{\rm eff}}g_{\rm max}$, replacing g in the latter case. When the collective coupling is strong, with $\sqrt{\bar{N}_{\rm eff}}g_{\rm max} \gg \max(\kappa, \gamma/2)$, the transmission spectrum takes the form of a vacuum Rabi doublet (Eqs. 13.179a and 13.179b); explicitly, for $C\kappa\gamma_h = \bar{N}_{\rm eff}g^2_{\rm max} > \frac{1}{4}(\kappa - \gamma_h/2)^2$, (14.112) reduces to

$$\frac{X^2(\Delta\omega)}{Y^2} = \frac{\kappa^2[(\gamma_h/2)^2 + \Delta\omega^2]}{\left[\frac{1}{4}(\kappa + \gamma_h/2)^2 + (\Delta\omega + G)^2\right]\left[\frac{1}{4}(\kappa + \gamma_h/2)^2 + (\Delta\omega - G)^2\right]}, \tag{14.115}$$

where

$$G \equiv \sqrt{\bar{N}_{\rm eff}g^2_{\rm max} - \tfrac{1}{4}(\kappa - \gamma_h/2)^2}. \tag{14.116}$$

Note how this spectrum vanishes on resonance for $\gamma_h = 0$, a counterpart to the quantum interference effect of Note 13.14.

Given the equivalence of (14.113) and (14.114), there are also many-atom versions of the cavity-enhanced emission rate (13.172a) (bad-cavity limit) and the spoiled-cavity decay rate (13.176b) (good-cavity limit), with associated frequency shifts when the atoms and cavity are detuned. The transmission spectrum is plotted for the three different regimes in Fig. 14.6.

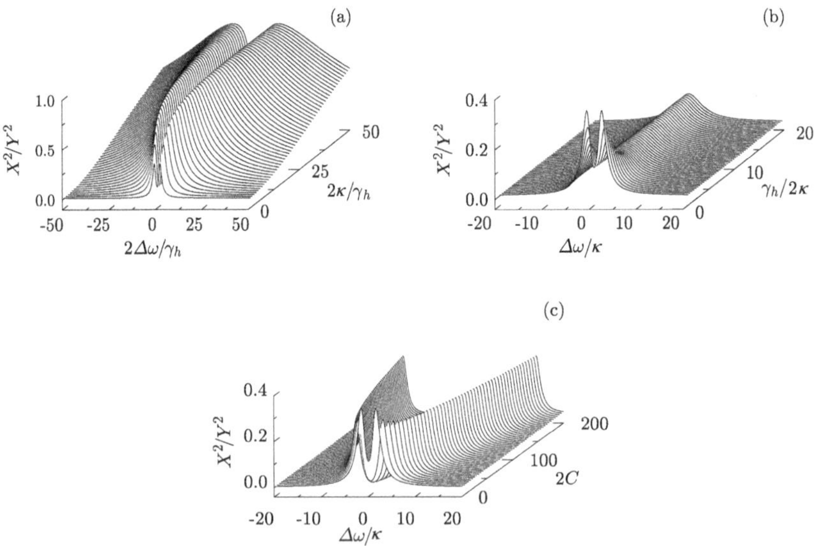

Fig. 14.6. Weak-probe transmission spectra (Eq. 14.112) showing: (**a**) development of the cavity-enhanced absorption dip, for $2C = 2$ and $1 \leq 2\kappa/\gamma_h \leq 50$ (bad-cavity limit); (**b**) development of the spoiled-cavity transmission resonance, for $2C = 2$ and $1 \leq \gamma_h/2\kappa \leq 20$ (good-cavity limit); (**c**) development of the many-atom vacuum Rabi doublet, for $\gamma_h/2\kappa = 1$ and $2 \leq 2C \leq 200$ (strong coupling)

While it is legitimate to regard the condition $\sqrt{\bar{N}_{\text{eff}}}g_{\max} \gg \max(\kappa, \gamma/2)$ as a special case of strong coupling, we should bear in mind that if $\bar{N}_{\text{eff}} \gg 1$, the coupling is not strong in the sense of a small system size, hence large quantum noise. The doublet transmission spectrum has been observed in experiments with one atom [14.30], a few atoms [14.31, 14.32, 14.33], many atoms [14.34, 14.35], and very many atoms [14.36]; in the latter case, at least, the linearized treatment of quantum fluctuations will hold.

In fact, the doublet may be understood from very general physics, without making reference to quantum energy levels at all. It is explained by the interplay of strong absorption, which suppresses transmission on the empty cavity resonance [the first term inside the bracket in (14.105)] and strong dispersion; the dispersion shifts the cavity resonance by a sufficiently large amount (beyond the anomalous dispersion regime) that resonance with the driving field is achieved by the shifted cavity resonance in the normal dispersion regime [the second term inside the bracket in (14.105)]. Given the right combination of absorption and dispersion, any linear medium can duplicate the effect. The simplicity and generality of the physics is perhaps even more apparent in a time domain picture. Here we meet, once again, with coupled harmonic oscillators, one oscillator for the field mode and the other for the polarization of the medium. The equations of motion are equivalent to (13.152a) and (13.152b), and the new resonances are the normal mode frequencies of the coupled harmonic oscillators, or, in the language of solid state physics, the polariton frequencies (Exercise 14.4).

Spectrum (14.112) is obtained in the weak-field limit of the more general input–output relationship provided by the optical bistability state equation. Figure 14.7 shows how the weak-probe, linear response connects with the nonlinear regime for the state equation (14.53). Gripp and coworkers [14.35] have made experimental observations of the nonlinear extension of the vacuum Rabi doublet.

Exercise 14.4. For weak excitation, the Maxwell–Bloch equations (14.85), (14.75a), and (14.75b) take the linear form

$$\frac{d}{dt}\begin{pmatrix}\mathcal{E}\\ \mathcal{P}\end{pmatrix} = \boldsymbol{M}\begin{pmatrix}\mathcal{E}\\ \mathcal{P}\end{pmatrix} + e^{i[\phi_0+\phi_T+\frac{1}{2}(\phi_f-\phi_b)]}(c/2L)\sqrt{T_1}\mathcal{E}_0\begin{pmatrix}1\\ 0\end{pmatrix}, \tag{14.117}$$

with

$$\boldsymbol{M} \equiv \begin{pmatrix}-\kappa(1-i\Phi) & i\omega_0\ell/2\epsilon_0 L\\ id^2D/\hbar & -(\gamma_h/2)(1-i\Delta)\end{pmatrix}, \tag{14.118}$$

where it has been assumed all atoms remain in the lower state ($\mathcal{M} = -D$). Diagonalize the matrix $\boldsymbol{M}$; hence find the *normal modes of the coupled field and polarization*, the so-called *cavity polaritons* [14.37,14.38]. Use the diagonal form of $\boldsymbol{M}$ to solve for the steady state field amplitude in the presence of $\mathcal{E}_0$ and thus verify the spectrum (14.112).

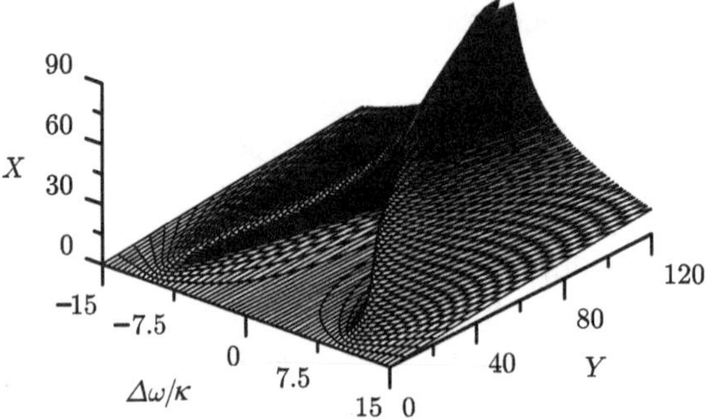

Fig. 14.7. Relationship between the vacuum Rabi transmission spectrum (14.115) and the optical bistability state equation (14.53). The output field amplitude X is plotted as a function of the input field amplitude and detuning, Y and $\Delta\omega/\kappa$, for $2C = 100$, $\gamma_h/2\kappa = 1$, and $\omega_A = \omega_C$. Lines of constant $\Delta\omega/\kappa$ correspond to the input–output curves plotted in Fig. 14.4 [where curve (ii) is the line $\Delta\omega/\kappa = 0$]; they are crossed by contours of constant X. The vacuum Rabi resonances appear for $Y \to 0$ at $\Delta\omega/\kappa = \pm\sqrt{2C} = \pm 10$

14.4.2 A Comment on Spatial Effects

Throughout this chapter we have taken the spatial distribution of the atoms into account. There is no particular difficulty in doing this when working at the level of the Maxwell–Bloch equations with the atoms treated as a continuous medium; we simply integrate the medium polarization against the cavity mode function, as in Sect. 14.2.1. In the nonlinear regime spatial effects are very important. They bring about the differences between the various optical bistability state equations, equation (14.53) versus (14.58)–(14.60). They are apparently unimportant in the linear regime, however, where every state equation reduces to (14.105). Of course, at low densities the assumption of a continuous medium does not hold, and the spatial distribution of atoms is important even in the linear regime. In the limit of just one atom, for example, it certainly matters where in the cavity the atom is placed, as the dipole coupling strength is position-dependent; thus, the parameter $2C$, defined for a continuous medium in terms of absorption length (Eq. 14.50), must be defined for one atom as in (13.36), by $2C = (\gamma/\gamma_h)2C_1 = 2g^2(\boldsymbol{r}_A)/\gamma_h\kappa$, where $\boldsymbol{r}_A$ is the position of the atom.

In the following chapter we begin our analysis of quantum fluctuations for many atoms. Spatial effects create a real difficulty in this context. Ultimately one is led towards a treatment in terms of coupled quantum fields—one field for the upper atomic state and one for the lower—under conditions where high excitation of the fields must be considered. The direction goes beyond the scope of the book; therefore, for the most part, we neglect spatial effects in what follows (Sect. 16.2 excepted). It is useful, nonetheless, to say a little about how the problem with spatial effects arises; in particular, to pinpoint the role of nonlinearity. We do this here, and at the same time fill in some detail about how the spatial distribution of atoms is taken into account in the linear regime.

Consider first a distribution of exactly N atoms located at positions $\boldsymbol{r}_j$, $j = 1, \ldots, N$. Hamiltonian (14.102) leads to a set of mean-value equations that are a straightforward generalization of (13.39a)–(13.39e). Factorizing the averages of operator products, we arrive at the Maxwell–Bloch equations (14.99a)–(14.99e). Under weak excitation, we may then make the replacement $\sigma_{jz} \to -1$, or in the Maxwell–Bloch equations, $m_j \to -1$, which yields a set of linear equations coupling the amplitude $\tilde{\alpha}$ of the cavity field to the polarization amplitudes $\tilde{v}_j$, $j = 1, \ldots, N$. There are $N + 1$ equations; how then do we make a connection to the pair of oscillator equations that produce the vacuum Rabi doublet?

The clue is to note that the field amplitude actually couples to just one variable, the collective polarization $\sum_{j=1}^{N} g(\boldsymbol{r}_j)\tilde{v}_j$. Thus, if we multiply the equation of motion for $\tilde{v}_j$ by the coupling constant g_j and sum over j, we obtain the pair of coupled equations

$$\frac{d\tilde{\alpha}}{dt} = -\kappa(1 - i\Phi)\tilde{\alpha} + \left[\sum_{j=1}^{N} g(\boldsymbol{r}_j)\tilde{v}_j\right] - i\bar{\mathcal{E}}_0, \tag{14.119a}$$

$$\frac{d\left[\sum_{j=1}^{N} g(\boldsymbol{r}_j)\tilde{v}_j\right]}{dt} = -\frac{\gamma_h}{2}(1 - i\Delta)\left[\sum_{j=1}^{N} g(\boldsymbol{r}_j)\tilde{v}_j\right] - \left[\sum_{j=1}^{N} g^2(\boldsymbol{r}_j)\right]\tilde{\alpha}. \tag{14.119b}$$

The effective number of interacting atoms, N_{eff}, is then defined through the relationship

$$\sum_{j=1}^{N} g^2(\boldsymbol{r}_j) = N_{\text{eff}} g_{\text{max}}^2, \tag{14.120}$$

and if we introduce the collective variable

$$\tilde{v} \equiv \frac{1}{\sqrt{N_{\text{eff}}}\, g_{\text{max}}} \sum_{j=1}^{N} g(\boldsymbol{r}_j)\tilde{v}_j, \tag{14.121}$$

the *equations of motion for coupled oscillator amplitudes in the linear regime of many-atom cavity QED* are

$$\frac{d\tilde{\alpha}}{dt} = -\kappa(1 - i\Phi)\tilde{\alpha} + \sqrt{N_{\text{eff}}}\, g_{\text{max}}\tilde{v} - i\bar{\mathcal{E}}_0, \tag{14.122a}$$

$$\frac{d\tilde{v}}{dt} = -\frac{\gamma_h}{2}(1 - i\Delta)\tilde{v} - \sqrt{N_{\text{eff}}}\, g_{\text{max}}\tilde{\alpha}, \tag{14.122b}$$

where the detunings $\kappa\Phi$ and $\gamma_h\Delta/2$ are defined, respectively, by (14.51) and (14.30).

The more likely scenario is that the number of interacting atoms is not definite and their spatial configuration is not fixed. This is the situation for cavity QED with atomic beams, for example [14.31, 14.32, 14.34, 14.35, 14.36]. In this case, assuming the fluctuations due to atomic motion are sufficiently slow, the results obtained for a definite number and configuration of atoms may be averaged. As an example, the weak-probe transmission spectrum may be obtained as an average of spectra calculated from (14.122a) and (14.122b), the approach taken by Thompson and coworkers [14.32]. In the high-density limit, the treatment of Sect. 14.4.1 assumes that averaging spectra gives the same result as calculating a single spectrum for an averaged effective dipole coupling strength. To recover this point of view from (14.119a) and (14.119b), we define $\bar{N}_{\mathrm{eff}}$ through the relationship

$$\overline{\sum_{j=1}^{N} g^2(\boldsymbol{r}_j)} = \bar{N}_{\mathrm{eff}} g_{\mathrm{max}}^2, \tag{14.123}$$

where the overbar denotes an average over the number and spatial configuration of the atoms. We also introduce the averaged collective variable

$$\bar{\tilde{v}} \equiv \frac{1}{\sqrt{\bar{N}_{\mathrm{eff}}} g_{\mathrm{max}}} \overline{\sum_{j=1}^{N} g(\boldsymbol{r}_j)\tilde{v}_j}. \tag{14.124}$$

Then averaging the equations of motion term by term, and adopting the factorization $\overline{\sum_{j=1}^{N} g_j^2(\boldsymbol{r}_j)\tilde{\alpha}} = \overline{\sum_{j=1}^{N} g_j^2(\boldsymbol{r}_j)}\bar{\tilde{\alpha}}$, we arrive at the *equations of motion for coupled oscillator amplitudes in the linear regime of many-atom cavity QED averaged over atomic configuration*:

$$\frac{d\bar{\tilde{\alpha}}}{dt} = -\kappa(1 - i\Phi)\bar{\tilde{\alpha}} + \sqrt{\bar{N}_{\mathrm{eff}}} g_{\mathrm{max}} \bar{\tilde{v}} - i\bar{\mathcal{E}}_0, \tag{14.125a}$$

$$\frac{d\bar{\tilde{v}}}{dt} = -\frac{\gamma_h}{2}(1 - i\Delta)\bar{\tilde{v}} - \sqrt{\bar{N}_{\mathrm{eff}}} g_{\mathrm{max}} \bar{\tilde{\alpha}}. \tag{14.125b}$$

These equations are related to the Maxwell–Bloch equations (14.117) through the scaling (14.91b) and (14.91c).

The difficulty introduced by nonlinearity becomes apparent if we attempt to apply the same averaging procedure to the full set of Maxwell–Bloch equations (14.99a)–(14.99e). Allowing that $\bar{\tilde{\alpha}}$ may be factorized from the average at high densities, there is still another difficulty to consider; namely, the sum over j can no longer be accounted for through the two equations (14.123) and (14.124). In place of (14.124), we find a dependence on the infinite hierarchy of collective variables

$$\overline{\sum_{j=1}^{N} g(\boldsymbol{r}_j)\tilde{v}_j}, \ \overline{\sum_{j=1}^{N} g^2(\boldsymbol{r}_j)\tilde{m}_j},$$
$$\overline{\sum_{j=1}^{N} g^3(\boldsymbol{r}_j)\tilde{v}_j}, \ \overline{\sum_{j=1}^{N} g^4(\boldsymbol{r}_j)\tilde{m}_j},$$
$$\vdots \qquad \vdots$$

As it happens, the resulting hierarchy of equations may be solved in the steady state, where the calculation is equivalent to carrying out the spatial average over the saturation term (14.35). Things become more complicated, however, with the quantum fluctuations included.

For an introduction to the quantum treatment, let us consider once more a distribution of exactly N atoms with fixed locations. For weak excitation, corresponding to the regime of the linear equations (14.122a) and (14.122b), we expand the state up to the level of one quantum of excitation:

$$|\psi\rangle = |0\rangle_a|0\rangle^{(N)} + \tilde{\alpha}e^{-i\omega_0 t}|0\rangle^{(N)}|1\rangle_a + \tilde{v}e^{-i\omega_0 t}|1\rangle^{(N)}|0\rangle_a, \tag{14.126}$$

where

$$|0\rangle^{(N)} \equiv \prod_{j=1}^{N} |1\rangle_j \tag{14.127}$$

is the ground state of the collection of N atoms, and

$$|1\rangle^{(N)} \equiv \frac{1}{\sqrt{N}} J_+|0\rangle^{(N)}, \tag{14.128}$$

where we introduce the ladder operators

$$J_\pm \equiv \left[\frac{N}{\sum_{j=1}^{N} g^2(\boldsymbol{r}_j)}\right]^{1/2} \sum_{j=1}^{N} g(\boldsymbol{r}_j)\sigma_{j\pm}. \tag{14.129}$$

It is perhaps unclear why the state may be taken as pure, but this is justified in Sects. 16.1.1 and 16.1.2. It is essentially because coherent excitation from the ground state to the one-quantum manifold dominates at weak excitation, since the time between incoherent scattering events (fluorescence or photon loss from the cavity) is far greater than the coherence time $[\frac{1}{2}(\kappa + \gamma_h/2)]^{-1}$. It then follows from the master equation that the one-quantum Schrödinger amplitudes, $\tilde{\alpha}$ and $\tilde{v}$ in (14.126), obey the coupled oscillator equations (14.122a) and (14.122b).

The ladder operators (14.129) look very much like the collective operators introduced in Sect. 6.2.1 (Eq. 6.44). Indeed, if the dipole coupling constants $g(\boldsymbol{r}_j)$ are all equal, they are the same collective operators. An important difference arises for unequal coupling constants, though. The collective operators defined by (6.44) obey a closed algebra, with commutation relations (6.45). For comparison, from (14.129), we have

$$[J_+, J_-] = J_z, \tag{14.130}$$

with

$$J_z \equiv \left[\frac{N}{\sum_{j=1}^N g^2(\boldsymbol{r}_j)}\right] \sum_{j=1}^N g^2(\boldsymbol{r}_j)\sigma_{jz}. \tag{14.131}$$

At first sight the parallel appears to continue. But now the algebra does not close. The commutator of $J_\pm$ and J_z is

$$[J_\pm, J_z] = \mp 2\left[\frac{N}{\sum_{j=1}^N g^2(\boldsymbol{r}_j)}\right]^{3/2} \sum_{j=1}^N g^3(\boldsymbol{r}_j)\sigma_{j\pm}. \tag{14.132}$$

When the $g(\boldsymbol{r}_j)$ are all equal the right-hand side is $\mp 2J_\pm$; more generally, (14.129), (14.131), and (14.132) begin the hierarchy we just met, which appears now as a consequence of the operator algebra.

Exercise 14.5. Consider the many-atom Hamiltonian

$$H_S \equiv \tfrac{1}{2}\hbar\omega_A \sum_{j=1}^N \sigma_{jz} + \hbar\omega_C a^\dagger a + i\hbar \sum_{j=1}^N g(\boldsymbol{r}_j)(a^\dagger\sigma_{j-} - a\sigma_{j+}). \tag{14.133}$$

At the one-quantum level a basis is provided by the states $|\{\}; -N/2\rangle|1\rangle_a$ and $|\{j\}; -N/2+1\rangle|0\rangle_a$, $j = 1, \ldots, N$ [the notation for atomic states is taken from (6.60)]. Diagonalize the Hamiltonian in the manifold of one-quantum states for $\omega_A = \omega_C$; hence show that there exists a pair of eigenstates,

$$|1, U\rangle = \frac{1}{\sqrt{2}}\left[|1\rangle^{(N)}|0\rangle_a + i|0\rangle^{(N)}|1\rangle_a\right], \tag{14.134a}$$

$$|1, L\rangle = \frac{1}{\sqrt{2}}\left[|1\rangle^{(N)}|0\rangle_a - i|0\rangle^{(N)}|1\rangle_a\right], \tag{14.134b}$$

with energies corresponding to the vacuum Rabi resonances,

$$E_{1,U} = \tfrac{1}{2}\hbar\omega_A + \hbar\sqrt{N_{\text{eff}}}g_{\max}, \tag{14.135a}$$

$$E_{1,L} = \tfrac{1}{2}\hbar\omega_A - \hbar\sqrt{N_{\text{eff}}}g_{\max}, \tag{14.135b}$$

where N_{eff} is given by (14.120), and $N-2$ degenerate eigenstates with energy $E_1 = \frac{1}{2}\hbar\omega_A$.

15

Many Atoms in a Cavity II: Quantum Fluctuations in the Small-Noise Limit

We proceed now from the macroscopic theory of the previous chapter to the analysis of quantum fluctuations for many atoms in a cavity. The program is similar to that of Chap. 8, where the phase-space approach to the quantum theory of the laser was developed; we work from a master equation to a Fokker–Planck equation with the help of the system size expansion. The treatment therefore is restricted to the small-noise limit. Recall, though, that the collective dipole coupling may be strong—$\sqrt{\bar{N}_{\rm eff}}g_{\rm max} \gg \frac{1}{2}(\kappa + \gamma_h/2)$— while the dipole coupling of the individual atoms is weak—$g_{\rm max} \ll \gamma_h/2, \kappa$; thus, collective vacuum Rabi oscillations may appear, even for a system size parameter $n_{\rm sat} = \gamma\gamma_h/8g_{\rm max}^2 \gg 1$.

The analysis of the present chapter goes beyond that for the laser in so far as the fluctuations are nonclassical. Unlike laser light, with its fluctuations dominated by spontaneous emission, the forward-scattered light in many-atom cavity QED may be antibunched and squeezed. We return to the theme of Chaps. 10–12, to nonclassical noise in the phase-space approach—i.e., to the vagaries of non-positive-semidefinite diffusion.

A great deal has been written on the topic of quantum fluctuations for many atoms in cavity in the small-noise limit, much of it during the 1970s and 1980s when optical bistability was widely studied. The article by Lugiato [15.1] is an excellent place to start for a review of the entire field. We will be satisfied with a few selected examples from the wide range of things accessible to the phase-space analysis. We concentrate on the limit of weak excitation, where nonclassical features are strongest and appear in their most interesting form: atom–atom correlations are discussed in Sect. 15.2.4, since they are not addressed explicitly anywhere else in the book; squeezing and photon antibunching are discussed in Sects. 15.2.5– 15.2.7. The small noise analysis provides an introduction and background for Chap. 16, where the treatment of quantum fluctuations in many-atom cavity QED is taken beyond the small-noise limit.

15.1 Microscopic Model

15.1.1 Master Equation for Optical Bistability

The microscopic model underlying our macroscopic treatment of an optical cavity filled with a homogeneously broadened two-level medium (Chap. 14) is defined by a straightforward generalization of the master equation for single-atom cavity QED (Eq. 13.14). Adopting the simplification that every atom couples to the cavity mode with the same coupling strength (see Sect. 14.4.2), and neglecting thermal fluctuations, we generalize (13.14) to N atoms and add an atomic dephasing term (Sect. 2.2.4) to obtain what is commonly known as the *master equation for optical bistability (with $\bar{n} = 0$)*:

$$\begin{aligned}\dot{\rho} &= -i\tfrac{1}{2}\omega_A[J_z, \rho] - i\omega_C[a^\dagger a, \rho] \\ &\quad + g[a^\dagger J_- - aJ_+, \rho] - i[\bar{\mathcal{E}}_0 e^{-i\omega_0 t}a^\dagger + \bar{\mathcal{E}}_0^* e^{i\omega_0 t}a, \rho] \\ &\quad + \frac{\gamma}{2}\left(\sum_{j=1}^N 2\sigma_{j-}\rho\sigma_{j+} - \tfrac{1}{2}J_z\rho - \tfrac{1}{2}\rho J_z - N\rho\right) + \frac{\gamma_p}{2}\left(\sum_{j=1}^N \sigma_{jz}\rho\sigma_{jz} - N\rho\right) \\ &\quad + \kappa(2a\rho a^\dagger - a^\dagger a\rho - \rho a^\dagger a),\end{aligned} \tag{15.1}$$

where J_-, J_+, and J_z are collective atomic operators,

$$J_\pm \equiv \sum_{j=1}^N \sigma_{j\pm}, \qquad J_z \equiv \sum_{j=1}^N \sigma_{jz}, \tag{15.2}$$

which satisfy angular momentum commutation relations

$$[J_+, J_-] = J_z, \qquad [J_\pm, J_z] = \mp 2J_\pm. \tag{15.3}$$

The cavity output fields are defined in Chap. 13 (Eqs. 13.9 and 13.20), and we note here, for completeness, that there is also an output field generated as fluorescence from the atoms. For this, we generalize the expression from Sect. 2.3.1 to a collection of atoms located within the cavity at positions $\boldsymbol{r}_j$, $j = 1, \ldots, N$; thus, we write

$$\hat{\boldsymbol{E}}^{(+)}(\boldsymbol{r}, t) = \hat{\boldsymbol{E}}_f^{(+)}(\boldsymbol{r}, t) - \frac{\omega_A^2}{4\pi\epsilon_0 c^2 r}(\boldsymbol{d}_{12} \times \hat{r}) \times \hat{r}\sum_{j=1}^N \sigma_{j-}(t - |\boldsymbol{r} - \boldsymbol{r}_j|/c), \tag{15.4}$$

where the free field is (Eq. 2.77)

$$\hat{\boldsymbol{E}}^{(+)}(\boldsymbol{r}, t) = i\sum_{\boldsymbol{k},\lambda}\sqrt{\frac{\hbar\omega_k}{2\epsilon_0 V}}\hat{e}_{\boldsymbol{k},\lambda} r_{\boldsymbol{k},\lambda}(0)e^{-i(\omega_k t - \boldsymbol{k}\cdot\boldsymbol{r})}. \tag{15.5}$$

To be in accord with our habit, the field must be reexpressed in photon flux units. In this instance, it is necessary to consider the direction of the emission and state that the considered flux is into the solid angle $d\Omega$ in the direction

$\hat{r} \equiv (\theta, \phi)$; then, for the directed fluorescence, we have the output field operator in photon flux units

$$\hat{\mathcal{E}}_{\hat{r}}(t) = \sqrt{\frac{cr^2 d\Omega}{V}} r_{Af}(t') + \sqrt{\frac{3d\Omega}{8\pi}} \sin\theta \sqrt{\gamma} \sum_{j=1}^{N} e^{-i\omega_A \hat{r}\cdot\boldsymbol{r}_j/c} \sigma_{j-}(t'), \tag{15.6}$$

where $t' = t - r/c$, and the free field is

$$r_{Af}(t') \equiv -i \sum_k \sqrt{\frac{\omega_k}{\omega_A}} r_{k\hat{r},\lambda_2}(0) e^{-i\omega_k t'}. \tag{15.7}$$

In order to stay with a scalar notion, we include only those free-field modes that have a nonvanishing projection onto the atomic dipole moment (polarization $\hat{e}_{\boldsymbol{k},\lambda_2}$ in Fig. 2.2); in fact, there is an additional free-field contribution polarized perpendicular to both $\hat{r}$ and the dipole moment (polarization $\hat{e}_{\boldsymbol{k},\lambda_1}$ in Fig. 2.2).

15.1.2 Fokker–Planck Equation in the *P* Representation

Let us begin by deriving the phase-space equations of motion corresponding to master equation (15.1) in the three representations we have met—the Glauber–Sudarshan P representation, the Q representation, and the Wigner representation. From these equations and the system size expansion, Fokker–Planck equations may be derived. It turns out that the diffusion matrix is not generally positive semi-definite in this case; it therefore helps to see the Fokker–Planck equations in all three representations so that comparisons can be made.

We consider the P representation first, where the appropriate characteristic function,

$$\chi_N(z, z^*, \xi, \xi^*, \eta) \equiv \mathrm{tr}\big(\rho e^{iz^* a^\dagger} e^{iza} e^{i\xi^* J_+} e^{i\eta J_z} e^{i\xi J_-}\big), \tag{15.8}$$

generates the hierarchy of normal-ordered operator averages (Eq. 7.96). The associated quasi-distribution function is the Fourier transform

$$\begin{aligned} &P(\alpha, \alpha^*, v, v^*, m) \qquad\qquad (15.9)\\ &\equiv \frac{1}{2\pi^5} \int d^2 z \int d^2\xi \int d\eta\, \chi_N(z, z^*, \xi, \xi^*, \eta) e^{-iz^*\alpha^*} e^{-iz\alpha} e^{-i\xi^* v^*} e^{-i\xi v} e^{-i\eta m}. \end{aligned}$$

Now, the phase-space equation of motion can be pieced together from things we already know. It is changed only slightly from the laser phase-space equation of motion (Sect. 7.2.4); hence, we go immediately from master equation (15.1) to the *phase-space equation of motion for optical bistability in the Glauber–Sudarshan P representation (with $\bar{n} = 0$):*

$$\begin{aligned} \frac{\partial P}{\partial t} = \bigg[&L_{+1}^{A}\left(v, v^*, m, \frac{\partial}{\partial v}, \frac{\partial}{\partial v^*}, \frac{\partial}{\partial m}\right) + L_{+1}^{F}\left(\alpha, \alpha^*, \frac{\partial}{\partial \alpha}, \frac{\partial}{\partial \alpha^*}\right) \\ &+ L_{+1}^{AF}\left(\alpha, \alpha^*, v, v^*, m, \frac{\partial}{\partial \alpha}, \frac{\partial}{\partial \alpha^*}, \frac{\partial}{\partial v}, \frac{\partial}{\partial v^*}, \frac{\partial}{\partial m}\right)\bigg] P, \end{aligned} \tag{15.10}$$

where the first term on the right-hand side is taken from (7.101a), with the replacements $\gamma_\downarrow \to \gamma$, $\gamma_\uparrow \to 0$ and the addition of the dephasing term (Eq. 6.158)

$$
\begin{aligned}
&L^A_{+1}\left(v, v^*, m, \frac{\partial}{\partial v}, \frac{\partial}{\partial v^*}, \frac{\partial}{\partial m}\right) \\
&\equiv i\omega_C\left(\frac{\partial}{\partial v}v - \frac{\partial}{\partial v^*}v^*\right) \\
&\quad + \frac{\gamma}{2}\left[\left(e^{2\frac{\partial}{\partial m}} - 1\right)(N+m) + \frac{\partial}{\partial v}v + \frac{\partial}{\partial v^*}v^*\right] \\
&\quad + \gamma_p\left[\frac{\partial}{\partial v}v + \frac{\partial}{\partial v^*}v^* + \frac{\partial^2}{\partial v \partial v^*}e^{-2\frac{\partial}{\partial m}}(N+m)\right]; \qquad (15.11a)
\end{aligned}
$$

the second term is taken from (7.101b) ($\bar{n} = 0$), with the addition of a term to account for the coherent driving of the cavity mode (Eq. 10.7),

$$
\begin{aligned}
L^F_{+1}\left(\alpha, \alpha^*, \frac{\partial}{\partial \alpha}, \frac{\partial}{\partial \alpha^*}\right) &\equiv (\kappa + i\omega_C)\frac{\partial}{\partial \alpha}\alpha + (\kappa - i\omega_C)\frac{\partial}{\partial \alpha^*}\alpha^* \\
&\quad + i\left(\frac{\partial}{\partial \alpha}\bar{\mathcal{E}}_0 e^{-i\omega_0 t} - \frac{\partial}{\partial \alpha^*}\bar{\mathcal{E}}_0^* e^{i\omega_0 t}\right); \qquad (15.11b)
\end{aligned}
$$

and the third term is exactly the interaction term from the laser phase-space equation of motion (Eq. 7.101c),

$$
\begin{aligned}
&L^{AF}_{+1}\left(\alpha, \alpha^*, v, v^*, m, \frac{\partial}{\partial \alpha}, \frac{\partial}{\partial \alpha^*}, \frac{\partial}{\partial v}, \frac{\partial}{\partial v^*}, \frac{\partial}{\partial m}\right) \\
&\qquad \equiv -g\left\{\left[\left(e^{-2\frac{\partial}{\partial m}} - 1\right)v^* + \frac{\partial}{\partial v}m - \frac{\partial^2}{\partial v^2}v\right]\alpha + \frac{\partial}{\partial \alpha}v \right. \\
&\qquad\quad \left. + \left[\left(e^{-2\frac{\partial}{\partial m}} - 1\right)v + \frac{\partial}{\partial v^*}m - \frac{\partial^2}{\partial v^{*2}}v^*\right]\alpha^* + \frac{\partial}{\partial \alpha^*}v^*\right\}. \qquad (15.11c)
\end{aligned}
$$

Note that the subscript $\sigma = -1, 0, +1$ (here $+1$) follows the notation of (10.32).

When developing the phase-space treatment of the laser in Chap. 8, we were careful to apply the system size expansion in a systematic way (Sect. 5.1.3): we passed directly from the full phase-space equation of motion to a linearized equation below threshold, a quasi-linearized equation above threshold, and to an equation one step beyond linearization at threshold which retained the lowest nonvanishing order of the nonlinearity. Treatments found in the literature are rarely concerned with such niceties. It is more usual to find the authors passing directly to a phase-space equation of motion truncated at second derivatives and retaining all orders of the nonlinearity. Of course, when this truncated equation is linearized, which it usually is, the end point is just the same. Since we have seen how the systematic approach works and since there is nothing to be learned from its repetition, we follow the common practice here. Thus, keeping only first and second derivatives in (15.11a)–(15.11c), we replace (15.10) by the *Fokker–Planck truncation of the phase-space equation of motion for optical bistability in the Glauber–Sudarshan P representation (with $\bar{n}=0$)*:

$$
\begin{aligned}
\frac{\partial P}{\partial t} = \bigg\{ & \frac{\partial}{\partial \alpha}\big[(\kappa + i\omega_C)\alpha - gv + i\bar{\mathcal{E}}_0 e^{-i\omega_0 t}\big] \\
&+ \frac{\partial}{\partial \alpha^*}\big[(\kappa - i\omega_C)\alpha^* - gv^* - i\bar{\mathcal{E}}_0^* e^{i\omega_0 t}\big] \\
&+ \frac{\partial}{\partial v}\Big[\Big(\frac{\gamma_h}{2} + i\omega_A\Big)v - gm\alpha\Big] \\
&+ \frac{\partial}{\partial v^*}\Big[\Big(\frac{\gamma_h}{2} - i\omega_A\Big)v^* - gm\alpha^*\Big] \\
&+ \frac{\partial}{\partial m}\big[\gamma(m+N) + 2g(v^*\alpha + v\alpha^*)\big] \\
&+ \frac{\gamma_h - \gamma}{2}(N+m)\frac{\partial^2}{\partial v \partial v^*} \\
&+ g\frac{\partial^2}{\partial v^2}v\alpha + g\frac{\partial^2}{\partial v^{*2}}v^*\alpha^* \\
&+ \frac{\partial^2}{\partial m^2}\big[\gamma(m+N) - 2g(v^*\alpha + v\alpha^*)\big]\bigg\}P.
\end{aligned} \tag{15.12}
$$

Note 15.1. When considering the Wigner representation in Sect. 15.1.4, we find it much easier to derive the Fokker–Planck truncation corresponding to (15.12) than the full phase-space equation of motion corresponding to (15.10). The difficulty is that the operator disentangling theorem used to manipulate the exponentials in the characteristic function—the Baker–Hausdorff theorem (4.8)—is considerably more difficult to apply when the commutator on the right-hand side produces an operator rather than a constant (see, for example, [15.2]).

15.1.3 Fokker–Planck Equation in the Q Representation

Considering now the Q representation, the characteristic function generates operator averages in antinormal order. We have

$$
\chi_A(z, z^*, \xi, \xi^*, \eta) \equiv \mathrm{tr}\big(\rho e^{iza} e^{iz^* a^\dagger} e^{i\xi J_-} e^{i\eta J_z} e^{i\xi^* J_+}\big), \tag{15.13}
$$

and, from its Fourier transform,

$$Q(\alpha,\alpha^*,v,v^*,m)$$
$$\equiv \frac{1}{2\pi^5}\int d^2z\int d^2\xi\int d\eta\,\chi_A(z,z^*,\xi,\xi^*,\eta)e^{-iz^*\alpha^*}e^{-iz\alpha}e^{-i\xi^*v^*}e^{-i\xi v}e^{-i\eta m}. \tag{15.14}$$

The derivation of the phase-space equation of motion requires a little extra work as we have not considered atomic operators in antinormal order before. In place of (15.10), we write

$$\frac{\partial Q}{\partial t}=\left[L_{-1}^A\left(v,v^*,m,\frac{\partial}{\partial v},\frac{\partial}{\partial v^*},\frac{\partial}{\partial m}\right)+L_{-1}^F\left(\alpha,\alpha^*,\frac{\partial}{\partial\alpha},\frac{\partial}{\partial\alpha^*}\right)\right.$$
$$\left.+L_{-1}^{AF}\left(\alpha,\alpha^*,v,v^*,m,\frac{\partial}{\partial\alpha},\frac{\partial}{\partial\alpha^*},\frac{\partial}{\partial v},\frac{\partial}{\partial v^*},\frac{\partial}{\partial m}\right)\right]Q. \tag{15.15}$$

From here, the explicit terms—specifically those involving the atoms—are not so easily constructed from equations we have written down before. Their derivation follows, however, from an obvious generalization of the methods described in Sects. 3.2.2, 6.1.3, and 6.3.4. In fact, once we have the results, it will be clear that a direct connection with the previously derived phase-space equation of motion for the laser exists; it is a useful exercise, nevertheless, to carry through the derivation from first principles, using the methods of the quantum-classical correspondence.

Exercise 15.1. Extend the methods of Sects. 3.2.2, 6.1.3, and 6.3.4 to the Q representation; hence show that in place of (15.11a), we have

$$L_{-1}^A\left(v,v^*,m,\frac{\partial}{\partial v},\frac{\partial}{\partial v^*},\frac{\partial}{\partial m}\right)$$
$$\equiv i\omega_C\left(\frac{\partial}{\partial v}v-\frac{\partial}{\partial v^*}v^*\right)$$
$$+\frac{\gamma}{2}\left[(e^{2\frac{\partial}{\partial m}}-1)(N+m)+\frac{\partial^4}{\partial v^2\partial v^{*2}}e^{-2\frac{\partial}{\partial m}}(N-m)\right.$$
$$\left.+2\left(e^{2\frac{\partial}{\partial m}}+\frac{\partial^2}{\partial v\partial v^*}-\frac{1}{2}\right)\left(\frac{\partial}{\partial v}v+\frac{\partial}{\partial v^*}v^*\right)+2N\frac{\partial^2}{\partial v\partial v^*}\right]$$
$$+\gamma_p\left[\frac{\partial}{\partial v}v+\frac{\partial}{\partial v^*}v^*+\frac{\partial^2}{\partial v\partial v^*}e^{2\frac{\partial}{\partial m}}(N-m)\right], \tag{15.16a}$$

in place of (15.11b), we have

$$L_{-1}^F\left(\alpha,\alpha^*,\frac{\partial}{\partial\alpha},\frac{\partial}{\partial\alpha^*}\right)\equiv(\kappa+i\omega_C)\frac{\partial}{\partial\alpha}\alpha+(\kappa-i\omega_C)\frac{\partial}{\partial\alpha^*}\alpha^*$$
$$+i\left(\frac{\partial}{\partial\alpha}\bar{\mathcal{E}}_0e^{-i\omega_0t}-\frac{\partial}{\partial\alpha^*}\bar{\mathcal{E}}_0^*e^{i\omega_0t}\right)+2\kappa\frac{\partial^2}{\partial\alpha\partial\alpha^*}, \tag{15.16b}$$

and in place of (15.11c), we have

$$
\begin{aligned}
L_{-1}^{AF}&\left(\alpha,\alpha^*,v,v^*,m,\frac{\partial}{\partial\alpha},\frac{\partial}{\partial\alpha^*},\frac{\partial}{\partial v},\frac{\partial}{\partial v^*},\frac{\partial}{\partial m}\right)\\
&\equiv g\Bigg\{\left[\left(e^{2\frac{\partial}{\partial m}}-1\right)v^*-\frac{\partial}{\partial v}m-\frac{\partial^2}{\partial v^2}v\right]\alpha-\frac{\partial}{\partial\alpha}v\\
&\quad+\left[\left(e^{2\frac{\partial}{\partial m}}-1\right)v-\frac{\partial}{\partial v^*}m-\frac{\partial^2}{\partial v^{*2}}v^*\right]\alpha^*-\frac{\partial}{\partial\alpha^*}v^*\Bigg\}.
\end{aligned}
\tag{15.16c}
$$

Note the similarity to the phase-space equation of motion in the Glauber–Sudarshan P representation.

On comparing (15.16a)–(15.16c) with (15.11a)–(15.11c), we note first that the diffusion term added to the right-hand side of (15.16b) is to be expected on the basis of (3.47) and (4.14); it follows from the relationship between the normal- and antinormal-ordered characteristic functions (Eq. 4.14). So far as atomic variables go, (15.16c) differs from (15.11c) only through a change of sign on all even-order derivatives, while (15.16a) may actually be taken from the term describing *upwards* transitions in the laser phase-space equation of motion (Eq. 7.100); the sign of m also changes in the dephasing term of (15.16a)—the term proportional to γ_p (Eq. 6.158).

Now keeping derivatives up to second order in (15.15), we arrive at the *Fokker–Planck truncation of the phase-space equation of motion for optical bistability in the Q representation (with $\bar{n}=0$)*:

$$
\begin{aligned}
\frac{\partial Q}{\partial t}=\Bigg\{&\frac{\partial}{\partial\alpha}\left[(\kappa+i\omega_C)\alpha-gv+i\bar{\mathcal{E}}_0e^{-i\omega_0t}\right]\\
&+\frac{\partial}{\partial\alpha^*}\left[(\kappa-i\omega_C)\alpha^*-gv^*-i\bar{\mathcal{E}}_0^*e^{i\omega_0t}\right]\\
&+\frac{\partial}{\partial v}\left[\left(\frac{\gamma_h}{2}+i\omega_A\right)v-gm\alpha\right]\\
&+\frac{\partial}{\partial v^*}\left[\left(\frac{\gamma_h}{2}-i\omega_A\right)v^*-gm\alpha^*\right]\\
&+\frac{\partial}{\partial m}\left[\gamma(m+N)+2g(v^*\alpha+v\alpha^*)\right]\\
&+2\kappa\frac{\partial^2}{\partial\alpha\partial\alpha^*}+\left(\frac{\gamma_h+\gamma}{2}N-\frac{\gamma_h-\gamma}{2}m\right)\frac{\partial^2}{\partial v\partial v^*}\\
&-g\frac{\partial^2}{\partial v^2}v\alpha-g\frac{\partial^2}{\partial v^{*2}}v^*\alpha^*+2\gamma\frac{\partial}{\partial m}\left(\frac{\partial}{\partial v}v+\frac{\partial}{\partial v^*}v^*\right)\\
&+\frac{\partial^2}{\partial m^2}\left[\gamma(m+N)+2g(v^*\alpha+v\alpha^*)\right]\Bigg\}Q.
\end{aligned}
\tag{15.17}
$$

15.1.4 Fokker–Planck Equation in the Wigner Representation

Turning now to the Wigner representation, we finally meet with an opportunity to do something entirely new. It is possible, in principle, to derive an exact phase-space equation of motion for the Wigner distribution as well. The calculations, however, are significantly more complicated, and since we plan to truncate the equation at second-order derivatives in any case, it hardly seems worth the effort to labor through a lot of algebra if it is possible to avoid it. Can we, then, go directly from the master equation to the Fokker–Planck truncation of the phase-space equation of motion? It turns out that this is possible. We could, in fact, follow this alternative approach in any of the representations. Let us therefore see how it works as a general method. The truncated equation of motion in the Wigner representation will then follow as a special case at the end of the calculations.

In any representation, we may start out by proposing a Fokker–Planck truncation of the phase-space equation of motion in the form

$$\frac{\partial F_\sigma}{\partial t} = L_\sigma(\boldsymbol{X},\boldsymbol{X}')F_\sigma, \tag{15.18}$$

with

$$L_\sigma(\boldsymbol{X},\boldsymbol{X}') \equiv -\sum_i \frac{\partial}{\partial X_i} A_\sigma^i(\boldsymbol{X}) + \frac{1}{2}\sum_{i,j} \frac{\partial^2}{\partial X_i \partial X_j} D_\sigma^{ij}(\boldsymbol{X}), \tag{15.19}$$

where $\sigma = +1$, 0, and -1 signifies the P, Wigner, and Q representations, respectively (Eq. 10.32). For the present application the vectors of phase-space variables and the corresponding derivatives with respect to these variables are

$$\boldsymbol{X} \equiv \begin{pmatrix} \alpha \\ \alpha^* \\ v \\ v^* \\ m \end{pmatrix}, \qquad \boldsymbol{X}' \equiv \begin{pmatrix} \partial/\partial\alpha \\ \partial/\partial\alpha^* \\ \partial/\partial v \\ \partial/\partial v^* \\ \partial/\partial m \end{pmatrix}. \tag{15.20}$$

Our task is to determine the explicit functions $A_\sigma^i(\boldsymbol{X})$ and $D_\sigma^{ij}(\boldsymbol{X})$ to be substituted as components of the drift vector,

$$\boldsymbol{A}_\sigma(\boldsymbol{X}) \equiv \begin{pmatrix} A_\sigma^\alpha(\boldsymbol{X}) \\ A_\sigma^{\alpha^*}(\boldsymbol{X}) \\ A_\sigma^v(\boldsymbol{X}) \\ A_\sigma^{v^*}(\boldsymbol{X}) \\ A_\sigma^m(\boldsymbol{X}) \end{pmatrix}, \tag{15.21}$$

and the elements of the diffusion matrix

$$
\boldsymbol{D}_\sigma(\boldsymbol{X}) \equiv \begin{pmatrix} D_\sigma^{\alpha\alpha}(\boldsymbol{X}) & D_\sigma^{\alpha\alpha^*}(\boldsymbol{X}) & D_\sigma^{\alpha v}(\boldsymbol{X}) & D_\sigma^{\alpha v^*}(\boldsymbol{X}) & D_\sigma^{\alpha m}(\boldsymbol{X}) \\ D_\sigma^{\alpha^*\alpha}(\boldsymbol{X}) & D_\sigma^{\alpha^*\alpha^*}(\boldsymbol{X}) & D_\sigma^{\alpha^* v}(\boldsymbol{X}) & D_\sigma^{\alpha^* v^*}(\boldsymbol{X}) & D_\sigma^{\alpha^* m}(\boldsymbol{X}) \\ D_\sigma^{v\alpha}(\boldsymbol{X}) & D_\sigma^{v\alpha^*}(\boldsymbol{X}) & D_\sigma^{vv}(\boldsymbol{X}) & D_\sigma^{vv^*}(\boldsymbol{X}) & D_\sigma^{vm}(\boldsymbol{X}) \\ D_\sigma^{v^*\alpha}(\boldsymbol{X}) & D_\sigma^{v^*\alpha^*}(\boldsymbol{X}) & D_\sigma^{v^* v}(\boldsymbol{X}) & D_\sigma^{v^* v^*}(\boldsymbol{X}) & D_\sigma^{v^* m}(\boldsymbol{X}) \\ D_\sigma^{m\alpha}(\boldsymbol{X}) & D_\sigma^{m\alpha^*}(\boldsymbol{X}) & D_\sigma^{mv}(\boldsymbol{X}) & D_\sigma^{mv^*}(\boldsymbol{X}) & D_\sigma^{mm}(\boldsymbol{X}) \end{pmatrix}. \tag{15.22}
$$

The determination is to be made by comparing the equations of motion for phase-space averages that follow from (15.18) with the equations of motion for operator averages derived directly from the master equation.

The equations of motion for phase-space averages are obtained by generalizing the calculation found in Sect. 5.1.1 to many dimensions. For the phase-space variable means, i.e., first-order moments, this yields

$$
\frac{d(\overline{X_i})_{F_\sigma}}{dt} = (\overline{A_\sigma^i(\boldsymbol{X})})_{F_\sigma}, \tag{15.23}
$$

while for the second-order moments,

$$
\begin{aligned}
\frac{d(\overline{X_i X_j})_{F_\sigma}}{dt} &= (\overline{X_i A_\sigma^j(\boldsymbol{X})})_{F_\sigma} + (\overline{X_j A_\sigma^i \boldsymbol{X})})_{F_\sigma} \\
&\quad + \tfrac{1}{2}\big[(\overline{D_\sigma^{ij}(\boldsymbol{X})})_{F_\sigma} + (\overline{D_\sigma^{ji}(\boldsymbol{X})})_{F_\sigma}\big].
\end{aligned} \tag{15.24}
$$

These are to be compared with the corresponding equations of motion for operator averages, with the correspondence defined by using the appropriate operator order for the three cases: $\sigma = +1$ (normal order), $\sigma = 0$ (symmetric order), and $\sigma = -1$ (antinormal order). Thus, each phase-space average appearing on the left-hand side of either (15.23) or (15.24) corresponds to an operator average in a particular order. The master equation tells us what the equation of motion for that operator average is. To connect that equation with (15.23) or (15.24), every term in it is to be written with its operators in the appropriate order. The functions $A_\sigma^i(\boldsymbol{X})$ and $D_\sigma^{ij}(\boldsymbol{X})$ are then determined by noting that once the operator ordering is correct, there must be a one-to-one correspondence between the averages appearing term by term throughout the two sets of equations.

The procedure is largely trivial to carry through for first-order moments. Corresponding to the vector of phase-space variables, we define the vector of associated operators

$$
\hat{\boldsymbol{X}} \equiv \begin{pmatrix} a \\ a^\dagger \\ J_- \\ J_+ \\ J_z \end{pmatrix}. \tag{15.25}
$$

Then, from master equation (15.1), the equations of motion for the operator means are given by

$$\frac{d\langle \hat{X}_i \rangle}{dt} = \langle \hat{A}_i \rangle, \tag{15.26}$$

$\hat{A}_i \equiv \hat{A}_{\hat{X}_i}$, with

$$\hat{\boldsymbol{A}} \equiv \begin{pmatrix} \hat{A}_a \\ \hat{A}_{a^\dagger} \\ \hat{A}_{J_-} \\ \hat{A}_{J_+} \\ \hat{A}_{J_z} \end{pmatrix} = \begin{pmatrix} -(\kappa + i\omega_C)a + gJ_- - i\bar{\mathcal{E}}_0 e^{-i\omega_0 t} \\ -(\kappa - i\omega_C)a^\dagger + gJ_+ + i\bar{\mathcal{E}}_0^* e^{i\omega_0 t} \\ -(\gamma_h/2 + i\omega_A)J_- + gJ_z a \\ -(\gamma_h/2 - i\omega_A)J_+ + gJ_z a^\dagger \\ -\gamma(J_z + N) - 2g(J_+ a + J_- a^\dagger) \end{pmatrix}. \tag{15.27}$$

Note, then, that the $\hat{A}_i$ contain no noncommuting operator products. Thus, the operator ordering issue is irrelevant and we may simply replace operators by associated phase-space variables to obtain the drift vector

$$\boldsymbol{A}_\sigma(\boldsymbol{X}) = \begin{pmatrix} -(\kappa + i\omega_C)\alpha + gv - i\bar{\mathcal{E}}_0 e^{-i\omega_0 t} \\ -(\kappa - i\omega_C)\alpha^* + gv^* + i\bar{\mathcal{E}}_0^* e^{i\omega_0 t} \\ -(\gamma_h/2 + i\omega_A)v + gm\alpha \\ -(\gamma_h/2 - i\omega_A)v^* + gm\alpha^* \\ -\gamma(m + N) - 2g(v^*\alpha + v\alpha^*) \end{pmatrix}, \tag{15.28}$$

which is independent of representation. From (15.13) and (15.28), the phase-space equations of motion for first-order moments lie in one-to-one correspondence with the Maxwell–Bloch equations (14.99a)–(14.99e).

There is rather more work to be done in determining the three diffusion matrices. Let us consider the Glauber–Sudarshan P representation first. The equations of motion for second-order moments are to be written in *normal order* and matched to (15.24). The tedious algebra is left as an exercise.

Exercise 15.2. For master equation (15.1) and $\hat{X}_i = a, a^\dagger, J_-, J_+, J_z$, show that the normal-ordered equations of motion for second-order moments may be written in the form

$$\frac{d\langle (\hat{X}_i \hat{X}_j)_N \rangle}{dt} = \langle (\hat{X}_i \hat{A}_j)_N \rangle + \langle (\hat{X}_j \hat{A}_i)_N \rangle + \tfrac{1}{2}\left(\langle \hat{D}_{+1}^{ij} \rangle + \langle \hat{D}_{+1}^{ji} \rangle\right), \tag{15.29}$$

where all $\hat{D}_{+1}^{ij} \equiv \hat{D}_{+1}^{\hat{X}_i \hat{X}_j}$ vanish, except for

$$\hat{D}_{+1}^{J_- J_-} = 2gJ_- a, \tag{15.30a}$$

$$\hat{D}_{+1}^{J_+ J_+} = 2gJ_+ a^\dagger, \tag{15.30b}$$

$$\hat{D}_{+1}^{J_z J_z} = 2\gamma(J_z + N) - 4g(J_+ a + J_- a^\dagger), \tag{15.30c}$$

$$\hat{D}_{+1}^{J_- J_+} = \hat{D}_{+1}^{J_+ J_-} = \frac{\gamma_h - \gamma}{2}(N + J_z). \tag{15.30d}$$

Hence find the diffusion matrix in the Fokker–Planck truncation of the phase-space equation of motion for optical bistability in the Glauber–Sudarshan P representation:

$$\boldsymbol{D}_{+1} = \begin{pmatrix} 0 & 0 & 0 & 0 & 0 \\ 0 & 0 & 0 & 0 & 0 \\ 0 & 0 & D_{+1}^{vv} & D_{+1}^{vv^*} & 0 \\ 0 & 0 & D_{+1}^{v^*v} & D_{+1}^{v^*v^*} & 0 \\ 0 & 0 & 0 & 0 & D_{+1}^{mm} \end{pmatrix}, \tag{15.31}$$

with nonzero elements

$$D_{+1}^{vv}(\boldsymbol{X}) = 2gv\alpha, \tag{15.32a}$$

$$D_{+1}^{v^*v^*}(\boldsymbol{X}) = 2gv^*\alpha^*, \tag{15.32b}$$

$$D_{+1}^{mm}(\boldsymbol{X}) = 2\gamma(m+N) - 4g(v^*\alpha + v\alpha^*), \tag{15.32c}$$

$$D_{+1}^{vv^*}(\boldsymbol{X}) = D_{+1}^{v^*v}(\boldsymbol{X}) = \frac{\gamma_h - \gamma}{2}(N+m). \tag{15.32d}$$

Note the agreement with (15.12).

Considering the Q representation next, equations of motion for second-order moments are to be written in *antinormal order* and matched to (15.24). Equation 15.30 must be rewritten as

$$\frac{d\langle(\hat{X}_i\hat{X}_j)_A\rangle}{dt} = \langle(\hat{X}_i\hat{A}_j)_A\rangle + \langle(\hat{X}_j\hat{A}_i)_A\rangle + \tfrac{1}{2}(\langle\hat{D}_{-1}^{ij}\rangle + \langle\hat{D}_{-1}^{ji}\rangle). \tag{15.33}$$

As an illustration of how to proceed, let us follow the details through for one operator product. We consider the case $\hat{X}_i = J_+$ and $\hat{X}_j = J_-$.

To begin, from the commutation relations for collective atomic operators (Eqs. 15.3), we have

$$\frac{d\langle(J_+J_-)_A\rangle}{dt} = \frac{d\langle(J_+J_-)_N\rangle}{dt} - \frac{d\langle J_z\rangle}{dt}. \tag{15.34}$$

Then substituting (15.33) and (15.29) on the left- and right-hand sides of this equation, respectively, and substituting for $d\langle J_z\rangle/dt$ using (15.26), we find the relationship

$$\begin{aligned}\hat{D}_{-1}^{J_+J_-} &= \hat{D}_{+1}^{J_+J_-} + \left[(J_+\hat{A}_{J_-})_N - (J_+\hat{A}_{J_-})_A\right] \\ &\quad + \left[(J_-\hat{A}_{J_+})_N - (J_-\hat{A}_{J_+})_A\right] - \hat{A}_z.\end{aligned} \tag{15.35}$$

On substituting the explicit expression for the drift operators $\hat{A}_{J_-}$ and $\hat{A}_{J_+}$ (Eq. 15.27) we obtain

$$\begin{aligned}\left[(J_+\hat{A}_{J_-})_N - (J_+\hat{A}_{J_-})_A\right] &= -\left(\frac{\gamma_h}{2} + i\omega_A\right)[J_+, J_-] + g[J_+, J_z]a \\ &= -\left(\frac{\gamma_h}{2} + i\omega_A\right)J_z - 2gJ_+a,\end{aligned} \tag{15.36a}$$

and

$$\left[\left(J_-\hat{A}_{J_+}\right)_N - \left(J_-\hat{A}_{J_+}\right)_A\right] = -\left(\frac{\gamma_h}{2} - i\omega_A\right)J_z - 2gJ_-a^\dagger. \tag{15.36b}$$

Finally, combining (15.35) with (15.30d), (15.36a), and (15.36b) yields the antinormal-ordered diffusion operator

$$\begin{aligned}\hat{D}_{-1}^{J_+J_-} &= \frac{\gamma_h - \gamma}{2}(N + J_z) - \gamma_h J_z - 2g(J_+a + J_-a^\dagger) \\ &\quad + \gamma(N + J_z) + 2g(J_+a + J_-a^\dagger) \\ &= \frac{\gamma_h + \gamma}{2}N - \frac{\gamma_h - \gamma}{2}J_z. \end{aligned} \tag{15.37}$$

A calculation along similar lines must be carried out for each operator product. The details are left, once again, as an exercise.

Exercise 15.3. For master equation (15.1) and $\hat{X}_i = a, a^\dagger, J_-, J_+, J_z$, use the operator moment equations (15.29) and (15.33) as above to show that all $\hat{D}_{-1}^{ij} \equiv \hat{D}_{-1}^{\hat{X}_i\hat{X}_j}$ vanish, except for

$$\hat{D}_{-1}^{aa^\dagger} = \hat{D}_{-1}^{a^\dagger a} = 2\kappa, \tag{15.38a}$$

$$\hat{D}_{-1}^{J_-J_-} = -2gJ_-a, \tag{15.38b}$$

$$\hat{D}_{-1}^{J_+J_+} = -2gJ_+a^\dagger, \tag{15.38c}$$

$$\hat{D}_{-1}^{J_zJ_z} = 2\gamma(J_z + N) + 4g(J_+a + J_-a^\dagger), \tag{15.38d}$$

$$\hat{D}_{-1}^{J_-J_+} = \hat{D}_{-1}^{J_+J_-} = \frac{\gamma_h + \gamma}{2}N - \frac{\gamma_h - \gamma}{2}J_z, \tag{15.38e}$$

$$\hat{D}_{-1}^{J_-J_z} = \hat{D}_{-1}^{J_zJ_-} = 2\gamma J_-, \tag{15.38f}$$

$$\hat{D}_{-1}^{J_+J_z} = \hat{D}_{-1}^{J_zJ_+} = 2\gamma J_+. \tag{15.38g}$$

Hence find the diffusion matrix in the Fokker–Planck truncation of the phase-space equation of motion for optical bistability in the Q representation:

$$\boldsymbol{D}_{-1} = \begin{pmatrix} 0 & D_{-1}^{\alpha\alpha^*} & 0 & 0 & 0 \\ D_{-1}^{\alpha^*\alpha} & 0 & 0 & 0 & 0 \\ 0 & 0 & D_{-1}^{vv} & D_{-1}^{vv^*} & D_{-1}^{vm} \\ 0 & 0 & D_{-1}^{v^*v} & D_{-1}^{v^*v^*} & D_{-1}^{v^*m} \\ 0 & 0 & D_{-1}^{mv} & D_{-1}^{mv^*} & D_{-1}^{mm} \end{pmatrix}, \tag{15.39}$$

with nonzero elements

$$D_{-1}^{\alpha\alpha^*} = D_{-1}^{\alpha^*\alpha} = 2\kappa, \tag{15.40a}$$

$$D_{-1}^{vv} = -2gv\alpha, \tag{15.40b}$$

$$D_{-1}^{v^*v^*} = -2gv^*\alpha^*, \tag{15.40c}$$

$$D_{-1}^{mm} = 2\gamma(m+N) + 4g(v^*\alpha + v\alpha^*), \tag{15.40d}$$

$$D_{-1}^{vv^*} = D_{-1}^{v^*v} = \frac{\gamma_h + \gamma}{2}N - \frac{\gamma_h - \gamma}{2}m, \tag{15.40e}$$

$$D_{-1}^{vm} = D_{-1}^{mv} = 2\gamma v, \tag{15.40f}$$

$$D_{-1}^{v^*m} = D_{-1}^{mv^*} = 2\gamma v^*. \tag{15.40g}$$

Note the agreement with (15.17).

This brings us, finally, to the Wigner representation, our principal objective in this section. In this case, results with the second-order moments written in *symmetric order* may be reached by averaging the normal- and antinormal-ordered results. In line with (15.29) and (15.33), we write the equations of motion for second-order moments in the symmetric-ordered form

$$\frac{d\langle(\hat{X}_i\hat{X}_j)_S\rangle}{dt} = \langle(\hat{X}_i\hat{A}_j)_S\rangle + \langle(\hat{X}_j\hat{A}_i)_S\rangle + \tfrac{1}{2}\left(\langle\hat{D}_0^{ij}\rangle + \langle\hat{D}_0^{ji}\rangle\right), \tag{15.41}$$

where, term by term, the symmetric-ordered equation is the average of the other two. Thus, by averaging diffusion matrices (15.31) and (15.39), we find that all $\hat{D}_0^{ij} \equiv \hat{D}_0^{\hat{X}_i\hat{X}_j}$ vanish, except for

$$\hat{D}_0^{aa^\dagger} = \hat{D}_0^{a^\dagger a} = \kappa, \tag{15.42a}$$

$$\hat{D}_0^{J_zJ_z} = 2\gamma(J_z + N), \tag{15.42b}$$

$$\hat{D}_0^{J_-J_+} = \hat{D}_0^{J_+J_-} = \frac{\gamma_h}{2}N, \tag{15.42c}$$

$$\hat{D}_0^{J_-J_z} = \hat{D}_{-1}^{J_zJ_-} = \gamma J_-, \tag{15.42d}$$

$$\hat{D}_0^{J_+J_z} = \hat{D}_0^{J_zJ_+} = \gamma J_+. \tag{15.42e}$$

We arrive at the diffusion matrix in the Fokker–Planck truncation of the phase-space equation of motion for optical bistability in the Wigner representation:

$$\boldsymbol{D}_0 = \begin{pmatrix} 0 & D_0^{\alpha\alpha^*} & 0 & 0 & 0 \\ D_0^{\alpha^*\alpha} & 0 & 0 & 0 & 0 \\ 0 & 0 & 0 & D_0^{vv^*} & D_0^{vm} \\ 0 & 0 & D_0^{v^*v} & 0 & D_0^{v^*m} \\ 0 & 0 & D_0^{mv} & D_0^{mv^*} & D_0^{mm} \end{pmatrix}, \tag{15.43}$$

with nonzero elements

$$D_0^{\alpha\alpha^*} = D_0^{\alpha^*\alpha} = \kappa, \tag{15.44a}$$

$$D_0^{mm} = 2\gamma(m+N), \tag{15.44b}$$

$$D_0^{vv^*} = D_0^{v^*v} = \frac{\gamma_h}{2}N, \tag{15.44c}$$

$$D_0^{vm} = D_0^{mv} = \gamma v, \tag{15.44d}$$

$$D_0^{v^*m} = D_0^{mv^*} = \gamma v^*. \tag{15.44e}$$

Now using (15.18)–(15.22), (15.28), (15.43), and (15.44a)–(15.44e), we construct the *Fokker–Planck truncation of the phase-space equation of motion for optical bistability in the Wigner representation (with $\bar{n} = 0$)*:

$$\begin{aligned}\frac{\partial W}{\partial t} = \Bigg\{ & \frac{\partial}{\partial \alpha}\left[(\kappa + i\omega_C)\alpha - gv + i\bar{\mathcal{E}}_0 e^{-i\omega_0 t}\right] \\ & + \frac{\partial}{\partial \alpha^*}\left[(\kappa - i\omega_C)\alpha^* - gv^* - i\bar{\mathcal{E}}_0^* e^{i\omega_0 t}\right] \\ & + \frac{\partial}{\partial v}\left[\left(\frac{\gamma_h}{2} + i\omega_A\right)v - gm\alpha\right] \\ & + \frac{\partial}{\partial v^*}\left[\left(\frac{\gamma_h}{2} - i\omega_A\right)v^* - gm\alpha^*\right] \\ & + \frac{\partial}{\partial m}\left[\gamma(m+N) + 2g(v^*\alpha + v\alpha^*)\right] \\ & + \kappa\frac{\partial^2}{\partial\alpha\partial\alpha^*} + \frac{\gamma_h}{2}N\frac{\partial^2}{\partial v\partial v^*} \\ & + \gamma\frac{\partial}{\partial m}\left(\frac{\partial}{\partial v}v + \frac{\partial}{\partial v^*}v^*\right) + \gamma\frac{\partial^2}{\partial m^2}(m+N)\Bigg\}W. \end{aligned} \tag{15.45}$$

The drift terms are the same as in the Glauber–Sudarshan P and Q representations, while the diffusion terms are the averages of those in the other two representations.

15.2 Linear Theory of Quantum Fluctuations

Much of the interesting physics exhibited by a system of many atoms in a cavity arises from the system nonlinearity. Most notably, we have the analogy with a first-order phase transition—the optical bistability of Sects. 14.1 and 14.2. Beyond this there are the numerous dynamical instabilities briefly mentioned at the beginning of Sect. 14.2.3. All of these things carry us into the interesting territory of nonlinear dynamics and chaos. They raise questions about global fluctuations, which might be addressed by retaining the full nonlinearity of the drift and diffusion in Fokker–Planck equations (15.12), (15.17), and (15.45), or in some instances, in the multimode versions of these equations [15.3, 15.4]. Although there has been some work in this direction, caution is called for when assessing the results, since they can be infected by inconsistencies introduced by the arbitrary truncation of derivatives at the second order.

To expand just a little on the point, we have already noted (Sect. 5.1.3) that retaining nonlinear terms in the drift and diffusion after truncating derivatives is an inconsistent procedure; it is not a systematic truncation at a particular order of the system size parameter. It is possible, though, that we might accept the "small" inaccuracy involved. But considering applications in quantum optics, there is something of a more fundamental nature to note. Different phase-space representations give different phase-space equations of motion. The moments obtained by solving these equations should nevertheless be related to one another through the operator ordering conventions—normal, antinormal, and symmetric—i.e., through the commutation relations. The expected relationships are, indeed, faithfully preserved by the Fokker–Planck truncation for moments up to second order, essentially by construction, as we can see from the derivation of the Fokker–Planck truncation in Sect. 15.1.4. Noting, then, that the linearized Fokker–Planck equations have Gaussian solutions—with all moments given in terms of first- and second-order moments—consistency between the P, Q, and Wigner representations is guaranteed under linearization. There is no guarantee, on the other hand, when nonlinearities in the drift and diffusion are kept. Equations 15.12, 15.17, and 15.45 are approximate equations, and it is important to realize that, without linearization, the approximations made in the different representations are not the same; they differ in just such a way that some disagreement is to be expected at the level of the commutation relations. Explicit examples of the kinds of inconsistency that result can be found in the work of Drummond and collaborators [15.5, 15.6].

For the most part we will not concern ourselves with global fluctuations, except for a brief consideration of the topic in Sect. 16.3.6. In this section we work from the linearized Fokker–Planck equations to develop a treatment of quantum fluctuations for many atoms in a cavity which follows a path we have traveled before.

15.2.1 System Size Expansion for Optical Bistability

The most suitable scaling of phase-space variables has become fairly standard for us now. We model what is to be done in the present case on (8.17) and (8.20), introducing scaled variables through the relationships

$$ie^{-i(\phi_T+\phi_0-\phi_C-\phi_C')}\alpha = n_{\rm sat}^{1/2}\bar{\alpha}, \tag{15.46a}$$

$$-ie^{i(\phi_T+\phi_0-\phi_C-\phi_C')}\alpha^* = n_{\rm sat}^{1/2}\bar{\alpha}^*, \tag{15.46b}$$

$$i\sqrt{2\gamma_h/\gamma}e^{-i(\phi_T+\phi_0-\phi_C-\phi_C')}v = N\bar{v}, \tag{15.46c}$$

$$-i\sqrt{2\gamma_h/\gamma}e^{i(\phi_T+\phi_0-\phi_C-\phi_C')}v^* = N\bar{v}^*, \tag{15.46d}$$

$$m = N\bar{m}, \tag{15.46e}$$

where $n_{\rm sat}$ is the saturation photon number [(14.104) with $g_{\rm max} \to g$]; the factor $\sqrt{2\gamma_h/\gamma}$ is introduced in (15.46c) and (16.46d) to simplify the equations and the phase factors [taken from (13.58)] ensure that the driving field amplitude will appear in the equations as a real number. It is convenient to also transform to a frame rotating at the driving field frequency, writing

$$\bar{\alpha} = e^{-i\omega_0 t}\tilde{\bar{\alpha}}, \qquad \bar{\alpha}^* = e^{i\omega_0 t}\tilde{\bar{\alpha}}^*, \tag{15.47a}$$

$$\bar{v} = e^{-i\omega_0 t}\tilde{\bar{v}}, \qquad \bar{v}^* = e^{i\omega_0 t}\tilde{\bar{v}}^*, \tag{15.47b}$$

and to combine the three Fokker–Planck equations, (15.12), (15.17), and (15.45), as a single equation using the index $\sigma = +1$, 0, or -1 in our usual way (Eq. 10.32). The distribution function is then

$$\begin{aligned}&\tilde{\bar{F}}'_\sigma(\tilde{\bar{\alpha}}, \tilde{\bar{\alpha}}^*, \tilde{\bar{v}}, \tilde{\bar{v}}^*, \bar{m}, t)\\ &\quad \equiv N^3 n_{\rm sat}\frac{\gamma}{2\gamma_h}F_\sigma\big(\alpha(\tilde{\bar{\alpha}}, t), \alpha^*(\tilde{\bar{\alpha}}^*, t), v(\tilde{\bar{v}}, t), v^*(\tilde{\bar{v}}^*, t), m(\bar{m}), t\big),\end{aligned} \tag{15.48}$$

and satisfies the Fokker–Planck equation

$$\begin{aligned}\frac{\partial \tilde{\bar{F}}'_\sigma}{\partial t} = \bigg\{&\kappa\frac{\partial}{\partial\tilde{\bar{\alpha}}}\big[(1-i\Phi)\tilde{\bar{\alpha}} - 2C\tilde{\bar{v}} - Y\big] + \kappa\frac{\partial}{\partial\tilde{\bar{\alpha}}^*}\big[(1+i\Phi)\tilde{\bar{\alpha}}^* - 2C\tilde{\bar{v}}^* - Y\big]\\ &+\frac{\gamma_h}{2}\frac{\partial}{\partial\tilde{\bar{v}}}\big[(1-i\Delta)\tilde{\bar{v}} - \bar{m}\tilde{\bar{\alpha}} + \frac{\gamma_h}{2}\frac{\partial}{\partial\tilde{\bar{v}}^*}\big[(1+i\Delta)\tilde{\bar{v}}^* - \bar{m}\tilde{\bar{\alpha}}^*\big]\\ &+\gamma\frac{\partial}{\partial\bar{m}}\big[\bar{m} + 1 + \tfrac{1}{2}(\tilde{\bar{v}}^*\tilde{\bar{\alpha}} + \tilde{\bar{v}}\tilde{\bar{\alpha}}^*)\big]\\ &+n_{\rm sat}^{-1}(1-\sigma)\kappa\frac{\partial^2}{\partial\tilde{\bar{\alpha}}\partial\tilde{\bar{\alpha}}^*} + N^{-1}\left(\frac{\gamma_h - \sigma\gamma}{\gamma} + \sigma\frac{\gamma_h-\gamma}{\gamma}\bar{m}\right)\gamma_h\frac{\partial^2}{\partial\tilde{\bar{v}}\partial\tilde{\bar{v}}^*}\\ &+N^{-1}\sigma\frac{\gamma_h}{2}\left(\frac{\partial^2}{\partial\tilde{\bar{v}}^2}\tilde{\bar{v}}\tilde{\bar{\alpha}} + \frac{\partial^2}{\partial\tilde{\bar{v}}^{*2}}\tilde{\bar{v}}^*\tilde{\bar{\alpha}}^*\right)\\ &+N^{-1}\gamma\frac{\partial^2}{\partial\bar{m}^2}\big[\bar{m} + 1 - \sigma\tfrac{1}{2}(\tilde{\bar{v}}^*\tilde{\bar{\alpha}} + \tilde{\bar{v}}\tilde{\bar{\alpha}}^*)\big]\bigg\}\tilde{\bar{F}}'_\sigma,\end{aligned} \tag{15.49}$$

where Δ and Φ are the dimensionless detunings (14.30) and (14.51), respectively, C is the cooperativity parameter [(Eq. 14.103) with $\bar{N}_{\rm eff} \to N$, $g_{\max} \to g$], and Y is the dimensionless driving field amplitude (13.88).

Now we pass from (15.49) to the linearized Fokker–Planck equation by way of the system size expansion. As in the case of the laser, we might adopt either $n_{\rm sat}$ or N as the system size parameter. For the laser we chose $n_{\rm sat}$. Here choosing N leads to equations in a slightly simpler form. We therefore make the expansions

$$\bar{\alpha} = \langle\bar{a}(t)\rangle + N^{-1/2}z, \tag{15.50a}$$

$$\bar{\alpha}^* = \langle\bar{a}^\dagger(t)\rangle + N^{-1/2}z^*, \tag{15.50b}$$

$$\bar{v} = \langle \bar{J}_-(t) \rangle + N^{-1/2}\nu, \tag{15.50c}$$

$$\bar{v}^* = \langle \bar{J}_+(t) \rangle + N^{-1/2}\nu^*, \tag{15.50d}$$

$$\bar{m} = \langle \bar{J}_z(t) \rangle + N^{-1/2}\mu, \tag{15.50e}$$

where the scaled operator averages are defined in parallel with (15.46a)–(15.46e),

$$ie^{-i(\phi_T+\phi_0-\phi_C-\phi'_C)}a = n_{\text{sat}}^{1/2}\bar{a}, \tag{15.51a}$$

$$-ie^{i(\phi_T+\phi_0-\phi_C-\phi'_C)}a^\dagger = n_{\text{sat}}^{1/2}\bar{a}^\dagger, \tag{15.51b}$$

$$i\sqrt{2\gamma_h/\gamma}e^{-i(\phi_T+\phi_0-\phi_C-\phi'_C)}J_- = N\bar{J}_-, \tag{15.51c}$$

$$-i\sqrt{2\gamma_h/\gamma}e^{i(\phi_T+\phi_0-\phi_C-\phi'_C)}J_+ = N\bar{J}_+, \tag{15.51d}$$

$$J_z = N\bar{J}_z. \tag{15.51e}$$

Then, corresponding to (15.48), the distribution function for the fluctuations is

$$\tilde{\bar{F}}_\sigma(\tilde{z},\tilde{z}^*,\tilde{\nu},\tilde{\nu}^*,\mu,t) \equiv N^{-5/2}\tilde{\bar{F}}'_\sigma\big(\tilde{\bar{\alpha}}(\tilde{z},t),\tilde{\bar{\alpha}}^*(\tilde{z}^*,t),\tilde{\bar{v}}(\tilde{\nu},t),\tilde{\bar{v}}^*(\tilde{\nu}^*,t),\bar{m}(\mu,t),t\big), \tag{15.52}$$

with

$$z = e^{-i\omega_0 t}\tilde{z}, \qquad z^* = e^{i\omega_0 t}\tilde{z}^*, \tag{15.53a}$$

$$\nu = e^{-i\omega_0 t}\tilde{\nu}, \qquad \nu^* = e^{i\omega_0 t}\tilde{\nu}^*, \tag{15.53b}$$

and it satisfies the equation of motion

$$\begin{aligned}
\frac{\partial\tilde{\bar{F}}_\sigma}{\partial t} &= N^{-5/2}\Bigg(\frac{\partial\tilde{\bar{F}}'_\sigma}{\partial\tilde{\bar{\alpha}}}\frac{\partial\tilde{\bar{\alpha}}}{\partial t} + \frac{\partial\tilde{\bar{F}}'_\sigma}{\partial\tilde{\bar{\alpha}}^*}\frac{\partial\tilde{\bar{\alpha}}^*}{\partial t} + \frac{\partial\tilde{\bar{F}}'_\sigma}{\partial\tilde{\bar{v}}}\frac{\partial\tilde{\bar{v}}}{\partial t} + \frac{\partial\tilde{\bar{F}}'_\sigma}{\partial\tilde{\bar{v}}^*}\frac{\partial\tilde{\bar{v}}^*}{\partial t} \\
&\quad + \frac{\partial\tilde{\bar{F}}'_\sigma}{\partial\bar{m}}\frac{\partial\bar{m}}{\partial t} + \frac{\partial\tilde{\bar{F}}'_\sigma}{\partial t}\Bigg) \\
&= N^{1/2}\left(\frac{\partial\tilde{\bar{F}}_\sigma}{\partial\tilde{z}}\frac{d\langle\tilde{\bar{a}}(t)\rangle}{dt} + \text{c.c.}\right) + N^{1/2}\left(\frac{\partial\tilde{\bar{F}}_\sigma}{\partial\tilde{\nu}}\frac{d\langle\tilde{\bar{J}}_-(t)\rangle}{dt} + \text{c.c.}\right) \\
&\quad + N^{1/2}\frac{\partial\tilde{\bar{F}}_\sigma}{\partial\mu}\frac{d\langle\bar{J}_z(t)\rangle}{dt} + \frac{\partial}{\partial t}(N^{-5/2}\tilde{\bar{F}}'_\sigma),
\end{aligned} \tag{15.54}$$

where the last term on the right-hand side is to be substituted from the Fokker–Planck equation (15.49). The substitution yields

$$
\begin{aligned}
\frac{\partial \tilde{\bar{F}}_\sigma}{\partial t} = N^{1/2}\Bigg(& \frac{\partial \tilde{\bar{F}}_\sigma}{\partial \tilde{z}}\left\{\frac{d\langle\tilde{\bar{a}}(t)\rangle}{dt} + \kappa\big[(1-i\Phi)\langle\tilde{\bar{a}}(t)\rangle - 2C\langle\tilde{\bar{J}}_-(t)\rangle - Y\big]\right\} + \text{c.c.} \\
& + \frac{\partial \tilde{\bar{F}}_\sigma}{\partial \tilde{\nu}}\left\{\frac{d\langle\tilde{\bar{J}}_-(t)\rangle}{dt} + \frac{\gamma_h}{2}\big[(1-i\Delta)\langle\tilde{\bar{J}}_-(t)\rangle - \langle\bar{J}_z(t)\rangle\langle\tilde{\bar{a}}(t)\rangle\big]\right\} + \text{c.c.} \\
& + \frac{\partial \tilde{\bar{F}}_\sigma}{\partial \mu}\left\{\frac{d\langle\bar{J}_z(t)\rangle}{dt} + \gamma\Big[\big(\langle\bar{J}_z(t)\rangle + 1\big) - \tfrac{1}{2}\big(\langle\tilde{\bar{J}}_+(t)\rangle\langle\tilde{\bar{a}}(t)\rangle + \text{c.c.}\big)\Big]\right\}\Bigg) \\
& + \Bigg\{\kappa\frac{\partial}{\partial \tilde{z}}[(1-i\Phi)\tilde{z} - 2C\tilde{\nu}] + \text{c.c.} \\
& + \frac{\gamma_h}{2}\frac{\partial}{\partial \tilde{\nu}}\big[(1-i\Delta)\tilde{\nu} - \langle\bar{J}_z(t)\rangle\tilde{z} - \langle\tilde{\bar{a}}(t)\rangle\mu\big] + \text{c.c.} \\
& + \gamma\frac{\partial}{\partial \mu}\Big[\mu + \tfrac{1}{2}\big(\langle\tilde{\bar{J}}_+(t)\rangle\tilde{z} + \langle\tilde{\bar{a}}(t)\rangle\tilde{\nu}^* + \text{c.c.}\big)\Big] \\
& + (1-\sigma)\kappa\xi 4C\frac{\partial^2}{\partial\tilde{z}\partial\tilde{z}^*} + \left(\frac{\gamma_h - \sigma\gamma}{\gamma} + \sigma\frac{\gamma_h - \gamma}{\gamma}\langle\bar{J}_z(t)\rangle\right)\gamma_h\frac{\partial^2}{\partial\tilde{\nu}\partial\tilde{\nu}^*} \\
& + \sigma\frac{\gamma_h}{2}\left(\langle\tilde{\bar{J}}_-(t)\rangle\langle\tilde{\bar{a}}(t)\rangle\frac{\partial^2}{\partial\tilde{\nu}^2} + \text{c.c.}\right) + (1-\sigma)\gamma\left(\langle\tilde{\bar{J}}_-(t)\rangle\frac{\partial}{\partial\nu} + \text{c.c.}\right)\frac{\partial}{\partial\mu} \\
& + \gamma\Big[\langle\bar{J}_z(t)\rangle + 1 - \sigma\tfrac{1}{2}\big(\langle\tilde{\bar{J}}_+(t)\rangle\langle\tilde{\bar{a}}(t)\rangle + \text{c.c.}\big)\Big]\frac{\partial^2}{\partial\mu^2} + O(N^{-1/2})\Bigg\}\tilde{\bar{F}}_\sigma,
\end{aligned}
\tag{15.55}
$$

where in order to eliminate n_{sat} in favor of N, we have used (14.103) and (14.104) (with $\bar{N}_{\text{eff}} \to N$ and $g_{\max} \to g$) to establish the relationship [compare (8.1)],

$$
N = n_{\text{sat}}\xi 4C, \qquad \xi \equiv \frac{2\kappa}{\gamma}. \tag{15.56}
$$

Now, in the limit $N \to \infty$, the terms of order $N^{1/2}$ in (15.55) must vanish. This yields the macroscopic law, the *Maxwell–Bloch equations for a collection of homogeneously broadened two-level atoms and one driven mode of the field (with equal coupling strengths)*:

$$
\kappa^{-1}\frac{d\langle\tilde{\bar{a}}\rangle}{dt} = -(1-i\Phi)\langle\tilde{\bar{a}}\rangle + 2C\langle\tilde{\bar{J}}_-\rangle + Y, \tag{15.57a}
$$

$$
\kappa^{-1}\frac{d\langle\tilde{\bar{a}}^\dagger\rangle}{dt} = -(1+i\Phi)\langle\tilde{\bar{a}}^\dagger\rangle + 2C\langle\tilde{\bar{J}}_+\rangle + Y, \tag{15.57b}
$$

$$\left(\frac{\gamma_h}{2}\right)^{-1}\frac{d\langle\tilde{\bar{J}}_-\rangle}{dt} = -(1-i\Delta)\langle\tilde{\bar{J}}_-\rangle + \langle\bar{J}_z\rangle\langle\tilde{\bar{a}}\rangle, \tag{15.57c}$$

$$\left(\frac{\gamma_h}{2}\right)^{-1}\frac{d\langle\tilde{\bar{J}}_+\rangle}{dt} = -(1+i\Delta)\langle\tilde{\bar{J}}_+\rangle + \langle\bar{J}_z\rangle\langle\tilde{\bar{a}}^\dagger\rangle, \tag{15.57d}$$

$$\gamma^{-1}\frac{d\langle\bar{J}_z\rangle}{dt} = -(\langle\bar{J}_z\rangle+1) - \tfrac{1}{2}\big(\langle\tilde{\bar{J}}_+\rangle\langle\tilde{\bar{a}}\rangle + \langle\tilde{\bar{J}}_-\rangle\langle\tilde{\bar{a}}^\dagger\rangle\big). \tag{15.57e}$$

Dropping the term of order $N^{-1/2}$, the fluctuations about the macroscopic state are governed by the *linearized Fokker–Planck equation for optical bistability (with $\bar{n}=0$)*,

$$\begin{aligned}
\frac{\partial\tilde{\bar{F}}_\sigma}{\partial t} = \bigg\{&\kappa\frac{\partial}{\partial\tilde{z}}[(1-i\Phi)\tilde{z}-2C\tilde{\nu}] + \kappa\frac{\partial}{\partial\tilde{z}^*}[(1+i\Phi)\tilde{z}^*-2C\tilde{\nu}^*] \\
&+\frac{\gamma_h}{2}\frac{\partial}{\partial\tilde{\nu}}\big[(1-i\Delta)\tilde{\nu} - \langle\bar{J}_z(t)\rangle\tilde{z} - \langle\tilde{\bar{a}}(t)\rangle\mu\big] \\
&+\frac{\gamma_h}{2}\frac{\partial}{\partial\tilde{\nu}^*}\big[(1+i\Delta)\tilde{\nu}^* - \langle\bar{J}_z(t)\rangle\tilde{z}^* - \langle\tilde{\bar{a}}^\dagger(t)\rangle\mu\big] \\
&+\gamma\frac{\partial}{\partial\mu}\Big[\mu - \tfrac{1}{2}\big(\langle\tilde{\bar{J}}_+(t)\rangle\tilde{z} + \langle\tilde{\bar{a}}(t)\rangle\tilde{\nu}^* + \langle\tilde{\bar{J}}_-(t)\rangle\tilde{z}^* + \langle\tilde{\bar{a}}^\dagger(t)\rangle\tilde{\nu}\big)\Big] \\
&+(1-\sigma)\kappa\xi 4C\frac{\partial^2}{\partial\tilde{z}\partial\tilde{z}^*} + \left(\frac{\gamma_h-\sigma\gamma}{\gamma} + \sigma\frac{\gamma_h-\gamma}{\gamma}\langle\bar{J}_z(t)\rangle\right)\gamma_h\frac{\partial^2}{\partial\tilde{\nu}\partial\tilde{\nu}^*} \\
&+\sigma\frac{\gamma_h}{2}\left(\langle\tilde{\bar{J}}_-(t)\rangle\langle\tilde{\bar{a}}(t)\rangle\frac{\partial^2}{\partial\tilde{\nu}^2} + \langle\tilde{\bar{J}}_+(t)\rangle\langle\tilde{\bar{a}}^\dagger(t)\rangle\frac{\partial^2}{\partial\tilde{\nu}^{*2}}\right) \\
&+(1-\sigma)\gamma\left(\langle\tilde{\bar{J}}_-(t)\rangle\frac{\partial}{\partial\nu} + \langle\tilde{\bar{J}}_+(t)\rangle\frac{\partial}{\partial\nu^*}\right)\frac{\partial}{\partial\mu} \\
&+\gamma\Big[\langle\bar{J}_z(t)\rangle + 1 - \sigma\tfrac{1}{2}\big(\langle\tilde{\bar{J}}_+(t)\rangle\langle\tilde{\bar{a}}(t)\rangle + \langle\tilde{\bar{J}}_-(t)\rangle\langle\tilde{\bar{a}}^\dagger(t)\rangle\big)\Big]\frac{\partial^2}{\partial\mu^2}\bigg\}\tilde{\bar{F}}_\sigma.
\end{aligned} \tag{15.58}$$

Exercise 15.4. Verify that the Maxwell–Bloch equations (15.57a)–(15.57e) follow from (14.99a)–(14.99e) when the dipole coupling constants $g(\boldsymbol{r}_j)$ are assumed to be equal to $g_{\max}\to g$ for all atoms.

Our first interest is to ask whether or not the linearized Fokker–Planck equation has positive semidefinite diffusion. To this end, it is helpful to introduce the real and imaginary parts of phase-space variables; we write

$$\tilde{z} = \tilde{z}_1 + i\tilde{z}_2, \qquad \tilde{z}^* = \tilde{z}_1 - i\tilde{z}_2, \tag{15.59a}$$

$$\tilde{\nu} = \tilde{\nu}_1 + i\tilde{\nu}_2, \qquad \tilde{\nu}^* = \tilde{\nu}_1 - i\tilde{\nu}_2. \tag{15.59b}$$

Then, from the coefficients of second-order derivatives in (15.58), the diffusion matrix is

$$\bar{\boldsymbol{D}}_\sigma = \begin{pmatrix} \bar{D}_\sigma^{\tilde{z}_1\tilde{z}_1} & 0 & 0 & 0 & 0 \\ 0 & \bar{D}_\sigma^{\tilde{z}_2\tilde{z}_2} & 0 & 0 & 0 \\ 0 & 0 & \bar{D}_\sigma^{\tilde{\nu}_1\tilde{\nu}_1} & \bar{D}_\sigma^{\tilde{\nu}_1\tilde{\nu}_2} & \bar{D}_\sigma^{\tilde{\nu}_1\mu} \\ 0 & 0 & \bar{D}_\sigma^{\tilde{\nu}_2\tilde{\nu}_1} & \bar{D}_\sigma^{\tilde{\nu}_2\tilde{\nu}_2} & \bar{D}_\sigma^{\tilde{\nu}_2\mu} \\ 0 & 0 & \bar{D}_\sigma^{\mu\tilde{\nu}_1} & \bar{D}_\sigma^{\mu\tilde{\nu}_2} & \bar{D}_\sigma^{\mu\mu} \end{pmatrix}, \tag{15.60}$$

with nonzero elements

$$\bar{D}_\sigma^{\tilde{z}_1\tilde{z}_1} = \bar{D}_\sigma^{\tilde{z}_2\tilde{z}_2} = (1-\sigma)\kappa\xi 2C, \tag{15.61a}$$

$$\bar{D}_\sigma^{\tilde{\nu}_1\tilde{\nu}_1} = \frac{\gamma_h}{2}\left[\frac{\gamma_h - \sigma\gamma}{\gamma} + \sigma\frac{\gamma_h-\gamma}{\gamma}\langle\bar{J}_z\rangle + \sigma\tfrac{1}{2}\big(\langle\tilde{\bar{J}}_-\rangle\langle\tilde{\bar{a}}\rangle + \text{c.c.}\big)\right], \tag{15.61b}$$

$$\bar{D}_\sigma^{\tilde{\nu}_2\tilde{\nu}_2} = \frac{\gamma_h}{2}\left[\frac{\gamma_h - \sigma\gamma}{\gamma} + \sigma\frac{\gamma_h-\gamma}{\gamma}\langle\bar{J}_z\rangle - \sigma\tfrac{1}{2}\big(\langle\tilde{\bar{J}}_-\rangle\langle\tilde{\bar{a}}\rangle + \text{c.c.}\big)\right], \tag{15.61c}$$

$$\bar{D}_\sigma^{\mu\mu} = 2\gamma\big[\langle\bar{J}_z\rangle + 1 - \sigma\tfrac{1}{2}\big(\langle\tilde{\bar{J}}_+\rangle\langle\tilde{\bar{a}}\rangle + \text{c.c.}\big)\big], \tag{15.61d}$$

$$\bar{D}_\sigma^{\tilde{\nu}_1\tilde{\nu}_2} = \bar{D}_\sigma^{\tilde{\nu}_2\tilde{\nu}_1} = \sigma\frac{\gamma_h}{2}\tfrac{1}{2}\big(-i\langle\tilde{\bar{J}}_-\rangle\langle\tilde{\bar{a}}\rangle + \text{c.c.}\big), \tag{15.61e}$$

$$\bar{D}_\sigma^{\tilde{\nu}_1\mu} = \bar{D}_\sigma^{\mu\tilde{\nu}_1} = (1-\sigma)\gamma\tfrac{1}{2}\big(\langle\tilde{\bar{J}}_-\rangle + \text{c.c.}\big), \tag{15.61f}$$

$$\bar{D}_\sigma^{\tilde{\nu}_2\mu} = \bar{D}_\sigma^{\mu\tilde{\nu}_2} = (1-\sigma)\gamma\tfrac{1}{2}\big(-i\langle\tilde{\bar{J}}_-\rangle + \text{c.c.}\big). \tag{15.61g}$$

The matrix is block diagonal, with diagonal diffusion in each field quadrature—i.e., the diffusion of the field variables is decoupled from the atoms. The diffusion constants for the field are positive definite in the Q and Wigner representations ($\sigma = -1$ and 0) and vanish in the Glauber–Sudarshan P representation ($\sigma = +1$). The question, then, is whether or not the 3×3 atomic diffusion matrix has nonnegative eigenvalues. Since it depends on σ, there is a possibility that its eigenvalues are nonnegative in one representation and not in another.

Let us consider the Wigner representation first, where for $\sigma = 0$, from (15.61b)–(15.61g), the atomic diffusion matrix is

$$\begin{pmatrix} \bar{D}_0^{\tilde{\nu}_1\tilde{\nu}_1} & \bar{D}_0^{\tilde{\nu}_1\tilde{\nu}_2} & \bar{D}_0^{\tilde{\nu}_1\mu} \\ \bar{D}_0^{\tilde{\nu}_2\tilde{\nu}_1} & \bar{D}_0^{\tilde{\nu}_2\tilde{\nu}_2} & \bar{D}_0^{\tilde{\nu}_2\mu} \\ \bar{D}_0^{\mu\tilde{\nu}_1} & \bar{D}_0^{\mu\tilde{\nu}_2} & \bar{D}_0^{\mu\mu} \end{pmatrix} = \frac{\gamma}{2}\begin{pmatrix} (\gamma_h/\gamma)^2 & 0 & \langle\tilde{\bar{J}}_-\rangle + \text{c.c.} \\ 0 & (\gamma_h/\gamma)^2 & -i\langle\tilde{\bar{J}}_-\rangle + \text{c.c.} \\ \langle\tilde{\bar{J}}_-\rangle + \text{c.c.} & -i\langle\tilde{\bar{J}}_-\rangle + \text{c.c.} & 4(\langle\bar{J}_z\rangle + 1) \end{pmatrix}. \tag{15.62}$$

The eigenvalues are

$$\gamma^{-1}\lambda_1 = \frac{1}{2}\left(\frac{\gamma_h}{\gamma}\right)^2, \tag{15.63a}$$

$$\gamma^{-1}\lambda_{2,3} = \frac{1}{4}\left(\frac{\gamma_h}{\gamma}\right)^2 + \langle\bar{J}_z\rangle + 1 \pm \left\{\left[\frac{1}{4}\left(\frac{\gamma_h}{\gamma}\right)^2 - \langle\bar{J}_z\rangle - 1\right]^2 + \langle\bar{\tilde{J}}_+\rangle\langle\bar{\tilde{J}}_-\rangle\right\}^{1/2}, \tag{15.63b}$$

from which the requirement for positive semidefinite diffusion is given by the inequality

$$\left(\frac{\gamma_h}{\gamma}\right)^2 \left(\langle\bar{J}_z\rangle + 1\right) - \langle\bar{\tilde{J}}_+\rangle\langle\bar{\tilde{J}}_-\rangle \geq 0. \tag{15.64}$$

Solutions to the optical Bloch equations must lie inside the Bloch sphere (see the discussion below Fig 2.3). It follows that $4\langle J_+\rangle\langle J_-\rangle + \langle J_z\rangle^2 \leq N^2$, which with the scaling (15.51c)–(15.51e) yields the inequality

$$4\langle\bar{\tilde{J}}_+\rangle\langle\bar{\tilde{J}}_-\rangle \leq \frac{2\gamma_h}{\gamma}\left(1 - \langle\bar{J}_z\rangle^2\right). \tag{15.65}$$

We are also assured that $\gamma_h/\gamma \geq 1$. Thus, noting that on the right-hand side of (15.65) we may write $2(1 - \langle\bar{J}_z\rangle^2) = 4(1 + \langle\bar{J}_z\rangle) - 2(1 + \langle\bar{J}_z\rangle)^2$, from the constraint that solutions to the Bloch equations lie within the Bloch sphere we prove that

$$\begin{aligned}\left(\frac{\gamma_h}{\gamma}\right)^2 \left(\langle\bar{J}_z\rangle + 1\right) - \langle\bar{\tilde{J}}_+\rangle\langle\bar{\tilde{J}}_-\rangle &\geq \tfrac{1}{2}\left(\langle\bar{J}_z\rangle + 1\right)^2 \\ &\geq 0.\end{aligned} \tag{15.66}$$

We conclude that the diffusion matrix in the Wigner representation is always positive semidefinite.

Note 15.2. The guarantee of positive semidefinite diffusion comes from the requirement that solutions to the optical Bloch equations lie within the Bloch sphere. On the other hand, the optical Bloch equations emerge as the macroscopic equations of motion as a consequence of linearization. There is, then, no assurance that Fokker–Planck equation (15.45), with its nonlinear drift and diffusion, has positive semidefinite diffusion. Indeed, there are certainly regions of phase space where this Fokker–Planck equation has non-positive-semidefinite diffusion—eigenvalue λ_3 can be negative if $(\langle\bar{\tilde{J}}_-\rangle, \langle\bar{\tilde{J}}_+\rangle, \langle\bar{J}_z\rangle) \to (\bar{\tilde{v}}, \bar{\tilde{v}}^*, \bar{m})$ is permitted to venture outside the Bloch sphere. This is hardly surprising: a Wigner function that satisfies a Fokker–Planck equation with positive semidefinite diffusion is itself everywhere positive; it can represent only a limited range of the possible quantum states.

We turn now to the Glauber–Sudarshan P representation, with $\sigma = +1$. To simplify the general case, let us take $\langle\tilde{\bar{a}}\rangle$ and $\langle\tilde{\bar{J}}_-\rangle$ to be real. This yields a diagonal atomic diffusion matrix with diffusion constants

$$\bar{D}_{+1}^{\tilde{\nu}_1\tilde{\nu}_1} = \frac{\gamma_h}{2}\left[\frac{\gamma_h-\gamma}{\gamma}\big(\langle\bar{J}_z\rangle+1\big)+\tfrac{1}{2}\big(\langle\tilde{\bar{J}}_-\rangle\langle\tilde{\bar{a}}\rangle+\text{c.c.}\big)\right], \tag{15.67a}$$

$$\bar{D}_{+1}^{\tilde{\nu}_2\tilde{\nu}_2} = \frac{\gamma_h}{2}\left[\frac{\gamma_h-\gamma}{\gamma}\big(\langle\bar{J}_z\rangle+1\big)-\tfrac{1}{2}\big(\langle\tilde{\bar{J}}_-\rangle\langle\tilde{\bar{a}}\rangle+\text{c.c.}\big)\right], \tag{15.67b}$$

$$\bar{D}_{+1}^{\mu\mu} = 2\gamma\big[\langle\bar{J}_z\rangle+1-\tfrac{1}{2}\big(\langle\tilde{\bar{J}}_+\rangle\langle\tilde{\bar{a}}\rangle+\text{c.c.}\big)\big]. \tag{15.67c}$$

We conclude that diffusion in the Glauber–Sudarshan P representation is definitely not positive semidefinite if, for example, $\gamma_h = \gamma$ and $\langle\tilde{\bar{J}}_-\rangle\langle\tilde{\bar{a}}\rangle \neq 0$; then either $\bar{D}_{+1}^{\tilde{\nu}_1\tilde{\nu}_1}$ or $\bar{D}_{+1}^{\tilde{\nu}_2\tilde{\nu}_2}$ is negative.

Finally, in the Q representation ($\sigma = -1$), although the atomic diffusion matrix is not diagonal under the same conditions, the $\tilde{\nu}_2$-diffusion decouples from the other two dimensions, with diffusion constant

$$\bar{D}_{-1}^{\tilde{\nu}_2\tilde{\nu}_2} = \frac{\gamma_h}{2}\left[\frac{\gamma_h+\gamma}{\gamma}-\frac{\gamma_h-\gamma}{\gamma}\langle\bar{J}_z\rangle+\tfrac{1}{2}\big(\langle\tilde{\bar{J}}_-\rangle\langle\tilde{\bar{a}}\rangle+\text{c.c.}\big)\right]. \tag{15.68}$$

It is straightforward to prove that this also gives non-positive-semidefinite diffusion in accessible regions of the phase space.

In summary, diffusion in our phase-space treatment of many atoms in a cavity is positive semidefinite in the Wigner representation so long as the Fokker–Planck equation is linearized. It is non-positive-semidefinite in the Glauber–Sudarshan P and Q representations, even under linearization. Both of the latter yield positive semidefinite diffusion in some regions of the phase space, though. The next task to check is whether or not the steady state lies within these regions.

15.2.2 Linearization About the Steady State

Introducing the dimensionless amplitude and phase defined in (14.52) and (14.56) [see also (13.58), (14.91), and (15.51) for the definition of the phase], we write the steady-state intracavity field amplitude as

$$\langle\tilde{\bar{a}}\rangle_{\text{ss}} = e^{i\phi_X}X, \tag{15.69a}$$

$$\langle\tilde{\bar{a}}^\dagger\rangle_{\text{ss}} = e^{-i\phi_X}X, \tag{15.69b}$$

and solve the Maxwell–Bloch equations (15.57a)–(15.57e) in steady state to obtain

$$\langle\tilde{\bar{J}}_-\rangle_{\text{ss}} = -\frac{1+i\Delta}{1+\Delta^2+X^2}e^{i\phi_X}X, \tag{15.70a}$$

$$\langle\tilde{\bar{J}}_+\rangle_{\text{ss}} = -\frac{1-i\Delta}{1+\Delta^2+X^2}e^{-i\phi_X}X, \tag{15.70b}$$

$$\langle\bar{J}_z\rangle_{\text{ss}} = -\frac{1+\Delta^2}{1+\Delta^2+X^2}, \tag{15.70c}$$

where X and ϕ_X satisfy the optical bistability state equation

$$Y = e^{i\phi_X} X \left[\left(1 + 2C \frac{1}{1 + \Delta^2 + X^2} \right) - i \left(\Phi - 2C\Delta \frac{1}{1 + \Delta^2 + X^2} \right) \right]. \quad (15.71)$$

With the adoption of equal dipole coupling strengths for all atoms, the state equation for a ring cavity is obtained. Now, to simplify the otherwise tedious algebra, let us specialize to the absorptive case ($\Delta = \Phi = 0$): from (15.71), the *state equation of absorptive optical bistability for a homogeneously broadened two-level medium in a plane-wave ring cavity* is

$$Y = X \left(1 + 2C \frac{1}{1 + X^2} \right), \qquad \phi_X = 0, \quad (15.72)$$

with

$$\frac{dY}{dX} = 1 + 2C \frac{1 - X^2}{(1 + X^2)^2}. \quad (15.73)$$

It is readily checked that dY/dX vanishes for $X_\pm^2 = C - 1 \pm \sqrt{C(C-4)}$, verifying that (15.72) possesses three real solutions for X whenever $2C > 8$ (Exercise 14.3).

We wish to determine whether or not diffusion matrix (15.60) is positive semidefinite when the linearization is made around these steady states. Let us restrict ourselves to radiative damping ($\gamma_h = \gamma$) to further simplify the calculation. Then, substituting from (15.69a)–(15.70c) into (15.61a)–(15.61g), the diffusion matrix has nonzero elements

$$(\bar{D}_\sigma^{\mathrm{ss}})_{\tilde{z}_1\tilde{z}_1} = (\bar{D}_\sigma^{\mathrm{ss}})_{\tilde{z}_2\tilde{z}_2} = \gamma(1 - \sigma)\xi^2 C, \quad (15.74a)$$

$$(\bar{D}_\sigma^{\mathrm{ss}})_{\tilde{\nu}_1\tilde{\nu}_1} = \frac{\gamma}{2} \left(1 - \sigma - \sigma \frac{X^2}{1 + X^2} \right), \quad (15.74b)$$

$$(\bar{D}_\sigma^{\mathrm{ss}})_{\tilde{\nu}_2\tilde{\nu}_2} = \frac{\gamma}{2} \left(1 - \sigma + \sigma \frac{X^2}{1 + X^2} \right), \quad (15.74c)$$

$$(\bar{D}_\sigma^{\mathrm{ss}})_{\mu\mu} = 2\gamma(1 + \sigma) \frac{X^2}{1 + X^2}, \quad (15.74d)$$

$$(\bar{D}_\sigma^{\mathrm{ss}})_{\nu_1\mu} = (\bar{D}_\sigma^{\mathrm{ss}})_{\mu\nu_1} = -\gamma(1 - \sigma) \frac{X}{1 + X^2}. \quad (15.74e)$$

As expected from our general proof (Eq. 15.66), it is positive semidefinite in the Wigner representation ($\sigma = 0$). In the Q representation ($\sigma = -1$), however, it is not, since one eigenvalue of the coupled (ν_1, μ)-diffusion is always negative. Also, in the Glauber–Sudarshan P representation ($\sigma = +1$) it is not; the diffusion matrix is diagonal with $(\bar{D}_\sigma^{\mathrm{ss}})_{\tilde{\nu}_1\tilde{\nu}_1} < 0$. Thus, both the Q and Glauber–Sudarshan P representations yield non-positive-semidefinite diffusion in the vicinity of the steady state.

We could carry through the linearized analysis of fluctuations in any one of these representations; alternatively, we might leave the representation unspecified, as we did for the degenerate parametric oscillator (Sect. 10.2). The simplicity of diagonal diffusion is appealing, though; moreover, since we can carry through the calculations formally, ignoring the intricacies of negative diffusion in practice, we proceed with the Glauber–Sudarshan P representation—seen now, however, to require a reinterpretation as the positive P representation.

Strictly, the positive P distribution is a function in ten dimensions. Instead of writing out the Fokker–Planck equation in the phase space of double dimensions, though, it is convenient to write it in the common complex notation, a notation mirroring that of the standard Glauber–Sudarshan P representation. Thus, we begin our linearized analysis of fluctuations from the *Fokker–Planck equation in the positive P representation for absorptive optical bistability linearized about the steady state (with $\bar{n} = 0$)*:

$$\frac{\partial \bar{\tilde{P}}}{\partial t} = \left(- \tilde{\boldsymbol{Z}}'^{T} \bar{\boldsymbol{J}}_{\mathrm{ss}} \tilde{\boldsymbol{Z}} + \tfrac{1}{2} \tilde{\boldsymbol{Z}}'^{T} \bar{\boldsymbol{D}}_{\mathrm{ss}} \tilde{\boldsymbol{Z}}' \right) \bar{\tilde{P}}, \tag{15.75}$$

with

$$\tilde{\boldsymbol{Z}} \equiv \begin{pmatrix} \tilde{z} \\ \tilde{z}_* \\ \tilde{\nu} \\ \tilde{\nu}_* \\ \mu \end{pmatrix}, \qquad \tilde{\boldsymbol{Z}}' \equiv \begin{pmatrix} \partial/\partial\tilde{z} \\ \partial/\partial\tilde{z}_* \\ \partial/\partial\tilde{\nu} \\ \partial/\partial\tilde{\nu}_* \\ \partial/\partial\mu \end{pmatrix}, \tag{15.76}$$

where, substituting (15.69a)–(15.70c) into (15.58) ($\Delta = \Phi = 0$ and $\gamma_h = \gamma$), the Jacobian matrix is

$$\bar{\boldsymbol{J}}_{\mathrm{ss}} = \frac{\gamma}{2} \begin{pmatrix} -\xi & 0 & \xi 2C & 0 & 0 \\ 0 & -\xi & 0 & \xi 2C & 0 \\ -1/(1+X^2) & 0 & -1 & 0 & X \\ 0 & -1/(1+X^2) & 0 & -1 & X \\ X/(1+X^2) & X/(1+X^2) & -X & -X & -2 \end{pmatrix}, \tag{15.77}$$

and from (15.74a)–(15.74e) with $\sigma = +1$ we obtain the diagonal diffusion

$$\bar{\boldsymbol{D}}_{\mathrm{ss}} = \gamma \frac{X^2}{1+X^2} \begin{pmatrix} 0 & 0 & 0 & 0 & 0 \\ 0 & 0 & 0 & 0 & 0 \\ 0 & 0 & -1 & 0 & 0 \\ 0 & 0 & 0 & 1 & 0 \\ 0 & 0 & 0 & 0 & 4 \end{pmatrix}. \tag{15.78}$$

Here $\tilde{Z}$ is a vector of five *independent* complex variables. The positive P interpretation of (15.75) amounts to a definition of how derivatives with respect to these variables are to be taken (see the discussion below Note 11.3); but we need not be concerned with these formal issues. All we need to perform the calculations is contained in the matrices $\bar{\boldsymbol{J}}_{\mathrm{ss}}$ and $\bar{\boldsymbol{D}}_{\mathrm{ss}}$. We work with them in exactly the way we would if the diffusion matrix were positive semidefinite (Sect. 11.1.3).

Note 15.3. Having discovered that we need to reinterpret Fokker–Planck equation (15.12) in the positive P representation, we should be aware that this equation, if accepted along with its nonlinear drift and diffusion, leads to difficulties similar to those discussed in Sect. 12.2. The first hint of this comes with the observation that, in addition to the physical steady states identified in (15.69)–(15.71), the extended phase space admits nonphysical steady states. Specifically, the state equation (Eq. 15.72) which we present as an equation for the intracavity intensity $X^2 \equiv \langle\tilde{\bar{a}}^\dagger\rangle_{\rm ss}\langle\tilde{\bar{a}}\rangle_{\rm ss}$, is now generalized to allow X^2 to be complex; variables $\bar{\alpha}$ and $\bar{\alpha}^*$, expanded in (15.50a) and (15.50b) about $\langle\bar{a}(t)\rangle$ and $\langle\bar{a}^\dagger(t)\rangle$—i.e., interpreted as complex conjugates—are replaced by $\bar{\alpha}$ and $\bar{\alpha}_*$, which need not be conjugates of one another. Looking beyond the steady states themselves, there is a reorganization of the deterministic flow analogous to that described for the degenerate parametric oscillator below (12.65). When the quantum fluctuations are sufficiently large, bringing the full nonlinearity into play, suspicious "spiking" occurs in the positive P stochastic simulations. Indeed, Carmichael and coworkers first discovered the problem of divergent trajectories in positive P stochastic simulations of absorptive optical bistability [15.7]. Of course, within the linear treatment of fluctuations no such anomalies arise.

We aim to calculate the covariance matrix for absorptive optical bistability. Before doing so, it is useful to consider the eigenvalues of Jacobian matrix (15.77). These govern the regression of fluctuations, and we can discover quite a bit of physics directly from them and the characteristic polynomial they solve. To this end, it is convenient to adopt a new set of variables,

$$\tilde{\boldsymbol{Z}}_1 \equiv \begin{pmatrix} -i\frac{1}{2}(\tilde{z}-\tilde{z}_*) \\ -i\frac{1}{2}(\tilde{\nu}-\tilde{\nu}_*) \\ \frac{1}{2}(\tilde{z}+\tilde{z}_*) \\ \frac{1}{2}(\tilde{\nu}+\tilde{\nu}_*) \\ \mu \end{pmatrix}. \tag{15.79}$$

In these variables the rows and columns of the Jacobian matrix are rearranged into block diagonal form:

$$\bar{\boldsymbol{J}}^1_{\rm ss} = \frac{\gamma}{2}\begin{pmatrix} -\xi & \xi 2C & 0 & 0 & 0 \\ -1/(1+X^2) & -1 & 0 & 0 & 0 \\ 0 & 0 & -\xi & \xi 2C & 0 \\ 0 & 0 & -1/(1+X^2) & -1 & X \\ 0 & 0 & 2X/(1+X^2) & -2X & -2 \end{pmatrix}. \tag{15.80}$$

The eigenvalues of the matrix then separate into a pair governing the evolution of phase-space variables $-i\frac{1}{2}(\tilde{z}-\tilde{z}_*)$ and $-i\frac{1}{2}(\tilde{\nu}-\tilde{\nu}_*)$, and a set of three governing the evolution of phase-space variables $\frac{1}{2}(\tilde{z}+\tilde{z}_*)$, $\frac{1}{2}(\tilde{\nu}+\tilde{\nu}_*)$, and μ.

Defining dimensionless eigenvalues

$$\bar{\lambda}_i = (2/\gamma)\lambda_i, \tag{15.81}$$

$i = 1, \cdots, 5$, the first pair, $\bar{\lambda}_1$ and $\bar{\lambda}_2$, are the roots of the quadratic

$$(\bar{\lambda} + \xi)(\bar{\lambda} + 1) + \xi 2C \frac{1}{1 + X^2}. \tag{15.82}$$

Explicitly, we have

$$\bar{\lambda}_{1,2} = -\tfrac{1}{2}(\xi + 1) \pm \sqrt{\tfrac{1}{4}(\xi - 1)^2 - \xi 2C \frac{1}{1 + X^2}}. \tag{15.83}$$

These eigenvalues are essentially Λ_+ and Λ_- from (14.114). They are responsible for the many-atom vacuum Rabi doublet and arise from the coupling of one quadrature of the intracavity field amplitude—phase-space variable $-i\frac{1}{2}(\tilde{z} - \tilde{z}_*)$—to the same quadrature of the atomic polarization—phase-space variable $-i\frac{1}{2}(\tilde{\nu} - \tilde{\nu}_*)$. Here, with $2C = 2Ng^2/\gamma\kappa$, they appear for a fixed number of atoms rather than the effective number $\bar{N}_{\text{eff}}$, and with γ in place of γ_h, as we are considering radiative damping. The factor $1/(1 + X^2)$ in (15.83) is an important nonlinear extension that takes us beyond the weak-excitation limit. It accounts for saturation of the atomic inversion. Because of it the vacuum Rabi peaks move together as the intracavity field amplitude is increased; saturation of the inversion effectively reduces the dipole coupling strength. Eventually, in the limit of strong excitation, eigenvalues λ_1 and λ_2 are transformed into the independent damping rates κ and $\gamma/2$. This "closing up" of the vacuum Rabi doublet is seen in experiments and was reported on in some detail by Gripp and coworkers [15.8].

Explicit forms for $\bar{\lambda}_3$, $\bar{\lambda}_4$, and $\bar{\lambda}_5$ do not teach us very much. We can learn something, though, from the cubic polynomial they solve. In fact, it is helpful to write this cubic in three different forms:

$$\left[(\bar{\lambda} + \xi)(\bar{\lambda} + 1) + \xi 2C \frac{1}{1 + X^2}\right](\bar{\lambda} + 2) = -2X^2\left(\bar{\lambda} + \xi - \xi 2C \frac{1}{1 + X^2}\right), \tag{15.84a}$$

or equivalently,

$$(\bar{\lambda} + \xi)[(\bar{\lambda} + 1)(\bar{\lambda} + 2) + 2X^2] = -\xi 2C \frac{1}{1 + X^2}(\bar{\lambda} + 2 - 2X^2), \tag{15.84b}$$

or

$$\bar{\lambda}\left[\bar{\lambda}^2 + (3 + \xi)\lambda + 3\xi + 2(1 + X^2) + \xi 2C \frac{1}{1 + X^2}\right] = -2\xi(1 + X^2)\frac{dY}{dX}, \tag{15.84c}$$

where dY/dX in the last equation is given by (15.73). Something can be learned from each of these equations.

The cubic polynomial in version (15.84a) provides another view of vacuum Rabi splitting. If we set the right-hand side to zero, we recover two eigenvalues, $\bar{\lambda}_3$ and $\bar{\lambda}_4$, degenerate with $\bar{\lambda}_1$ and $\bar{\lambda}_2$, plus the eigenvalue $\bar{\lambda}_5 = 2$. The degeneracy reflects the fact that the field and polarization amplitudes are complex, so vacuum Rabi oscillation occurs in both their real and imaginary parts. Eigenvalue $\lambda_5 = \gamma$ describes the decay of fluctuations in the atomic inversion. Generally, for nonvanishing X, coupled field and polarization variables $\frac{1}{2}(\tilde{z} + \tilde{z}_*)$ and $\frac{1}{2}(\tilde{\nu} + \tilde{\nu}_*)$ also couple to the inversion variable μ; thus, there is an interplay of three eigenvalues at moderate excitation levels.

The ultimate development of this three-way coupling is revealed by (15.84b), where in the strong-excitation limit the right-hand side of the equation may be set to zero. We see then that two of the eigenvalues satisfying this equation describe the evolution of the coupled polarization and inversion in free-space resonance fluorescence (ordinary Rabi oscillation). The third reverts to the decay rate κ for the quadrature $\frac{1}{2}(\tilde{z} + \tilde{z}_*)$ of the cavity field amplitude.

The third version of the cubic polynomial, (15.84c), reveals local (linearized) dynamical features of the optical bistability phenomenon. From it we see that one of the eigenvalues vanishes at each turning point of the input–output curves (Fig. 14.4), whenever the slope dY/dX is zero. Furthermore, since the right-hand side of (15.84c) is the product $\bar{\lambda}_3\bar{\lambda}_4\bar{\lambda}_5$, we may conclude that the eigenvalue that vanishes at the turning point is positive when $dY/dX < 0$—i.e., the steady state is unstable in regions of negative slope. We have in (15.84c) an explicit expression of the stability issues discussed more generally in Sect. (14.2.3).

Exercise 15.5. Find the roots of the cubic characteristic polynomial numerically and determine how the eigenvalues λ_i, $i = 3, 4, 5$, vary as a function of the intracavity field amplitude X. Observe how the properties identified in (15.84a), (15.84b), and (15.84c) are carried continuously from the region of weak excitation to the region of strong excitation, passing through the range of unstable steady states.

Many things can be calculated from the linearized Fokker–Planck equation: optical spectra, spectra of squeezing, intensity correlation functions, to mention some things we have already met. In addition to the number of correlation functions and spectra that might be of interest, there are a wide range of operating conditions to consider. Results depend on the steady state about which the linearization is made, and as our observations about eigenvalues show, the behavior of the quantum fluctuations can be expected to change significantly with the steady state—the steady-state field amplitude X. The main distinction is between the *lower branch*,

$$X^2 < X_-^2 \equiv C - 1 - \sqrt{C(C-4)}, \tag{15.85a}$$

and the *upper branch*,

$$X^2 > X_+^2 \equiv C - 1 + \sqrt{C(C-4)}, \tag{15.85b}$$

of the input–output curve (Fig. 14.4). The lower branch is sometimes called the *cooperative branch*, because it is here that coherence induced by the collective interaction of the N atoms with the cavity mode beats out the spontaneous emission; the latter exerts its influence through individual atoms and tends to destroy both collectivity and coherence. Along the upper branch the spontaneous emission wins. The upper branch is therefore referred to as the *independent-atom branch.*

A sense of this physics is made available by the roots of (15.84b) and (15.84c) with the right-hand sides of the equations set to zero. The first equation (weak-excitation limit or lower branch) produces a pair of eigenvalues which depend on the collectivity parameter $2C$, while the second (strong-excitation limit or upper branch) produces the eigenvalues of single-atom resonance fluorescence. Considering the first in the *bad-cavity limit*, $\kappa \gg \gamma/2, \sqrt{N}g$, we find that one of the eigenvalues corresponds to the many-atom version of the cavity-enhanced emission rate; the enhancement factor $2C_1$ of (13.50) is replaced by $2C = N2C_1$, the decay rate of cavity-assisted superradiance [15.9,15.10,15.11]; thus, fluctuations along the lower branch of absorptive optical bistability exhibit features associated with both collective and cavity QED effects.

Another aspect of the variety of operating conditions comes from the time scales set by damping rates κ and $\gamma/2$, and the coupling strength $\sqrt{N}g$; we might make a distinction between this collective coupling strength and the strength of the single-atom coupling, but so long as we stay in the small-noise regime the single-atom coupling is necessarily weak.

Most of what is found in the literature deals with either the bad-cavity limit, just mentioned, or the *good-cavity limit*, $\kappa \ll \gamma/2, \sqrt{N}g$. Perhaps the more interesting results are those for the bad-cavity limit; specifically, along the lower, or cooperative branch, where the connections with perturbative cavity QED and superradiance show up [15.12,15.13,15.14]. We cannot review everything, however, so let us limit ourselves to a few calculations that are not covered so widely in the literature. We begin with a calculation of the covariance matrix, carried through generally, without any restriction on the operating conditions. We then look explicitly at the atom–atom correlations associated with collective effects (Sect. 15.2.4), and finally at some results relating to nonperturbative cavity QED (Sects. 15.2.6 and 15.2.7).

15.2.3 Covariance Matrix for Absorptive Bistability

We aim to calculate the steady-state covariance matrix for absorptive optical bistability. According to (11.69), we must solve the matrix equation

$$\bar{\boldsymbol{J}}_{\mathrm{ss}}\boldsymbol{C}_\infty + \boldsymbol{C}_\infty\bar{\boldsymbol{J}}_{\mathrm{ss}} = -\bar{\boldsymbol{D}}_{\mathrm{ss}}, \tag{15.86}$$

where $\bar{\boldsymbol{J}}_{\mathrm{ss}}$ and $\bar{\boldsymbol{D}}_{\mathrm{ss}}$ are the Jacobian and diffusion matrices defined in (15.77) and (15.78). The covariance matrix is symmetric, and for purely absorptive bistability all of its matrix elements are real. We therefore write it as

$$\boldsymbol{C}_\infty \equiv \left(\overline{\tilde{Z}\tilde{Z}^T}\right)_{\tilde{\bar{P}}_{\mathrm{ss}}} \equiv \begin{pmatrix} A & B & U & V & W \\ \cdots & A & V & U & W \\ \cdots & \cdots & P & Q & R \\ \cdots & \cdots & \cdots & P & R \\ \cdots & \cdots & \cdots & \cdots & S \end{pmatrix}. \tag{15.87}$$

Then, from the quantum–classical correspondence, and using the scaling relationships (15.50a)–(15.50e), we have

$$\begin{pmatrix} \langle\Delta\tilde{\bar{a}}\Delta\tilde{\bar{a}}\rangle_{\mathrm{ss}} & \langle\Delta\tilde{\bar{a}}^\dagger\Delta\tilde{\bar{a}}\rangle_{\mathrm{ss}} & \langle\Delta\tilde{\bar{a}}\Delta\tilde{\bar{J}}_-\rangle_{\mathrm{ss}} & \langle\Delta\tilde{\bar{a}}\Delta\tilde{\bar{J}}_+\rangle_{\mathrm{ss}} & \langle\Delta\tilde{\bar{a}}\Delta\bar{J}_z\rangle_{\mathrm{ss}} \\ \cdots & \langle\Delta\tilde{\bar{a}}^\dagger\Delta\tilde{\bar{a}}^\dagger\rangle_{\mathrm{ss}} & \langle\Delta\tilde{\bar{a}}^\dagger\Delta\tilde{\bar{J}}_-\rangle_{\mathrm{ss}} & \langle\Delta\tilde{\bar{a}}^\dagger\Delta\tilde{\bar{J}}_+\rangle_{\mathrm{ss}} & \langle\Delta\tilde{\bar{a}}^\dagger\Delta\bar{J}_z\rangle_{\mathrm{ss}} \\ \cdots & \cdots & \langle\Delta\tilde{\bar{J}}_-\Delta\tilde{\bar{J}}_-\rangle_{\mathrm{ss}} & \langle\Delta\tilde{\bar{J}}_+\Delta\tilde{\bar{J}}_-\rangle_{\mathrm{ss}} & \langle\Delta\bar{J}_z\Delta\tilde{\bar{J}}_-\rangle_{\mathrm{ss}} \\ \cdots & \cdots & \cdots & \langle\Delta\tilde{\bar{J}}_+\Delta\tilde{\bar{J}}_+\rangle_{\mathrm{ss}} & \langle\Delta\tilde{\bar{J}}_+\Delta\bar{J}_z\rangle_{\mathrm{ss}} \\ \cdots & \cdots & \cdots & \cdots & \langle\Delta\bar{J}_z\Delta\bar{J}_z\rangle_{\mathrm{ss}} \end{pmatrix}$$

$$= \frac{1}{N}\begin{pmatrix} A & B & U & V & W \\ \cdots & A & V & U & W \\ \cdots & \cdots & P & Q & R \\ \cdots & \cdots & \cdots & P & R \\ \cdots & \cdots & \cdots & \cdots & S \end{pmatrix}, \tag{15.88}$$

where the fluctuation operators are

$$\Delta\bar{a} \equiv \bar{a} - \langle\bar{a}\rangle_{\mathrm{ss}}, \tag{15.89a}$$

$$\Delta\bar{a}^\dagger \equiv \bar{a}^\dagger - \langle\bar{a}^\dagger\rangle_{\mathrm{ss}}, \tag{15.89b}$$

$$\Delta\bar{J}_- \equiv \bar{J}_- - \langle\bar{J}_-\rangle_{\mathrm{ss}}, \tag{15.89c}$$

$$\Delta\bar{J}_+ \equiv \bar{J}_+ - \langle\bar{J}_+\rangle_{\mathrm{ss}}, \tag{15.89d}$$

$$\Delta\bar{J}_z \equiv \bar{J}_z - \langle\bar{J}_z\rangle_{\mathrm{ss}}. \tag{15.89e}$$

One approach to the solution of (15.86) is to diagonalize the drift matrix, $\bar{\boldsymbol{J}}_{\mathrm{ss}}$, in the manner of (5.62), and obtain as in (5.81)

$$C_\infty = -\boldsymbol{S}^{-1}\left(\frac{\left(\boldsymbol{S}\bar{\boldsymbol{D}}_{\mathrm{ss}}\boldsymbol{S}^{-1}\right)_{ij}}{\lambda_i+\lambda_j}\right)(\boldsymbol{S}^{-1})^T. \tag{15.90}$$

Although possible in principle, this is not a practical approach here, since even the eigenvalues of $\bar{\boldsymbol{J}}_{\mathrm{ss}}$ are not given by closed expressions, let alone the eigenvectors. We are nevertheless able to find a closed form solution by directly solving the simultaneous equations. There are nine equations to solve: the first row of (15.86) gives the five equations

$$A - 2CU = 0, \tag{15.91a}$$

$$B - 2CV = 0, \tag{15.91b}$$

$$(1+\xi)U - \xi 2CP + \frac{1}{1+X^2}A - XW = 0, \tag{15.91c}$$

$$(1+\xi)V - \xi 2CQ + \frac{1}{1+X^2}B - XW = 0, \tag{15.91d}$$

$$(2+\xi)W - \xi 2CR - \frac{X}{1+X^2}(A+B) + X(U+V) = 0; \tag{15.91e}$$

the third row (columns three, four, and five) gives,

$$\frac{1}{1+X^2}U + P - XR = -\frac{X^2}{1+X^2}, \tag{15.91f}$$

$$\frac{1}{1+X^2}V + Q - XR = 0, \tag{15.91g}$$

$$\frac{1}{1+X^2}W + 3R - XS - \frac{X}{1+X^2}(U+V) + X(P+Q) = 0; \tag{15.91h}$$

and the fifth row (column five) gives

$$\frac{X}{1+X^2}W - XR - S = -\frac{2X^2}{1+X^2}. \tag{15.91i}$$

From this point the algebraic manipulations are left to the reader to explore as an exercise.

Exercise 15.6. Solve the set of nine simultaneous equations (15.91a)–(15.91i). The following four steps provide a possible avenue to the solution:

(i) From (15.91a)–(15.91d), show that

$$A = 2CV - 2CX^2\frac{\xi}{1+\xi}\left(1 - \frac{X}{Y}\right), \tag{15.92a}$$

$$B = 2CV, \tag{15.92b}$$

$$U = V - X^2\frac{\xi}{1+\xi}\left(1 - \frac{X}{Y}\right). \tag{15.92c}$$

(ii) From (15.91f) and (15.91g), show that

$$P = Q - \frac{X^2}{1+X^2}\frac{1+\xi X/Y}{1+\xi}. \tag{15.92d}$$

(iii) Now express W and R in terms of V and Q, with

$$W = \frac{1}{X}\left[\left(1+\xi+2C\frac{1}{1+X^2}\right)V - \xi 2CQ\right], \tag{15.92e}$$

$$R = \frac{1}{X}\left(\frac{1}{1+X^2}V + Q\right). \tag{15.92f}$$

(iv) Finally, solve (15.91e), (15.91h), and (15.91i) to obtain

$$S = \frac{1}{1+X^2}\left(\xi + 2C\frac{1}{1+X^2}\right)V - \left(1 + \xi 2C\frac{1}{1+X^2}\right)Q + \frac{2X^2}{1+X^2}, \tag{15.92g}$$

$$Q = \frac{\xi(3+\xi) + 2(1+X^2)dY/dX}{\xi(3+\xi)}\frac{1}{2C}V + \frac{X^4}{1+X^2}\frac{1-2X/Y}{(1+\xi)(3+\xi)}, \tag{15.92h}$$

and

$$V = 2C\left(\frac{X^2}{1+X^2}\right)^2 \frac{\xi}{1+\xi}\left(\frac{dY^2}{dX^2}\right)^{-1} \times \frac{(1+X^2)(2-Y/X)(3+\xi-\xi dY/dX) + (3+\xi)[2\xi + (3+\xi)Y/X]}{2(1+X^2)(3+\xi-\xi dY/dX) + \xi(3+\xi)(2+Y/X)}. \tag{15.92i}$$

Through back-substitution, starting from (15.92i), all elements of the steady-state covariance matrix are obtained.

This solution provides the input needed for the calculations of the next three sections. As a first indication of what it can tell us, Fig. 15.1 shows how the fluctuations in the intracavity field amplitude change as a function of the mean amplitude X; matrix elements A and B are plotted for three values of ξ ranging from the bad-cavity limit ($\xi \gg 1$) to the good-cavity limit ($\xi \ll 1$). The most obvious feature of the plots is the divergence at the turning points of the input–output curve, a repetition of behavior we have seen before—for the degenerate parametric oscillator in Fig. 10.2, and in our treatment of the laser threshold in Sect. 8.2. The magnitude of the amplitude variance $A = N\langle\Delta\tilde{a}\Delta\tilde{a}\rangle_{\rm ss}$ is also something to note. Along much of the lower branch [frames (a), (b), and (c)] it is larger than the fluctuation intensity $B = N\langle\Delta\tilde{a}^\dagger\Delta\tilde{a}\rangle_{\rm ss}$. This is not possible for any classical stochastic field since the fluctuation intensity is the sum of two quadrature amplitude variances, both *positive* numbers (see Sect. 15.2.5). We can deduce from this observation that the field fluctuations are squeezed. To explicitly demonstrate the presence of squeezing, we use the scaling (15.51a) and (15.51b), and the relation (15.56), to calculate the quadrature phase variances

$$\begin{aligned}(\Delta X)^2_{\rm ss} &\equiv \tfrac{1}{4}\big\langle\big[-i\Delta\tilde{a}e^{-i(\phi_T+\phi_0-\phi_c-\phi'_c)} + i\Delta\tilde{a}^\dagger e^{i(\phi_T+\phi_0-\phi_c-\phi'_c)}\big]^2\big\rangle \\ &= n_{\rm sat}\tfrac{1}{2}\big(\langle\Delta\tilde{a}^\dagger\Delta\tilde{a}\rangle_{\rm ss} + \langle\Delta\tilde{a}\Delta\tilde{a}\rangle_{\rm ss}\big) + \tfrac{1}{4} \\ &= (\xi 4C)^{-1}\tfrac{1}{2}(B+A) + \tfrac{1}{4},\end{aligned} \tag{15.93a}$$

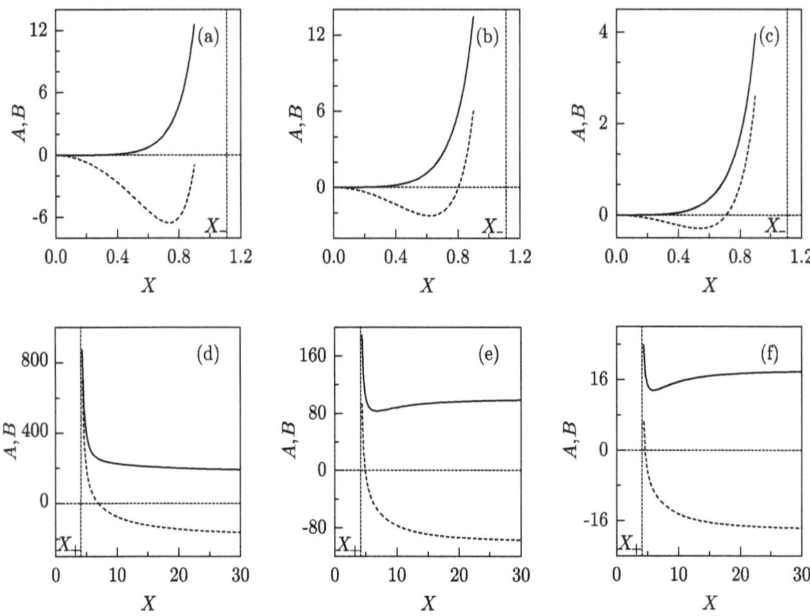

Fig. 15.1. Steady-state fluctuations in absorptive optical bistability as a function of the mean intracavity field amplitude. Quadrature amplitude fluctuations $A = N\langle\Delta\tilde{\bar{a}}\Delta\tilde{\bar{a}}\rangle_{\rm ss}$ (*dashed lines*) and intensity fluctuations $B = N\langle\Delta\tilde{\bar{a}}^\dagger\Delta\tilde{\bar{a}}\rangle_{\rm ss}$ (*solid lines*) are plotted as a function of X for $2C = 20$ and $\xi = 10.0$ [(**a**) and (**d**)], $\xi = 1.0$ [(**b**) and (**e**)], and $\xi = 0.1$ [(**c**) and (**f**)]

and

$$\begin{aligned}(\Delta Y)^2_{\rm ss} &\equiv \tfrac{1}{4}\big\langle\big[\Delta\tilde{a}e^{-i(\phi_T+\phi_0-\phi_c-\phi_c')} + \Delta\tilde{a}^\dagger e^{i(\phi_T+\phi_0-\phi_c-\phi_c')}\big]^2\big\rangle \\ &= n_{\rm sat}\tfrac{1}{2}\big(\langle\Delta\tilde{\bar{a}}^\dagger\Delta\tilde{\bar{a}}\rangle_{\rm ss} - \langle\Delta\tilde{\bar{a}}\Delta\tilde{\bar{a}}\rangle_{\rm ss}\big) + \tfrac{1}{4} \\ &= (\xi 4C)^{-1}\tfrac{1}{2}(B - A) + \tfrac{1}{4}. \end{aligned} \tag{15.93b}$$

When $-A > B$, which is true along much of the lower branch, the variance $(\Delta X)^2$ falls below the vacuum fluctuation level of $\frac{1}{4}$; thus, the normal-ordered quadrature amplitude variance is negative and the intracavity field quadrature in phase with the driving field is squeezed.

Squeezing for a system of many atoms in a cavity was observed in experiments by Raizen and coworkers [15.15], who used balanced homodyne detection to measure a narrow frequency component within the squeezing spectrum [the variance given in (15.93a) is the integral over the squeezing spectrum]. The normal-ordered variance—$(\Delta X)^2 - \frac{1}{4}$—can also be measured through conditional homodyne detection [15.16]. Foster and coworkers [15.17] have demonstrated squeezing for atoms in a cavity using this method.

15.2.4 Atom–Atom Correlations

The elements of the covariance matrix (15.92) provide the starting point for numerous calculations—of optical spectra, spectra of squeezing, intensity correlation functions, to name just some of the possibilities. Here, and in the following two sections, we make a brief survey of what can be done, restricting our attention to examples where analytical calculations can be carried through. To begin with, in the present section, we explore the question of correlations between atoms.

In the paragraph below (15.85b), we noted that the lower and upper branches of the steady-state input–output curve are sometimes referred to, respectively, as the *cooperative branch* (or the *collective branch*) and the *independent atom branch.* Invoking the word "cooperative" again, we denoted C the *cooperativity parameter* (Eqs. 14.50 and 15.56). Very little in the way of motivation was given for the "cooperative" terminology, though clearly it has something to do with correlations between atoms. The reasoning behind the choice of names is made clearer by calculating the atom–atom correlation functions explicitly. Atom–atom correlations come in two types. Consider, for example, the collective operator correlation function

$$\begin{aligned}
\langle \Delta\tilde{\tilde{J}}_{+}\Delta\tilde{\tilde{J}}_{-}\rangle &= \langle \tilde{\tilde{J}}_{+}\tilde{\tilde{J}}_{-}\rangle - \langle \tilde{\tilde{J}}_{+}\rangle\langle \tilde{\tilde{J}}_{-}\rangle \\
&= N^{-2}\frac{2\gamma_h}{\gamma}\Bigg[\sum_{j=1}^{N}\big(\langle\tilde{\sigma}_{j+}\tilde{\sigma}_{j-}\rangle - \langle\tilde{\sigma}_{j+}\rangle\langle\tilde{\sigma}_{j-}\rangle\big) \\
&\qquad + \sum_{\substack{j,k=1\\ j\neq k}}^{N}\big(\langle\tilde{\sigma}_{j+}\tilde{\sigma}_{-}a_{k-}\rangle - \langle\tilde{\sigma}_{j+}\rangle\langle\tilde{\sigma}_{-}a_{k-}\rangle\big)\Bigg] \\
&= N^{-1}\frac{2\gamma_h}{\gamma}\big[\langle\Delta\tilde{\sigma}_{+}\Delta\tilde{\sigma}_{-}\rangle_{\rm like} + (N-1)\langle\Delta\tilde{\sigma}_{+}\Delta\tilde{\sigma}_{-}\rangle_{\rm unlike}\big].
\end{aligned} \tag{15.94}$$

Now, an explicit expansion of the summation over atoms gives N *like-atom correlations*,

$$\langle\Delta\tilde{\sigma}_{+}\Delta\tilde{\sigma}_{-}\rangle_{\rm like} \equiv \langle\tilde{\sigma}_{j+}\tilde{\sigma}_{j-}\rangle - \langle\tilde{\sigma}_{j+}\rangle\langle\tilde{\sigma}_{j-}\rangle, \qquad \text{any } j, \tag{15.95}$$

and $N(N-1)$ *unlike-atom correlations*,

$$\langle\Delta\tilde{\sigma}_{+}\Delta\tilde{\sigma}_{-}\rangle_{\rm unlike} \equiv \langle\tilde{\sigma}_{j+}\tilde{\sigma}_{-}a_{k-}\rangle - \langle\tilde{\sigma}_{j+}\rangle\langle\tilde{\sigma}_{-}a_{k-}\rangle, \qquad \text{any } j\neq k. \tag{15.96}$$

There is only one like-atom correlation, the same for each atom, because we take the dipole coupling constants for all atoms to be equal; for the same reason, there is only one unlike-atom correlation, which holds for all pairs of different atoms. Like- and unlike-atom correlations must be distinguished, though, since they take on different values as we will see.

Note 15.4. The equality of all like-atom and all unlike-atom correlations suggests that we are dealing with identical atoms. Indeed, the atoms are identical in a certain sense, since they couple to the cavity mode with the same strength and decay spontaneously at the same rate; it follows that they are interchangeable in any operator average. It would be a mistake to think of the atoms as indistinguishable particles in the conventional sense, however. They are, in fact, distinguishable, due to their independent spontaneous emission output channels. Thus, so long as the atoms remain far apart compared with a resonant wavelength, the fluorescence from each atom may, at least in principle, be imaged separately. One can in principle assign the emission of a detected side-scattered photon to a particular atom and know that it was this atom, and no other, that just returned to its ground state (also see Note 6.3).

The like-atom correlation (15.95) is readily evaluated from the single-operator expectation values. Using (2.25a), and (15.89c) and (15.89d), the result is

$$\langle \Delta\tilde{\sigma}_+ \Delta\tilde{\sigma}_- \rangle_{\text{like}} = \tfrac{1}{2}(\langle \sigma_{jz} \rangle + 1) - \langle \tilde{\sigma}_{j+} \rangle \langle \tilde{\sigma}_{j-} \rangle, \qquad \text{any } j, \tag{15.97}$$

and then the unlike-atom correlation (15.96) is obtained by subtracting this from the collective operator correlation function:

$$\langle \Delta\tilde{\sigma}_+ \Delta\tilde{\sigma}_- \rangle_{\text{unlike}} = (N-1)^{-1} \left[\frac{\gamma}{2\gamma_h} N \langle \Delta\tilde{\tilde{J}}_+ \Delta\tilde{\tilde{J}}_- \rangle - \langle \Delta\tilde{\sigma}_+ \Delta\tilde{\sigma}_- \rangle_{\text{like}} \right]. \tag{15.98}$$

As one would expect, the collective atom correlation functions tell us about unlike-atom correlations.

Equations 15.97 and 15.98 hold generally for any collection of atoms interacting with a cavity mode, accepting the restriction to equal coupling strengths. In the specific case of absorptive optical bistability ($\varDelta = \varPhi = 0$) in the steady state, and for radiative damping, from (15.97), using (15.70a)–(15.70c), we obtain

$$\begin{aligned} \langle \Delta\tilde{\sigma}_+ \Delta\tilde{\sigma}_- \rangle_{\text{like}}^{\text{ss}} &= \tfrac{1}{2}(\langle \sigma_{jz} \rangle_{\text{ss}} + 1) - \langle \tilde{\sigma}_{j+} \rangle_{\text{ss}} \langle \tilde{\sigma}_{j-} \rangle_{\text{ss}} \\ &= \frac{1}{2}\frac{X^2}{1+X^2} - \frac{1}{2}\frac{X^2}{(1+X^2)^2} \\ &= \frac{1}{2}\left(\frac{X^2}{1+X^2}\right)^2, \end{aligned} \tag{15.99}$$

while from (15.98), using the expression (15.92h), and noting $Q = N\langle \Delta\tilde{\tilde{J}}_+ \Delta\tilde{\tilde{J}}_- \rangle$, we obtain

$$\langle \varDelta\tilde{\sigma}_+ \varDelta\tilde{\sigma}_- \rangle_{\text{unlike}}^{\text{ss}} = N^{-1}\Bigg[\frac{\xi(3+\xi)+2(1+X^2)dY/dX}{\xi(3+\xi)}\frac{1}{4C}V + \frac{1}{2}\left(\frac{X^2}{1+X^2}\right)^2\frac{(1+X^2)(1-2X/Y)-(1+\xi)(3+\xi)}{(1+\xi)(3+\xi)}\Bigg], \tag{15.100}$$

where we are satisfied that $N-1$ can be replaced by N on the right-hand side of (15.98) since terms of order N^{-1} are already neglected when performing the system size expansion; in fact, the additional term of order N^{-1} should be dropped to retain a consistent expansion.

It is instructive now to evaluate the expression for the unlike-atom correlation in both good- and bad-cavity limits. To this end, we substitute for V (Eq. 15.92i) in (15.100) and first take the limit $\xi \to 0$. The result is

$$\begin{aligned}\langle \varDelta\tilde{\sigma}_+ \varDelta\tilde{\sigma}_- \rangle_{\text{unlike}}^{\text{ss}} \underset{\xi\to 0}{\longrightarrow} & N^{-1}\frac{1}{2}\left(\frac{X^2}{1+X^2}\right)^2 \\ & \times\Bigg[\frac{(1+X^2)X/Y}{3}\frac{(1+X^2)(2-Y/X)+3Y/X}{1+X^2} \\ & +\frac{(1+X^2)(1-2X/Y)-3}{3}\Bigg] \\ \underset{\xi\to 0}{\longrightarrow} & \; 0. \end{aligned} \tag{15.101}$$

Thus, the unlike-atom correlation vanishes in the good cavity limit—at least this is so to dominant order in N^{-1}. In contrast, the bad-cavity limit, $\xi \to \infty$, yields a nonvanishing correlation:

$$\langle \varDelta\tilde{\sigma}_+ \varDelta\tilde{\sigma}_- \rangle_{\text{unlike}}^{\text{ss}} \underset{\xi\to\infty}{\longrightarrow} N^{-1}\frac{1}{2}\left(\frac{X^2}{1+X^2}\right)^2\left(\frac{1}{dY^2/dX^2}-1\right). \tag{15.102}$$

Note now that there are $N(N-1)$ unlike-atom correlations contributing to the collective operator correlation function, which is to be compared with just N like-atom correlations. It follows that although like- and unlike-atom correlations are individually of order N^0 and N^{-1}, respectively, they contribute overall at the same order to the collective operator average.

These results teach two pieces of physics. First, that there is a decrease in unlike-atom correlation as the good-cavity limit is approached. This may be seen to result from a filtering of the atomic fluctuations by the cavity mode. The fluctuation bandwidth of an individual atom is determined by its spontaneous emission linewidth γ. To set up an unlike-atom correlation, there must be a communication from one atom to another, which takes place via the cavity mode. The latter provides a communication channel of finite bandwidth 2κ. Thus, communication is increasingly restricted as $\xi = 2\kappa/\gamma \to 0$, and the correlation between unlike atoms is consequently reduced. The same filtering effect in the good-cavity limit protects the bimodality of the steady state from destruction by atomic fluctuations, even when the atom number is reduced to $N = 1$. Indeed, Savage and Carmichael [15.18] have demonstrated the existence of single-atom optical bistability by numerically solving master equation (13.57).

Note 15.5. At first sight it might appear that $N = 1$ must surely invalidate any treatment of fluctuations based on the system size expansion. Recall, however, that we may choose to use either N or $n_{\rm sat}$ as the system size parameter; we chose $n_{\rm sat}$ in Chap. 8, while in the present chapter we have chosen N. The two alternatives are related through (15.56). They remain interchangeable so long as N^{-1} (alternatively $n_{\rm sat}^{-1}$) is genuinely the smallest parameter in the system. This is not so when the good-cavity limit is taken, since then ξ is a small parameter as well. Clearly, in the good-cavity limit, $n_{\rm sat} = N/\xi 4C$, and not N, is the system size parameter; as $\xi \to 0$, $n_{\rm sat}$ goes to infinity for fixed $2C$ and N, even if $N = 1$.

The second lesson is taken from (15.102), where we see that while the unlike-atom correlation is generally nonzero in the bad-cavity limit, it approaches zero along the upper branch of the input–output curve—i.e., for $X^2 \gtrsim 2C \gg 1$ (Eq. 15.85b) and $dY^2/dX^2 \to 1$ (Eq. 15.73). By comparison, dY^2/dX^2 is significantly larger than unity along the lower branch, where like- and unlike-atom correlations contribute to the collective operator correlation function at the same level. It is here that we find justification for the common designations of the upper branch as the *independent atom branch* and the lower branch as the *cooperative*, or *collective branch*. The parameter $2C$ determines just how far the slope dY^2/dX^2 deviates from unity. As this slope determines the magnitude of the unlike-atom correlation, the parameter C is termed the *cooperativity parameter*.

Exercise 15.7. Show that one obtains similar behavior for the other unlike-atom correlations in the good- and bad-cavity limits. There are three correlations to consider: $\langle \Delta\sigma_- \Delta\tilde{\sigma}_- \rangle_{\rm unlike}^{\rm ss}$, $\langle \Delta\sigma_z \Delta\tilde{\sigma}_- \rangle_{\rm unlike}^{\rm ss}$, and $\langle \Delta\sigma_z \Delta\sigma_z \rangle_{\rm unlike}^{\rm ss}$; two others, $\langle \Delta\sigma_+ \Delta\tilde{\sigma}_+ \rangle_{\rm unlike}^{\rm ss}$ and $\langle \Delta\tilde{\sigma}_+ \Delta\sigma_z \rangle_{\rm unlike}^{\rm ss}$, are conjugates of these.

15.2.5 A Comment on Measures of Squeezing

We briefly mentioned squeezing at the end of Sect. 15.2.3. Squeezing occurs for many atoms in a cavity along the lower branch of the input–output curve. If the aim is to maximize the degree of squeezing, it is necessary to go beyond the expressions (15.93) for intracavity quadrature phase variances; spectra of squeezing must be calculated from (9.144a) and (9.144b). We will not concern ourselves with this exercise here. Surveys showing how the magnitude of the squeezing changes under different operating conditions can be found in the literature; for example, Raizen and coworkers [15.15] carried out a search for the optimum squeezing in preparation for their experiments on squeezed-state generation by the normal modes of a coupled atom-cavity system (squeezing at the vacuum-Rabi resonances).

When speaking here of the degree of squeezing, we refer, of course, to a standard squeezing measurement based on balanced homodyne detection. In that scheme the degree of squeezing is determined by the relative drop in the homodyne current noise level below the shot noise level (Sect. 9.3.3). More simply, considering the integrated squeezing spectrum, the measure is provided by the magnitude of the term $(\xi 4C)^{-1}\frac{1}{2}(B+A)$ in (15.93a) relative to the vacuum fluctuation term—the additional $+\frac{1}{4}$. By this measure, the squeezing can never be large for very weak fields. Significant squeezing requires photon numbers per mode that approach the $\frac{1}{2}$-photon characterizing the zero-point energy—an output photon flux that approaches $\frac{1}{2}$-photon per inverse bandwidth.

The nonclassicality of squeezed light need not be characterized in this standard way, though. An alternate measure in the case of (15.93a) is simply the ratio $|A|/B$. Consider the values of A and B when the excitation is weak. For $X^2 \ll 1$, (15.92a) and (15.92b) yield

$$A \xrightarrow[X^2 \ll 1]{} -X^2 \frac{\xi}{1+\xi} 4C^2, \tag{15.103a}$$

and

$$B \xrightarrow[X^2 \ll 1]{} X^4 \frac{\xi}{1+\xi} \frac{4C^2(2+\xi+2C)}{(1+2C)^2(1+\xi)}. \tag{15.103b}$$

As we noted already below (15.93b), these results imply squeezing because $B+A$ is negative—i.e., because $-A > B$. Certainly, the degree of squeezing is negligible by the standard measure, since $-A \sim X^2 \ll 1$. On the other hand, (15.103a) and (15.103b) reveal something entirely new. Suppose we adopt the relative measure $|A|/B$ as suggested. In these terms, $-A$ exceeds B by the greatest amount precisely in the weak-excitation limit; indeed, $|A|/B$ diverges as $X^2 \to 0$.

To see what this means physically, consider a classical field of complex amplitude $\tilde{z} = \tilde{z}_1 + i\tilde{z}_2$. Assuming $\langle \Delta\tilde{a}\Delta\tilde{a}\rangle_{\mathrm{ss}}$ to be real, as it is in the present case, the ratio

$$\frac{|A|}{B} \equiv \frac{|\langle \Delta\tilde{a}\Delta\tilde{a}\rangle_{\mathrm{ss}}|}{\langle \Delta\tilde{a}^\dagger\Delta\tilde{a}\rangle_{\mathrm{ss}}} \tag{15.104}$$

is given for the classical field by

$$\frac{|A|}{B} = \frac{\left|\left(\overline{\tilde{z}_1^2}\right)_{P_{\mathrm{ss}}} - \left(\overline{\tilde{z}_2^2}\right)_{P_{\mathrm{ss}}}\right|}{\left(\overline{\tilde{z}_1^2}\right)_{P_{\mathrm{ss}}} + \left(\overline{\tilde{z}_2^2}\right)_{P_{\mathrm{ss}}}}, \tag{15.105}$$

where P_{ss} is the steady-state Glauber–Sudarshan P distribution (assumed to be a nonsingular and positive semidefinite probability density). The ratio, therefore, has an upper bound of unity, since the quadrature amplitude variances are nonnegative.

How then can the expressions (15.103a) and (15.103b) produce the result $|A|/B \sim X^{-2} \gg 1$? This can happen because, for a general quantum field, the phase-space averages $\left(\overline{\tilde{z}_1^2}\right)_{P_{ss}}$ and $\left(\overline{\tilde{z}_2^2}\right)_{P_{ss}}$ evaluate *normal-ordered* variances of quadrature phase amplitudes, and these are allowed to be negative. The variances are, from (15.93a) and (15.93b),

$$\begin{aligned}\langle :(\Delta\hat{X})^2: \rangle_{ss} &= (\Delta X)^2_{ss} - \tfrac{1}{4} \\ &= (\xi 4C)^{-1}\tfrac{1}{2}(B+A), \end{aligned} \tag{15.106a}$$

$$\begin{aligned}\langle :(\Delta\hat{Y})^2: \rangle_{ss} &= (\Delta Y)^2_{ss} - \tfrac{1}{4} \\ &= (\xi 4C)^{-1}\tfrac{1}{2}(B-A), \end{aligned} \tag{15.106b}$$

with a sum proportional to B, as indicated by the denominator in (15.105). Equations 15.103 and 15.106 show that in the weak-excitation limit, both normal-ordered variances are of order X^2, while they sum to something of order X^4. This classically disallowed result is possible because one of the normal-ordered variances is negative. The ratio $|A|/B$ dramatically captures this particular nonclassical result in the divergence as $X^2 \to 0$.

The measure $|A|/B$ contrasts the standard squeezing measure by comparing the quadrature phase amplitude variance with the light intensity rather than the vacuum fluctuation level. It is a particularly sensitive detector of squeezing when the light intensity is weak and may be implemented experimentally through conditional homodyne detection [15.16]. Measurements have been performed for a system of many atoms in a cavity [15.17], though violation of the classical bound on $|A|/B$ has not been observed to date.

Note 15.6. Conditional homodyne detection separately measures the ratios $\langle :(\Delta\hat{X})^2: \rangle_{ss}/\langle \Delta\tilde{a}^\dagger \Delta\tilde{a}\rangle_{ss}$ and $\langle :(\Delta\hat{Y})^2: \rangle_{ss}/\langle \Delta\tilde{a}^\dagger \Delta\tilde{a}\rangle_{ss}$. It therefore explicitly reveals the minus sign in the normal-ordered variance of the squeezed quadrature phase amplitude. It is also worth noting that the conditional measurement is independent of detection efficiency, unlike the standard squeezing measurement, which is degraded by lower efficiency (Eq. 9.154 and Sect. 9.3.4).

15.2.6 Spectrum of the Transmitted Light in the Weak-Excitation Limit

The incoherent spectrum of the transmitted light provides yet another measure of squeezing in the weak-excitation limit. Recall what we observed for the degenerate parametric oscillator by writing the optical spectrum as a sum of squeezing spectra (Eqs. 10.65 and 10.66). When one of the squeezing spectra appears with negative weight the sum is actually a subtraction. Then, if dominant terms cancel, as they do to produce (9.94b) from (15.106a) and (15.106b), Lorentzian spectra of squeezing can yield an optical spectrum that is a Lorentzian squared. Considering single-atom cavity QED, Rice and

Carmichael showed that this mechanism produces a vacuum Rabi doublet of squared Lorentzians [15.19]. Carmichael and coworkers found a similar result for many-atom cavity QED [15.20]. We reproduce their result here. The single-atom calculation is discussed in Sect. 16.3.4.

The optical spectrum is given by the Fourier transform of the first-order correlation function of the transmitted light. For the output field (13.19), we write

$$T_{\rightarrow}(\omega) = \frac{\gamma_{a2}}{2\pi}\int_{-\infty}^{\infty} d\tau e^{i\omega\tau}\langle a^\dagger(0)a(\tau)\rangle_{\rm ss}, \tag{15.107}$$

where $\langle a^\dagger(0)a(\tau)\rangle_{\rm ss} \equiv \lim_{t\to\infty}\langle a^\dagger(t)a(t+\tau)\rangle$ and the normalization is in units of photon flux. Dividing the spectrum into its coherent and incoherent parts, we have

$$T_{\rightarrow}(\omega) = T_{\rightarrow}^{\rm coh}(\omega) + T_{\rightarrow}^{\rm inc}(\omega), \tag{15.108}$$

where the coherent part is defined by the mean output field,

$$\begin{aligned} T_{\rightarrow}^{\rm coh}(\omega) &= \frac{\gamma_{a2}}{2\pi}\int_{-\infty}^{\infty} d\tau e^{i(\omega-\omega_0)\tau}\langle\tilde{a}^\dagger\rangle_{\rm ss}\langle\tilde{a}\rangle_{\rm ss} \\ &= \gamma_{a2}n_{\rm sat}X^2\delta(\omega-\omega_0), \end{aligned} \tag{15.109}$$

and the incoherent part accounts for the field fluctuations,

$$T_{\rightarrow}^{\rm inc}(\omega) = \frac{\gamma_{a2}}{2\pi}\int_{-\infty}^{\infty} d\tau e^{i(\omega-\omega_0)\tau}\langle\Delta\tilde{a}^\dagger(0)\Delta\tilde{a}(\tau)\rangle_{\rm ss}. \tag{15.110}$$

The integral over frequency gives the coherent transmitted photon flux,

$$\begin{aligned} F_{\rightarrow}^{\rm coh} &= \gamma_{a2}\langle\tilde{a}^\dagger\tilde{a}\rangle_{\rm ss} \\ &= \gamma_{a2}n_{\rm sat}X^2, \end{aligned} \tag{15.111}$$

and the incoherent photon flux

$$\begin{aligned} F_{\rightarrow}^{\rm inc} &= \gamma_{a2}\langle\Delta\tilde{a}^\dagger\Delta\tilde{a}\rangle_{\rm ss} \\ &= \gamma_{a2}n_{\rm sat}\langle\Delta\tilde{\tilde{a}}^\dagger\Delta\tilde{\tilde{a}}\rangle_{\rm ss} \\ &= \gamma_{a2}n_{\rm sat}N^{-1}4C^2\left(\frac{X^2}{1+X^2}\right)^2\frac{\xi}{1+\xi}\left(\frac{dY^2}{dX^2}\right)^{-1} \\ &\quad\times\frac{(1+X^2)(2-Y/X)(3+\xi-\xi dY/dX)+(3+\xi)[2\xi+(3+\xi)Y/X]}{2(1+X^2)(3+\xi-\xi dY/dX)+\xi(3+\xi)(2+Y/X)}. \end{aligned} \tag{15.112}$$

Since a calculation for arbitrary X is too difficult to carry through analytically, we confine our attention to the weak-excitation limit, where, from (15.112), we have

$$F_{\rightarrow}^{\mathrm{inc}} \underset{X^2 \ll 1}{\longrightarrow} \gamma_{a2} n_{\mathrm{sat}} N^{-1} X^4 \frac{\xi}{1+\xi} \frac{4C^2(2+\xi+2C)}{(1+2C)^2(1+\xi)}. \tag{15.113}$$

Aside from an overall scaling factor, $F_{\rightarrow}^{\mathrm{inc}}$ is equal to the quantity B from the previous section; it shows the X^4 dependence of (15.103b), which arises from the cancelation of the dominant X^2 dependence of each of the quadrature phase variances.

To calculate the incoherent spectrum (15.110), we must determine the first-order correlation function of the fluctuations. Thus, we set out to solve the equation of motion (Eq. 11.68)

$$\frac{d\boldsymbol{C}_{\mathrm{ss}}}{d\tau} = \begin{cases} \boldsymbol{C}_{\mathrm{ss}} \bar{\boldsymbol{J}}_{\mathrm{ss}}^T & \tau \geq 0 \\ \bar{\boldsymbol{J}}_{\mathrm{ss}} \boldsymbol{C}_{\mathrm{ss}} & \tau \leq 0 \end{cases}, \tag{15.114}$$

where $\bar{\boldsymbol{J}}_{\mathrm{ss}}$ is the Jacobian matrix (11.62) and

$$\boldsymbol{C}_{\mathrm{ss}}(\tau) \equiv \lim_{t\to\infty} \overline{\left(\tilde{Z}(t) \tilde{Z}(t+\tau) \right)}_{\tilde{P}}. \tag{15.115}$$

The spectrum is obtained from the element $C_{\mathrm{ss}}^{\tilde{z}_*\tilde{z}}(\tau)$ of the correlation matrix, where, from (15.110), (15.115), and the quantum-classical correspondence for two-time operator averages in normal order (Sect. 4.3.3),

$$\begin{aligned} T_{\rightarrow}^{\mathrm{inc}}(\omega) &= \frac{\gamma_{a2}}{2\pi} n_{\mathrm{sat}} \frac{1}{N} \int_{-\infty}^{\infty} d\tau e^{i(\omega-\omega_0)\tau} C_{\mathrm{ss}}^{\tilde{z}_*\tilde{z}}(\tau) \\ &= \frac{\gamma_{a2}}{\pi} n_{\mathrm{sat}} \frac{1}{N} \mathrm{Re}\left[\int_{0}^{\infty} d\tau e^{i(\omega-\omega_0)\tau} C_{\mathrm{ss}}^{\tilde{z}_*\tilde{z}}(\tau) \right] \\ &= \frac{\gamma_{a2}}{\pi} n_{\mathrm{sat}} \frac{1}{N} \mathrm{Re}\{ C_{\mathrm{ss}}^{\tilde{z}_*\tilde{z}}[-i(\omega-\omega_0)] \}, \end{aligned} \tag{15.116}$$

where $C_{\mathrm{ss}}^{\tilde{z}_*\tilde{z}}(s)$ and $C_{\mathrm{ss}}^{\tilde{z}_*\tilde{z}}(\tau)$ form a Laplace transform pair.

The correlation function $C_{\mathrm{ss}}^{\tilde{z}_*\tilde{z}}(\tau)$ and four others make up the second row of the autocorrelation matrix. Forming the five correlation functions as a column vector, we have

$$\begin{pmatrix} C_{\mathrm{ss}}^{\tilde{z}_*\tilde{z}}(\tau) \\ C_{\mathrm{ss}}^{\tilde{z}_*\tilde{z}_*}(\tau) \\ C_{\mathrm{ss}}^{\tilde{z}_*\tilde{\nu}}(\tau) \\ C_{\mathrm{ss}}^{\tilde{z}_*\tilde{\nu}_*}(\tau) \\ C_{\mathrm{ss}}^{\tilde{z}_*\mu}(\tau) \end{pmatrix} = N \lim_{t\to\infty} \begin{pmatrix} \langle \Delta\tilde{\bar{a}}^\dagger(t) \Delta\tilde{\bar{a}}(t+\tau) \rangle \\ \langle \Delta\tilde{\bar{a}}^\dagger(t) \Delta\tilde{\bar{a}}^\dagger(t+\tau) \rangle \\ \langle \Delta\tilde{\bar{a}}^\dagger(t) \Delta\tilde{\bar{J}}_-(t+\tau) \rangle \\ \langle \Delta\tilde{\bar{a}}^\dagger(t) \Delta\tilde{\bar{J}}_+(t+\tau) \rangle \\ \langle \Delta\tilde{\bar{a}}^\dagger(t) \Delta\bar{J}_z(t+\tau) \rangle \end{pmatrix}, \tag{15.117}$$

with the equation of motion

$$\frac{d}{d\tau}\begin{pmatrix} C_{\rm ss}^{\tilde{z}_*\tilde{z}} \\ C_{\rm ss}^{\tilde{z}_*\tilde{z}_*} \\ C_{\rm ss}^{\tilde{z}_*\tilde{\nu}} \\ C_{\rm ss}^{\tilde{z}_*\tilde{\nu}_*} \\ C_{\rm ss}^{\tilde{z}_*\mu} \end{pmatrix} = \bar{\boldsymbol{J}}_{\rm ss}\begin{pmatrix} C_{\rm ss}^{\tilde{z}_*\tilde{z}} \\ C_{\rm ss}^{\tilde{z}_*\tilde{z}_*} \\ C_{\rm ss}^{\tilde{z}_*\tilde{\nu}} \\ C_{\rm ss}^{\tilde{z}_*\tilde{\nu}_*} \\ C_{\rm ss}^{\tilde{z}_*\mu} \end{pmatrix}, \tag{15.118}$$

where $\bar{\boldsymbol{J}}_{\rm ss}$ is given in explicit form in (15.77), and the initial conditions are, in the notation of (15.87),

$$\begin{pmatrix} C_{\rm ss}^{\tilde{z}_*\tilde{z}}(0) \\ C_{\rm ss}^{\tilde{z}_*\tilde{z}_*}(0) \\ C_{\rm ss}^{\tilde{z}_*\tilde{\nu}}(0) \\ C_{\rm ss}^{\tilde{z}_*\tilde{\nu}_*}(0) \\ C_{\rm ss}^{\tilde{z}_*\mu}(0) \end{pmatrix} = \begin{pmatrix} B \\ A \\ U \\ V \\ W \end{pmatrix}. \tag{15.119}$$

Explicitly, the initial conditions are given by (15.92a)–(15.92c) and (15.92e).

Note 15.7. Equation of motion (15.118) is simply a phase-space version of the equations of motion obtained by applying the quantum regression formula directly to the linearized Maxwell–Bloch equations (15.57).

Since we are interested in the weak-excitation limit, we retain only dominant terms in the intracavity field amplitude X. The initial conditions are then given by

$$C_{\rm ss}^{\tilde{z}_*\tilde{z}}(0) = X^4\frac{\xi}{1+\xi}\frac{4C^2(2+\xi+2C)}{(1+2C)^2(1+\xi)}, \tag{15.120a}$$

$$C_{\rm ss}^{\tilde{z}_*\tilde{z}_*}(0) = -X^2\frac{\xi}{1+\xi}\frac{4C^2}{1+2C}, \tag{15.120b}$$

$$C_{\rm ss}^{\tilde{z}_*\tilde{\nu}}(0) = X^4\frac{\xi}{1+\xi}\frac{2C(2+\xi+2C)}{(1+2C)^2(1+\xi)}, \tag{15.120c}$$

$$C_{\rm ss}^{\tilde{z}_*\tilde{\nu}_*}(0) = -X^2\frac{\xi}{1+\xi}\frac{2C}{1+2C}, \tag{15.120d}$$

$$C_{\rm ss}^{\tilde{z}_*\mu}(0) = X^3\frac{\xi}{1+\xi}\frac{2C}{1+2C}, \tag{15.120e}$$

and also expanding $\bar{\boldsymbol{J}}_{\mathrm{ss}}$ to lowest order, from (15.118) and (15.77), the equations of motion are

$$\frac{dC_{\mathrm{ss}}^{\tilde{z}_*\tilde{z}}}{d\bar{\tau}} = -\xi C_{\mathrm{ss}}^{\tilde{z}_*\tilde{z}} + \xi 2C C_{\mathrm{ss}}^{\tilde{z}_*\tilde{\nu}}, \tag{15.121a}$$

$$\frac{dC_{\mathrm{ss}}^{\tilde{z}_*\tilde{z}_*}}{d\bar{\tau}} = -\xi C_{\mathrm{ss}}^{\tilde{z}_*\tilde{z}_*} + \xi 2C C_{\mathrm{ss}}^{\tilde{z}_*\tilde{\nu}_*}, \tag{15.121b}$$

$$\frac{dC_{\mathrm{ss}}^{\tilde{z}_*\tilde{\nu}}}{d\bar{\tau}} = -C_{\mathrm{ss}}^{\tilde{z}_*\tilde{\nu}} - C_{\mathrm{ss}}^{\tilde{z}_*\tilde{z}} + XC_{\mathrm{ss}}^{\tilde{z}_*\mu}, \tag{15.121c}$$

$$\frac{dC_{\mathrm{ss}}^{\tilde{z}_*\tilde{\nu}_*}}{d\bar{\tau}} = -C_{\mathrm{ss}}^{\tilde{z}_*\tilde{\nu}_*} - C_{\mathrm{ss}}^{\tilde{z}_*\tilde{z}_*} + XC_{\mathrm{ss}}^{\tilde{z}_*\mu}, \tag{15.121d}$$

$$\frac{dC_{\mathrm{ss}}^{\tilde{z}_*\mu}}{d\bar{\tau}} = -2C_{\mathrm{ss}}^{\tilde{z}_*\mu} + X(C_{\mathrm{ss}}^{\tilde{z}_*\tilde{z}} + C_{\mathrm{ss}}^{\tilde{z}_*\tilde{z}_*} - C_{\mathrm{ss}}^{\tilde{z}_*\tilde{\nu}} - C_{\mathrm{ss}}^{\tilde{z}_*\tilde{\nu}_*}), \tag{15.121e}$$

with dimensionless time

$$\bar{\tau} \equiv \gamma\tau/2. \tag{15.122}$$

Note now that terms on the right-hand side of each equation need be kept only where they are of the same order, in X, as the correlation function on the left-hand side. Thus, reading the orders of the correlation functions from the initial conditions, the term $XC_{\mathrm{ss}}^{\tilde{z}_*\mu}$ may be dropped from (15.121d), and the terms $XC_{\mathrm{ss}}^{\tilde{z}_*\tilde{z}}$ and $XC_{\mathrm{ss}}^{\tilde{z}_*\tilde{\nu}}$ from (15.121e). With this simplification, the five equations of motion are conveniently arranged as a set of three matrix equations which are readily solved in sequence:

$$\frac{d}{d\bar{\tau}}\begin{pmatrix} C_{\mathrm{ss}}^{\tilde{z}_*\tilde{z}} \\ C_{\mathrm{ss}}^{\tilde{z}_*\tilde{\nu}} \end{pmatrix} = \begin{pmatrix} -\xi & \xi 2C \\ -1 & -1 \end{pmatrix}\begin{pmatrix} C_{\mathrm{ss}}^{\tilde{z}_*\tilde{z}} \\ C_{\mathrm{ss}}^{\tilde{z}_*\tilde{\nu}} \end{pmatrix} + XC_{\mathrm{ss}}^{\tilde{z}_*\mu}\begin{pmatrix} 0 \\ 1 \end{pmatrix}, \tag{15.123a}$$

$$\frac{d}{d\bar{\tau}}\begin{pmatrix} C_{\mathrm{ss}}^{\tilde{z}_*\tilde{z}_*} \\ C_{\mathrm{ss}}^{\tilde{z}_*\tilde{\nu}_*} \end{pmatrix} = \begin{pmatrix} -\xi & \xi 2C \\ -1 & -1 \end{pmatrix}\begin{pmatrix} C_{\mathrm{ss}}^{\tilde{z}_*\tilde{z}_*} \\ C_{\mathrm{ss}}^{\tilde{z}_*\tilde{\nu}_*} \end{pmatrix}, \tag{15.123b}$$

$$\frac{dC_{\mathrm{ss}}^{\tilde{z}_*\mu}}{d\bar{\tau}} = -2C_{\mathrm{ss}}^{\tilde{z}_*\mu} + X(C_{\mathrm{ss}}^{\tilde{z}_*\tilde{z}_*} - C_{\mathrm{ss}}^{\tilde{z}_*\tilde{\nu}_*}). \tag{15.123c}$$

To solve the equations, we first take Laplace transforms to obtain a corresponding set of algebraic equations. The transformed equations are

$$\begin{pmatrix} \xi+\bar{s} & -\xi 2C \\ 1 & 1+\bar{s} \end{pmatrix}\begin{pmatrix} \bar{C}_{\mathrm{ss}}^{\tilde{z}_*\tilde{z}} \\ \bar{C}_{\mathrm{ss}}^{\tilde{z}_*\tilde{\nu}} \end{pmatrix} = X^4\frac{\xi}{1+\xi}\frac{2C(2+\xi+2C)}{(1+2C)^2(1+\xi)}\begin{pmatrix} 2C \\ 1 \end{pmatrix} + X\bar{C}_{\mathrm{ss}}^{\tilde{z}_*\mu}\begin{pmatrix} 0 \\ 1 \end{pmatrix}, \tag{15.124a}$$

$$\begin{pmatrix} \xi+\bar{s} & -\xi 2C \\ 1 & 1+\bar{s} \end{pmatrix}\begin{pmatrix} \bar{C}_{\mathrm{ss}}^{\tilde{z}_*\tilde{z}_*} \\ \bar{C}_{\mathrm{ss}}^{\tilde{z}_*\tilde{\nu}_*} \end{pmatrix} = -X^2\frac{\xi}{1+\xi}\frac{2C}{1+2C}\begin{pmatrix} 2C \\ 1 \end{pmatrix}, \tag{15.124b}$$

$$(2+\bar{s})\bar{C}_{\mathrm{ss}}^{\tilde{z}_*\mu} = X^3\frac{\xi}{1+\xi}\frac{2C}{1+2C} + X(\bar{C}_{\mathrm{ss}}^{\tilde{z}_*\tilde{z}_*} - \bar{C}_{\mathrm{ss}}^{\tilde{z}_*\tilde{\nu}_*}), \tag{15.124c}$$

where scaled quantities have been introduced, with

$$\bar{s} \equiv 2s/\gamma, \qquad C_{\rm ss}^{ij}(s) = \frac{2}{\gamma}\bar{C}_{\rm ss}^{ij}(\bar{s}), \tag{15.125}$$

where $C_{\rm ss}^{ij}(s)$ is the Laplace transform of $C_{\rm ss}^{ij}(\tau)$. Now invert the 2×2 matrix appearing on the left-hand side of (15.124a) and (15.124b). The inverse is

$$\begin{pmatrix} \xi+\bar{s} & -\xi 2C \\ 1 & 1+\bar{s} \end{pmatrix}^{-1} = \frac{1}{(\xi+\bar{s})(1+\bar{s})+\xi 2C}\begin{pmatrix} 1+\bar{s} & \xi 2C \\ -1 & \xi+\bar{s} \end{pmatrix}, \tag{15.126}$$

from which the solution to (15.124b) is

$$\bar{C}_{\rm ss}^{\tilde{z}_*\tilde{z}_*}(\bar{s}) = -X^2\frac{\xi}{1+\xi}\frac{4C^2}{1+2C}\frac{\xi+1+\bar{s}}{(\xi+\bar{s})(1+\bar{s})+\xi 2C}, \tag{15.127}$$

and

$$\bar{C}_{\rm ss}^{\tilde{z}_*\tilde{\nu}_*}(\bar{s}) = -X^2\frac{\xi}{1+\xi}\frac{2C}{1+2C}\frac{\xi+\bar{s}-2C}{(\xi+\bar{s})(1+\bar{s})+\xi 2C}; \tag{15.128}$$

hence, from (15.124c),

$$\bar{C}_{\rm ss}^{\tilde{z}_*\mu}(\bar{s}) = X^3\frac{\xi}{1+\xi}\frac{2C}{1+2C}\frac{\xi+\bar{s}-2C}{(\xi+\bar{s})(1+\bar{s})+\xi 2C}. \tag{15.129}$$

Finally, substituting the latter expression into (15.124a) and using the matrix inverse once again, we obtain the Laplace transform of the correlation function we seek:

$$\begin{aligned}\bar{C}_{\rm ss}^{\tilde{z}_*\tilde{z}}(\bar{s}) = X^4\frac{\xi}{1+\xi}\frac{4C^2}{1+2C}\bigg\{&\frac{2+\xi+2C}{(1+2C)(1+\xi)}\frac{1+\xi+\bar{s}}{(\xi+\bar{s})(1+\bar{s})+\xi 2C} \\ &+\frac{\xi(\xi-2C+\bar{s})}{[(\xi+\bar{s})(1+\bar{s})+\xi 2C]^2}\bigg\}.\end{aligned} \tag{15.130}$$

We may now combine (15.113), (15.116), (15.125), and (15.130), to arrive at the *incoherent spectrum of the transmitted light for a system of many atoms in a cavity, on resonance ($\Delta=\Phi=0$), and in the weak-excitation limit (with $\bar{n}=0$):*

$$T_{\rightarrow}^{\rm inc}(\omega) = F_{\rightarrow}^{\rm inc}\bar{T}_{\rm inc}[2(\omega-\omega_0)/\gamma], \tag{15.131}$$

with

$$\begin{aligned}\bar{T}_{\rm inc}(y) = \frac{2}{\pi\gamma}{\rm Re}\bigg\{&\frac{1+\xi-iy}{(\xi-iy)(1-iy)+\xi 2C} \\ &+\frac{(1+2C)(1+\xi)}{2+\xi+2C}\frac{\xi(\xi-2C-iy)}{[(\xi-iy)(1-iy)+\xi 2C]^2}\bigg\}.\end{aligned} \tag{15.132}$$

The spectrum comprises two pieces: one, the first term inside the curly brackets, is of the usual Lorentzian form, while the other, the second term, takes the promised squared-Lorentzian form. By considering some limiting cases we can reveal the squared Lorentzians more clearly.

Bad-cavity limit ($\xi \gg 1, 2C$): Taking $\xi \to \infty$ in (15.132) yields

$$\begin{aligned}\bar{T}_{\mathrm{inc}}(y) &= \frac{2}{\pi\gamma}\mathrm{Re}\left[\frac{1}{1+2C-iy} + (1+2C)\left(\frac{1}{1+2C-iy}\right)^2\right] \\ &= \frac{2}{\pi\gamma}\frac{2(1+2C)^3}{[(1+2C)^2+y^2]^2},\end{aligned} \tag{15.133}$$

and the result from (15.131) is *the spectrum of collective cavity-enhanced resonance fluorescence in the weak-excitation limit*:

$$T_{\rightarrow}^{\mathrm{inc}}(\omega) = F_{\rightarrow}^{\mathrm{inc}}\frac{2}{\pi}\frac{[\gamma(1+2C)/2]^3}{\{[\gamma(1+2C)/2]^2+(\omega-\omega_0)^2\}^2}. \tag{15.134}$$

The form as a Lorentzian squared is the same as that identified by Rice and Carmichael [15.19] for resonance fluorescence in free space. The interaction with the cavity mode adds the linewidth enhancement factor $2C = N2C_1$; the factor $2C_1$ is the spontaneous emission enhancement factor (13.36), which applies to each atom individually, and, through atom–atom correlations, the overall enhancement is raised by the additional factor N. Thus, the linewidth is collectively enhanced; it corresponds to the familiar superradiant decay rate for a collection of atoms in a cavity [15.9, 15.10, 15.11]. Working against the enhancement, the width of the Lorentzian squared is less than the width of the Lorentzian itself. Since the square arises from squeezing, Rice and Carmichael [15.19] speak of *squeezing-induced linewidth narrowing.*

Good-cavity limit ($\xi \ll 1$): The limiting form of (15.132) in the good-cavity limit is only a little more difficult to obtain. We first note that the function $\bar{T}_{\mathrm{inc}}(y)$ is peaked around $y \sim \xi \ll 1$. Then we may write

$$\begin{aligned}\bar{T}_{\mathrm{inc}}(y) &= \frac{2}{\pi\gamma}\mathrm{Re}\left\{\frac{1}{\xi(1+2C)-iy} - \xi(1+2C)\frac{C}{1+C}\left[\frac{1}{\xi(1+2C)-iy}\right]^2\right\} \\ &= \frac{2}{\pi\gamma}\left\{\frac{\xi(1+2C)}{[\xi(1+2C)]^2+y^2} - \frac{C}{1+C}\frac{[\xi(1+2C)]^3-\xi(1+2C)y^2}{\{[\xi(1+2C)]^2+y^2\}^2}\right\},\end{aligned} \tag{15.135}$$

and the spectrum is

$$\begin{aligned}T_{\rightarrow}^{\mathrm{inc}}(\omega) = F_{\rightarrow}^{\mathrm{inc}}\frac{1}{\pi}\Bigg\{&\frac{\kappa(1+2C)}{[\kappa(1+2C)]^2+(\omega-\omega_0)^2} \\ &-\frac{C}{1+C}\frac{[\kappa(1+2C)]^3-\kappa(1+2C)(\omega-\omega_0)^2}{\{[\kappa(1+2C)]^2+(\omega-\omega_0)^2\}^2}\Bigg\}.\end{aligned} \tag{15.136}$$

The spectrum in this limit is the sum of a Lorentzian and a Lorentzian squared. The Lorentzian squared is subtracted from the Lorentzian, which produces a prominent dip in the center of the spectrum. A two-peaked structure results, though the resulting "doublet" has nothing to do with vacuum Rabi splitting. Rice and Carmichael [15.19] speak of a *squeezing-induced spectral hole.*

Many-atom strong-coupling limit ($\sqrt{\xi 2C} \gg \frac{1}{2}(\xi+1)$): Here we recover a connection with the vacuum Rabi doublet. The spectrum develops two peaks in the manner of the weak-probe transmission spectrum derived in Sect. 14.4.1, and the approximations we make are targeted towards the two-peaked form. First, the denominator in (15.132) is factorized and written as

$$(\xi - iy)(1 - i\xi) + \xi 2C = (\bar{\Lambda}_+ + iy)(\bar{\Lambda}_- + iy), \tag{15.137}$$

with

$$\bar{\Lambda}_\pm = -\tfrac{1}{2}(\xi+1) \pm i\sqrt{\xi 2C - \tfrac{1}{4}(\xi-1)^2}. \tag{15.138}$$

We then make the approximations for large $\xi 2C$:

$$\begin{aligned}&(\xi - iy)(1-iy) + \xi 2C \\ &\approx \left[\tfrac{1}{2}(\xi+1) - i(y + \sqrt{\xi 2C})\right]\left[\tfrac{1}{2}(\xi+1) - i(y - \sqrt{\xi 2C})\right],\end{aligned} \tag{15.139a}$$

$$\begin{aligned}&\frac{1+\xi-iy}{(\xi-iy)(1-iy)+\xi 2C} \\ &\approx \frac{1}{2}\left[\frac{1}{\frac{1}{2}(\xi+1) - i(y+\sqrt{\xi 2C})} + \frac{1}{\frac{1}{2}(\xi+1) - i(y-\sqrt{\xi 2C})}\right],\end{aligned} \tag{15.139b}$$

and

$$\begin{aligned}&\frac{(1+2C)(1+\xi)}{2+\xi+2C}\,\frac{\xi(\xi-2C-iy)}{[(\xi-iy)(1-iy)+\xi 2C]^2} \\ &\approx \frac{1}{4}\left\{\frac{1}{\left[\frac{1}{2}(\xi+1) - i(y+\sqrt{\xi 2C})\right]^2} + \frac{1}{\left[\frac{1}{2}(\xi+1) - i(y-\sqrt{\xi 2C})\right]^2}\right\},\end{aligned} \tag{15.139c}$$

where in (15.139b) and (15.139c), close to the resonances at $y = +\sqrt{\xi 2C}$ and $y = -\sqrt{\xi 2C}$, we replace y in the nonresonant term by its on-resonance value. From (15.132), and (15.139b) and (15.139c), we obtain

$$\begin{aligned}&\bar{T}_{\rm inc}(y) \\ &= \frac{1}{\pi\gamma}\left\{\frac{2\left[\frac{1}{2}(\xi+1)\right]^3}{\left\{\left[\frac{1}{2}(\xi+1)\right]^2 + (y+\sqrt{\xi 2C})^2\right\}^2} + \frac{2\left[\frac{1}{2}(\xi+1)\right]^3}{\left\{\left[\frac{1}{2}(\xi+1)\right]^2 + (y-\sqrt{\xi 2C})^2\right\}^2}\right\}.\end{aligned} \tag{15.140}$$

The resulting spectrum, from (15.131), is *the many-atom vacuum Rabi doublet with squeezing-induced linewidth narrowing*:

$$T^{\rm inc}_{\rightarrow}(\omega) = F^{\rm inc}_{\rightarrow}\frac{1}{2}\left\{\frac{2\left[\frac{1}{2}(\kappa+\gamma/2)\right]^3/\pi}{\left\{\left[\frac{1}{2}(\kappa+\gamma/2)\right]^2+(\omega-\omega_0+\sqrt{N}g)^2\right\}^2} + \frac{2\left[\frac{1}{2}(\kappa+\gamma/2)\right]^3/\pi}{\left\{\left[\frac{1}{2}(\kappa+\gamma/2)\right]^2+(\omega-\omega_0-\sqrt{N}g)^2\right\}^2}\right\}. \qquad (15.141)$$

Each vacuum Rabi peak is a Lorentzian squared.

Note 15.8. In the bad-cavity limit, the Lorentzian squared (Eq. 15.134) approaches zero as $|\omega-\omega_0|^{-4}$ as $|\omega-\omega_0|\rightarrow\infty$, instead of following the usual $|\omega-\omega_0|^{-2}$. Mollow noted this interesting fact in his original paper on the spectrum of resonance fluorescence [15.21]. He also identified the subtraction responsible for the behavior, though he had nothing to say about squeezing, which was unknown at the time. The relevant passage in the Mollow paper starts below Eq. 4.21: "It is important to note, however, that one of the parameters D_0, D_+, or D_- (the one associated with the root of intermediate magnitude) is negative, and hence the Lorentzian in which it appears has negative weight. It is an interesting feature of the model we are considering that, in the limit $|\nu-\omega|\rightarrow\infty$, the spectral density falls to zero as $|\nu-\omega|^{-4}$, rather than as $|\nu-\omega|^{-2}$, as it would for positive Lorentzian functions." The noted property can be traced to the short-time behavior of the first-order correlation function [15.22]; if the derivative of the correlation function vanishes at zero delay, the spectrum shows the $|\omega-\omega_0|^{-4}$ behavior. It is easily verified, from (15.121a), and (15.120a) and (15.120c), that the derivative of $C_{\rm ss}^{\tilde{z}_*\tilde{z}}$ vanishes at $\bar{\tau}=0$.

Exercise 15.8. Show that the general result (15.132) for the spectrum approaches zero as $|y|^{-4}$ as $|y|\rightarrow\infty$. Why then does (15.136) approach zero as $|\omega-\omega_0|^{-2}$?

15.2.7 Forwards Photon Scattering in the Weak-Excitation Limit

As a last application of our solution for the steady-state covariance matrix, let us calculate the second-order correlation function of the forward-scattered light. We once again restrict our attention to the weak-excitation limit where it is possible to obtain a closed expression. Forwards photon scattering for one atom in a cavity was treated in Sect. 13.2.3, though only for the bad-cavity limit [however see (13.75)]. Here, a linearized treatment of the fluctuations is made and the number of atoms is assumed to be large. We return to the topic of forwards photon scattering in Sects. 16.1.3 and 16.2, where a treatment for arbitrary numbers of atoms is developed in which no restriction is imposed other than the requirement that the excitation be weak. The result from Sect. 16.1.3 is compared with the present calculation in Sect. 16.1.4.

The second-order correlation function of the forward-scattered light is given by

$$g_{\rightarrow}^{(2)}(\tau) = \frac{\langle \tilde{\bar{a}}^{\dagger}(0)\tilde{\bar{a}}^{\dagger}(\tau)\tilde{\bar{a}}(\tau)\tilde{\bar{a}}(0)\rangle_{\mathrm{ss}}}{\left(\langle \tilde{\bar{a}}^{\dagger}\tilde{\bar{a}}\rangle_{\mathrm{ss}}\right)^2}, \tag{15.142}$$

where $\langle \tilde{\bar{a}}^{\dagger}(0)\tilde{\bar{a}}^{\dagger}(\tau)\tilde{\bar{a}}(\tau)\tilde{\bar{a}}(0)\rangle_{\mathrm{ss}} \equiv \lim_{t\to\infty}\langle \tilde{\bar{a}}^{\dagger}(t)\tilde{\bar{a}}^{\dagger}(t+\tau)\tilde{\bar{a}}(t+\tau)\tilde{\bar{a}}(t)\rangle_{\mathrm{ss}}$. Expanding the annihilation and creation operators as sums of a mean and a fluctuation (Eqs. 15.89a and 15.89b), we write

$$\begin{aligned}\langle \tilde{\bar{a}}^{\dagger}\tilde{\bar{a}}\rangle_{\mathrm{ss}} &= \langle \tilde{\bar{a}}^{\dagger}\rangle_{\mathrm{ss}}\langle \tilde{\bar{a}}\rangle_{\mathrm{ss}} + \langle \Delta\tilde{\bar{a}}^{\dagger}\Delta\tilde{\bar{a}}\rangle_{\mathrm{ss}} \\ &= X^2 + N^{-1}B,\end{aligned} \tag{15.143}$$

where we use (15.69) and (15.88), and

$$\begin{aligned}&\langle \tilde{\bar{a}}^{\dagger}(0)\tilde{\bar{a}}^{\dagger}(\tau)\tilde{\bar{a}}(\tau)\tilde{\bar{a}}(0)\rangle_{\mathrm{ss}} \\ &= \left(\langle \tilde{\bar{a}}^{\dagger}\rangle_{\mathrm{ss}}\langle \tilde{\bar{a}}\rangle_{\mathrm{ss}} + \langle \Delta\tilde{\bar{a}}^{\dagger}\Delta\tilde{\bar{a}}\rangle_{\mathrm{ss}}\right)^2 + \langle \tilde{\bar{a}}^{\dagger}\rangle_{\mathrm{ss}}\langle \tilde{\bar{a}}\rangle_{\mathrm{ss}}\left[\langle \Delta\tilde{\bar{a}}^{\dagger}(0)\Delta\tilde{\bar{a}}(\tau)\rangle_{\mathrm{ss}} + \mathrm{c.c.}\right] \\ &\quad + \left(\langle \tilde{\bar{a}}\rangle_{\mathrm{ss}}\right)^2\langle \Delta\tilde{\bar{a}}^{\dagger}(0)\Delta\tilde{\bar{a}}^{\dagger}(\tau)\rangle_{\mathrm{ss}} + \mathrm{c.c.} \\ &= (X^2 + N^{-1}B)^2 + 2N^{-1}X^2[C_{\mathrm{ss}}^{\tilde{z}_*\tilde{z}}(\tau) + C_{\mathrm{ss}}^{\tilde{z}_*\tilde{z}_*}(\tau)],\end{aligned} \tag{15.144}$$

where we make use of (15.117). In order to remain consistent with the assumptions of the linearized treatment of fluctuations, in (15.144) we neglect $\langle \Delta\tilde{\bar{a}}^{\dagger}(0)\Delta\tilde{\bar{a}}^{\dagger}(\tau)\Delta\tilde{\bar{a}}(\tau)\Delta\tilde{\bar{a}}(0)\rangle_{\mathrm{ss}}$ as a term of second-order in N^{-1}; of course, third-order terms in $\Delta\tilde{\bar{a}}^{\dagger}$ and $\Delta\tilde{\bar{a}}$ vanish because the linearized fluctuations are Gaussian. Now, from (15.142)–(15.144), using (15.103b), (15.127) and (15.130), we obtain

$$g^{(2)}(\tau) = 1 + 2N^{-1}\frac{C_{\mathrm{ss}}^{\tilde{z}_*\tilde{z}_*}(\tau)}{X^2}, \tag{15.145}$$

where the standard correlation $C_{\mathrm{ss}}^{\tilde{z}_*\tilde{z}}(\tau) \sim X^4$ is neglected in the weak-excitation limit compared with the anomalous correlation $C_{\mathrm{ss}}^{\tilde{z}_*\tilde{z}_*}(\tau) \sim X^2$; thus, the source of nonclassicality discussed in Sect. 15.2.5 enters the picture once again.

One further step takes us to the final result. The Laplace transform of $C_{\mathrm{ss}}^{\tilde{z}_*\tilde{z}_*}(\tau)$ is given as (15.127). Making the factorization (15.137) in its denominator, the transform is

$$\begin{aligned}\bar{C}_{\mathrm{ss}}^{\tilde{z}_*\tilde{z}_*}(\bar{s}) &= -X^2\frac{\xi}{1+\xi}\frac{4C^2}{1+2C}\frac{\xi+1+\bar{s}}{(\bar{\Lambda}_+ - \bar{s})(\bar{\Lambda}_- - \bar{s})} \\ &= -X^2\frac{\xi}{1+\xi}\frac{4C^2}{1+2C}\frac{1}{\bar{\Lambda}_+ - \bar{\Lambda}_-}\left(\frac{\bar{\Lambda}_-}{\bar{\Lambda}_+ - \bar{s}} - \frac{\bar{\Lambda}_+}{\bar{\Lambda}_- - \bar{s}}\right),\end{aligned} \tag{15.146}$$

with inverse

$$C_{\rm ss}^{\tilde{z}_*\tilde{z}_*}(\tau) = -X^2\frac{\xi}{1+\xi}\frac{4C^2}{1+2C}e^{-\frac{1}{2}(\kappa+\gamma/2)\tau}\left[\cos(G\tau) + \frac{\frac{1}{2}(\kappa+\gamma/2)}{G}\sin(G\tau)\right]. \tag{15.147}$$

Thus, substituting this result into (15.145), *the second-order correlation function of forwards scattering for many atoms in a cavity in the weak-excitation limit (within the linearized treatment of fluctuations)* is given by

$$g^{(2)}_{\rightarrow}(\tau) - 1$$
$$= -2N^{-1}\frac{\xi}{1+\xi}\frac{4C^2}{1+2C}e^{-\frac{1}{2}(\kappa+\gamma/2)\tau}\left[\cos(G\tau) + \frac{\frac{1}{2}(\kappa+\gamma/2)}{G}\sin(G\tau)\right], \tag{15.148}$$

where

$$G \equiv \sqrt{Ng^2 - \tfrac{1}{4}(\kappa-\gamma/2)^2}. \tag{15.149}$$

The principal feature of interest is that the forwards photon scattering is antibunched. The size of the antibunching effect, for $2C \gg 1$, is determined by $N^{-1}2C = 2C_1$, the enhancement factor of cavity-enhanced spontaneous emission from a single atom (Eq. 13.36). In the bad-cavity limit, the regression of fluctuations from the steady state damps at the collectively enhanced rate $\gamma(1+2C)$ [compare the spectrum (15.134)]. More generally, it is oscillatory, with frequency determined by the splitting of the many-atom vacuum Rabi doublet (Sect. 14.4.1, with $g_{\rm max}$ replaced by g and $\bar{N}_{\rm eff}$ replaced by N). This is the so-called *vacuum Rabi oscillation*, a *collective vacuum Rabi oscillation* in this case.

Examples of oscillatory correlation functions are plotted in Fig. 15.2. The chosen parameters trace something of the history of experiments in optical bistability and cavity QED. Figure 15.2a corresponds to an optical bistability experiment by Kimble and coworkers [15.23], which, although it realizes $2C \sim 30$, uses a large cavity and realizes a very small dipole coupling constant; the high value of $2C$ results from the large number of cooperating atoms. Under these conditions, the sizes of the fluctuations and the photon antibunching effect are small. For comparison, a photon antibunching effect some $40\times$ larger is displayed in Fig. 15.2b, where the parameters correspond to the measurement of the many-atom vacuum Rabi doublet by Zhu and coworkers [15.24]. Finally, Figs. 15.2c and d use parameters from the experiment by Raizen and coworkers [15.25], who also measured the vacuum Rabi doublet. In this case, the photon antibunching effect is raised to a few percent. Of course, all plots overlook the degradation of the photon antibunching effect by the average over dipole coupling strengths for spatially distributed atoms (Sect. 16.2).

Finally, returning to the comment below (15.145), we note that $g^{(2)}_{\rightarrow}(\tau) - 1$ provides a measurement of the anomalous correlation $C_{\rm ss}^{\tilde{z}_*\tilde{z}_*}(\tau) = C_{\rm ss}^{\tilde{z}\tilde{z}}(\tau)$,

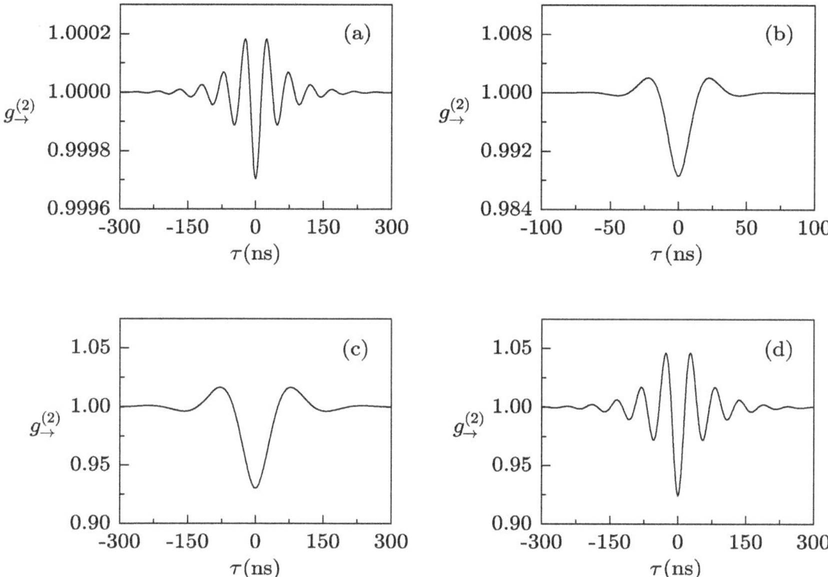

Fig. 15.2. Antibunching of forwards photon scattering for many atoms in a cavity in the weak-excitation limit. Intensity correlation function (15.148) is plotted for (**a**) $(g, \kappa, \gamma)/2\pi = (0.07, 1.45, 10)$ MHz, $N = 90,000$ [15.23]; (**b**) $(g, \kappa, \gamma)/2\pi = (1.3, 15, 19)$ MHz, $N = 300$ [15.24]; and (**c**) $(g, \kappa, \gamma)/2\pi = (1.06, 0.88, 10)$ MHz, $N = 40$, and (**d**) $(g, \kappa, \gamma)/2\pi = (1.06, 0.88, 10)$ MHz, $N = 310$ [15.25]

which is responsible for squeezing in the weak-excitation limit; the negative sign on the right-hand side of (15.147) produces antibunching in Fig. 15.2 and a negative spectrum of squeezing.

Note 15.9. Equation 15.148 is based upon the small-noise approximation, which assumes $N^{-1} \ll 1$. If it is used where the approximation is not valid, it can produce a negative $g^{(2)}(0)$, which is of course nonphysical since $g^{(2)}(0)$ is nonnegative by definition. The possibility of a negative result is introduced by dropping the term $\langle \Delta\tilde{\bar{a}}^\dagger(0)\Delta\tilde{\bar{a}}^\dagger(\tau)\Delta\tilde{\bar{a}}(\tau)\Delta\tilde{\bar{a}}(0)\rangle_{\rm ss}$ from (15.144). By keeping it, negative values of $g^{(2)}(0)$ can be avoided (see Sect. 16.1.4). The resulting expression for $g^{(2)}_{\rightarrow}(\tau) - 1$ is certainly, in one sense, a better approximation, though it is still in error at order N^{-2}, and the error can be very large if $N \sim 1$ (see, for example, Fig. 16.3).

16

Cavity QED II: Quantum Fluctuations

In Chap. 15 we made some progress applying linear fluctuation theory to the system of many atoms in a cavity. Conditions of strong dipole coupling were included, assuming the coupling was strong only in the many-atom sense. Ultimately, cavity QED is concerned with strong coupling for single atoms, though—i.e., with $g \gg \kappa, \gamma/2$ (alternatively $2C_1 \gg 1$), conditions that invalidate the small-noise assumption upon which the linear theory of fluctuations rests (Sect. 13.1).

The remaining chapters of the book deal with alternatives to the linearized treatment of fluctuations. The present chapter deals specifically with quantum fluctuations in cavity QED, while Chaps. 17– 19 take up the issue from a broader point of view, reformulating our whole approach to the treatment of fluctuations in open quantum systems.

Much of the present chapter stays with the weak-excitation limit. Sections 16.1 and 16.2 tackle the limit by making a perturbation expansion in the driving field strength, an approach that allows us to generalize a number of earlier results. We then sketch the behavior that unfolds as the excitation strength is increased (Sect. 16.3). Though analytical solutions cannot be found here, we are able to demonstrate the breakdown of the small-noise picture of deterministic (semiclassical) evolution plus quantum "fuzz"; thus, we return to the theme of the degenerate parametric oscillator example treated in Chap. 12.

16.1 Density Matrix Expansion for the Weak-Excitation Limit

When treating forwards photon scattering for one atom in a cavity in Sect. 13.2.3, we took the weak-excitation limit by factorizing the steady-state density operator in pure state form (Eqs. 13.81, 13.82 and 13.89). It turns out that this factorization has rather wide application, requiring only that the excitation be sufficiently weak that over periods on the order of the correlation time (the larger of κ^{-1} and $2\gamma^{-1}$) the probability of irreversible photon emission be much less than one. Then, the scattering processes that destroy the coherence of the quantum evolution are weak, in so far as the regression of a typical fluctuation is unlikely to be interrupted by a scattering event (an irreversible photon emission); thus, the purity of the quantum state is preserved during the regression of fluctuations—at least to lowest order in the driving field strength. A rigorous justification for the factorization follows by expanding the density operator, and its master equation evolution, in powers of the driving field amplitude $|\bar{\mathcal{E}}_0|$. Let us see how this is done, first for one atom, then generalized to the many-atom case.

16.1.1 Pure-State Factorization of the Density Operator for One Atom

The master equation for single-atom cavity QED with coherent driving is given as an operator equation in (13.57). For the present purposes we rewrite it as a density matrix equation, specializing, for simplicity, to the case of resonant excitation ($\omega_A = \omega_C = \omega_0$). Then, working in the interaction picture, with $\tilde{\rho}(t) = e^{i\omega_0(\sigma_z/2+a^\dagger a)t}\rho(t)e^{-i\omega_0(\sigma_z/2+a^\dagger a)t}$, the *density matrix equation for single-atom cavity QED with coherent driving of the cavity mode* is

$$
\begin{aligned}
&\begin{pmatrix} \dot{\tilde{\rho}}_{2,n;2,m} & \dot{\tilde{\rho}}_{2,n;1,m} \\ \dot{\tilde{\rho}}_{1,n;2,m} & \dot{\tilde{\rho}}_{1,n:1,m} \end{pmatrix} \\
&= g\left[\sqrt{n}\begin{pmatrix} 0 & 0 \\ \tilde{\rho}_{2,n-1;2,m} & \tilde{\rho}_{2,n-1;1,m} \end{pmatrix} - \sqrt{n+1}\begin{pmatrix} \tilde{\rho}_{1,n+1;2,m} & \tilde{\rho}_{1,n+1;1,m} \\ 0 & 0 \end{pmatrix}\right. \\
&\qquad \left. +\sqrt{m}\begin{pmatrix} 0 & \tilde{\rho}_{2,n;2,m-1} \\ 0 & \tilde{\rho}_{1,n;2,m-1} \end{pmatrix} - \sqrt{m+1}\begin{pmatrix} \tilde{\rho}_{2,n;1,m+1} & 0 \\ \tilde{\rho}_{1,n;1,m+1} & 0 \end{pmatrix}\right] \\
&\quad + i\bar{\mathcal{E}}_0\left[\sqrt{m+1}\begin{pmatrix} \tilde{\rho}_{2,n;2,m+1} & \tilde{\rho}_{2,n;1,m+1} \\ \tilde{\rho}_{1,n;2,m+1} & \tilde{\rho}_{1,n;1,m+1} \end{pmatrix} - \sqrt{n}\begin{pmatrix} \tilde{\rho}_{2,n-1;2,m} & \tilde{\rho}_{2,n-1;1,m} \\ \tilde{\rho}_{1,n-1;2,m} & \tilde{\rho}_{1,n-1;1,m} \end{pmatrix}\right] \\
&\quad - i\bar{\mathcal{E}}_0^*\left[\sqrt{n+1}\begin{pmatrix} \tilde{\rho}_{2,n+1;2,m} & \tilde{\rho}_{2,n+1;1,m} \\ \tilde{\rho}_{1,n+1;2,m} & \tilde{\rho}_{1,n+1;1,m} \end{pmatrix} - \sqrt{m}\begin{pmatrix} \tilde{\rho}_{2,n;2,m-1} & \tilde{\rho}_{2,n;1,m-1} \\ \tilde{\rho}_{1,n;2,m-1} & \tilde{\rho}_{1,n;1,m-1} \end{pmatrix}\right] \\
&\quad + 2\kappa\sqrt{(n+1)(m+1)}\begin{pmatrix} \tilde{\rho}_{2,n+1;2,m+1} & \tilde{\rho}_{2,n+1;1,m+1} \\ \tilde{\rho}_{1,n+1;2,m+1} & \tilde{\rho}_{1,n+1;1,m+1} \end{pmatrix} \\
&\quad - \kappa(n+m)\begin{pmatrix} \tilde{\rho}_{2,n;2,m} & \tilde{\rho}_{2,n;1,m} \\ \tilde{\rho}_{1,n;2,m} & \tilde{\rho}_{1,n;1,m} \end{pmatrix} - \frac{\gamma}{2}\begin{pmatrix} 2\tilde{\rho}_{2,n;2,m} & \tilde{\rho}_{2,n;1,m} \\ \tilde{\rho}_{1,n;2,m} & -2\tilde{\rho}_{2,n;2,m} \end{pmatrix}, \qquad (16.1)
\end{aligned}
$$

where $\tilde{\rho}_{\eta,n;\xi,m} \equiv \langle\eta,n|\tilde{\rho}|\xi,m\rangle$; $n,m = 1,2,\ldots$; $\eta,\xi = 1$ (atomic lower state) and 2 (atomic upper state).

We propose to make a self-consistent truncation of the matrix element equations based on an expansion in powers of $|\bar{\mathcal{E}}_0|$. Since the equations themselves determine the solution for $\tilde{\rho}$, and hence the correct powers to use in the expansion, the best we can do is start out with an *ansatz*. For this, we choose

$$\tilde{\rho}_{\eta,n;\xi,m} \sim |\bar{\mathcal{E}}_0|^{n+m+\eta+\xi-2}. \tag{16.2}$$

Verification will come if we can show that the *ansatz* is consistent with the explicit terms in $|\bar{\mathcal{E}}_0|$ appearing in the equations of motion for density matrix elements. The *ansatz* does make physical sense, though, since the exponent $n+m+\eta+\xi-2$ simply counts the total number of quanta in the states $|\eta,n\rangle$ and $|\xi,m\rangle$.

Let us work through the truncation of the first two matrix element equations explicitly. Treatment of the rest follows by a straightforward generalization of the procedure illustrated by these examples. Consider the equation

$$\dot{\tilde{\rho}}_{1,0;1,0} = i\bar{\mathcal{E}}_0\tilde{\rho}_{1,0;1,1} - i\bar{\mathcal{E}}_0^*\tilde{\rho}_{1,1;1,0} + \gamma\tilde{\rho}_{2,0;2,0} + 2\kappa\tilde{\rho}_{1,1;1,1}$$

According to the *ansatz*, to lowest order $\tilde{\rho}_{1,0;1,0} \sim 1$. On the other hand, every term on the right-hand side of the equation is of order $|\bar{\mathcal{E}}_0|^2$. For consistency, we therefore write

$$\dot{\tilde{\rho}}_{1,0;1,0} = 0. \tag{16.3a}$$

For a second example, we begin with

$$\begin{aligned}\dot{\tilde{\rho}}_{2,0;2,0} = &-g(\tilde{\rho}_{2,0;1,1} + \tilde{\rho}_{1,1;2,0}) + i\bar{\mathcal{E}}_0\tilde{\rho}_{2,0;2,1} - i\bar{\mathcal{E}}_0^*\tilde{\rho}_{2,1;2,0}\\ &- \gamma\tilde{\rho}_{2,0;2,0} + 2\kappa\tilde{\rho}_{2,1;2,1}.\end{aligned}$$

In this case, the matrix elements $\tilde{\rho}_{2,0;2,0}$, $\tilde{\rho}_{2,0;1,1}$, and $\tilde{\rho}_{1,1;2,0}$ have the same dominant order, $|\bar{\mathcal{E}}_0|^2$; but $\tilde{\rho}_{2,0;2,1}$ and $\tilde{\rho}_{2,1;2,0}$ appear multiplied by $|\bar{\mathcal{E}}_0|$, which gives terms of order $|\bar{\mathcal{E}}_0|^3$, and $\tilde{\rho}_{2,1;2,1}$ is of order $|\bar{\mathcal{E}}_0|^4$. We therefore drop the latter terms and write

$$\dot{\tilde{\rho}}_{2,0;2,0} = -g(\tilde{\rho}_{2,0;1,1} + \tilde{\rho}_{1,1;2,0}) - \gamma\tilde{\rho}_{2,0;2,0}. \tag{16.3b}$$

Proceeding in this way yields the self-consistent truncation we are after. The process could be carried through for all matrix elements. We will be satisfied, however, with a solution for the density matrix taking into account up to a maximum of two energy quanta; this is sufficient to extend our earlier results on forwards photon scattering for an atom in a cavity (Sect. 13.2.3). With this restriction, three diagonal matrix elements remain, for which the truncated equations are

$$\dot{\tilde{\rho}}_{1,1;1,1} = g(\tilde{\rho}_{2,0;1,1} + \tilde{\rho}_{1,1;2,0}) - i\bar{\mathcal{E}}_0\tilde{\rho}_{1,0;1,1} + i\bar{\mathcal{E}}_0^*\tilde{\rho}_{1,1;1,0} - 2\kappa\tilde{\rho}_{1,1;1,1}, \tag{16.3c}$$

$$\begin{aligned}\dot{\tilde{\rho}}_{2,1;2,1} = &-\sqrt{2}g(\tilde{\rho}_{1,2;2,1} + \tilde{\rho}_{2,1;1,2}) - i\bar{\mathcal{E}}_0\tilde{\rho}_{2,0;2,1} + i\bar{\mathcal{E}}_0^*\tilde{\rho}_{2,1;2,0}\\ &- (2\kappa+\gamma)\tilde{\rho}_{2,1;2,1},\end{aligned} \tag{16.3d}$$

$$\begin{aligned}\dot{\tilde{\rho}}_{1,2;1,2} = &\sqrt{2}g(\tilde{\rho}_{2,1;1,2} + \tilde{\rho}_{1,2;2,1}) - i\sqrt{2}\bar{\mathcal{E}}_0\tilde{\rho}_{1,1;1,2} + i\sqrt{2}\bar{\mathcal{E}}_0^*\tilde{\rho}_{1,2;1,1}\\ &- 4\kappa\tilde{\rho}_{1,2;1,2}.\end{aligned} \tag{16.3e}$$

Then, for the twenty off-diagonal matrix elements, we obtain

$$\dot{\tilde{\rho}}_{1,0;2,0} = -g\tilde{\rho}_{1,0;1,1} - (\gamma/2)\tilde{\rho}_{1,0;2,0}, \tag{16.4a}$$

$$\dot{\tilde{\rho}}_{1,0;1,1} = g\tilde{\rho}_{1,0;2,0} + i\bar{\mathcal{E}}_0^*\tilde{\rho}_{1,0;1,0} - \kappa\tilde{\rho}_{1,0;1,1}, \tag{16.4b}$$

$$\dot{\tilde{\rho}}_{1,0;2,1} = -\sqrt{2}g\tilde{\rho}_{1,0;1,2} + i\bar{\mathcal{E}}_0^*\tilde{\rho}_{1,0;2,1} - (\kappa + \gamma/2)\tilde{\rho}_{1,0;2,1}, \tag{16.4c}$$

$$\dot{\tilde{\rho}}_{1,0;1,2} = \sqrt{2}g\tilde{\rho}_{2,1;1,0} + i\sqrt{2}\bar{\mathcal{E}}_0^*\tilde{\rho}_{1,0;1,1} - 2\kappa\tilde{\rho}_{1,0;1,2}, \tag{16.4d}$$

$$\dot{\tilde{\rho}}_{2,0;1,1} = -g(\tilde{\rho}_{1,1;1,1} - \tilde{\rho}_{2,0;2,0}) + i\bar{\mathcal{E}}_0^*\tilde{\rho}_{1,0;2,0} - (\kappa + \gamma/2)\tilde{\rho}_{2,0;1,1}, \tag{16.4e}$$

$$\dot{\tilde{\rho}}_{2,0;2,1} = -g\big(\sqrt{2}\tilde{\rho}_{2,0;1,2} + \tilde{\rho}_{1,1;2,1}\big) + i\bar{\mathcal{E}}_0^*\tilde{\rho}_{2,0;2,0} - (\kappa + \gamma)\tilde{\rho}_{2,0;2,1}, \tag{16.4f}$$

$$\begin{aligned}\dot{\tilde{\rho}}_{2,0;1,2} = &-g\big(\tilde{\rho}_{1,1;1,2} - \sqrt{2}\tilde{\rho}_{2,0;2,1}\big) + i\sqrt{2}\bar{\mathcal{E}}_0^*\tilde{\rho}_{2,0;2,0} \\ &- (2\kappa + \gamma/2)\tilde{\rho}_{2,0;1,2},\end{aligned} \tag{16.4g}$$

$$\begin{aligned}\dot{\tilde{\rho}}_{1,1;2,1} = &-g\big(\sqrt{2}\tilde{\rho}_{1,1;1,2} - \tilde{\rho}_{2,0;2,1}\big) - i\bar{\mathcal{E}}_0\tilde{\rho}_{1,0;2,1} + i\bar{\mathcal{E}}_0^*\tilde{\rho}_{1,1;2,0} \\ &- (2\kappa + \gamma/2)\tilde{\rho}_{1,1;2,1},\end{aligned} \tag{16.4h}$$

$$\begin{aligned}\dot{\tilde{\rho}}_{1,1;1,2} = &\, g\big(\tilde{\rho}_{2,0;1,2} + \sqrt{2}\tilde{\rho}_{1,1;2,1}\big) - i\bar{\mathcal{E}}_0\tilde{\rho}_{1,0;1,2} + i\sqrt{2}\bar{\mathcal{E}}_0^*\tilde{\rho}_{1,1;1,1} \\ &- 3\kappa\tilde{\rho}_{1,1;1,2},\end{aligned} \tag{16.4i}$$

$$\begin{aligned}\dot{\tilde{\rho}}_{2,1;1,2} = &-\sqrt{2}g(\tilde{\rho}_{1,2;1,2} - \tilde{\rho}_{2,1;2,1}) - i\bar{\mathcal{E}}_0\tilde{\rho}_{2,0;1,2} + i\sqrt{2}\bar{\mathcal{E}}_0^*\tilde{\rho}_{2,1;1,1} \\ &- (3\kappa + \gamma/2)\tilde{\rho}_{2,1;1,2},\end{aligned} \tag{16.4j}$$

and the ten complex conjugates of (16.4a)–(16.4j).

The self-consistency of the truncation is apparent from the fact that each equation of motion is of uniform order in $|\bar{\mathcal{E}}_0|$. The truncation is helpful because it removes unnecessary terms from the equations, considering our goal of solving the master equation to dominant order only in $|\bar{\mathcal{E}}_0|$; each of (16.4a)–(16.4j) retains the minimum number of terms needed to obtain the correct matrix elements to dominant order. Of course we must remember the spirit of the procedure when looking at an equation like (16.3a). Starting from the ground state, the solution to this equation is $\tilde{\rho}_{1,0;1,0}(t) = 1$, which clearly violates the normalization requirement for the density matrix as a whole; nonetheless, it is correct to dominant order, which is all that is claimed and all that is required.

Having obtained the truncated matrix element equations, we are now in a position to follow through with the factorization of the density operator. We propose that (16.3a)–(16.3e) and (16.4a)–(16.4j) are consistent with a solution for the density operator in the form

$$\tilde{\rho}(t) = |\tilde{\psi}(t)\rangle\langle\tilde{\psi}(t)|, \tag{16.5}$$

with $|\tilde{\psi}(t)\rangle$ expanded in the *two-quanta basis for one atom in a cavity*,

$$\left.\begin{array}{l}|1\rangle_A|0\rangle_a\\|1\rangle_A|1\rangle_a\\|2\rangle_A|0\rangle_a\\|1\rangle_A|2\rangle_a\\|2\rangle_A|1\rangle_a\end{array}\right\}, \tag{16.6}$$

as

$$\begin{aligned}|\tilde{\psi}(t)\rangle = |1\rangle_A|0\rangle_a + \tilde{\alpha}(t)|1\rangle_A|1\rangle_a + \tilde{\beta}(t)|2\rangle_A|0\rangle_a \\ + \tilde{\eta}(t)|1\rangle_A|2\rangle_a + \tilde{\zeta}(t)|2\rangle_A|1\rangle_a,\end{aligned} \tag{16.7}$$

where $|n\rangle_a$, $n = 0, 1, 2$, are Fock states for the cavity mode, and $|1\rangle_A$ and $|2\rangle_A$ are the lower and upper states, respectively, of the atom. With the proposed factorization, each matrix element is to be expressed as a product of two of the five state amplitudes—1, $\tilde{\alpha}$, $\tilde{\beta}$, $\tilde{\eta}$, and $\tilde{\zeta}$—and there is of course no guarantee that a consistent set of equations of motion for the amplitudes will emerge when the products are substituted into (16.3a)–(16.3e) and (16.4a)–(16.4j)—the substitution gives an overdetermined set of twenty-five equations for only five state amplitudes. It turns out that the overdetermined set *is* consistent, though: the four time-varying state amplitudes must satisfy the *equations of motion in the two-quanta truncation of the master equation for single-atom cavity QED*,

$$\dot{\tilde{\alpha}} = -\kappa\tilde{\alpha} + g\tilde{\beta} - i\bar{\mathcal{E}}_0, \tag{16.8a}$$

$$\dot{\tilde{\beta}} = -\frac{\gamma}{2}\tilde{\beta} - g\tilde{\alpha}, \tag{16.8b}$$

$$\dot{\tilde{\eta}} = -2\kappa\tilde{\eta} + \sqrt{2}g\tilde{\zeta} - i\sqrt{2}\bar{\mathcal{E}}_0\tilde{\alpha}, \tag{16.8c}$$

$$\dot{\tilde{\zeta}} = -(\kappa + \gamma/2)\tilde{\zeta} - \sqrt{2}g\tilde{\eta} - i\bar{\mathcal{E}}_0\tilde{\beta}. \tag{16.8d}$$

Equations 16.8a and 16.8b are the coupled oscillator equations whose normal modes explain the phenomenology of the vacuum Rabi doublet (compare Eqs. 13.152, 14.117, and 14.125). The doublet is a consequence of the vacuum Rabi oscillation of the one-quantum amplitudes (Sect. 13.3.2). By adding the two-quanta amplitudes, we are able to describe certain features of the quantum fluctuations. Before setting out in this direction, though, let us first set up the pure-state factorization in the many-atom case.

16.1.2 Pure-State Factorization of the Density Operator for Many Atoms

The two-quanta truncation for many atoms must consider a basis of six, rather than five states, since we must now account for the possibility that two atoms are excited simultaneously. The *two-quanta basis for N atoms in a cavity* is denoted by

$$\left.\begin{array}{l} |0\rangle^{(N)}|0\rangle_a \\ |0\rangle^{(N)}|1\rangle_a \\ |1\rangle^{(N)}|0\rangle_a \\ |0\rangle^{(N)}|2\rangle_a \\ |1\rangle^{(N)}|1\rangle_a \\ |2\rangle^{(N)}|0\rangle_a \end{array}\right\}, \tag{16.9}$$

where the states for the atoms are the collective, or symmetrized, atomic states

$$|0\rangle^{(N)} \equiv \prod_{j=1}^{N} |1\rangle_j = |1, N/2, -N/2\rangle, \tag{16.10a}$$

$$|1\rangle^{(N)} \equiv \frac{1}{\sqrt{N}} J_+ |0\rangle^{(N)} = |1, N/2, -N/2+1\rangle, \tag{16.10b}$$

$$|2\rangle^{(N)} \equiv \sqrt{\frac{2}{N-1}} J_+ |1\rangle^{(N)} = |1, N/2, -N/2+2\rangle, \tag{16.10c}$$

with the notation on the right-hand sides taken from Sect. 6.2.2.

A comment is in order regarding the choice of symmetrized atomic states; their use might seem inappropriate as it suggests the atoms are indistinguishable, which is definitely not the case. In the first place, in a standing-wave cavity, atoms at different locations experience different coupling strengths; although, this source of distinguishability is ignored in master equation (15.1), and we continue with the simplification throughout the present chapter (with the exception of Sect. 16.2). Of more fundamental concern is the role of spontaneous emission. Each emission event may be assigned to a particular atom (Note 15.4); surely this counters any argument for using symmetrized atomic states. Contrary to appearances, though, there is a justification for our choice of basis. It follows from the limited goal of solving for each density matrix element to dominant order in $|\bar{\mathcal{E}}_0|$. To show that our basis is, indeed, appropriate and consistent, let us consider the various processes that cause transitions between the basis states.

It is clear, first, that symmetrized states couple self-consistently through the interaction between the atoms and the cavity mode; the Hamiltonian $g(a^\dagger J_- - a\, J_+)$ generates transitions up and down the truncated ladder of states: $|0\rangle^{(N)}|1\rangle_a \to |1\rangle^{(N)}|0\rangle_a \to |0\rangle^{(N)}|1\rangle_a$ and $|0\rangle^{(N)}|2\rangle_a \to |1\rangle^{(N)}|1\rangle_a \to |2\rangle^{(N)}|0\rangle_a \to |1\rangle^{(N)}|1\rangle_a \to |0\rangle^{(N)}|2\rangle_a$. Spontaneous emission, on the other hand, should work to destroy the coherence. Emission from a particular atom j induces two kinds of transitions. For a transition out of state $|1\rangle^{(N)}$, we obtain the final state

$$\begin{aligned}\sigma_{j-}|1\rangle^{(N)} &= \sigma_{j-}\frac{1}{\sqrt{N}}\sum_{k=1}^{N}\sigma_{k+}|0\rangle^{(N)} \\ &= \frac{1}{\sqrt{N}}\sigma_{j-}\sigma_{j+}|0\rangle^{(N)} \\ &= \frac{1}{\sqrt{N}}|0\rangle^{(N)}. \end{aligned} \tag{16.11}$$

This state is trivially one of the symmetrized states and lies within the basis (16.9). It is the second kind of transition, out of state $|2\rangle^{(N)}$, that is potentially problematic. In this case we find

$$\begin{aligned}\sigma_{j-}|2\rangle^{(N)} &= \sigma_{j-}\sqrt{\frac{2}{N(N-1)}}\sum_{\substack{k,l=1\\k>l}}^{N}\sigma_{k+}\sigma_{l+}|0\rangle^{(N)} \\ &= \sqrt{\frac{2}{N(N-1)}}|1\rangle_j\left[\sum_{\substack{k=1\\k\neq j}}^{N}|2\rangle_k\left(\prod_{\substack{l=1\\l\neq j,k}}^{N}|1\rangle_l\right)\right] \\ &= \sqrt{\frac{2}{N}}|1\rangle_j|1\rangle_{\{j\}}^{(N-1)}, \end{aligned} \tag{16.12}$$

where $|m\rangle_{\{j_1,\ldots,j_n\}}^{(N-n)}$ is the symmetrized state of m excited atoms amongst $N-n$ atoms, excluding specifically atoms $j_1,\ldots,j_n$. Here the final state lies outside the basis (16.9); it is not a symmetrized state, since a particular atom—atom j—is definitely in its lower state (see Note 15.4). To clarify the point, we might write the symmetrized one-quantum state as

$$|1\rangle^{(N)} = \frac{1}{\sqrt{N}}\left[\sqrt{N-1}|1\rangle_j|1\rangle_{\{j\}}^{(N-1)} + |2\rangle_j|0\rangle_{\{j\}}^{(N-1)}\right], \tag{16.13}$$

where the transition (16.12) produces only the first term on the right-hand side. While this is the dominant term when the number of atoms is large, spontaneous emission from state $|2\rangle^{(N)}$ nevertheless degrades the coherence, as we would expect, and the basis (16.9) appears to eliminate any possibility of accounting for this process.

There can be no argument about the omission. The relevant point for our program, though, is that while transitions of the latter type threaten to expand the basis at the one-quantum level where diagonal matrix elements are of order $|\bar{\mathcal{E}}_0|^2$, they contribute themselves at order $|\bar{\mathcal{E}}_0|^4$; this is the order of the probability for a two-quanta excitation. The threatened expansion of the basis may therefore be neglected within the framework of our adopted dominant-order truncation.

Accepting, then, the basis defined by (16.9) and (16.10), we propose a factorization with pure-state expansion

$$|\tilde{\psi}(t)\rangle = |0\rangle^{(N)}|0\rangle_a + \tilde{\alpha}(t)|0\rangle^{(N)}|1\rangle_a + \tilde{\beta}(t)|1\rangle^{(N)}|0\rangle_a \\ + \tilde{\eta}(t)|0\rangle^{(N)}|2\rangle_a + \tilde{\zeta}(t)|1\rangle^{(N)}|1\rangle_a + \tilde{\theta}(t)|2\rangle^{(N)}|0\rangle_a. \quad (16.14)$$

Bypassing the tedious details involved in the explicit verification of the factorization, we go directly to the *equations of motion in the two-quanta truncation of the master equation for many-atom cavity QED (with equal coupling strengths)* first reported by Carmichael, Rice, and Brecha [16.1]:

$$\dot{\tilde{\alpha}} = -\kappa\tilde{\alpha} + \sqrt{N}g\tilde{\beta} - i\bar{\mathcal{E}}_0, \quad (16.15a)$$

$$\dot{\tilde{\beta}} = -\frac{\gamma}{2}\tilde{\beta} - \sqrt{N}g\tilde{\alpha}, \quad (16.15b)$$

$$\dot{\tilde{\eta}} = -2\kappa\tilde{\eta} + \sqrt{2}\sqrt{N}g\tilde{\zeta} - i\sqrt{2}\bar{\mathcal{E}}_0\tilde{\alpha}, \quad (16.15c)$$

$$\dot{\tilde{\zeta}} = -(\kappa + \gamma/2)\tilde{\zeta} - \sqrt{2}\sqrt{N}g\tilde{\eta} + \sqrt{2}\sqrt{N-1}g\tilde{\theta} - i\bar{\mathcal{E}}_0\tilde{\beta}, \quad (16.15d)$$

$$\dot{\tilde{\theta}} = -\gamma\tilde{\theta} - \sqrt{2}\sqrt{N-1}g\tilde{\zeta}. \quad (16.15e)$$

Note 16.1. It is possible to develop the pure-state expansion from a self-consistent truncation of the matrix element equations of motion with the density matrix written in the product state basis—i.e., with all atoms explicitly labeled. In this approach we would initially distinguish between the like-atom matrix elements, $\langle\{j\}, -N/2+1|\tilde{\rho}|\{j\}, -N/2+1\rangle$, and the unlike-atom matrix elements $\langle\{j\}, -N/2+1|\tilde{\rho}|\{k\}, -N/2+1\rangle$, $k \neq j$; the former are equal for all j, by symmetry, and the latter are equal for all $j \neq k$. It is then shown, on the basis of the self-consistent truncation, that like- and unlike-atom matrix elements are also equal. The approach provides an alternative path to the conclusion that process (16.12) may be neglected.

It may be helpful in conclusion to look at what we have done from a more formal point of view. Formally, we may write the *master equation for many-atom cavity QED*—alternatively, the master equation for optical bistability (Eq. 15.1)—as

$$\dot{\tilde{\rho}} = \tilde{\mathcal{L}}\tilde{\rho}, \quad (16.16)$$

with the superoperator $\tilde{\mathcal{L}}$ defined by

$$\tilde{\mathcal{L}} \equiv g[a^\dagger J_- - aJ_+, \cdot\,] - i[\bar{\mathcal{E}}_0 a^\dagger + \bar{\mathcal{E}}_0^* a, \cdot\,] \\ + \frac{\gamma}{2}\left(\sum_{j=1}^{N} 2\sigma_{j-}\cdot\sigma_{j+} - \tfrac{1}{2}J_z\cdot - \cdot\tfrac{1}{2}J_z - N\cdot\right) + \frac{\gamma_p}{2}\left(\sum_{j=1}^{N}\sigma_{jz}\cdot\sigma_{jz} - N\cdot\right) \\ + \kappa(2a\cdot a^\dagger - a^\dagger a\cdot - \cdot a^\dagger a), \quad (16.17)$$

generalizing (13.89) to many atoms. The pure-state factorization follows from these equations by making the approximation ($\gamma_p = 0$)

$$\tilde{\mathcal{L}} \approx g[a^\dagger J_- - aJ_+, \cdot\,] - i[\bar{\mathcal{E}}_0 a^\dagger, \cdot\,] - \frac{\gamma}{4}[J_z + N, \cdot\,]_+ - \kappa[a^\dagger a, \cdot\,]_+, \tag{16.18}$$

where $[\ ,\ ]_+$ denotes the anticommutator. Then, substituting the factorized form $\tilde{\rho} \approx |\tilde{\psi}\rangle\langle\tilde{\psi}|$ into the approximate equation, the pure state $|\tilde{\psi}\rangle$ satisfies the Schrödinger equation

$$\frac{d|\tilde{\psi}\rangle}{dt} = \left[g(a^\dagger J_- - aJ_+) - i\bar{\mathcal{E}}_0 a^\dagger - \frac{\gamma}{4}(J_z + N) - \kappa a^\dagger a\right]|\tilde{\psi}\rangle. \tag{16.19}$$

The equations of motion for state amplitudes (16.15) are readily recovered from this equation.

Note that the factorized solution is permitted because after making the approximation (16.18) $\tilde{\mathcal{L}}$ is a sum of commutators and anticommutators. It is specifically the terms $\sum_{j=1}^{N} 2\sigma_{j-} \cdot \sigma_{j+}$ and $2a \cdot a^\dagger$ in the master equations—the terms omitted under the approximation—that impose the usual mixed-state character on the density matrix. We have also dropped the term $\bar{\mathcal{E}}_0^* a$, as it only contributes at higher-order in $|\bar{\mathcal{E}}_0|$; but this plays no role in justifying the factorization.

Note 16.2. Equation 16.19 is a Schrödinger equation with a non-Hermitian Hamiltonian. It describes a nonunitary evolution which does not preserve the state norm. The equation is nevertheless acceptable, given our aim of solving for density matrix elements to dominant order in $|\bar{\mathcal{E}}_0|$. Thus the comment in the paragraph below (16.4) applies here as well.

Note 16.3. Approximation (16.18) omits nonradiative dephasing. It should be noted that if dephasing is included, the pure-state factorization can no longer be consistently carried through. Thus there is a fundamental difference between radiative and nonradiative damping. This is apparent from the decay (without coherent driving) of a single atom. The matrix element equations are

$$\dot{\tilde{\rho}}_{12} = -\left(\frac{\gamma}{2} + \gamma_p\right)\tilde{\rho}_{12}, \qquad \dot{\tilde{\rho}}_{22} = -\gamma\tilde{\rho}_{22}, \quad \text{and} \quad \dot{\tilde{\rho}}_{11} = \gamma\tilde{\rho}_{22}.$$

For weak excitation ($\tilde{\rho}_{22} \ll \tilde{\rho}_{11}$), we might propose a pure-state solution

$$\tilde{\rho}(t) \approx |\tilde{\psi}(t)\rangle\langle\tilde{\psi}(t)|, \quad \text{with} \quad \tilde{\psi}(t) = |1\rangle + \tilde{\beta}(t)|2\rangle.$$

This proposal is consistent with the matrix element equations when $\gamma_p = 0$—one need only require that $\dot{\tilde{\beta}} = -(\gamma/2)\tilde{\beta}$. On the other hand, there exists no consistent equation of motion for $\tilde{\beta}$ when $\gamma_p \neq 0$.

In the introduction to Sect. 16.1 we gave a physical argument for the pure-state factorization. To reiterate, the basic requirement is that photon emission events are rare on the scale of the system correlation time—both emission through the cavity mirrors and spontaneous emission. The correlation time sets the scale because it governs how long it takes for the system to relax to a stationary state; thus, if we associate a photon emission with a fluctuation away from the stationary state, it must be rare that a second emission interrupts the regression of the fluctuation.

We are now in a position to be a little more precise about this picture. The stationary state referred to is the stationary solution to Schrödinger equation (16.19) [equivalently (16.8)]; this is the equation that governs the regression of the fluctuations after an emission. In the master equation the emission events are accounted for by terms

$$\mathcal{J}_A = \sum_{j=1}^{N} \mathcal{J}_A^j \equiv \sum_{j=1}^{N} \gamma\sigma_{j-} \cdot \sigma_{j+}, \tag{16.20a}$$

$$\mathcal{J}_a \equiv 2\kappa a \cdot a^\dagger, \tag{16.20b}$$

the terms that are dropped in making the approximation (16.18); specifically, spontaneous emission from atom j is expressed through the transition

$$|\tilde{\psi}\rangle\langle\tilde{\psi}| \rightarrow \mathcal{J}_A^j|\tilde{\psi}\rangle\langle\tilde{\psi}| = \gamma\big(\sigma_{j-}|\tilde{\psi}\rangle\big)\big(\langle\tilde{\psi}|\sigma_{j+}\big), \tag{16.21a}$$

and emission of a photon from the cavity through

$$|\tilde{\psi}\rangle\langle\tilde{\psi}| \rightarrow \mathcal{J}_a|\tilde{\psi}\rangle\langle\tilde{\psi}| = 2\kappa\big(a|\tilde{\psi}\rangle\big)\big(\langle\tilde{\psi}|a^\dagger\big). \tag{16.21b}$$

By dropping the terms (16.20), we assume that the emission events are rare: (i) that they are separated by long periods in the stationary state given by the stationary solution to (16.19), and (ii) that they rarely interrupt the otherwise coherent evolution into the stationary state.

Note 16.4. The approximation leading to the pure-state factorization can be extremely restrictive. It assumes that typically no photon emissions, cavity or spontaneous, take place during a randomly selected interval of duration $\tau_{\rm coh} \equiv \max(\kappa^{-1}, 2\gamma^{-1})$. Taking $\kappa \sim \gamma/2$, the condition on the rate of cavity emissions requires that the intracavity photon number be small. Even $\langle a^\dagger a\rangle_{\rm ss} \sim 10^{-2}$ may not be small enough, however, when the additional constraint on spontaneous emission is considered. Compare the spontaneous emission rate for weak excitation—given by $\gamma N\frac{1}{2}(\langle\bar{J}_z\rangle_{\rm ss}+1) = \gamma N X^2/2$—with the rate of cavity emissions $2\kappa\langle a^\dagger a\rangle_{\rm ss} = 2\kappa n_{\rm sat}X^2$; the estimates are taken from (15.69) and (15.70c) using the scaling (15.51). The spontaneous emission rate exceeds the rate of cavity emissions by a factor $N/2n_{\rm sat}\xi = 2C$ (Eq. 15.56). Since experiments realize values of $2C$ on the order of 10^2 [16.2,16.3], intracavity photon numbers as low as 10^{-4}, or less, may be needed to strictly satisfy the requirements of the pure-state factorization.

16.1.3 Forwards Photon Scattering for N Atoms in a Cavity

Let us now apply the pure-state factorization to calculate the second-order correlation function of the forward-scattered light. The calculation generalizes the result of Sect. 13.2.3 to an arbitrary number of atoms and arbitrary $\xi = 2\kappa/\gamma$. We compare the result obtained with that obtained from the linear theory of fluctuations in Sect. 15.2.7; thus, we are able to perform an explicit check on the limitations of the linearization.

The calculation begins, as in (13.77), from the expression for the second-order correlation function,

$$g_{\rightarrow}^{(2)}(\tau) = \frac{\langle(\tilde{a}^{\dagger}\tilde{a})(\tau)\rangle_{\tilde{\rho}(0)=\tilde{\rho}'_{\mathrm{ss}}}}{\langle\tilde{a}^{\dagger}\tilde{a}\rangle_{\mathrm{ss}}}, \tag{16.22}$$

with

$$\tilde{\rho}'_{\mathrm{ss}} \equiv \frac{\tilde{a}\tilde{\rho}_{\mathrm{ss}}\tilde{a}^{\dagger}}{\mathrm{tr}(\tilde{a}\rho_{\mathrm{ss}}\tilde{a}^{\dagger})}. \tag{16.23}$$

Making use of the pure-state factorization (16.5), this expression is recast in simpler form as

$$g_{\rightarrow}^{(2)}(\tau) = \frac{\langle\tilde{\psi}(\tau)|\tilde{a}^{\dagger}\tilde{a}|\tilde{\psi}(\tau)\rangle}{\langle\tilde{\psi}_{\mathrm{ss}}|\tilde{a}^{\dagger}\tilde{a}|\tilde{\psi}_{\mathrm{ss}}\rangle}, \tag{16.24}$$

where

$$|\tilde{\psi}(\tau)\rangle = |0\rangle^{(N)}|0\rangle_a + \tilde{\alpha}(\tau)|0\rangle^{(N)}|1\rangle_a + \tilde{\beta}(\tau)|1\rangle^{(N)}|0\rangle_a, \tag{16.25}$$

with initial conditions, $\tilde{\alpha}(0)$ and $\tilde{\beta}(0)$, defined by

$$|\tilde{\psi}(0)\rangle \equiv \frac{\tilde{a}|\tilde{\psi}_{\mathrm{ss}}\rangle}{\sqrt{\langle\tilde{\psi}_{\mathrm{ss}}|\tilde{a}^{\dagger}\tilde{a}|\tilde{\psi}_{\mathrm{ss}}\rangle}} e^{-i\arg\left({}_a\langle 0|^{(N)}\langle 0|\tilde{a}|\tilde{\psi}_{\mathrm{ss}}\rangle\right)}, \tag{16.26}$$

where $|\tilde{\psi}_{\mathrm{ss}}\rangle$ is the stationary state obtained from the steady-state solutions to (16.15a)–(16.15e); the phase factor $e^{-i\arg\left({}_a\langle 0|^{(N)}\langle 0|\tilde{a}|\tilde{\psi}_{\mathrm{ss}}\rangle\right)}$ is inserted in order to make the coefficient of the ground state in the expansion of $|\tilde{\psi}(\tau)\rangle$ real. From (16.24), (16.25), and (16.14), we arrive at the simple expression

$$g_{\rightarrow}^{(2)}(\tau) = \frac{|\tilde{\alpha}(\tau)|^2}{|\tilde{\alpha}_{\mathrm{ss}}|^2}, \tag{16.27}$$

where $\tilde{\alpha}(\tau)$, the conditional one-photon amplitude, is the solution to the coupled oscillator equations (16.15a) and (16.16b) with initial conditions

$$\tilde{\alpha}(0) = \frac{\sqrt{2}\tilde{\eta}_{\mathrm{ss}}}{\tilde{\alpha}_{\mathrm{ss}}}, \qquad \tilde{\beta}(0) = \frac{\tilde{\zeta}_{\mathrm{ss}}}{\tilde{\alpha}_{\mathrm{ss}}}; \tag{16.28}$$

$\tilde{\alpha}_{\mathrm{ss}}$, $\tilde{\eta}_{\mathrm{ss}}$, and $\tilde{\zeta}_{\mathrm{ss}}$ satisfy (16.15a)–(16.15e) in the steady state.

Calculation of the correlation function has been reduced to two straightforward tasks: first, find the steady-state solutions to (16.15a)–(16.15e), then solve for $\tilde{\alpha}(\tau)$ with initial condition (16.28). Considering the steady state, it is trivial to solve for the one-quantum amplitudes

$$\tilde{\alpha}_{\rm ss} = -i\frac{\bar{\mathcal{E}}_0/\kappa}{1+2C}, \tag{16.29a}$$

$$\tilde{\beta}_{\rm ss} = i\frac{\sqrt{N}g}{\gamma/2}\frac{\bar{\mathcal{E}}_0/\kappa}{1+2C}, \tag{16.29b}$$

and also, from (16.15e), to obtain

$$\tilde{\theta}_{\rm ss} = -\sqrt{2}\frac{\sqrt{N-1}g}{\gamma}\tilde{\zeta}_{\rm ss}. \tag{16.30}$$

Substituting these results into (16.15c) and (16.15d), we then obtain a pair of coupled equations for the two-quanta amplitudes $\tilde{\eta}_{\rm ss}$ and $\tilde{\zeta}_{\rm ss}$:

$$\tilde{\eta}_{\rm ss} - \frac{1}{\sqrt{2}}\frac{\sqrt{N}g}{\kappa}\tilde{\zeta}_{\rm ss} = -\frac{1}{\sqrt{2}}\frac{(\bar{\mathcal{E}}_0/\kappa)^2}{1+2C}, \tag{16.31a}$$

$$\sqrt{2}\frac{\sqrt{N}g}{\gamma/2}\tilde{\eta}_{\rm ss} + [1+\xi(1+2C-2C_1)]\tilde{\zeta}_{\rm ss} = \xi\frac{\sqrt{N}g}{\gamma/2}\frac{(\bar{\mathcal{E}}_0/\kappa)^2}{1+2C}. \tag{16.31b}$$

The solution to the pair of equations is

$$\tilde{\zeta}_{\rm ss} = \frac{\sqrt{N}g}{\gamma/2}\frac{(\bar{\mathcal{E}}_0/\kappa)^2}{1+2C}\frac{1}{1+2C-2C_1\xi/(1+\xi)}, \tag{16.32a}$$

$$\tilde{\eta}_{\rm ss} = -\frac{1}{\sqrt{2}}\frac{(\bar{\mathcal{E}}_0/\kappa)^2}{1+2C}\left[1-2C\frac{1}{1+2C-2C_1\xi/(1+\xi)}\right]. \tag{16.32b}$$

Thus, (16.29), (16.30), and (16.32) provide the required stationary state $|\tilde{\psi}\rangle_{\rm ss}$.

Turning now to the second task, the general solution to (16.15a) and (16.15b) is governed by eigenvalues $\Lambda_\pm = \gamma\bar{\Lambda}_\pm$, with $\bar{\Lambda}_\pm$ given in (15.138); we may write

$$\tilde{\alpha}(\tau) = -i\frac{\bar{\mathcal{E}}_0/\kappa}{1+2C} + Ae^{\Lambda_+\tau} + Be^{\Lambda_-\tau}, \tag{16.33}$$

where the first term is the steady-state solution (16.29a), and the constants A and B are to be determined from the initial conditions. Matching (16.33) with the expression for $\tilde{\alpha}(0)$ using (16.28), (16.29a), and (16.32b), we have

$$\begin{aligned} A+B &= i\frac{\bar{\mathcal{E}}_0/\kappa}{1+2C} - i(\bar{\mathcal{E}}_0/\kappa)\left[1-\frac{2C}{1+2C-2C_1\xi/(1+\xi)}\right] \\ &= -i(\bar{\mathcal{E}}_0/\kappa)\left[\frac{2C}{1+2C}-\frac{2C}{1+2C-2C_1\xi/(1+\xi)}\right] \\ &= i\frac{\bar{\mathcal{E}}_0/\kappa}{1+2C}\frac{2C_1\xi}{1+\xi}\frac{2C}{1+2C-2C_1\xi/(1+\xi)}, \end{aligned} \tag{16.34}$$

and matching the derivative of $\tilde{\alpha}(\tau)$ at $\tau = 0$, we have

$$(\Lambda_+ + \kappa)A + (\Lambda_- + \kappa)B = i\bar{\mathcal{E}}_0 \frac{1}{1+2C} + i\bar{\mathcal{E}}_0 \frac{2C}{1+2C-2C_1\xi/(1+\xi)} - i\bar{\mathcal{E}}_0$$
$$= i\kappa \frac{\bar{\mathcal{E}}_0/\kappa}{1+2C} \frac{2C_1\xi}{1+\xi} \frac{2C}{1+2C-2C_1\xi/(1+\xi)}, \tag{16.35}$$

where we use also (16.15a) and (16.32a). Subtracting $\kappa(A+B)$ from both sides of this expression and using (16.34) yields

$$\Lambda_+ A + \Lambda_- B = 0; \tag{16.36}$$

hence, on substituting $\Lambda_\pm = \gamma\bar{\Lambda}_\pm = \frac{1}{2}(\kappa + \gamma/2) \pm G$ (Eqs. 15.138 and 15.149),

$$A - B = -i\frac{\frac{1}{2}(\kappa+\gamma/2)}{G}(A+B). \tag{16.37}$$

If follows from (16.33), (16.34), and (16.37) that the solution for the conditional one-photon amplitude is

$$\tilde{\alpha}(\tau) = -i\frac{\bar{\mathcal{E}}_0/\kappa}{1+2C}\left\{1 - 2C_1\frac{\xi}{1+\xi}\frac{2C}{1+2C-2C_1\xi/(1+\xi)}\right.$$
$$\left.\times e^{-\frac{1}{2}(\kappa+\gamma/2)\tau}\left[\cos(G\tau) + \frac{\frac{1}{2}(\kappa+\gamma/2)}{G}\sin(G\tau)\right]\right\}. \tag{16.38}$$

Finally, (16.27), (16.29a), and (16.38) yield the *second-order correlation function of forwards scattering in cavity QED in the weak-excitation limit (equal coupling strengths)*:

$$g^{(2)}_{\rightarrow}(\tau) = \left\{1 - 2C_1\frac{\xi}{1+\xi}\frac{2C}{1+2C-2C_1\xi/(1+\xi)}\right.$$
$$\left.\times e^{-\frac{1}{2}(\kappa+\gamma/2)\tau}\left[\cos(G\tau) + \frac{\frac{1}{2}(\kappa+\gamma/2)}{G}\sin(G\tau)\right]\right\}^2, \tag{16.39}$$

where $G \equiv \sqrt{Ng^2 - \frac{1}{4}(\kappa - \gamma/2)^2}$. This result holds for any number of atoms (even $N = 1$) and reproduces (13.96) in the bad-cavity limit ($\xi \gg 1$).

It is notable that (16.39) is a perfect square, the square of the one-photon amplitude conditioned upon the detection of a forward-scattered photon at $\tau = 0$. Qualitatively, the regression of the amplitude is simple and unremarkable—merely a damped oscillation, whose damping rate and frequency are determined by the eigenvalues of the coupled oscillator equations (16.15a) and (16.15b). A number of interesting features appear, however, when the inequalities constraining $g^{(2)}_{\rightarrow}(\tau)$ for a classical noise process are considered. These follow, in the first place, from the *negative* deviation of the one-photon amplitude from its steady-state value at $\tau = 0$. The minus sign produces the photon antibunching illustrated by Fig. 15.2, and when using (16.39), rather than (15.148), there is no longer any requirement that the magnitude of the deviation be small. Thus, in Fig. 16.1a we see an example of photon antibunching with $g^{(2)}(0) = 0$—antibunching of the same order as in free-space resonance fluorescence (Sect. 2.3.5). Although $N = 1$ in this example, it is clear that $g^{(2)}(0) = 0$ can be achieved for a large number of atoms as well; the requirement for large $2C$ is only that $2C_1\xi/(1+\xi) = 1$.

With a moderate increase in the dipole coupling strength, the correlation function changes rather dramatically, from Fig. 16.1a to Fig. 16.1b. All that happens to the underlying state amplitude is that the magnitude and frequency of its excursion increases. After the square is taken, however, we obtain the interesting result shown in the figure, with its new violation of classicality. The correlation function for a classical intensity is constrained by the Schwartz inequality [16.4]

$$|g^{(2)}(\tau) - 1| \leq g^{(2)}(0) - 1 \geq 0, \tag{16.40}$$

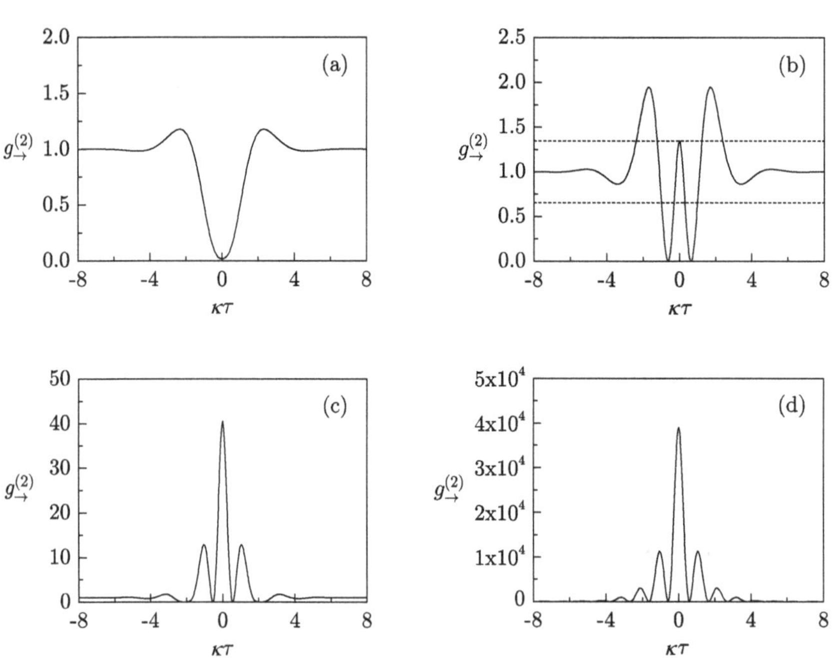

Fig. 16.1. Photon correlations in forwards scattering for single-atom cavity QED. The second-order correlation function (16.39) is plotted for $N = 1$ and (**a**) $g/\kappa = 1.35$, $\xi = 1$; (**b**) $g/\kappa = 1.85$, $\xi = 1$; (**c**) $g/\kappa = 3.0$, $\xi = 1$; (**d**) $g/\kappa = 3.0$, $\xi = 5$. The *dashed lines* in frame (**b**) mark upper and lower bounds derived from the Schwartz inequality (16.40)

which generalizes the usual definition of photon antibunching (second inequality only) to give the upper and lower bounds depicted in Fig. 16.1b. Violation of the upper bound has been observed by Mielke and coworkers [16.5], though in a many-atom system rather than for a single atom, as shown [see Fig. 16.5b]. On increasing the coupling strength still further, the violation of inequality (16.40) disappears and the correlation function looks like Fig. 16.1c. In a bad cavity and for a single atom, truly enormous intensity fluctuations are possible, for relatively moderate values of $2C_1$; $g_{\rightarrow}^{(2)}(0)$ approaches a limiting value of $(1-4C_1^2)^2$. Figure 16.1d shows an example of this extreme regime.

Note 16.5. Although the correlation functions plotted in Figs. 16.1c and d do not violate any classical inequality, in both cases the underlying field amplitude, (16.38), does. The amplitude rather than the intensity may be measured using conditional homodyne detection. It violates the classical constraint on the ratio $|A|/B$ discussed in Sect. 15.2.5 [16.6].

Figure 16.2 shows something of how the intensity fluctuations are affected by an increase in the number of atoms. Qualitatively, similar phenomenology is seen regardless of the number of atoms. On the other hand, the extreme size of the fluctuations shown by Fig. 16.1d is largely a single-atom feature. Considering the bad-cavity limit and $2C \gg 1$, for one atom, we have $g^{(2)}(0) = (1-2C_1^2)^2$, while $g^{(2)}(0) = (1-2C_1)^2$ in the many-atom case. Increasing the number of atoms with $2C_1$ fixed can change the magnitude of the fluctuations by orders of magnitude. Figure 16.2a illustrates the change from Fig. 16.1d, where the number of atoms is changed from $N = 1$ to $N = 15$. If the vacuum Rabi frequency is held fixed, rather than $2C_1$, the change can be even more dramatic; Fig. 16.2b replaces Fig. 16.1d, for example.

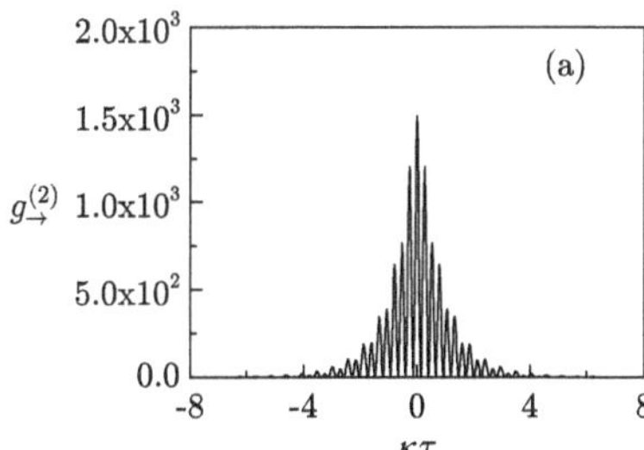

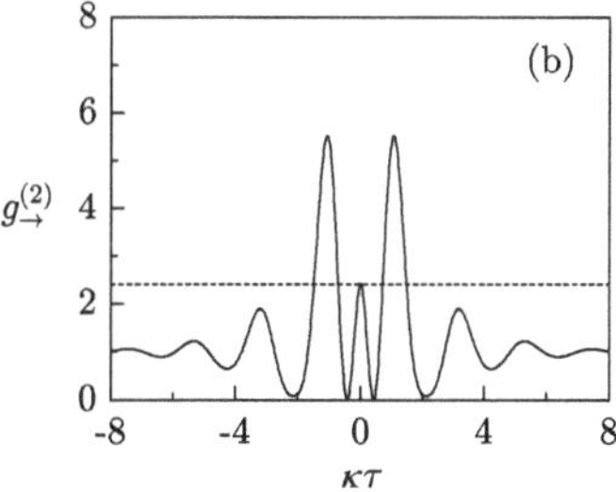

Fig. 16.2. Photon correlations in forwards scattering for many-atom cavity QED. The second-order correlation function (16.39) is plotted for $N = 15$, $\xi = 5$, and (**a**) $g/\kappa = 3.0$, as in Fig. 16.1(**d**); (**b**) $g/\kappa = 0.77$, giving the same vacuum Rabi frequency as in Fig. 16.1(**d**). The *dashed line* in frame (**b**) marks the upper bound derived from the Schwartz inequality (16.40)

16.1.4 Corrections to the Small-Noise Approximation

The linearized treatment of fluctuations led us to the result (15.148) for the second-order correlation function of forwards scattering, which is in general disagreement with (16.39). A comparison of the two expressions provides a quantitative illustration of the breakdown of linear fluctuation theory under strong-coupling conditions. For the purposes of this comparison, it is helpful to rewrite (16.39) in the form

$$g_{\rightarrow}^{(2)}(\tau) = [1 + \lambda \Gamma(\tau)]^2 \quad \text{(pure-state factorization)}, \tag{16.41}$$

with

$$\Gamma(\tau) \equiv e^{-\frac{1}{2}(\kappa+\gamma/2)\tau}\left[\cos(G\tau) + \frac{\frac{1}{2}(\kappa+\gamma/2)}{G}\sin(G\tau)\right], \tag{16.42}$$

$$\lambda \equiv -2C_1 \frac{\xi}{1+\xi}\frac{2C}{1+2C-2C_1\xi/(1+\xi)}. \tag{16.43}$$

Equation 15.148 may be similarly rewritten as

$$g_{\rightarrow}^{(2)}(\tau) = 1 + 2\lambda' \Gamma(\tau) \quad \text{(linear fluctuation theory)}, \tag{16.44}$$

with

$$\lambda' \equiv -2C_1 \frac{\xi}{1+\xi}\frac{2C}{1+2C}. \tag{16.45}$$

There are then two distinct differences between these expressions. First, after expanding the square on the right-hand side of (16.41), the term $\lambda^2\Gamma^2(\tau)$ must be dropped in order to recover something similar to (16.44). Then there is the difference between the prefactors λ and λ'.

The missing term, $\lambda'^2\Gamma^2(\tau)$, can in fact be recovered working entirely within the small noise approximation. The omission occured when we dropped $\langle \Delta\tilde{\bar{a}}^\dagger(0)\Delta\tilde{\bar{a}}^\dagger(\tau)\Delta\tilde{\bar{a}}(\tau)\Delta\tilde{\bar{a}}(0)\rangle_{\text{ss}}$ from the expansion (15.144). To show this, let us return briefly to the calculation of Sect. 15.2.7, where now, for simplicity, we consider only $g_{\rightarrow}^{(2)}(0)$; extending the calculation to include time dependence follows in a straightforward way from the $\tau = 0$ case.

Working from (15.142), introducing the decomposition of field operators into a mean value and fluctuation (Eqs. 15.89a and 15.89b), the second-order correlation function at zero delay is

$$g_{\rightarrow}^{(2)}(0) = \frac{\left\langle \left(\langle\tilde{\bar{a}}^\dagger\rangle_{\text{ss}} + \Delta\tilde{\bar{a}}^\dagger\right)^2\left(\langle\tilde{\bar{a}}\rangle_{\text{ss}} + \Delta\tilde{\bar{a}}\right)^2\right\rangle_{\text{ss}}}{\left\langle\left(\langle\tilde{\bar{a}}^\dagger\rangle_{\text{ss}} + \Delta\tilde{\bar{a}}^\dagger\right)\left(\langle\tilde{\bar{a}}\rangle_{\text{ss}} + \Delta\tilde{\bar{a}}\right)\right\rangle_{\text{ss}}^2}. \tag{16.46}$$

Since the fluctuations are Gaussian in the small noise approximation, we may expand the numerator and denominator setting all moments of odd order to zero. The numerator expands as

$$\begin{aligned}&\Big\langle\big(\langle\tilde{\bar{a}}^\dagger\rangle_{ss}+\Delta\tilde{\bar{a}}^\dagger\big)^2\big(\langle\tilde{\bar{a}}\rangle_{ss}+\Delta\tilde{\bar{a}}\big)^2\Big\rangle_{ss}\\&=\big(\langle\tilde{\bar{a}}^\dagger\rangle_{ss}\langle\tilde{\bar{a}}\rangle_{ss}\big)^2\Bigg[\left|1+\frac{\langle(\Delta\tilde{\bar{a}})^2\rangle_{ss}}{\langle\tilde{\bar{a}}\rangle_{ss}^2}\right|^2\\&\quad+\frac{\langle(\Delta\tilde{\bar{a}}^\dagger)^2(\Delta\tilde{\bar{a}})^2\rangle_{ss}-\langle(\Delta\tilde{\bar{a}}^\dagger)^2\rangle_{ss}\langle(\Delta\tilde{\bar{a}})^2\rangle_{ss}}{\big(\langle\tilde{\bar{a}}^\dagger\rangle_{ss}\langle\tilde{\bar{a}}\rangle_{ss}\big)^2}+4\frac{\langle\Delta\tilde{\bar{a}}^\dagger\Delta\tilde{\bar{a}}\rangle_{ss}}{\langle\tilde{\bar{a}}^\dagger\rangle_{ss}\langle\tilde{\bar{a}}\rangle_{ss}}\Bigg],\end{aligned}\tag{16.47}$$

while for the denominator we have

$$\Big\langle\big(\langle\tilde{\bar{a}}^\dagger\rangle_{ss}+\Delta\tilde{\bar{a}}^\dagger\big)\big(\langle\tilde{\bar{a}}\rangle_{ss}+\Delta\tilde{\bar{a}}\big)\Big\rangle_{ss}^2=\big(\langle\tilde{\bar{a}}^\dagger\rangle_{ss}\langle\tilde{\bar{a}}\rangle_{ss}\big)^2\left(1+\frac{\langle\Delta\tilde{\bar{a}}^\dagger\Delta\tilde{\bar{a}}\rangle_{ss}}{\langle\tilde{\bar{a}}^\dagger\rangle_{ss}\langle\tilde{\bar{a}}\rangle_{ss}}\right)^2.\tag{16.48}$$

Previously we neglected the second term inside the square bracket in (16.44); we argued that it is a term of second order in the expansion parameter N^{-1} and should therefore be neglected for self-consistency—terms of the same order are already dropped in setting up the linearized treatment of fluctuations. On the other hand, one might argue in favor of keeping the term, which is, at least, an approximate representation of the higher-order corrections to linear theory. We may do so by evaluating it using the Gaussian moment theorem [16.7]; thus, we write

$$\langle(\Delta\tilde{\bar{a}}^\dagger)^2(\Delta\tilde{\bar{a}})^2\rangle_{ss}=\langle(\Delta\tilde{\bar{a}}^\dagger)^2\rangle_{ss}\langle(\Delta\tilde{\bar{a}})^2\rangle_{ss}+2\big(\langle\Delta\tilde{\bar{a}}^\dagger\Delta\tilde{\bar{a}}\rangle_{ss}\big)^2.\tag{16.49}$$

The steps taken so far are completely general with no limitation placed on the excitation strength. Specializing now to weak excitation, we make use of the inequality

$$\frac{\langle\Delta\tilde{\bar{a}}^\dagger\Delta\tilde{\bar{a}}\rangle_{ss}}{\langle\tilde{\bar{a}}^\dagger\tilde{\bar{a}}\rangle_{ss}}\ll\left|\frac{\langle(\Delta\tilde{\bar{a}})^2\rangle_{ss}}{\big(\langle\tilde{\bar{a}}\rangle_{ss}\big)^2}\right|,\tag{16.50}$$

which holds on the basis of (15.120a) and (15.120b). The inequality shows that the first term on the right-hand side of (16.47) is dominant; therefore, retaining this term only, we arrive at the result

$$g_{\rightarrow}^{(2)}(0)=\left|1+\frac{\langle(\Delta\tilde{\bar{a}})^2\rangle_{ss}}{\langle\tilde{\bar{a}}\rangle_{ss}^2}\right|^2.\tag{16.51}$$

The additional term in this expression, $|\langle(\Delta\tilde{\bar{a}}\rangle_{ss})^2/\langle\tilde{\bar{a}}\rangle_{ss}^2|^2$, finds its origin in the previously omitted moment $\langle(\Delta\tilde{\bar{a}}^\dagger)^2(\Delta\tilde{\bar{a}})^2\rangle_{ss}\approx\langle(\Delta\tilde{\bar{a}}^\dagger)^2\rangle_{ss}\langle(\Delta\tilde{\bar{a}})^2\rangle_{ss}$.

Inclusion of time dependence calls for a somewhat more involved calculation, but leads in the end to the natural generalization of (16.51); it yields the *second-order correlation function of forwards scattering for many atoms in a cavity including second-order corrections within the linearized treatment of fluctuations*:

$$g^{(2)}_{\rightarrow}(\tau) = \left|1 + \frac{\langle \Delta\tilde{\bar{a}}(0)\Delta\tilde{\bar{a}}(\tau)\rangle_{ss}}{\langle\tilde{\bar{a}}\rangle^2_{ss}}\right|^2$$

$$= \left\{1 - N^{-1}\frac{\xi}{1+\xi}\frac{4C^2}{1+2C}\right.$$

$$\left.\times e^{-\frac{1}{2}(\kappa+\gamma/2)\tau}\left[\cos(G\tau) + \frac{\frac{1}{2}(\kappa+\gamma/2)}{G}\sin(G\tau)\right]\right\}^2, \quad (16.52a)$$

or

$$g^{(2)}_{\rightarrow}(\tau) = [1 + \lambda' \Gamma(\tau)]^2, \quad (16.52b)$$

where the explicit form of the correlation function $\langle \Delta\tilde{\bar{a}}(0)\Delta\tilde{\bar{a}}(\tau)\rangle_{ss}$ has been taken from (15.147).

The end result shows that retaining the second-order moment is, indeed, inconsistent, since the difference between λ in (16.41) and λ' in (16.52b) makes the new result incorrect at the order N^{-2}; specifically, the term $-2C_1\xi/(1+\xi) = -N^{-1}2C\xi/(1+\xi)$ appearing in the denominator of λ is absent from λ'. This term affects the overall magnitude of the fluctuation $g^{(2)}_{\rightarrow}(\tau)-1$. Large errors can arise if (16.52a) is applied to a single atom under conditions of strong coupling to a single atom. Figure 16.3 illustrates this breakdown of linear fluctuation theory under cavity QED conditions. If the number of atoms is sufficiently large, the agreement between (16.39) and (16.52a) can be rather good, as shown in Fig. 16.3a. On the other hand, for $N = 1$ and moderately strong coupling, significant errors are made by (16.52a), at the level shown by Fig. 16.3b. In extreme cases, the error can underestimate the size of the fluctuation by more than an order of magnitude; an example is presented in Fig. 16.3c.

Note 16.6. The missing factor $-2C_1\xi/(1+\xi)$ illustrates, once again, how different system size parameters and scalings of the quantum fluctuations apply in the good- and bad-cavity limits [see the discussion below (15.102)]. While this factor is negligible in the good-cavity limit, even for $N = 1$, it requires $N \gg 1$ to be neglected in the bad-cavity limit.

16.1.5 Antibunching of Fluorescence for One Atom in a Cavity

Having considered the forwards scattering, it is now natural to ask about the fluorescence, about the side-scattered light. Early reports for many atoms in a cavity predicted interesting modifications compared with resonance fluorescence in free space [16.8, 16.9, 16.10, 16.11]. These reports were eventually shown to be incorrect. It was shown by Carmichael [16.12] and Lugiato [16.13] (and also Agarwal [16.14]) that there are no significant cavity-induced modifications to the fluorescence if the small-noise approximation holds. The fluorescence exhibits the Mollow spectrum [Sect. 2.3.4] and the usual photon bunching of resonant scattering from many independent sources [16.15]. Strictly, of

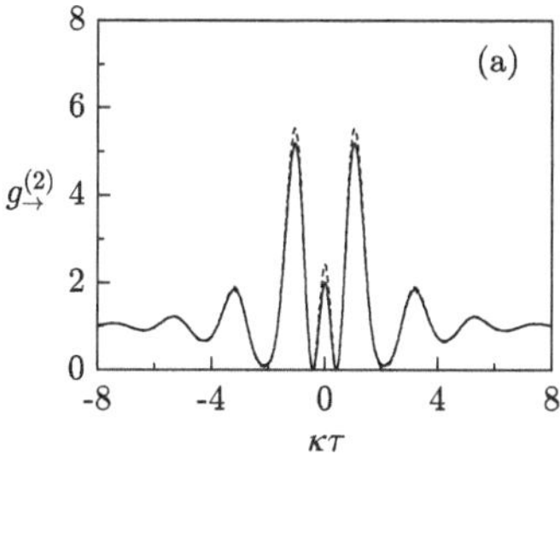

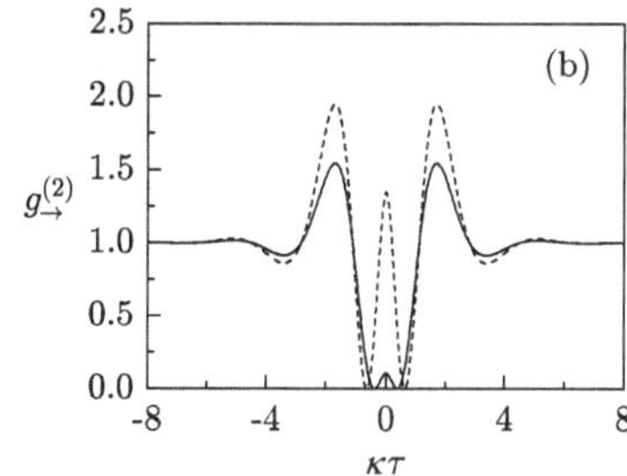

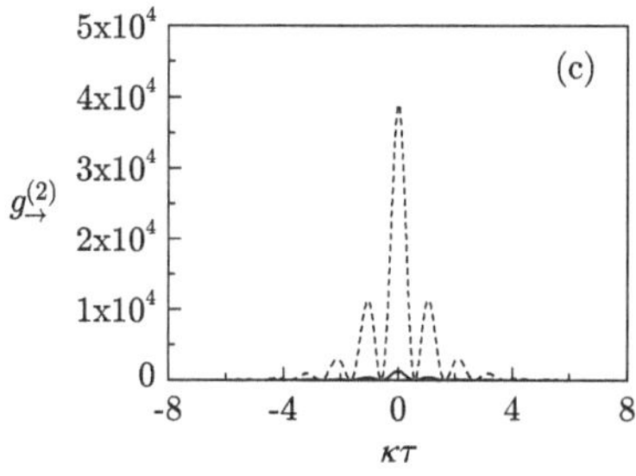

Fig. 16.3. Comparison of the correlation function with second-order corrections (16.52a) (*solid curves*) and the exact correlation function (16.39) (*dashed curves*): for (**a**) $g/\kappa = 0.77$, $\xi = 5$, and $N = 15$ [Fig. 16.2(**b**)]; (**b**) $g/\kappa = 1.85$, $\xi = 1$, and $N = 1$ [Fig. 16.1(**b**)]; (**c**) $g/\kappa = 3.0$, $\xi = 5$, and $N = 1$ [Fig. 15.1(**d**)]

course, the sources are not independent due to their coupling though the intracavity field (Sect. 15.2.4). This brings only very small corrections, though, of order N^{-1}. We note then that there is a fundamental difference between scattering in the forwards direction and scattering from the sides of the cavity: the latter does not take place into a single collective mode; thus, while there are nonvanishing correlations between pairs of atoms (Sect. 15.2.4), these unlike-atom correlations do not add coherently (constructively) to correlation functions measured on the side-scattered light as in an expression like (15.94).

From what we have just seen, however, the situation should be very different under cavity QED conditions—for strong coupling in the single-atom sense. In that case, corrections of order N^{-1} will appear at dominant order, and we might anticipate the side-scattered light to differ significantly from the scattering of ordinary resonance fluorescence. To illustrate the point, let us compute the second-order correlation function of the side-scattered light for one atom in a cavity. We specialize again to the limit of weak excitation, so the calculation runs parallel to that of Sect. 16.1.3.

We begin from the expression for the second-order correlation function (16.22) with $\tilde{a} \to \tilde{\sigma}_-$,

$$g^{(2)}_{\text{side}}(\tau) = \frac{\langle(\tilde{\sigma}_+\tilde{\sigma}_-)(\tau)\rangle_{\tilde{\rho}(0)=\tilde{\rho}'_{\text{ss}}}}{\langle\tilde{\sigma}_+\tilde{\sigma}_-\rangle_{\text{ss}}}, \tag{16.53}$$

where

$$\tilde{\rho}'_{\text{ss}} \equiv \frac{\tilde{\sigma}_-\tilde{\rho}_{\text{ss}}\tilde{\sigma}_+}{\text{tr}(\tilde{\sigma}_-\rho_{\text{ss}}\tilde{\sigma}_+)}. \tag{16.54}$$

Using the pure-state factorization for one atom (Eqs. 16.5 and 16.7), this may be recast as

$$g^{(2)}_{\text{side}}(\tau) = \frac{\langle\tilde{\psi}(\tau)|\tilde{\sigma}_+\tilde{\sigma}_-|\tilde{\psi}(\tau)\rangle}{\langle\tilde{\psi}_{\text{ss}}|\tilde{\sigma}_+\tilde{\sigma}_-|\tilde{\psi}_{\text{ss}}\rangle}, \tag{16.55}$$

with the one-quantum expansion of the state

$$|\tilde{\psi}(\tau)\rangle = |1\rangle_A|0\rangle_a + \tilde{\alpha}(\tau)|1\rangle_A|1\rangle_a + \tilde{\beta}(\tau)|2\rangle_A|0\rangle_a, \tag{16.56}$$

and initial condition

$$|\tilde{\psi}(0)\rangle \equiv \frac{\tilde{\sigma}_-|\tilde{\psi}_{\text{ss}}\rangle}{\sqrt{\langle\tilde{\psi}_{\text{ss}}|\tilde{\sigma}_+\tilde{\sigma}_-|\tilde{\psi}_{\text{ss}}\rangle}} e^{-i\arg\left({}_a\langle 0|_A\langle 1|\tilde{a}|\tilde{\psi}_{\text{ss}}\rangle\right)}. \tag{16.57}$$

We obtain

$$g^{(2)}_{\text{side}}(\tau) = \frac{|\tilde{\beta}(\tau)|^2}{|\tilde{\beta}_{\text{ss}}|^2}, \tag{16.58}$$

where, using (16.56), (16.57), and the two-quanta expansion of the steady state (Eq. 16.7), the initial values of the conditional one-quantum amplitudes are

$$\tilde{\alpha}(0) = \frac{\tilde{\zeta}_{\text{ss}}}{\tilde{\beta}_{\text{ss}}}, \qquad \tilde{\beta}(0) = 0. \tag{16.59}$$

The steady-state amplitudes $\tilde{\beta}_{\text{ss}}$ and $\tilde{\zeta}_{\text{ss}}$ may be taken from (16.29a)–(16.32b) with the number of atoms set to $N = 1$.

The solution for the time-dependent amplitude $\tilde{\beta}(\tau)$ follows the steps below (16.32b). We write

$$\tilde{\beta}(\tau) = i\frac{g}{\gamma/2}\frac{\bar{\mathcal{E}}_0/\kappa}{1+2C_1} + Ae^{\Lambda_+\tau} + Be^{\Lambda_-\tau}, \tag{16.60}$$

where the initial condition requires

$$A + B = -i\frac{g}{\gamma/2}\frac{\bar{\mathcal{E}}_0/\kappa}{1+2C_1}. \tag{16.61}$$

Then, using (16.8b) and (16.29a) to match the initial value of the derivative,

$$\Lambda_+ A + \Lambda_- B = -\frac{g}{\gamma/2}(\bar{\mathcal{E}}_0/\kappa)\frac{\kappa + \gamma/2}{1 + \xi + 2C_1}, \tag{16.62}$$

or with $\Lambda_\pm = \gamma\bar{\Lambda}_\pm = -\frac{1}{2}(\kappa + \gamma/2) \pm g'$ (Eqs. 15.138, 15.149, and 13.180),

$$\begin{aligned} A - B &= -i\frac{\frac{1}{2}(\kappa + \gamma/2)}{g'}\left[2i\frac{g}{\gamma/2}\frac{\bar{\mathcal{E}}_0/\kappa}{1 + \xi + 2C_1} + (A + B)\right] \\ &= -i\frac{\frac{1}{2}(\kappa + \gamma/2)}{g'}(A + B)\left(-2\frac{1 + 2C_1}{1 + \xi + 2C_1} + 1\right) \\ &= -\frac{\frac{1}{2}(\kappa + \gamma/2)}{g'}(A + B)\frac{1 + 2C_1 - \xi}{1 + 2C_1 + \xi}. \end{aligned} \tag{16.63}$$

Thus, from (16.60), (16.61), and (16.63), the solution for the conditional dipole amplitude is

$$\begin{aligned} \tilde{\beta}(\tau) = i\frac{g}{\gamma/2}\frac{\bar{\mathcal{E}}_0/\kappa}{1 + 2C_1}\Bigg\{&1 - e^{-\frac{1}{2}(\kappa + \gamma/2)\tau} \\ &\times\left[\cos(g'\tau) + \frac{\kappa - \gamma'/2}{\kappa + \gamma'/2}\frac{\frac{1}{2}(\kappa + \gamma/2)}{g'}\sin(g'\tau)\right]\Bigg\}, \end{aligned} \tag{16.64}$$

where $\gamma' = \gamma(1 + 2C_1)$ is the cavity-enhanced emission rate. Finally, (16.58), (16.64), and (16.29b) yield the *second-order correlation function of the side-scattered light in cavity QED in the weak-excitation limit (single-atom case)*:

$$g^{(2)}_{\text{side}}(\tau) = \left\{1 - e^{-\frac{1}{2}(\kappa + \gamma/2)\tau}\left[\cos(g'\tau) + \frac{\kappa - \gamma'/2}{\kappa + \gamma'/2}\frac{\frac{1}{2}(\kappa + \gamma/2)}{g'}\sin(g'\tau)\right]\right\}^2. \tag{16.65}$$

The corresponding result for free-space resonance fluorescence in the weak-excitation limit is (Eq. 2.152)

$$g^{(2)}_{\text{side}}(\tau) = (1 - e^{-(\gamma/2)\tau})^2. \tag{16.66}$$

Both results show photon antibunching with $g^{(2)}_{\text{side}}(0) = 0$—formally, a simple consequence of the identity $\sigma_+^2 = \sigma_-^2 = 0$. Beyond this, there is little similarity between the two expressions. Equation 16.65 shows the cavity-modified dynamics seen in spontaneous emission (Sects. 13.2.1, 13.3.1, and 13.3.2)—a cavity-enhanced emission rate in the perturbative regime (bad-

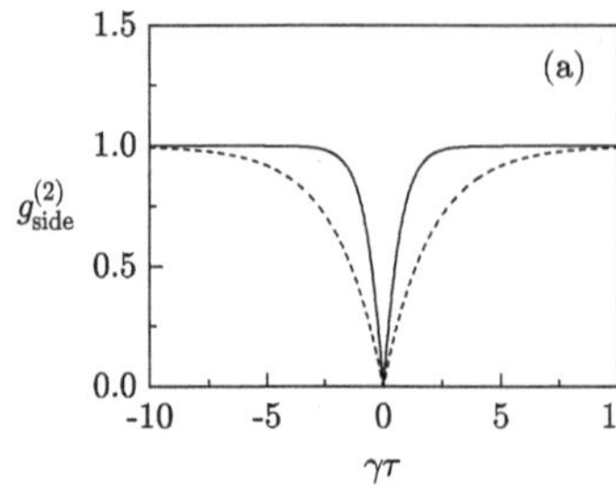

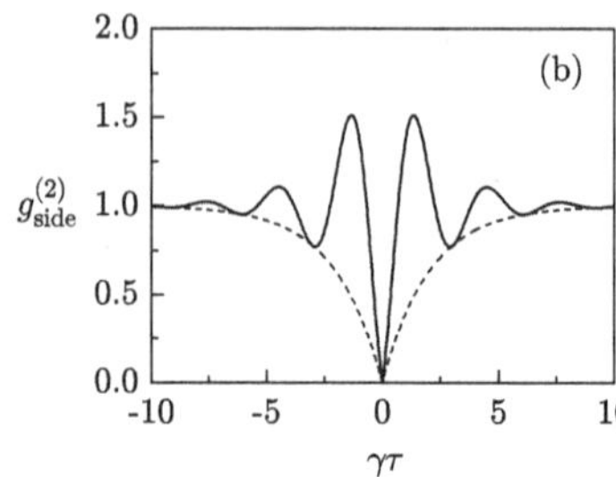

Fig. 16.4. Second-order correlation function of the side-scattered light in single-atom cavity QED (*solid curves*) compared with the correlation function for free-space resonance fluorescence (*dashed curves*): for $g/\gamma = 2$ and (**a**) $\kappa/\gamma = 5$ (perturbative regime), (**b**) $\kappa/\gamma = 0.5$ (nonperturbative regime)

cavity limit) and vacuum Rabi oscillations in the nonperturbative regime (strong-coupling limit). The free-space expression shows neither of these effects (Fig. 16.4).

Returning to the comment at the beginning of the section, we might also contrast strong coupling for a single atom with strong coupling in the many-atom sense. In the latter case, $2C_1$ is negligible, although $2C = N2C_1$ is large; setting side corrections of order N^{-1}, the correlation function for the side-scattered light is to be computed from

$$g^{(2)}_{\rm side}(\tau) = 1 + \frac{|\langle\tilde{\sigma}_+(0)\tilde{\sigma}_-(\tau)\rangle_{\rm ss}|^2 - \left(\langle\tilde{\sigma}_+\rangle_{\rm ss}\langle\tilde{\sigma}_-\rangle_{\rm ss}\right)^2}{\langle\tilde{\sigma}_+\tilde{\sigma}_-\rangle_{\rm ss}^2 + \left(\langle\tilde{\sigma}_+\rangle_{\rm ss}\langle\tilde{\sigma}_-\rangle_{\rm ss}\right)^2}, \tag{16.67}$$

where the operator moments are calculated from (2.120a), (2.121a), and (2.127) using the intracavity field amplitude X in place of the input field amplitude Y. The result for weak excitation is

$$g^{(2)}_{\rm side}(\tau) = 1 + X^2(2e^{-(\gamma/2)\tau} - e^{-\gamma\tau}). \tag{16.68}$$

This is again a very different expression from (16.65).

Forwards photon scattering is described by (16.39) whether or not the coupling for a single atom is strong; thus, forwards scattering does not distinguish, qualitatively (through the time dependence of the regression), between a $2C$ that is large because many atoms cooperate and a situation where $2C_1$ is large and the number of atoms is small. Sideways scattering does make this distinction.

Exercise 16.1. Derive (16.67) making use of the following two observations: (i) the side-scattered light is a sum of fields scattered by many different atoms and (ii) the ratio of unlike-atom to like-atom correlations is of order N^{-1}.

16.1.6 Spectra of Squeezing in the Weak-Excitation Limit

We have now seen that an expansion of the density operator and master equation in powers of $|\bar{\mathcal{E}}_0|$ can take us away from the small-noise limit. The approach has its limitations, though; for example, when applied to calculate the optical spectrum, where the calculation, even to lowest order, becomes unexpectedly involved.

The difficulty here is easy enough to find. To lowest order, the pure state (16.14) yields the expectations $\langle\tilde{a}\rangle_{\rm ss} = \tilde{\alpha}_{\rm ss}$, $\langle\tilde{a}^\dagger\tilde{a}\rangle_{\rm ss} = |\tilde{\alpha}_{\rm ss}|^2$, and $\langle\tilde{a}^2\rangle_{\rm ss} = \sqrt{2}\tilde{\eta}_{\rm ss}$. Thus, while the anomalous moment is of order $\langle(\Delta\tilde{a})^2\rangle_{\rm ss} = \langle\tilde{a}^2\rangle_{\rm ss} - \langle\tilde{a}\rangle_{\rm ss}^2 \sim |\bar{\mathcal{E}}_0|^2$, consistent with (15.120b), in this approximation the ordinary moment vanishes: $\langle\Delta\tilde{a}^\dagger\Delta\tilde{a}\rangle_{\rm ss} = \langle\tilde{a}^\dagger\tilde{a}\rangle_{\rm ss} - \langle\tilde{a}^\dagger\rangle_{\rm ss}\langle\tilde{a}\rangle_{\rm ss} = 0$. To obtain a nonzero result for $\langle\Delta\tilde{a}^\dagger\Delta\tilde{a}\rangle_{\rm ss}$ the calculation must be carried to higher order. This presents no difficulty so long as we may continue to work with the pure-state factorization. Doing so we keep two-quanta contributions to obtain

$$\langle\tilde{a}\rangle_{\rm ss} = \tilde{\alpha}_{\rm ss} + \sqrt{2}\tilde{\alpha}_{\rm ss}\tilde{\eta}_{\rm ss} + \tilde{\beta}_{\rm ss}\tilde{\zeta}_{\rm ss}, \tag{16.69a}$$

$$\langle\tilde{a}^\dagger\tilde{a}\rangle_{\rm ss} = |\tilde{\alpha}_{\rm ss}|^2 + 2|\tilde{\eta}_{\rm ss}|^2 + |\tilde{\zeta}_{\rm ss}|^2, \tag{16.69b}$$

with nonvanishing $\langle\Delta\tilde{a}^\dagger\Delta\tilde{a}\rangle_{\rm ss} \sim |\bar{\mathcal{E}}_0|^4$, just as expected from (15.120a). The result obtained in this way cannot be correct, however, beyond its giving a result of the correct order. Note that $\tilde{\alpha}_{\rm ss}$ on the right-hand side of (16.69a) is calculated as the dominant contribution to the matrix element ${}_a\langle 0|^{(N)}\langle 0|\tilde{\rho}_{\rm ss}|0\rangle^{(N)}|1\rangle_a$, and $|\tilde{\alpha}_{\rm ss}|^2$ on the right-hand side of (16.69b) as the dominant contribution to the matrix element ${}_a\langle 1|^{(N)}\langle 0|\tilde{\rho}_{\rm ss}|0\rangle^{(N)}|1\rangle_a$. However, these dominant terms cancel when $\langle\Delta\tilde{a}^\dagger\Delta\tilde{a}\rangle_{\rm ss}$ is computed. The matrix elements clearly must be determined to higher order, a task for which the pure-state factorization does not hold. Thus, in order to compute the optical spectrum, one must return to the equations of motion for the density matrix elements themselves and develop the perturbative approach at that level. This laborious task has been carried through by Rice [16.16] for single-atom cavity QED.

Fortunately, there is an alternative path to the optical spectrum, which we develop in Sect. (16.3.4). For now we simply avoid the difficulties by limiting our attention to the spectra of squeezing. These are determined from the anomalous correlation $\langle\Delta\tilde{a}(\tau)\Delta\tilde{a}(0)\rangle_{\rm ss}$, which evaluates in the weak-excitation limit without a cancellation of dominant terms. Working from (9.156), noting that $\langle\Delta\tilde{a}^\dagger(0)\Delta\tilde{a}(\tau)\rangle_{\rm ss} \ll \langle\Delta\tilde{a}(\tau)\Delta\tilde{a}(0)\rangle_{\rm ss}$, the spectrum of squeezing is

$$\begin{aligned}\bar{S}^\theta_{\rightarrow}(\omega) &= \eta\gamma_{a2}2\int_0^\infty d\tau\cos\omega\tau\left[e^{-2i\theta}\langle\Delta\tilde{a}(\tau)\Delta\tilde{a}(0)\rangle_{\rm ss} + \text{c.c.}\right]\\ &= \eta\gamma_{a2}2\,\text{Re}\left\{\int_0^\infty d\tau e^{i\omega\tau}\left[e^{-2i\theta}\langle\Delta\tilde{a}(\tau)\Delta\tilde{a}(0)\rangle_{\rm ss} + \text{c.c.}\right]\right\}.\end{aligned} \tag{16.70}$$

Of course it is not possible to recover the optical spectrum from a sum of spectra of squeezing calculated within this approximation.

Let us turn now to the evaluation of (16.70) using the quantum regression formula (1.98) and the pure-state factorization for many atoms (16.14). Having the pure-state factorization, the expression for the anomalous correlation is recast as

$$\begin{aligned}\langle\Delta\tilde{a}(\tau)\Delta\tilde{a}(0)\rangle_{\mathrm{ss}} &= \mathrm{tr}\{(\tilde{a}-\tilde{\alpha}_{\mathrm{ss}})e^{\tilde{\mathcal{L}}\tau}[(\tilde{a}-\tilde{\alpha}_{\mathrm{ss}})\tilde{\rho}_{\mathrm{ss}}]\}\\ &= \langle\tilde{\psi}_{\mathrm{ss}}|(\tilde{a}-\tilde{\alpha}_{\mathrm{ss}})|\bar{\tilde{\psi}}(\tau)\rangle,\end{aligned} \tag{16.71}$$

with the one-quantum expansion of the state

$$|\bar{\tilde{\psi}}(\tau)\rangle = \Delta\tilde{\alpha}(\tau)|0\rangle^{(N)}|1\rangle_a + \Delta\tilde{\beta}(\tau)|1\rangle^{(N)}|0\rangle_a, \tag{16.72}$$

where $\Delta\tilde{\alpha}$ and $\Delta\tilde{\beta}$ obey equations of motion (16.15a) and (16.15b), for initial condition

$$|\bar{\tilde{\psi}}(0)\rangle = (\tilde{a}-\tilde{\alpha}_{\mathrm{ss}})|\tilde{\psi}_{\mathrm{ss}}\rangle. \tag{16.73}$$

The right-hand side of (16.71) has been developed by substituting $\tilde{\mathcal{L}}$ in the approximate form (16.18) and using the fact that $|\tilde{\psi}_{\mathrm{ss}}\rangle$ is the stationary solution to (16.19); the overbar on $|\bar{\tilde{\psi}}(\tau)\rangle$ indicates that the state is not normalized, in this case even at dominant order—the argument of $e^{\tilde{\mathcal{L}}\tau}$ in (16.71) is not a density operator. From (16.71), (16.72), and (16.14), the correlation function is given by the one-photon amplitude of the time-dependent state,

$$\langle\Delta\tilde{a}(\tau)\Delta\tilde{a}(0)\rangle_{\mathrm{ss}} = \Delta\tilde{\alpha}(\tau), \tag{16.74}$$

where the dominant term only has been kept: $\langle\tilde{\psi}_{\mathrm{ss}}|(\tilde{a}-\tilde{\alpha}_{\mathrm{ss}})|\bar{\tilde{\psi}}(\tau)\rangle$ is replaced by ${}_a\langle 0|^{(N)}\langle 0|a|\bar{\tilde{\psi}}(\tau)\rangle$.

The initial values $\Delta\tilde{\alpha}(0)$ and $\Delta\tilde{\beta}(0)$ are evaluated using (16.72), (16.73), and the solution (16.29a)–(16.32b) for the two-quanta expansion in the steady state. We find

$$\begin{aligned}\Delta\tilde{\alpha}(0) &= \sqrt{2}\tilde{\eta}_{\mathrm{ss}} - \tilde{\alpha}_{\mathrm{ss}}^2\\ &= -\frac{(\bar{\mathcal{E}}_0/\kappa)^2}{1+2C}\left[1-\frac{2C}{1+2C-2C_1\xi/(1+\xi)}\right]+\left(\frac{\bar{\mathcal{E}}_0/\kappa}{1+2C}\right)^2\\ &= \left(\frac{\bar{\mathcal{E}}_0/\kappa}{1+2C}\right)^2 2C_1\frac{\xi}{1+\xi}\frac{2C}{1+2C-2C_1\xi/(1+\xi)},\end{aligned} \tag{16.75}$$

and

$$\begin{aligned}\Delta\tilde{\beta}(0) &= \tilde{\zeta}_{\mathrm{ss}} - \tilde{\alpha}_{\mathrm{ss}}\tilde{\beta}_{\mathrm{ss}} \\ &= \frac{\sqrt{N}g}{\gamma/2}\frac{(\bar{\mathcal{E}}_0/\kappa)^2}{1+2C}\frac{1}{1+2C-2C_1\xi/(1+\xi)} - \frac{\sqrt{N}g}{\gamma/2}\left(\frac{\bar{\mathcal{E}}_0/\kappa}{1+2C}\right)^2 \\ &= \frac{\sqrt{N}g}{\gamma/2}\left(\frac{\bar{\mathcal{E}}_0/\kappa}{1+2C}\right)^2 2C_1\frac{\xi}{1+\xi}\frac{1}{1+2C-2C_1\xi/(1+\xi)}.\end{aligned} \tag{16.76}$$

Then the one-photon amplitude is written as

$$\Delta\tilde{\alpha}(\tau) = Ae^{\Lambda_+\tau} + Be^{\Lambda_-\tau}, \tag{16.77}$$

which, from (16.75), yields

$$A + B = \left(\frac{\bar{\mathcal{E}}_0/\kappa}{1+2C}\right)^2 2C_1\frac{\xi}{1+\xi}\frac{1}{1+2C-2C_1\xi/(1+\xi)}, \tag{16.78}$$

and by matching the initial derivative in (16.15a),

$$(\Lambda_+ + \kappa)A + (\Lambda_- + \kappa)B = \kappa\left(\frac{\bar{\mathcal{E}}_0/\kappa}{1+2C}\right)^2 2C_1\frac{\xi}{1+\xi}\frac{1}{1+2C-2C_1\xi/(1+\xi)}, \tag{16.79}$$

where we use (16.75) and (16.76). Substituting the eigenvalues $\Lambda_\pm = \gamma\bar{\Lambda}_\pm = -\frac{1}{2}(\kappa+\gamma/2) \pm G$ (Eqs. 15.138 and 15.149) into the latter expression yields

$$A - B = -i\frac{\frac{1}{2}(\kappa+\gamma/2)}{G}(A+B); \tag{16.80}$$

hence, from (16.74), (16.75), (16.77), and (16.80), the anomalous correlation function is

$$\langle\Delta\tilde{a}(0)\Delta\tilde{a}(\tau)\rangle_{\mathrm{ss}} = \langle(\Delta\tilde{a})^2\rangle_{\mathrm{ss}}e^{-\frac{1}{2}(\kappa+\gamma/2)\tau}\left[\cos(G\tau) + \frac{\frac{1}{2}(\kappa+\gamma/2)}{G}\sin(G\tau)\right], \tag{16.81}$$

with $\langle(\Delta\tilde{a})^2\rangle_{\mathrm{ss}} = \Delta\tilde{\alpha}(0)$ given in (16.75). Thus, expression (16.70) for the spectrum of squeezing is now evaluated as

$$\begin{aligned}\bar{S}^{\theta}_{\rightarrow}(\omega) &= \eta\gamma_{a2}4\cos[2(\theta - \arg(\bar{\mathcal{E}}_0))]|\langle(\Delta\tilde{a})^2\rangle_{\mathrm{ss}}| \\ &\quad\times \mathrm{Re}\left\{\int_0^\infty d\tau e^{i\omega\tau}e^{-\frac{1}{2}(\kappa+\gamma/2)\tau}\left[\cos(G\tau) + \frac{\frac{1}{2}(\kappa+\gamma/2)}{G}\sin(G\tau)\right]\right\}.\end{aligned} \tag{16.82}$$

Finally, after carrying out the integration we obtain the *spectrum of squeezing for the forwards scattered light in cavity QED in the weak-excitation limit*:

$$\bar{S}^{\theta}_{\rightarrow}(\omega) = \eta\gamma_{a2} 4\cos[2(\theta - \arg(\bar{\mathcal{E}}_0))] \frac{|\langle(\Delta\tilde{a})^2\rangle_{ss}|}{2|G|}$$
$$\times \begin{cases} \dfrac{|G|^2 - [\frac{1}{2}(\kappa+\gamma/2)]^2}{\left[\frac{1}{2}(\kappa+\gamma/2) + |G|\right]^2 + \omega^2} - \dfrac{|G|^2 - [\frac{1}{2}(\kappa+\gamma/2)]^2}{\left[\frac{1}{2}(\kappa+\gamma/2) - |G|\right]^2 + \omega^2} & \text{(i)} \\ \dfrac{\frac{1}{2}(\kappa+\gamma/2)(2G+\omega)}{\left[\frac{1}{2}(\kappa+\gamma/2)\right]^2 + (\omega+G)^2} + \dfrac{\frac{1}{2}(\kappa+\gamma/2)(2G-\omega)}{\left[\frac{1}{2}(\kappa+\gamma/2)\right]^2 + (\omega-G)^2} & \text{(ii)} \end{cases} \tag{16.83}$$

where cases (i) and (ii) apply, respectively, for $Ng^2 \leq \frac{1}{4}(\kappa-\gamma/2)^2$ and $Ng^2 \geq \frac{1}{4}(\kappa-\gamma/2)^2$. The squeezing is in phase with the mean intracavity field, i.e., the spectrum is negative for $\theta = -\pi/2 + \arg(\bar{\mathcal{E}}_0)$.

The result should be compared with the Fourier transform of (15.147), which, leaving aside an overall scale factor, is the spectrum of squeezing in the small noise approximation. The difference is again what we noted at the end of Sect. 16.1.4—the factor $2C_1\xi/(1+\xi)$, which appears in (16.75) but not in (13.138).

16.2 Spatial Effects

Taking spatial effects into account in a general treatment of quantum fluctuations in cavity QED is a difficult task. Some background to the problems one faces is provided by the discussion in Sect. 14.4.2. The principal difficulty is that the atoms are no longer indistinguishable (overlooking the distinguishability due to spontaneous emission); one cannot account for the coherent interaction with the cavity mode using symmetrized atomic states. More formally, as explained in Sect. 14.4.2, we are no longer able to find a set of collective atomic operators like J_-, J_+, and J_z that obey a closed commutator algebra.

It is, however, possible to account for spatial effects in the weak-excitation limit, where the pure-state factorization may be used. One must simply drop the symmetrized atomic states in favor of a direct product state basis and labor through the algebra required to solve the resulting equations of motion for state amplitudes. At a quantitative level, spatial effects can change things rather dramatically, though there is no qualitative change in the character of the fluctuations. We illustrate this by revisiting the treatment of forwards photon scattering in Sect. 16.1.3.

Consider N atoms located at positions $\boldsymbol{r}_j$, $j = 1, \ldots, N$ within the standing-wave cavity mode shown in Fig. 13.2. The dipole coupling constants vary with position according to (14.94a). To treat this situation, in place of the two-quanta expansion (16.14), we expand the atomic part of state $|\tilde{\psi}(t)\rangle$

in the direct product basis (6.60), writing

$$
\begin{aligned}
|\tilde{\psi}(t)\rangle = {} & |\{\}, -N/2\rangle_A |0\rangle_a \\
& + \tilde{\alpha}(t)|\{\}, -N/2\rangle_A |1\rangle_a \\
& + \sum_{j=1}^{N} \tilde{\beta}_j(t)|\{j\}, -N/2+1\rangle_A |0\rangle_a \\
& + \tilde{\eta}(t)|\{\}, -N/2\rangle_A |2\rangle_a \\
& + \sum_{j=1}^{N} \tilde{\zeta}_j(t)|\{j\}, -N/2+1\rangle_A |1\rangle_a \\
& + \sum_{\substack{j,k=1 \\ j>k}}^{N} \tilde{\theta}_{(j,k)}(t)|\{j,k\}, -N/2+2\rangle_A |0\rangle_a .
\end{aligned} \tag{16.84}
$$

The bracketed notation (j, k) for the subscript on $\theta_{(j,k)}(t)$ is intended to indicate that the order of the indices is unimportant. The Schrödinger equation replacing (16.19) is

$$
\frac{d|\tilde{\psi}\rangle}{dt} = \left[\sum_{j=1}^{N} g(\boldsymbol{r}_j)(a^\dagger \sigma_{j-} - a\sigma_{j+}) - i\bar{\mathcal{E}}_0 a^\dagger - \frac{\gamma}{4}\sum_{j=1}^{N}(\sigma_{jz}+1) - \kappa a^\dagger a \right] |\tilde{\psi}\rangle. \tag{16.85}
$$

From the Schrödinger equation and state expansion we obtain the *equations of motion in the two-quanta truncation of the master equation for many atoms in a cavity, with unequal coupling strengths*:

$$
\dot{\tilde{\alpha}} = -\kappa\tilde{\alpha} + \sum_{j=1}^{N} g(\boldsymbol{r}_j)\tilde{\beta}_j - i\bar{\mathcal{E}}_0, \tag{16.86a}
$$

$$
\dot{\tilde{\beta}}_j = -\frac{\gamma}{2}\tilde{\beta}_j - g(\boldsymbol{r}_j)\tilde{\alpha}, \tag{16.86b}
$$

$$
\dot{\tilde{\eta}} = -2\kappa\tilde{\eta} + \sqrt{2}\sum_{j=1}^{N} g(\boldsymbol{r}_j)\tilde{\zeta}_j - i\sqrt{2}\bar{\mathcal{E}}_0\tilde{\alpha}, \tag{16.86c}
$$

$$
\dot{\tilde{\zeta}}_j = -(\kappa + \gamma/2)\tilde{\zeta}_j - \sqrt{2}g(\boldsymbol{r}_j)\tilde{\eta} + \sum_{\substack{k=1 \\ k\neq j}}^{N} g(\boldsymbol{r}_k)\tilde{\theta}_{(j,k)} - i\bar{\mathcal{E}}_0\tilde{\beta}_j, \tag{16.86d}
$$

$$
\dot{\tilde{\theta}}_{(j,k)} = -\gamma\tilde{\theta}_{(j,k)} - [g(\boldsymbol{r}_j)\tilde{\zeta}_k + g(\boldsymbol{r}_k)\tilde{\zeta}_j]. \tag{16.86e}
$$

These equations are the generalization of (16.15a)–(16.15e).

The equations of motion for the one-quantum state amplitudes, (16.86a) and (16.86b), may be written in a form resembling (14.119a) and (14.119b). Multiplying the latter by $g(\boldsymbol{r}_j)$ and summing over j yields the pair of equations

$$\frac{d\tilde{\alpha}}{dt} = -\kappa\tilde{\alpha} + \left[\sum_{j=1}^{N} g(\boldsymbol{r}_j)\tilde{\beta}_j\right] - i\bar{\mathcal{E}}_0, \tag{16.87a}$$

$$\frac{d\left[\sum_{j=1}^{N} g(\boldsymbol{r}_j)\tilde{\beta}_j\right]}{dt} = -\frac{\gamma}{2}\left[\sum_{j=1}^{N} g(\boldsymbol{r}_j)\tilde{\beta}_j\right] - \left[\sum_{j=1}^{N} g^2(\boldsymbol{r}_j)\right]\tilde{\alpha}. \tag{16.87b}$$

These are again equivalent (apart from the added driving field) to the oscillator equations underlying the treatment of spontaneous emission in single-atom cavity QED (Eqs. 13.152). They have the steady-state solution

$$\tilde{\alpha}_{\mathrm{ss}} = -i\frac{\bar{\mathcal{E}}_0/\kappa}{1+2C}, \tag{16.88a}$$

$$\frac{1}{\sqrt{N_{\mathrm{eff}}}g_{\mathrm{max}}}\left[\sum_{j=1}^{N} g(\boldsymbol{r}_j)\tilde{\beta}_j\right]_{\mathrm{ss}} = i\frac{\sqrt{N_{\mathrm{eff}}}g_{\mathrm{max}}}{\gamma/2}\frac{\bar{\mathcal{E}}_0/\kappa}{1+2C}, \tag{16.88b}$$

which generalizes (16.29a) and (16.29b), with the effective number of atoms N_{eff} (Eq. 14.120) in place of N, and

$$2C \equiv 2\frac{\sum_{j=1}^{N} g^2(\boldsymbol{r}_j)}{\gamma\kappa} = 2\frac{N_{\mathrm{eff}}g_{\mathrm{max}}^2}{\gamma\kappa} = N_{\mathrm{eff}}2C_1^{\mathrm{max}}, \tag{16.89}$$

in agreement with definition (14.103) of the cooperativity parameter in optical bistability. Proceeding now as in Sect 16.1.3, the second-order correlation function is to be calculated by solving the oscillator equations (16.87) for the conditional one-photon amplitude $\tilde{\alpha}(\tau)$, with initial conditions

$$\tilde{\alpha}(0) = \frac{\sqrt{2}\tilde{\eta}_{\mathrm{ss}}}{\tilde{\alpha}_{\mathrm{ss}}}, \qquad \left[\sum_{j=1}^{N} g(\boldsymbol{r}_j)\tilde{\beta}_j\right](0) = \frac{\left[\sum_{j=1}^{N} g(\boldsymbol{r}_j)\tilde{\zeta}_j\right]_{\mathrm{ss}}}{\tilde{\alpha}_{\mathrm{ss}}}, \tag{16.90}$$

where $\tilde{\eta}_{\mathrm{ss}}$ and $\left[\sum_{j=1}^{N} g(\boldsymbol{r}_j)\tilde{\zeta}_j\right]_{\mathrm{ss}}$ are found by solving (16.86c)–(16.86e) in the steady state. As in (16.27), the correlation function is computed as

$$g_{\rightarrow}^{(2)}(\tau) = \frac{|\tilde{\alpha}(\tau)|^2}{|\tilde{\alpha}_{\mathrm{ss}}|^2}. \tag{16.91}$$

Thus, the only remaining task is to find the steady-state solution to (16.86c)–(16.86e).

The first steps towards the solution may be followed in similar fashion to those leading from (16.15c)–(16.15e) to (16.30) and (16.31). Starting with (16.86e), we obtain

$$\tilde{\theta}_{(j,k)} = -\frac{g(\boldsymbol{r}_j)\tilde{\zeta}_k^{\mathrm{ss}} + g(\boldsymbol{r}_k)\tilde{\zeta}_j^{\mathrm{ss}}}{\gamma}, \tag{16.92}$$

and from (16.86c) and (16.86d), the amplitudes $\tilde{\eta}_{\rm ss}$ and $\left[\sum_{j=1}^N g(\boldsymbol{r}_j)\tilde{\zeta}_j\right]_{\rm ss}$ then satisfy simultaneous equations

$$\sqrt{2}\kappa\tilde{\eta}_{\rm ss} - \left[\sum_{j=1}^N g(\boldsymbol{r}_j)\tilde{\zeta}_j\right]_{\rm ss} = -\kappa\frac{(\bar{\mathcal{E}}_0/\kappa)}{1+2C}, \tag{16.93a}$$

and

$$2\sqrt{2}\left[\sum_{j=1}^N g^2(\boldsymbol{r}_j)\right]\tilde{\eta}_{\rm ss} + \gamma(1+\xi)\left[\sum_{j=1}^N g(\boldsymbol{r}_j)\tilde{\zeta}_j\right]_{\rm ss} + \frac{4}{\gamma}\sum_{\substack{j,k=1\\k\neq j}}^N g^2(\boldsymbol{r}_j)g(\boldsymbol{r}_k)\tilde{\zeta}_k^{\rm ss}$$

$$= 2\xi\left[\sum_{j=1}^N g^2(\boldsymbol{r}_j)\right]\frac{(\bar{\mathcal{E}}_0/\kappa)^2}{1+2C}. \tag{16.93b}$$

From this point something new is needed, since, while (16.92) is a straightforward generalization of (16.30), the system of simultaneous equations is not closed—unlike (16.30) and (16.31). To verify this, consider the third term on the left-hand side of (16.93b) rewritten as

$$\sum_{\substack{j,k=1\\k\neq j}}^N g^2(\boldsymbol{r}_j)g(\boldsymbol{r}_k)\tilde{\zeta}_k^{\rm ss} = \left[\sum_{j=1}^N g^2(\boldsymbol{r}_j)\right]\left[\sum_{k=1}^N g(\boldsymbol{r}_k)\tilde{\zeta}_k\right]_{\rm ss} - \left[\sum_{j=1}^N g^3(\boldsymbol{r}_j)\tilde{\zeta}_j\right]_{\rm ss}. \tag{16.94}$$

A new collective variable, $\left[\sum_{j=1}^N g^3(\boldsymbol{r}_j)\tilde{\zeta}_j\right]_{\rm ss}$, has been introduced; we encounter the start of the hierarchy of equations discussed in Sect. 14.4.2.

Having recognized the hierarchy, we can now tackle it head on. First, generalizing the definitions of $\sum_{j=1}^N g(\boldsymbol{r}_j)\tilde{\zeta}_j$ and $\sum_{j=1}^N g^2(\boldsymbol{r}_j)$, we define the infinite set of collective variables

$$Z^{(2n-1)} \equiv \sum_{j=1}^N g^{2n-1}(\boldsymbol{r}_j)\tilde{\zeta}_j, \tag{16.95}$$

and

$$\Gamma^{(2n)} \equiv \sum_{j=1}^N g^{2n}(\boldsymbol{r}_j). \tag{16.96}$$

Equation 16.93a is transcribed as

$$\sqrt{2}\kappa\tilde{\eta}_{\rm ss} - Z_{\rm ss}^{(1)} = -\kappa\frac{(\bar{\mathcal{E}}_0/\kappa)^2}{1+2C}. \tag{16.97}$$

The process of arriving at the replacement for (16.93b) is just a little more involved. Multiplying (16.86d) by $g^{2n-1}(\boldsymbol{r}_j)$ and summing over j yields

$$2\sqrt{2}\Gamma^{(2n)}\tilde{\eta}_{\rm ss} + \gamma(1+\xi)Z_{\rm ss}^{(2n-1)} + \frac{2}{\gamma}\sum_{\substack{j,k=1\\k\neq j}}^{N}\left[g^{2n}(\boldsymbol{r}_j)g(\boldsymbol{r}_k)\tilde{\zeta}_k^{\rm ss}\right.$$

$$\left.+\, g^{2n-1}(\boldsymbol{r}_j)g^2(\boldsymbol{r}_k)\tilde{\zeta}_j^{\rm ss}\right] = 2i\bar{\mathcal{E}}_0\sum_{j=1}^{N} g^{2n-1}(\boldsymbol{r}_j)\tilde{\beta}_j^{\rm ss}. \qquad (16.98)$$

We then simplify the third term on the left-hand side, by writing

$$\sum_{\substack{j,k=1\\k\neq j}}^{N}\left[g^{2n}(\boldsymbol{r}_j)g(\boldsymbol{r}_k)\tilde{\zeta}_k^{\rm ss} + g^{2n-1}(\boldsymbol{r}_j)g^2(\boldsymbol{r}_k)\tilde{\zeta}_j^{\rm ss}\right]$$

$$= \Gamma^{(2n)}Z_{\rm ss}^{(1)} + \Gamma^{(2)}Z_{\rm ss}^{(2n-1)} - 2Z_{\rm ss}^{(2n+1)}, \qquad (16.99)$$

and on the right-hand side, using (16.86b) and (16.88a), have

$$2i\bar{\mathcal{E}}_0\sum_{j=1}^{N} g^{2n-1}(\boldsymbol{r}_j)\tilde{\beta}_j^{\rm ss} = 2\xi\Gamma^{(2n)}\frac{(\bar{\mathcal{E}}_0/\kappa)^2}{1+2C}. \qquad (16.100)$$

Equation 16.98 is rewritten as

$$2\sqrt{2}\Gamma^{(2n)}\tilde{\eta}_{\rm ss} + \gamma[1+\xi(1+C)]Z_{\rm ss}^{(2n-1)} + \frac{2}{\gamma}\left(\Gamma^{(2n)}Z_{\rm ss}^{(1)} - 2Z_{\rm ss}^{(2n+1)}\right)$$

$$= 2\xi\Gamma^{(2n)}\frac{(\bar{\mathcal{E}}_0/\kappa)^2}{1+2C}. \qquad (16.101)$$

Our task is reduced to solving the hierarchy of equations (19.97) and (16.101) for the two-quanta amplitudes $\tilde{\eta}_{\rm ss}$ and $Z_{\rm ss}^{(1)} \equiv \left[\sum_{j=1}^{N} g(\boldsymbol{r}_j)\tilde{\zeta}_j\right]_{\rm ss}$. We first eliminate $\tilde{\eta}_{\rm ss}$ from (16.101), where, using (16.97),

$$\sqrt{2}\tilde{\eta}_{\rm ss} = \frac{1}{\kappa}Z_{\rm ss}^{(1)} - \frac{(\bar{\mathcal{E}}_0/\kappa)^2}{1+2C}. \qquad (16.102)$$

This yields the recurrence relation

$$\gamma[1+\xi(1+C)]Z_{\rm ss}^{(2n-1)} - \frac{4}{\gamma}Z_{\rm ss}^{(2n+1)}$$

$$= 2\left[(1+\xi)\frac{(\bar{\mathcal{E}}_0/\kappa)^2}{1+2C} - (1+\xi/2)\frac{1}{\kappa}Z_{\rm ss}^{(1)}\right]\Gamma^{(2n)}, \qquad (16.103)$$

or, after a little rearrangement,

$$Z_{\rm ss}^{(2n-1)} = \frac{4}{\gamma^2}\frac{1}{1+\xi(1+C)}\left\{Z_{\rm ss}^{(2n+1)} + \frac{\gamma}{2}\left[(1+\xi)\frac{(\bar{\mathcal{E}}_0/\kappa)^2}{1+2C} - (1+\xi/2)\frac{1}{\kappa}Z_{\rm ss}^{(1)}\right]\Gamma^{(2n)}\right\}. \quad (16.104)$$

To solve the recurrence relation, we iterate the substitution, from the left to the right, of $Z_{\rm ss}^{(2n+1)}$, $Z_{\rm ss}^{(2n+3)}$, etc. In this way an expansion for $Z_{\rm ss}^{(2n-1)}$ in terms of $Z_{\rm ss}^{(1)}$ is developed. Using (16.96), we find

$$\begin{aligned} Z_{\rm ss}^{(2n-1)} &= \frac{2}{\gamma}\frac{1}{1+\xi(1+C)}\left[(1+\xi)\frac{(\bar{\mathcal{E}}_0/\kappa)^2}{1+2C} - (1+\xi/2)\frac{1}{\kappa}Z_{\rm ss}^{(1)}\right] \\ &\quad\times\left\{\Gamma^{(2n)} + \frac{4/\gamma^2}{1+\xi(1+C)}\Gamma^{(2n+2)} + \left[\frac{4/\gamma^2}{1+\xi(1+C)}\right]^2\Gamma^{(2n+4)} + \cdots\right\} \\ &= \left[(1+\xi)\frac{(\bar{\mathcal{E}}_0/\kappa)^2}{1+2C} - (1+\xi/2)\frac{1}{\kappa}Z_{\rm ss}^{(1)}\right]\sum_{j=1}^{N}\frac{2g^{2n}(\boldsymbol{r}_j)/\gamma}{1+\xi(1+C)} \\ &\quad\times\sum_{p=0}^{\infty}\left[\frac{4/\gamma^2}{1+\xi(1+C)}\right]^p g^{2p}(\boldsymbol{r}_j). \end{aligned} \quad (16.105)$$

The case $n=1$ gives

$$\frac{1}{\kappa}Z_{\rm ss}^{(1)} = \left[(1+\xi)\frac{(\bar{\mathcal{E}}_0/\kappa)^2}{1+2C} - (1+\xi/2)\frac{1}{\kappa}Z_{\rm ss}^{(1)}\right]S, \quad (16.106)$$

with

$$S \equiv \sum_{j=1}^{N}\frac{2C_{1j}}{1+\xi(1+C)-2C_{1j}\xi}, \quad (16.107)$$

where (Eq. 13.36)

$$2C_{1j} \equiv 2\frac{g^2(\boldsymbol{r}_j)}{\gamma\kappa} \quad (16.108)$$

is the spontaneous emission enhancement factor for an atom located at $\boldsymbol{r}_j$. The solution for $Z_{\rm ss}^{(1)}$ is

$$Z_{\rm ss}^{(1)} \equiv \left[\sum_{j=1}^{N} g(\boldsymbol{r}_j)\tilde{\zeta}_j\right]_{\rm ss} = \kappa\frac{(\bar{\mathcal{E}}_0/\kappa)^2}{1+2C}\frac{(1+\xi)S}{1+(1+\xi/2)S}, \quad (16.109)$$

and for $\tilde{\eta}_{\rm ss}$, from (16.102),

$$\tilde{\eta}_{\rm ss} = -\frac{1}{\sqrt{2}}\frac{(\bar{\mathcal{E}}_0/\kappa)^2}{1+2C}\left[1 - \frac{(1+\xi)S}{1+(1+\xi/2)S}\right]. \quad (16.110)$$

Note that while for some choices of parameters the sum over p in (16.105) is not guaranteed to converge—e.g. when $N = 1$ and $g^2(\boldsymbol{r}_1)/\gamma \gg 1+\xi$—we can extend our results to these cases by analytic continuation.

We are finally in a position to construct the solution for the second-order correlation function (16.91). Following (16.33) we write

$$\tilde{\alpha}(\tau) = -i\frac{\bar{\mathcal{E}}_0/\kappa}{1+2C} + Ae^{\Lambda_+\tau} + Be^{\Lambda_-\tau}, \tag{16.111}$$

where the eigenvalues $\Lambda_\pm = \gamma\bar{\Lambda}_\pm$ are still defined by (15.138), even with unequal coupling constants for the atoms. Then, with the help of (16.90), (16.88a), and (16.110), initial conditions require

$$\begin{aligned} A+B &= i\frac{\bar{\mathcal{E}}_0/\kappa}{1+2C} - i\frac{\bar{\mathcal{E}}_0}{\kappa}\left[1 - \frac{(1+\xi)S}{1+(1+\xi/2)S}\right] \\ &= -i\frac{\bar{\mathcal{E}}_0}{\kappa}\left[\frac{2C}{1+2C} - \frac{(1+\xi)S}{1+(1+\xi/2)S}\right] \\ &= i\frac{\bar{\mathcal{E}}_0/\kappa}{1+2C}\frac{[1+\xi(1+C)]S-2C}{1+(1+\xi/2)S}, \end{aligned} \tag{16.112}$$

and, using also (16.87a) and (16.109),

$$\begin{aligned} (\Lambda_+ + \kappa)A + (\Lambda_- + \kappa)B &= i\bar{\mathcal{E}}_0\frac{1}{1+2C} + i\bar{\mathcal{E}}_0\frac{(1+\xi)S}{1+(1+\xi/2)S} - i\bar{\mathcal{E}}_0 \\ &= i\kappa\frac{\bar{\mathcal{E}}_0/\kappa}{1+2C}\frac{[1+\xi(1+C)]S-2C}{1+(1+\xi/2)S}. \end{aligned} \tag{16.113}$$

From (16.112) and (16.113), we find $\Lambda_+A + \Lambda_-B = 0$; hence

$$A - B = -i\frac{\frac{1}{2}(\kappa+\gamma/2)}{G}(A+B). \tag{16.114}$$

Substituting (16.112) and (16.114) into (16.111), we arrive at the desired generalization of (16.39), the *second-order correlation function of forwards scattering in cavity QED in the weak-excitation limit with unequal coupling strengths*,

$$\begin{aligned} g^{(2)}_{\rightarrow}(\tau) = \Bigg\{1 &- \frac{[1+\xi(1+C)]S-2C}{1+(1+\xi/2)S} \\ &\times e^{-\frac{1}{2}(\kappa+\gamma/2)\tau}\left[\cos(G\tau) + \frac{\frac{1}{2}(\kappa+\gamma/2)}{G}\sin(G\tau)\right]\Bigg\}^2, \end{aligned} \tag{16.115}$$

where, to replace (15.149), $G \equiv \sqrt{N_{\text{eff}}g_{\max}^2 - \frac{1}{4}(\kappa-\gamma/2)^2}$.

The new expression has the same structure as (16.39). The differences are quantitative: the prefactor that determines the amplitude of the damped oscillation, and the use of N_{eff} rather than N in the frequency G of the vacuum

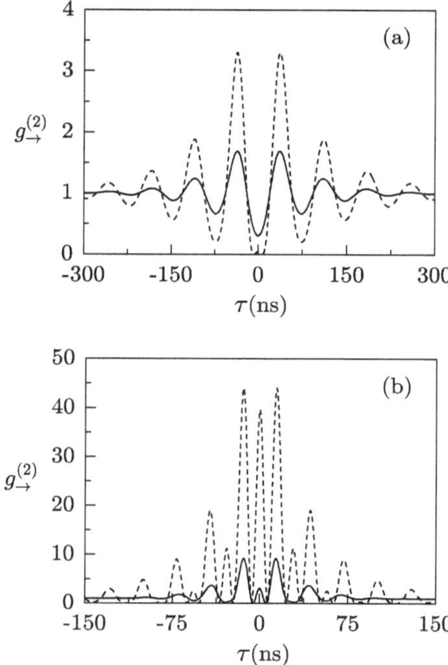

Fig. 16.5. Comparison of (16.39) evaluated for $\bar{N}_{\rm eff}$ optimally coupled atoms (*dashed curves*) and the configuration average of (16.115), with mean number density $D = \bar{N}_{\rm eff}/[\pi(w_0/2)^2\ell]$ (*solid curves*): for (**a**) $g_{\rm max}/2\pi = 3.2$MHz, $\kappa/2\pi = 5.0$MHz, $\gamma/2\pi = 0.9$MHz, and $\bar{N}_{\rm eff} = 18$ (Fig. 4a of [16.2]); (**b**) $g_{\rm max}/2\pi = 11.1$MHz, $\kappa/2\pi = 2.73$MHz, $\gamma/2\pi = 6.07$MHz, and $\bar{N}_{\rm eff} = 10$ (Fig. 2b of [16.5])

Rabi oscillation. The differences are significant though. With regard to the amplitude of oscillation, using (16.39) and $N_{\rm eff}$ atoms each with the optimal coupling $g = g_{\rm max}$ can yield a result that differs from (16.115) by something greater than an order of magnitude. To consider an example, Rempe and coworkers [16.2] derived (16.115) to improve the fit of the theory to their experimental results. Since an atomic beam is used in the experiment, an average over an ensemble of atomic configurations, $\boldsymbol{r}_j$, $j = 1, \dots, N$, must be taken; it is taken at fixed atomic density D with each atom randomly distributed in some suitably large interaction volume. The atomic density determines the effective number of atoms to be used: for the standing-wave mode function (14.94a), the configuration average of $N_{\rm eff}$ is $\bar{N}_{\rm eff} = D\pi(w_0/2)^2\ell$, where ℓ is the atomic beam width along the cavity axis [16.17].

Using the parameters of the experiment, Fig. 16.5a demonstrates the difference between the Monte Carlo average of (16.115) and the equal-coupling-constant formula (16.39). Figure 16.5b makes the same comparison for a later experiment by Mielke and coworkers [16.5].

Exercise 16.2. Show that the expression (16.115) reduces to (16.39) when $g(\boldsymbol{r}_j) = g$, $j = 1, \ldots, N$.

Exercise 16.3. Consider the continuous medium limit defined by $\bar{N}_{\text{eff}} \to \infty$, $g_{\text{max}} \to 0$, with $2C = 2\bar{N}_{\text{eff}} g_{\text{eff}}^2 / \gamma\kappa$ constant. Show that the limit is approached with

$$g_{\to}^{(2)}(\tau) = \left\{ 1 - \bar{N}_{\text{eff}}^{-1} \frac{3}{8} \frac{\xi}{1+\xi} \frac{4C^2}{1+2C} \right.$$
$$\left. \times e^{-\frac{1}{2}(\kappa+\gamma/2)\tau} \left[\cos(G\tau) + \frac{\frac{1}{2}(\kappa+\gamma/2)}{G} \sin(G\tau) \right] \right\}^2 , \quad (16.116)$$

where $G = \sqrt{\bar{N}_{\text{eff}} g_{\text{max}}^2 - \frac{1}{4}(\kappa - \gamma/2)^2}$. Compare this result with (16.52a).

Exercise 16.4. Devise a Monte Carlo implementation of the average over configurations described in the last paragraph. Compare the Monte Carlo average with the continuum result (16.116) for the parameters of Fig. 16.5. Explain any differences observed.

Exercise 16.5. It is possible to solve for the steady state of the coupled equations (18.86a)–(18.86e) more directly, without introducing the hierarchy of equations (16.97) and (16.101). Develop the alternate method of solution.

16.3 Beyond Classical Trajectories plus "Fuzz": Spontaneous Dressed-State Polarization

Earlier chapters have introduced us to three major applications of linear fluctuation theory—(i) to the homogeneously broadened laser (Chaps. 7 and 8), the paradigm of a quantum noise process, but one where the noise is actually classical; (ii) to the degenerate parametric oscillator (Chaps. 10 and 11) with its squeezed and nonclassical fluctuations; and (iii) to a system of many atoms in a driven cavity, or optical bistability (Chap. 15), where again the fluctuations are nonclassical and exhibit both photon antibunching and squeezing.

Quantum fluctuations were added in each of these examples as a perturbation upon a background of classical nonlinear physics. For the system of many atoms in a cavity, to pick a particular case, the basic features of the classical physics are summarized in the optical bistability state equations (14.53) and (14.58)–(14.60). So long as the fluctuations remain small, the hysteresis cycle deduced from these equations is expected to remain intact. Fluctuations simply add a "fuzz-ball" of noise around a chosen operating point—except, of course, close to the cycle boundaries, where even small fluctuations can precipitate switching from one operating branch to the other. Certainly, if the fluctuations are nonclassical, the "fuzz-ball" lives within the double-dimension phase-space of the positive P representation. Nevertheless, the background of classical nonlinear physics is still present in exactly the same way.

Contrast cavity QED conditions: the system size is small, the quantum fluctuations large, and the underlying classical physics is expected to break down. We expect to see a disintegration of the hysteresis cycle accompanied by a transformation of the noise process similar to that illustrated for the degenerate parametric oscillator in Figs. 12.6–12.8. For an illustration, let us take the parameters of Fig. 16.5b and imagine they are realized for a single atom localized at an antinode of the standing-wave cavity mode. We find a saturation photon number $n_{\rm sat} = 0.037$ (Eq. 14.104), which should be compared with the threshold photon number, $\langle a^\dagger a\rangle_{\rm thr} = 0.025$, of Fig. 12.8b. For the latter example, the noise process is entirely changed from the picture of local fluctuations about classical steady states shown in Figs. 12.6 and 12.7a; its interpretation involves transitions between even and odd superpositions of coherent states (Sect. 12.1.8)—a decidedly quantum-mechanical evolution.

Unfortunately, the assumed conditions are very difficult to realize in the laboratory. The experiment to which the parameters of Fig. 16.5b apply [16.5] does not work with a single localized atom, but with a thermal atomic beam, with $\bar{N}_{\rm eff} = 10$, and is strongly affected by spatial effects; Horvath and Carmichael have shown just how troublesome the atomic motion and density fluctuations in an atomic beam can be [16.18]. In order to overcome these difficulties, experiments with single trapped atoms are performed [16.19, 16.20, 16.21].

There are complementary difficulties on the theoretical side. It is not possible to arbitrarily extend computations with many-atoms, particularly when including spatial effects, as the size of the basis required grows as 2^N—a very large number, even for $N = 100$. While Horvath and Carmichael [16.18] treat systems with $\bar{N}_{\rm eff} = 13$ and 18 at weak excitation, it is not feasible to extend their numerical methods to high excitation.

We can carry our treatment of the *single-atom* system further, though, thus moving away from the weak-excitation limit and its reliance on the pure-state factorization. This is the goal for the rest of the chapter. A number of miscellaneous calculations are presented. Although a general analytical solution like the one underlying the progression followed through Sect. 12.1.7 cannot be given, the accumulated results of the next few sections trace a clear outline of the quantum noise process which replaces the picture of classical trajectories plus "fuzz" at intermediate and strong excitation for single-atom cavity QED.

16.3.1 Maxwell–Bloch Equations for "Zero System Size"

We are interested in the limit of small system size. A good way to start out is to set the system size parameter to zero. This can be done by turning off the spontaneous emission; thus, we let $\gamma_h = \gamma \to 0$, and in this way reach the limit $n_{\rm sat} = \gamma^2/8g^2 \to 0$. We refer to the limit as the limit of "zero system size." Of course, $n_{\rm sat}$ rather than N is taken as the system size parameter, since there is nothing useful to be learned from setting $N=0$.

A quick path to the basic physics of this limit is provided by the Maxwell–Bloch equations. Perhaps this appears to be a contradiction: the fluctuations scale inversely with system size and the Maxwell–Bloch equations neglect fluctuations altogether; what then can they have to say about a limit in which fluctuations diverge? In fact there is no contradiction, as we soon see. The essential point is that from the Maxwell–Bloch equations we are able to identify new stationary states that play a central role in organizing the dynamics once the fluctuations are reintroduced. These states are complementary to the steady states identified in (15.69)–(15.71). The latter are steady states only when γ_h and γ are nonzero. The new states are strictly stationary only when $\gamma_h=\gamma=0$. Through the Maxwell–Bloch equations, we first familiarize ourselves with these states in their semiclassical version. Their quantum-mechanical character is explored in the following section.

The Maxwell–Bloch equations for scaled variables are given as (15.57a)–(15.57e). We first remove the scaling (Eqs. 15.51) then set $\gamma_h=\gamma=0$. The resulting equations are the *Maxwell–Bloch equations for "zero system size"*:

$$\frac{d\langle\tilde{a}\rangle}{dt}=-\kappa\langle\tilde{a}\rangle+g\langle\tilde{J}_-\rangle-i\bar{\mathcal{E}}_0, \tag{16.117a}$$

$$\frac{d\langle\tilde{a}^\dagger\rangle}{dt}=-\kappa\langle\tilde{a}^\dagger\rangle+g\langle\tilde{J}_+\rangle+i\bar{\mathcal{E}}_0, \tag{16.117b}$$

$$\frac{d\langle\tilde{J}_-\rangle}{dt}=g\langle J_z\rangle\langle\tilde{a}\rangle, \tag{16.117c}$$

$$\frac{d\langle\tilde{J}_+\rangle}{dt}=g\langle J_z\rangle\langle\tilde{a}^\dagger\rangle, \tag{16.117d}$$

$$\frac{d\langle J_z\rangle}{dt}=-2g\big(\langle\tilde{J}_+\rangle\langle\tilde{a}\rangle+\langle\tilde{J}_-\rangle\langle\tilde{a}^\dagger\rangle\big), \tag{16.117e}$$

where we specialize to the resonant case ($\Delta=\Phi=0$). Note now that with $\gamma_h=\gamma=0$, the Bloch equations, (16.117c)–(16.117e), are undamped, and conserve the length of the Bloch vector, which for atoms prepared in the ground state is given by the number of atoms. The five-dimensional state vector $\big(\langle\tilde{a}\rangle,\langle\tilde{a}^\dagger\rangle,\langle\tilde{J}_-\rangle,\langle\tilde{J}_+\rangle,\langle J_z\rangle\big)$ therefore moves on a four-dimensional surface: two dimensions—the $\big(\langle\tilde{a}\rangle,\langle\tilde{a}^\dagger\rangle\big)$-plane—specify the cavity field amplitude, and the other two identify the state of the atoms with a point on the surface of the *collective Bloch sphere*,

$$4\langle\tilde{J}_+\rangle\langle\tilde{J}_-\rangle+\langle J_z\rangle^2=N^2. \tag{16.118}$$

Although in future sections it will be necessary to limit our attention to a single atom, here we encounter no difficulty retaining the full dependence on N.

Our task is to find stationary solutions to (16.117a)–(16.117e) subject to the constraint (16.118). It is clear from (16.117c) and (16.117d) that the solutions fall into two classes: those with vanishing field amplitude and those with vanishing inversion. On looking more carefully at the equations, the classification is to be made according to whether the driving field amplitude lies below, at, or above the threshold value $|\bar{\mathcal{E}}_0| = Ng/2$:

Below threshold ($|\bar{\mathcal{E}}_0|/g < N/2$): there is a single stationary solution with vanishing field amplitude,

$$\langle \tilde{a} \rangle_< \equiv \langle \tilde{a} \rangle_{\mathrm{ss}}^< = 0. \tag{16.119a}$$

The vanishing of the field amplitude guarantees that (16.117c), (16.117d), and (16.117e) are all satisfied. The polarization is then determined by the requirement that it cancel the driving field exactly; from (16.117a), (16.117b), and (16.118), the collective state of the atoms is

$$\langle \tilde{J}_\mp \rangle_< \equiv \langle \tilde{J}_\mp \rangle_{\mathrm{ss}}^< = \pm i e^{\pm i \arg(\bar{\mathcal{E}}_0)} |\bar{\mathcal{E}}_0|/g, \tag{16.119b}$$

$$\langle J_z \rangle_< \equiv \langle J_z \rangle_{\mathrm{ss}}^< = -N\sqrt{1 - (2|\bar{\mathcal{E}}_0|/Ng)^2}. \tag{16.119c}$$

Since $\langle J_z \rangle_<$ is real, the solution is unacceptable above the defined threshold, $|\bar{\mathcal{E}}_0|/g = N/2$.

At threshold ($|\bar{\mathcal{E}}_0|/g = N/2$): the field amplitude and inversion both vanish. We have

$$\langle \tilde{a} \rangle_{\mathrm{thr}} \equiv \langle \tilde{a} \rangle_{\mathrm{ss}}^{\mathrm{thr}} = 0, \tag{16.120a}$$

$$\langle \tilde{J}_\mp \rangle_{\mathrm{thr}} \equiv \langle \tilde{J}_\mp \rangle^{\mathrm{thr}} = \pm i e^{\pm i \arg(\bar{\mathcal{E}}_0)} N/2, \tag{16.120b}$$

$$\langle J_z \rangle_{\mathrm{thr}} \equiv \langle J_z \rangle_{\mathrm{ss}}^{\mathrm{thr}} = 0, \tag{16.120c}$$

and the solution is doubly degenerate.

Above threshold ($|\bar{\mathcal{E}}_0|/g > N/2$): there are two solutions with vanishing inversion. The field amplitude is nonzero and, from (16.117e),

$$\frac{\langle \tilde{J}_- \rangle_{\mathrm{ss}}}{\langle \tilde{J}_+ \rangle_{\mathrm{ss}}} = -\frac{\langle \tilde{a} \rangle_{\mathrm{ss}}}{\langle \tilde{a}^\dagger \rangle_{\mathrm{ss}}}. \tag{16.121}$$

The solutions are of the form

$$\begin{aligned} \langle \tilde{a} \rangle_>^\pm &= A e^{i\phi_\pm}, & \langle \tilde{a}^\dagger \rangle_>^\pm &= A e^{-i\phi_\pm}, \\ \langle \tilde{J}_- \rangle_>^\pm &= \pm i e^{i\phi_\pm} N/2, & \langle \tilde{J}_+ \rangle_>^\pm &= \mp i e^{-i\phi_\pm} N/2, \end{aligned} \tag{16.122a}$$

$\langle \tilde{a} \rangle_> \equiv \langle \tilde{a} \rangle_{\mathrm{ss}}^>$, $\langle \tilde{a}^\dagger \rangle_> \equiv \langle \tilde{a}^\dagger \rangle_{\mathrm{ss}}^>$, and $\langle \tilde{J}_\pm \rangle_> \equiv \langle \tilde{J}_\pm \rangle_{\mathrm{ss}}^>$, with

$$\langle J_z \rangle_> \equiv \langle J_z \rangle_{\mathrm{ss}} = 0. \tag{16.122b}$$

The magnitude of the polarization amplitude is set by (16.118), while the magnitude and phase of the field amplitude are determined by (16.117a) and (16.117b), which require

$$-\kappa A e^{i\phi_\pm} \pm i(Ng/2)e^{i\phi_\pm} - i\bar{\mathcal{E}}_0 = 0. \tag{16.123}$$

We obtain

$$A = (|\bar{\mathcal{E}}_0|/\kappa)\sqrt{1-(Ng/2|\bar{\mathcal{E}}_0|)^2}, \tag{16.124}$$

and

$$\begin{aligned} e^{i\phi_\pm} &= \frac{-i\bar{\mathcal{E}}_0/\kappa}{A \mp i(Ng/2\kappa)} \\ &= -ie^{i\arg(\bar{\mathcal{E}}_0)}\left[\sqrt{1-(Ng/2|\bar{\mathcal{E}}_0|)^2} \pm i(Ng/2|\bar{\mathcal{E}}_0|)\right]. \end{aligned} \tag{16.125}$$

It is clear that the stationary solutions (16.119), (16.120), and (16.122) are not those recovered from the $\gamma_h = \gamma \to 0$ limit of the steady states in absorptive optical bistability (Eqs. 15.69–15.71). This follows, in particular, because according to (16.125) the phases ϕ_+ and ϕ_- change with driving field strength, while the phases in absorptive bistability are fixed by the driving field. The new transition, passing through threshold, is therefore a *symmetry-breaking* transition: the phases of the individual field states, $\big(\langle\tilde{a}\rangle_>^+, \langle\tilde{a}^\dagger\rangle_>^+\big)$ and $\big(\langle\tilde{a}\rangle_>^-, \langle\tilde{a}^\dagger\rangle_>^-\big)$, do not reflect the phase of the external field, though the phase of an equally weighted average of them does; a similar comment holds for the states on the Bloch sphere, $\big(\langle\tilde{J}_-\rangle_>^+, \langle\tilde{J}_+\rangle_>^+, 0\big)$ and $\big(\langle\tilde{J}_-\rangle_>^-, \langle\tilde{J}_+\rangle_>^-, 0\big)$.

It is useful to make an explicit comparison between the new stationary states and the "zero system size" limit of absorptive optical bistability. To do this we must first determine precisely what that limit is:

Exercise 16.6. Show that in the limit $\gamma_h = \gamma \to 0$ the stable steady states determined from (15.69)–(16.71) (with $\Delta = \Phi = 0$) approach, for $\sqrt{2}|\bar{\mathcal{E}}_0|/g \leq N/2$,

$$\langle\tilde{a}\rangle_{\rm ss} = \langle\tilde{a}^\dagger\rangle_{\rm ss} = 0, \tag{16.126a}$$

$$\langle\tilde{J}_\mp\rangle_{\rm ss} = ie^{\pm i\arg\bar{\mathcal{E}}_0}|\bar{\mathcal{E}}_0|/g, \tag{16.126b}$$

$$\langle J_z\rangle_{\rm ss} = -\frac{N}{2}\left[1+\sqrt{1-8(|\bar{\mathcal{E}}_0|/Ng)^2}\right], \tag{16.126c}$$

which defines the limit of the lower branch of the absorptive bistability input-output curve, and for all $|\bar{\mathcal{E}}_0|/g \geq 0$,

$$\langle\tilde{a}\rangle_{\rm ss} = i\bar{\mathcal{E}}_0/\kappa, \qquad \langle\tilde{a}^\dagger\rangle_{\rm ss} = -i\bar{\mathcal{E}}_0^*/\kappa, \tag{16.127a}$$

$$\langle\tilde{J}_-\rangle_{\rm ss} = \langle\tilde{J}_+\rangle_{\rm ss} = \langle J_z\rangle_{\rm ss} = 0, \tag{16.127b}$$

which defines the limit of the upper branch of the input–output curve.

Figure 16.6 shows how the intracavity photon number changes with driving field strength according to the optical bistability state equation and compares the change according to the new stationary states. There appears to be very little relationship between the two curves. The steady states of absorptive optical bistability, (16.126) and (16.127), are, of course, also stationary solutions to (16.117a)–(16.117e). But they do not respect the conservation law (16.118), and in the limit, when $\gamma_h = \gamma = 0$, they cannot be reached from the initial state $\langle \tilde{J}_- \rangle = \langle \tilde{J}_+ \rangle = 0$, $\langle J_z \rangle = -N/2$. Moreover, close to the limit, when $\gamma_h = \gamma$ is close to zero, the relaxation time to these steady states becomes extremely long. The limit of "zero system size" is in this sense structurally unstable. It follows that near to the limit all of the mentioned states are *quasi-stationary* and fragile to fluctuations. This is borne out in Sect. 16.3.6, where we see that quantum fluctuations in single-atom cavity QED are strongly influenced by all four quasi-stationary states, and, at least in a qualitative way, merge the contrasting pictures depicted in Fig. 16.6.

To help clarify the physics behind the new stationary states, let us define real and imaginary parts of the complex cavity field amplitude by

$$x + iy \equiv N^{-1/2}\big[ie^{-i\arg(\bar{\mathcal{E}}_0)}\langle \tilde{a} \rangle\big], \tag{16.128}$$

and the collective atomic state Bloch vector

$$\boldsymbol{m} \equiv N^{-1}\{2\mathrm{Re}\big[ie^{-i\arg(\bar{\mathcal{E}}_0)}\langle \tilde{J}_- \rangle\big]\hat{x} + 2\mathrm{Im}\big[ie^{-i\arg(\bar{\mathcal{E}}_0)}\langle \tilde{J}_- \rangle\big]\hat{y} + \langle J_z \rangle \hat{z}\}, \tag{16.129}$$

which represents a point on the unit sphere. The Maxwell–Bloch equations for "zero system size," (16.117a)–(16.117e), may then be written more compactly as

$$\dot{x} = -\kappa x + \tfrac{1}{2}\sqrt{N} g m_x + |\bar{\mathcal{E}}_0|/\sqrt{N}, \tag{16.130a}$$

$$\dot{y} = -\kappa y + \tfrac{1}{2}\sqrt{N} g m_y, \tag{16.130b}$$

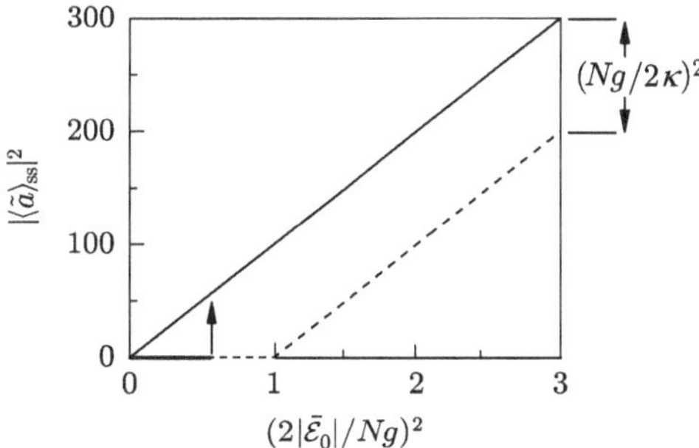

Fig. 16.6. Intracavity photon number as a function of input intensity for the stable steady states, (16.126a) and (16.127a), of absorptive optical bistability (*solid lines*) and the stationary states (16.119), (16.120), and (16.122) (*dashed lines*); for $Ng/\kappa = 20$

and

$$\dot{\boldsymbol{m}} = \boldsymbol{B} \times \boldsymbol{m}, \qquad \boldsymbol{m} \cdot \boldsymbol{m} = 1, \tag{16.130c}$$

where we introduce the fictitious magnetic field

$$\boldsymbol{B} \equiv 2\sqrt{N}g \begin{pmatrix} -y \\ x \\ 0 \end{pmatrix}. \tag{16.131}$$

We may now discuss the physics in terms of the magnetic analogy introduced in Sect. 2.3.3.

Equations 16.129, 16.130c, and 16.131 correspond to (2.105), (2.106), and (2.107) of Sect. 2.3.3, respectively. According to the analogy, they describe a magnetic moment $\boldsymbol{m}$ moving under the influence of the magnetic field $\boldsymbol{B}$. In the present case there is a significant difference from the standard picture, though; the magnetic field $\boldsymbol{B}$ is not a fixed external field, but a field that changes through its dynamical coupling to the magnetic moment. In terms of the true physical variables, the induced collective dipole changes under the influence of the intracavity field, while at the same time radiating into the cavity mode and thus, in turn, changing the intracavity field.

It is seen from this construction that the stationary states above threshold, where $\boldsymbol{B}_{\mathrm{ss}}$ is nonzero, are *self-consistent dressed states*—i.e., the steady state is established by both $\boldsymbol{m}_{\mathrm{ss}}$ and $\boldsymbol{B}_{\mathrm{ss}}$ adjusting their values so that the pair either align or anti-align with one another. Alsing and Carmichael call this phenomenon *spontaneous dressed-state polarization* [16.22]. This particular strategy for reaching a stationary state is only followed above threshold, where it is no longer possible for the polarization to cancel the driving field, as it does below threshold, since that would require $|\langle\tilde{J}_{\mp}\rangle|$ to be larger than the maximum value of $N/2$ permitted by (16.118).

Figure 16.7 illustrates the progression of the new stationary states from below threshold, where the polarization grows steadily to cancel the increasing driving field, to above threshold, where the stationary state is reached through spontaneous dressed-state polarization.

Before leaving our discussion of the Maxwell–Bloch equations, we should say something about the stability of the stationary states. Consider, first, the bad-cavity limit, where κ is much larger than $\sqrt{N}g$ and $|\bar{\mathcal{E}}_0|/\sqrt{N}$. In this case we can adiabatically eliminate the cavity field and, from (16.130a) and (16.130b), write

$$x = |\bar{\mathcal{E}}_0|/\sqrt{N}\kappa + \big(\sqrt{N}g/2\kappa\big)m_x, \tag{16.132a}$$

$$y = \big(\sqrt{N}g/2\kappa\big)m_y. \tag{16.132b}$$

After substituting these expressions into (16.130c), the resulting equations are the nonlinear Bloch equations of collective resonance fluorescence—i.e., the semiclassical equations of the single-mode, or Dicke model of superradiance with an added external driving field. Drummond and Carmichael [16.26]

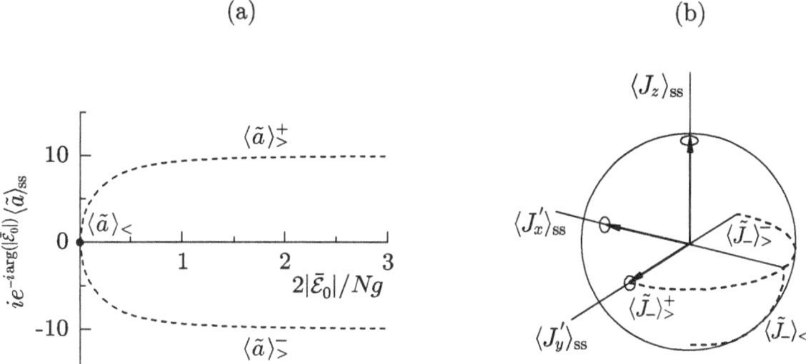

Fig. 16.7. Stationary states (16.119a)–(16.122c) plotted as a function of $2|\bar{\mathcal{E}}_0|/Ng$ (**a**) in the phase-plane of the cavity field and (**b**) on the Bloch sphere. Bloch-sphere axes are defined with $J_-' \equiv ie^{-i\arg(\bar{\mathcal{E}}_0)}J_-$

studied these equations and identify the same threshold behavior. They also solved the general time-dependent problem. The time dependence shows that while, below threshold, the stationary state (16.119a)–(16.119c) is a globally attracting fixed point, the state above threshold, (16.122a)–(16.122c), is nonstable; here the two fixed points lie at the foci of families of cyclic solutions, with frequency $\omega = (4g|\bar{\mathcal{E}}_0|/\kappa)\sqrt{1-(Ng/2|\bar{\mathcal{E}}_0|)^2}$. By solving the full set of Maxwell–Bloch equations numerically, we can show that similar behavior holds above threshold to that observed in the bad-cavity limit (Fig. 16.8).

This completes the semiclassical background against which quantum fluctuations will be added in Sect. 16.3.6. The next few sections make a diversion into the spectroscopy of single-atom cavity QED in the presence of an external field.

Note 16.7. The Hamiltonian for a collection of two-state atoms interacting with a single mode of the electromagnetic field may be mapped onto an analogous system of ions interacting with a single vibrational mode in an RF trap. The Jaynes–Cummings Hamiltonian maps onto the single ion case. The trapped ion analogy may be carried through to realize various cavity QED phenomena. Meekhof and coworkers [16.23], for example, performed an ion experiment to extend the time-domain measurement of few-quanta Rabi oscillations by Brune and coworkers [16.24] to nonclassical states of the vibrational mode (field). Spontaneous dressed-state polarization also exists in a trapped ion version, and has been discussed in this context by Milburn and Alsing [16.25]. We should perhaps note that the analogy is not complete, though; in particular, because of the different loss mechanisms that come into play. The results of this and the next section find a direct parallel in the ion system, but those on quantum fluctuations (Sect. 16.3.6) do not transfer directly.

Note 16.8. When considering many atoms, the limit of "zero system size" ($\gamma_h = \gamma \to 0$) is not necessarily a limit of large fluctuations. We might let

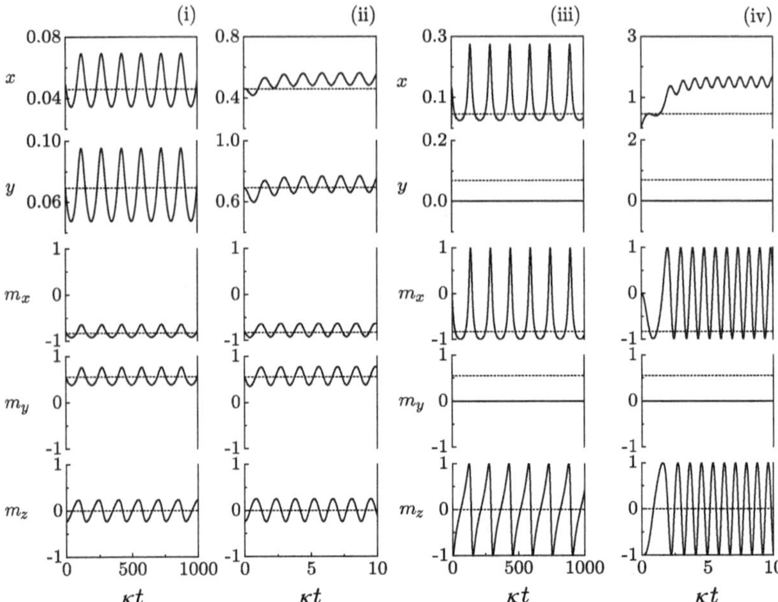

Fig. 16.8. Solutions to the Maxwell–Bloch equations (16.130) above threshold: for $\sqrt{N}g/\kappa = 0.25$, $|\bar{\mathcal{E}}_0|/\sqrt{N}\kappa = 0.15$ [columns (i) and (iii)] and $\sqrt{N}g/\kappa = 2.5$, $|\bar{\mathcal{E}}_0|/\sqrt{N}\kappa = 1.5$ [columns (ii) and (iv)]; the first case is a good approximation to the bad-cavity limit. Columns (i) and (ii) are for an initial condition close to the stationary state (*dashed lines*). Columns (iii) and (iv) show the evolution from the ground state

$g/|\bar{\mathcal{E}}_0| \to 0$, $N \to \infty$, with $Ng/2|\bar{\mathcal{E}}_0|$ finite to reach the small noise regime analyzed (bad-cavity case) by Drummond [16.27] and Carmichael [16.28].

16.3.2 Dressed Jaynes–Cummings Eigenstates

With the elimination of spontaneous emission, the limit of "zero system size" would appear to bring us closer to the physics of the pure Jaynes–Cummings model. This, however, is not really the case, because we are still concerned with the asymptotic dynamics of an open system—a system driven by an external field and damped by cavity loss. Even with the loss set aside as a small perturbation, we are not concerned with the Jaynes–Cummings Hamiltonian but with the Hamiltonian of the *driven Jaynes–Cummings model*:

$$\begin{aligned} H(t) \equiv & \tfrac{1}{2}\hbar\omega_A\sigma_z + \hbar\omega_C a^\dagger a + i\hbar g(a^\dagger\sigma_- - a\sigma_+) \\ & + \hbar(\bar{\mathcal{E}}_0 e^{-i\omega_0 t} a^\dagger + \bar{\mathcal{E}}_0^* e^{i\omega_0 t} a). \end{aligned} \tag{16.133}$$

There are two competing interactions: first the Jaynes–Cummings interaction between the atom and cavity mode, and then the interaction of the cavity mode with the driving (external) field. We should note that the results of the previous section assume that they have similar strength, since the threshold for spontaneous dressed-state polarization occurs at $2|\bar{\mathcal{E}}_0| = g$.

In this section we calculate the energy spectrum and stationary states of Hamiltonian (16.133) for the case of resonant excitation, $\omega_0 = \omega_C = \omega_A$. For an idea of where we are headed, we might compare (16.133) with the Hamiltonian of resonance fluorescence (Eq. 2.68a). In resonance fluorescence, the external field shifts, or splits, the bare energy levels of the atom to produce the so-called dressed atomic energies. We might anticipate similar level shifts here, but where the bare energies are now those of the Jaynes–Cummings Hamiltonian; thus, we can anticipate a *dressing of the dressed states*—if we care to speak in that way.

We meet an example of such double dressing in Sect. 13.1. There, with frequencies satisfying $\omega_0 + g = \omega_A = \omega_C$, we made a two-state approximation to predict a splitting of the lower vacuum Rabi resonance at intermediate driving field strengths—for $\frac{1}{2}(\kappa + \gamma/2) < |\bar{\mathcal{E}}_0| \ll g$. In the present situation, with $\omega_0 = \omega_C = \omega_A$, we can expect the spectrum for weak driving fields, $2|\bar{\mathcal{E}}_0| \ll g$, to be the Jaynes–Cummings spectrum plus a perturbative level shift of order $(2|\bar{\mathcal{E}}_0|/g)^2$. More generally, the perturbed-dressed-state picture must be qualified, though, since in the opposite extreme, for strong driving fields, $2|\bar{\mathcal{E}}_0| \gg g$, the roles of the interactions is reversed; the Jaynes–Cummings interaction becomes the perturbation. The dominant interaction is the interaction with the external field, which, written in the interaction picture as $\hbar(\bar{\mathcal{E}}_0 a^\dagger + \bar{\mathcal{E}}_0^* a)$, is formally a potential energy proportional to the position (or momentum) of a harmonic oscillator. It has a continuous spectrum. It seems then that the spectrum of (16.133) must change from the discrete Jaynes–Cummings spectrum to a continuous spectrum with increasing $(2|\bar{\mathcal{E}}_0|/g)$. Considering what we learned in the previous section, it is reasonable to anticipate a transition from a discrete spectrum to a continuous one at $2|\bar{\mathcal{E}}_0|/g = 1$.

As it stands, Hamiltonian (16.133) is explicitly time-dependent. It does not define an eigenvalue problem of the standard sort. Instead of looking for energy eigenvalues and eigenstate, we follow the method outlined in Note 2.5 to find periodic solutions, $|\psi_E(t)\rangle$, to the Schrödinger equation

$$\frac{d|\psi\rangle}{dt} = \frac{1}{i\hbar} H(t)|\psi\rangle, \tag{16.134}$$

and their associated quasienergies E. The procedure is rather straightforward in this instance because the time-dependence can be removed by working in the interaction picture; thus, setting $\omega_0 = \omega_C = \omega_A$, we write

$$|\psi\rangle = e^{-i\omega_A(\frac{1}{2}\sigma_z + a^\dagger a)t}|\tilde{\psi}\rangle, \tag{16.135}$$

to obtain

$$\frac{d|\tilde{\psi}\rangle}{dt} = \frac{1}{i\hbar}\tilde{H}|\tilde{\psi}\rangle, \tag{16.136}$$

with time-independent Hamiltonian

$$\tilde{H} = i\hbar g(a^\dagger\sigma_- - a\sigma_+) + \hbar(\bar{\mathcal{E}}_0 a^\dagger + \bar{\mathcal{E}}_0^* a). \tag{16.137}$$

Solutions to the time-independent Schrödinger equation

$$\tilde{H}|\tilde{\psi}_E\rangle = E|\tilde{\psi}_E\rangle \tag{16.138}$$

yield the desired periodic solutions to Schrödinger equation (16.134),

$$|\psi_E(t)\rangle = e^{-i(E/\hbar)t}e^{-i\omega_A(\frac{1}{2}\sigma_z + a^\dagger a)t}|\tilde{\psi}_E\rangle. \tag{16.139}$$

The rotation operator, $e^{-i\omega_A(\frac{1}{2}\sigma_z + a^\dagger a)t}$, which enters the transformation into the interaction picture (16.135), generates periodicity at the frequency of the driving field, $\omega_0 = \omega_A$.

The substance of the calculation involves finding solutions to the time-independent Schrödinger equation (16.138). We follow the solution scheme devised by Alsing and coworkers [16.29]. Expanding the eigenstate as

$$|\tilde{\psi}_E\rangle = |2\rangle_A|\tilde{\psi}_E^2\rangle_a + |1\rangle_A|\tilde{\psi}_E^1\rangle_a, \tag{16.140}$$

from (16.137), (16.138), and (16.140), we obtain coupled equations for $|\tilde{\psi}_E^2\rangle_a$ and $|\tilde{\psi}_E^1\rangle_a$,

$$[\hbar(\bar{\mathcal{E}}_0 a^\dagger + \bar{\mathcal{E}}_0^* a) - E]|\tilde{\psi}_E^2\rangle_a = i\hbar g a|\tilde{\psi}_E^1\rangle_a, \tag{16.141a}$$

$$[\hbar(\bar{\mathcal{E}}_0 a^\dagger + \bar{\mathcal{E}}_0^* a) - E]|\tilde{\psi}_E^1\rangle_a = -i\hbar g a^\dagger|\tilde{\psi}_E^2\rangle_a. \tag{16.141b}$$

The goal is to first decouple these equations and then solve separately for $|\tilde{\psi}_E^1\rangle_a$ and $|\tilde{\psi}_E^1\rangle_a$. Decoupling the equations is straightforward. We multiply (16.141a) and (16.141b) on the left by $a^\dagger$ and a, respectively, to obtain

$$\{[\hbar(\bar{\mathcal{E}}_0 a^\dagger + \bar{\mathcal{E}}_0^* a) - E]a^\dagger - \hbar\bar{\mathcal{E}}_0^*\}|\tilde{\psi}_E^2\rangle_a = i\hbar g a^\dagger a|\tilde{\psi}_E^1\rangle_a, \tag{16.142a}$$

$$\{[\hbar(\bar{\mathcal{E}}_0 a^\dagger + \bar{\mathcal{E}}_0^* a) - E]a + \hbar\bar{\mathcal{E}}_0\}|\tilde{\psi}_E^1\rangle_a = -i\hbar g a a^\dagger|\tilde{\psi}_E^2\rangle_a. \tag{16.142b}$$

Then, substituting for $a^\dagger|\tilde{\psi}_E^2\rangle_a$ and $a|\tilde{\psi}_E^1\rangle_a$, respectively, on the left-hand sides of (16.142a) and (16.142b), using (16.141b) and (16.141a), we obtain the new set of coupled equations

$$[\hat{O}_E + \tfrac{1}{2}(\hbar g)^2]|\tilde{\psi}_E^1\rangle_a = -i\hbar^2\bar{\mathcal{E}}_0^* g|\tilde{\psi}_E^2\rangle_a, \tag{16.143a}$$

$$[\hat{O}_E - \tfrac{1}{2}(\hbar g)^2]|\tilde{\psi}_E^2\rangle_a = -i\hbar^2\bar{\mathcal{E}}_0 g|\tilde{\psi}_E^1\rangle_a, \tag{16.143b}$$

with

$$\hat{O}_E \equiv [E - \hbar(\bar{\mathcal{E}}_0 a^\dagger + \bar{\mathcal{E}}_0^* a)]^2 - (\hbar g)^2\tfrac{1}{2}(a^\dagger a + a a^\dagger). \tag{16.144}$$

These have constant coefficients on the right-hand side and are therefore readily decoupled; $|\tilde{\psi}_E^1\rangle$ and $|\tilde{\psi}_E^2\rangle$ are both solutions to the equation

$$[\hat{O}_E + \tfrac{1}{2}(\hbar g)^2 e^{-2r}][\hat{O}_E - \tfrac{1}{2}(\hbar g)^2 e^{-2r}]|\tilde{\psi}_E^{1,2}\rangle_a = 0, \tag{16.145}$$

where

$$e^{-2r} \equiv \sqrt{1 - (2|\bar{\mathcal{E}}_0|/g)^2}. \tag{16.146}$$

Thus, general solutions for $|\tilde{\psi}_E^1\rangle$ and $|\tilde{\psi}_E^2\rangle$ may be written as linear combinations

$$|\tilde{\psi}_E^1\rangle_a = c_P|P\rangle_a + c_M|M\rangle_a, \tag{16.147a}$$
$$|\tilde{\psi}_E^2\rangle_a = d_P|P\rangle_a + d_M|M\rangle_a, \tag{16.147b}$$

where $c_{P,M}$ and $d_{P,M}$ are complex constants, and $|P\rangle_a$ and $|M\rangle_a$ vanish when operated on by one or other of the prefactors in (16.145),

$$[\hat{O}_E + \tfrac{1}{2}(\hbar g)^2 e^{-2r}]|P\rangle_a = 0, \tag{16.148a}$$
$$[\hat{O}_E - \tfrac{1}{2}(\hbar g)^2 e^{-2r}]|M\rangle_a = 0. \tag{16.148b}$$

It remains for us to solve these equations and determine the coefficients $c_{P,M}$ and $d_{P,M}$.

The equations have solutions for particular values of E only. We must determine both these values of E and the corresponding solutions. To this end, we note, first, that $\hat{O}_E$ is quadratic in a and $a^\dagger$; it can be diagonalized by the combination of a displacement and squeeze transformation. Thus, we write

$$|P\rangle_a = D(\alpha)S(\eta)|P\rangle'_a, \qquad |M\rangle_a = D(\alpha)S(\eta)|M\rangle'_a, \tag{16.149}$$

where α and η are complex parameters, and the squeeze operator, $S(\eta)$, is defined by (9.39) and the displacement operator, $D(\alpha)$, by (9.41). Under the displacement and squeeze, (16.148a) and (16.148b) transform as

$$[\hat{O}'_E(\alpha,\eta) + \tfrac{1}{2}(\hbar g)^2 e^{-2r}]|P\rangle'_a = 0, \tag{16.150a}$$
$$[\hat{O}'_E(\alpha,\eta) - \tfrac{1}{2}(\hbar g)^2 e^{-2r}]|M\rangle'_a = 0, \tag{16.150b}$$

with

$$\hat{O}'_E(\alpha,\eta) \equiv D(\alpha)S(\eta)\hat{O}_E S^\dagger(\eta)D^\dagger(\alpha). \tag{16.151}$$

We must now choose the adjustable parameters α and η so that the transformed operator $\hat{O}'_E$ takes on the diagonal form

$$\hat{O}'_E = A(E)\tfrac{1}{2}(a^\dagger a + aa^\dagger) + B(E), \tag{16.152}$$

where $A(E)$ and $B(E)$ are some undetermined functions of E. With this done, substituting (16.152) into (16.150a) and (16.150b), the latter are rewritten as

$$\big[A(E)(a^\dagger a + \tfrac{1}{2}) + B(E) + \tfrac{1}{2}(\hbar g)^2 e^{-2r}\big]|P\rangle'_a = 0, \tag{16.153a}$$

$$\big[A(E)(a^\dagger a + \tfrac{1}{2}) + B(E) - \tfrac{1}{2}(\hbar g)^2 e^{-2r}\big]|M\rangle'_a = 0, \tag{16.153b}$$

whose solutions are photon number states: we have $|P\rangle'_a = |n\rangle_a$, $n \geq 0$ an integer, which holds for quasienergies satisfying

$$(n + \tfrac{1}{2})A(E) + B(E) + \tfrac{1}{2}(\hbar g)^2 e^{-2r} = 0, \tag{16.154a}$$

and $|M\rangle'_a = |n\rangle_a$, $n \geq 0$ an integer, which holds for quasienergies satisfying

$$(n + \tfrac{1}{2})A(E) + B(E) - \tfrac{1}{2}(\hbar g)^2 e^{-2r} = 0. \tag{16.154b}$$

The unknown quantities α, η, $A(E)$, and $B(E)$ are all fixed by the diagonalization procedure. Their determination is left as an exercise.

Exercise 16.7. Using (16.144) and (16.151), write the displacement and squeeze transformation (16.152) in terms of quadrature phase operators as

$$\begin{aligned} &D(\alpha)S(\eta)\big[(E - \hbar 2|\bar{\mathcal{E}}_0|\hat{X})^2 + (\hbar g)^2(\hat{X}^2 + \hat{Y}^2)\big]S^\dagger(\eta)D^\dagger(\alpha) \\ &= A(E)(\hat{X}^2 + \hat{Y}^2) + B(E), \end{aligned} \tag{16.155}$$

with

$$\hat{X} \equiv \tfrac{1}{2}(ae^{-i\theta} + a^\dagger e^{i\theta}), \qquad \hat{Y} \equiv -i\tfrac{1}{2}(ae^{-i\theta} - a^\dagger e^{i\theta}), \tag{16.156}$$

where $\theta \equiv \arg(\bar{\mathcal{E}}_0)$. Show that the equality holds for the choice of parameters

$$\eta = -re^{i2\arg(\bar{\mathcal{E}}_0)}, \qquad \alpha(E) = -e^{i\arg(\bar{\mathcal{E}}_0)}\frac{E/\hbar}{g}e^{4r}\sqrt{1 - e^{-4r}}, \tag{16.157}$$

and functions of the quasienergy

$$A(E) = -(\hbar g)^2 e^{-2r}, \qquad B(E) = E^2 e^{4r}. \tag{16.158}$$

Note that the squeeze parameter r (degree of squeezing) is determined by the strength of the driving field relative to the dipole coupling strength (Eq. 16.146) and goes to infinity for $2|\bar{\mathcal{E}}_0|/g = 1$.

We are now in a position to determine the allowed quasienergies E. These follow from (16.154a) and (16.154b), with the functions $A(E)$ and $B(E)$ taken from (16.158): from (16.154a) and (16.158), the quasienergy satisfies

$$n + \tfrac{1}{2} = \left(\frac{E/\hbar}{g}\right)^2 e^{6r} + \tfrac{1}{2}, \tag{16.159}$$

with corresponding solution to (16.148a)

$$|P\rangle_a = D[\alpha(E)]S(\eta)|n\rangle_a, \tag{16.160}$$

and from (16.154b) and (16.158) the quasienergy satisfies

$$n + \tfrac{1}{2} = \left(\frac{E/\hbar}{g}\right)^2 e^{6r} - \tfrac{1}{2}, \tag{16.161}$$

with corresponding solution to (16.148b)

$$|M\rangle_a = D[\alpha(E)]S(\eta)|n\rangle_a. \tag{16.162}$$

Thus, in summary, the quasienergies and corresponding states solving (16.145) as elementary solutions are, for $n = 0$,

$$E = 0, \qquad |P\rangle_a = S(\eta)|0\rangle_a, \tag{16.163}$$

[note that (16.161) is not satisfied by any $n \geq 0$ if $E = 0$] and, for $n = 1, 2, \ldots$,

$$E = \pm e^{-3r}\sqrt{n}\hbar g, \qquad \begin{cases} |P\rangle_a &= D[\alpha(E)]S(\eta)|n\rangle_a \\ |M\rangle_a &= D[\alpha(E)]S(\eta)|n-1\rangle_a \end{cases}, \tag{16.164}$$

where both of the bracketed solutions are admitted for each value (positive or negative) of quasienergy E.

As a final step, the coefficients $c_{P,M}$ and $d_{P,M}$ in the expansions (16.147a) and (16.147b) of $|\tilde{\psi}_E^1\rangle_a$ and $|\tilde{\psi}_E^2\rangle_a$ must be determined. First, coefficients d_P and d_M are written in terms of c_P and c_M by substituting (16.147a) and (16.147b) into (16.141a) and equating the coefficients of $|P\rangle_a$ and $|M\rangle_a$ on either side of the equation. Then two more relationships are needed. One is the normalization condition $\langle\tilde{\psi}_E|\tilde{\psi}_E\rangle = 1$, and for the second, we can return to the step taking us from (16.141a) and (16.141b) to (16.142a) and (16.142b). Consider, specifically, that we multiplied (16.141b) by a to obtain (16.142b). It follows that a solution to (16.142b) need not necessarily satisfy (16.141b); we can add a multiple of the vacuum state on the right-hand side of the latter without altering the former. Thus, we must now impose the requirement that (16.141b) be satisfied. This provides the fourth relationship needed to fix the four expansion coefficients. The details of the derivation are left as an exercise.

Exercise 16.8. Show that for the quasienergy $E = 0$ the expansion coefficients in (16.147a) and (16.147b) are

$$c_P = \frac{1}{\sqrt{2}}\sqrt{1 + e^{-2r}}, \qquad d_P = ie^{i\arg(\bar{\mathcal{E}}_0)}\frac{1}{\sqrt{2}}\sqrt{1 - e^{-2r}}, \tag{16.165}$$

$c_M = d_M = 0$, and that otherwise, for quasienergies $E \neq 0$, the expansion coefficients depend only on the sign of the quasienergy, not on n, with

$$c_P = \pm i\frac{1}{2}\sqrt{1 + e^{-2r}}, \qquad d_P = \mp e^{i\arg(\bar{\mathcal{E}}_0)}\frac{1}{2}\sqrt{1 - e^{-2r}}, \tag{16.166}$$

and

$$c_M = -ie^{-i\arg(\bar{\mathcal{E}}_0)}\frac{1}{2}\sqrt{1-e^{-2r}}, \qquad d_M = \frac{1}{2}\sqrt{1+e^{-2r}}. \tag{16.167}$$

In writing (16.166) and (16.167), a convenient choice has been made for the arbitrary phase factor.

According to (16.163) and (16.164), for quasienergy $E = 0$ the state of the intracavity field is a squeezed vacuum state, while for nonzero quasienergies it is formed as a superposition of squeezed and displaced Fock states—$|n\rangle_a$ and $|n-1\rangle_a$—entangled with the state of the atom. The threshold from Sect. 16.3.1 emerges through the squeeze parameter r (Eq. 16.146), which tends to infinity (perfect squeezing) as $2|\bar{\mathcal{E}}_0|/g \to 1$. The squeeze parameter also determines the nonzero quasienergies, through (16.171a) and (16.171b), where the semiclassical threshold emerges once again. All quasienergies collapse to zero at $2|\bar{\mathcal{E}}_0|/g = 1$, and the above-threshold spectrum is continuous (Fig. 16.9). The constructed stationary states no longer hold above threshold.

Finally, putting all the pieces together, the *stationary states and quasienergies of the Jaynes–Cummings model driven on resonance [interaction picture Hamiltonian (16.137)] and for* $2|\bar{\mathcal{E}}_0|/g \leq 1$, are the "ground" state

$$|\tilde{\psi}_G\rangle = S(\eta)|O(r)\rangle_A|0\rangle_a, \tag{16.168}$$

quasienergy

$$E_G = 0, \tag{16.169}$$

and the "excited" state doublets ($n = 1, 2, \ldots$)

$$|\tilde{\psi}_{n,U}\rangle = D[\alpha(E_{n,U})]S(\eta)\frac{1}{\sqrt{2}}\big[|T(r)\rangle_A|n-1\rangle_a + i|O(r)\rangle_A|n\rangle_a\big], \tag{16.170a}$$

$$|\tilde{\psi}_{n,L}\rangle = D[\alpha(E_{n,L})]S(\eta)\frac{1}{\sqrt{2}}\big[|T(r)\rangle_A|n-1\rangle_a - i|O(r)\rangle_A|n\rangle_a\big], \tag{16.170b}$$

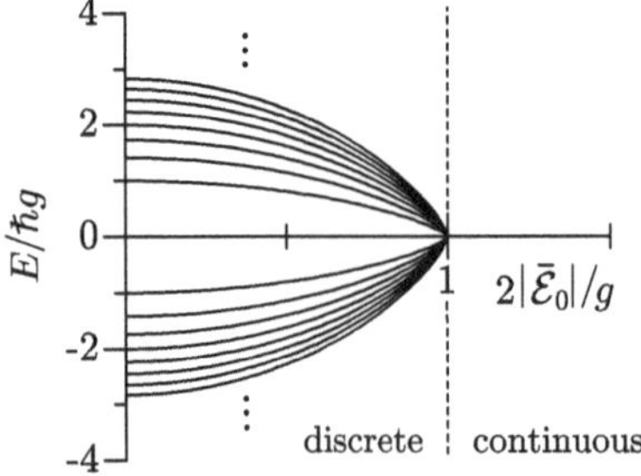

Fig. 16.9. First eight quasienergies plotted as a function of the driving field strength showing the transition from a discrete to a continuous spectrum at $2|\bar{\mathcal{E}}_0|/g = 1$

with quasienergies

$$E_{n,U} = +e^{-3r}\sqrt{n}\hbar g, \tag{16.171a}$$
$$E_{n,L} = -e^{-3r}\sqrt{n}\hbar g, \tag{16.171b}$$

where

$$|O(r)\rangle_A \equiv \frac{1}{\sqrt{2}}\Big(\sqrt{1+e^{-2r}}|1\rangle_A + ie^{i\arg(\bar{\mathcal{E}}_0)}\sqrt{1-e^{-2r}}|2\rangle_A\Big), \tag{16.172}$$

$$|T(r)\rangle_A \equiv \frac{1}{\sqrt{2}}\Big(\sqrt{1+e^{-2r}}|2\rangle_A - ie^{-i\arg(\bar{\mathcal{E}}_0)}\sqrt{1-e^{-2r}}|1\rangle_A\Big), \tag{16.173}$$

and r, η, and $\alpha(E)$ are defined in (16.146) and (16.157). We refer to the stationary states as the *dressed Jaynes–Cummings eigenstates.* Note that the words "ground" and "excited" have been placed in quotes. This is because they refer to quasienergies, not energies, and the associated states are stationary in the interaction but not the Schrödinger picture. Indeed, through (16.139), each stationary state defines a periodic solution to the Schrödinger equation, expanded in a superposition of an infinite number of terms oscillating at frequencies $(m-\frac{1}{2})\omega_A+E/\hbar$, $m=0,1,2,\ldots$; thus, each dressed Jaynes–Cummings eigenstate—even the "ground" state—is expanded over the entire spectrum of bare states—the bare ground state $|1\rangle_A|0\rangle_a$ and all excited states $|1\rangle_A|m\rangle_a$, $|2\rangle_A|m-1\rangle_a$, $m=1,2,\ldots$.

Exercise 16.9. The threshold condition is not the only trace of the semiclassical analysis to surface in the quantum solution. Show that for the "ground" state (16.168) the state of the atom is the same as that given by the semiclassical solution (16.119b) and (16.119c).

Note 16.9. In the general case of arbitrary detunings between the driving field and the cavity, and the cavity and the atom, the interaction picture Hamiltonian (16.137) acquires the additional term $\frac{1}{2}\Delta\omega_A\sigma_z + \hbar\Delta\omega_C a^\dagger a$, with $\Delta\omega_A \equiv \omega_A - \omega_0$ and $\Delta\omega_C \equiv \omega_C - \omega_0$ [replace ω_A by ω_0 in (16.135)]. If $\Delta\omega_A \neq 0$ and $\Delta\omega_C = 0$, the solution scheme carries through as before. The only change is that the operator $\hat{O}(E)$ (Eq. 16.144) acquires the extra term $-(\hbar\Delta\omega_A/2)^2$. Since the operator is still quadratic in a and $a^\dagger$, it can still be diagonalized by a suitable choice of displacement and squeeze operators. For nonzero cavity detuning, the solution scheme fails, as $\hat{O}(E)$ is then quartic in a and $a^\dagger$.

16.3.3 Secular Approximation in the Basis of Dressed Jaynes–Cummings Eigenstates

The dressed Jaynes–Cummings eigenstates are nondegenerate eigenstates of the Hamiltonian (16.137). They form an orthonormal basis and may be used in place of the bare states or ordinary dressed states (Eqs. 13.189 and 13.190) to write the master equation in matrix element form. Thus, in the interaction picture and for the resonant case ($\omega_0 = \omega_C = \omega_A$), we may write master equation (13.57) as

$$\begin{aligned}\dot{\tilde{\rho}}_{EE'} &= \frac{1}{i\hbar}(E - E')\tilde{\rho}_{EE'} \\ &+ \sum_{\epsilon,\epsilon'} \left(\gamma\langle E|\sigma_-|\epsilon\rangle\langle\epsilon'|\sigma_+|E'\rangle + 2\kappa\langle E|a|\epsilon\rangle\langle\epsilon'|a^\dagger|E'\rangle\right)\tilde{\rho}_{\epsilon\epsilon'} \\ &- \sum_{\epsilon}\left(\frac{\gamma}{2}\langle E|\sigma_+\sigma_-|\epsilon\rangle + \kappa\langle E|a^\dagger a|\epsilon\rangle\right)\tilde{\rho}_{\epsilon E'} \\ &- \sum_{\epsilon}\left(\frac{\gamma}{2}\langle \epsilon|\sigma_+\sigma_-|E'\rangle + \kappa\langle \epsilon|a^\dagger a|E'\rangle\right)\tilde{\rho}_{E\epsilon},\end{aligned} \tag{16.174}$$

where

$$\tilde{\rho}_{EE'} \equiv \langle\tilde{\psi}_E|\tilde{\rho}|\tilde{\psi}_{E'}\rangle. \tag{16.175}$$

We may then remove the trivial time dependence, writing

$$\tilde{\rho}_{EE'} = e^{-(i/\hbar)(E-E')t}\bar{\rho}_{EE'}, \tag{16.176}$$

with $\bar{\rho}_{EE'} \equiv \langle\psi_E(t)|\rho|\psi_{E'}(t)\rangle$ (Eq. 16.139). Then the evolution of matrix elements $\bar{\rho}_{EE'}(t)$ is due to the cavity and atomic damping alone.

Most generally, the matrix elements obey rather complicated equations of motion. First, an explicit time dependence is introduced by the transformation (16.176), and, although only damping terms remain, these terms are far from simple when written in the basis of dressed Jaynes–Cummings eigenstates; diagonal and off-diagonal matrix elements are coupled, and in place of the relatively simple coupling in the bare-state basis (Eq. 16.1), each state, $|\tilde{\psi}_E\rangle$, is coupled to every other state, $|\tilde{\psi}_{E'}\rangle$. Nevertheless, the $\bar{\rho}_{EE'}$-representation can still be useful if a secular approximation is made.

The secular approximation was popularized by Cohen–Tannoudji and Reynaud [16.30] who applied it to resonance fluorescence, where it provides an appealing understanding of the Mollow spectrum, especially in the strong-excitation limit. A similar approximation holds here for the limit of strong nonperturbative coupling—$g \gg \kappa, \gamma/2$—and sufficiently weak excitation; under these conditions the quasienergy splittings of Fig. 16.9 may be assumed to be much larger than the level widths. Consider then the equation of motion for a diagonal density matrix element $\bar{\rho}_{EE}$. Terms in the equation that couple to the off-diagonal elements, $\bar{\rho}_{EE'}$, oscillate with frequencies $(E - E')/\hbar$. Assuming a slow time variation for $\bar{\rho}_{EE'}$, of order $(2\kappa)^{-1}$ or γ^{-1}, and integrating over an interval short compared with this timescale but much longer than the period of oscillation, the time-averaged oscillatory terms contribute at the order $2\kappa/|E - E'|$ or $\gamma/|E - E'|$. Similar comments apply to nonresonant terms in the equation of motion for an off-diagonal matrix element.

In summary, the *secular approximation* neglects the time-averaged terms of order $2\kappa/|E-E'|$ or $\gamma/|E-E'|$ to describe the dynamics on a course-grained timescale. It drops off-diagonal matrix elements from the equations of motion for diagonal matrix elements, and keeps only resonant terms [terms that also oscillate as $e^{-(i/\hbar)(E-E')t}$] in the equations of motion for off-diagonal matrix elements. Thus, coming back to master equation (16.174), for the diagonal matrix elements we obtain the *rate equations in the basis of dressed Jaynes–Cummings eigenstates*

$$\dot{\bar{\rho}}_{EE} = \sum_{\epsilon}(\gamma_{\epsilon,E}\bar{\rho}_{\epsilon\epsilon} - \gamma_{E,\epsilon}\bar{\rho}_{EE}), \tag{16.177}$$

with

$$\gamma_{\epsilon,E} = \gamma|\langle E|\sigma_-|\epsilon\rangle|^2 + 2\kappa|\langle E|a|\epsilon\rangle|, \tag{16.178}$$

where we have used $\langle E|(\gamma\sigma_+\sigma_- + 2\kappa a^\dagger a)|E\rangle = \sum_\epsilon \gamma_{E,\epsilon}$, which is easily proved using the completeness relation $\sum_\epsilon |\epsilon\rangle\langle\epsilon| = 1$. For the off-diagonal elements, note that resonance between quasienergies occurs with $\epsilon = E$ and $\epsilon' = E'$, and also with $\epsilon = -E'$ and $\epsilon' = -E$; thus, we obtain ($E \neq E'$)

$$\dot{\bar{\rho}}_{EE'} = -\left(\sum_{\epsilon}\frac{\gamma_{E,\epsilon} + \gamma_{E',\epsilon}}{2} - K^{(1)}_{E,E'}\right)\bar{\rho}_{EE'} + K^{(2)}_{E,E'}\bar{\rho}_{-E'-E}, \tag{16.179}$$

with

$$K^{(1)}_{E,E'} \equiv \gamma\langle E|\sigma_-|E\rangle\langle E'|\sigma_+|E'\rangle + 2\kappa\langle E|a|E\rangle\langle E'|a^\dagger|E'\rangle, \tag{16.180a}$$

$$K^{(2)}_{E,E'} \equiv \gamma\langle E|\sigma_-|-E'\rangle\langle -E|\sigma_+|E'\rangle + 2\kappa\langle E|a|-E'\rangle\langle -E|a^\dagger|E'\rangle. \tag{16.180b}$$

Note 16.10. Frequencies $|E-E'|/\hbar$ may be classified according to whether E and E' have the same or opposite signs. For opposite signs, these frequencies are all larger than $g \gg \kappa, \gamma/2$. For the same sign, on the other hand, they can be quite small—for example, choose $E = E_{n,U}$ and $E' = E_{n-1,U}$ with $n \gg 1$. It might appear that the latter case creates a problem for the secular approximation, as matrix elements between basis states having quasienergy of the same sign are not identically zero. These matrix elements approach zero, though, as $2|\bar{\mathcal{E}}_0|/g \to 0$. Therefore, so long as the excitation is not too strong, the seemingly troublesome terms may still be neglected.

16.3.4 Spectrum of the Transmitted Light in the Strong-Coupling and Weak-Excitation Limits

Although the density matrix element equations in the secular approximation are considerably simpler than for the full master equation, they are still difficult to work with; evaluating the matrix elements (16.178) and (16.180) is a daunting task in itself. Noting, however, that the approximation is only valid for weak excitation in any case, we may further simplify calculations by making a quasienergy truncation. In this section we employ a truncation similar to the two-quanta truncation of Sect. 16.1.1 and calculate the spectrum of the transmitted light in the strong-coupling and weak-excitation limits.

The resulting spectrum is the squared-Lorentzian doublet obtained from the linear theory of fluctuations in Sect. 15.2.6. The $\bar{\rho}_{EE'}$-representation is particularly suited to this derivation as the basis states already incorporate the source of the squared Lorentzians—the squeezing of Sects. 15.2.5 and 16.1.6. A quasienergy truncation at the first "excited" state is adequate and provides a particularly quick route to the spectrum. For comparison, the alternative derivation carried through by Rice [16.16] adds higher-order terms to the two-quanta truncation in the bare-state basis. It is considerably more tedious, and less transparent concerning the role of the squeezing of the intracavity field. The following calculation expands upon the sketch of Carmichael and coworkers [16.31].

Truncated at the first "excited" state, the density matrix equations in the secular approximation, (16.177) and (16.179), describe a birth–death process with transitions between the three states $|\tilde{\psi}_G\rangle$, $|\tilde{\psi}_{1,U}\rangle$, and $|\tilde{\psi}_{1,L}\rangle$. In steady state the off-diagonal matrix elements vanish, since they are uncoupled from the diagonal elements and obey a homogeneous set of equations. The steady state is the mixed state

$$\bar{\rho}_{\mathrm{ss}} = |\tilde{\psi}_G\rangle\langle\tilde{\psi}_G| + p_1^{\mathrm{ss}}\left(|\tilde{\psi}_{1,U}\rangle\langle\tilde{\psi}_{1,U}| + |\tilde{\psi}_{1,L}\rangle\langle\tilde{\psi}_{1,L}|\right), \tag{16.181}$$

where $p_1^{\mathrm{ss}} \ll 1$ is the occupation probability of the first "excited" state doublet. From Sect. 16.1.1 we know that the steady state is pure to lowest order in $|\bar{\mathcal{E}}_0|/g$, with $|\tilde{\psi}_{\mathrm{ss}}\rangle$ given by the two-quanta truncation in the bare-state basis. This requires $|\tilde{\psi}_G\rangle = |\tilde{\psi}_{\mathrm{ss}}\rangle$, which we verify below. To this lowest-order approximation, the birth–death process adds the higher orders needed to go beyond the pure-state factorization and obtain the optical spectrum at dominant order (see the opening paragraphs of Sect. 16.1.6).

Before the calculation of the spectrum can begin, we must find explicit expressions for the basis states $|\tilde{\psi}_G\rangle$, $|\tilde{\psi}_{1,U}\rangle$, and $|\tilde{\psi}_{1,L}\rangle$, compute the transition rates $\gamma_{\epsilon,E}$, and solve the rate equations (16.177). The parameters entering the formulae for the states are given by (16.146) and (16.157). For weak excitation, $2|\bar{\mathcal{E}}_0|/g \ll 1$, we may write

$$r = \left(\frac{|\bar{\mathcal{E}}_0|}{g}\right)^2, \tag{16.182}$$

and

$$\eta = -\left(\frac{\bar{\mathcal{E}}_0}{g}\right)^2, \qquad \alpha(E) = -\mathrm{sgn}(E)\sqrt{n}\frac{2\bar{\mathcal{E}}_0}{g}. \tag{16.183}$$

Then, substituting for r in (16.172) and (16.173), and expanding the exponentials to lowest order in $|\bar{\mathcal{E}}_0|/g \ll 1$, the atomic part of the "excited" states

$|\tilde{\psi}_{1,U}\rangle$ and $|\tilde{\psi}_{1,L}\rangle$ evaluates as

$$|O(r)\rangle_A = |1\rangle_A + i(\bar{\mathcal{E}}_0/g)|2\rangle_A, \tag{16.184a}$$
$$|T(r)\rangle_A = |2\rangle_A - i(\bar{\mathcal{E}}_0^*/g)|1\rangle_A. \tag{16.184b}$$

Similarly, from the definitions (9.39) and (9.41) of the squeeze and displacement operators, to lowest order we have

$$S(\eta)|0\rangle_a = |0\rangle_a + \frac{1}{\sqrt{2}}\left(\frac{\bar{\mathcal{E}}_0}{g}\right)^2|2\rangle_a, \tag{16.185a}$$
$$S(\eta)|1\rangle_a = |1\rangle_a, \tag{16.185b}$$

and

$$D[\alpha(E)]S(\eta)|0\rangle_a = |0\rangle_a - \mathrm{sgn}(E)2\frac{\bar{\mathcal{E}}_0}{g}|1\rangle_a + \frac{1}{\sqrt{2}}\left(\frac{\bar{\mathcal{E}}_0}{g}\right)^2|2\rangle_a, \tag{16.186a}$$
$$D[\alpha(E)]S(\eta)|1\rangle_a = \mathrm{sgn}(E)2\frac{\bar{\mathcal{E}}_0^*}{g}|0\rangle_a + |1\rangle_a - \mathrm{sgn}(E)2\sqrt{2}\frac{\bar{\mathcal{E}}_0}{g}|2\rangle_a. \tag{16.186b}$$

Thus, from (16.168), (16.184a), and (16.185a), we obtain the *dressed Jaynes–Cummings "ground" state in the limit of weak excitation,*

$$|\tilde{\psi}_G\rangle = |1\rangle_A|0\rangle_a + i\frac{\bar{\mathcal{E}}_0}{g}|2\rangle_A|0\rangle_a + \frac{1}{\sqrt{2}}\left(\frac{\bar{\mathcal{E}}_0}{g}\right)^2|1\rangle_A|2\rangle_a, \tag{16.187}$$

and from (16.170a), (16.170b), and (16.184a)–(16.186b), the *dressed Jaynes–Cummings first "excited" state doublet,*

$$|\tilde{\psi}_{1,U}\rangle = |1,U\rangle + i\frac{1}{\sqrt{2}}\frac{\bar{\mathcal{E}}_0^*}{g}|1\rangle_A|0\rangle_a, \tag{16.188a}$$
$$|\tilde{\psi}_{1,L}\rangle = |1,L\rangle + i\frac{1}{\sqrt{2}}\frac{\bar{\mathcal{E}}_0^*}{g}|1\rangle_A|0\rangle_a. \tag{16.188b}$$

Exercise 16.10. Show that in the strong-coupling limit, $g \gg \frac{1}{2}(\kappa+\gamma/2)$, the "ground" state (16.187) agrees with the pure state defined by (16.7) and the steady-state Schrödinger amplitudes (16.29a), (16.29b), (16.32a), and (16.32b).

Expressions for the transition rates $\gamma_{\epsilon,E}$ follow in a straightforward manner. Using (16.178), (16.187), (16.188a), and (16.188b), the transition rates into and out of the "ground" state are

$$\gamma_{G,G} = \gamma\left(\frac{|\bar{\mathcal{E}}_0|}{g}\right)^2, \tag{16.189a}$$
$$\gamma_{G;1,U} = \gamma_{G;1,L} = (\kappa+\gamma/2)\left(\frac{|\bar{\mathcal{E}}_0|}{g}\right)^4, \tag{16.189b}$$
$$\gamma_{1,U;G} = \gamma_{1,L;G} = (\kappa+\gamma/2), \tag{16.189c}$$

while the rates for transitions within the "excited" state doublet are

$$\gamma_{1,U;1,U} = \gamma_{1,L;1,L} = (\kappa + \gamma/2)\frac{1}{2}\left(\frac{|\bar{\mathcal{E}}_0|}{g}\right)^2, \tag{16.189d}$$

$$\gamma_{1,U;1,L} = \gamma_{1,L;1,U} = (\kappa + \gamma/2)\frac{1}{2}\left(\frac{|\bar{\mathcal{E}}_0|}{g}\right)^2. \tag{16.189e}$$

We can now write out the rate equations explicitly. We may ignore the equation for $\bar{\rho}_{GG}$ since it only adds higher-order corrections to the *ansatz* $\bar{\rho}_{GG} = 1$. Transition rates (16.189d) and (16.189e) may be ignored as well; they contribute further higher-order corrections. The only equations of importance are those accounting for the excitation and de-excitation of the first "excited" states. Thus, we obtain the *rate equations in the limit of weak excitation within the secular approximation in the basis of dressed Jaynes–Cummings eigenstates*:

$$\dot{\bar{\rho}}_{1,U;1,U} = -(\kappa + \gamma/2)\bar{\rho}_{1,U;1,U} + (\kappa + \gamma/2)\left(\frac{|\bar{\mathcal{E}}_0|}{g}\right)^4 \bar{\rho}_{GG}, \tag{16.190a}$$

$$\dot{\bar{\rho}}_{1,L;1,L} = -(\kappa + \gamma/2)\bar{\rho}_{1,L;1,L} + (\kappa + \gamma/2)\left(\frac{|\bar{\mathcal{E}}_0|}{g}\right)^4 \bar{\rho}_{GG}, \tag{16.190b}$$

with steady-state solution $\bar{\rho}_{1,U;1,U} = \bar{\rho}_{1,L;1,L} = (|\bar{\mathcal{E}}_0|/g)^4\bar{\rho}_{GG}$, or steady-state density matrix (setting $\bar{\rho}_{GG} = 1$)

$$\bar{\rho}_{\rm ss} = |\tilde{\psi}_G\rangle\langle\tilde{\psi}_G| + \left(\frac{|\bar{\mathcal{E}}_0|}{g}\right)^4 \Big(|\tilde{\psi}_{1,U}\rangle\langle\tilde{\psi}_{1,U}| + |\tilde{\psi}_{1,L}\rangle\langle\tilde{\psi}_{1,L}|\Big). \tag{16.191}$$

Note how the probability of occupation of the first "excited" state doublet is of order $(|\bar{\mathcal{E}}_0|/g)^4$. This probability sets the order of the incoherent part of the optical spectrum, which we are now in a position to calculate.

The spectrum of the transmitted light is calculated from the Fourier transform of the first-order correlation function of the cavity output field:

$$\begin{aligned} T_{\rightarrow}(\omega) &= \frac{\gamma_{a2}}{2\pi}\int_{-\infty}^{\infty} d\tau e^{-i\omega\tau}\langle a^\dagger(0)a(\tau)\rangle_{\rm ss}, \\ &= \frac{\gamma_{a2}}{\pi}{\rm Re}\left[\int_0^{\infty} d\tau e^{-i\omega\tau}\langle a^\dagger(0)a(\tau)\rangle_{\rm ss}\right], \end{aligned} \tag{16.192}$$

where we use $\langle a^\dagger(0)a(-\tau)\rangle_{\rm ss} = \langle a^\dagger(\tau)a(0)\rangle_{\rm ss} = \langle a^\dagger(0)a(\tau)\rangle_{\rm ss}^*$. From the quantum regression formula (1.97), the required correlation function is calculated as ($\tau \geq 0$)

$$\begin{aligned} \langle a^\dagger(0)a(\tau)\rangle_{\rm ss} &= e^{-i\omega_0\tau}{\rm tr}[ae^{\tilde{\mathcal{L}}\tau}(\tilde{\rho}_{\rm ss}a^\dagger)] \\ &= e^{-i\omega_0\tau}{\rm tr}[a\tilde{R}(\tau)], \end{aligned} \tag{16.193}$$

with

$$\tilde{R}(\tau) \equiv e^{\tilde{\mathcal{L}}\tau}(\tilde{\rho}_{ss}a^\dagger). \tag{16.194}$$

Like the density matrix elements themselves, the matrix elements of $\tilde{R}(\tau)$ obey the equations of motion (16.174). Our plan is to solve these matrix element equations in the secular approximation, with the important change from the above that now a transient solution for the off-diagonal matrix elements is required.

Continuing with the weak-excitation limit, each matrix element has a dominant order determined by its initial value $\tilde{R}_{EE'}(0) = \langle E|\tilde{\rho}_{ss}a^\dagger|E'\rangle$. From (16.187), (16.188a), (16.188b), and (16.191), we have

$$\left.\begin{aligned} (\tilde{\rho}_{ss}a^\dagger)_{G;1,U} &= i\frac{1}{\sqrt{2}}\left(\frac{\bar{\mathcal{E}}_0^*}{g}\right)^2, \qquad (\tilde{\rho}_{ss}a^\dagger)_{1,U;G} = -i\frac{1}{\sqrt{2}}\left(\frac{|\bar{\mathcal{E}}_0|}{g}\right)^4 \\ (\tilde{\rho}_{ss}a^\dagger)_{G;1,L} &= -i\frac{1}{\sqrt{2}}\left(\frac{\bar{\mathcal{E}}_0^*}{g}\right)^2, \qquad (\tilde{\rho}_{ss}a^\dagger)_{1,L;G} = i\frac{1}{\sqrt{2}}\left(\frac{|\bar{\mathcal{E}}_0|}{g}\right)^4 \end{aligned}\right\}, \tag{16.195}$$

with all other matrix elements of higher order than $(|\bar{\mathcal{E}}_0|/g)^4$. We therefore limit our attention to the four equations (Eqs. 16.179):

$$\dot{\bar{R}}_{G;1,U} = -\left[\tfrac{1}{2}(\kappa+\gamma/2) - K^{(1)}_{G;1,U}\right]\bar{R}_{G;1,U} + K^{(2)}_{G;1,U}\bar{R}_{1,L;G}, \tag{16.196a}$$

$$\dot{\bar{R}}_{G;1,L} = -\left[\tfrac{1}{2}(\kappa+\gamma/2) - K^{(1)}_{G;1,L}\right]\bar{R}_{G;1,L} + K^{(2)}_{G;1,L}\bar{R}_{1,U;G}, \tag{16.196b}$$

$$\dot{\bar{R}}_{1,U;G} = -\left[\tfrac{1}{2}(\kappa+\gamma/2) - K^{(1)}_{1,U;G}\right]\bar{R}_{1,U;G} + K^{(2)}_{1,U;G}\bar{R}_{G;1,L}, \tag{16.196c}$$

$$\dot{\bar{R}}_{1,L;G} = -\left[\tfrac{1}{2}(\kappa+\gamma/2) - K^{(1)}_{1,L;G}\right]\bar{R}_{1,L;G} + K^{(2)}_{1,L;G}\bar{R}_{G;1,U}, \tag{16.196d}$$

where

$$\bar{R}_{EE'} \equiv e^{(i/\hbar)(E-E')t}\tilde{R}_{EE'}. \tag{16.197}$$

$K^{(1)}_{E,E'}$ and $K^{(2)}_{E,E'}$ are to be evaluated from (16.180a) and (16.180b), using the dressed Jaynes–Cummings eigenstates (16.187), (16.188a), and (16.188b). The first coefficient is of order $(|\bar{\mathcal{E}}_0|/g)^2$ for all relevant E and E'; thus, it makes negligible corrections to the damping rate $\frac{1}{2}(\kappa+\gamma/2)$. Furthermore, the second coefficient need not be evaluated in (16.196a) and (16.196b), since the terms proportional to this coefficient may be dropped; according to (16.195), the dominant order of the two equations is $(|\bar{\mathcal{E}}_0|/g)^2$, while the dropped terms involve matrix elements of order $(|\bar{\mathcal{E}}_0|/g)^4$. There are then just two coefficients to evaluate, for which we find

$$K^{(2)}_{1;U;G} = K^{(2)}_{1;L;G} = (\kappa+\gamma/2)\left(\frac{\bar{\mathcal{E}}_0}{g}\right)^2. \tag{16.198}$$

The resulting equations of motion are

$$\dot{\bar{R}}_{G;1,U} = -\tfrac{1}{2}(\kappa+\gamma/2)\bar{R}_{G;1,U}, \tag{16.199a}$$

$$\dot{\bar{R}}_{G;1,L} = -\tfrac{1}{2}(\kappa+\gamma/2)\bar{R}_{G;1,L}, \tag{16.199b}$$

and

$$\dot{\bar{R}}_{1,U;G} = -\tfrac{1}{2}(\kappa + \gamma/2)\bar{R}_{1,U;G} + (\kappa + \gamma/2)\left(\frac{\bar{\mathcal{E}}_0}{g}\right)^2 \bar{R}_{G;1,L}, \tag{16.199c}$$

$$\dot{\bar{R}}_{1,L;G} = -\tfrac{1}{2}(\kappa + \gamma/2)\bar{R}_{1,L;G} + (\kappa + \gamma/2)\left(\frac{\bar{\mathcal{E}}_0}{g}\right)^2 \bar{R}_{G;1,U}, \tag{16.199d}$$

with solutions

$$\bar{R}_{G;1,U}(\tau) = i\frac{1}{\sqrt{2}}\left(\frac{\bar{\mathcal{E}}_0^*}{g}\right)^2 e^{-\frac{1}{2}(\kappa+\gamma/2)\tau}, \tag{16.200a}$$

$$\bar{R}_{G;1,L}(\tau) = -i\frac{1}{\sqrt{2}}\left(\frac{\bar{\mathcal{E}}_0^*}{g}\right)^2 e^{-\frac{1}{2}(\kappa+\gamma/2)\tau}, \tag{16.200b}$$

and

$$\bar{R}_{1,U;G}(\tau) = -i\frac{1}{\sqrt{2}}\left(\frac{|\bar{\mathcal{E}}_0|}{g}\right)^4 \{-1 + 2[1 + \tfrac{1}{2}(\kappa + \gamma/2)\tau]\}e^{-\frac{1}{2}(\kappa+\gamma/2)\tau}, \tag{16.200c}$$

$$\bar{R}_{1,L;G}(\tau) = i\frac{1}{\sqrt{2}}\left(\frac{|\bar{\mathcal{E}}_0|}{g}\right)^4 \{-1 + 2[1 + \tfrac{1}{2}(\kappa + \gamma/2)\tau]\}e^{-\frac{1}{2}(\kappa+\gamma/2)\tau}. \tag{16.200d}$$

With these we can evaluate the trace in (16.193) and the optical spectrum from (16.192).

The effect of squeezing on the spectrum enters through the equations of motion (16.199c) and (16.199d), where each matrix element, $\bar{R}_{1,U;G}$ or $\bar{R}_{1,L;G}$, couples to a conjugate element—respectively to $\bar{R}_{G;1,L}$ and $\bar{R}_{G;1,U}$; compare the two-mode coupling in (9.183a) and (9.183b). As a result of this coupling and the degeneracy of the decay rates, solutions (16.200c) and (16.200d) do not exhibit the usual exponential decay, but include terms decaying as $(1+\bar{\tau})e^{-\bar{\tau}}$, where $\bar{\tau} \equiv \frac{1}{2}(\kappa+\gamma/2)\tau$. The Fourier transform of such a decay is a Lorentzian squared.

To complete the derivation of the spectrum and explicitly reveal its squared-Lorentzian form, from (16.193), and employing the truncation at the first "excited" state doublet, we have

$$\begin{aligned} e^{i\omega_0\tau}\langle a^\dagger(0)a(\tau)\rangle_{\rm ss} = {}& \langle\tilde{\psi}_G|a|\tilde{\psi}_{1,U}\rangle\bar{R}_{1,U;G}(\tau) + \langle\tilde{\psi}_G|a|\tilde{\psi}_{1,L}\rangle\bar{R}_{1,L;G}(\tau) \\ & + \langle\tilde{\psi}_{1,U}|a|\tilde{\psi}_G\rangle\bar{R}_{G;1,U}(\tau) + \langle\tilde{\psi}_{1,L}|a|\tilde{\psi}_G\rangle\bar{R}_{G;1,L}(\tau). \end{aligned} \tag{16.201}$$

Then, using (16.187), (16.188a), and (16.188b), the matrix elements of a in the dressed Jaynes–Cummings basis are

$$\left.\begin{array}{ll}\langle\tilde{\psi}_G|a|\tilde{\psi}_{1,U}\rangle = i\frac{1}{\sqrt{2}}, & \langle\tilde{\psi}_{1,U}|a|\tilde{\psi}_G\rangle = -i\frac{1}{\sqrt{2}}(\bar{\mathcal{E}}_0/g)^2 \\ \langle\tilde{\psi}_G|a|\tilde{\psi}_{1,L}\rangle = -i\frac{1}{\sqrt{2}}, & \langle\tilde{\psi}_{1,L}|a|\tilde{\psi}_G\rangle = i\frac{1}{\sqrt{2}}(\bar{\mathcal{E}}_0/g)^2\end{array}\right\}. \tag{16.202}$$

It follows that ($\tau \geq 0$)

$$\langle a^\dagger(0)a(\tau)\rangle_{\rm ss} = 2\left(\frac{|\bar{\mathcal{E}}_0|}{g}\right)^4 e^{-[\frac{1}{2}(\kappa+\gamma/2)+i\omega_0]\tau}\left[1+\tfrac{1}{2}(\kappa+\gamma/2)\tau\right]\cos(g\tau), \tag{16.203}$$

where we have used (16.197) with $E_{1,U} = \hbar g$ and $E_{1,L} = -\hbar g$. Substituting (16.203) into (16.192) and carrying out the Fourier transform, we arrive at the spectrum of the transmitted light, the *single-atom vacuum Rabi doublet with squeezing-induced linewidth narrowing*,

$$\begin{aligned}T_{\rightarrow}(\omega) = \gamma_{a2}\left(\frac{|\bar{\mathcal{E}}_0|}{g}\right)^4\Bigg\{&\frac{2\left[\frac{1}{2}(\kappa+\gamma/2)\right]^3/\pi}{\{\left[\frac{1}{2}(\kappa+\gamma/2)\right]^2+(\omega-\omega_0+g)^2\}^2} \\ &+\frac{2\left[\frac{1}{2}(\kappa+\gamma/2)\right]^3/\pi}{\{\left[\frac{1}{2}(\kappa+\gamma/2)\right]^2+(\omega-\omega_0-g)^2\}^2}\Bigg\}.\end{aligned} \tag{16.204}$$

This spectrum should be compared with the many-atom result (15.141).

Note 16.11. The Fourier transform yielding the Lorentzian squared may be performed using

$$\begin{aligned}\frac{1}{\pi}\mathrm{Re}\int_0^\infty d\tau e^{-i\omega\tau}(1+K\tau)e^{-K\tau} &= \frac{1}{\pi}\mathrm{Re}\left(1-K\frac{d}{dK}\right)\int_0^\infty d\tau e^{-(K+i\omega)\tau} \\ &= \frac{1}{\pi}\mathrm{Re}\left[\frac{1}{K+i\omega}+\frac{K}{(K+i\omega)^2}\right] \\ &= \frac{2K^3/\pi}{(K^2+\omega^2)^2}.\end{aligned} \tag{16.205}$$

16.3.5 The $\sqrt{n}$ Anharmonic Oscillator

The dressed Jaynes–Cummings eigenstates of Sect. 16.3.2 show a number of features that may be connected with the phenomenon of spontaneous dressed-state polarization (Sect. 16.3.1): the threshold at $2|\bar{\mathcal{E}}_0|/g = 1.0$ and a "ground" state (Eq. 16.168) that is a fully quantum mechanical version of the below-threshold semiclassical steady state (Eqs. 16.119b and 16.119c). We would now like to say something more about spontaneous dressed-state polarization; in particular, we wish to identify the broken-symmetry steady states above threshold within a fully quantum-mechanical analysis. To do so, in this section we exploit a relationship between the driven Jaynes–Cummings model and the $\sqrt{n}$ anharmonic oscillator model introduced by Chough and Carmichael [16.32]. The two models agree closely far above threshold, and even show similar behavior throughout the threshold region.

We return to the standard Jaynes–Cummings eigenstates (Eqs. 13.189a and 13.189b) and introduce the ladder operators

$$\hat{U} \equiv |G\rangle\langle 1, U| + \sum_{n=1}^{\infty} \sqrt{n+1}|n, U\rangle\langle n+1, U|, \tag{16.206a}$$

$$\hat{L} \equiv |G\rangle\langle 1, L| + \sum_{n=1}^{\infty} \sqrt{n+1}|n, L\rangle\langle n+1, L|. \tag{16.206b}$$

These operators are similar to annihilation operators for independent Boson modes, but the identification is not exact because the two ladders of states, one connecting the upper dressed states and the other the lower, begin out of the same ground state $|G\rangle$. The split ladder is illustrated in Fig. 16.10. The one common state means that the modes are not strictly independent, and satisfy commutation relations

$$[\hat{U}, \hat{U}^\dagger] = \hat{P}_U + |G\rangle\langle G|, \qquad [\hat{L}, \hat{L}^\dagger] = \hat{P}_L + |G\rangle\langle G|, \tag{16.207a}$$

and

$$[\hat{U}, \hat{L}^\dagger] = |1, L\rangle\langle 1, U|, \tag{16.207b}$$

where $\hat{P}_U$ and $\hat{P}_L$ are projectors onto the two branches of excited states:

$$\hat{P}_U \equiv \sum_{n=1}^{\infty} |n, U\rangle\langle n, U|, \qquad \hat{P}_L \equiv \sum_{n=1}^{\infty} |n, L\rangle\langle n, L|. \tag{16.208}$$

Independence of the modes is prevented by the appearance of the projector $|G\rangle\langle G|$ in (16.207a), and the nonvanishing of the commutator (16.207b). Note, however, that for excitations far above the ground state, the commutators might be replaced by $[\hat{U}, \hat{L}^\dagger] = 0$ and $[\hat{U}, \hat{U}^\dagger] = [\hat{L}, \hat{L}^\dagger] = 1$.

We now build upon this picture of the Jaynes–Cummings spectrum as two loosely joined energy-level ladders by writing the Jaynes–Cummings Hamiltonian in terms of the defined ladder operators and expanding the field annihilation operator in the *driven* Jaynes–Cummings Hamiltonian as far as possible in the same way. The details are left as an exercise:

Exercise 16.11. Show that the Hamiltonian (16.133) of the driven Jaynes–Cummings model can be written in the resonant case ($\omega_0 = \omega_C = \omega_A$) as

$$\begin{aligned} &H_S + \tfrac{1}{2}\hbar\omega_A \\ &= 0|G\rangle\langle G| + \left(\hbar\omega_A \hat{U}^\dagger \hat{U} + \hbar g \sqrt{\hat{U}^\dagger \hat{U}}\right) + \left(\hbar\omega_A \hat{L}^\dagger \hat{L} - \hbar g \sqrt{\hat{L}^\dagger \hat{L}}\right) \\ &\quad + \hbar(\bar{\mathcal{E}}_0 e^{-i\omega_A t} a^\dagger + \bar{\mathcal{E}}_0^* e^{i\omega_A t} a). \end{aligned} \tag{16.209}$$

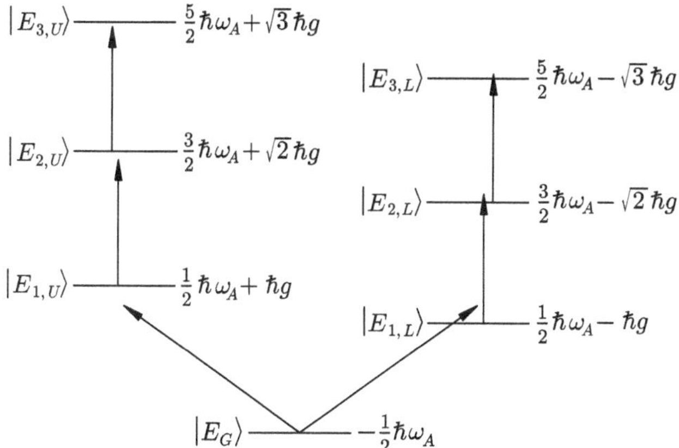

Fig. 16.10. The split ladder of Jaynes–Cummings eigenstates. Arrows show how the detuning of resonant excitation ($\omega_0=\omega_C=\omega_A$) decreases with increasing level of excitation

Show also that the annihilation operator for the cavity mode may be written in the form

$$\begin{aligned}a=\hat{U}+\hat{L}-|G\rangle\Big({}_a\langle 1|{}_A\langle 1|+\sqrt{2}{}_a\langle 0|{}_A\langle 2|\Big)\\-\sum_{n=1}^{\infty}\big(\sqrt{n+1}-\sqrt{n}\big)\big(|2\rangle\langle 2|\big)_A\big(|n-1\rangle\langle n|\big)_a.\end{aligned}\tag{16.210}$$

At high levels of excitation, the expansion (16.210) of the annihilation operator may be simplified as $a=\hat{U}+\hat{L}$, since the factors $\sqrt{n+1}-\sqrt{n}$ are all small; hence, for high excitation, we may write Hamiltonian (16.209) as a sum of two parts,

$$H_S+\frac{1}{2}\hbar\omega_A=H^{+}_{\sqrt{n}}+H^{-}_{\sqrt{n}},\tag{16.211}$$

with

$$H^{+}_{\sqrt{n}}\equiv\hbar\omega_A\hat{U}^\dagger\hat{U}+\hbar g\sqrt{\hat{U}^\dagger\hat{U}}+\hbar(\bar{\mathcal{E}}_0e^{-i\omega_At}\hat{U}^\dagger+\bar{\mathcal{E}}_0^*e^{i\omega_At}\hat{U}),\tag{16.212a}$$

$$H^{-}_{\sqrt{n}}\equiv\hbar\omega_A\hat{L}^\dagger\hat{L}-\hbar g\sqrt{\hat{L}^\dagger\hat{L}}+\hbar(\bar{\mathcal{E}}_0e^{-i\omega_At}\hat{L}^\dagger+\bar{\mathcal{E}}_0^*e^{i\omega_At}\hat{L}).\tag{16.212b}$$

Now if we regard U and L as independent bosons, $H^{+}_{\sqrt{n}}$ and $H^{-}_{\sqrt{n}}$ each define a $\sqrt{n}$ anharmonic oscillator driven away from resonance. The anharmonicity arises from an excitation-dependent detuning: detunings of $\pm\hbar g$ for transitions from the ground to the first excited state, evolve with increasing n to $\pm\hbar g\big(\sqrt{n+1}-\sqrt{n}\big)\approx\pm\hbar g/2\sqrt{n}$ for transitions between highly excited states.

Let us now add a small cavity damping, $\kappa \ll g$, and consider, for example, excitation of the U-oscillator using master equation

$$\dot{\rho} = \frac{1}{i\hbar}[H^{+}_{\sqrt{n}}, \rho] + \kappa(2\hat{U}^{\dagger}\rho\hat{U} - \hat{U}^{\dagger}\hat{U}\rho - \rho\hat{U}^{\dagger}\hat{U}). \tag{16.213}$$

Two conclusions readily follow: (i) for weak driving fields, since single-photon excitation is far from resonance, a small steady-state mean photon number will result,

$$\langle a^{\dagger}a\rangle_{\rm ss} \approx \left|\frac{\bar{\mathcal{E}}_0}{\kappa + ig}\right| \approx \left(\frac{|\bar{\mathcal{E}}_0|}{g}\right)^2; \tag{16.214}$$

and (ii) for driving fields sufficiently strong to drive multiphoton transitions past the first few excited states, high-level excitation becomes quasi-resonant (detuning $g/2\sqrt{n}$), and

$$\langle a^{\dagger}a\rangle_{\rm ss} \approx \left|\frac{\bar{\mathcal{E}}_0}{\kappa + ig/2\sqrt{\langle a^{\dagger}a\rangle_{\rm ss}}}\right|^2 \Rightarrow \langle a^{\dagger}a\rangle_{\rm ss} \approx \left(\frac{\bar{\mathcal{E}}_0}{\kappa}\right)^2 - \left(\frac{g}{2\kappa}\right)^2. \tag{16.215}$$

Equation 16.215 is in agreement with the square of the semiclassical intracavity field amplitude (16.124).

The steady-state photon number computed for the $\sqrt{n}$ anharmonic oscillator model with damping is plotted in Fig. 16.11. It agrees well with our two conclusions.

Of course, details given by the $\sqrt{n}$ anharmonic oscillator model are a rather poor approximation to the real thing. We know, for example, that the mean photon number below threshold is of order $(|\bar{\mathcal{E}}_0|/g)^4$ (Eqs. 15.113 and 16.203), not of order $(|\bar{\mathcal{E}}_0|/g)^2$ as given by (16.214). This difference comes about from an interference between the two possible pathways for excitation out of the ground state. In fact, quantum fluctuations associated with the competition between pathways are missing entirely—below threshold, at threshold, and even well above threshold. Nevertheless, Fig. 16.11 mirrors Fig. 16.6 in its

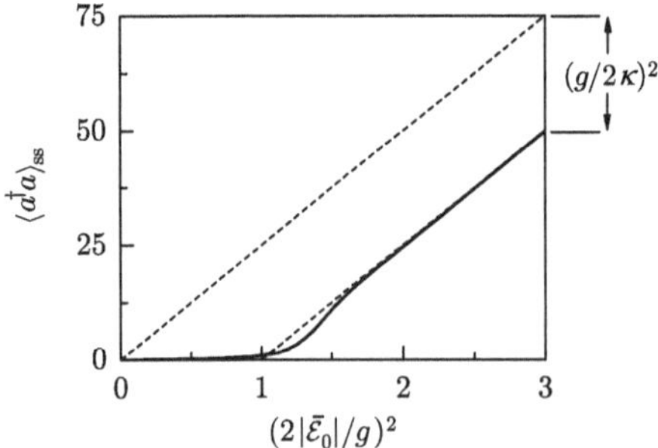

Fig. 16.11. Mean photon number as a function of driving field strength for the $\sqrt{n}$ anharmonic oscillator with $g/\kappa = 10$

essential features: it shows a smoothed out transition at $2|\bar{\mathcal{E}}_0|/g = 1$, and the same offset from $\langle a^\dagger a \rangle_{\rm ss} = (|\bar{\mathcal{E}}_0|/\kappa)^2$ in the strong-excitation limit arising from quasiresonant excitation up one or the other Jaynes–Cummings ladder. Thus the phenomenon of spontaneous dressed-state polarization is captured in broad outline, with the shortcoming that the symmetry breaking has been introduced by hand. To conclude the chapter, let us briefly review what a direct numerical solution of the master equation reveals about how quantum fluctuations, in fact, execute the symmetry breaking.

16.3.6 Quantum Fluctuations for Strong Excitation

The limit of "zero system size" takes $\gamma = \gamma_h \to 0$ with $n_{\rm sat} = \gamma^2/8g^2 \to 0$. We investigated the limit in Sect. 16.3.1 by neglecting quantum fluctuations altogether. There we uncovered the new semiclassical stationary states (16.119a)–(16.122c). In Sects. 16.3.2 and 16.3.5 we discovered that the main features of the semiclassical analysis are recovered from a fully quantum-mechanical treatment: the threshold at $2|\bar{\mathcal{E}}_0|/g = 1$ and the above-threshold dressed-state polarization. It remains to put the pieces together by solving the master equation for the limit of "zero system size."

The numerical results presented in this section reproduce those reported by Alsing and Carmichael [16.22]. They illustrate the behavior of the steady-state solution to master equation (13.57) for the case of resonant dipole coupling ($\omega_A = \omega_C$) and resonant excitation of the cavity mode ($\omega_0 = \omega_C$). Figure 16.12 corresponds to the limit of "zero system size." Here, with $\gamma = 0$, the steady state is established through cavity damping alone.

A number of features are worthy of note. First, the mean photon number, shown in frame (a), reproduces the threshold behavior deduced from the Maxwell–Bloch equations (Fig. 16.6), with the transition region smoothed out by fluctuations as we would expect. Note again how much the behavior differs from that deduced from a small-noise analysis of absorptive optical bistability (the light dashed line in the figure). Second, if we compare the $\sqrt{n}$ anharmonic oscillator (Fig. 16.11), the plot of Fig. 16.12a is similar, but the fluctuations clearly behave differently in the threshold region. The importance of fluctuations near threshold is illustrated, for example, by Fig. 16.12c.

Finally, the most significant thing to note does not concern the stationary states themselves, but rather their stability. Recall from Sect. 16.3.1 (below Fig. 16.8) that the semiclassical stationary states above threshold are nonstable—they are the foci of an infinite family of cyclic asymptotic solutions (nonstable limit cycles or Volterra-Lotka cycles). As such, we might expect fluctuations to smear the asymptotic distribution across the whole family of solutions—across the entire surface of the Bloch sphere and throughout a corresponding contiguous region in the phase plane of the field. This is indeed what happens for many atoms and weak coupling, where a small-noise analysis is justified (Note 16.8). In contrast to the expectation, though,

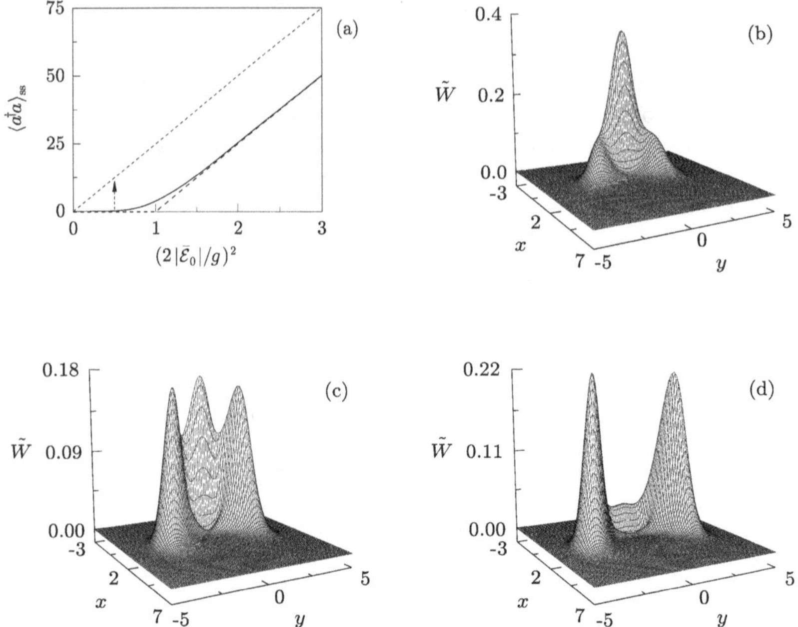

Fig. 16.12. Quantum fluctuations in spontaneous dressed-state polarization for $g/\kappa = 10$ and $\gamma/2\kappa = 0$. (**a**) Steady-state photon number (*solid line*) compared with the semiclassical stationary states (16.119a)–(16.122c) (*dashed line*) and the stable steady states of optical bistability (16.126a)–(16.127b) (*light dashed line*). The steady-state Wigner distribution is plotted for (**b**) $2|\bar{\mathcal{E}}_0|/g = 0.9$, (**c**) $2|\bar{\mathcal{E}}_0|/g = 1.0$, and (**d**) $2|\bar{\mathcal{E}}_0|/g = 1.1$

Fig. 16.12d shows that for one atom and strong coupling the semiclassical stationary states are "attractors" in the full quantum-mechanical treatment; for stronger excitation than that used in the figure the steady-state Wigner distribution approaches two isolated Gaussians, the quantum state corresponding to a mixture of coherent states with amplitudes (16.122a). In short, as with the degenerate parametric oscillator (Sect. 12.1.8), in the limit of small system size a situation is encountered where the notion of semiclassical dynamics plus "fuzz" does not provide even an approximate starting point for understanding the quantum dynamics. In this limit, we must turn to the spectrum of the Hamiltonian H_S, as we did in Sects. 16.3.2 and 16.3.5, and rethink the physics for a situation where the internal coupling is sufficiently strong to produce (few-quantum) energy-level shifts and splittings in excess of the level widths.

Note 16.12. Coherent driving of the cavity mode can be replaced by coherent driving of the atom without fundamentally altering the behavior. Density operators satisfying the different master equations are related through a displacement of the cavity field. In particular, the solution to (13.57) with driving field amplitude $\bar{\mathcal{E}}_0$ is recovered from the solution to the driven-atom master equation with driving field amplitude $-g\bar{\mathcal{E}}_0/\kappa$; the former solution is obtained from the latter through a displacement of magnitude $\bar{\mathcal{E}}_0/\kappa$ [16.22]. This flexibility was used when computing the Wigner distributions displayed in Figs. 16.12 and 16.13, where improved numerical accuracy was achieved by driving the atom and cavity mode simultaneously with parameters chosen so that the mean steady-state amplitude of the field was zero; the Wigner distribution for pure cavity driving was then recovered by appropriately displacing the computed distribution.

The story remains unfinished so long as we have considered only the strict "zero system size" limit. The limit helped us uncover the phenomenon of spontaneous dressed-state polarization, but it yields a structurally unstable model, since the behavior with $\gamma = 0$ is constrained by the conservation law (16.118)—or, more precisely, when quantum fluctuations are included by $4\langle \tilde{J}_+\tilde{J}_-\rangle + \langle J_z\rangle^2 = N(N+2)$. The conservation law is broken for nonzero γ. To complete the story, then, we must consider how spontaneous emission changes the picture presented in Fig. 16.12

Results for $\gamma/2\kappa = 1$ are displayed in Fig. 16.13. The behavior of the mean photon number shows a significant change, with movement in the direction of what might be predicted from the steady states of absorptive optical bistability. In fact, as Fig. 16.13c shows, the threshold region is now an amalgamation of the bimodality of absorptive bistability—the double-peaked character in the x-dimension—and the bimodality associated with spontaneous dressed-state polarization—the double peak in the y-dimension. Only the latter remains in evidence above threshold, where the general shape of the Wigner distribution shown in Fig. 16.13d persists as the strength of the excitation is increased; the distribution is simply displaced further and further along the x-axis.

What, then, is happening above threshold to explain the difference between Figs. 16.12d and 16.13d? Observe first that a spontaneous emission puts the atom into its ground state $|1\rangle_A$—i.e., into a superposition of the dressed states $|U\rangle = \big(|2\rangle_A + i|1\rangle_A\big)/\sqrt{2}$ and $|L\rangle = \big(|2\rangle_A - i|1\rangle_A\big)/\sqrt{2}$. We might now start from the picture of spontaneous dress-state polarization without spontaneous emission and assume that the system is polarized along the U-branch (Fig. 16.10) prior to the emission. Thus, prior to the emission the atomic state is $|U\rangle$, to a good approximation, and factorized from the field state (assuming a position well up the ladder of states). Now after the spontaneous emission, with the transition to atomic state $|1\rangle_A$, the system is no longer localized on the U-branch; in fact it is no longer on either branch. The field state remains localized on the U-branch, but the atomic state is a superposition of one state lying on the U-branch and another lying on the L-branch.

The subsequent evolution may be unclear. We must expect, though, that it attempts to return the system to one or other of the stationary states, to one or other of the self-consistent dressed states. Either the atomic state returns to $|U\rangle$ and the system remains polarized on the U-branch, or the system

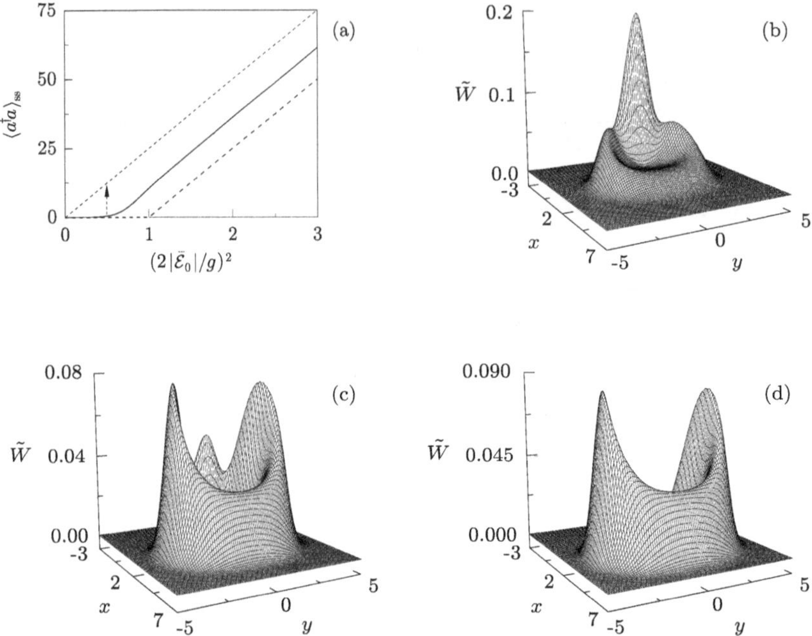

Fig. 16.13. Quantum fluctuations in spontaneous dressed-state polarization for $g/\kappa = 10$ and $\gamma/2\kappa = 1$. (**a**) Steady-state photon number (*solid line*) compared with the semiclassical stationary states (16.119a)–(16.122c) (*dashed line*) and the stable steady states of optical bistability (16.126a)–(16.127b) (*light dashed line*). The steady-state Wigner distribution is plotted for (**b**) $2|\bar{\mathcal{E}}_0|/g = 0.9$, (**c**) $2|\bar{\mathcal{E}}_0|/g = 1.0$, and (**d**) $2|\bar{\mathcal{E}}_0|/g = 1.1$

polarization switches to localize on the L-branch. If the second option is taken, the phase of the intracavity field must switch. The "skirt" connecting the peaks in Fig. 16.13d [compare Fig. 16.12d] is a result of the phase switching.

Alsing and Carmichael [16.22] have provided a mathematical formulation to verify this picture. Kilin and Krinitskaya [16.33] provide another treatment. Summarizing the model of Alsing and Carmichael, well above threshold, the state of the cavity field at time t is to a good approximation a coherent state $|\tilde{\alpha}(t)\rangle$ whose complex amplitude obeys the stochastic equation

$$\frac{d\tilde{\alpha}}{dt} = -(\kappa + i\epsilon g/2|\tilde{\alpha}|)\alpha + i\bar{\mathcal{E}}_0, \tag{16.216}$$

where $\epsilon = \pm 1$ is a random number that either switches sign or does not switch sign (with 50/50 probability) whenever there is a spontaneous emission; spontaneous emissions occur randomly at rate $\gamma/2$. The term $i\epsilon g/2|\tilde{\alpha}|$ in (16.215) represents the detunings, either $+g/2\sqrt{n}$ or $-g/2\sqrt{n}$, of the two ladders of states in Fig. 16.10. A typical phase-space trajectory of the field amplitude is shown in Fig. 16.14. If we now imagine the Wigner distribution that represents $|\tilde{\alpha}(t)\rangle$—a stochastic Wigner distribution—and average it over time, we recover the Wigner distribution of a mixed state, one looking just like the distribution plotted in Fig. 16.13d.

The main point for us to appreciate is that the phase switching, which is macroscopic, is initiated by a single quantum event, a single spontaneous emission. Consequently, the fluctuations have nothing in common with those envisaged by the system size expansion, where the outcomes of single-quantum events are assumed microscopic with only their accumulated effect appearing as a diffusion process at the macroscopic level.

There is a second, less obvious point to note. With the introduction of equation (16.216), we have slipped from our earlier discussions of stochastic processes in phase space into a discussion of a stochastic process in Hilbert space, which is correlated with (conditioned on) detectable scattering events: a stochastic process in Hilbert space; one conditioned on a record of scattering events. These are the themes of quantum trajectory theory; in fact, Fig. 16.14 is a reproduction of one of the earliest applications of quantum trajectories [16.22]. The final three chapters of the book will teach us how a representation of quantum fluctuations like that depicted in Fig. 16.14 can be formulated in a systematic way.

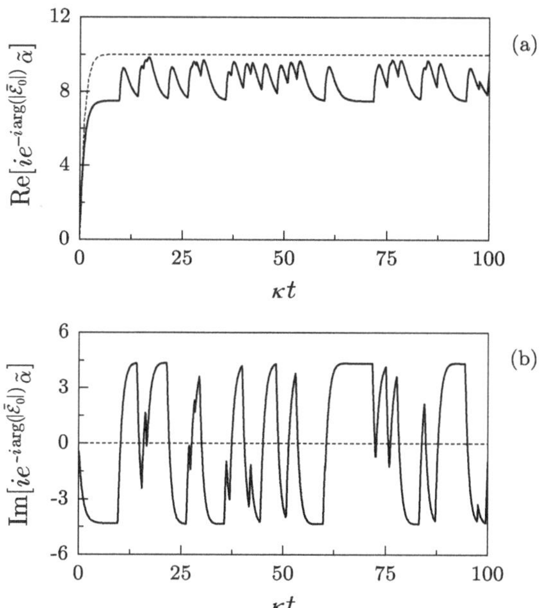

Fig. 16.14. A typical switching trajectory for spontaneous dressed-state polarization: (**a**) real part of the field amplitude, (**b**) imaginary part of the field amplitude. The parameters are $g/\kappa = 10$, $\gamma/2\kappa = 1$, and $2|\bar{\mathcal{E}}_0|/g = 2.0$. The *dashed line* shows the trajectory in the absence of the atom

Note 16.13. The polarization of a pseudo-spin in the presence of an external laser field and a quantized cavity field can be seen as an analog of the polarization of a real spin in a Stern–Gerlach apparatus. The analogy has been worked out by Carmichael and coworkers [16.34]. From the analogy, it is clear that the ideas carry over to any number of atoms; the only change is the increase in the number of available spin states. See Kochan and coworkers [16.35] for a discussion of the two-atom generalization.

17

Quantum Trajectories I: Background and Interpretation

We have discussed many things in the last eight chapters, but there have been just two overriding themes. Both arise from a consideration of the phase-space methods which account so effectively for the quantum fluctuations of laser light (Chap. 8). First, there is the issue of nonclassical fluctuations (Chaps. 9, 10) which led us to an investigation of the positive P representation (Chaps. 11 and 12). The second, interrelated theme, concerns the scaling of fluctuations with system size. After introducing the topic by way of the positive P representation (Chap. 12), we were led to a study of cavity QED (Chaps. 13–16) which provides an experimentally accessible realization of the limit of small system size (large quantum noise).

Our grand conclusion from all of this is that the methods which provide a satisfying visualization of quantum fluctuations for the laser fail more generally. In fact, it is not only the visualization that is lost, but also the utility of the phase-space approach for calculating correlation functions and moments. The foundation upon which the phase-space methods rest is not simply the existence of a representation for quantum states, but the existence of an accessible and analyzable classical stochastic process—specifically, a phase-space diffusion process. While the positive P representation might extend the phase space to accommodate nonclassical noise, its target remains unaltered; it aims to find a classical stochastic process whose correlation functions and moments are mapped, via some ordering rule, to those of the quantum mechanical problem. For all but a few model examples, the system size expansion is needed to access such a pseudo-classical accounting of quantum fluctuations. Ultimately, then, nonclassicality remains as the central issue. The phase-space paradigm is a classical noise paradigm. Reasonably, we can expect a quantum stochastic process to differ from it in fundamental ways. Quantum statistics is founded on probability amplitudes, not probabilities, and surely this introduces subtleties. One might wonder, for example, how phase-space diffusion is to account for the intricacies of entangled states.

We have set no firm formal bounds on what can be achieved with the phase-space approach. Precisely circumscribing the limits is not so easy and leads us into questions of interpretation; in particular, the question of a definition for what is, and is not, "nonclassical" (see Note 9.19). We have seen, for example, that both the Wigner representation and the Q representation make the squeezed fluctuations of the degenerate parametric oscillator appear classical (Sect. 10.1.2); also, the positive P representation exists for entangled states of the electromagnetic field and has been used to treat correlations of the Einstein–Podolsky–Rosen type [17.1]. Against this, we have seen that nonlinearity is a decisive factor. Nonlinearities cause the Wigner and Q representations to lose their ability to make quantum fluctuations look classical (Sect. 10.1.2, Note 15.2). Strong nonlinearity presents the positive P representation with profound difficulties (Sect. 12.2).

In this chapter and what follows we explore a new approach to the formulation of a quantum stochastic process. We use the name *quantum trajectories* or *quantum trajectory theory* (the terminology originates in Sect. 5 of [17.2]); others speak of the *quantum Monte Carlo wavefunction method* [17.3] or simply Monte Carlo wavefunction simulations of the master equation [17.4,17.5]. The approach is not founded upon a particular representation of the density operator. It sets up a stochastic process that is fully equivalent to the master equation plus the regression formula for time- and normal-ordered correlation functions. It is a *quantum* stochastic process in its conception. Of course, we can work directly with the master equation and regression formula themselves, as we have done in Chap. 16. When doing this we are surely working, formally, with a quantum stochastic process; nevertheless, the central feature of a stochastic evolution, the feature that yields visualizable realizations or "trajectories"—i.e., the playing out of probabilistic decisions over time—is missing if we limit ourselves to this type of activity. The master equation and regression formula constitute a mathematical apparatus to tell us about ensemble-averaged quantities. They reveal nothing of the stochastic realizations from which the ensemble is made up, and cannot, for example, provide a Fig. 16.14 to illuminate Fig. 16.13. We wish to do more than calculate ensemble-averaged quantities. We wish to regain the appeal of the phase-space methods in providing a path to an explicit stochastic process with visualizable realizations of the quantum fluctuations.

A glimpse of the stochastic process underlying a quantum master equation first crept into quantum optics from mathematical physics, principally through the work Davies [17.6] and Lindblad [17.7], and the quantum dynamical semigroup approach to open quantum systems. The connection between a Davies process, a trajectory of realized events (photocounts in this case), and a quantum optics master equation was, however, unclear for a long time, and the early ideas from mathematical physics had little impact on mainstream quantum optics. Looking around more widely, there is now an extensive mathematical physics literature relevant to stochastic processes in quantum optics. Much of it is focused on technical questions of a kind swept aside by quantum opticians, in their informal recourse to the Markov approximation. There is a considerable amount of work on the *quantum stochastic calculus* [17.8,17.9], for example, which provides a stochastic formulation within the Heisenberg picture, at the level of $S \otimes R$. Quantum Langevin equations, like those met in Sects. 7.3.3 and 9.3.1, fall within its scope, as does the input–output theory of Gardiner and Collet [17.10]. The theory of *dilations of a quantum dynamical semigroup* [17.8, 17.9, 17.11, 17.12, 17.13] is also relevant to quantum optics. Interestingly, a dilation moves in the opposite direction to the derivations of the master equation in Chaps. 1 and 2; it takes us from a master equation for S (from the generator of a quantum dynamical semigroup) to a unitary evolution in $S \otimes R$, or to a realization of the evolution in the language of the quantum stochastic calculus.

Concrete connections between these and related interests in mathematical physics and applications in quantum optics are, even today, rather slight. The work of Barchielli is notable. Barchielli discusses photoelectron counting in quantum optics [17.14,17.15], and with Lupieri [17.16] has treated the problem of resonance fluorescence in great detail. Bouten and coworkers [17.17] also treat resonance fluorescence. Their work is of particular interest as it has the explicit objective of showing how quantum trajectory theory may be fitted into the rigorous framework developed in mathematical physics. In the following we make no attempt to match the rigor of the mathematical work. We keep to our customary physics style.

Quantum trajectory theory sets out in a new direction, but the direction is not so far removed from things we have already met. In fact, we stumbled upon elements of the approach in Sect. 12.1.8; specifically, through the Dyson expansion (12.102) and the subsequent discussion of switching between coherent state superpositions, $|A_{\text{even}}\rangle$ and $|A_{\text{odd}}\rangle$.

Working ahead from this point, quantum trajectory theory is formulated as a photoelectron counting theory for photoemissive sources. It was developed as such by Carmichael and coworkers [17.2,17.18,17.19]. The Dyson expansion is the quickest path to a connection with photoelectron counts and, indeed, Davies processes (see Exercise 12.10). Building the topic in this direction is perhaps not the best way for us to start out, though. Although quite informal compared with the style in mathematical physics, the typical treatment of photoelectron counting in quantum optics is sufficiently encumbered, by such things as quantum fields and superoperators, to obscure the simplicity of the central ideas upon which quantum trajectory theory is based. There is, then, some advantage to introducing these ideas without attempting to make any explicit connections to photoelectron counting theory. We do this in the present chapter. The systematic development of quantum trajectories from photoelectron counting is put off until Chap. 18.

17.1 Density Operators and Scattering Records

Considered most broadly, quantum optics deals with the interactions of light with matter, laying its emphasis on the fluctuations of the light and questions about how these fluctuations might be understood in terms of a physical process unfolding in time—i.e., the focus is on quantum dynamics. The special characteristics of laser light are important, because it is the laser, ultimately, that makes most things of interest in quantum optics possible. Two related characteristics of laser light are central: (i) the high single-mode photon flux, which makes it possible to excite a material system far from thermal equilibrium where nonlinearities set in, and (ii) that the excitation is coherent, as implied by the word "single-mode," or more accurately "single quasimode;" thus it is possible to induce *coherences* between highly excited material states, states reached through the absorption of many quanta. Along with these characteristics comes the important role of radiative decay or dissipation; far from equilibrium the system maintains its steady state by balancing excitation with some form of decay.

Throughout this book we have seen many examples that fit this paradigm and learned a lot about how to analyze their quantum fluctuations. It is useful before moving on to step back and summarize what, in the big picture, our approach has been. It starts from the system plus reservoir point of view introduced in Sect. 1.3, illustrated for specific examples by Figs. 7.7 and 9.4. From here the development is made in three steps. In the first, attention is focused on the system; we trace over the reservoir, or what might be called the system's environment, and work with the reduced density operator

$$\rho(t) \equiv \mathrm{tr}_R[\chi(t)] = \sum_n \langle E_n|\chi(t)|E_n\rangle, \tag{17.1}$$

where $|E_n\rangle$ denotes an energy eigenstate of the reservoir (environment). We derive an equation of motion for $\rho(t)$ in the Born–Markov approximation (Sects. 1.4, 2.2.1, and 2.2.4), which takes the generic *Lindblad form* [17.7]

$$\dot{\rho} = \mathcal{L}\rho, \tag{17.2}$$

with

$$\mathcal{L} = \frac{1}{i\hbar}[H_S, \cdot\,] + \sum_j \frac{\gamma_j}{2}(2\hat{O}_j \cdot \hat{O}_j^\dagger - \hat{O}_j^\dagger\hat{O}_j \cdot - \cdot\, \hat{O}_j^\dagger\hat{O}_j). \tag{17.3}$$

Most of our time has been spent in analyzing this equation. In particular, phase-space methods were introduced to help with this analysis. Analyzing the master equation is certainly an important issue, but a largely technical one. It can be approached in many ways, in many representations, and it is not a part of our general approach.

The second step takes us back to the level of system plus reservoir. While derivation of the master equation was carried through in the Schrödinger picture, we now switch and adopt the Heisenberg picture. We solve Heisenberg equations of motion for output fields (Sects. 2.3.1 and 7.3.1). Again, the Born–Markov approximation is used. A typical field operator is expanded as

$$\hat{\boldsymbol{E}}^{(+)}(\boldsymbol{r},t) = \hat{\boldsymbol{E}}_f^{(+)}(\boldsymbol{r},t) + \hat{\boldsymbol{E}}_s^{(+)}(\boldsymbol{r},t), \tag{17.4}$$

where $\hat{\boldsymbol{E}}_f^{(+)}(\boldsymbol{r},t)$ is the free-field operator—the field of the reservoir, or environment, with the coupling to the system turned off—and $\hat{\boldsymbol{E}}_s^{(+)}(\boldsymbol{r},t)$ is the source-field operator. The goal and main achievement of the second step is the derivation of an explicit expression for $\hat{\boldsymbol{E}}_s^{(+)}(\boldsymbol{r},t)$ in terms of retarded system operators; specific examples appear in (2.83), (7.111), (9.116), and (9.117).

The third step makes the connection between steps one and two. Continuing to work in the Heisenberg picture, a regression formula for time-ordered and normal-ordered correlation functions is derived (Sect. 1.5). When the reservoir (the input field to the system) is in the vacuum state, the correlation function may be expressed in terms of source fields alone; for example, we may write

$$\langle \hat{\boldsymbol{E}}_s^{(-)}(\boldsymbol{r},t)\hat{\boldsymbol{E}}_s^{(+)}(\boldsymbol{r},t+\tau)\rangle = \mathrm{tr}_S\{\hat{\boldsymbol{E}}_s^{(+)}(\boldsymbol{r})e^{\mathcal{L}\tau}\big[\rho(t-r/c)\hat{\boldsymbol{E}}_s^{(-)}(\boldsymbol{r})\big]\}. \quad (17.5)$$

The right-hand side of this expression depends on operators and superoperators of the system only: operators $\rho(t-r/c)$ and $\hat{\boldsymbol{E}}_s^{(+)}(\boldsymbol{r}) = [\hat{\boldsymbol{E}}_s^{(-)}(\boldsymbol{r})]^\dagger$, and superoperator $\mathcal{L}$. Returning to the Schrödinger picture, the correlation function can be evaluated by solving master equation (17.2) twice—assuming, of course, that we know the explicit form of $\hat{\boldsymbol{E}}_s^{(+)}(\boldsymbol{r})$.

Quantum trajectory theory makes two changes to this picture. First, in place of the *trace* over the environment, it *disentangles the system from its environment*, and second, rather than approaching the master equation and output field calculations as two separate exercises, and being satisfied with the resulting *local* description of the system, it lets the output fields—more precisely the detection of these fields—tell us how the disentanglement should be made. Through the latter device, quantum trajectory theory provides a *nonlocal description of the system and its environment* taken as a whole, or of the system and certain classically described elements of the environment (recording or measuring devices).

These few words are hardly sufficient to fully define the altered approach. In order to clarify it, we might start by recognizing that quantum optics experiments are scattering experiments. We should not be thinking of a localized system, with internal fluctuations, as the usual focus of attention on the master equation tends to imply. We should think of a scattering scenario, where the system S replaces the potential, $V(\boldsymbol{r})$, of elementary scattering theory, and the reservoir, or environment, carries the incoming and outgoing fields—generally photons in quantum optics, though they might be phonons, or even atoms. Perhaps the optical tables in a quantum optics laboratory do not seem to fit the standard picture of a scattering experiment. Nevertheless, the role of the experimenter is fundamentally the same as in any scattering scenario: the experimenter, in essence, controls the inputs applied and detects the generated outputs.

Figure 17.1 illustrates a general scattering scenario in quantum optics. The figure is schematic. In practice, a lot of nontrivial apparatus might be involved: optical cavities, beam splitters, polarizers, traps for atoms or ions

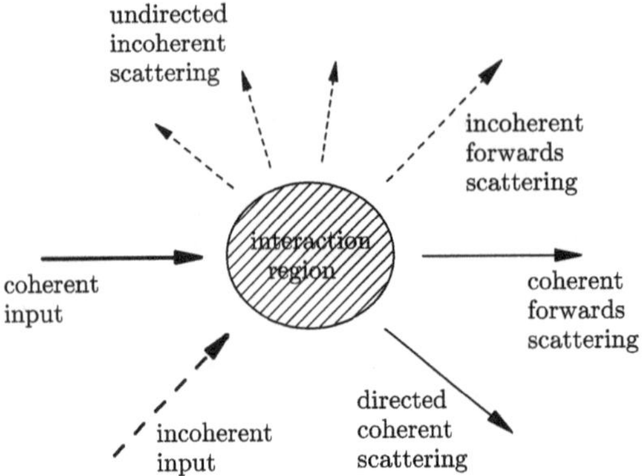

Fig. 17.1. Schematic of a typical scattering scenario in quantum optics

and the like. Once everything is set up, however, the experimenter sits outside the interaction region (the system S) controlling inputs and detecting outputs. Certainly the interaction region differs from that of elementary scattering theory. Generally, it must account for a complex process of excitation and deexcitation, with the excitation possibly reaching highly excited states. Most importantly, at least a part of this process is coherent. The interactions are strong and quasiresonant, and the scattering is not simply a matter of bouncing a particle off a potential. The scattering of inputs into outputs is nevertheless what quantum optics experiments are about.

As alluded to above, we tend to lose sight of the scattering scenario once we get tied up with solving the master equation. The master equation makes no reference to scattering events. Detail of this kind is averaged away by the trace (17.1). Quantum trajectory theory aims to bring it back. To arrive at the quantum trajectory point of view, we recognize that one might construct a *scattering record*—let us denote it by REC—by counting ingoing and outgoing particles and noting the times and locations at which the particle numbers change. This may clearly be done for all outputs by surrounding the interaction region with detectors. It is also allowed for incoherent inputs, because incoherent fields admit particle number as a good quantum number (the phase noise induced by monitoring particle number does not upset the coherence of an already incoherent field). Coherent light inputs are a different matter, though. They are, by definition, in a state of uncertain particle number and well-defined phase. For them, the tracking of particle number is not applicable, in principle. This presents no problem, however, since coherent inputs are boson quantum fields taken to the classical limit, and as such, may be described by classical waves; they are *external fields* and may be introduced as parameters of the system Hamiltonian H_S.

Note 17.1. If one insists on not admitting the notion of an external field [17.20] and asks for coherent inputs (laser inputs) to be described in a fully quantum-mechanical way, this may be done by coupling the laser source to the target system S using the cascaded system version of quantum trajectory theory (Sect. 19.2). The coherence of the input is then accounted for through an entanglement of the source and target system states. It is expressed as a preserved relative phase between entangled states, rather than as a preserved absolute phase of an external field. The situation is similar to that discussed for recent BEC experiments [17.21,17.22,17.23]. As our aim is to disentangle—to eliminate all operationally meaningless and cumbersome entanglement—we accept the notion of an external field for coherent inputs as a complement to the particle counting strategy for outputs and incoherent inputs.

Many-atom cavity QED provides an example of a quantum optics scattering scenario. The master equation appears in (15.1). Inputs and outputs corresponding to this model are illustrated in Fig. 17.2 [except for the elastic scattering represented in the master equation by the term proportional to γ_p (Sect. 2.2.4), which we neglect]. The notation $\gamma_A^j = \gamma$, $j = 1, \ldots, N$, distinguishes the fluorescence output channels of different atoms; the form of the master equation assumes the atoms are sufficiently far apart that the scattering from individual atoms can be resolved (see Notes 6.3 and 15.4). A particular *scattering record* is written in this notation as

$$\mathrm{REC} \equiv \left\{ {}_0\emptyset, \frac{\gamma_{a2}}{T_1}, \emptyset, \frac{\gamma_A^{j_2}}{T_2}, \emptyset, \frac{\gamma_A^{j_3}}{T_3}, \emptyset, \frac{\gamma_{a2}}{T_4}, \emptyset, \frac{\gamma_{a1}}{T_5}, \emptyset, \frac{\gamma_A^{j_6}}{T_6}, \emptyset, \frac{\cdots}{\cdots} \right\}, \tag{17.6}$$

with

$$T_k \equiv [t_k, t_k + dt_k). \tag{17.7}$$

The sequence of events is time-ordered, $0 < t_1 < t_2 \ldots$, and the entries indicate the times and channels of photon detections in the far field, along with the intervals without detection; the symbol $\emptyset$ indicates that no photons are detected between the events to the left and the right in the sequence, while ${}_0\emptyset$ denotes an interval without detection beginning at $t = 0$. The record (17.6) is simpler than what we might write down more generally, since the photon detections are not assigned spatial locations. It is important to state that it is not *the* scattering record, but a feasible scattering record. We return to this point, which is intimately tied up with the interpretation of quantum trajectories, in Sect. 18.3.1.

Note 17.2. If the scattering from individual atoms is not resolvable, then a simple record like that of (17.6) cannot be given and the location of the detection would replace the assignment to a particular atom. An example like this can be found in the work of Carmichael and coworkers [17.24, 17.25] who consider how quantum trajectory theory applies to the superradiance master equation. Spatially resolved detections are also important when the atomic center-of-mass motion is quantized. In this situation the direction of the scattering is correlated with the direction of the momentum recoil [17.26].

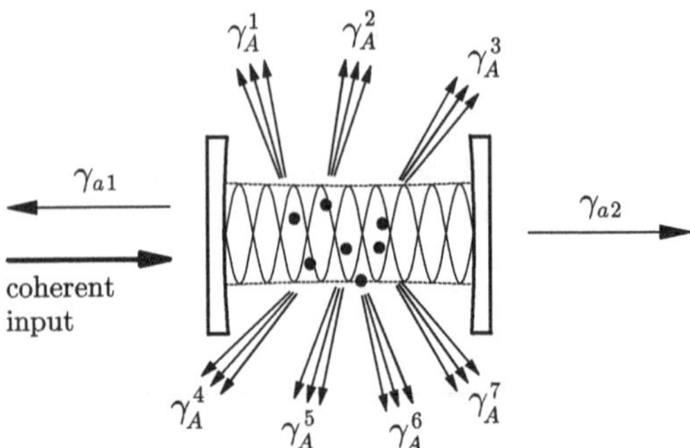

Fig. 17.2. A specific scattering scenario in many-atom cavity QED. Output channels γ_{a2} and γ_{a1} carry the forwards and backwards scattering, respectively, of the coherent input. Undirected scattering (fluorescence) from individual atoms is carried by output channels $\gamma_A^1, \gamma_A^2, \ldots, \gamma_A^7$

Having recognized the possibility of making scattering records, the strategy now is to construct an *unraveling (decomposition) of the reduced density operator* to replace the trace (17.1). The reduced density operator is expanded in the form

$$\rho(t) = \sum_{\mathrm{REC}} P(\mathrm{REC})|\psi_{\mathrm{REC}}(t)\rangle\langle\psi_{\mathrm{REC}}(t)|. \tag{17.8}$$

The unraveling disentangles the state of the system from that of its environment. The state $|\psi_{\mathrm{REC}}(t)\rangle$ is a *conditional state* of the system S, the state of the system given a particular scattering record. If the record is complete, in so far as it accounts for every scattered particle, the conditional state is pure. $P(\mathrm{REC})$ is the probability (or probability density, Sect. 18.1.1) for the particular scattering record to occur. Taking the trace of both sides of (17.8) yields

$$\sum_{\mathrm{REC}} P(\mathrm{REC}) = 1, \tag{17.9}$$

which must hold since scattering records are comprehensive and mutually exclusive: some particular record must occur, and the probabilities of mutually exclusive events add to unity in the usual way. Although $|\psi_{\mathrm{REC}}(t)\rangle$ refers to the Hilbert space of S alone, having added the label REC it provides a nonlocal description of S plus R.

The question now is how to construct $|\psi_{\mathrm{REC}}(t)\rangle$ and $P(\mathrm{REC})$ so that the expansion (17.8) holds—with $\rho(t)$ satisfying master equation (17.2)—and $P(\mathrm{REC})$ gives the correct frequency of occurrence for every conceivable record. The second point is important, because it is the records, or contaminated versions of them (finite detection efficiency, technical noise, etc.) that must account for the experimenter's observations. Of course, agreement with observations can only be checked in a statistical sense. The statistical agreement should be a sophisticated one, though, and go beyond the trivial statement that average quantities should agree with mean values computed from $\rho(t)$. The records define a stochastic process in their own right, a process to account for data sets recorded by an experimenter monitoring the environment. This process, which is implicit in (17.8), should agree with actual data sets—certainly with their means, but also with all their correlations. Thus, the aims that are achieved through the three steps summarized by (17.1)–(17.5) are, in a sense, all rolled into one when we build our treatment around $P(\mathrm{REC})$ and $|\psi_{\mathrm{REC}}(t)\rangle$. Assuming we find a way to simulate the scattering records, we can then, for example, calculate temporal correlation functions by directly correlating simulated data sets, imitating what is done in the laboratory.

We should perhaps consider an obvious question at this point. Taking a particular system observable $\hat{O}$, we are familiar with its usual quantum mechanical mean $\langle\hat{O}(t)\rangle = \mathrm{tr}_S[\hat{O}\rho(t)]$, but how should we refer to the quantity $\langle\psi_{\mathrm{REC}}(t)|\hat{O}|\psi_{\mathrm{REC}}(t)\rangle$? Clearly it is a conditional average, contrasting with $\langle\hat{O}(t)\rangle$, which is an unconditional average. We will refer to it as a *conditional expectation*, where the word "expectation" is specifically appropriate, because $\langle\psi_{\mathrm{REC}}(t)|\hat{O}|\psi_{\mathrm{REC}}(t)\rangle$ is an inference; it tells us what we might *expect*—what we might expect the value of $\hat{O}$ to be under the condition that the particular scattering record labeling the state has occurred.

Since the unraveling (17.8) hinges on $P(\mathrm{REC})$, the natural place to start is with the theory of photoelectron counting; scattering records are photoelectron counting records. We discuss the photoelectron counting approach in Sects. 18.1 and 18.2; but there are some new ideas to familiarize ourselves with first. Although simple enough in conception, formulating a statistical description in terms of conditional states is not usual in physics. Certainly, in quantum optics, the traditional emphasis is on unconditional quantities like $\rho(t)$. It is therefore helpful to see how a simple stochastic evolution appears using conditional states. It is also helpful to begin with an example that makes it clear that the basic idea behind quantum trajectories is a shift in the mode of description, and that this shift can be made for a classical process just as well; specifically, it does not originate in technical manipulations involving superoperators and quantum states.

We start from the old-fashioned idea of Bohr–Einstein quantum jumps. These certainly provide a stochastic account that includes scattering events. Their principle shortcoming is that they are unable to deal with coherence, with superpositions of quantum states, though they work just fine for incoherent scattering. Let us see how these elementary quantum jump ideas can be formulated in the language of scattering records and conditional states. We will find that the language permits a generalization to include the missing coherence.

17.2 Generalizing the Bohr–Einstein Quantum Jump

The earliest version of quantum dynamics made no reference to deBroglie waves, the Schrödinger equation, or Feynman diagrams. It grew out of a semi-literal acceptance of Planck's energy quanta [17.27], supported by Bohr's model of the hydrogen atom [17.28] and Einstein's treatment of the photoelectric effect [17.29]. The so-called quantum jump associated with the exchange of an energy quantum between two systems captures the essence of everything discrete and particle-like in quantum physics. So far as interpretation goes, a dynamics of quantum jumps and quantum jumps alone, even if it were stochastic, would be far less troubling than the quantum mechanics we actually have. The earliest useful quantum dynamics was precisely something of this sort, proposed by Einstein to explain the approach to thermal equilibrium of radiation interacting with matter [17.30].

17.2.1 The Einstein Stochastic Process

We are already familiar with Einstein A and B theory from Sects. 2.2.3 and 7.1. Consider a material resonance, a two-state system, coming to equilibrium with blackbody radiation. The Einstein stochastic process proposes an evolution based on the quantum jump scheme illustrated in Fig. 17.3. In our notation, the Einstein A coefficient is γ (Sect. 2.2.2) and $\bar{n}$ is the mean photon number for a mode of the electromagnetic field resonant with the $|1\rangle \rightarrow |2\rangle$ transition. The stochastic process is a random telegraph process [17.31] with transitions between states $|1\rangle$ and $|2\rangle$ mediated by the absorption and re-emission (resonant scattering) of thermal quanta. It is usual to ignore the explicit stochastic character of Einstein theory and reduce the description to a set of rate equations for state occupation probabilities. Denoting these probabilities by p_1 and p_2, they obey the equations of motion

$$\dot{p}_2 = -\gamma_{\downarrow} p_2 + \gamma_{\uparrow} p_1, \tag{17.10a}$$

$$\dot{p}_1 = -\gamma_{\uparrow} p_1 + \gamma_{\downarrow} p_2, \tag{17.10b}$$

with

$$\gamma_{\downarrow} \equiv \gamma(\bar{n}+1), \qquad \gamma_{\uparrow} \equiv \gamma\bar{n}. \tag{17.11}$$

These ideas have remarkable power considering their simplicity. From them alone, for example, one can write down a statistical theory of the laser, fully equivalent to the Scully–Lamb master equation (Sect. 7.1.3). A first-principles derivation of the master equation does little more than elaborate the physical conditions required for Einstein theory to be valid; we do get derived expressions for jump rates, like γ, but nothing new emerges in the dynamics.

Filled out as a stochastic process, Einstein theory gives a detailed elaboration of the absorptions and emissions, which may be summarized in a scattering record of the sort defined above. It is straightforward to generate realizations of *quantum trajectories* consisting of a quantum state evolution and

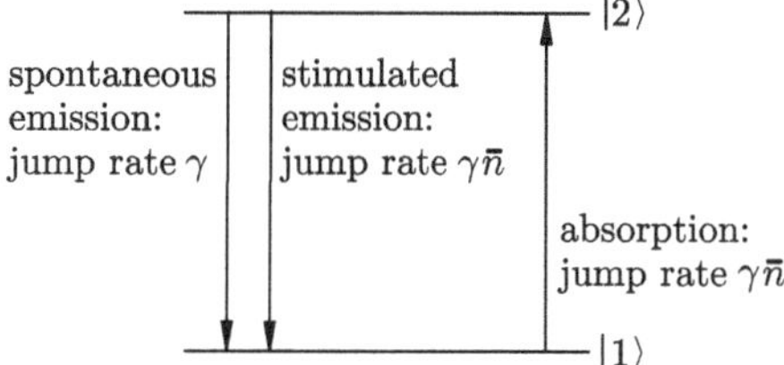

Fig. 17.3. Quantum jump scheme for the Einstein stochastic process

associated scattering record. We use Monte Carlo rules to decide whether or not jumps take place within each of a series of time steps of length Δt. At time $t = k\Delta t$, the rules for advancing the state turn on the value of a uniformly distributed random number r_k, $0 \leq r_k \leq 1$:

$$\left.\begin{array}{rl} \textbf{if} \text{ state is } |2\rangle \textbf{ then:} & \textbf{1. compute } p_\downarrow = \gamma_\downarrow \Delta t \\ & \textbf{2. if } p_\downarrow > r_k, \textbf{ then } |2\rangle \to |1\rangle \\ & \quad\ \textbf{else } |2\rangle \to |2\rangle \\ \textbf{else} \text{ (state is } |1\rangle\text{):} & \textbf{1. compute } p_\uparrow = \gamma_\uparrow \Delta t \\ & \textbf{2. if } p_\uparrow > r_k, \textbf{ then } |1\rangle \to |2\rangle \\ & \quad\ \textbf{else } |1\rangle \to |1\rangle \end{array}\right\}, \tag{17.12}$$

where $p_\downarrow$ and $p_\uparrow$ are the probabilities, respectively, for downward and upwards jumps to take place. It is assumed that Δt is sufficiently small that both $p_\downarrow$ and $p_\uparrow$ are much less than unity. A single realization of this process is an ongoing random telegraph signal, jumping between states $|1\rangle$ and $|2\rangle$, with associated scattering record

$$\mathrm{REC} \equiv \left\{ \begin{matrix} \cdots \\ \cdots \end{matrix}, \emptyset, \frac{\gamma_\uparrow}{T_{k-1}}, \emptyset, \frac{\gamma_\downarrow}{T_k}, \emptyset, \frac{\gamma_\uparrow}{T_{k+1}}, \emptyset, \frac{\gamma_\downarrow}{T_{k+2}}, \emptyset, \begin{matrix} \cdots \\ \cdots \end{matrix} \right\}. \tag{17.13}$$

The record is rather monotonous and trivial; only the times of the jumps contain interesting information.

Bohr–Einstein quantum jumps are appealing because they describe realized events occurring in time. Stochasticity might be unwelcome—"God does not play dice"—but its status is surely debatable, and probabilities are definitely not the most curious thing about quantum mechanics. Its deep puzzles stem from the inadequacy of any description that is either entirely discontinuous (as here) or entirely continuous. Bohr, in particular, was a strong opponent of the "energy quanta bookkeeping" offered by quantum jumps. His view was that so much in optics calls for wave interference that one cannot dispense with the idea of light as an electromagnetic wave. Where, though, in the Einstein stochastic process is there room for electromagnetic waves? Together with Kramers and Slater, Bohr attempted to make room by splicing the electromagnetic waves onto Einstein A and B theory [17.32]. The proposal was inadequate and quickly abandoned, although in retrospect it seems to have had more than a little in common with quantum trajectory theory [17.33, 17.34]. We need not retrace this old ground, however. We simply note that the limitation of Bohr–Einstein quantum jumps, as Bohr himself recognized, is the commitment to transitions between stationary states and the consequent exclusion of any notion of coherence; there is no place for a phase, of an electromagnetic wave or of anything else. From a modern point of view, there is no way to retain the appeal of quantum jumps when the atom starts out in a superposition of states $|1\rangle$ and $|2\rangle$—when it starts out with a non-vanishing mean dipole moment. Quantum trajectory theory generalizes the Bohr–Einstein jump so as to overcome this deficiency.

Before introducing anything new, we need to say something about the contrasting languages that can be used when describing a stochastic process. The probabilities $p_1(t)$ and $p_2(t)$ define the density operator $\rho(t)$. For the jump scheme of Fig. 17.3, we have

$$\rho(t) = p_2(t)|2\rangle\langle 2| + p_1(t)|1\rangle\langle 1|. \tag{17.14}$$

This is not, however, a decomposition of the kind envisioned in (17.8). The labels, 1 and 2, refer to stationary states—energy eigenstates—not to a scattering record, and the probabilities $p_1(t)$ and $p_2(t)$ describe the unconditional dynamics. We need to see how a description in terms of conditional dynamics can be made. In order to keep the records as simple as possible, we set $\bar{n} = 0$; thus, we specialize to the almost trivial example of spontaneous emission.

To illustrate the basic ideas, we have in mind an ensemble of atoms prepared in state $|2\rangle$ with probability $p_2(0)$, and in state $|1\rangle$ with probability $p_1(0) = 1 - p_2(0)$. There are clearly two types of trajectories: a trajectory of the first type (initial state $|2\rangle$) executes a quantum jump at some unpredictable time; a trajectory of the second type (initial state $|1\rangle$) never executes a quantum jump. We aim to find a decomposition of the unconditional probabilities, $p_1(t)$ and $p_2(t)$, in terms of conditional probabilities $p_{2|\mathrm{REC}}(t)$ and $p_{1|\mathrm{REC}}(t)$, where the sum over records covers realizations of both trajectory types. As we will see, records can be defined in two ways: the first leads to something familiar and serves only to start us thinking about conditional dynamics; the second leads to something unfamiliar—a flexible language that can accommodate the generalization of the Bohr–Einstein jump to quantum state superpositions.

17.2.2 Conditional Evolution: Trajectories for Known Initial States

The first definition considers the initial state to be included in the record. The conditional probabilities are trivial since they are always 0 or 1; nevertheless, we may still decompose the unconditional probabilities as a sum over these trivial conditional probabilities. To this end, in the spirit of (17.8), we write

$$p_2(t) = P(\{A\})p_{2|\{A\}}(t) + P(\{B\})p_{2|\{B\}}(t) + \int_0^t dt' P(\{C_{T'}\})p_{2|\{C_{T'}\}}(t), \quad (17.15a)$$

and

$$p_1(t) = P(\{A\})p_{1|\{A\}}(t) + P(\{B\})p_{1|\{B\}}(t) + \int_0^t dt' P(\{C_{T'}\})p_{1|\{C_{T'}\}}(t),. \quad (17.15b)$$

where the records appearing in these expressions are

$$\{A\} \equiv \left\{ \begin{matrix} |2\rangle \\ 0 \end{matrix}, \emptyset_t \right\}, \quad (17.16a)$$

$$\{B\} \equiv \left\{ \begin{matrix} |1\rangle \\ 0 \end{matrix}, \emptyset_t \right\}, \quad (17.16b)$$

$$\{C_{T'}\} \equiv \left\{ \begin{matrix} |2\rangle \\ 0 \end{matrix}, \emptyset, \frac{\gamma}{T'}, \emptyset_t \right\}. \quad (17.16c)$$

The first entry of each record specifies the initial state, otherwise the notation is the same as in (17.6). The record $\{C_{T'}\}$, for example, says the atom begins at $t = 0$ in state $|2\rangle$, no photon is detected in the interval $[0, t')$, a photon is detected at $t' \leq t$—in the interval $T' \equiv [t', t' + dt')$, and no photon is detected in the interval $[t' + dt', t)$. Note that $P(\{A\})$ and $P(\{B\})$ are probabilities, while $P(\{C_{T'}\})$ is a probability density.

Note 17.3. We consider scattering records to be made by detecting photons in the far field. The time of a detection is therefore retarded relative to the time of the quantum jump inferred from that detection. For simplicity, we neglect this detail here, though the retardation is included in Sects. 18.1 and 18.2.

The goal now is to write explicit expressions for all of the pieces appearing in expansions (17.15a) and (17.15b). The example is so trivial that for the most part what is needed is obvious. Clearly, the conditional probabilities are

$$p_{2|\{A\}}(t) = 1, \qquad p_{2|\{B\}}(t) = 0, \qquad p_{2|\{C_{T'}\}}(t) = 0, \quad (17.17a)$$

and

$$p_{1|\{A\}}(t) = 0, \qquad p_{1|\{B\}}(t) = 1, \qquad p_{1|\{C_{T'}\}}(t) = 1. \quad (17.17b)$$

This leaves the record probabilities and probability densities. It is obvious that

$$P(\{A\}) = p_2(t), \quad (17.18)$$

which is consistent with (17.15a) and the conditional probabilities (17.17a). Then, since rate equations (17.10a) and (17.10b) (with $\bar{n} = 0$) have solution

$p_2(t) = p_2(0)e^{-\gamma t}$, record $\{A\}$ has probability

$$P(\{A\}) = p_2(0)e^{-\gamma t}. \tag{17.19a}$$

It is also obvious that

$$P(\{B\}) = p_1(0) = 1 - p_2(0). \tag{17.19b}$$

What we should write down for the probability density of record $\{C_{t'}\}$ is possibly not so obvious. A simple way to arrive at the correct result is to consider the sum over record probabilities (Eq. 17.9). The sum yields

$$P(\{A\}) + P(\{B\}) + \int_0^t dt' P(\{C_{T'}\}) = 1. \tag{17.20}$$

So, differentiating with respect to t and making use of (17.19a) and (17.19b), the probability density is

$$P(\{C_{T'}\}) = P(\{C_T\})|_{t=t'} = \gamma e^{-\gamma t'} p_2(0). \tag{17.21}$$

In fact $P(\{C'_T\})/p_2(0)$ is simply the waiting-time distribution for a spontaneous emission, given that the atom was prepared in state $|2\rangle$ (see Sects. 2.3.6 and 17.3.5).

Let us now put all these pieces together. Introducing the expansions (17.15a) and (17.15b) into (17.14), the density operator is written as

$$\begin{aligned} \rho(t) = {} & P(\{A\})|\psi_{\{A\}}(t)\rangle\langle\psi_{\{A\}}(t)| + P(\{B\})|\psi_{\{B\}}(t)\rangle\langle\psi_{\{B\}}(t)| \\ & + \int_0^t dt' P(\{C_{T'}\})|\psi_{\{C_{T'}\}}(t)\rangle\langle\psi_{\{C_{T'}\}}(t)|, \end{aligned} \tag{17.22}$$

where the conditional probabilities have been made redundant by defining conditional states

$$|\psi_{\{A\}}(t)\rangle = |2\rangle, \qquad |\psi_{\{B\}}(t)\rangle = |1\rangle, \qquad |\psi_{\{C_{T'}\}}(t)\rangle = |1\rangle. \tag{17.23}$$

Equation 17.22 is an unraveling of the density operator of the sort envisaged in (17.8). In this case the unraveling is rather obvious and may appear as little more than so much confusing notation. It does add something to the unconditional expansion (17.14), though, by explicitly separating out the two kinds of paths, or trajectories, that may be taken to reach state $|1\rangle$: the atom might be prepared in state $|2\rangle$ and at some time t' jump to state $|1\rangle$, or the atom might be prepared in state $|1\rangle$ and remain there forever. The paths are illustrated in Fig. 17.4. The density operator is expanded as a sum over record probabilities and conditional states that refer to these paths.

To include the initial state of the atom in the record is not strictly keeping to the spirit of the scattering point of view. The initial state is a local property of the atom and not something that can be directly detected in the environment. Certainly it might be inferred through some form of scattering, but it is

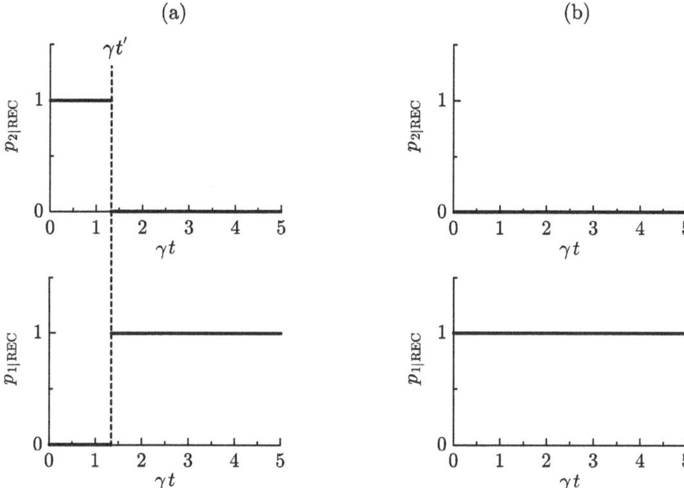

Fig. 17.4. Two kinds of trajectories followed by the conditional state of a spontaneously emitting atom that has been selected from an ensemble with a fraction $p_2(0)$ of the atoms prepared in the excited state. The conditioning is with respect to records that specify the initial state of the atom (compare Fig. 17.5)

not the kind of primitive information from which scattering records are made up. Nonetheless, this first way of defining records is useful as it reflects the way in which we conventionally think; it is more natural to think in terms of properties that *are so* than properties to be inferred. Of course, there is a huge difference between the two, because inferences are rarely definite—they generally come only with graded *likelihoods* that "this may be so" or "that may be so."

Thus, with our first way of defining records, having introduced definiteness about the state at the start, it is translated throughout the entire trajectory; in Fig. 17.4 the state of the atom, $|1\rangle$ or $|2\rangle$, is definite at every point in time. In fact, Fig. 17.4a is just the picture we would draw naively to illustrate a quantum jump between discrete states. Given, then, this built-in commitment to a definite state at all times, this way of defining records leaves no room at all for talking about superpositions of stationary states. Preparation of a superposition requires a different way of talking, one that allows the state, $|1\rangle$ or $|2\rangle$, to be left uncertain even as we follow the conditional evolution of a particular atom.

17.2.3 Conditional Evolution: Trajectories for "Blind" Realizations

To set up this "different way of talking," we remove the initial state from the record and condition our description on the detection or nondetection of an emitted photon alone. Consider the following scenario. An ensemble of many realizations of the kind illustrated in Fig. 17.4 is laid out *a priori*, from which we pick one realization at random. We are not to look to see which kind of realization we chose—an example like in Fig. 17.4a or the nonjumping case of Fig. 17.4b. We are, however, told the probability, $p_2(0)$, that our choice begins in state $|2\rangle$. We therefore start out with a measure of the likelihood that the state is $|2\rangle$, but we do not definitely know the state.

There is a friend who does know the realization we chose, and he/she steps through time, in steps of length Δt, telling us at each step whether or not a spontaneous photon is detected (whether or not a quantum jump of the atomic state occurs). Our job is to construct the conditional probabilities $p_{2|\mathrm{REC}}(t)$ and $p_{1|\mathrm{REC}}(t)$, where in this case the record includes only the information provided by the friend. As a rule, the definite state of the chosen realization remains unknown to us; although, perhaps we will come to know it over time. In this scenario the conditional probabilities quite literally express our evolving *expectation* of what the state of the atom is. The immediate question is: how do these probabilities evolve over time? They surely differ from those plotted in Fig. 17.4.

It is still possible to write the unconditional probabilities as a sum over conditional probabilities, as in (17.15a) and (17.15b). We write

$$p_2(t) = P(\{D\})p_{2|\{D\}}(t) + \int_0^t dt' P(\{E_{T'}\})p_{2|\{E_{T'}\}}(t), \tag{17.24a}$$

$$p_1(t) = P(\{D\})p_{1|\{D\}}(t) + \int_0^t dt' P(\{E_{T'}\})p_{1|\{E_{T'}\}}(t), \tag{17.24b}$$

where the records are now

$$\{D\} \equiv \{{}_0\emptyset_t\}, \tag{17.25a}$$

$$\{E_{T'}\} \equiv \left\{{}_0\emptyset, \frac{\gamma}{T'}, \emptyset_t\right\}. \tag{17.25b}$$

The two records (17.16a) and (17.16b) have collapsed into the single record, (17.25a), and record (17.25b) is essentially equivalent to (17.16c). The notation ${}_0\emptyset_t$ indicates an interval $[0, t)$ without a photon detection.

Our goal once again is to write explicit expressions for all the pieces in the expansion. It is obvious that

$$p_{2|\{E_{T'}\}}(t) = 0, \tag{17.26a}$$

$$p_{1|\{E_{T'}\}}(t) = 1, \tag{17.26b}$$

since as soon as our friend tells us a photon has been detected we know to set $p_{2|\mathrm{REC}} = 0$; thus, in this trivial example, one piece of information makes the state of the chosen realization definite. What we should write for the probabilities conditioned on $\{D\}$ is less clear. Let us come back to this question after writing down the two record probabilities.

These are straightforward to determine. The probability for no photon being detected up to time t is the sum of two probabilities, one associated with each of the initial states. Using (17.19a) and (17.19b), we have the probability for no photon detection

$$P(\{D\}) = P(\{A\}) + P(\{B\}) = e^{-\gamma t} p_2(0) + p_1(0). \tag{17.27}$$

It is also clear that if a photon is detected at some time t', the initial state must have been $|2\rangle$; therefore, according to (17.21),

$$P(\{E_{T'}\}) = P(\{C_{T'}\}) = \gamma e^{-\gamma t'} p_2(0). \tag{17.28}$$

The sum over record probabilities is of course unity once again:

$$P(\{D\}) + \int_0^t dt' P(\{E_{T'}\}) = 1, \tag{17.29}$$

where the first term on the left-hand side accounts for the first two terms of the sum (17.20).

We now come back to the conditional probabilities $p_{2|\{D\}}(t)$ and $p_{1|\{D\}}(t)$. How should we update our expectation about the atomic state when we continue to hear that no spontaneous photon is detected—when apparently nothing has happened? In quantum mechanics, a "nothing has happened" situation like this is called a *null measurement*. It would be incorrect to say that nothing is learned from hearing that nothing has happened. Something is learned, and exactly *what* constitutes the null measurement conditioning to be built into probabilities $p_{2|\{D\}}(t)$ and $p_{1|\{D\}}(t)$. In fact, this null measurement conditioning is going to develop into the core idea of quantum trajectory theory. It is the one thing that makes the "blind" realizations we are considering different from the conventional way of looking at the Einstein stochastic process, and it is the essential ingredient that allows the Bohr–Einstein quantum jump to be generalized to incorporate coherence.

Null measurements have a special place in quantum mechanics, but the basic ideas are not uniquely quantum mechanical. Perhaps the words "null measurement" are misleading. More correctly, it is the *result* of the measurement that is null; the physical conditions to make a spontaneous emission possible are in place whether or not an emission actually occurs. The occurrence of no emission in a given interval of time is one of two possible results—the null result. Conditional evolution given the null result is a statistical notion arising from the way in which statistical inferences are made. The evolution is determined by *Bayesian inference.* There are two alternative ways to deduce it. The first goes straight to the conditional probability, while the second goes to an equation of motion for the probability. The first is quicker so we will use it here; derivation of the equation of motion is postponed until Sect. 17.3.1.

In order to deduce the conditional probability $p_{2|\{D\}}(t)$, all we need do is relate it to the joint probability $p_{2\wedge\{D\}}(t)$ using Bayes theorem [17.35]. We write

$$p_{2\wedge\{D\}}(t) = p_{2|\{D\}}(t) P(\{D\}). \tag{17.30}$$

Then clearly $p_{2\wedge\{D\}}(t) = p_2(t)$, since $p_{2\wedge\{E_{T'}\}}(t) = 0$; thus,

$$p_{2|\{D\}}(t) = \frac{e^{-\gamma t} p_2(0)}{e^{-\gamma t} p_2(0) + p_1(0)}, \tag{17.31a}$$

and since the probabilities sum to unity,

$$p_{1|\{D\}}(t) = \frac{p_1(0)}{e^{-\gamma t} p_2(0) + p_1(0)}. \tag{17.31b}$$

Notice how $p_{2|\{D\}}(t)$ decays to zero for $t \to \infty$ in all cases except when $p_2(0) = 0$. This goes to the heart of the inference we are making; it expresses the quite reasonable conclusion that the longer our friend continues to say "no spontaneous emission is detected," the more convinced we become that the chosen realization started in state $|1\rangle$ and is never going to jump. Since a fraction $p_1(0)$ of the atoms begin in state $|1\rangle$, for this fraction of our "blind" realizations there will be no jump. Instead, our expectation that the atom is in state $|1\rangle$ increases continuously, to eventually reach unity. For the remaining fraction $p_2(0)$ of the realizations, our friend informs us that a spontaneous photon is detected at some time t', at which time our expectation that the atom is in state $|1\rangle$ jumps discontinuously to unity. Thus, as before, there are two kinds of paths, or trajectories, for the conditional probabilities, but now they appear as in Fig. 17.5.

Note 17.4. The jumps in quantum trajectory theory are not strictly discontinuous. There is an implied course-graining in time. Monte Carlo simulations (Sect. 17.3.4) use a nonzero time step set by practical considerations. It would normally be maximized, keeping in mind the constraint that the net jump probability per time step (all outputs) be much less than unity. So far as physics is concerned, however, the more important point is that there is a loosely defined lower bound on the time step, which is assumed large compared to the period of the transition resonance. This follows from the separation of timescales assumed by the Markov approximation (Sect. 1.3.3).

Putting the pieces together for our "blind" realizations, we arrive at an alternative unraveling of the density operator. In this instance, we cannot write it down in quite the same form as in (17.22) because the initial condition is not known as a definite pure state, $|1\rangle$ or $|2\rangle$, but only as the probabilities $p_1(0)$ and $p_2(0)$. The unraveling is made in terms of mixed states; thus, in place of (17.8) we write

$$\rho(t) = \sum_{\mathrm{REC}} P(\mathrm{REC}) \rho_{\mathrm{REC}}(t). \tag{17.32}$$

Its explicit form for "blind" realizations of spontaneous emission is

$$\rho(t) = P(\{D\}) \rho_{\{D\}}(t) + \int_0^t dt' P(\{E_{T'}\}) \rho_{\{E_{T'}\}}(t), \tag{17.33}$$

with

$$\rho_{\{D\}}(t) = p_{2|\{D\}}(t)|2\rangle\langle 2| + p_{1|\{D\}}(t)|1\rangle\langle 1|, \tag{17.34a}$$

and

$$\rho_{\{E_{T'}\}}(t) = |1\rangle\langle 1|, \tag{17.34b}$$

where the various probabilities and probability densities are given in (17.27), (17.28), (17.31a), and (17.31b).

The principal difference between trajectories in the sense of Fig. 17.5 and those in Fig. 17.4 is that in the former the atom is not assigned a definite stationary state, so that a central physical property of the atom—its energy—remains uncertain. The description is therefore given in terms of an expectation for the energy instead of a definite energy value. Of course, for the ensemble as constructed, the uncertainty arises by denying ourselves information that could, in principle, be had.

The next step is to recognize that in quantum mechanics uncertainty is intrinsic; denied information cannot always, in principle, be had. If, for example, we were to prepare an ensemble of atoms in a superposition of states,

$$|\psi(0)\rangle = c_1(0)|1\rangle + c_2(0)|2\rangle, \tag{17.35}$$

these atoms would be polarized. Such an ensemble is not equivalent to one with a fraction $p_1(0) = |c_1(0)|^2$ of the atoms in state $|1\rangle$ and the others in

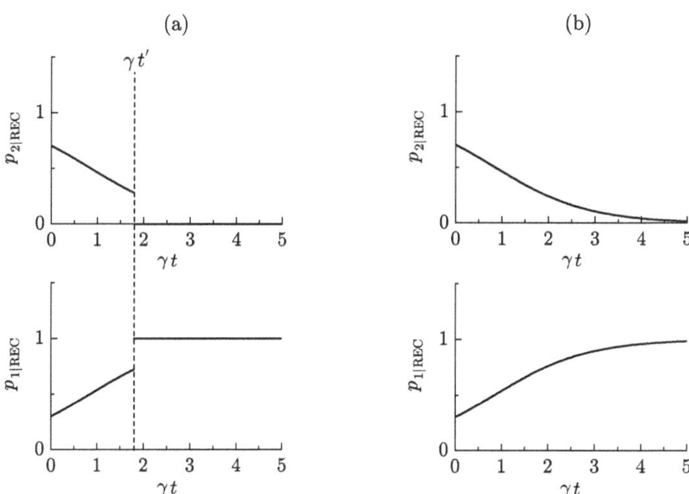

Fig. 17.5. Two kinds of trajectories followed by the conditional state of a spontaneously emitting atom that has been selected from an ensemble with a fraction $p_2(0) = 0.7$ of the atoms prepared in the excited state. The conditioning is with respect to records that do not specify the initial state of the atom—i.e., for "blind" realizations (compare Fig. 17.4). Note that trajectories of kind (**b**) are a limiting case ($\gamma t' \to \infty$) of those of kind (**a**)

state $|2\rangle$. Polarization amplitude is another physical property, and in quantum mechanics it and the energy cannot be assigned definite values at the same time; indeed, if the energy is assigned a definite value the atom may not be polarized at all. In this setting, there is a break between the notions of definite property and definite state. We can imagine making a conditional description in terms of a definite state

$$|\psi_{\mathrm{REC}}(t)\rangle = c_{1|\mathrm{REC}}(t)|1\rangle + c_{2|\mathrm{REC}}(t)|2\rangle, \tag{17.36}$$

where neither the energy nor the polarization amplitude take a definite value, though we are in possession of conditional expectations for both. We aim now for a generalization of the Bohr–Einstein quantum jump that is able to accommodate such a description. Our "blind" realizations are just a short step away from what we want. How that final step should be made will made apparent by formalizing what we have just done.

17.2.4 The Master Equation

Spontaneous emission may be described by a master equation (Sect. 2.2), and the unraveling of the density operator (17.33) can be arrived at by developing its solution in a Dyson expansion. We met an example of this expansion in Sect. 12.1.8, where the solution to master equation (12.97) was expanded as (12.102). The expansion takes a far simpler form when there can be only a single photon emission.

Introducing notation similar to that of (12.97)–(12.98c), let us write the spontaneous emission master equation (Eq. 2.26 with $\bar{n} = 0$) as

$$\dot{\rho} = (\mathcal{L}_B + \mathcal{L}_\gamma)\rho, \tag{17.37}$$

with

$$\mathcal{L}_B \equiv -i[\tfrac{1}{2}\omega_A\sigma_z, \cdot] - \frac{\gamma}{2}(\sigma_+\sigma_- \cdot + \cdot \sigma_+\sigma_-), \tag{17.38a}$$

$$\mathcal{L}_\gamma \equiv \gamma\sigma_- \cdot \sigma_+. \tag{17.38b}$$

By separating the right-hand side into two pieces, we are able to develop the solution for $\rho(t)$ in the manner of time-dependent perturbation theory. The difference is that instead of evolving a pure state $|\psi(t)\rangle$, with Hamiltonian $H = H_0 + H_I$, here we evolve a mixed state $\rho(t)$ with the superoperator $\mathcal{L} = \mathcal{L}_B + \mathcal{L}_\gamma$. The subscript B indicates the superoperator governing the *between jump evolution*, while the superoperator that executes the quantum jump carries the label of the associated output channel. The meaning will become clearer as we proceed.

We now transform to the equivalent of the interaction picture, by writing

$$\rho = e^{\mathcal{L}_B t}\bar{\rho}, \tag{17.39}$$

where $\bar{\rho}$ satisfies the equation of motion

$$\dot{\bar{\rho}} = e^{-\mathcal{L}_B t}\mathcal{L}_\gamma e^{\mathcal{L}_B t}\bar{\rho}. \tag{17.40}$$

Then, integrating this equation formally and inverting the transformation (17.39), the Dyson expansion of the density operator is

$$\rho(t) = e^{\mathcal{L}_B t}\rho(0) + \int_0^t dt' e^{\mathcal{L}_B (t-t')}\mathcal{L}_\gamma e^{\mathcal{L}_B t'}\rho(0). \tag{17.41}$$

In this case there is no need for the usual iteration to obtain terms of higher order in $\mathcal{L}_\gamma$, since the expansion truncates after just one photon emission—$(\mathcal{L}_\gamma)^2 = 0$.

It is straightforward to show that (17.41) is, in fact, precisely the quantum trajectory unraveling of the density operator (17.33). Note first that the elementary superoperator operations are trivial:

$$\begin{aligned} \mathcal{L}_B|2\rangle\langle 2| &= -\gamma|2\rangle\langle 2|, \qquad & \mathcal{L}_B|1\rangle\langle 1| &= 0, \\ \mathcal{L}_\gamma|2\rangle\langle 2| &= \gamma|1\rangle\langle 1|, \qquad & \mathcal{L}_\gamma|1\rangle\langle 1| &= 0. \end{aligned} \tag{17.42a}$$

From these,

$$e^{\mathcal{L}_B t}\rho(0) = e^{-\gamma t}p_2(0)|2\rangle\langle 2| + p_1(0)|1\rangle\langle 1|, \tag{17.43a}$$

and

$$e^{\mathcal{L}_B (t-t')}\mathcal{L}_\gamma e^{\mathcal{L}_B t'}\rho(0) = \gamma e^{-\gamma t'}p_2(0)|1\rangle\langle 1|. \tag{17.43b}$$

Thus, on comparing these expressions with (17.27) and (17.28), apparently the record probabilities and probability densities for "blind" realizations may be written, more formally, as

$$P(\{D\}) = \mathrm{tr}[e^{\mathcal{L}_B t}\rho(0)], \tag{17.44a}$$

$$P(\{E_{T'}\}) = \mathrm{tr}[e^{\mathcal{L}_B (t-t')}\mathcal{L}_\gamma e^{\mathcal{L}_B t'}\rho(0)]. \tag{17.44b}$$

The Dyson expansion then takes the form (17.33) if the conditional states, $\rho_{\{D\}}(t)$ and $\rho_{\{E_{T'}\}}(t)$, are written as

$$\rho_{\{D\}}(t) = \frac{e^{\mathcal{L}_B t}\rho(0)}{\mathrm{tr}[e^{\mathcal{L}_B t}\rho(0)]}, \tag{17.45a}$$

$$\rho_{\{E_{T'}\}}(t) = \frac{e^{\mathcal{L}_B (t-t')}\mathcal{L}_\gamma e^{\mathcal{L}_B t'}\rho(0)}{\mathrm{tr}[e^{\mathcal{L}_B (t-t')}\mathcal{L}_\gamma e^{\mathcal{L}_B t'}\rho(0)]}. \tag{17.45b}$$

Using (17.43a) and (17.43b), these expressions do indeed agree, respectively, with (17.34a) and (17.34b).

In summary, the summation over records of (17.32) [equivalently (17.8)] has emerged from the Dyson expansion through the simple expedient of normalizing the state entering into each term of the expansion; the raw expansion

(17.41) is simply rewritten as

$$\rho(t) = \mathrm{tr}[e^{\mathcal{L}_B t}\rho(0)]\frac{e^{\mathcal{L}_B t}\rho(0)}{\mathrm{tr}[e^{\mathcal{L}_B t}\rho(0)]} + \int_0^t dt' \mathrm{tr}[e^{\mathcal{L}_B(t-t')}\mathcal{L}_\gamma e^{\mathcal{L}_B t'}\rho(0)]\frac{e^{\mathcal{L}_B(t-t')}\mathcal{L}_\gamma e^{\mathcal{L}_B t'}\rho(0)}{\mathrm{tr}[e^{\mathcal{L}_B(t-t')}\mathcal{L}_\gamma e^{\mathcal{L}_B t'}\rho(0)]}. \tag{17.46}$$

The result separates the physically distinct roles played by the superoperators $\mathcal{L}_B$ and $\mathcal{L}_\gamma$. The quantum jump, $|2\rangle \to |1\rangle$, is executed formally by $\mathcal{L}_\gamma$, and the probability for the record $\{E_{t'}\}$ containing the jump depends on $\mathcal{L}_\gamma$. The superoperator $\mathcal{L}_B$, on the other hand, generates the time evolution before and after the jump—more generally, between jumps in a sequence, if many photons are scattered. The evolution generated by $\mathcal{L}_B$, together with the renormalization of the state, accounts for the null measurement conditioning; thus, $\mathcal{L}_B$ alone enters the expression for the record-$\{D\}$ probability.

The formal structure will be clarified further when we develop unravelings of the master equation and its density operator from photoelectron counting theory in Sects. 18.1 and 18.2. For the time being we simply use it as an aid to extend our quantum jump description to the more interesting case where quantum coherence is present.

17.2.5 Quantum Jumps in the Presence of Coherence

Master equation (17.37) is valid whether the initial state is mixed or pure. We may therefore take over (17.44a)–(17.46) more or less directly to arrive at the desired generalization of the Bohr–Einstein quantum jump. There is, however, one simplification we can make.

As we have already said, superposition states like (17.35) and (17.36) are definite states, even though they assign an indefinite energy to the atom. For a definite state we may get away from speaking about density operators and superoperators, even if the realizations are "blind" so far as properties like the energy are concerned. Consider how $\mathcal{L}_B$ and $\mathcal{L}_\gamma$ act upon an arbitrary pure state $\rho = |\Psi\rangle\langle\Psi|$. From (17.38a), we may write

$$\mathcal{L}_B(|\Psi\rangle\langle\Psi|) = \frac{1}{i\hbar}\Big(H_B|\Psi\rangle\langle\Psi| - |\Psi\rangle\langle\Psi|H_B^\dagger\Big), \tag{17.47}$$

where on the right-hand side we introduce the *non-Hermitian Hamiltonian*

$$H_B \equiv \tfrac{1}{2}\hbar\omega_A\sigma_z - i\hbar\frac{\gamma}{2}\sigma_+\sigma_-. \tag{17.48}$$

It is then straightforward to show that

$$e^{\mathcal{L}_B t}(|\Psi\rangle\langle\Psi|) = |\Psi_t\rangle\langle\Psi_t|, \qquad |\Psi_t\rangle \equiv e^{-(i/\hbar)H_B t}|\Psi\rangle, \tag{17.49}$$

or, equivalently, that the time evolving state between jumps, $|\Psi_t\rangle$, satisfies the *nonunitary Schrödinger equation*

$$\frac{d|\Psi_t\rangle}{dt} = \frac{1}{i\hbar} H_B|\Psi_t\rangle. \tag{17.50}$$

In addition, following from the definition (17.38b), we may expand the action of the jump superoperator as

$$\mathcal{L}_\gamma(|\Psi\rangle\langle\Psi|) = |\Psi_\gamma\rangle\langle\Psi_\gamma|, \qquad |\Psi_\gamma\rangle = J_\gamma|\Psi\rangle, \tag{17.51}$$

with the introduction of the *jump operator*

$$J_\gamma \equiv \sqrt{\gamma}\sigma_-. \tag{17.52}$$

We see from (17.47)–(17.52) that if the initial state is pure, the actions of both superoperators, $\mathcal{L}_B$ and $\mathcal{L}_\gamma$, may be replaced by the action of an ordinary operator on a pure state. Thus, the pure state factorization of the initial density operator is preserved in each term of the Dyson expansion (17.41), even though $\rho(t)$ does not factorize itself.

Note 17.5. The non-Hermitian Hamiltonian (17.48) is not as strange or unfamiliar as it might seem. We may alternatively write it as

$$H_B = \left(\tfrac{1}{2}\hbar\omega_A - i\hbar\gamma/2\right)|2\rangle\langle 2| - \tfrac{1}{2}\hbar\omega_A|1\rangle\langle 1|. \tag{17.53}$$

Thus, the non-Hermitian Hamiltonian is obtained by adding an imaginary term, $-i\hbar\gamma/2$, to the energy of the unstable excited state. Such an addition is a fairly common practice in the treatment of damped quantum systems. Note that quantum trajectory theory shows the practice to be strictly valid only if quantum jumps are also added to the description. In many applications, however, the unstable states are excited so weakly that the jumps are rare and may be omitted—at least to a good approximation. The pure-state expansion of Sect. 16.1 provides an example of such an application (see Note 16.1).

Consider now the operations appearing in the record probabilities and probability densities, (17.44a) and (17.44b), and the conditional states, (17.45a) and (17.45b). Using (17.49) and (17.51), for the initial superposition state (17.35) the probabilities and probability densities are given in terms of pure state norms:

$$P(\{D\}) = \langle\bar{\psi}_{\{D\}}|\bar{\psi}_{\{D\}}\rangle, \tag{17.54a}$$
$$P(\{E_{T'}\}) = \langle\bar{\psi}_{\{E_{T'}\}}|\bar{\psi}_{\{E_{T'}\}}\rangle, \tag{17.54b}$$

where the time dependence of the null measurement record follows from

$$\begin{aligned}|\bar{\psi}_{\{D\}}\rangle &= e^{-(i/\hbar)H_Bt}|\psi(0)\rangle \\ &= e^{i(\omega_A/2)t}c_1(0)|1\rangle + e^{-i(\omega_A/2)t}e^{-(\gamma/2)t}c_2(0)|2\rangle,\end{aligned} \tag{17.55}$$

and that of record $\{E_{T'}\}$ from

$$\begin{aligned}|\bar{\psi}_{\{E_{T'}\}}\rangle &= e^{-(i/\hbar)H_B(t-t')}J_\gamma|\Psi_t\rangle \\ &= e^{i(\omega_A/2)(t-2t')}\sqrt{\gamma}e^{-(\gamma/2)t'}c_2(0)|1\rangle. \end{aligned} \tag{17.56}$$

Equations 17.54a and 17.54b reproduce (17.27) and (17.28) with $p_1(0) = |c_1(0)|^2$ and $p_2(0) = |c_2(0)|^2$; thus, the record probabilities are not changed by the presence of coherence.

The conditional states do change, however, qualitatively, since now they are pure rather than mixed. Specifically, from (17.45a), (17.45b), (17.49), and (17.51), the conditional states in the presence of coherence are normalized versions of (17.55) and (17.56):

$$|\psi_{\{D\}}\rangle = \frac{|\bar{\psi}_{\{D\}}\rangle}{\sqrt{\langle\bar{\psi}_{\{D\}}|\bar{\psi}_{\{D\}}\rangle}}, \tag{17.57a}$$

$$|\psi_{\{E_{T'}\}}\rangle = \frac{|\bar{\psi}_{\{E_{T'}\}}\rangle}{\sqrt{\langle\bar{\psi}_{\{E_{T'}\}}|\bar{\psi}_{\{E_{T'}\}}\rangle}}. \tag{17.57b}$$

Putting the pieces together, using (17.46), the unraveling of the density operator for an initial pure state superposition takes the form

$$\rho(t) = P(\{D\})|\psi_{\{D\}}(t)\rangle\langle\psi_{\{D\}}(t)| + \int_0^t dt' P(\{E_{T'}\})|\psi_{\{E_{T'}\}}(t)\rangle\langle\psi_{\{E_{T'}\}}(t)|, \tag{17.58}$$

with conditional states

$$|\psi_{\{D\}}(t)\rangle = c_{2|\{D\}}(t)|2\rangle + c_{1|\{D\}}(t)|1\rangle, \tag{17.59a}$$

and

$$|\psi_{\{E_{T'}\}}\rangle = e^{i(\omega_A/2)(t-2t')}|1\rangle, \tag{17.59b}$$

where $c_{2|\{D\}}(t)$ and $c_{1|\{D\}}(t)$ are *conditional probability amplitudes*; from (17.55), (17.56), (17.57a), and (17.57b) they are given by

$$c_{2|\{D\}}(t) = \frac{e^{-i(\omega_A/2)t}e^{-(\gamma/2)t}c_2(0)}{\sqrt{e^{-\gamma t}|c_2(0)|^2 + |c_1(0)|^2}}, \tag{17.60a}$$

$$c_{1|\{D\}}(t) = \frac{e^{i(\omega_A/2)t}c_1(0)}{\sqrt{e^{-\gamma t}|c_2(0)|^2 + |c_1(0)|^2}}. \tag{17.60b}$$

We have thus arrived at our goal of unraveling the density operator into a sum over scattering records for the simple example of spontaneous emission. The unraveling is a straightforward generalization of the conditional evolution for "blind" realizations of Bohr–Einstein quantum jumps. The generalization does little more than accept the notion of a superposition state—the notion of a pure-state ensemble—and apply the idea of a stochastic evolution conditioned on the detection of a scattered photon, just as it was applied to the Einstein stochastic process (with its more conventional notion of ensemble). There are the same two kinds of trajectories for conditional probabilities, exactly as in Fig. 17.5.

A little extra is added, however, since the fundamental quantities are not conditional probabilities but conditional probability amplitudes. These amplitudes provide us with expectations for physical attributes other than the energy. In the present example, there is the *conditional dipole amplitude expectation*,

$$\langle\psi_{\{D\}}(t)|\sigma_-|\psi_{\{D\}}(t)\rangle = \frac{e^{-(\gamma/2)t}c_1^*(0)c_2(0)}{e^{-\gamma t}|c_2(0)|^2 + |c_1(0)|^2}. \tag{17.61}$$

It is important to note that the null measurement conditioning takes on a new physical significance here. When we met it in Sect. 17.2.3, it merely resolved our ignorance about the actual initial state of the selected "blind" realization —a state our friend knew about all along. In the present situation, the atom is initially polarized, and for a fraction $|c_1(0)|^2$ of the realizations, under the null-measurement evolution the polarization decays continuously all the way to zero. Now an ensemble of polarized atoms is physically different from an ensemble of unpolarized atoms, some initially in state $|2\rangle$ and some in state $|1\rangle$. The quantum jumps of Bohr and Einstein can speak about the latter but not the former. So we may summarize our progress as follows: *by moving away from the idea of states with definite properties and adopting the notion of a conditional evolution for expectations, we have constructed a jump process that can speak about both.*

Note 17.6. The proposal of Bohr, Kramers, and Slater (BKS) [17.32] comes very close to making the quantum trajectory generalization of Bohr–Einstein quantum jumps. Motivated by Bohr's insistence that coherence must have a place in the interaction of light with matter, it also proposes a stochastic process to generalize the quantum jump idea. Coherence is accounted for by introducing so-called "virtual atomic oscillators" whose action is turned on and off by jumps between stationary states. Virtual oscillator amplitudes are similar to conditional dipole amplitude expectations in quantum trajectory theory. Carmichael [17.33, 17.34] has discussed the similarities as they apply to the process of amplification without inversion.

In the absence of coherence, a quantum trajectory reverts to a sequence of jumps between stationary states. Thus, the *only* new part to the evolution is the null-measurement conditioning in between the jumps, which attempts, through Bayesian inference, to resolve the ambiguity of a superposition state. Spontaneous emission is of course a trivial example; the initially prepared coherence disappears as soon as the photon is emitted. Alternatively, we might consider the Einstein stochastic process with $\bar{n} \neq 0$ to keep the jumps going on forever; any initially prepared coherence is still, however, eliminated at the first jump.

Initially prepared coherence in a multilevel system need not disappear in such a trivial way. More interesting, however, are the examples of ongoing scattering processes driven by a coherent external field, examples such as resonance fluorescence or the cavity QED system shown in Fig. 17.2. In these cases, coherence is continuously recreated to oppose the ongoing sequence of quantum jumps which destroy it [17.36, 17.37]. Coherence can also enter in other ways, without being prepared initially or induced by an external field. It can take the form of entanglement between the interacting parts of a composite system, or arise spontaneously from the quantum jumps themselves—when an output channel collects, and thus interferes, the fields scattered by different sources [17.22, 17.23]. In examples of collective emission like superradiance, such interference is expressed through collective jump operators which leave the individual atoms in entangled states [17.24, 17.25].

At this stage it is undoubtedly difficult to see how the quantum trajectory idea is to be applied to these various situations. We need a more systematic development. Before tackling that, though, it might be helpful to focus on some observations arising out of what we have already done.

17.3 Miscellaneous Observations

17.3.1 Time Evolution Under Null Measurements

There is an alternative way to deduce the null-measurement conditioning expressed by (17.31a) and (17.31b). The idea is to find an equation of motion for the conditional probabilities, $p_{2|\{D\}}(t)$ and $p_{1|\{D\}}(t)$. We consider $p_{2|\{D\}}(t)$ explicitly; having treated it, the equation of motion for $p_{1|\{D\}}(t)$ will also be known.

We assume that at time t we know the conditional probability $p_{2|\{D\}}(t)$ and that at $t+\Delta t$ we receive the null result "no jump occurred between t and $t+\Delta t$." From this information we aim to construct the conditional probability $p_{2|\{D\}}(t+\Delta t)$. The desired equation of motion is then determined. In order to be clear about how this works, it is helpful to introduce a slightly more elaborate notation. Let us define the event

$$e_t \equiv \text{atom in state } |2\rangle \text{ at time } t, \tag{17.62}$$

and write out the conditional probabilities in full as

$$p_{2|\{D\}}(t) = P[e_t|\{_0\emptyset_t\}], \tag{17.63a}$$

$$p_{2|\{D\}}(t+\Delta t) = P[e_{t+\Delta t}|\{_0\emptyset_{t+\Delta t}\}], \tag{17.63b}$$

where $P[e_t|\{_0\emptyset_t\}]$ is the probability that event e_t is true given that the record up to time t is $\{_0\emptyset_t\}$ (Eq. 17.25a), and $P[e_{t+\Delta t}|\{_0\emptyset_{t+\Delta t}\}]$ is similarly defined for the record up to time $t+\Delta t$.

To begin, we use Bayesian inference to write the conditional probability $P[e_{t+\Delta t}|\{_0\emptyset_{t+\Delta t}\}]$ in terms of the joint probability $P[e_{t+\Delta t}\wedge\{_0\emptyset_{t+\Delta t}\}]$; thus, we write

$$P[e_{t+\Delta t}|\{_0\emptyset_{t+\Delta t}\}]=\frac{P[e_{t+\Delta t}\wedge\{_0\emptyset_{t+\Delta t}\}]}{P[\{_0\emptyset_{t+\Delta t}\}]}. \tag{17.64}$$

Clearly $e_{t+\Delta t}$ may be replaced by e_t on the right-hand side of this expression, yielding

$$P[e_{t+\Delta t}|\{_0\emptyset_{t+\Delta t}\}]=\frac{P[e_t\wedge\{_0\emptyset_{t+\Delta t}\}]}{P[\{_0\emptyset_{t+\Delta t}\}]}=\frac{P[\{_0\emptyset_{t+\Delta t}\}|e_t]P[e_t]}{P[\{_0\emptyset_{t+\Delta t}\}]}, \tag{17.65}$$

where Bayesian inference is used for a second time to move the conditioning from the record $\{_0\emptyset_{t+\Delta t}\}$ to the event e_t. Three more results allow for a rewriting of the right-hand side: the pair of results

$$P[\{_0\emptyset_{t+\Delta t}\}|e_t]=P[\{_0\emptyset_t\}|e_t]P[\{_t\emptyset_{t+\Delta t}\}|e_t], \tag{17.66a}$$

$$P[\{_0\emptyset_{t+\Delta t}\}]=P[\{_t\emptyset_{t+\Delta t}\}|\{_0\emptyset_t\}]P[\{_0\emptyset_t\}], \tag{17.66b}$$

the first of which follows from $P[A\wedge B]=P[A]P[B]$ for independent events A and B, and

$$P[\{_0\emptyset_t\}|e_t]P[e_t]=P[\{_0\emptyset_t\}\wedge e_t]=P[e_t|\{_0\emptyset_t\}]P[\{_0\emptyset_t\}], \tag{17.66c}$$

which follows by Bayesian inference, where we move the conditioning from the event e_t back to the record $\{_0\emptyset_t\}$. Now, substituting (17.66a) and (17.66b) on the right-hand side of (17.65), and using (17.66c), we arrive at the expression

$$P[e_{t+\Delta t}|\{_0\emptyset_{t+\Delta t}\}]=\frac{P[e_t|\{_0\emptyset_t\}]P[\{_t\emptyset_{t+\Delta t}\}|e_t]}{P[\{_t\emptyset_{t+\Delta t}\}|\{_0\emptyset_t\}]}. \tag{17.67}$$

Finally, the explicit probabilities are

$$P[\{_t\emptyset_{t+\Delta t}\}|e_t]=(1-\gamma\Delta t), \tag{17.68a}$$

$$P[\{_t\emptyset_{t+\Delta t}\}|\{_0\emptyset_t\}]=(1-\gamma\Delta t)p_{2|\{D\}}(t)+p_{1|\{D\}}(t). \tag{17.68b}$$

The first of these is the probability for the excited atom not to emit a photon in time step Δt, and the second composes the full null-measurement record probability out of probabilities for the two conditions that yield a null result. Thus, using (17.68a) and (17.68b) in (17.67), and returning to the simpler notation (Eq. 17.63), we reach the result

$$p_{2|\{D\}}(t+\Delta t)=\frac{p_{2|\{D\}}(t)(1-\gamma\Delta t)}{1-\gamma\Delta t\,p_{2|\{D\}}(t)}. \tag{17.69}$$

The desired equation of motion follows by expanding the right-hand side of (17.69) to first order in Δt. This yields

$$\dot{p}_{2|\{D\}} = -\gamma p_{2|\{D\}}\left(1 - p_{2|\{D\}}\right), \tag{17.70}$$

or, alternatively, since $p_{1|\{D\}}(t) + p_{2|\{D\}}(t) = 1$, the coupled *equations of motion for null-measurement conditional probabilities*

$$\dot{p}_{2|\{D\}} = -\gamma p_{2|\{D\}} p_{1|\{D\}}, \tag{17.71a}$$

$$\dot{p}_{1|\{D\}} = \gamma p_{2|\{D\}} p_{1|\{D\}}. \tag{17.71b}$$

When superposition states are considered, one can also write equations of motion in the fancier Hilbert space language of Sect. 17.2.4. In place of the explicit solution (17.55), we can clearly substitute the equation of motion

$$\frac{d|\bar{\psi}_{\{D\}}\rangle}{dt} = \frac{1}{i\hbar} H_B |\bar{\psi}_{\{D\}}\rangle, \tag{17.72}$$

and the state $|\psi_{\{D\}}\rangle$ conditioned on the null-measurement record is just the normalized version of this. An equation of motion for the normalized state may be derived from (17.72). Recall first that the norm, $\langle\bar{\psi}_{\{D\}}|\bar{\psi}_{\{D\}}\rangle$, is the probability for the null-measurement record (Eq. 17.54a). The equation of motion for this probability is

$$\frac{d\langle\bar{\psi}_{\{D\}}|\bar{\psi}_{\{D\}}\rangle}{dt} = \frac{1}{i\hbar}\langle\bar{\psi}_{\{D\}}|(H_B - H_B^\dagger)|\bar{\psi}_{\{D\}}\rangle. \tag{17.73}$$

Then, for the normalized state, we may write

$$\frac{d|\psi_{\{D\}}\rangle}{dt} = \frac{1}{(\langle\bar{\psi}_{\{D\}}|\bar{\psi}_{\{D\}}|\rangle^{1/2}} \frac{d|\bar{\psi}_{\{D\}}\rangle}{dt} - \frac{1}{2}\frac{d\langle\bar{\psi}_{\{D\}}|\bar{\psi}_{\{D\}}\rangle/dt}{\langle\bar{\psi}_{\{D\}}|\bar{\psi}_{\{D\}}\rangle^{3/2}}|\bar{\psi}_{\{D\}}\rangle. \tag{17.74}$$

Thus, we arrive at the *nonlinear Schrödinger equation for null-measurement conditioning*,

$$\frac{d|\psi_{\{D\}}\rangle}{dt} = \frac{1}{i\hbar}\left[H_B - \tfrac{1}{2}\langle\psi_{\{D\}}|(H_B - H_B^\dagger)|\psi_{\{D\}}\rangle\right]|\psi_{\{D\}}\rangle. \tag{17.75}$$

This equation generalizes (17.71a) and (17.71b), replacing the equations of motion for conditional probabilities by an equation of motion for conditional probability amplitudes.

Exercise 17.1. Show that with the non-Hermitian Hamiltonian given by (17.48) and for a conditional state in the form (17.59a), the Schrödinger equation (17.75) yields the same equations, (17.71a) and (17.71b), for the conditional probabilities.

17.3.2 Conditional States and Nonlinearity

When we compare (17.71a) and (17.71b) with the equations of motion for the unconditional probabilities (Eqs. 17.10 with $\bar{n}=0$), the most noticeable difference is that the former are *nonlinear* while the more familiar equations are linear. The form of the nonlinearity ensures that the conditional probabilities do not change if the energy expectation is definite—i.e., if either $p_{1|\{D\}}(t)$ or $p_{2|\{D\}}(t)$ is zero. Thus, as we might expect, the null-measurement evolution is trivial if there is no uncertainty to be resolved; the function of this evolution is to update our expectation and through the continual updating eventually resolve an uncertainty—was the atom at time t in fact in state $|2\rangle$, from which it simply failed to emit up to time $t+\Delta t$, or was it in state $|1\rangle$ from which it cannot emit? Strictly, we should avoid saying "in fact," though, because if the underlying state is a superposition state, the energy of the atom is not to be seen as a matter of *fact* that is simply unknown to us; the energy uncertainty is an intrinsic and necessary corollary to physical conditions that assign another particular attribute to the atom—the atom is polarized.

The nonlinear equations (17.71a) and (17.71b) seem very foreign and certainly something new. Surprisingly though, they are not new to quantum optics. They are precisely the equations of motion obtained for the radiative damping of a two-state system in the so-called *neoclassical radiation theory*, which was studied by Jaynes and coworkers in the late 1960s and early 1970s [17.38, 17.39]. Bouwmeester and coworkers [17.40] were the first to notice this curious fact. In hindsight, at a mathematical level, the reason for the correspondence is clear. Neoclassical theory starts out by factorizing expectations involving products of radiation field and atomic operators in the Heisenberg equations of motion describing the atom–field interaction. Alternatively, one might say that the radiation field operators are replaced by a classical c-number field; this is an equivalent statement, since the factored Heisenberg equations involve only the mean value of the radiation field. With this background, one might ask: what property must the quantum state of the coupled atom and radiation field have to justify such a factorization? The answer is that it must be a product state, a state for the atom multiplied by a state for the field; the atom and its environment —the surrounding radiation field—must not be entangled.

Now to make the connection with quantum trajectory theory. We saw in Sect. 17.1 that disentanglement is precisely what quantum trajectory theory sets out to achieve; in place of the usual trace over the environment, it aims to disentangle the system from its environment. Thus, the product state required by neoclassical radiation theory is satisfied throughout a quantum trajectory: during the null-measurement evolution, the conditioning continuously verifies that the state of the radiation field is the vacuum state, while once a scattered photon is detected (counted) the vacuum state is replaced by a one-photon state. The scattering record is nothing but a time-dependent label for the disentangled state of the radiation field. Of course, this interpretation envisages photon detection, or counting, in the nondemolition sense.

Note 17.7. The nonlinearity introduced by conditioning can be used to construct a continuous stochastic evolution that takes an initial superposition of states to a state localized on one component of the superposition. To consider the most famous example, in each realization an initial Schrödinger cat state [17.41] localizes onto a state describing *either* a dead cat *or* an alive cat [17.42,17.43]. More generally, one arrives at a dynamical description of the state selection (decoherence) that in the standard treatment of quantum measurements is accounted for by the formal projection hypothesis (Sect. 18.3.2). With regard to these topics, the interpretation of the nonlinearity in quantum trajectory theory differs from the proposal of others to modify the fundamental Schrödinger equation by adding nonlinear and stochastic terms tailored to achieve localization [17.44,17.45,17.46,17.47,17.48,17.49]. The nonlinearity in quantum trajectory theory is a straightforward consequence of conditioning and is not considered to arise from any previously unknown intrinsic stochasticity. It is perhaps a little surprising that the stochastic Schrödinger equations reached from the two points of view can have identical form (see Note 18.9). On the other hand, the convergence is no doubt imposed by the requirement that any modification of the Schrödinger equation should not lead to a gross disagreement with standard quantum mechanics—i.e., mean values should remain unchanged.

17.3.3 Record Probabilities and Norms

The notation of (17.8) is schematic and as a consequence a little loose. It should be clear that the sum over records is a generalized sum—i.e., when written out explicitly, $\sum_{\mathrm{REC}}$ expands as a series of sums and integrals. With regard to the integrals, this leaves us with a choice as to where to place the integration measure. If $P(\mathrm{REC})$ is to be an actual probability, then the measure must be included there, in the record probability. Alternatively, we can display the integration measure explicitly inside the integral, in which case $P(\mathrm{REC})$ is a probability density. We opted for the latter choice when writing out the unravelings of the density operator (17.22), (17.33), and (17.58). With this convention in mind, the normalization of the states $|\bar{\psi}_{\mathrm{REC}}(t)\rangle$ (Eqs. 17.55 and 17.56) is chosen so that

$$P(\mathrm{REC}) = \langle\bar{\psi}_{\mathrm{REC}}(t)|\bar{\psi}_{\mathrm{REC}}(t)\rangle. \tag{17.76}$$

Note that $|\bar{\psi}_{\mathrm{REC}}(t)\rangle$ has units of $t^{-n/2}$, where n is the number of photon detection events in the record. Only for $n = 0$—the special case of the null-measurement record—is $P(\mathrm{REC})$ an actual probability. For simplicity, we will refer to $P(\mathrm{REC})$ as the *record probability density*.

It follows from (17.76) that as an alternative to the unraveling of the density operator (17.8), we might sum over unnormalized states, writing

$$\rho(t) = \sum_{\mathrm{REC}} |\bar{\psi}_{\mathrm{REC}}(t)\rangle\langle\bar{\psi}_{\mathrm{REC}}(t)|, \tag{17.77}$$

where

$$|\bar{\psi}_{\mathrm{REC}}(t)\rangle = \sqrt{P(\mathrm{REC})}|\psi_{\mathrm{REC}}(t)\rangle. \tag{17.78}$$

From a physical point of view, the normalized state $|\psi_{\mathrm{REC}}(t)\rangle$ and the unnormalized state $|\bar{\psi}_{\mathrm{REC}}(t)\rangle$ differ in a rather elementary way. As a conditional state, $|\psi_{\mathrm{REC}}(t)\rangle$ provides us with conditional probabilities; the probability $|\langle 2|\psi_{\mathrm{REC}}(t)\rangle|^2$, for example, is the probability of finding the atom in state $|2\rangle$ at time t, *given* the record REC. In contrast, the unnormalized state $|\bar{\psi}_{\mathrm{REC}}(t)\rangle$ yields joint probability densities; $|\langle 2|\bar{\psi}_{\mathrm{REC}}(t)\rangle|^2$ is the probability density of finding the atom in state $|2\rangle$ at time t *and* the record REC. The distinction, then, is one between "conditional" and "joint."

To reiterate the comment at the beginning of Sect. (17.3.2), this difference carries over into a different character for the time evolutions: a nonlinear evolution for the normalized state, $|\psi_{\mathrm{REC}}(t)\rangle$, and its conditional probability densities (Eq. 17.75), and a linear evolution for the unnormalized state, $|\bar{\psi}_{\mathrm{REC}}(t)\rangle$, and its joint probability densities (Eq. 17.72).

17.3.4 Monte Carlo Simulations

The fundamental content of the quantum trajectory approach is provided by the record probability densities, $P(\mathrm{REC})$, and conditional states, $|\psi_{\mathrm{REC}}(t)\rangle$, or equivalently the unnormalized states, $|\bar{\psi}_{\mathrm{REC}}(t)\rangle$. To be useful, these quantities must be evaluated explicitly, which for all but the simplest examples is not possible analytically. At this point Monte Carlo simulations provide the most useful implementation of quantum trajectory ideas.

Monte Carlo rules for simulating standard quantum jumps were given in (17.12). Quantum trajectories for known initial states (Sect. 17.2.2) are simulated in precisely the same way. For the spontaneous emission example only a single jump ever occurs and the simulation algorithm is trivial. Considering the evolution of $p_{2|\mathrm{REC}}(t)$ (Fig. 17.4) over a sequence of discrete times $t_k = k\Delta t$, we follow the rules

$$\left.\begin{array}{ll}\textbf{if } p_{2|\mathrm{REC}}(t_k) = 1 \textbf{ then:} & \textbf{1. compute } p_\downarrow = \gamma\Delta t \\ & \textbf{2. if } p_\downarrow > r_k, \textbf{ then } p_{2|\mathrm{REC}}(t_{k+1}) = 0 \\ & \quad\ \textbf{else } p_{2|\mathrm{REC}}(t_{k+1}) = 1 \\ \textbf{else } [p_{2|\mathrm{REC}}(t_k) = 0]\textbf{:} & \textbf{1. compute } p_\uparrow = 0 \\ & \textbf{2. } p_{2|\mathrm{REC}}(t_{k+1}) = 0 \end{array}\right\}, \tag{17.79}$$

where r_k is a uniformly distributed random number, $0 \le r_k \le 1$.

Quantum trajectories for "blind" realizations (Sect. 17.2.3) may be simulated in similar fashion, with the change that in this case the initial state, $|1\rangle$ or $|2\rangle$, is not known, so the mutually exclusive sets of rules for $p_{2|\mathrm{REC}(t_k)} = 1$ or 0 are combined:

$$\left.\begin{array}{l} \textbf{1. compute } p_\downarrow = \gamma\Delta t p_{2|\mathrm{REC}}(t_k) \\ \textbf{2. if } p_\downarrow > r_k, \textbf{ then } p_{2|\mathrm{REC}}(t_{k+1}) = 0 \\ \quad \textbf{else } p_{2|\mathrm{REC}}(t_{k+1}) = p_{2|\mathrm{REC}}(t_k)\{1-\gamma\Delta t[1-p_{2|\mathrm{REC}}(t_k)]\} \end{array}\right\}. \tag{17.80}$$

The changed rules still resolve the **if/else** of (17.79), but they do so through the conditioning of the probabilities on the record, which plays out over an extended period of time: in some realizations the uncertainty about the initial state is resolved by a jump, as in Fig. 17.5a, in others through the continuous decay of Fig. 17.5b.

Exercise 17.2. It may not be obvious from the algorithm (17.80) alone that the two kinds of realizations occur in the correct proportion, $p_2(0) : p_1(0)$, for initial state $p_{2|\mathrm{REC}}(0) = p_2(0)$. Prove from the simulation rules alone that this is indeed the case.

Exercise 17.3. Write a computer program to implement both algorithms, (17.79) and (17.80). Show for both that a large ensemble of realizations gives the correct exponential distribution of jump (spontaneous emission) times. Show also that (17.80) realizes the correct proportion of discontinuous to continuous trajectories, hence verifying your answer to Exercise 17.2.

Algorithm (17.80) holds in the absence of coherence. It is readily generalized to account for an initial superposition of states, though. We use the results of Sect. 17.2.4 and carry out the evolution in the interaction picture, first making the transformation

$$|\psi_{\mathrm{REC}}(t)\rangle = e^{-i\frac{1}{2}\omega_A\sigma_z t}|\tilde{\psi}_{\mathrm{REC}}(t)\rangle, \tag{17.81}$$

introducing the transformed non-Hermitian Hamiltonian

$$\tilde{H}_B \equiv -i\hbar\frac{\gamma}{2}\sigma_+\sigma_-. \tag{17.82}$$

The Monte Carlo rules replacing those of (17.80) are:

$$\left.\begin{array}{l} \textbf{1. compute } p_\downarrow = \Delta t\langle\tilde{\psi}_{\mathrm{REC}}(t_k)|J_\gamma^\dagger J_\gamma|\tilde{\psi}_{\mathrm{REC}}(t_k)\rangle \\ \textbf{2. if } p_\downarrow > r_k, \textbf{ then } |\tilde{\psi}_{\mathrm{REC}}(t_{k+1})\rangle = \dfrac{J_\gamma|\tilde{\psi}_{\mathrm{REC}}(t_k)\rangle}{\sqrt{\langle\tilde{\psi}_{\mathrm{REC}}(t_k)|J_\gamma^\dagger J_\gamma|\tilde{\psi}_{\mathrm{REC}}(t_k)\rangle}} \\ \quad \textbf{else} \\ \quad |\tilde{\psi}_{\mathrm{REC}}(t_{k+1})\rangle \\ \quad = \left\{1+\Delta t\dfrac{1}{i\hbar}\Big[\tilde{H}_B - \tfrac{1}{2}\langle\tilde{\psi}_{\mathrm{REC}}(t_k)|(\tilde{H}_B-\tilde{H}_B^\dagger)|\tilde{\psi}_{\mathrm{REC}}(t_k)\rangle\Big]|\tilde{\psi}_{\mathrm{REC}}(t_k)\rangle\right\} \end{array}\right\}. \tag{17.83}$$

As given they make explicit use of the nonlinear Schrödinger evolution under null-measurement conditioning (17.75). It is also possible, and often convenient, to specify the rules in terms of the unnormalized state

$$|\bar{\psi}_{\mathrm{REC}}(t)\rangle = e^{-i\frac{1}{2}\omega_A\sigma_z t}|\tilde{\bar{\psi}}_{\mathrm{REC}}(t)\rangle, \tag{17.84}$$

which obeys the linear evolution equation (17.72). When written in terms of $|\tilde{\bar{\psi}}_{\mathrm{REC}}(t)\rangle$, the alternate Monte Carlo rules for spontaneous emission are:

$$\left.\begin{aligned} &\textbf{1. compute } p_{\downarrow} = \Delta t\frac{\langle\tilde{\bar{\psi}}_{\mathrm{REC}}(t_k)|J_\gamma^\dagger J_\gamma|\tilde{\bar{\psi}}_{\mathrm{REC}}(t_k)\rangle}{\langle\tilde{\bar{\psi}}_{\mathrm{REC}}(t_k)|\tilde{\bar{\psi}}_{\mathrm{REC}}(t_k)\rangle} \\ &\textbf{2. if } p_{\downarrow} > r_k, \textbf{ then } |\tilde{\bar{\psi}}_{\mathrm{REC}}(t_{k+1})\rangle = J_\gamma|\tilde{\bar{\psi}}_{\mathrm{REC}}(t_k)\rangle \\ &\qquad\textbf{else } |\tilde{\bar{\psi}}_{\mathrm{REC}}(t_{k+1})\rangle = \left(1+\Delta t\frac{1}{i\hbar}\tilde{H}_B\right)|\tilde{\bar{\psi}}_{\mathrm{REC}}(t_k)\rangle \end{aligned}\right\}. \tag{17.85}$$

Exercise 17.4. Write a computer program to implement algorithms (17.83) and (17.84), and with it study the decay of initial superposition states under spontaneous emission. Verify that the properties mentioned in Exercise 17.3 still hold. Show that the average over realizations reproduces the correct result for the decay of the coherence between states $|1\rangle$ and $|2\rangle$.

A single quantum trajectory simulated using algorithm (17.85) provides us with three things: (i) a realization of the scattering record, in this case the time of at most one photon emission; (ii) the corresponding conditional state $|\psi_{\mathrm{REC}}(t)\rangle$; and (iii) the probability density, $\langle\bar{\psi}_{\mathrm{REC}}(t)|\bar{\psi}_{\mathrm{REC}}(t)\rangle$, for the realized scattering record.

We might expect that the record probability densities must be known in advance for each record to be realized with the correct frequency. This is not so, however. The record probability densities are recovered "on the fly," as it were, as part of the simulation procedure. More to the point, explicit record probability densities are used by neither algorithm, not by (17.83) nor (17.85); the jump probability is calculated from the normalized conditional state. It follows that in practice any normalization at all is permitted for the state used in the second and third lines of (17.85). It is important to emphasize that the different records are automatically realized at the relative frequencies dictated by the probability density $P(\mathrm{REC})$.

17.3.5 The Waiting-Time Distribution

When discussing photon antibunching in resonance fluorescence (Sect. 2.3.6), we introduced the photoelectron waiting-time distribution, $w_{\mathrm{ss}}(\tau)$, the distribution of time intervals τ between two successive photoelectric detections for an atom in steady state. If the photoelectron detection efficiency is unity, this may be read as the distribution of time intervals between two successive photon emissions. For resonance fluorescence, the waiting-time distribution defines all record probabilities. Consider, for example, the record

$$\text{REC} \equiv \left\{ \genfrac{}{}{0pt}{}{\cdots}{\cdots}, \emptyset, \genfrac{}{}{0pt}{}{\gamma}{T_{k-1}}, \emptyset, \genfrac{}{}{0pt}{}{\gamma}{T_k}, \emptyset, \genfrac{}{}{0pt}{}{\gamma}{T_{k+1}}, \emptyset, \genfrac{}{}{0pt}{}{\gamma}{T_{k+2}}, \emptyset, \genfrac{}{}{0pt}{}{\cdots}{\cdots} \right\}. \tag{17.86}$$

The record probability density is given by the product of waiting-time distributions

$$P(\text{REC}) = \ldots w_{\text{ss}}(t_k - t_{k-1}) w_{\text{ss}}(t_{k+1} - t_k) w_{\text{ss}}(t_{k+2} - t_{k+1}) \ldots ; \tag{17.87}$$

thus, the scattering records are defined by the distribution of null measurement intervals. Carmichael and coworkers [17.18] have analyzed resonance fluorescence in some detail from this point of view.

Clearly, the record probability density might be built up in a similar way for any scattering process. In general, however, we would find that the waiting-time distribution does not depend only on the interval since the last scattering event; the whole history of past events can affect the distribution of times to be waited before the next. One way to say this is that, taken on their own, set apart from the conditional quantum state, the probability densities for scattering records do not generally define a Markov stochastic process. Of course, Markov behavior holds for the *complete* process defined by the scattering records *and* the conditional states. In the case of resonance fluorescence, the records taken alone *are* Markov.

We have encountered two waiting-time distributions in our treatment of spontaneous emission. When the initial state is known, and known to be the upper state $|2\rangle$, the distribution of elapsed times before the photon emission is

$$w_{\text{known}}(t_1) \equiv P(\{C_t\})|_{t=t_1} = \gamma e^{-\gamma t_1}. \tag{17.88}$$

On the other hand, for "blind" realizations the distribution is

$$w_{\text{“blind”}}(t_1) \equiv P(\{E_t\})|_{t=t_1} = \gamma e^{-\gamma t_1} p_2(0). \tag{17.89}$$

In the latter case there is a probability, $p_1(0) = 1 - p_2(0)$, that no photon will be emitted; the waiting-time distribution therefore integrates to $p_2(0)$ rather than to unity.

If we were to directly sample the distributions (17.88) and (17.89), we would obtain realizations of the spontaneous emission time t_1 in a single step. Thus there is no need for the evolution, time-step by time-step, of the previous algorithms. Of course, we are able do this only because the waiting-time distributions are known in closed form. In most situations of interest this will not be the case. It is nonetheless always possible to build a Monte Carlo algorithm based on waiting times by solving for the unknown waiting-time distributions as the simulated trajectory unfolds. Let us assume we do not have access to the explicit expressions (17.88) and (17.89). Note then that

$$w_{\text{known}}(t_1) = \langle \tilde{\bar{\psi}}_{\{C_t\}} | \tilde{\bar{\psi}}_{\{C_t\}} \rangle |_{t=t_1} = \langle \tilde{\bar{\psi}}_{\{A\}} | J_\gamma^\dagger J_\gamma | \tilde{\bar{\psi}}_{\{A\}} \rangle |_{t=t_1}, \tag{17.90}$$

and

$$w_{\text{“blind”}}(t_1) = \langle \tilde{\bar{\psi}}_{\{E_t\}} | \tilde{\bar{\psi}}_{\{E_t\}} \rangle |_{t=t_1} = \langle \tilde{\bar{\psi}}_{\{D\}} | J_\gamma^\dagger J_\gamma | \tilde{\bar{\psi}}_{\{D\}} \rangle |_{t=t_1}, \tag{17.91}$$

and introduce the cumulative distribution

$$C_w(t) = \int_0^t dt_1 w(t_1), \tag{17.92}$$

where either $w_{\text{known}}(t_1)$ or $w_{\text{“blind”}}(t_1)$ is to be substituted for $w(t_1)$. Using the cumulative waiting-time distribution, we may generate realizations of the trajectories from a single random number r_1 and the following rules:

$$\left.\begin{array}{l}\textbf{1. compute } C_w(t_{k+1}) = C_w(t_k) + \Delta t\langle\bar{\tilde{\psi}}_{\text{REC}}(t_k)|J_\gamma^\dagger J_\gamma|\bar{\tilde{\psi}}_{\text{REC}}(t_k)\rangle \\ \textbf{2. if } C_w(t_{k+1}) > r_1, \textbf{ then } |\bar{\tilde{\psi}}_{\text{REC}}(t_{k+1})\rangle = J_\gamma|\bar{\tilde{\psi}}_{\text{REC}}(t_k)\rangle \\ \quad\textbf{else } |\bar{\tilde{\psi}}_{\text{REC}}(t_{k+1})\rangle = \left(1 + \Delta t\dfrac{1}{i\hbar}\tilde{H}_B\right)|\bar{\tilde{\psi}}_{\text{REC}}(t_k)\rangle\end{array}\right\}. \tag{17.93}$$

The rules as written assume that the initial state is pure, so they apply to “blind” realizations in the sense of the initial superposition state (17.35). If the initial state is mixed, the same rules are used, restated in terms of the conditional density operator.

To see how algorithm (17.93) works, consider the random number r_1 to be a function of the time t_1 at which the emission is to take place, or more precisely, consider dr_1 to represent the fraction of realizations in which the photon emission occurs in the interval $[t_1, t_1 + dt_1)$. From the definition of the waiting time, it is then required that

$$dr_1 = w(t_1)dt_1 \Rightarrow r_1(t_1) = C_w(t_1). \tag{17.94}$$

Algorithm (17.93) looks for the crossing of r_1 by $C_w(t_1)$ while solving for the distributions $w(t_1)$ and $C_w(t_1)$. Figure 17.6 illustrates the procedure for choices of r_1 that produce the realizations presented in Figs. 17.4 and 17.5.

Exercise 17.5. Write a computer program to implement (17.93), and use it to repeat the calculations of Exercise 17.4.

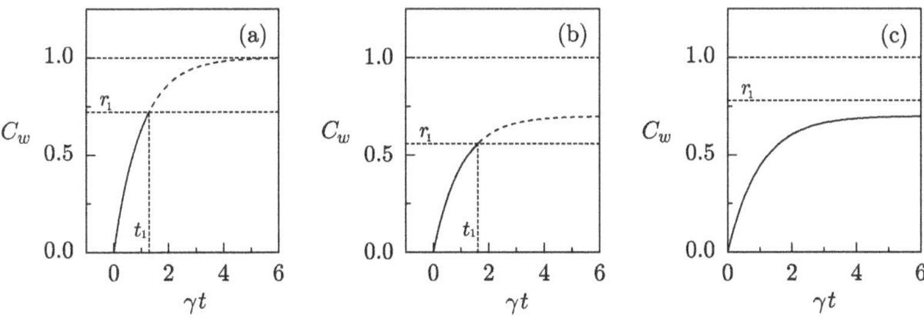

Fig. 17.6. Implementation of the simulation algorithm (17.93). Crossing (or non-crossing) of the cumulative waiting-time distribution, $C_w(t_1)$, with random number r_1 for: (**a**) a realization with known initial state $|2\rangle$ (Fig. 17.4), (**b**) a “blind” realization with $p_2(0) = 0.7$ [Fig. 17.5a], (**c**) a “blind” realization showing no quantum jump [Fig. 17.5b]

18

Quantum Trajectories II: The Degenerate Parametric Oscillator

The last chapter has provided us with a general background to quantum trajectory ideas. The central notion is to set up a conditional evolution for an open system (the interaction region of Fig. 17.1) with the state of the system conditioned on the record of scattered photons detected in the reservoir. The scattering record may be made in practice by monitoring the incoherent inputs and outputs to the system, and takes the form of a photoelectron counting sequence. This counting sequence characterizes the inputs and outputs as a classical stochastic process (Eq. 17.6 for example).

Up to this point our development of these ideas has been motivated more by physics than mathematics and is perhaps not as systematic as one might wish. The goal, formally, is to expand, or unravel, the reduced density of the system in the form (17.8), where conditional states $|\psi_{\mathrm{REC}}\rangle$ are labeled by a scattering record with probability (density) $P(\mathrm{REC})$. We demonstrated such an unraveling by construction in Sect. 17.2.4, but only for the simplest example of spontaneous emission. In this section we generalize the demonstration, starting with an expression for scattering record probability densities taken from photoelectron counting theory.

18.1 Scattering Records and Photoelectron Counting

It is again helpful to keep a specific example in view. We will work for definiteness with the degenerate parametric oscillator model of Chap. 9. It contains sufficient generality to illustrate the principal ideas. To simplify just a little, however, let us omit the thermal photons and crystal loss. Then, master equation (9.97) is taken over in the form ($\bar{n} = \bar{n}_p = 0$, $\gamma_{a\alpha} = \gamma_{b\alpha} = 0$)

$$\dot{\rho} = \mathcal{L}\rho, \tag{18.1}$$

with

$$\begin{aligned}\mathcal{L} \equiv & -i\omega_C[a^\dagger a, \cdot] - i2\omega_C[b^\dagger b, \cdot] \\ & + (g/2)[a^{\dagger 2}b - a^2 b^\dagger, \cdot] - i[\bar{\mathcal{E}}_0 e^{-i2\omega_C t} b^\dagger + \bar{\mathcal{E}}_0^* e^{i2\omega_C t} b, \cdot] \\ & + \tfrac{1}{2}(\gamma_{a1} + \gamma_{a2})(2a \cdot a^\dagger - a^\dagger a \cdot - \cdot a^\dagger a) \\ & + \tfrac{1}{2}(\gamma_{b1} + \gamma_{b2})(2b \cdot b^\dagger - b^\dagger b \cdot - \cdot b^\dagger b),\end{aligned} \tag{18.2}$$

or in the interaction picture,

$$\dot{\tilde{\rho}} = \tilde{\mathcal{L}}\tilde{\rho}, \tag{18.3}$$

with

$$\begin{aligned}\tilde{\mathcal{L}} = & (g/2)[a^{\dagger 2}b - a^2 b^\dagger, \cdot] - i[\bar{\mathcal{E}}_0 b^\dagger + \bar{\mathcal{E}}_0^* b, \cdot] \\ & + \tfrac{1}{2}(\gamma_{a1} + \gamma_{a2})(2a \cdot a^\dagger - a^\dagger a \cdot - \cdot a^\dagger a) \\ & + \tfrac{1}{2}(\gamma_{b1} + \gamma_{b2})(2b \cdot b^\dagger - b^\dagger b \cdot - \cdot b^\dagger b),\end{aligned} \tag{18.4}$$

where the parameters g and $\bar{\mathcal{E}}_0$ are defined by (9.80) and (9.81), and there are loss rates $\gamma_{a\mu}$ and $\gamma_{b\mu}$, $\mu = 1, 2$, arising from the coupling of the intracavity fields to vacuum-state reservoirs through the mirrors at either end of the cavity; the mirrors are labeled mirror 1 and mirror 2 as in Figs. 9.1 and 9.4. The output fields are specified in Sect. 9.2.5.

The scattering scenario corresponding to master equation (18.3) and (18.4) is depicted in Fig. 18.1. There is a single coherent input—the field driving the pump mode (the source of energy to the system)—and four output channels, each labeled by a cavity loss rate. Scattering records are made by four ideal (unit efficiency) photodetectors, as shown in the figure. A typical record is written as

$$\text{REC} \equiv \left\{ {}_0\emptyset, \frac{\gamma_1}{T_1}, \emptyset, \frac{\gamma_2}{T_2}, \ldots, \frac{\gamma_n}{T_n}, \emptyset_t \right\}, \tag{18.5}$$

with

$$\gamma_k \in \{\gamma_{a1}, \gamma_{a2}, \gamma_{b1}, \gamma_{b2}\}. \tag{18.6}$$

Note that in order to identify photodetections labeled by γ_{b1}, a coherent displacement is introduced, whose magnitude is adjusted to cancel the back-reflected external field. This strategy yields the simplest unraveling of the density operator and its master equation evolution. With the displacement the detector monitoring channel γ_{b1} sees only the field *transmitted* at mirror 1; the discarded coherent amplitude is omitted from the record-keeping (see the discussion above Note 18.4).

18.1.1 Record Probabilities

When discussing photoelectron counting records we divide the time line into intervals of equal length, T_k (Eq. 17.7), and seek the probability for a specific

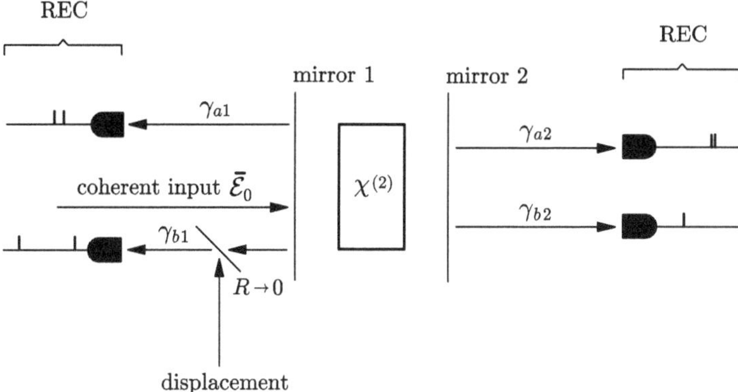

Fig. 18.1. Scattering scenario for the degenerate parametric oscillator model of Figs. 9.1 and 9.4 with scattering records made by direct photoelectron counting

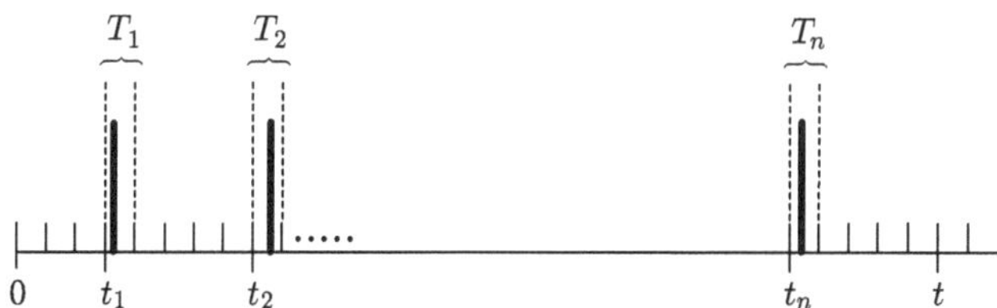

Fig. 18.2. Subdivision of the time line used in the construction of photoelectron counting probabilities. Each interval T_k, $k = 1, \dots, n$, contains either no count or at the most one count

sequence of counts within a subset of these intervals—for example at the times $t_1, t_2, \dots, t_n$, as shown in Fig. 18.2. The interval length is taken to be sufficiently short that there is negligible probability for two or more counts to occur in any one interval. Thus, the probability for one count to occur within any interval is itself very small, and in an overwhelming majority of the intervals there is no count at all.

Let us start out by considering the intracavity fields to be classical, with complex amplitudes $\alpha(t)$ and $\beta(t)$ for the subharmonic and pump modes, respectively. Then the output channel photon fluxes are calculated as

$$F_{\leftarrow}(t) = \gamma_{a1}|\alpha(t-\tau_R)|^2, \qquad F_{\rightarrow}(t) = \gamma_{a2}|\alpha(t-\tau_R)|^2, \tag{18.7a}$$

$$F_{p\leftarrow}(t) = \gamma_{b1}|\beta(t-\tau_R)|^2, \qquad F_{p\rightarrow}(t) = \gamma_{b2}|\beta(t-\tau_R)|^2, \tag{18.7b}$$

where τ_R is the retardation time from the mirrors to the detectors. In the limit of infinitesimal interval length (Eq. 17.7), the record probability is

$$\begin{aligned} P(\mathrm{REC})dt_1dt_2\cdots dt_n = W(t-t_n)[F_n(t_n)dt_n]W(t_n-t_{n-1}) \\ \cdots [F_2(t_2)dt_2]W(t_2-t_1)[F_1(t_1)dt_1]W(t_1), \end{aligned} \tag{18.8}$$

where $W(t_k - t_{k-1})$ is the no-count probability for the continuous sequence of intervals between count times t_{k-1} and t_k (see Exercise 13.6), and

$$F_k(t) \in \{F_{\leftarrow}(t), F_{\rightarrow}(t), F_{p\leftarrow}, F_{p\rightarrow}(t)\}. \tag{18.9}$$

Thus, the record probability is a product of probabilities for no count from $t = 0$ to t_1, a count in interval T_1, no counts from $t = t_1 + dt_1$ to t_2, a count in interval T_2, and so on. Alternatively, $P(\text{REC})$ may be subdivided as a product of waiting-time distributions, $w(t_k - t_{k-1}) = F_k(t_k)W(t_k - t_{k-1})$. The no-count or null-measurement probability, $W(t_k - t_{k-1})$, is determined from the total output flux

$$F(t) = F_{\leftarrow}(t) + F_{\rightarrow}(t) + F_{p\leftarrow}(t) + F_{p\rightarrow}(t). \tag{18.10}$$

It is constructed as a product of elementary no-count probabilities, one for each interval of length Δt containing no count. Following the derivation of the waiting-time distribution in Note 2.9, we find

$$\begin{aligned} W(t_k - t_{k-1}) &= \lim_{\substack{N\to\infty,\Delta t\to 0\\ N\Delta t=(t_k-t_{k-1})}} \prod_{m=0}^{N-1} \left[1 - \Delta t F(t_{k-1} + m\Delta t)\right] \\ &= \exp\left[-\int_{t_{k-1}}^{t_k} dt' F(t')\right]. \end{aligned} \tag{18.11}$$

Of course, in a more general case, the field amplitudes $\alpha(t)$ and $\beta(t)$ might be stochastic. This requires an average over stochastic realizations to be added. Record probabilities for classical intracavity fields are then given by

$$\begin{aligned} P(\text{REC})dt_1dt_2\cdots dt_n = &\overline{W(t-t_n)[F_n(t_n)dt_n]W(t_n - t_{n-1})} \\ &\overline{\cdots[F_2(t_2)dt_2]W(t_2 - t_1)[F_1(t_1)dt_1]W(t_1)}. \end{aligned} \tag{18.12}$$

Consider now the extension to quantized fields. Expression (18.12) may hold in the quantum domain, but only if is possible to represent the system state (at all times) by a Glauber–Sudarshan P distribution that is positive-definite. Even then, the classical expression is the phase-space form of an underlying expression involving field *operators* and a *quantum* (rather than stochastic) average. The extension to the quantum form is straightforward since the structure of the expression remains the same; thus, record probabilities for quantized fields are given by

$$\begin{aligned} P(\text{REC})dt_1dt_2\cdots dt_n = \Big\langle &: e^{-\hat{\Omega}(t-t_n)}[\hat{F}_n(t_n)dt_n]e^{-\hat{\Omega}(t_n-t_{n-1})} \\ &\cdots[\hat{F}_2(t_2)dt_2]e^{-\hat{\Omega}(t_2-t_1)}[\hat{F}_1(t_1)dt_1]e^{-\hat{\Omega}(t_1)} :\Big\rangle, \end{aligned} \tag{18.13}$$

with

$$\hat{\Omega}(t_k - t_{k-1}) \equiv \int_{t_{k-1}}^{t_k} dt' \hat{F}(t'), \tag{18.14}$$

where, reexpressed in operator form, (18.9) and (18.10) carry over as

$$\hat{F}_k(t) \in \{\hat{F}_{\leftarrow}(t), \hat{F}_{\rightarrow}(t), \hat{F}_{p\leftarrow}(t), \hat{F}_{p\rightarrow}(t)\}, \tag{18.15}$$

and

$$\hat{F}(t) = \hat{F}_{\leftarrow}(t) + \hat{F}_{\rightarrow}(t) + \hat{F}_{p\leftarrow}(t) + \hat{F}_{p\rightarrow}(t), \tag{18.16}$$

and in place of (18.7a) and (18.7b), we now have photon flux operators

$$\begin{aligned} \hat{F}_{\leftarrow}(t) &= \hat{\mathcal{E}}_{\leftarrow}^{\dagger}(t)\hat{\mathcal{E}}_{\leftarrow}(t), \qquad & \hat{F}_{\rightarrow}(t) &= \hat{\mathcal{E}}_{\rightarrow}^{\dagger}(t)\hat{\mathcal{E}}_{\rightarrow}(t), \\ \hat{F}_{p\leftarrow}(t) &= \hat{\mathcal{E}}_{p\leftarrow}^{\dagger}(t)\hat{\mathcal{E}}_{p\leftarrow}(t), \qquad & \hat{F}_{p\rightarrow}(t) &= \hat{\mathcal{E}}_{p\rightarrow}^{\dagger}(t)\hat{\mathcal{E}}_{p\rightarrow}(t), \end{aligned} \tag{18.17a}$$

with cavity output fields defined by (9.120a), (9.120b), (9.122a), and (9.122b). Note that free fields enter as well as source fields into these expressions. Also, like all other outputs, the free-field part of $\hat{\mathcal{E}}_{p\leftarrow}$ is a vacuum field, the backscattered driving field amplitude having been removed—the displacement in Fig. 18.1.

Passage from record probability (18.12) to record probability (18.13) has been made by introducing two rather obvious changes: (i) field operators replace classical field amplitudes and (ii) an average over a classical random process has been replaced by a quantum-mechanical average. We should note that along with these changes comes the requirement for normal-ordering and time-ordering of the operators; this is indicated in (18.13) by the two dots. Actually, time-ordering is not strictly required at this stage, because the total output field operators (free-field part plus source-field part) commute at different times. Time-ordering is going to become important for us, though, as we plan to manipulate (18.13) in a way that works only for time-ordered operators. After the manipulation time-ordering will be locked in. This, along with a few other formal steps, will cast the record probability into a form that allows us to see the connection with an unraveling of master equation (18.3) and (18.4).

Note 18.1. We have not derived the photoelectron counting probability (18.13), only motivated its form. A few words about the derivation are in order. The standard reference is to the work of Kelley and Kleiner [18.1]. The probability on the left-hand side of (18.13) is an example of what is called an *exclusive counting probability*; it corresponds to the quantity denoted $\mathcal{P}'_K(i_1, i_2, \cdots, i_K)$ in that work (taken here with $K = n$). It is to be contrasted with the *nonexclusive counting probability*, denoted there by $\mathcal{P}_K(i_1, i_2, \cdots, i_K)$. The nonexclusive probability is the probability that photoelectron counts occur in subintervals $T_1, T_2, \ldots, T_n$, without regard to the occurrence or not of counts at other times; it is precisely the multicoincidence probability that forms the basis of the coherence theory of Glauber [18.2]. The two kinds of probability can be expressed in terms of one another, and relationships between them are given in [18.1], as well as in books on photoelectron counting statistics like the one by Saleh [18.3]. Kelley and Kleiner are principally interested in the photoelectron count probability, $P_n(t, t+T)$—the probability that some fixed number, n, of photoionizations occurs during the interval from t to $t+T$. For this quantity they ultimately find the formula (9.146). They comment that "the natural definition of $P_n(t, t+T)$ is in terms of the exclusive probability function (2.7) rather than in terms of the nonexclusive probability function (2.4)." Nevertheless, as nonexclusive probabilities—the multicoincidence probabilities of Glauber's theory—are the easiest to calculate, Kelley and Kleiner aimed to express $P_n(t, t+T)$ in terms of them. The core of their work is a derivation from perturbation theory of a quantum-mechanical expression for the nonexclusive counting probabilities; thus, (18.13) does not appear explicitly in their paper, although it is implied, through the stated relationship between exclusive and nonexclusive probabilities.

Note 18.2. A certain amount of confusion has arisen around one aspect of Kelley and Kleiner's work. Its source is the assumption referenced in their comment [18.1]: "With regard to assumption (2), our treatment does not explicitly take account of modification of the field distribution engendered by the presence of the detector, particularly in the region of the detector where, for example, one may expect attenuation by the detector." It was noted at that time [18.4], and has been reemphasized more recently [18.5,18.6], that as a consequence of the referenced assumption, formula (9.146) can yield unphysical results if applied naively to situations where attenuation does in fact occur. A single-mode field in Fock state $|N\rangle$ illustrates the point clearly. Picture exactly N photons trapped inside an optical cavity with perfectly reflecting walls. If a photoelectric detector, a device that absorbs photons, is introduced into the cavity, clearly the number of undetected photons decays as the number of photoelectron counts grows. Formula (9.146) does not account for this, and, in fact, after a time produces negative probabilities and a mean number of counts that exceeds the initial photon number. Recent comments on this feature by Srinivas and Davies [18.5,18.6] and Srinivas [18.7] draw the Kelley–Kleiner approach into question; but they can too easily mislead. It is too easy to come away from the Srinvias and Davies papers with the impression that the Kelley–Kleiner formula (also called the Mandel formula) is invalid at a fundamental level because it leaves out the backaction of a quantum measuring device (the photoelectric detector) on the system measured (the detected field). We know, however, that the photoelectron counting theory developed by Glauber, Kelley–Kleiner, and Mandel accounts successfully for numerous experiments in quantum optics, including photon antibunching in resonance fluorescence. The latter occurs precisely because the first detected photon is no longer present after detection, to be detected a second time (Sect. 2.3.6). How can a treatment that omits field attenuation by the detector get such an effect right? Part of the answer is that Kelley and Kleiner do actually account for field attenuation, but only to lowest order in perturbation theory; in the above example, formula (9.146) gives zero probability to count more than N photoelectrons. But the more important part of the answer was given by Mandel [18.8] in his reply to Srinivas and Davies [18.5]: formula (9.146) [by implication (18.13)] *assumes an open system situation*, one in which the field from a photoemissive source propagates towards the detector, and if undetected, away to infinity never to return. If one inserts correlation functions derived for the output field of a photoemissive source into (9.146) [or (18.13)], the abovementioned problems will not occur. This assertion is demonstrated explicitly by the development below, where the semi-group treatment of photoelectron counting advocated by Srinivas and Davies emerges from (18.13) [implicitly from the treatment of Glauber, Kelley–Kleiner, and Mandel]. In order to see why everything should fit together so nicely, one need only think about the physics. Are the fields impinging on the detectors in Fig. 18.1 altered if the detectors are removed? Surely not. These are outgoing fields, destined to be absorbed in the environment, if not by a detector, then elsewhere. Indeed, a photoelectric detector is nothing but a part of the absorbing environment that remembers what it absorbs and when. So we might return to the single mode prepared in state $|N\rangle$ and ask how to treat it correctly using the Kelley–Kleiner formula. Clearly one must recognize the presence of an absorber in the cavity, write down a master equation to account for the absorption, solve for correlation functions using it, and insert these correlation functions into (9.146). This self-consistent program yields not only the answer, but also the mathematical form of the Srinivas and Davies treatment.

Returning now to the development of the record probability density from (18.13), we aim to cast this expression into a form that makes contact with master equation (18.3) and (18.4). Our first task is to deal with the operator ordering. To this end, we expand the exponentials. We write the expansion out explicitly up to time t_1 only, leaving the rest implicit. This yields

$$
\begin{aligned}
P(\text{REC}) = {} & \not{\Sigma}\sum_{m=0}^{\infty}(-1)^m\int_0^{t_1}d\xi_m\cdots\int_0^{\xi_3}d\xi_2\int_0^{\xi_2}d\xi_1\sum_{\mu_m=1}^{4}\cdots\sum_{\mu_1=1}^{4} \\
& \operatorname{tr}\Big[\hat{\mathcal{E}}^\dagger_{\mu_1}(\xi_1)\hat{\mathcal{E}}^\dagger_{\mu_2}(\xi_2)\cdots\hat{\mathcal{E}}^\dagger_{\mu_m}(\xi_m)\hat{\mathcal{E}}^\dagger_1(t_1)\prod\nolimits_{\hat{\mathcal{E}}^\dagger} \\
& \times\prod\nolimits_{\hat{\mathcal{E}}}\hat{\mathcal{E}}_1(t_1)\hat{\mathcal{E}}_{\mu_m}(\xi_m)\cdots\hat{\mathcal{E}}_{\mu_2}(\xi_2)\hat{\mathcal{E}}_{\mu_1}(\xi_1)\chi(0)\Big],
\end{aligned}
\tag{18.18}
$$

with

$$
\hat{\mathcal{E}}_k(t)\in\{\hat{\mathcal{E}}_{\leftarrow}(t),\hat{\mathcal{E}}_{\rightarrow}(t),\hat{\mathcal{E}}_{p\leftarrow}(t),\hat{\mathcal{E}}_{p\rightarrow}(t)\}.
\tag{18.19}
$$

The symbols $\prod_{\hat{\mathcal{E}}^\dagger}$ and $\prod_{\hat{\mathcal{E}}}$ stand in for all operators $\hat{\mathcal{E}}^\dagger_k(t_k)$ and $\hat{\mathcal{E}}_k(t_k)$ not shown explicitly—those whose time arguments are larger than t_1—and $\not{\Sigma}$ represents the sums and integrals over these operators and their arguments, similar to those shown. From the cyclic property of the trace, we then have

$$P(\text{REC}) = \not\sum \sum_{m=0}^{\infty} (-1)^m \int_0^{t_1} d\xi_m \cdots \int_0^{\xi_3} d\xi_2 \int_0^{\xi_2} d\xi_1 \sum_{\mu_m=1}^{4} \cdots \sum_{\mu_1=1}^{4}$$
$$\text{tr}\Big[\prod_{\hat{\mathcal{E}}} \hat{\mathcal{E}}_1(t_1)\hat{\mathcal{E}}_{\mu_m}(\xi_m)\cdots\hat{\mathcal{E}}_{\mu_2}(\xi_2)\hat{\mathcal{E}}_{\mu_1}(\xi_1)\chi(0)$$
$$\times\hat{\mathcal{E}}^\dagger_{\mu_1}(\xi_1)\hat{\mathcal{E}}^\dagger_{\mu_2}(\xi_2)\cdots\hat{\mathcal{E}}^\dagger_{\mu_m}(\xi_m)\hat{\mathcal{E}}^\dagger_1(t_1)\prod_{\hat{\mathcal{E}}^\dagger}\Big]. \tag{18.20}$$

From this form it is clear that if the initial reservoir state is the vacuum, the free fields may be dropped in the definitions (9.120a), (9.120b), (9.122a), and (9.122b); thus, we may make the replacement

$$\hat{\mathcal{E}}_k(t) \rightarrow \hat{S}_k(t-\tau_R), \tag{18.21}$$

where in place of (18.19) we now have

$$\hat{S}_k(t) \in \{\sqrt{\gamma_{a1}}a(t), \sqrt{\gamma_{a2}}a(t), \sqrt{\gamma_{b1}}b(t), \sqrt{\gamma_{b2}}b(t)\}. \tag{18.22}$$

The replacement is justified because, although the free fields and source fields do not commute at all times, they do commute when the source field is evaluated at an *earlier* time then the free field; at the level of commutator expectations, this is shown explicitly by (9.124a)–(9.125b). With this commutation rule, and explicit normal- and time-ordering of (18.20), it is possible to move all free-field operators $\sqrt{c/2L'}r_{kf}(t_k)$ to the immediate left of $\chi(0)$, and all conjugate operators $\sqrt{c/2L'}r^\dagger_{kf}(t_k)$ to the right. The free-field operators then act on $\chi(0)$ to produce zero when the reservoir is in the vacuum state.

We now make the substitution (18.21), and at the same time introduce a more compact superoperator notation, with

$$\mathcal{S}_{\leftarrow} \equiv \gamma_{a1}a\cdot a^\dagger, \qquad \mathcal{S}_{\rightarrow} \equiv \gamma_{a2}a\cdot a^\dagger, \tag{18.23a}$$
$$\mathcal{S}_{p\leftarrow} \equiv \gamma_{b1}b\cdot b^\dagger, \qquad \mathcal{S}_{p\rightarrow} \equiv \gamma_{b2}b\cdot b^\dagger, \tag{18.23b}$$

and $L \equiv (1/i\hbar)[H,\cdot]$, where H is the Hamiltonian of the system in interaction with the reservoir (Sect. 9.2.4). Record probability density (18.20) is then written as

$$P(\text{REC})$$
$$= \not\sum \sum_{m=0}^{\infty} (-1)^m \int_0^{t_1} d\xi_m \cdots \int_0^{\xi_3} d\xi_2 \int_0^{\xi_2} d\xi_1 \sum_{\mu_m=1}^{4} \cdots \sum_{\mu_1=1}^{4}$$
$$\text{tr}\Big\{\prod_{\mathcal{S}}\Big[\mathcal{S}_1 e^{L(t_1-\xi_m)}\mathcal{S}_{\mu_m}e^{L(\xi_m-\xi_{m-1})}\cdots\mathcal{S}_{\mu_2}e^{L(\xi_2-\xi_1)}\mathcal{S}_{\mu_1}e^{L\xi_1}\chi(-\tau_R)\Big]\Big\}, \tag{18.24}$$

with

$$\mathcal{S}_k \in \{\mathcal{S}_{\leftarrow}, \mathcal{S}_{\rightarrow}, \mathcal{S}_{p\leftarrow}, \mathcal{S}_{p\rightarrow}\}, \tag{18.25}$$

where $\prod_{\mathcal{S}}$ denotes the product of source superoperators and propagators obtained from $\prod_{\hat{\mathcal{E}}} \cdot \prod_{\hat{\mathcal{E}}^\dagger}$ through the substitution (18.21).

Two further steps take us to our final result. First, in the Born–Markov approximation we may replace Liouvillian L by superoperator $\mathcal{L}$, or, after transforming to the interaction picture, by superoperator $\tilde{\mathcal{L}}$; at the same time $\chi(-\tau_R) \to \tilde{\rho}(-\tau_R)$ and $\mathrm{tr}\{\} \to \mathrm{tr}_S\{\}$. Second, the expanded exponentials are resummed. Clearly, the sums over $\mu_1, \ldots, \mu_m$ simply result in each $\mathcal{S}_k$ being replaced by the summed source superoperator

$$\mathcal{S} \equiv \mathcal{S}_{\leftarrow} + \mathcal{S}_{\rightarrow} + \mathcal{S}_{p\leftarrow} + \mathcal{S}_{p\rightarrow}. \tag{18.26}$$

The sum over m is then seen to be a Dyson expansion, which when resummed gives

$$P(\text{REC}) = \not{\sum} \mathrm{tr}_S\Big\{\prod\nolimits_{\mathcal{S}} \Big[\mathcal{S}_1 e^{(\tilde{\mathcal{L}}-\mathcal{S})(t_1'+\tau_R)} \tilde{\rho}(-\tau_R)\Big]\Big\}, \tag{18.27}$$

where we introduce the retarded time $t_1' = t_1 - \tau_R$. There are n similar expansions hidden in the generalized sum over $\prod_{\mathcal{S}}$ which are treated in the same way. Thus, after resumming all of these terms, we arrive at our final expression for the *probability density for scattering (photoelectron counting) records in superoperator and density matrix form*,

$$P(\text{REC}) = \mathrm{tr}_S\big[\tilde{\mathcal{K}}_{\text{REC}}(t')\tilde{\rho}(-\tau_R)\big], \tag{18.28}$$

with

$$\tilde{\mathcal{K}}_{\text{REC}}(t') \equiv e^{(\tilde{\mathcal{L}}-\mathcal{S})(t'-t_n')} \mathcal{S}_n e^{(\tilde{\mathcal{L}}-\mathcal{S})(t_n'-t_{n-1}')} \cdots \mathcal{S}_2 e^{(\tilde{\mathcal{L}}-\mathcal{S})(t_2'-t_1')} \mathcal{S}_1 e^{(\tilde{\mathcal{L}}-\mathcal{S})(t_1'+\tau_R)}, \tag{18.29}$$

where $t_k' \equiv t_k - \tau_R$.

Note 18.3. Time ordering is not strictly necessary in the starting expression (18.13) because output field operators commute at different times. After substitution (18.21) is made, however, the time ordering is locked in; this substitution is only possible after the field operators have been written out explicitly in time order. The end result, record probability density (18.28) and (18.29), is given in terms of source-field operators only, and clearly the quantum trajectory approach envisages the detection of output fields that are coupled to their sources. On the other hand, nothing in principle prevents us from moving the detectors (Fig. 18.1) so far away from the cavity that the retardation time τ_R is larger than the time for which the source is turned on. Then, when the outputs are eventually detected, they will be freely propagating fields. Another comment of Kelley and Kleiner's is pertinent here [18.1]: "We now discuss an assumption made implicitly by other authors and which is of far reaching consequence. The time evolution of the field operators under the full Hamiltonian of field-plus-sources is replaced by the evolution under the field Hamiltonian alone. ... The correlation functions used by Glauber,[11–13] Sudarshan,[14,15] and others are of this type. A question which remains unanswered is under what circumstances and how this simplification can be justified." In response, we can say that the detected fields may indeed be considered freely propagating ("evolution under the field Hamiltonian alone")—at least this is so within the Born–Markov framework—so long as an appropriate description is used for the state of the field: the correlation functions inserted into (18.13) should describe a field that, sometime in the past (possibly the distant past), was produced by a photoemissive source. This is consistent with the open system aspect of Kelley–Kleiner theory commented upon by Mandel [18.8].

18.1.2 Factorization for Pure Initial States

Superoperators are convenient for deriving a result like (18.28), but now that we have this expression, it is helpful to rewrite it in its factorized, pure-state form. The factorization is made possible by the structure of the superoperator $\tilde{\mathcal{L}} - \mathcal{S}$, which defines the propagator entering (18.28); this structure follows from the Lindblad form of the master equation (Eq. 18.4) and its relationship to the source superoperators (Eqs. 18.23a, 18.23b, and 18.26). The central point is that $\tilde{\mathcal{L}} - \mathcal{S}$ decomposes as a sum of commutators and anticommutators. Specifically, in the case of the degenerate parametric oscillator example,

$$\tilde{\mathcal{L}} - \mathcal{S} = (g/2)[a^{\dagger 2}b - a^2 b^\dagger, \cdot\,] - i[\bar{\mathcal{E}}_0 b^\dagger + \bar{\mathcal{E}}_0^* b, \cdot\,] \\ - \tfrac{1}{2}(\gamma_{a1} + \gamma_{a2})[a^\dagger a, \cdot\,]_+ - \tfrac{1}{2}(\gamma_{b1} + \gamma_{b2})[b^\dagger b, \cdot\,]_+. \tag{18.30}$$

where $[\,\cdot\,,\cdot\,]_+$ denotes the anticommutator. It is thus possible to write

$$\tilde{\mathcal{L}} - \mathcal{S} = \frac{1}{i\hbar}(\tilde{H}_B \,\cdot - \cdot\, \tilde{H}_B^\dagger), \tag{18.31}$$

with *non-Hermitian Hamiltonian*

$$\tilde{H}_B \equiv -i\hbar\tfrac{1}{2}(\gamma_{a1} + \gamma_{a2})a^\dagger a - i\hbar\tfrac{1}{2}(\gamma_{b1} + \gamma_{b2})b^\dagger b \\ + i\hbar(g/2)(a^{\dagger 2}b - a^2 b^\dagger) + \hbar(\bar{\mathcal{E}}_0 b^\dagger + \bar{\mathcal{E}}_0^* b). \tag{18.32}$$

Evolution under the propagator $e^{(\tilde{\mathcal{L}}-\mathcal{S})t}$ in-between the photoelectric counts is then governed by the *nonunitary Schrödinger equation* with Hamiltonian $\tilde{H}_B$ [generalizing (17.50)]. We write

$$e^{(\tilde{\mathcal{L}}-\mathcal{S})t} = \tilde{B}(t) \cdot \tilde{B}^\dagger(t), \tag{18.33}$$

with

$$\tilde{B}(t) \equiv \exp[-(i/\hbar)\tilde{H}_B t]. \tag{18.34}$$

To complete the rewriting, the source superoperators, (18.23a) and (18.23b), are replaced by their corresponding *jump operators* [the generalization of (17.52)],

$$J_{\leftarrow} \equiv \sqrt{\gamma_{a1}}a, \qquad J_{\rightarrow} \equiv \sqrt{\gamma_{a2}}a, \tag{18.35a}$$

$$J_{p\leftarrow} \equiv \sqrt{\gamma_{b1}}b, \qquad J_{p\rightarrow} \equiv \sqrt{\gamma_{b2}}b. \tag{18.35b}$$

Then with

$$\tilde{K}_{\mathrm{REC}}(t') \equiv \tilde{B}(t'-t'_n)J_n\tilde{B}(t'_n-t'_{n-1})\cdots J_2\tilde{B}(t'_2-t'_1)J_1\tilde{B}(t'_1+\tau_R), \tag{18.36}$$

where $J_k \in \{J_{\leftarrow}, J_{\rightarrow}, J_{p\leftarrow}, J_{p\rightarrow}\}$, superoperator (18.29), which defines the record probability density (18.28), takes the simpler factorized form

$$\tilde{\mathcal{K}}_{\mathrm{REC}}(t') = \tilde{K}_{\mathrm{REC}}(t') \cdot \tilde{K}^{\dagger}_{\mathrm{REC}}(t'). \tag{18.37}$$

Thus, for any pure initial state,

$$\tilde{\rho}(-\tau_R) = |\tilde{\psi}_{\mathrm{REC}}(-\tau_R)\rangle\langle\tilde{\psi}_{\mathrm{REC}}(-\tau_R)|, \tag{18.38}$$

the record probability density (18.28) and (18.29) may be rewritten as a conditional state norm, the generalization of (17.76); we arrive at the *probability density for scattering (photoelectron counting) records in operator and pure-state form*:

$$P(\mathrm{REC}) = \langle\bar{\tilde{\psi}}_{\mathrm{REC}}(t')|\bar{\tilde{\psi}}_{\mathrm{REC}}(t')\rangle, \tag{18.39}$$

with *unnormalized conditional state*

$$|\bar{\tilde{\psi}}_{\mathrm{REC}}(t')\rangle = \tilde{K}_{\mathrm{REC}}(t')|\tilde{\psi}(-\tau_R)\rangle. \tag{18.40}$$

18.2 Unraveling the Density Operator

18.2.1 Photoelectron Counting Records

It remains for us to demonstrate that the derived record probability density and associated conditional state do, indeed, provide an unraveling of the density operator in the form (17.8) [equivalently (17.77)]. To show that this is so we generalize the Dyson expansion approach applied to the example of spontaneous emission in Sect. 17.2.4.

On the right-hand side of master equation (18.3) we write $\tilde{\mathcal{L}} = (\tilde{\mathcal{L}}-\mathcal{S})+\mathcal{S}$, viewing $\tilde{\mathcal{L}}-\mathcal{S}$ as the generator of "free" evolution and $\mathcal{S}$ as an interaction. For the degenerate parametric oscillator, $\mathcal{S}$ is the summed source superoperator (18.26) and $\tilde{\mathcal{L}} - \mathcal{S}$ is given by (18.30). The Dyson expansion is

$$\begin{aligned}
&\tilde{\rho}(t') \\
&= e^{[(\tilde{\mathcal{L}}-\mathcal{S})+\mathcal{S}]t}\tilde{\rho}(-\tau_R) \\
&= \sum_{n=0}^{\infty}\int_{-\tau_R}^{t'} dt'_n \cdots \int_{-\tau_R}^{t'_3} dt'_2 \int_{-\tau_R}^{t'_2} dt'_1 \\
&\quad e^{(\tilde{\mathcal{L}}-\mathcal{S})(t'-t'_n)}\mathcal{S}e^{(\tilde{\mathcal{L}}-\mathcal{S})(t'_n-t'_{n-1})}\ldots\mathcal{S}e^{(\tilde{\mathcal{L}}-\mathcal{S})(t'_2-t'_1)}\mathcal{S}e^{(\tilde{\mathcal{L}}-\mathcal{S})(t'_1+\tau_R)}\tilde{\rho}(-\tau_R) \\
&= \sum_{n=0}^{\infty}\sum_{\nu_n=1}^{4}\cdots\sum_{\nu_1=1}^{4}\int_{-\tau_R}^{t'} dt'_n \cdots \int_{-\tau_R}^{t'_3} dt'_2 \int_{-\tau_R}^{t'_2} dt'_1 \\
&\quad e^{(\tilde{\mathcal{L}}-\mathcal{S})(t'-t'_n)}\mathcal{S}_{\nu_n}e^{(\tilde{\mathcal{L}}-\mathcal{S})(t'_n-t'_{n-1})}\ldots\mathcal{S}_{\nu_2}e^{(\tilde{\mathcal{L}}-\mathcal{S})(t'_2-t'_1)}\mathcal{S}_{\nu_1}e^{(\tilde{\mathcal{L}}-\mathcal{S})(t'_1+\tau_R)}\tilde{\rho}(-\tau_R),
\end{aligned} \tag{18.41}$$

with

$$\mathcal{S}_{\nu_k} \in \{\mathcal{S}_{\leftarrow}, \mathcal{S}_{\rightarrow}, \mathcal{S}_{p\leftarrow}, \mathcal{S}_{p\rightarrow}\}. \tag{18.42}$$

The quantity inside the sums and integrals is $\tilde{\mathcal{K}}_{\mathrm{REC}}(t')\tilde{\rho}(-\tau_R)$, the argument of the trace in the expression for the record probability (18.28). Thus, using its factorized form (18.37), and assuming a pure initial pure state, the Dyson expansion takes on the form

$$\tilde{\rho}(t') = \sum_{\mathrm{REC}} |\tilde{\bar{\psi}}_{\mathrm{REC}}(t')\rangle\langle\tilde{\bar{\psi}}_{\mathrm{REC}}(t')|, \tag{18.43}$$

where $|\tilde{\bar{\psi}}_{\mathrm{REC}}(t')\rangle$ is the unnormalized conditional state (18.40). The explicit sums and integrals in (18.41) make up the generalized sum over records: the sum over n is the sum over the number of photons emitted, sums over $\nu_1, \nu_2, \ldots, \nu_n$ cover the possible output channels for each emission, while the integrals sum over possible emission times. The Dyson expansion is in effect an unraveling of the density operator in the form (17.77). After normalizing the conditional state, and using (18.39), the unraveling is converted to the originally proposed form:

$$\tilde{\rho}(t') = \sum_{\mathrm{REC}} P(\mathrm{REC})|\tilde{\psi}_{\mathrm{REC}}(t')\rangle\langle\tilde{\psi}_{\mathrm{REC}}(t')|, \tag{18.44}$$

with *normalized conditional state*

$$|\tilde{\psi}_{\mathrm{REC}}(t')\rangle = \frac{|\tilde{\bar{\psi}}_{\mathrm{REC}}(t')\rangle}{\sqrt{\langle\tilde{\bar{\psi}}_{\mathrm{REC}}(t')|\tilde{\bar{\psi}}_{\mathrm{REC}}(t')\rangle}}. \tag{18.45}$$

Let us now address the practical matter of making use of such an unraveling. As noted in Sect. 17.3.4, in most applications $P(\text{REC})$ and $|\tilde{\psi}_{\text{REC}}(t')\rangle$ are difficult to determine analytically, and in order to put the unraveling to use it must be implemented numerically, as the basis of a Monte Carlo simulation scheme. The numerical approach generates stochastic realizations of both the conditional state and its associated scattering record. The different records are automatically realized in the correct proportions—with relative frequency dictated by the probability density $P(\text{REC})$. The expansion of the density operator is then evaluated as

$$\tilde{\rho}(t') = N^{-1} \sum_{\{\text{REC}\}_N} |\tilde{\psi}_{\text{REC}}(t')\rangle\langle\tilde{\psi}_{\text{REC}}(t')|, \tag{18.46}$$

where $\{\text{REC}\}_N$ denotes the realized ensemble of N simulated records.

In order to appreciate fully how the Monte Carlo evolution proceeds, let us assume the unnormalized state $|\tilde{\tilde{\psi}}_{\text{REC}}(t'_k)\rangle$ has been reached after the first k time steps ($t_k = k\Delta t$). In the next time step, there is a five-way splitting of the trajectory governed by the five probabilities for the next entry in the record. The branching ratio is determined from the general record probability density (18.39). With the help of (18.40) and (18.36), and (18.34)–(18.35b), the five record probabilities up to time $t_k + \Delta t$ are

$$P\left(\text{REC} \wedge \begin{Bmatrix} \gamma_{a1} \\ T_k \end{Bmatrix}\right) = \Delta t\langle\tilde{\tilde{\psi}}_{\text{REC}}(t'_k)|J^\dagger_\rightarrow J_\rightarrow|\tilde{\tilde{\psi}}_{\text{REC}}(t'_k)\rangle, \tag{18.47a}$$

$$P\left(\text{REC} \wedge \begin{Bmatrix} \gamma_{a2} \\ T_k \end{Bmatrix}\right) = \Delta t\langle\tilde{\tilde{\psi}}_{\text{REC}}(t'_k)|J^\dagger_\leftarrow J_\leftarrow|\tilde{\tilde{\psi}}_{\text{REC}}(t'_k)\rangle, \tag{18.47b}$$

$$P\left(\text{REC} \wedge \begin{Bmatrix} \gamma_{b1} \\ T_k \end{Bmatrix}\right) = \Delta t\langle\tilde{\tilde{\psi}}_{\text{REC}}(t'_k)|J^\dagger_{p\rightarrow} J_{p\rightarrow}|\tilde{\tilde{\psi}}_{\text{REC}}(t'_k)\rangle, \tag{18.47c}$$

$$P\left(\text{REC} \wedge \begin{Bmatrix} \gamma_{b2} \\ T_k \end{Bmatrix}\right) = \Delta t\langle\tilde{\tilde{\psi}}_{\text{REC}}(t'_k)|J^\dagger_{p\leftarrow} J_{p\leftarrow}|\tilde{\tilde{\psi}}_{\text{REC}}(t'_k)\rangle, \tag{18.47d}$$

and

$$P\left(\text{REC} \wedge \{\emptyset_{t_{k+1}}\}\right) = \langle\tilde{\tilde{\psi}}_{\text{REC}}(t'_k)|e^{-(i/\hbar)(\tilde{H}_B - \tilde{H}_B^\dagger)\Delta t}|\tilde{\tilde{\psi}}_{\text{REC}}(t'_k)\rangle. \tag{18.47e}$$

Then, using Bayesian inference, there are five conditional probabilities which govern the choice of which branch along the splitting trajectory to follow:

$$p_\rightarrow \equiv p\left(\begin{Bmatrix} \gamma_{a1} \\ T_k \end{Bmatrix}|\text{REC}\right) = \Delta t\langle\tilde{\psi}_{\text{REC}}(t'_k)|J^\dagger_\rightarrow J_\rightarrow|\tilde{\psi}_{\text{REC}}(t'_k)\rangle, \tag{18.48a}$$

$$p_\leftarrow \equiv p\left(\begin{Bmatrix} \gamma_{a2} \\ T_k \end{Bmatrix}|\text{REC}\right) = \Delta t\langle\tilde{\psi}_{\text{REC}}(t'_k)|J^\dagger_\leftarrow J_\leftarrow|\tilde{\psi}_{\text{REC}}(t'_k)\rangle, \tag{18.48b}$$

$$p_{p\rightarrow} \equiv p\left(\begin{Bmatrix} \gamma_{b1} \\ T_k \end{Bmatrix}|\text{REC}\right) = \Delta t\langle\tilde{\psi}_{\text{REC}}(t'_k)|J^\dagger_{p\rightarrow} J_{p\rightarrow}|\tilde{\psi}_{\text{REC}}(t'_k)\rangle, \tag{18.48c}$$

$$p_{p\leftarrow} \equiv p\left(\begin{Bmatrix} \gamma_{b2} \\ T_k \end{Bmatrix}|\text{REC}\right) = \Delta t\langle\tilde{\psi}_{\text{REC}}(t'_k)|J^\dagger_{p\leftarrow} J_{p\leftarrow}|\tilde{\psi}_{\text{REC}}(t'_k)\rangle, \tag{18.48d}$$

and

$$p\left(\{\emptyset_{t_{k+1}}\}|\mathrm{REC}\right) = \langle\tilde{\psi}_{\mathrm{REC}}(t'_k)|\left[1 + \Delta t\frac{1}{i\hbar}\left(\tilde{H}_B - \tilde{H}_B^\dagger\right)\right]|\tilde{\psi}_{\mathrm{REC}}(t'_k)\rangle. \quad (18.48e)$$

One readily checks that the sum of (18.48a)–(18.48e) is unity, guaranteeing that the probabilities sum to unity for all records extending over the entire simulation. Given the conditional probabilities, the *Monte Carlo algorithm for quantum trajectory simulations of the degenerate parametric oscillator with photoelectron counting records* follows in a natural way:

$$\left.\begin{array}{l}
\textbf{1. compute } p_{\rightarrow},\ p_{\leftarrow},\ p_{p\rightarrow},\ p_{p\leftarrow} \\
\qquad \text{and} \\
\qquad p_{\mathrm{jump}} = p_{\rightarrow} + p_{\leftarrow} + p_{p\rightarrow} + p_{p\leftarrow} \\
\textbf{2. if } p_{\mathrm{jump}} > r_k, \textbf{ then if } \dfrac{p_{\rightarrow}}{p_{\mathrm{jump}}} > r'_k \\
\qquad |\tilde{\psi}_{\mathrm{REC}}(t'_{k+1})\rangle = \dfrac{J_{\rightarrow}|\tilde{\psi}_{\mathrm{REC}}(t'_k)\rangle}{\sqrt{\langle\tilde{\psi}_{\mathrm{REC}}(t'_k)|J_{\rightarrow}^\dagger J_{\rightarrow}|\tilde{\psi}_{\mathrm{REC}}(t'_k)\rangle}} \\
\qquad \textbf{else if } \dfrac{p_{\rightarrow} + p_{\leftarrow}}{p_{\mathrm{jump}}} > r'_k \\
\qquad |\tilde{\psi}_{\mathrm{REC}}(t'_{k+1})\rangle = \dfrac{J_{\leftarrow}|\tilde{\psi}_{\mathrm{REC}}(t'_k)\rangle}{\sqrt{\langle\tilde{\psi}_{\mathrm{REC}}(t'_k)|J_{\leftarrow}^\dagger J_{\leftarrow}|\tilde{\psi}_{\mathrm{REC}}(t'_k)\rangle}} \\
\qquad \textbf{else if } \dfrac{p_{\rightarrow} + p_{\leftarrow} + p_{p\rightarrow}}{p_{\mathrm{jump}}} > r'_k \\
\qquad |\tilde{\psi}_{\mathrm{REC}}(t'_{k+1})\rangle = \dfrac{J_{p\rightarrow}|\tilde{\psi}_{\mathrm{REC}}(t'_k)\rangle}{\sqrt{\langle\tilde{\psi}_{\mathrm{REC}}(t'_k)|J_{p\rightarrow}^\dagger J_{p\rightarrow}|\tilde{\psi}_{\mathrm{REC}}(t'_k)\rangle}} \\
\qquad \textbf{else} \\
\qquad |\tilde{\psi}_{\mathrm{REC}}(t'_{k+1})\rangle = \dfrac{J_{p\leftarrow}|\tilde{\psi}_{\mathrm{REC}}(t'_k)\rangle}{\sqrt{\langle\tilde{\psi}_{\mathrm{REC}}(t'_k)|J_{p\leftarrow}^\dagger J_{p\leftarrow}|\tilde{\psi}_{\mathrm{REC}}(t'_k)\rangle}} \\
\quad \textbf{else} \\
\quad |\tilde{\psi}_{\mathrm{REC}}(t'_{k+1})\rangle \\
\quad = \left\{1 + \Delta t\dfrac{1}{i\hbar}\left[\tilde{H}_B - \tfrac{1}{2}\langle\tilde{\psi}_{\mathrm{REC}}(t'_k)|(\tilde{H}_B - \tilde{H}_B^\dagger)|\tilde{\psi}_{\mathrm{REC}}(t'_k)\rangle\right]\right\}|\tilde{\psi}_{\mathrm{REC}}(t'_k)\rangle
\end{array}\right\}. \quad (18.49)$$

This version of the algorithm uses two independent random numbers, r_k and r'_k, each uniformly distributed on the unit line. The first is employed to decide whether a jump or no-jump takes place and the second to decide which jump to make. Alternatively, a single random number can be used, with the decision between the five options made on the basis of where the random number falls within a unit interval divided into five subintervals in accordance with the conditional probabilities.

A selection of results obtained from an implementation of (18.49) is shown in Fig. 18.3. The steady-state Wigner distributions on the left are computed from the time-averaged density operator,

$$\tilde{\rho}_{ss} = \frac{1}{T}\int_0^T dt |\tilde{\psi}_{REC}(t)\rangle\langle\tilde{\psi}_{REC}(t)|, \tag{18.50}$$

for a single trajectory run continuously in time. The time average is equivalent to the ensemble average (18.46) since quantum trajectories are ergodic (with a few exceptions) [18.9,18.10]. The parameters chosen yield $n_p^{\rm thr} = \gamma_{a2}/2g = 2$ and $\langle a^\dagger a\rangle_{\rm thr} = 0.676$ [from (10.102)]. Threshold occurs at

$$\lambda = \frac{4g|\bar{\mathcal{E}}_0|}{\gamma_{a2}\gamma_{b2}} = 1 \quad \Rightarrow \quad \frac{2|\bar{\mathcal{E}}_0|}{\gamma_{a2}} = 2;$$

thus, Figs. 18.3a, b, and c correspond to conditions below threshold, at threshold, and above threshold, respectively. Although the system size is relatively large and the threshold is not sharp, we still see the progression predicted by the small-noise analysis of Sect. 10.2.4—from a single-peaked distribution below threshold to a two-peaked distribution above (compare Fig. 12.5 where $\langle a^\dagger a\rangle_{\rm thr} = 1.0$ and 0.5).

Short segments of the subharmonic-mode photoelectron count records are shown above each frame on the right in the figure. The records show intermittent bursts of photon emissions—"bright intervals"—separated by intervals where there are no emissions at all—"dark intervals;" the latter become less and less frequent as the oscillator passes above threshold.

The photon statistics within the bright intervals exhibit a feature that illustrates the kind of subtlety quantum trajectory simulations can extract. The plotted distributions, p_n, are probability distributions of bright-interval photon numbers, where for the purposes of the figure, a bright interval has been defined to be any continuous period of subharmonic-mode photoelectron counts beginning and ending with a dark interval of duration $2.5\gamma_{a2}^{-1}$ or more. Notice, now, that nonzero values of the probability fall almost entirely on the even integers; though more difficult to see at the resolution of the figure, this is true even for Fig. 18.3c, where typical bright intervals contain 50–100 counts. Thus, near threshold, the degenerate parametric oscillator is producing bursts of even numbers of photons, clearly a remnant of the pair production process. This is a subtle quantum correlation when the photon numbers are large. It would be very difficult to account for it using the phase-space approach of Chaps. 10–12. It is nevertheless readily accessible to quantum trajectory simulations.

18.2.2 Homodyne-Current Records

We have seen how the program of Sect. 17.1 can be formulated systematically from a photoelectron counting point of view. Thus, our primary aim has been

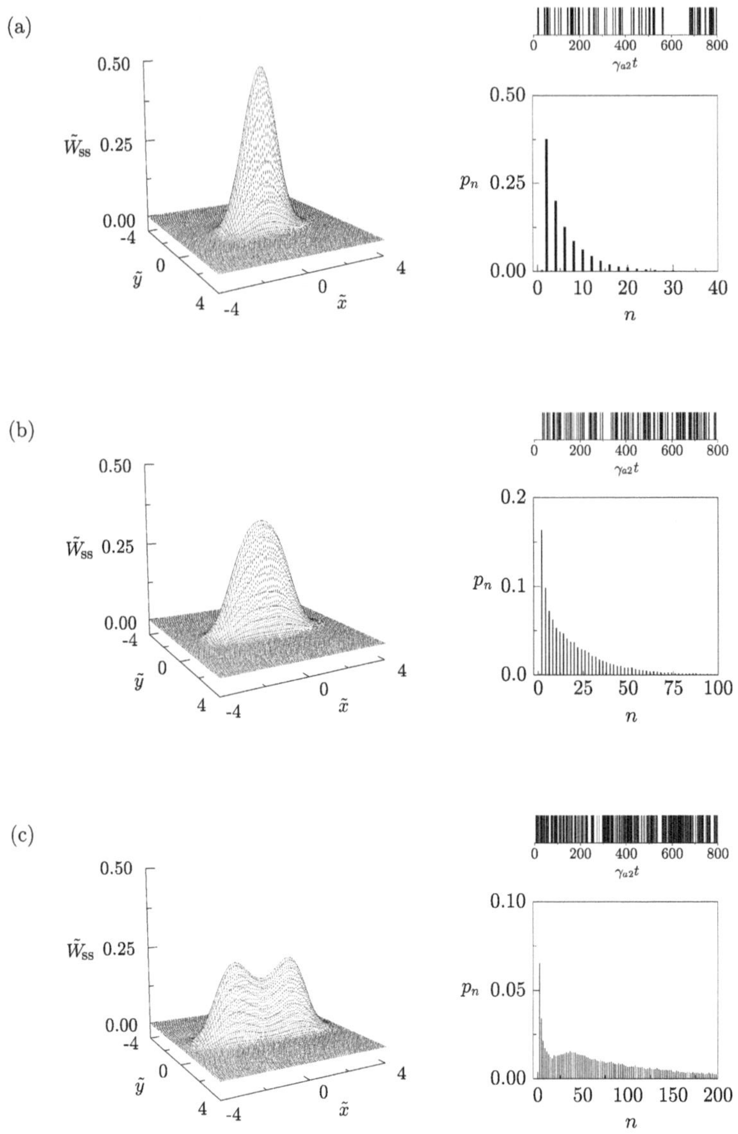

Fig. 18.3. Monte Carlo simulation of quantum trajectories with photoelectron counting records for the degenerate parametric oscillator near threshold in the strong coupling regime (for small system size); for $\gamma_{a1} = \gamma_{b1} = 0$, $\gamma_{a2} = \gamma_{b2} = 4g$, and (**a**) $2|\bar{\mathcal{E}}_0|/\gamma_{a2} = 1.4$, (**b**) $2|\bar{\mathcal{E}}_0|/\gamma_{a2} = 2.0$, and (**c**) $2|\bar{\mathcal{E}}_0|/\gamma_{a2} = 2.6$. The steady-state Wigner distribution is shown to the left and the probability distribution of "bright interval" counts (see text) to the right. A typical record of subharmonic counts appears above each frame on the right

achieved and there are now new issues to consider. Central amongst them is the observation that unravelings of the density operator are not unique—different unravelings are defined by different strategies for making scattering records. In all cases, ultimately the photoelectron counting scenario of the previous section is invoked—ultimately scattered photons are detected and recorded as photoelectron counts. The scattered field need not be directly detected, though; it might, for example, pass first through a filter cavity to effect a measurement of the optical spectrum, or be superposed with a local oscillator to accomplish homodyne or heterodyne detection. In this section we develop the unraveling of the density operator and its master equation evolution for the homodyne detection case.

Let us first consider a simple example that illustrates the possibilities. Returning to Fig. 18.1, we might ask what happens if the displacement which subtracts the backscattered field amplitude before detection is removed. In this case, the detector (lower left in the figure) sees output field $\sqrt{\gamma_{b1}}b$ superposed with a coherent amplitude $i\bar{\mathcal{E}}_0/\sqrt{\gamma_{b1}}$. The source superoperator in (18.23b), $\mathcal{S}_{p\leftarrow} \equiv \gamma_{b1} b \cdot b^\dagger$, is replaced by

$$\mathcal{S}'_{p\leftarrow} \equiv \left[i\frac{\bar{\mathcal{E}}_0}{\sqrt{\gamma_{b1}}} + \sqrt{\gamma_{b1}}b \right] \cdot \left[-i\frac{\bar{\mathcal{E}}_0^*}{\sqrt{\gamma_{b1}}} + \sqrt{\gamma_{b1}}b^\dagger \right]. \tag{18.51}$$

From here the mathematical development carries through as before, but with jump operator $J_{p\leftarrow} \equiv \sqrt{\gamma_{b1}}b$ (Eq. 18.35b) replaced by

$$J'_{p\leftarrow} \equiv i\frac{\bar{\mathcal{E}}_0}{\sqrt{\gamma_{b1}}} + \sqrt{\gamma_{b1}}b, \tag{18.52}$$

and altered non-Hermitian Hamiltonian [in place of (18.32)]

$$\tilde{H}'_B \equiv \tilde{H}_B - \hbar\bar{\mathcal{E}}_0^* b - i\hbar\frac{|\bar{\mathcal{E}}_0|^2}{2\gamma_{b1}}. \tag{18.53}$$

The Monte Carlo algorithm still holds, but with $J_{p\leftarrow} \to J'_{p\leftarrow}$ and $\tilde{H}_B \to \tilde{H}'_B$. Thus we arrive at a different unraveling. In fact, the unraveling arrived at in this way has much in common with the homodyne case. It differs in two small, though important, respects: in homodyne detection the superposed field is a local oscillator with (i) an adjustable phase and (ii) a very large amplitude (photon flux) compared to that of the cavity output field.

Note 18.4. The driving field enters non-Hermitian Hamiltonian (18.53) through a *unidirectional* coupling term, $\hbar(\bar{\mathcal{E}}_0 b^\dagger + \bar{\mathcal{E}}_0^* b) - \hbar\bar{\mathcal{E}}_0^* b = \hbar\bar{\mathcal{E}}_0 b^\dagger$; there is an operator to create, but not to annihilate, cavity photons. Such terms are met more generally in the cascaded open systems formalism of Sect. 18.2. Note also that the constant term, $-i\hbar|\bar{\mathcal{E}}_0|^2/2\gamma_{b1}$ in $\tilde{H}'_B$, has no effect on the normalized conditional state since it cancels out in the last line of (18.49). Indeed, in simulations, this term may be dropped; although if that is done the unnormalized state norm [defined via (18.40) and (18.36)] no longer gives the record probability density. We return to this point below (18.74).

The homodyne detection unraveling is derived by applying photoelectron counting theory to the scattering scenario depicted in Fig. 18.4. For simplicity, we retain direct photoelectron counting for output channels γ_{a1}, γ_{b1}, and γ_{b2}; only at the fourth output channel, γ_{a2}, are homodyne current records made. The records are made by a *balanced homodyne detector*, as depicted in the figure. Unbalanced detection was discussed in Sect 9.3.3, where it was used to derive an expression for the spectrum of squeezing. Balanced detection was proposed by Yuen and Chan [18.11] as the practical way to observe the reduced quadrature phase amplitude fluctuations of squeezed light. In this measurement, a strong local oscillator field is superposed at a 50/50 beam splitter with the signal field to be measured. The two beam splitter outputs are passed to photoelectric detectors whose photocurrents are subtracted and filtered to produce the measured signal—the difference current $i(t)$ in the figure.

The analysis of the scattering scenario of Fig. 18.4 is carried out as before (Sects. 18.1.1–18.2.1), but now there are five, rather than four, photoelectric detectors to consider. The two that form the balanced detector see fields that have been superposed with the local oscillator, so in place of source superoperators (18.25) and (18.42), we now have

$$\mathcal{S}_{\nu_k} \in \{\mathcal{S}_{\leftarrow}, \mathcal{S}_P, \mathcal{S}_M, \mathcal{S}_{p\leftarrow}, \mathcal{S}_{p\rightarrow}\}, \tag{18.54}$$

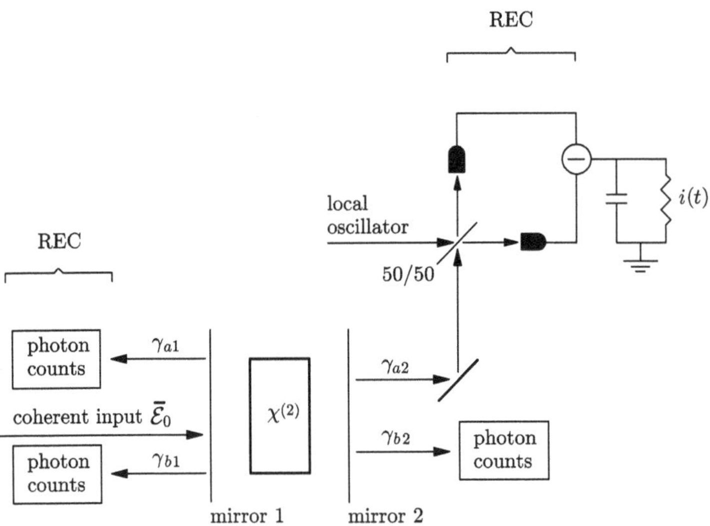

Fig. 18.4. Scattering scenario for the degenerate parametric oscillator model of Figs. 9.1 and 9.4 with scattering records made by direct photoelectron counting of output channels γ_{a1}, γ_{b1}, and γ_{b2}, and balanced homodyne detection of channel γ_{a2}

where superoperator $\mathcal{S}_{\rightarrow}$ is replaced by the two superoperators

$$\mathcal{S}_P \equiv \tfrac{1}{2}(\mathcal{E}_{\rm lo} + J_{\rightarrow}) \cdot (\mathcal{E}_{\rm lo}^* + J_{\rightarrow}^\dagger), \tag{18.55a}$$

$$\mathcal{S}_M \equiv \tfrac{1}{2}(\mathcal{E}_{\rm lo} - J_{\rightarrow}) \cdot (\mathcal{E}_{\rm lo}^* - J_{\rightarrow}^\dagger). \tag{18.55b}$$

The local oscillator field is written as $\mathcal{E}_{\rm lo}e^{-i\omega_C t}$, with frequency tuned to the cavity resonance and complex amplitude (in photon flux units)

$$\mathcal{E}_{\rm lo} = |\mathcal{E}_{\rm lo}|e^{i\theta}. \tag{18.56}$$

The local oscillator phase, θ, determines the quadrature phase amplitude to be measured. With the newly defined source superoperator sum, $\mathcal{S} \equiv \mathcal{S}_{\leftarrow} + \mathcal{S}_P + \mathcal{S}_M + \mathcal{S}_{p\leftarrow} + \mathcal{S}_{p\rightarrow}$, we may again write the propagator between quantum jumps as $e^{(\tilde{\mathcal{L}}-\mathcal{S})t} = \tilde{B}(t) \cdot \tilde{B}^\dagger(t)$, but with non-Hermitian Hamiltonian

$$\tilde{H}_B \rightarrow \tilde{H}_B - i\hbar\tfrac{1}{2}|\mathcal{E}_{\rm lo}|^2, \tag{18.57}$$

an analogous change to that made in (18.53). In the balanced detection scheme, the change arises from the detection of *two* superposed fields with a π phase difference between the superpositions; hence, the two terms corresponding to $-\hbar\bar{\mathcal{E}}_0^* b$ in (18.53) cancel. In place of the single jump operator, $J_{\rightarrow}$, there are now two,

$$J_P = \frac{1}{\sqrt{2}}(\mathcal{E}_{\rm lo} + J_{\rightarrow}), \tag{18.58a}$$

and

$$J_M = \frac{1}{\sqrt{2}}(\mathcal{E}_{\rm lo} - J_{\rightarrow}). \tag{18.58b}$$

Note 18.5. The local oscillator may be modeled as an additional source whose field is in a coherent state. The system state, including this additional source, is then written as $|\bar{\tilde{\psi}}_{\rm REC}(t)\rangle|\alpha_{\rm lo}\rangle$, where $\alpha_{\rm lo}$ is the complex amplitude of the local oscillator mode and the local oscillator source field is $\sqrt{\gamma_{\rm lo}}a_{\rm lo}$, with output coupling rate $\gamma_{\rm lo}$. Quantum jumps applied to the local oscillator mode yield $\sqrt{\gamma_{\rm lo}}a_{\rm lo}|\alpha_{\rm lo}\rangle \rightarrow \mathcal{E}_{\rm lo}|\alpha_{\rm lo}\rangle$, with $\mathcal{E}_{\rm lo} = \sqrt{\gamma_{\rm lo}}\alpha_{\rm lo}$. Thus, as an eigenstate of the jump operator, the coherent state $|\alpha_{\rm lo}\rangle$ is superfluous and might just as well be dropped.

We return now to the general quantum jump formalism of Sects. 18.1.1–18.2.1 and consider the propagator (18.36). In view of the assumed high local oscillator photon flux, the two homodyne photodetectors will fire at a much faster rate than the other three photodetectors, and the homodyne count records require special attention. To this end, we introduce notation to distinguish their counts from the rest. Consider a record that contains m nonhomodyne counts, at times $t_{d_1}, t_{d_2}, \ldots t_{d_m}$, with n_1 homodyne counts before the first, n_2 after the first and before the second, and so on. Firings of the three nonhomodyne detectors therefore have count indicators

$$\left.\begin{array}{ll} d_1 & = n_1 + 1 \\ d_2 & = n_1 + n_2 + 2 \\ \vdots & \quad \vdots \\ d_m & = n_1 + n_2 + \cdots n_m + m \end{array}\right\}. \tag{18.59}$$

For the propagator, (18.36), we then write

$$\begin{aligned} \tilde{K}_{\rm REC}(t') = {} & \tilde{K}_{\rm hom}(t' - t'_{d_m}) J_{d_m} \tilde{K}_{\rm hom}(t'_{d_m} - t'_{d_{m-1}}) \cdots \\ & \cdots J_{d_2} \tilde{K}_{\rm hom}(t'_{d_2} - t'_{d_1}) J_{d_1} \tilde{K}_{\rm hom}(t'_{d_1}), \end{aligned} \tag{18.60}$$

with jump operators for the direct photoelectric detections

$$J_{d_l} \in \{J_{\leftarrow}, J_{p\leftarrow}, J_{p\rightarrow}\}, \tag{18.61}$$

and with

$$\begin{aligned} \tilde{K}_{\rm hom}(t'_{d_{l+1}} - t'_{d_l}) \equiv {} & \tilde{B}(t'_{d_{l+1}} - t'_{d_l+n_{l+1}}) J_{d_l+n_{l+1}} \tilde{B}(t'_{d_l+n_{l+1}} - t'_{d_l+n_{l+1}-1}) \cdots \\ & \cdots J_{d_l+2} \tilde{B}(t'_{d_l+2} - t'_{d_l+1}) J_{d_l+1} \tilde{B}(t'_{d_l+1} - t'_{d_l}), \end{aligned} \tag{18.62}$$

where the homodyne detection jump operators are

$$J_{d_l+k_l} \in \{J_P, J_M\}. \tag{18.63}$$

Our goal now is to derive a stochastic Schrödinger equation for the conditional state that incorporates the effect of propagator $\tilde{K}_{\rm hom}(t'_{d_{l+1}} - t'_{d_l})$ at the level of a course-graining over time. The course-grained evolution becomes a continuous evolution in the limit of infinite local oscillator photon flux.

To begin, let us consider a time interval from t to $t + \Delta t$ (retarded time t' to $t' + \Delta t$) during which there are no nonhomodyne counts and $q_P + q_M$ homodyne counts occurring at times $t_{h_1}, t_{h_2}, \ldots t_{h_{q_P+q_M}}$; during this interval, the conditional state suffers q_P jumps of type J_P and q_M jumps of type J_M. We locate the latter within the record by the indicator

$$h_r = n + r, \qquad r = 1, 2, \ldots q_P + q_M, \tag{18.64}$$

where n is the total number of jumps prior to time t. The associated jump times are denoted by

$$t_{h_r} = t + \tau_1 + \tau_2 + \cdots \tau_r. \tag{18.65}$$

The length of the interval, Δt, is assumed to be small compared with the timescale for significant change of the conditional state, which is characterized by the various rates $|\bar{\mathcal{E}}_0|$, g, γ_{a1}, γ_{a2}, γ_{b1}, and γ_{b2} from master equation (18.3) and (18.4). The numbers q_P and q_M are nevertheless both extremely large, and approach infinity in the limit of infinite local oscillator photon flux—specifically, we have $q_P \sim q_M \sim \frac{1}{2}|\mathcal{E}_{\rm lo}|^2 \Delta t$. Under these conditions, we expand the propagator

$$\tilde{K}_{\mathrm{hom}}(\Delta t) = \tilde{B}\Big(\Delta t - \textstyle\sum_{s=1}^{q_P+q_M} \tau_s\Big) J_{h_{q_P+q_M}} \tilde{B}(\tau_{q_P+q_M}) \cdots J_{h_2}\tilde{B}(\tau_2) J_{h_1}\tilde{B}(\tau_1), \tag{18.66}$$

with

$$J_{h_r} \in \{J_P, J_M\}, \tag{18.67}$$

to dominant order in $\xi\Delta t$, where ξ stands in for the aforementioned rates, but including $\frac{1}{2}|\mathcal{E}_{\mathrm{lo}}|^2\Delta t$ to all orders. Thus, using (18.34) and (18.57)–(18.58b), we write

$$\tilde{K}_{\mathrm{hom}}(\Delta t) = \big(\mathcal{E}_{\mathrm{lo}}/\sqrt{2}\big)^{q_P+q_M} e^{-\frac{1}{2}|\mathcal{E}_{\mathrm{lo}}|^2\Delta t}\left[1 + \frac{1}{i\hbar}\tilde{H}_B\Delta t + \frac{q_P - q_M}{\mathcal{E}_{\mathrm{lo}}} J_{\rightarrow} + \frac{1}{2}\frac{(q_P - q_M)^2 - (q_P + q_M)}{\mathcal{E}_{\mathrm{lo}}^2} J_{\rightarrow}^2\right]. \tag{18.68}$$

The third and fourth terms in the square bracket come from an expansion to second order of the product of jump operators in (18.66); the fourth is of order $\gamma_{a2}\Delta t$ [note, for example, that $(q_P + q_M) \sim |\mathcal{E}_{\mathrm{lo}}|^2\Delta t$], while the third contains two contributions, one of order $\gamma_{a2}\Delta t$ and one of order $(\gamma_{a2}\Delta t)^{1/2}$. To replace them by explicit expressions, we write the photoelectron count numbers as sums of a mean and a fluctuation:

$$q_P = \frac{1}{2}\big\langle(\mathcal{E}_{\mathrm{lo}}^* + J_{\rightarrow}^\dagger)(\mathcal{E}_{\mathrm{lo}} + J_{\rightarrow})\big\rangle_{\mathrm{REC}}\Delta t + \frac{1}{\sqrt{2}}|\mathcal{E}_{\mathrm{lo}}|\Delta W_P, \tag{18.69a}$$

$$q_M = \frac{1}{2}\big\langle(\mathcal{E}_{\mathrm{lo}}^* - J_{\rightarrow}^\dagger)(\mathcal{E}_{\mathrm{lo}} - J_{\rightarrow})\big\rangle_{\mathrm{REC}}\Delta t + \frac{1}{\sqrt{2}}|\mathcal{E}_{\mathrm{lo}}|\Delta W_M. \tag{18.69b}$$

In these expressions photon flux conditional expectations are taken to determine the means—the operator averages $\langle\cdots\rangle_{\mathrm{REC}}$—while terms $|\mathcal{E}_{\mathrm{lo}}|\Delta W_P/\sqrt{2}$ and $|\mathcal{E}_{\mathrm{lo}}|\Delta W_M\sqrt{2}$ account for Poisson fluctuations in the count numbers (shot noise) for a local oscillator photon flux $\frac{1}{2}|\mathcal{E}_{\mathrm{lo}}|^2$ at each detector; the Wiener increments, ΔW_P and ΔW_M, are Gaussian-distributed random variables of zero mean and variance Δt.

The propagator depends on the three quantities $q_P - q_M$, $(q_P - q_M)^2$, and $q_P + q_M$. From (18.69a) and (18.69b), we obtain

$$q_P - q_M = |\mathcal{E}_{\mathrm{lo}}|\Big(\big\langle e^{i\theta} J_{\rightarrow}^\dagger + e^{-i\theta} J_{\rightarrow}\big\rangle_{\mathrm{REC}}\Delta t + \Delta W\Big), \tag{18.70}$$

where

$$\Delta W = \frac{1}{\sqrt{2}}(\Delta W_P - \Delta W_M) \tag{18.71}$$

is also Gaussian-distributed with zero mean and variance Δt. Then, since the dominant term in this expression for the photoelectron count difference is

$\Delta W \sim \sqrt{\Delta t}$, we use the Ito rule to write

$$(q_P - q_M)^2 = |\mathcal{E}_{\text{lo}}|^2 (\Delta W)^2 = |\mathcal{E}_{\text{lo}}|^2 \Delta t, \tag{18.72}$$

where we replace $(\Delta W)^2$ by its mean. Keeping only the dominant term also in the photoelectron count sum, from (18.69a) and (18.69b) we write

$$q_P + q_M = |\mathcal{E}_{\text{lo}}|^2 \Delta t. \tag{18.73}$$

Now, substituting (18.70), (18.72), and (18.73) into (18.68), our expansion of the propagator to dominant order in $\xi \Delta t$ takes the form

$$\begin{aligned}\tilde{K}_{\text{hom}}(\Delta t) = (\mathcal{E}_{\text{lo}}/\sqrt{2})^{q_P+q_M} \exp\left(-\tfrac{1}{2}|\mathcal{E}_{\text{lo}}|^2 \Delta t\right)\bigg\{1 + \frac{1}{i\hbar}\tilde{H}_B \Delta t \\ + \Big(\langle e^{i\theta} J_{\rightarrow}^{\dagger} + e^{-i\theta} J_{\rightarrow}\rangle_{\text{REC}} \Delta t + \Delta W\Big) e^{-i\theta} J_{\rightarrow}\bigg\}.\end{aligned} \tag{18.74}$$

Of course, the prefactor on the right-hand side does nothing more than change the normalization of the state. It may be removed by introducing a new unnormalized conditional state $|\bar{\tilde{\tilde{\psi}}}_{\text{REC}}(t)\rangle$, with

$$|\bar{\tilde{\psi}}_{\text{REC}}(t')\rangle = (\mathcal{E}_{\text{lo}}/\sqrt{2})^{q_{P,M}(t')} \exp\left(-\tfrac{1}{2}|\mathcal{E}_{\text{lo}}|^2 t'\right)|\bar{\tilde{\tilde{\psi}}}_{\text{REC}}(t')\rangle, \tag{18.75}$$

where $q_{P,M}(t')$ is the total number of homodyne counts up to time t (retarded time T'). While the new state serves our purpose of following the conditional evolution just as well, it must be remembered that its norm does not give the scattering record probability density. With the changed normalization, we write

$$\begin{aligned}&|\bar{\tilde{\tilde{\psi}}}_{\text{REC}}(t' + \Delta t)\rangle \\ &= (\mathcal{E}_{\text{lo}}/\sqrt{2})^{-[q_P+q_M+q_{P,M}(t')]} \exp\left[\tfrac{1}{2}|\mathcal{E}_{\text{lo}}|^2 (t' + \Delta t)\right]|\bar{\tilde{\psi}}_{\text{REC}}(t' + \Delta t)\rangle \\ &= (\mathcal{E}_{\text{lo}}/\sqrt{2})^{-[q_P+q_M+q_{P,M}(t')]} \exp\left[\tfrac{1}{2}|\mathcal{E}_{\text{lo}}|^2 (t' + \Delta t)\right]\left[\tilde{K}_{\text{hom}}(\Delta t)|\bar{\tilde{\psi}}_{\text{REC}}(t')\rangle\right] \\ &= \left[(\mathcal{E}_{\text{lo}}/\sqrt{2})^{-(q_P+q_M)} \exp\left(\tfrac{1}{2}|\mathcal{E}_{\text{lo}}|^2 \Delta t\right)\tilde{K}_{\text{hom}}(\Delta t)\right]|\bar{\tilde{\tilde{\psi}}}_{\text{REC}}(t')\rangle,\end{aligned} \tag{18.76}$$

and, using (18.74) and (18.76), arrive at the course-grained stochastic differential equation

$$\Delta|\bar{\tilde{\tilde{\psi}}}_{\text{REC}}\rangle = \left[\frac{1}{i\hbar}\tilde{H}_B \Delta t + \Big(\langle e^{i\theta} J_{\rightarrow}^{\dagger} + e^{-i\theta} J_{\rightarrow}\rangle_{\text{REC}} \Delta t + \Delta W\Big) e^{-i\theta} J_{\rightarrow}\right]|\bar{\tilde{\tilde{\psi}}}_{\text{REC}}\rangle. \tag{18.77}$$

Finally, taking the limit of infinite local oscillator photon flux allows us to adopt this equation at the level of infinitesimals, with $\Delta t \rightarrow dt$ and $\Delta W \rightarrow dW$. The result is the *stochastic Schrödinger equation within quan-*

tum trajectory theory for the degenerate parametric oscillator with homodyne-current records:

$$d|\bar{\bar{\tilde{\psi}}}_{\rm REC}\rangle = \left[\frac{1}{i\hbar}\tilde{H}_B dt + (Ge|\mathcal{E}_{\rm lo}|)^{-1}e^{-i\theta}J_{\rightarrow}dq\right]|\bar{\bar{\tilde{\psi}}}_{\rm REC}\rangle, \tag{18.78}$$

where

$$dq = Ge|\mathcal{E}_{\rm lo}|\Big(\langle e^{i\theta}J_{\rightarrow}^{\dagger} + e^{-i\theta}J_{\rightarrow}\rangle_{\rm REC}dt + dW\Big) \tag{18.79}$$

is the charge deposited in the detector circuit in interval t to $t+dt$; here e is the electronic charge and G the detector gain. The *filtered homodyne current*, $i(t)$ in Fig. 18.4, satisfies the stochastic differential equation

$$di = -\tau_d^{-1}(idt - dq), \tag{18.80}$$

where τ_d^{-1} is the detection bandwidth.

Note 18.6. The claim that the conditional state norm, $\langle\bar{\bar{\tilde{\psi}}}_{\rm REC}(t')|\bar{\bar{\tilde{\psi}}}_{\rm REC}(t')\rangle$, is the scattering record probability density (Eq. 18.39) is verified by noting that the scale factor in (18.75) yields the probability density for the record of local oscillator photoelectric counts (with the signal field in the vacuum state). It is a product of waiting-time distributions for coherent light:

$$\big(|\mathcal{E}_{\rm lo}|^2/2\big)^{(q_P+q_M)}e^{-|\mathcal{E}_{\rm lo}|^2\Delta t} = \exp\Big[-|\mathcal{E}_{\rm lo}|^2\Big(\Delta t - \textstyle\sum_{s=1}^{q_P+q_M}\tau_s\Big)\Big]\displaystyle\prod_{s=1}^{q_P+q_M}\tfrac{1}{2}W_{\rm lo}(\tau_s), \tag{18.81}$$

where $\frac{1}{2}W_{\rm lo}(\tau_s) = \frac{1}{2}|\mathcal{E}_{\rm lo}|^2\exp\big(-|\mathcal{E}_{\rm lo}|^2\tau_s\big)$ is the product of the no-count (at either detector) or null-measurement probability for an interval τ_s and the probability density for a count at one of the detectors at the end of the τ_s waiting time.

Note 18.7. The stochastic differential equation (18.78) propagates the conditional state at the retarded time $t' = t - \tau_R$, while dq and di refer to charge and current changes during the interval t to $t+dt$. The distinction is generally inconsequential and often overlooked.

Exercise 18.1. Corresponding to (18.78) and (18.79), show that the normalized conditional state satisfies the stochastic Schrödinger equation

$$\begin{aligned} d|\tilde{\psi}_{\rm REC}\rangle = \bigg\{&\frac{1}{i\hbar}\Big[\tilde{H}_B - \tfrac{1}{2}\langle\tilde{\psi}_{\rm REC}|\big(\tilde{H}_B - \tilde{H}_B^{\dagger} + i\hbar J_{\rightarrow}^{\dagger}J_{\rightarrow}\big)|\tilde{\psi}_{\rm REC}\rangle\Big]dt \\ &- \tfrac{1}{2}\langle\tilde{\psi}_{\rm REC}|J_{\rightarrow}^{\theta}|\tilde{\psi}_{\rm REC}\rangle^2 dt + \langle\tilde{\psi}_{\rm REC}|J_{\rightarrow}^{\theta}|\tilde{\psi}_{\rm REC}\rangle e^{-i\theta}J_{\rightarrow}dt \\ &+ dW\big(e^{-i\theta}J_{\rightarrow} - \langle\tilde{\psi}_{\rm REC}|J_{\rightarrow}^{\theta}|\tilde{\psi}_{\rm REC}\rangle\big)\bigg\}|\tilde{\psi}_{\rm REC}\rangle, \end{aligned} \tag{18.82}$$

with

$$J_{\rightarrow}^{\theta} \equiv \tfrac{1}{2}\big(e^{-i\theta}J_{\rightarrow} + e^{i\theta}J_{\rightarrow}^{\dagger}\big). \tag{18.83}$$

Fig. 18.5. Monte Carlo simulation of quantum trajectories with photoelectron counting (output channels γ_{a1}, γ_{b1}, and γ_{b2}) and homodyne-current (output channel γ_{a2}) records for the degenerate parametric oscillator near threshold in the strong coupling regime; for $\gamma_{a1} = \gamma_{b1} = 0$, $\gamma_{a2} = \gamma_{b2} = 4g$, and $2|\bar{\mathcal{E}}_0|/\gamma_{a2} = 1.4$ (**a** and **c**) and $2|\bar{\mathcal{E}}_0|/\gamma_{a2} = 2.6$ (**b** and **d**). Photocurrent correlation functions (to the left) and squeezing spectra (to the right) are shown for X-quadrature (**a** and **b**) and Y-quadrature (**c** and **d**) measurements, with detection bandwidth $2\tau_d^{-1}/\gamma_{a2} = 10$ [The definition of X and Y quadrature phase amplitudes appears above (10.54a).]

The stochastic Schrödinger equation propagates the conditional state in-between the quantum jumps associated with photoelectron counts at output channels γ_{a1}, γ_{b1}, and γ_{b2} [jump operators (18.61)]. The Monte Carlo algorithm (18.49) is therefore changed with the $J_{\rightarrow}$ jump removed and the evolution calculated from (18.78) (plus normalization) inserted in its place. Filtered homodyne-current records are simultaneously obtained from the integration of (18.80). They provide simulated data sets, in imitation of the data from a squeezing experiment. Thus, the spectrum of squeezing (Sects. 9.3.2–9.3.5) can be calculated by explicitly constructing the photocurrent autocorrelation $\overline{i(t)i(t+\tau)}$—as a time average for ergodic records [18.9, 18.10]—and taking the Fourier transform. This gives the spectrum $S_\theta(\omega)$ defined in (9.150) and (9.155); the collection efficiency is $\zeta = \gamma_{a2}/(\gamma_{a1} + \gamma_{a2})$ and we have assumed ideal detectors, detection efficiency $\eta = 1$. Non-unit detection efficiency can be treated by introducing a beam splitter with reflectivity $\sqrt{\eta}$ into the γ_{a2} output channel; one makes homodyne-current records for the reflected light and photoelectron counting records on the light transmitted.

Results are plotted in Fig. 18.5 for parameters corresponding to the below-threshold operating conditions of Fig. 18.3a (a and c) and the above-threshold conditions of Fig. 18.3c (b and d). The X-quadrature measurement (a and b) shows a narrowing of the spectrum above threshold and evidence of the developing bimodality in the displayed sample homodyne-current records [compare the Wigner function of Fig. 8.3c]. Here the fluctuations are amplified rather than squeezed; they rise noticeably above the background shot noise level. This is seen most clearly in the right frame of Fig. 18.5a, where the shot noise level is the broad Lorentzian pedestal upon which the central peak of amplified low-frequency fluctuations sits.

On the other hand, fluctuations of the Y-quadrature phase amplitude are squeezed. In place of the peak in Figs. 18.5a and b, the squeezing spectra of Figs. 18.5c and d show a clear central dip, while for long times the correlation function makes a negative, rather than positive excursion.

18.2.3 Heterodyne-Current Records

In homodyne detection the frequency ω of the local oscillator matches the central frequency of the source-field spectrum. Heterodyne detection uses a local

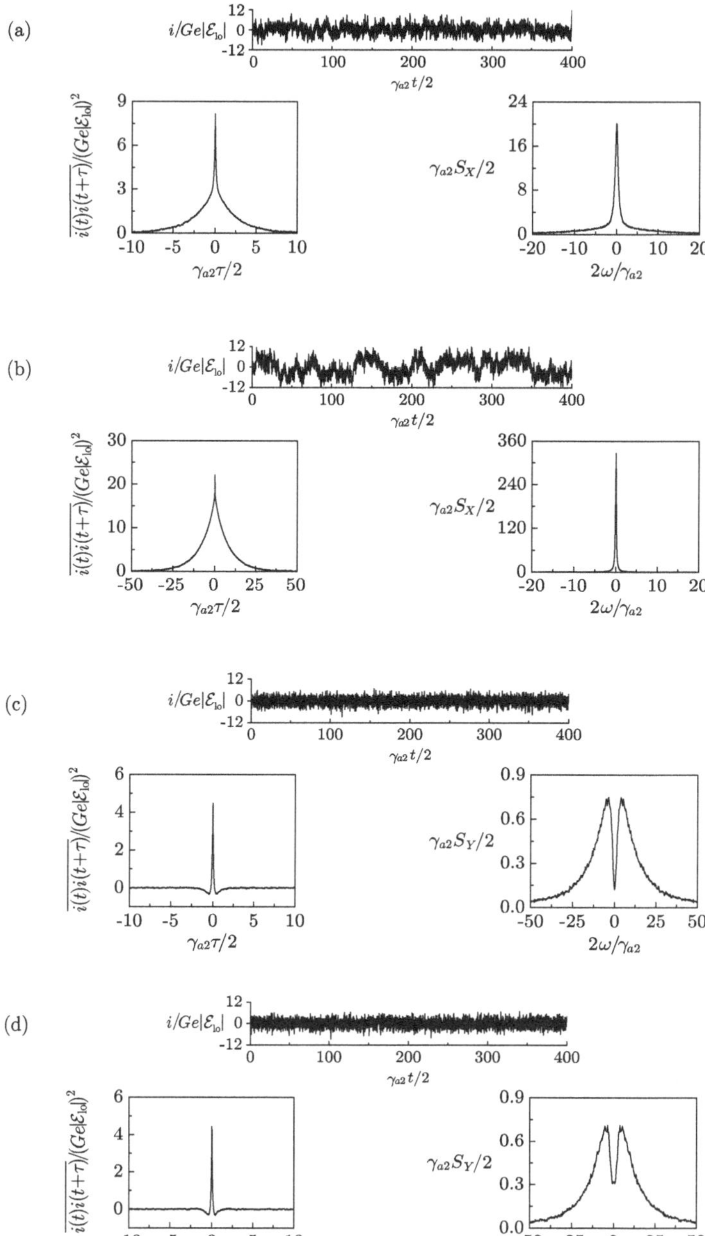
(a)
$i/Ge|\mathcal{E}_{lo}|$
12
0
-12
0 100 200 300 400
$\gamma_{a2}t/2$
$\overline{i(t)i(t+\tau)}/(Ge|\mathcal{E}_{lo}|)^2$
9 6 3 0
-10 -5 0 5 10
$\gamma_{a2}\tau/2$
$\gamma_{a2}S_X/2$
24 16 8 0
-20 -10 0 10 20
$2\omega/\gamma_{a2}$
(b)
$i/Ge|\mathcal{E}_{lo}|$
12
0
-12
0 100 200 300 400
$\gamma_{a2}t/2$
$\overline{i(t)i(t+\tau)}/(Ge|\mathcal{E}_{lo}|)^2$
30 20 10 0
-50 -25 0 25 50
$\gamma_{a2}\tau/2$
$\gamma_{a2}S_X/2$
360 240 120 0
-20 -10 0 10 20
$2\omega/\gamma_{a2}$
(c)
$i/Ge|\mathcal{E}_{lo}|$
12
0
-12
0 100 200 300 400
$\gamma_{a2}t/2$
$\overline{i(t)i(t+\tau)}/(Ge|\mathcal{E}_{lo}|)^2$
6 4 2 0
-10 -5 0 5 10
$\gamma_{a2}\tau/2$
$\gamma_{a2}S_Y/2$
0.9 0.6 0.3 0.0
-50 -25 0 25 50
$2\omega/\gamma_{a2}$
(d)
$i/Ge|\mathcal{E}_{lo}|$
12
0
-12
0 100 200 300 400
$\gamma_{a2}t/2$
$\overline{i(t)i(t+\tau)}/(Ge|\mathcal{E}_{lo}|)^2$
6 4 2 0
-10 -5 0 5 10
$\gamma_{a2}\tau/2$
$\gamma_{a2}S_Y/2$
0.9 0.6 0.3 0.0
-50 -25 0 25 50
$2\omega/\gamma_{a2}$

oscillator that is far detuned. Having noted this difference, the analysis of the last section then carries through with the addition of detuning factors $e^{\pm i\Delta\omega t}$, $\Delta\omega \equiv \omega - \omega_C$, in the interference between the local oscillator and the source field. Effectively, the local oscillator phase factor θ is replaced by $-\Delta\omega t$. In place of stochastic Schrödinger equation (18.78), we obtain

$$d|\tilde{\tilde{\tilde{\psi}}}_{\rm REC}\rangle = \left[\frac{1}{i\hbar}\tilde{H}_B dt + (Ge|\mathcal{E}_{\rm lo}|)^{-1}e^{i\Delta\omega t}J_{\rightarrow}dq\right]|\tilde{\tilde{\tilde{\psi}}}_{\rm REC}\rangle, \tag{18.84}$$

with

$$dq = Ge|\mathcal{E}_{\rm lo}|\Big(\langle e^{-i\Delta\omega t}J_{\rightarrow}^{\dagger} + e^{i\Delta\omega t}J_{\rightarrow}\rangle_{\rm REC}dt + dW\Big). \tag{18.85}$$

Assuming then that the frequency mismatch $\Delta\omega$ is very large compared with the bandwidth of the source-field fluctuations, we may introduce the slowly-varying incremental charge

$$d\tilde{q} \equiv e^{i\Delta\omega t}dq, \tag{18.86}$$

and neglect the rapidly oscillating term $e^{2i\Delta\omega t}\langle J_{\rightarrow}\rangle$ that enters on the right-hand side of (18.85). We also make the substitution

$$e^{i\Delta\omega t}dW \rightarrow dZ, \tag{18.87}$$

where $dZ = (dW_x + idW_y)/\sqrt{2}$ is a complex-valued Wiener increment with covariances

$$\overline{dZdZ} = \overline{dZ^*dZ^*} = 0, \qquad \overline{dZ^*dZ} = dt, \tag{18.88}$$

and dW_x and dW_y are statistically independent (real) Wiener increments, with covariances

$$\overline{dW_xdW_x} = \overline{dW_ydW_y} = dt, \qquad \overline{dW_xdW_y} = 0. \tag{18.89}$$

Thus, from (18.84) and (18.85), with the substitutions (18.86) and (18.87), we arrive at the *stochastic Schrödinger equation within quantum trajectory theory for the degenerate parametric oscillator with heterodyne-current records*:

$$d|\tilde{\tilde{\tilde{\psi}}}_{\rm REC}\rangle = \left[\frac{1}{i\hbar}\tilde{H}_B dt + (Ge|\mathcal{E}_{\rm lo}|)^{-1}J_{\rightarrow}d\tilde{q}\right]|\tilde{\tilde{\tilde{\psi}}}_{\rm REC}\rangle, \tag{18.90}$$

where

$$d\tilde{q} \equiv e^{i\Delta\omega t}q = Ge|\mathcal{E}_{\rm lo}|\Big(\langle J_{\rightarrow}^{\dagger}\rangle_{\rm REC}dt + dZ\Big). \tag{18.91}$$

There is now a complex-valued *filtered heterodyne current*, $\tilde{i}(t)$, satisfying the stochastic differential equation

$$d\tilde{i} = -\tau_d^{-1}(\tilde{i}dt - d\tilde{q}), \tag{18.92}$$

where τ_d^{-1} is the detection bandwidth. Of course the raw heterodyne current is real. It takes the form of a modulated rf carrier oscillation with carrier

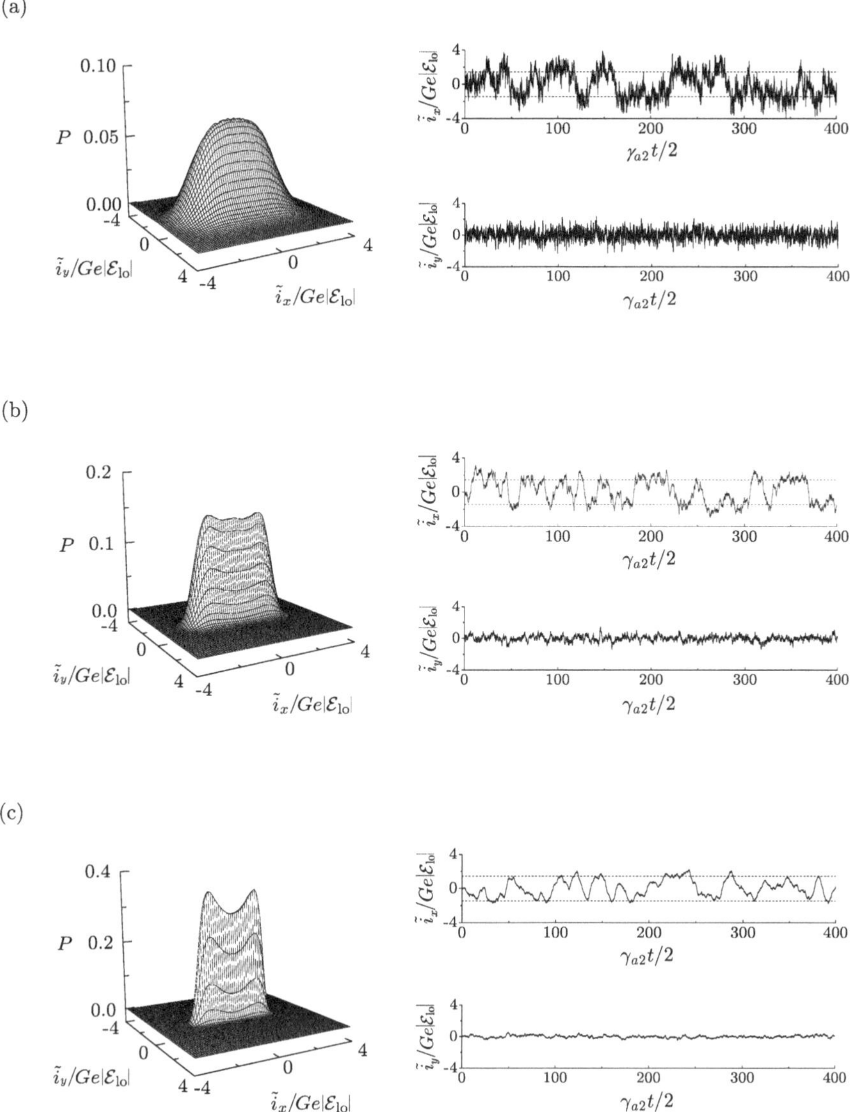

Fig. 18.6. Monte Carlo simulation of quantum trajectories with photoelectron counting (output channels γ_{a1}, γ_{b1}, and γ_{b2}) and heterodyne-current (output channel γ_{a2}) records for the degenerate parametric oscillator near threshold in the strong coupling regime; for $\gamma_{a1} = \gamma_{b1} = 0$, $\gamma_{a2} = \gamma_{b2} = 4g$, and $2|\bar{\mathcal{E}}_0|/\gamma_{a2} = 2.6$. Sample time series and probability distributions of the complex-valued heterodyne current are shown for three detection bandwidths: (**a**) $2\tau_d^{-1}/\gamma_{a2} = 4.0$, (**b**) $2\tau_d^{-1}/\gamma_{a2} = 1.0$, and (**c**) $2\tau_d^{-1}/\gamma_{a2} = 0.25$

frequency $\Delta\omega$. The complex-valued heterodyne current $\tilde{i}(t)$ accounts for the slowly-varying amplitude and phase of the modulation.

Through this complex-valued current, heterodyne detection provides us with a measurement of the complex amplitude of the intracavity field. In the presence of shot noise, dZ, the measurement is strongly affected by detection bandwidth. Figure 18.6 illustrates these bandwidth effects for the conditions of Fig. 18.3c (also Fig. 18.5b and d), where the degenerate parametric oscillator is just above threshold and a bimodal distribution for the real part of the subharmonic-mode field amplitude begins to appear. To produce the histograms shown to the left in the figure, ongoing time series, similar to the examples shown, were sampled at regular intervals of time. The beginning bimodality is clearly evident at the lowest detection bandwidth [Fig. 18.6c], though lost in the shot noise at the highest [Fig. 18.6a].

It is clear that none of the computed histograms correspond precisely to the Wigner distribution of Fig. 9.3c. In fact, heterodyne detection measures the Q distribution, not the Wigner distribution, but the histograms do not correspond to the Q distribution of the intracavity field either. This is because the measurement is not mode-matched to the cavity; hence an additional contamination by shot noise results [alternatively, contamination by the vacuum fluctuations of the free output field]. A mode-matched measurement of the intracavity Q distribution is possible, though, as shown in Sect. 18.3.2.

A final useful observation is that in the high-bandwidth limit, the Fourier transform of the photocurrent autocorrelation $\overline{\tilde{i}^*(t)\tilde{i}(t+\tau)}$ yields the optical spectrum from simulated data sets. The spectrum sits on a pedestal of shot noise just like the spectrum of squeezing in Fig. 18.5. Measurement of the optical spectrum through heterodyne detection is treated in Sect. 19.3.3.

Exercise 18.2. Show that the stochastic Schrödinger equation conditioned on heterodyne-current records may be derived from the equation conditioned on homodyne-current records (Eqs. 18.78 and 19.79) by dividing the γ_{a2} output field at a 50/50 beam splitter and measuring the X-quadrature phase amplitude at one beam splitter output and the Y-quadrature phase amplitude at the other.

Exercise 18.3. Use the stochastic Schrödinger equation conditioned on homodyne-current records (Eqs. 18.78 and 18.79) complemented by quantum jumps at the γ_{a1}, γ_{b1}, and γ_{b2} outputs to derive an expression for the net change, $d(|\tilde{\psi}_{\rm REC}\rangle\langle\tilde{\psi}_{\rm REC}|)$, of the conditional density operator in time interval dt. Hence show by taking the average over scattering records that the unconditional density operator satisfies master equation (18.3) and (18.4). Repeat the calculations for the stochastic Schrödinger equation conditioned on heterodyne-current records (Eqs. 18.90 and 18.91).

Note 18.8. A stochastic differential equation conditioned on heterodyne-current records and satisfied by the normalized conditional state follows from (18.82) by making the replacement $\theta \to -\Delta\omega t$ and neglecting rapidly oscillating terms. The resulting equation reads

$$
\begin{aligned}
&d|\tilde{\psi}_{\mathrm{REC}}\rangle \\
&=\bigg\{\frac{1}{i\hbar}\Big[\tilde{H}_B-\tfrac{1}{2}\langle\tilde{\psi}_{\mathrm{REC}}|\big(\tilde{H}_B-\tilde{H}_B^\dagger+i\hbar J_\rightarrow^\dagger J_\rightarrow\big)|\tilde{\psi}_{\mathrm{REC}}\rangle\Big]dt \\
&\quad-\tfrac{1}{4}\langle\tilde{\psi}_{\mathrm{REC}}|J_\rightarrow^\dagger|\tilde{\psi}_{\mathrm{REC}}\rangle\langle\tilde{\psi}_{\mathrm{REC}}|J_\rightarrow|\tilde{\psi}_{\mathrm{REC}}\rangle dt+\tfrac{1}{2}\langle\tilde{\psi}_{\mathrm{REC}}|J_\rightarrow^\dagger|\tilde{\psi}_{\mathrm{REC}}\rangle J_\rightarrow dt \\
&\quad+dZ\big(J_\rightarrow-\tfrac{1}{2}\langle\tilde{\psi}_{\mathrm{REC}}|J_\rightarrow|\tilde{\psi}_{\mathrm{REC}}\rangle\big)-dZ^*\tfrac{1}{2}\langle\tilde{\psi}_{\mathrm{REC}}|J_\rightarrow^\dagger|\tilde{\psi}_{\mathrm{REC}}\rangle\bigg\}|\tilde{\psi}_{\mathrm{REC}}\rangle,
\end{aligned}
\tag{18.93}
$$

and depends on both the complex Wiener increment dZ and its conjugate dZ^*. The latter dependence may be removed by introducing a time-dependent global phase shift of the conditional state. Thus, with

$$
|\tilde{\psi}'_{\mathrm{REC}}(t)\rangle\equiv e^{i\phi(t)}|\tilde{\psi}_{\mathrm{REC}}(t)\rangle, \tag{18.94a}
$$

$$
|\tilde{\psi}'_{\mathrm{REC}}\rangle+d|\tilde{\psi}'_{\mathrm{REC}}\rangle=e^{id\phi}e^{i\phi}(|\tilde{\psi}_{\mathrm{REC}}\rangle+d|\tilde{\psi}_{\mathrm{REC}}\rangle), \tag{18.94b}
$$

and infinitesimal phase change

$$
id\phi\equiv-dZ\tfrac{1}{2}\langle\tilde{\psi}_{\mathrm{REC}}|J_\rightarrow|\tilde{\psi}_{\mathrm{REC}}\rangle+dZ^*\tfrac{1}{2}\langle\tilde{\psi}_{\mathrm{REC}}|J_\rightarrow^\dagger|\tilde{\psi}_{\mathrm{REC}}\rangle, \tag{18.95}
$$

we expand (18.94b) to second order in $d\phi$ and $d|\tilde{\psi}_{\mathrm{REC}}\rangle$:

$$
\begin{aligned}
&|\tilde{\psi}'_{\mathrm{REC}}\rangle+d|\tilde{\psi}'_{\mathrm{REC}}\rangle \\
&\quad=\big[1+(id\phi)+\tfrac{1}{2}(id\phi)^2\big]|\tilde{\psi}'_{\mathrm{REC}}\rangle+e^{i\phi}\big[d|\tilde{\psi}_{\mathrm{REC}}\rangle+(id\phi)d|\tilde{\psi}_{\mathrm{REC}}\rangle\big].
\end{aligned}
\tag{18.96}
$$

Now, invoking the Ito rule, as in (18.72), we replace the second-order terms $\frac{1}{2}(id\phi)^2$ and $(id\phi)d|\tilde{\psi}_{\mathrm{REC}}\rangle$ by their means with, from (18.95) and (18.93),

$$
\overline{\tfrac{1}{2}(id\phi)^2}=-\tfrac{1}{4}\langle\tilde{\psi}_{\mathrm{REC}}|J_\rightarrow^\dagger|\tilde{\psi}_{\mathrm{REC}}\rangle\langle\tilde{\psi}_{\mathrm{REC}}|J_\rightarrow|\tilde{\psi}_{\mathrm{REC}}\rangle dt, \tag{18.97a}
$$

and

$$
\overline{(id\phi)d|\tilde{\psi}_{\mathrm{REC}}\rangle}=\tfrac{1}{2}\langle\tilde{\psi}_{\mathrm{REC}}|J_\rightarrow^\dagger|\tilde{\psi}_{\mathrm{REC}}\rangle J_\rightarrow dt. \tag{18.97b}
$$

Finally, using (18.96)–(18.97b) and (18.93), the stochastic differential equation conditioned on heterodyne-current records and satisfied by the normalized conditional state is

$$
\begin{aligned}
&d|\tilde{\psi}'_{\mathrm{REC}}\rangle \\
&=\bigg\{\frac{1}{i\hbar}\Big[\tilde{H}_B-\tfrac{1}{2}\langle\tilde{\psi}'_{\mathrm{REC}}|\big(\tilde{H}_B-\tilde{H}_B^\dagger+i\hbar J_\rightarrow^\dagger J_\rightarrow\big)|\tilde{\psi}'_{\mathrm{REC}}\rangle\Big]dt \\
&\quad-\tfrac{1}{2}\langle\tilde{\psi}'_{\mathrm{REC}}|J_\rightarrow^\dagger|\tilde{\psi}'_{\mathrm{REC}}\rangle\langle\tilde{\psi}'_{\mathrm{REC}}|J_\rightarrow|\tilde{\psi}'_{\mathrm{REC}}\rangle dt+\langle\tilde{\psi}'_{\mathrm{REC}}|J_\rightarrow^\dagger|\tilde{\psi}'_{\mathrm{REC}}\rangle J_\rightarrow dt \\
&\quad+dZ\big(J_\rightarrow-\langle\tilde{\psi}'_{\mathrm{REC}}|J_\rightarrow|\tilde{\psi}'_{\mathrm{REC}}\rangle\big)\bigg\}|\tilde{\psi}'_{\mathrm{REC}}\rangle.
\end{aligned}
\tag{18.98}
$$

Interestingly, this corresponds to the equation proposed by Gisin and Percival in their *quantum state diffusion* model [18.12, 18.13, 18.14]. The connection is notable, since these authors derived their equation from rather general considerations, aiming only to decompose the deterministic evolution of a quantum ensemble into a stochastic evolution obeyed by individual members of the ensemble (stochastic realizations). They aim to do this in such a way that each realization will localize, dynamically, onto an eigenstate of a specified observable (e.g., an energy eigenstate), thus resolving the dispersion within the ensemble; in particular, an initial superposition state should localize onto one state within the superposition. In essence, then, quantum state diffusion was designed to model a dynamical wavefunction collapse. In light of this, its connection with (18.98) represents a somewhat surprising convergence of ideas, since the starting point of quantum trajectory theory is very different. Our aim has been to set up a conditional dynamics, not to alter the Schrödinger evolution "in reality." The different starting points, and outlooks, become particularly clear when we observe that within quantum state diffusion the proposed stochastic Schrödinger equation is considered to be unique. In contrast, we have constructed three quite different unravelings of the same master equation.

18.3 Physical Interpretation

The ideas upon which quantum trajectory theory are based were discussed in Chap. 17. Nothing is said there, though, about the uniqueness of the proposed unraveling of the density operator and its master equation evolution. From the preceeding we have seen that it is not, in fact, unique. In light of this, some additional words are called for on the question of physical interpretation.

18.3.1 Systems, Environments, and Complementarity

The difference between quantum trajectory theory and the standard master equation point of view is summarized by the two observations made in Sect. 17.1 (below Eq. 17.5); these may be restated as follows: (i) while the standard master equation approach takes the *trace* over the reservoir (environment) R, quantum trajectory theory *disentangles* system S from R; and (ii) while the master equation is *local* in its dependence on R, quantum trajectory theory relies on a *nonlocal* description.

First let us expand on the second point. Nonlocality enters through conditioning: the system state is conditioned on a scattering record that is made by detectors which, importantly, do not upset the local interaction between S and R—i.e., are not placed within a wavelength or two of where S couples to R—although they alter R at some distance from S. The form of the master equation depends on the local interaction only of S and R. It follows that there are infinitely many conditional evolutions (unravelings of the density operator and its master equation evolution) because there are infinitely many actual environments that appear *locally* (where S and R couple) to be the same. The different arrays of detectors surrounding the cavity in Figs. 18.1 and 18.2 constitute different actual environments, and the change from the on-resonance local oscillator of homodyne detection (Sect. 18.2.2) to the detuned local oscillator of heterodyne detection (Sect. 18.2.3) is again a change in the actual environment.

In examples like these, upon which the most common quantum trajectory unravelings are based, the introduced detectors may be said to constitute *idealized environments*. An idealized environment could correspond to true experimental conditions; but more generally it should be viewed as a formal aid, helping us conceptualize some aspect of the system quantum dynamics—by putting the system output fields into specific action, by introducing a specific environment upon which the output fields can act. Of course, for any particular experiment there is ultimately only one relevant environment—the environment as it exists in fact.

While the quantum trajectory formalism provides us with ways to calculate things—often where alternate methods fail—its main distinction is the conceptual outlook just described—i.e., its many unravelings of an open system evolution, each one matched to the (nonlocal) action of the system upon a specific environment. No single unraveling suffices to account for every such action. This is the essence of Bohr's *complementarity* [18.15, 18.16, 18.17], or to invoke another word often used, the *contextuality* of quantum mechanics. A quantum trajectory does not tell us what the system actually does, as a matter of fact. We do not argue that the system actually jumps between stationary states as photoelectron counting records suggest, or, in opposition, that it evolves continuously in time as homodyne- or heterodyne-current records suggest. Each of these visualizations is a statistical inference, a *conditional* evolution. The scattering records, the measurement results, are the actual happening, and the inference works backwards from this. We might prefer to work forwards, from a stochastic description of the system state independent of all nonlocal monitoring, and adopt, in effect, a local hidden variables point of view. Quantum trajectories, however, do not provide us with that.

In the following few pages we explore some of the subtleties that different unravelings of the master equation reveal for the degenerate parametric oscillator example. We consider the one-mode master equation obtained with adiabatic elimination of the pump (Eq. 12.10). The simplification makes the computations much more manageable. Before considering our first unraveling, though, something needs to be said about how the adiabatic elimination of the pump affects the making of scattering records.

Let us return to the photoelectron counting scheme of Fig. 18.1 and set γ_{a1} and γ_{b1} to zero, thus simplifying our model by considering a single-ended cavity. Note, then, that the master equation with adiabatic elimination of the pump (Eq. 12.10) possesses one- and two-photon damping terms, and jump operators

$$J_{\text{one}} = \sqrt{2\kappa}a = \sqrt{\gamma_{a2}}a, \tag{18.99a}$$

$$J_{\text{two}} = \sqrt{g^2/2\kappa_p}a^2 = \sqrt{g^2/\gamma_{b2}}a^2. \tag{18.99b}$$

These terms account for losses from the γ_{a2} and γ_{b2} outputs, respectively. Note, in particular, that after adiabatic elimination of the pump, the two-photon loss from the subharmonic mode accounts for the loss of single pump-mode photons at the γ_{b2} output. The physical process accomplishing this is explained below (12.10). In the present situation we must recognize, however, that the entire γ_{b2} output cannot be assigned in this way, since pump photons are lost from the cavity even when the nonlinear crystal is absent—in the absence of any two-photon loss. In fact, the field at the γ_{b2} output is comprised of two pieces: a steady-state amplitude induced by coherent driving of the cavity at frequency $2\omega_C$, plus a term proportional to the square of the subharmonic field. The latter only accounts for the two-photon loss.

The explicit decomposition of the γ_{b2} output is shown at the input to the beam splitter in Fig. 18.7 [compare (12.2)]. As the figure illustrates, the beam splitter can be used to effect a displacement that removes the steady-state amplitude, separating off the two-photon term; the scheme is similar to the one used to remove the back-reflection in Fig. 18.1. The discussion that follows refers to quantum trajectories conditioned on direct photoelectron counting of the *displaced* γ_{b2} output [with jump operator (18.99b)] and various detection schemes for the γ_{a2} output.

We aim to explore some of the subtlety associated with the dynamics introduced in Sect. 12.1.8. The suggestion made there was that the state of the intracavity field switches back and forth between even and odd coherent state superpositions (Eqs. 18.90a and 18.90b) as successive photons are emitted by the cavity. If the suggestion is correct, quantum trajectories conditioned on photoelectron counting records should produce even and odd conditional states correlated with an even and odd cumulative photoelectron count. This

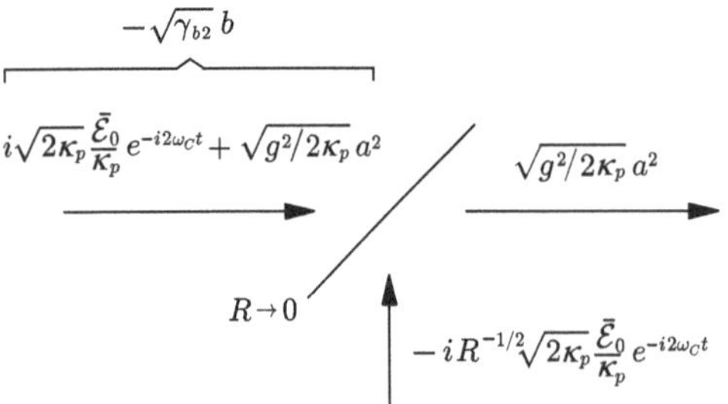

Fig. 18.7. Subtraction of the coherent amplitude of the γ_{b2} output in the limit $\gamma_{b2} \gg \gamma_{a2}$ (adiabatic elimination of the pump with $\gamma_{a1} = \gamma_{b1} = 0$)

is exactly what one finds. The result is shown in Fig. 18.8, which compares the time-averaged conditional density operator for a single long-running trajectory [frame (a)] with two partial time averages, one [frame (b)] averaging over all time intervals preceeded by an even number of photoelectron counts and the other [frame (c)] averaging over all time intervals preceeded by an odd number of photoelectron counts. The Wigner distributions of frames (b) and (c) exhibit phase-space interference, with different phases of interference. The two interference patterns evidence the proposed change from even to odd coherent state superposition—or vice versa—on each photon emission.

Plotting computed Wigner distributions tells us nothing about *observed*, or observable physics, however. A more interesting question is whether the scattering records themselves contain any indication of the even-odd switching. Certainly, photoelectron counting records will not show the interference fringes of Figs. 18.8b and c. They do contain indirect evidence of the even–odd dichotomy, however. One indication of this appears in Fig. 18.3, which shows that "bright-interval" photoelectron count numbers are predominantly even. Figure 18.9 looks a little more carefully at this feature. The distribution of waiting times (between one count and the next) is plotted in Fig. 18.9d for the two cases where the considered interval is preceeded by either an even or

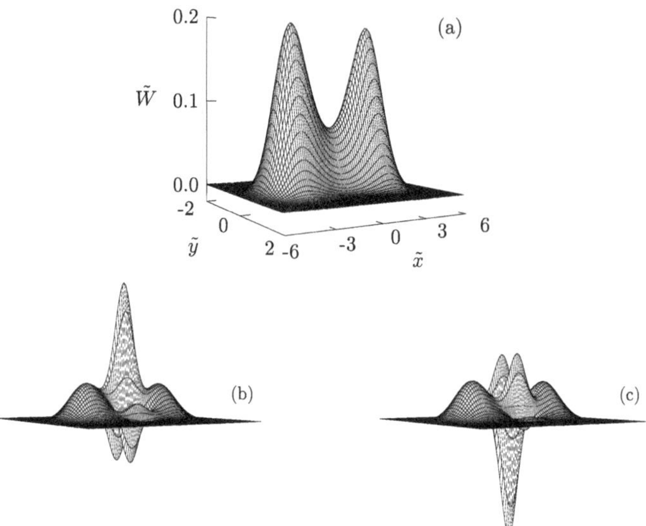

Fig. 18.8. Monte Carlo simulation of quantum trajectories with photoelectron counting records for the degenerate parametric oscillator with adiabatic elimination of the pump. Wigner distributions are computed from the time-averaged conditional density operator with the average taken over (**a**) all time intervals between photoelectron counts, and (**b**) and (**c**) all time intervals preceded, respectively, by even and odd numbers of photoelectron counts; for $g|\bar{\mathcal{E}}_0|/\kappa\kappa_p = 1.15$ and $g^2/4\kappa\kappa_p = 10^{-2}$ (mean intracavity photon number $\langle a^\dagger a\rangle_{\rm ss} = 6.6$)

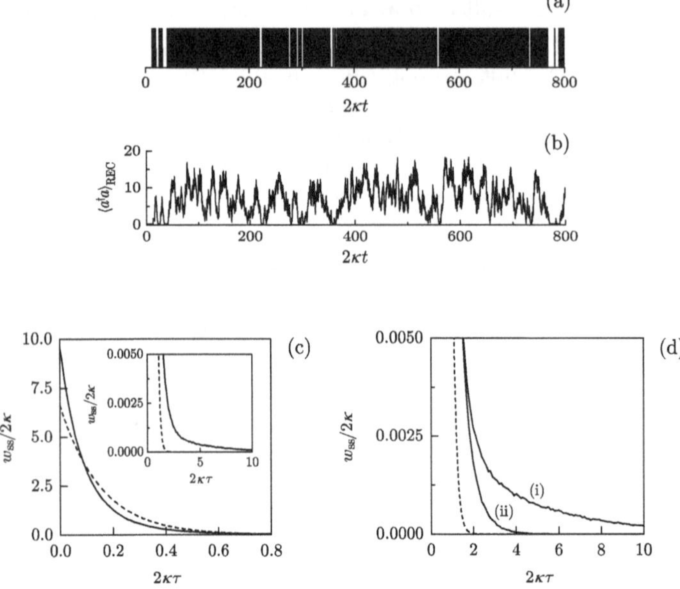

Fig. 18.9. Time series and photoelectron waiting-time distributions computed from the Monte Carlo simulations of Fig. 18.8: (**a**) sample photoelectron count record, (**b**) corresponding conditional intracavity photon number expectation, (**c**) photoelectron waiting-time distribution, and (**d**) distribution of waiting times for intervals preceeded by (i) an even number and (ii) an odd number of counts; the *dashed line* is the waiting-time distribution for coherent light of the same photon flux

odd number of counts. The distributions are not the same: very long waiting times are far more likely to be preceeded by an even number of counts. This occurs because, for the chosen parameters, gaps appear in the photoelectron count record, as seen in Fig. 18.9a, and they are always preceeded by an even number of counts (subharmonic photons are created in pairs). Of course, the idealized record, by assumption, includes a count for every photon.

To carry our question one step further, we might ask whether the interference fringes appearing in Fig. 18.8 can be recovered from some other type of scattering record. As noted, the phase of the interference fringe correlates with the parity of the cumulative emitted photon number. Certainly, then, a photoelectron counting record must be taken in order to sort the data sets, after which, a complementary unraveling might reveal the fringes of the separated even and odd subensembles. Indeed, this may be accomplished in the following way.

First, a photoelectron counting record is taken for set time T, and then the pump is turned off ($\lambda = 0$) and the subharmonic field allowed to freely decay. During the decay the record keeping is switched to homodyne detection, with local oscillator phase θ and an exponentially decaying (temporally mode-matched) local oscillator amplitude. Throughout the decay the net charge deposited in the detector is integrated over time, so that after all the light has left the cavity the homodyne record is the real number

$$Q = \sqrt{\kappa/2}(Ge|\mathcal{E}_{\mathrm{lo}}|)^{-1} \int_T^\infty e^{-\kappa(t-T)} dq, \tag{18.100}$$

with dq defined in (18.79). Carmichael [18.18] has shown that the probability distribution, $P_\theta(Q)$, obtained in this way measures a marginal of the Wigner distribution representing the state of the intracavity field immediately prior to the period of free decay. The similar result for temporally mode-matched heterodyne detection—which measures the Q distribution—is derived in Sect. 18.3.2.

Figure 8.10 shows the integrated-charge probability distributions, $P_X(Q)$ and $P_Y(Q)$, obtained from Monte Carlo simulations of X- and Y-quadrature homodyne detection. In particular, frames (e) and (f) display the Y-quadrature distributions for even and odd numbers, respectively, of prior photoelectron counts. They directly reveal the different phase of the interference fringe for even and odd coherent state superpositions (Figs. 18.8b and c). Notice that the presence of the fringes also effects the X-quadrature distributions presented in frames (b) and (c).

The point to note about this example is that three unravelings of master equation (12.10) are used to produce the simulated data sets: an unraveling for photoelectron counting record up to time T and an unraveling for either X- or Y-quadrature homodyne-current records after that. Although all three unravel master equation (12.10), each employs a distinct stochastic Schrödinger evolution to produce records of a distinct character: a sequence of integers for the cumulative photoelectron counts, and of real numbers for the charge Q. The distributions over Q correlate with the parity of the cumulative photoelectron count. Superficially, there appears to be nothing extraordinary about the records. Below the surface, however, there is. We encounter the issue of quantum contexuality. Within a classical measurement paradigm we could view the records as simply "read-off" properties of the evolving system. The suggestion from the quantum trajectory scheme, on the other hand, is that there is no single system dynamics, expressible in terms of these properties (and possibly others—hidden variables), not even a stochastic one, that is able to generate records with distributions as displayed in Fig. 18.10—including the subtle correlation distinguishing n from $n+1$ prior photoelectron counts. Instead, the dynamics must adjust to the context, to recognize which recording device exists (nonlocally) in the environment. It is not at all strange that a conditional dynamics should be contextual in this way; indeed, a conditional dynamics *must* be, since that is its very nature. What is strange, however, is that quantum mechanics appears to have no other way of speaking about the generation of data sets.

For completeness, Fig. 18.11 displays results for one final measurement scheme: the same procedure to time T, followed by heterodyne rather than

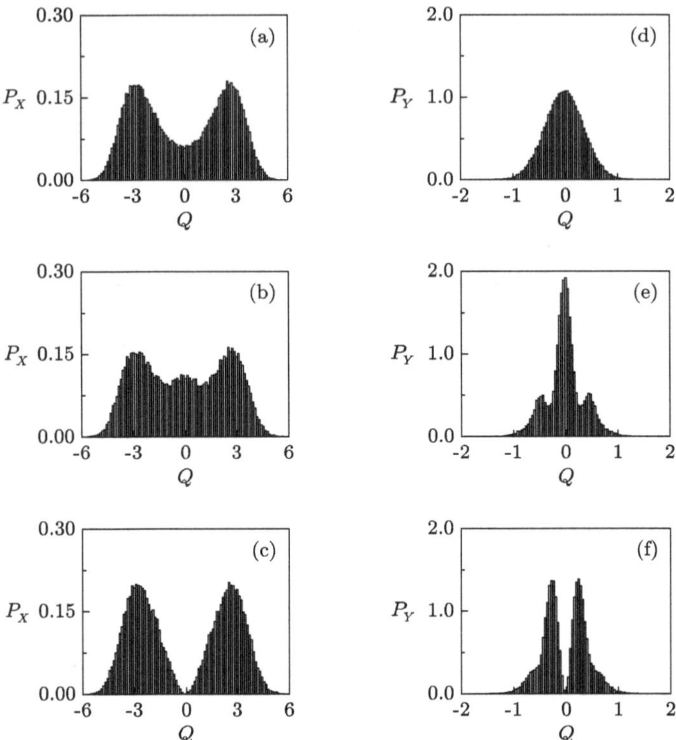

Fig. 18.10. Monte Carlo simulation of quantum trajectories with initial photoelectron counting records followed by temporally mode-matched homodyne detection records for the degenerate parametric oscillator with adiabatic elimination of the pump. Histograms of the integrated charge, Q, are shown for X-quadrature (**a**–**c**) and Y-quadrature (**d**–**e**) homodyne detection. A total of 10^5 samples are taken and histograms are taken for: (**a** and **d**) all samples; (**b** and **e**) samples with the homodyne detection preceeded by an even number of photoelectron counts (55% of samples); (**c** and **f**) samples with the homodyne detection preceeded by an odd number of photoelectric counts (45% of samples). Parameters are the same as for Figs. 18.8 and 18.9

homodyne detection. The integrated heterodyne-current record is the complex number

$$\tilde{Q} = \sqrt{\kappa/2}(Ge|\mathcal{E}_{\mathrm{lo}}|)^{-1}\int_T^\infty e^{-\kappa(t-T)}d\tilde{q}, \tag{18.101}$$

with $d\tilde{q}$ defined in (18.91). As shown in Sect. 18.3.2, the distribution $P(\tilde{Q})$ measures the Q distribution representing the state of the intracavity field immediately prior to the free decay. Again, the distributions displayed in Figs. 18.11b and c reveal a difference for even and odd numbers of prior photoelectron counts; although interference fringes are not seen due to the broad-

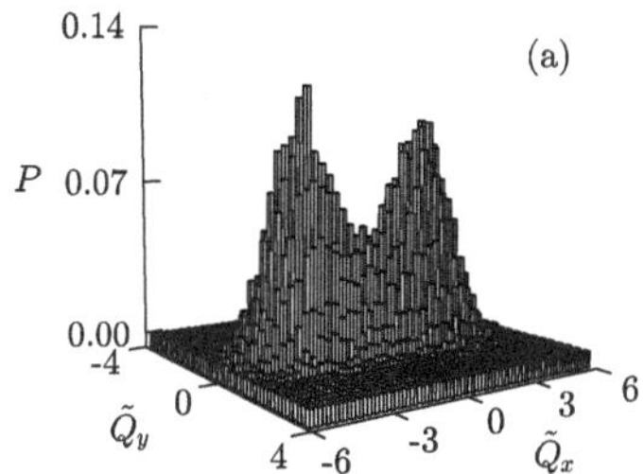

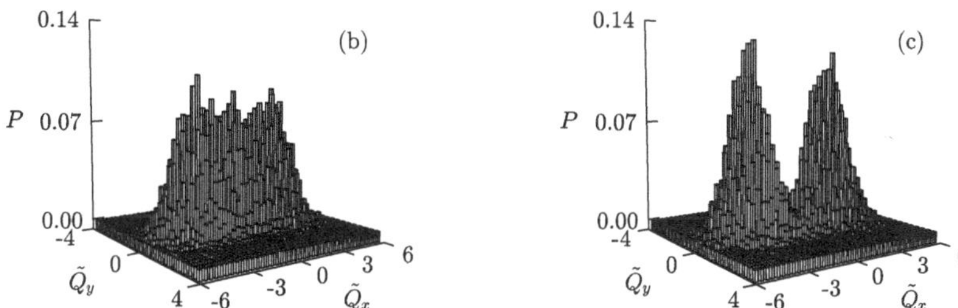

Fig. 18.11. Monte Carlo simulation of quantum trajectories with initial photoelectron counting records followed by temporally mode-matched heterodyne detection records for the degenerate parametric oscillator with adiabatic elimination of the pump. Histograms of the integrated charge, $\tilde{Q}$, are shown. A total of 10^6 samples were taken and histograms made for (**a**) all samples, (**b**) samples with the heterodyne detection preceeded by an even number of photoelectron counts (55% of samples), and (**c**) samples with the heterodyne detection preceeded by an odd number of photoelectric counts (45% of samples). Parameters are the same as for Figs. 18.8–18.10

ening of the Q relative to the Wigner distribution (Eq. 4.35b)—alternatively, the smaller signal-to-noise ratio of heterodyne detection.

18.3.2 Modeling Projective Measurements

As an approach to photoelectron counting, quantum trajectory theory appears as a quantum measurement theory. It deals, however, with destructive measurements, rather than the projective measurements of the standard von Neumann scheme. Whether or not the counted photons are in fact destroyed (absorbed) on detection is not so much the point; photons are merely *scattered* by system S, and while something might be learned about the system state from the record of photon counts, one certainly does not read off an eigenvalue a of observable $\hat{A}$, with the system state projected into the corresponding eigenstate $|a\rangle$.

Measurements of the von Neumann type can be constructed from a quantum trajectory evolution, though. In this section we explore one of the ways in which this can be done. The example suggests that we view the standard von Neumann measurement as a special case of the conditional evolution of dissipative quantum systems. It shows also how coherent states—as eigenstates of the photon annihilation operator—have a special role to play as the stationary states of the conditional evolution.

The idea is to use an optical cavity mode as a meter which having been entangled with the eigenstates $|a_j\rangle$, $j = 1, 2, \ldots$, of some observable of a second system (an atom for example) $\hat{A}$, reads out—through interaction with the environment—a particular eigenvalue $a_j = a$. The readout is performed by temporally mode-matched heterodyne detection of the cavity output. When complete, it leaves the second system in the eigenstate $|a_j\rangle = |a\rangle$. A full summary of how the scheme works is given further on. A required result to be proved first is the one claimed at the end of the last section: temporally mode-matched heterodyne detection of a freely decaying cavity mode yields records whose distribution of realizations $\tilde{Q}$ (Eq. 18.101) measures the Q distribution of the initial state of the intracavity field.

Consider an empty cavity prepared in state $|\psi(0)\rangle$. From the stochastic Schrödinger equation for heterodyne detection [Eqs. 18.90 and 18.91 with $J_{\rightarrow} = \sqrt{2\kappa}a$], free decay of the quantum state of the intracavity field conditioned on temporally mode-matched detection ($|\mathcal{E}_{\rm lo}| \rightarrow e^{-\kappa t}|\mathcal{E}_{\rm lo}|$) is governed by the equation

$$d|\bar{\bar{\bar{\psi}}}_{\rm REC}\rangle = \big(-\kappa a^\dagger a dt + \sqrt{2\kappa} a\, d\xi\big)|\bar{\bar{\bar{\psi}}}_{\rm REC}\rangle, \tag{18.102}$$

with

$$\begin{aligned} d\xi &\equiv e^{\kappa t}(Ge|\mathcal{E}_{\rm lo}|)^{-1} d\tilde{q} \\ &= \sqrt{2\kappa}\langle a^\dagger\rangle_{\rm REC} dt + dZ. \end{aligned} \tag{18.103}$$

We wish to determine the distribution, call it $P(\tilde{Q}, \tilde{Q}^*, t)$, of the cumulative complex charge

$$\tilde{Q} \equiv \sqrt{2\kappa}(Ge|\mathcal{E}_{\rm lo}|)^{-1} \int_0^t d\tilde{q} = \sqrt{2\kappa} \int_0^t e^{-\kappa t'} d\xi', \tag{18.104}$$

where the incremental charge deposited in the detector in time step dt is

$$\begin{aligned} d\tilde{Q} &= \sqrt{2\kappa} e^{-\kappa t} d\xi \\ &= e^{\kappa t}\langle a^\dagger\rangle_{\rm REC}(2\kappa e^{-2\kappa t} dt) + \sqrt{2\kappa} e^{-\kappa t} dZ. \end{aligned} \tag{18.105}$$

Our program is to solve the stochastic Schrödinger equation, (18.102), compute the conditional expectation $\langle a^\dagger\rangle_{\rm REC}$, substitute the result for $\langle a^\dagger\rangle_{\rm REC}$ into (18.105) to obtain a stochastic differential equation for $\tilde{Q}$, and finally to solve the stochastic differential equation for $\tilde{Q}$.

We begin with the transformation

$$|\bar{\bar{\bar{\psi}}}_{\rm REC}\rangle = e^{-\kappa a^\dagger a t}|\chi\rangle, \tag{18.106}$$

and, from (18.102), obtain the transformed equation

$$\begin{aligned} d|\chi\rangle &= \sqrt{2\kappa}e^{\kappa a^\dagger a t} a e^{-\kappa a^\dagger a t} d\xi|\chi\rangle \\ &= \sqrt{2\kappa}e^{-\kappa t} a d\xi|\chi\rangle \\ &= a d\tilde{Q}|\chi\rangle. \end{aligned} \tag{18.107}$$

Thus, the solution to stochastic Schrödinger equation (18.102) is obtained as

$$|\bar{\bar{\tilde{\psi}}}_{\mathrm{REC}}(t)\rangle = e^{-\kappa a^\dagger a t} e^{\tilde{Q}a}|\psi(0)\rangle. \tag{18.108}$$

Then the conditional expectation $\langle a^\dagger\rangle_{\mathrm{REC}}$ is evaluated as

$$\begin{aligned} \langle a^\dagger\rangle_{\mathrm{REC}} &= \frac{\langle\psi(0)|e^{\tilde{Q}^* a^\dagger} e^{-\kappa a^\dagger a t} a^\dagger e^{-\kappa a^\dagger a t} e^{\tilde{Q}a}|\psi(0)\rangle}{\langle\psi(0)|e^{\tilde{Q}^* a^\dagger} e^{-\kappa a^\dagger a t} e^{-\kappa a^\dagger a t} e^{\tilde{Q}a}|\psi(0)\rangle} \\ &= e^{-\kappa t}\frac{\partial}{\partial\tilde{Q}^*}\ln\left[\langle\psi(0)|e^{\tilde{Q}^* a^\dagger} e^{-2\kappa a^\dagger a t} e^{\tilde{Q}a}|\psi(0)\rangle\right]. \end{aligned} \tag{18.109}$$

Now, substituting result (18.109) into (18.105), the charge $\tilde{Q}$ deposited in the heterodyne detector satisfies the stochastic differential equation

$$d\tilde{Q} = -\frac{\partial}{\partial\tilde{Q}^*}V(\tilde{Q},\tilde{Q}^*,t)\left(2\kappa e^{-2\kappa t}dt\right) + \sqrt{2\kappa}e^{-\kappa t}dZ, \tag{18.110}$$

where we introduce the time-dependent "potential"

$$V(\tilde{Q},\tilde{Q}^*,t) \equiv -\ln\left[\langle\psi(0)|e^{\tilde{Q}^* a^\dagger} e^{-2\kappa a^\dagger a t} e^{\tilde{Q}a}|\psi(0)\rangle\right]. \tag{18.111}$$

Making the change of variable from t to

$$\eta \equiv 1 - e^{-2\kappa t}, \tag{18.112}$$

we arrive at the simpler equation for the deposited charge,

$$d\tilde{Q} = -\frac{\partial}{\partial\tilde{Q}^*}V(\tilde{Q},\tilde{Q}^*,\eta)d\eta + d\zeta, \tag{18.113}$$

where

$$V(\tilde{Q},\tilde{Q}^*,\eta) \equiv -\ln\left[\langle\psi(0)|e^{\tilde{Q}^* a^\dagger}(1-\eta)^{a^\dagger a} e^{\tilde{Q}a}|\psi(0)\rangle\right], \tag{18.114}$$

and $d\zeta$ is a Wiener increment with $\overline{d\zeta d\zeta} = \overline{d\zeta^* d\zeta^*} = 0$ and $\overline{d\zeta^* d\zeta} = d\eta$. It follows that the probability distribution over the cumulative complex charge,

$P'(\tilde{Q}, \tilde{Q}^*, \eta) \equiv P(\tilde{Q}, \tilde{Q}^*, t)$, satisfies the Fokker–Planck equation (Sect. 5.3.5)

$$\frac{\partial P'}{\partial \eta} = \left\{ \frac{\partial}{\partial \tilde{Q}} \left[\frac{\partial}{\partial \tilde{Q}^*} V(\tilde{Q}, \tilde{Q}^*, \eta) \right] + \frac{\partial}{\partial \tilde{Q}^*} \left[\frac{\partial}{\partial \tilde{Q}} V(\tilde{Q}, \tilde{Q}^*, \eta) \right] + \frac{\partial^2}{\partial \tilde{Q} \partial \tilde{Q}^*} \right\} P'. \tag{18.115}$$

We now require the solution for $t \to \infty$ (equivalently $\eta \to 1$). This last step is left as an exercise.

Exercise 18.4. Show by direct substitution or otherwise that the distribution of cumulative complex charge in temporally mode-matched heterodyne detection is given by

$$\begin{aligned} P'(\tilde{Q}, \tilde{Q}^*, \eta) &= \frac{1}{\pi\eta} e^{-\eta^{-1}|\tilde{Q}|^2} e^{-V(\tilde{Q}, \tilde{Q}^*, \eta)} \\ &= \frac{1}{\pi\eta} e^{-\eta^{-1}|\tilde{Q}|^2} \langle \psi(0)| e^{\tilde{Q}^* a^\dagger} (1-\eta)^{a^\dagger a} e^{\tilde{Q} a} |\psi(0)\rangle. \end{aligned} \tag{18.116}$$

The proposed connection with the Q distribution follows immediately from (18.116) by setting $\eta = 1$. We find

$$\begin{aligned} P(\tilde{Q}, \tilde{Q}^*, \infty) &= \frac{1}{\pi} e^{-|\tilde{Q}|^2} \langle \psi(0)| e^{\tilde{Q}^* a^\dagger} |0\rangle \langle 0| e^{\tilde{Q} a} |\psi(0)\rangle \\ &= \frac{1}{\pi} \langle \psi(0)|\tilde{Q}^*\rangle \langle \tilde{Q}^*|\psi(0)\rangle, \end{aligned} \tag{18.117}$$

where $|\tilde{Q}^*\rangle = e^{-\frac{1}{2}|\tilde{Q}|^2} e^{\tilde{Q}^* a^\dagger} |0\rangle$ is the coherent state of complex amplitude $\tilde{Q}^*$; thus,

$$P(\tilde{Q}, \tilde{Q}^*, \infty) = Q_0(\tilde{Q}^*, \tilde{Q}), \tag{18.118}$$

where the right-hand side is the Q distribution of the initial state of the intracavity field—i.e., representing density operator $\rho(0) = |\psi(0)\rangle\langle\psi(0)|$ (Eq. 4.6). Note that we have arrived at $Q_0(\tilde{Q}^*, \tilde{Q})$ rather than $Q_0(\tilde{Q}, \tilde{Q}^*)$ because the incremental charge (18.105) defines $\tilde{Q}$ in terms of the conditional expectation of $a^\dagger$ rather than of a.

The possibilities for modeling projective measurements follow directly from this result. Consider now the entangled initial state

$$|\chi(0)\rangle = \sum_{j=1}^{N} c_j |a_j\rangle |\psi_j(0)\rangle, \tag{18.119}$$

where $|a_j\rangle$, $j = 1, 2, \ldots, N$, are orthogonal eigenstates of an observable $\hat{A}$ of a second quantum system (a collection of atoms for example) and $|\psi_j(0)\rangle$ is a meter state—some state of the cavity mode that is correlated with $|a_j\rangle$. Three observations take us from initial state (18.119) to a modeling of the projective measurement of observable $\hat{A}$:

1. The derivation of $P(\tilde{Q},\tilde{Q}^*,\infty)$ carries through as before, leading us to the generalized result for the distribution of cumulative charge

$$P(\tilde{Q},\tilde{Q}^*,\infty)=\sum_{j=1}^{N}|c_j|^2Q_0^{(j)}(\tilde{Q}^*,\tilde{Q}), \tag{18.120}$$

where $Q_0^{(j)}(\tilde{Q}^*,\tilde{Q})$ is the Q distribution representing state $|\psi_j(0)\rangle$. Thus, assuming the distributions $Q_0^{(j)}(\tilde{Q}^*,\tilde{Q})$, $j=1,2,\ldots,N$, are localized in phase space, the realized records $\tilde{Q}$ fall into distinct sets, each coordinated with a particular eigenstate of $\hat{A}$. The probability of obtaining a record coordinated with $|a_j\rangle$ is as expected for a projective measurement, namely given by the coefficient $|c_j|^2$ in sum (18.120).

2. Coherent states are particularly well suited to serve as the initial meter states, $|\psi_j(0)\rangle$, $j=1,\ldots,N$, since, from (18.108), the meter states then remain coherent as the cavity mode decays. Choosing coherent states $|\alpha_j\rangle$, $j=1,2,\ldots,N$, we find

$$|\bar{\bar{\tilde{\psi}}}_{\mathrm{REC}}(t)\rangle=\sum_{j=1}^{N}c_j\exp\left[-\tfrac{1}{2}|\alpha_j|^2(1-e^{-\kappa t})\right]e^{\tilde{Q}\alpha_j}|a_j\rangle|e^{-\kappa t}\alpha_j\rangle|a_j\rangle. \tag{18.121}$$

3. For the given choice of initial coherent states, the unnormalized state in the long-time limit is

$$|\bar{\bar{\tilde{\psi}}}_{\mathrm{REC}}(\infty)\rangle=\sum_{j=1}^{N}c_je^{-\frac{1}{2}|\alpha_j|^2}e^{\tilde{Q}\alpha_j}|a_j\rangle|0\rangle, \tag{18.122}$$

where the relative weighting of the N components is determined by

$$\left|c_je^{-\frac{1}{2}|\alpha_j|^2}e^{\tilde{Q}\alpha_j}\right|^2=|c_j|^2e^{-|\alpha_j-\tilde{Q}^*|^2}e^{|\tilde{Q}|^2}. \tag{18.123}$$

Clearly, if the meter states $|\alpha_j\rangle$, $j=1,2,\ldots,N$, are well-localized in phase space (macroscopically distinct), then, from (18.120), $\tilde{Q}^*$ is close to one initial amplitude ($\tilde{Q}^*\approx\alpha_k$) and far from all others. It follows from (18.122) and (18.123), plus normalization, that

$$|\tilde{\psi}_{\mathrm{REC}}(\infty)\rangle=|a_k\rangle|0\rangle; \tag{18.124}$$

the standard state reduction (collapse of the wavefunction) has taken place.

Exercise 18.5. Design a numerical simulation to demonstrate the properties of the described modeling of a projective measurement. Investigate the dependence on the phase-space separation of the initial meter states.

19

Quantum Trajectories III: More Examples

The examples met in previous chapters could now be reanalyzed from a quantum trajectory point of view, opening up new perspectives on each—on the laser, the degenerate parametric oscillator, on optical bistability and cavity QED. In addition, much more could be said about methods of calculation and simulation within the quantum trajectory approach. There is therefore a great deal more that could be done. There is only one chapter left, however, in which to do it, and many things must be set aside.

In this final chapter we explore just a small selection of the possible extensions and applications of quantum trajectory theory. For the most part, topics are chosen not so much for their connection to examples treated in earlier chapters, but because each teaches us something new about what can be done with the quantum trajectory approach.

19.1 Photon Scattering in the Weak-Excitation Limit

We begin with the one topic that is, in fact, motivated by a calculation performed previously. Quantum trajectory theory is the natural language for developing the weak-excitation expansion of Sect. 16.1. There, assuming weak excitation, the density operator that satisfies the cavity QED master equation is factorized in pure state form—for one atom, as the pure state (16.7), and as (16.14) in the case of many atoms. While the factorization can be verified through tedious algebra—by following the sketch of Sects. 16.1.1 and 16.1.2—the physical basis and mathematical form of the approximation appear much more clearly within a quantum trajectory treatment.

Figure 19.1 uses quantum trajectories to illustrate the argument for factorization put forward in the opening paragraph of Sect. 16.1 (also see the paragraph below Note 16.1). It plots the photon number expectation, conditioned on photoelectron counting records, for a typical realization of the scattering process in single-atom cavity QED. The important point to note is that in the limit of weak excitation the quantum jumps are spaced very far apart in

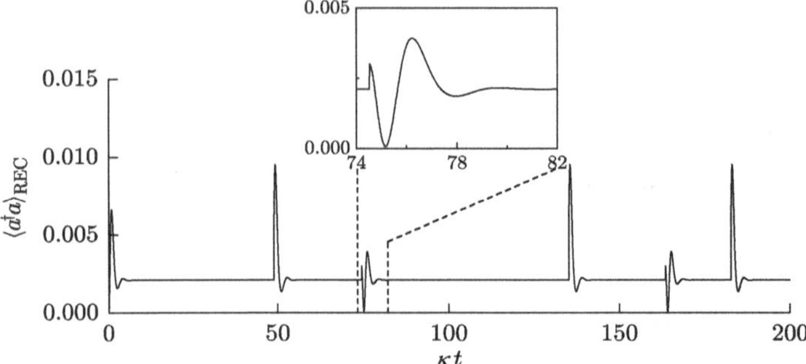

Fig. 19.1. Sample quantum trajectory with photoelectron counting records for the weak-excitation limit of single-atom cavity QED (Eq. 13.57 with $\omega_A = \omega_C = \omega_0$). The parameters are those of Fig. 16.1b: $g/\kappa = 1.85$, $\gamma/2\kappa = 1$, with driving field strength $|\bar{\mathcal{E}}_0|/\kappa = 0.2$. The smaller amplitude perturbations correspond to cavity emissions (given as the *inset*) and the larger to atomic emissions

comparison with the short time taken to recover the steady state following a quantum jump; thus, when averaging over scattering records, the dominant term—in the summation (17.8)—comes from the steady state reached under the nonunitary Schrödinger evolution between quantum jumps. This observation is verified by the equations for state amplitudes (Eqs. 16.8 and 16.15), which are, indeed, precisely those obtained from the nonunitary Schrödinger equation (Eq. 17.50). The conclusion is in fact apparent from the equation of motion (16.19) for the factorized state, which, in anticipation of quantum trajectory theory, is just the between-jump Schrödinger equation for master equation (16.16) with the second-order term $-i\bar{\mathcal{E}}_0^* a$ neglected.

The treatment in Sects. 16.1.3 and 16.2 of forwards photon scattering in cavity QED is also, in essence, a quantum trajectory calculation. The derived correlation function is simply the response of the photon number expectation to an isolated cavity emission—i.e., the response shown inset in Fig. 19.1. The physical content of the weak-excitation approximation is then as we just noted: one and only one term is kept in the sum over scattering records. The approximation is acceptable so long as the recovery of the steady state after a quantum jump (regression of the quantum fluctuation) is not interrupted—or more correctly rarely interrupted—by another quantum jump, either of cavity- or atomic-emission type. Of course, the experimental measurement of the second-order correlation function [19.1, 19.2, 19.3] records only those rare events where the first cavity emission *is* followed by a second; nevertheless, in the weak-excitation limit, the *probability* of the second is determined by the uninterrupted coherent evolution displayed by the inset in Fig. 19.1.

We see from this that our earlier calculation of $g^{(2)}(\tau)$ is, at its root, based upon a perturbation expansion, one in which the number of photon emissions (quantum jumps) sets the order of each term in the expansion. To formulate the approximation we used in a more precise way, we may start from the full quantum trajectory expansion of $g^{(2)}(\tau)$. Thus, we define photon emission events

$$\{A\} \equiv \begin{Bmatrix} \kappa \\ T \end{Bmatrix}, \qquad \{B\} \equiv \begin{Bmatrix} \kappa \\ T' \end{Bmatrix}, \tag{19.1}$$

to correspond to photoelectron counts in the respective intervals $T \equiv [t, t+dt)$ and $T' \equiv [t', t'+dt')$. Aside from an overall normalization, $g^{(2)}(\tau)$ is the joint probability $P(\{A\} \wedge \{B\})$—the probability for both events to occur. Now let REC1 denote the photoelectron counting record (all output channels) up to time t, and REC2 the record from $t + dt$ to t', $t' > t$. The required joint probability may then be expanded as

$$\begin{aligned}
&P(\{A\} \wedge \{B\}) \\
&= \sum_{\mathrm{REC1}} P(\mathrm{REC1} \wedge \{A\}) \sum_{\mathrm{REC2}} P(\mathrm{REC2} \wedge \{B\} | \mathrm{REC1} \wedge \{A\}) \\
&= \sum_{\mathrm{REC1}} P(\mathrm{REC1}) \Big[2\kappa \langle (a^\dagger a)(t) \rangle_{\mathrm{REC1}} dt \Big] \sum_{\mathrm{REC2}} P(\mathrm{REC2} | \mathrm{REC1} \wedge \{A\}) \\
&\quad \times \Big[2\kappa \langle \tilde{\psi}_{\mathrm{REC2}|\mathrm{REC1}\wedge\{A\}}(t') | a^\dagger a | \tilde{\psi}_{\mathrm{REC2}|\mathrm{REC1}\wedge\{A\}}(t') \rangle dt' \Big] \\
&= \sum_{\mathrm{REC1}} P(\mathrm{REC1}) \Big[2\kappa \langle (a^\dagger a)(t) \rangle_{\mathrm{REC1}} dt \Big] \sum_{\mathrm{REC2}} P(\mathrm{REC2} | \mathrm{REC1} \wedge \{A\}) \\
&\quad \times \mathrm{tr} \Big[a^\dagger a | \tilde{\psi}_{\mathrm{REC2}|\mathrm{REC1}\wedge\{A\}}(t') \rangle \langle \tilde{\psi}_{\mathrm{REC2}|\mathrm{REC1}\wedge\{A\}}(t') | \Big] 2\kappa dt'.
\end{aligned} \tag{19.2}$$

The sum over REC2 yields the density operator, $\tilde{\rho}_{\mathrm{REC1}\wedge\{A\}}(t')$, reached from the pure state (realized at time $t+dt$) with prior scattering record REC1$\wedge\{A\}$. The unnormalized state at this time is $a|\tilde{\psi}_{\mathrm{REC1}}(t)\rangle$. Therefore, the sum over REC2 yields

$$\begin{aligned}
P(\{A\} \wedge \{B\}) &= \sum_{\mathrm{REC1}} P(\mathrm{REC1}) \Big[2\kappa \langle (a^\dagger a)(t) \rangle_{\mathrm{REC1}} dt \Big] \\
&\quad \times \mathrm{tr} \left\{ a^\dagger a \, e^{\tilde{\mathcal{L}}(t'-t)} \left[\frac{a|\tilde{\psi}_{\mathrm{REC1}}(t)\rangle\langle\tilde{\psi}_{\mathrm{REC1}}(t)|a^\dagger}{\langle\tilde{\psi}_{\mathrm{REC1}}(t)|a^\dagger a|\tilde{\psi}_{\mathrm{REC1}}(t)\rangle} \right] \right\} 2\kappa dt' \\
&= \mathrm{tr}\{a^\dagger a \, e^{\tilde{\mathcal{L}}(t'-t)} [a\tilde{\rho}(t)a^\dagger]\} (2\kappa dt)(2\kappa dt'),
\end{aligned} \tag{19.3}$$

where we use

$$\tilde{\rho}(t) = \sum_{\mathrm{REC1}} P(\mathrm{REC1}) |\tilde{\psi}_{\mathrm{REC1}}(t)\rangle\langle\tilde{\psi}_{\mathrm{REC1}}(t)|. \tag{19.4}$$

The final result reproduces the expression for the joint photoelectron counting probability given by the quantum regression formula (Eq. 1.102).

Returning to the second line of (19.2), the approximation for weak excitation is made by retaining only dominant terms in the sums over scattering records. We assume first that for all records up to time t of non-negligible probability, the conditional state reached is

$$|\tilde{\psi}_{\mathrm{REC1}}(t)\rangle = |\tilde{\psi}_{\mathrm{ss}}\rangle, \tag{19.5}$$

where $|\tilde{\psi}_{\mathrm{ss}}\rangle$ denotes the between-jump steady state, satisfying

$$\tilde{H}_B|\tilde{\psi}_{\mathrm{ss}}\rangle = 0. \tag{19.6}$$

The approximation is a good one for a situation like that depicted Fig. 19.1, where the vast majority of quantum jumps do occur out of the between-jump steady state. Furthermore, we assume that between $t+dt$ to t' only the no-count record is realized with non-negligible probability. Under these two assumptions, working from the second line of (19.2) we may write ($\tau \equiv t'-t$)

$$\begin{aligned} g^{(2)}(\tau) &= (2\kappa dt)^{-1}(2\kappa dt')^{-1}\frac{P(\{A\}\wedge\{B\})}{(\langle a^\dagger a\rangle_{\mathrm{ss}})^2} \\ &= \frac{\langle\tilde{\psi}_{\{t\emptyset_{t+\tau}\}}(\tau)|a^\dagger a|\tilde{\psi}_{\{t\emptyset_{t+\tau}\}}(\tau)\rangle}{\langle a^\dagger a\rangle_{\mathrm{ss}}}, \end{aligned} \tag{19.7}$$

where $\langle a^\dagger a\rangle_{\mathrm{ss}}$ is the photon number expectation computed in state $|\tilde{\psi}_{\mathrm{ss}}\rangle$, and

$$|\tilde{\psi}_{\{t\emptyset_{t+\tau}\}}(\tau)\rangle = \frac{|\tilde{\bar{\psi}}_{\{t\emptyset_{t+\tau}\}}(\tau)\rangle}{\sqrt{\langle\tilde{\bar{\psi}}_{\{t\emptyset_{t+\tau}\}}(\tau)|\tilde{\bar{\psi}}_{\{t\emptyset_{t+\tau}\}}(\tau)\rangle}}, \tag{19.8}$$

with the unnormalized state at time $t' = t+\tau$, $|\tilde{\bar{\psi}}_{\{t\emptyset_{t+\tau}\}}(\tau)\rangle$, satisfying the between-jump nonunitary Schrödinger equation with initial condition

$$|\tilde{\bar{\psi}}_{\{t\emptyset_{t+\tau}\}}(0)\rangle = \frac{a|\tilde{\psi}_{\mathrm{ss}}\rangle}{\sqrt{\langle\tilde{\psi}_{\mathrm{ss}}|a^\dagger a|\tilde{\psi}_{\mathrm{ss}}\rangle}}. \tag{19.9}$$

Equations 19.7–19.9 correspond to (16.24)–(16.26).

The procedure followed might be taken a step further by retaining higher-order terms (records involving more quantum jumps) in the summations of (19.2). In particular, it is important to note that the scenario just presented does not always hold in the weak-excitation limit; for example, sometimes, even for weak excitation, an additional term must be kept in order to obtain the lowest-order nontrivial result. Such a situation is illustrated by Fig. 19.2, where a typical trajectory for the degenerate parametric oscillator operated well below threshold is plotted. In this case, because photons are created inside the cavity in pairs, a first photon emission out of the between-jump steady state is followed by a second photon emission before the steady state is recovered. The first assumption above (Eq. 19.5) still holds, but we must now

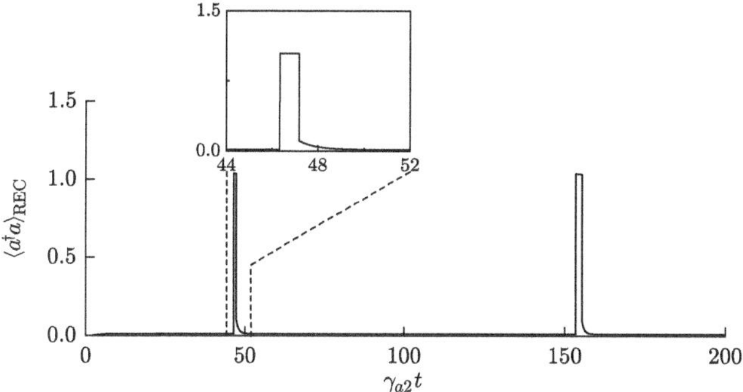

Fig. 19.2. Sample quantum trajectory with photoelectron counting records for the degenerate parametric oscillator well below threshold (Eqs. 18.3 and 18.4); for $\gamma_{a1} = \gamma_{b1} = 0$, $\gamma_{a2} = \gamma_{b2} = 4g$, and $2|\bar{\mathcal{E}}_0|/\gamma_{a2} = 0.5$. The quantum jumps occur in pairs, separated on average by the photon lifetime of the subharmonic mode

sum over two records between $t + dt$ and t': the no-count record, as before, and all records of the form

$$\{C_{T''}\} \equiv \left\{ {}_t\emptyset, \overset{\kappa}{T''}, \emptyset_{t+\tau} \right\} \tag{19.10}$$

that include an additional photon emission in the interval

$$T'' \equiv [t'', t'' + dt'') \equiv [t + \tau', t + \tau' + d\tau'). \tag{19.11}$$

Working as before from the second line of (19.2), the correlation function is written as

$$\begin{aligned} g^{(2)}(\tau) &= (2\kappa dt)^{-1}(2\kappa dt')^{-1}\frac{P(\{A\} \wedge \{B\})}{(\langle a^\dagger a\rangle_{\rm ss})^2} \\ &= P(\{{}_t\emptyset_{t+\tau}\})\frac{\langle\tilde{\psi}_{\{{}_t\emptyset_{t+\tau}\}}(\tau)|a^\dagger a|\tilde{\psi}_{\{{}_t\emptyset_{t+\tau}\}}(\tau)\rangle}{\langle a^\dagger a\rangle_{\rm ss}} \\ &\quad + \int_t^{t+\tau} d\tau' P(\{C_{T''}\})\frac{\langle\tilde{\psi}_{\{C_{T''}\}}(\tau)|a^\dagger a|\tilde{\psi}_{\{C_{T''}\}}(\tau)\rangle}{\langle a^\dagger a\rangle_{\rm ss}}, \end{aligned} \tag{19.12}$$

where the quantum state conditioned on the record $\{C_{T''}\}$,

$$|\tilde{\psi}_{\{C_{T''}\}}(\tau)\rangle = \frac{|\tilde{\bar{\psi}}_{\{C_{T''}\}}(\tau)\rangle}{\sqrt{\langle\tilde{\bar{\psi}}_{\{C_{T''}\}}(\tau)|\tilde{\bar{\psi}}_{\{C_{T''}\}}(\tau)\rangle}}, \tag{19.13}$$

solves the between-jump nonunitary Schrödinger equation with initial condition

$$|\tilde{\bar{\psi}}_{\{C_{T''}\}}(\tau')\rangle = 2\kappa a|\tilde{\bar{\psi}}_{\{{}_t\emptyset_{t+\tau'}\}}(\tau')\rangle. \tag{19.14}$$

Since record probabilities $P(\{{}_t\emptyset_{t+\tau}\})$ and $P(\{C_{T''}\})$ are given by the corresponding state norms, the correlation function is most readily calculated from

$$g^{(2)}(\tau) = \frac{\langle\bar{\bar{\psi}}_{\{{}_t\emptyset_{t+\tau}\}}(\tau)|a^\dagger a|\bar{\bar{\psi}}_{\{{}_t\emptyset_{t+\tau}\}}(\tau)\rangle}{\langle a^\dagger a\rangle_{\rm ss}} + \int_t^{t+\tau} d\tau' \frac{\langle\bar{\bar{\psi}}_{\{C_{T''}\}}(\tau)|a^\dagger a|\bar{\bar{\psi}}_{\{C_{T''}\}}(\tau)\rangle}{\langle a^\dagger a\rangle_{\rm ss}}. \tag{19.15}$$

Exercise 19.1. Show that for the degenerate parametric oscillator with weak coupling and adiabatic elimination of the pump (Eq. 12.10 with $g^2/\kappa\kappa_p \ll 1$)

$$\langle\bar{\bar{\psi}}_{\{{}_t\emptyset_{t+\tau}\}}(\tau)|a^\dagger a|\bar{\bar{\psi}}_{\{{}_t\emptyset_{t+\tau}\}}(\tau)\rangle = e^{-2\kappa\tau}, \tag{19.16a}$$

and

$$\langle\bar{\bar{\psi}}_{\{C_{T''}\}}(\tau)|a^\dagger a|\bar{\bar{\psi}}_{\{C_{T''}\}}(\tau)\rangle = 2\kappa(\lambda/2)^2[1 - e^{-2\kappa(\tau-\tau')}]^2 e^{-2\kappa\tau'}; \tag{19.16b}$$

hence show that in the weak-excitation limit $[\langle a^\dagger a\rangle_{\rm ss} = (\lambda/2)^2 \ll 1]$ the *second-order correlation function for the degenerate parametric oscillator* obtained from (19.15) is

$$g^{(2)}(\tau) = 1 + (2/\lambda)^2 e^{-2\kappa\tau}. \tag{19.17}$$

Note the extreme photon bunching as $\lambda \to 0$.

19.2 Unraveling the Density Operator: Cascaded Systems

In Chap. 18 we met the three principal unravelings of the density operator used in quantum optics, those based on photoelectron counting, homodyne and heterodyne detection. There are numerous other possibilities, many of which involve combinations of these three. A good example is conditional homodyne detection, which combines photoelectron counting with homodyne detection. The scheme, introduced by Carmichael and Orozco [19.4], effects a direct measurement of the spectrum of squeezing; it uncovers surprising aspects of the weak squeezing limit, where, by an appropriate measure, both squeezed *and* unsqueezed fluctuations are shown to be nonclassical—most surprisingly, the degree of nonclassicality *increases* as the squeezing is reduced. Conditional homodyne detection was implemented by Foster and coworkers [19.5, 19.6] for many-atom cavity QED, where it provides a sensitive probe of field fluctuations in forwards scattering, a complement to the intensity fluctuations discussed in Sects. 16.1.3 and 16.2. An example in a somewhat different vein

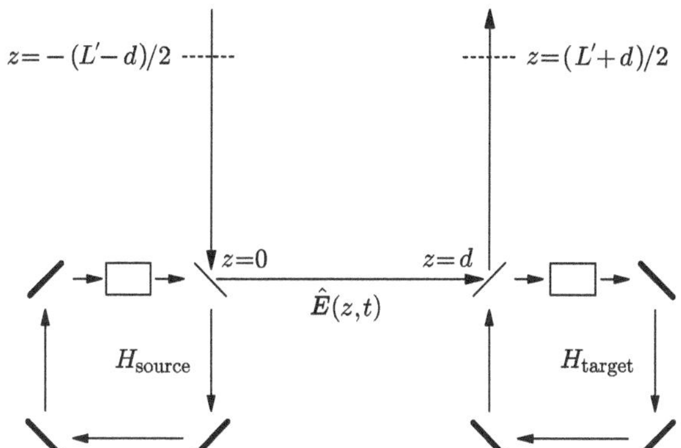

Fig. 19.3. A typical cascaded system. Light generated in the source subsystem, the cavity on the left, is coupled into the target subsystem, the cavity on the right. The coupling is unidirectional and mediated by the reservoir field $\hat{\boldsymbol{E}}(z,t)$

is provided by the quantum trajectory formalism for continuous variable quantum teleportation [19.7]. It also is built upon the three standard unravelings.

The possibilities are in fact endless. But our aim in this book is not simply to enumerate unravelings. We aim to gain an understanding of the principles upon which they are built. In this regard one more unraveling, in particular, should be mentioned, as it introduces principles for the treatment of a whole new class of systems—cascaded systems, i.e., systems of the kind sketched in Fig. 19.3.

The example sketched involves a source subsystem, represented by a ring-cavity mode, with possible intracavity interaction (Hamiltonian H_{source}), whose output couples to a target subsystem, represented by a second ring-cavity mode, also with possible intracavity interaction (Hamiltonian H_{target}). The source output field is carried to the target by the reservoir, which mediates a unidirectional coupling. The reservoir field is expanded as

$$\hat{\boldsymbol{E}}(z,t) = \hat{\boldsymbol{E}}^{(+)}(z,t) + \hat{\boldsymbol{E}}^{(-)}(z,t), \tag{19.18a}$$

with

$$\hat{\boldsymbol{E}}^{(+)}(z,t) = i\hat{e}_0 \sum_j \sqrt{\frac{\hbar\omega_j}{2\epsilon_0 AL'}}\, r_j(t) e^{i[(\omega_j/c)z+\phi(z)]}, \tag{19.18b}$$

$$\hat{\boldsymbol{E}}^{(-)}(z,t) = \hat{\boldsymbol{E}}^{(+)}(z,t)^\dagger, \tag{19.18c}$$

where

$$\phi(z) \equiv \begin{cases} \phi_R^a + \phi_R^b & z > d \\ \phi_R^a & 0 > z \le d \\ 0 & z \le 0; \end{cases} \tag{19.19}$$

ϕ_R^a and ϕ_R^b are the respective phase changes upon reflection from the source output mirror at $z = 0$ and the target input mirror at $z = d$.

19.2.1 System–Reservoir Interaction Hamiltonian

Our aim is to derive a master equation for the composite system, S, of source and target, and to decompose the field $\hat{\boldsymbol{E}}(z,t)$ into free- and source-field components, as was done in Sects. 7.3.1 and 9.2.5 [also see (13.19)–(13.21)]. We begin from Hamiltonian (1.16) for the entire system $S \otimes R$; the pieces of the Hamiltonian are

$$H_S = H_{\text{source}} + H_{\text{target}}, \tag{19.20a}$$

$$H_R = \sum_j \hbar\omega_j r_j^\dagger r_j, \tag{19.20b}$$

$$H_{SR} = H_{SR}^{\text{source}} + H_{SR}^{\text{target}}, \tag{19.20c}$$

with reservoir interactions

$$H_{SR}^{\text{source}} = \sum_j \hbar(\kappa_{aj}^* a r_j^\dagger + \text{H.c.}), \tag{19.21a}$$

$$H_{SR}^{\text{target}} = \sum_j \hbar\big(\kappa_{bj}^* b r_j^\dagger e^{-i[(\omega_j/c)d+\phi_R^a]} + \text{H.c.}\big), \tag{19.21b}$$

where a and b are photon annihilation operators for the source and target modes, respectively. The reservoir interactions account for transmission and reflection of the fields at $z = 0$ and $z = d$ (Fig. 19.3). Thus, if κ_a and κ_b denote the source and target mode decay rates, following (7.112), explicit expressions for the reservoir coupling coefficients may be written as

$$\kappa_{aj}^* = i\sqrt{2\kappa_a}\sqrt{\frac{c}{L'}}e^{i(\phi_T^a-\phi_R^a)} = \sqrt{2\kappa_a}\sqrt{\frac{c}{L'}}, \tag{19.22a}$$

$$\kappa_{bj}^* = i\sqrt{2\kappa_b}\sqrt{\frac{c}{L'}}e^{i(\phi_T^b-\phi_R^b)} = \sqrt{2\kappa_b}\sqrt{\frac{c}{L'}}, \tag{19.22b}$$

where, as usual, in the Markov approximation, the coupling coefficients are taken to be independent of the reservoir mode frequency ω_j. Note also that the phase changes on reflection and transmission satisfy $\phi_R^a - \phi_T^a = \phi_R^b - \phi_T^b = \pi/2$ (see Note 7.12). Then, expressing the reservoir field $\hat{\boldsymbol{E}}(z,t)$ in units of the square root of photon flux, with Heisenberg operator

$$\hat{\mathcal{E}}(z,t) \equiv -i\sqrt{\frac{2\epsilon_0 Ac}{\hbar\frac{1}{2}(\omega_a+\omega_b)}}\hat{e}_0 \cdot \hat{\boldsymbol{E}}^{(+)}(z,t), \tag{19.23}$$

and setting $\omega_j/\frac{1}{2}(\omega_a+\omega_b) \to 1$, the system–reservoir interaction Hamiltonians take the simple form

$$H_{SR}^{\text{source}} = \hbar\sqrt{2\kappa_a}\big[a\hat{\mathcal{E}}^\dagger(0) + \text{H.c.}\big], \tag{19.24a}$$

$$H_{SR}^{\text{target}} = \hbar\sqrt{2\kappa_b}\big[b\hat{\mathcal{E}}^\dagger(d) + \text{H.c.}\big], \tag{19.24b}$$

where we introduce the Schrödinger picture operator $\hat{\mathcal{E}}(z) \equiv \hat{\mathcal{E}}(z,0)$.

19.2.2 Reservoir Field

Solving Heisenberg equations of motion for the mode annihilation operators $r_j(t)$ yields a decomposition of the reservoir field into free- and source-field terms (Sect. 7.3.1). We write

$$\hat{\boldsymbol{E}}^{(+)}(z,t) = \hat{\boldsymbol{E}}_f^{(+)}(z,t) + \hat{\boldsymbol{E}}_{\text{source}}^{(+)}(z,t) + \hat{\boldsymbol{E}}_{\text{target}}^{(+)}(z,t), \tag{19.25}$$

where the free field is

$$\hat{\boldsymbol{E}}_f^{(+)}(z,t) = i\hat{e}_0\sum_j\sqrt{\frac{\hbar\omega_j}{2\epsilon_0 AL'}}r_j(0)e^{-i[\omega_j(t-z/c)-\phi(z)]}, \tag{19.26}$$

and the source fields are (Eq. 7.111)

$$\hat{\boldsymbol{E}}_{\text{source}}^{(+)}(z,t) = \begin{cases} \hat{e}_0\sqrt{\dfrac{\hbar\omega_a}{2\epsilon_0 Ac}}\sqrt{\dfrac{L'}{c}}\kappa_a^*(\omega_a)e^{i\phi(z)}a(t') & ct > z > 0 \\ \dfrac{1}{2}\hat{e}_0\sqrt{\dfrac{\hbar\omega_a}{2\epsilon_0 Ac}}\sqrt{\dfrac{L'}{c}}\kappa_a^*(\omega_a)a(t') & z = 0 \\ 0 & z < 0, \end{cases} \tag{19.27a}$$

for the source subsystem, with $t' = t - z/c$, and

$$\hat{\boldsymbol{E}}_{\text{target}}^{(+)}(z,t) = \begin{cases} \hat{e}_0\sqrt{\dfrac{\hbar\omega_b}{2\epsilon_0 Ac}}\sqrt{\dfrac{L'}{c}}\kappa_b^*(\omega_b)e^{i\phi_R^b}b(t'') & ct > z > d \\ \dfrac{1}{2}\hat{e}_0\sqrt{\dfrac{\hbar\omega_b}{2\epsilon_0 Ac}}\sqrt{\dfrac{L'}{c}}\kappa_b^*(\omega_b)b(t'') & z = d \\ 0 & z < d, \end{cases} \tag{19.27b}$$

for the target subsystem, with $t'' = t - (z-d)/c$. Note that $\hat{\boldsymbol{E}}_{\text{source}}^{(+)}(z,t)$ and $\hat{\boldsymbol{E}}_{\text{target}}^{(+)}(z,t)$ show strict discontinuities at $z = 0$ and $z = d$, respectively. This arises from the δ-function in the integrand of (7.111), a result of the Markov approximation; in this approximation, rapid changes of the field in space—changes over a distance of a few wavelengths—appear as discontinuities. What we have here is a spatial counterpart to the singular appearance of rapid changes (on the order of the optical period) of the reservoir correlation function in time (Sect. 1.3.3 and below Eq. 1.52).

Alternatively, using the definition (19.23), written in units of the square root of photon flux, the reservoir field is

$$\hat{\mathcal{E}}(z,t) = \hat{\mathcal{E}}_f(z,t) + \hat{\mathcal{E}}_{\text{source}}(z,t) + \hat{\mathcal{E}}_{\text{target}}(z,t), \tag{19.28}$$

with free field

$$\hat{\mathcal{E}}_f(z,t) = \sqrt{\frac{c}{L'}}\sum_j \sqrt{\frac{2\omega_j}{\omega_a+\omega_b}}\, r_j(0) e^{-i[\omega_j(t-z/c)-\phi(z)]}, \tag{19.29}$$

and source fields

$$\hat{\mathcal{E}}_{\text{source}}(z,t) = \begin{cases} -ie^{i\phi(z)}\sqrt{2\kappa_a}\,a(t') & ct > z > 0 \\ -i\frac{1}{2}\sqrt{2\kappa_a}\,a(t') & z = 0 \\ 0 & z < 0, \end{cases} \tag{19.30a}$$

with $t' = t - z/c$, and

$$\hat{\mathcal{E}}_{\text{target}}(z,t) = \begin{cases} -ie^{i\phi_R^b}\sqrt{2\kappa_b}\,b(t'') & ct > z > d \\ -i\frac{1}{2}\sqrt{2\kappa_b}\,b(t'') & z = d \\ 0 & z < d, \end{cases} \tag{19.30b}$$

with $t'' = t - (z-d)/c$. Here, once again, on the basis of the Markov approximation we can write $2\omega_a/(\omega_a+\omega_b) \approx 2\omega_b/(\omega_a+\omega_b) \approx 1$.

19.2.3 The Cascaded Systems Master Equation

Three derivations of the cascaded systems master equation have been given, the first by Kolobov and Sokolov [19.8], and the other two, independently and more-or-less simultaneously, by Gardiner [19.9] and Carmichael [19.10]. We follow the approach of Carmichael, starting, as in Sect. 1.3.1, with the Schrödinger equation for the complete system and reservoir $S \otimes R$; thus, with $H = H_S + H_R + H_{SR}$ defined by (19.20a)–(19.21b), we begin with the Schrödinger equation

$$\dot{\chi} = \frac{1}{i\hbar}[H,\chi]. \tag{19.31}$$

From this point, the important difference from the development in Sects. 1.3 and 1.4 is that while the reservoirs seen by the source and target differ—they couple to the different field operators $\hat{\mathcal{E}}(0)$ and $\hat{\mathcal{E}}(d)$—they are, nevertheless, clearly correlated, as the field at $z=0$ freely propagates to $z=d$. Moreover, the field seen by the target carries photons emitted by the source, so it hardly constitutes a thermal reservoir as was assumed before.

To cope with these differences, our first step is to introduce the time-retarded density operator

$$\chi_{\text{ret}} \equiv U_{\text{source}}(-d/c)\chi U^\dagger_{\text{source}}(-d/c), \tag{19.32}$$

with

$$U_{\rm source}(-d/c) \equiv \exp\left[\frac{1}{i\hbar}(H_{\rm source} + H_R + H_{SR}^{\rm source})(-d/c)\right]. \tag{19.33}$$

The retardation applies to the source subsystem and its interaction with the reservoir; it is introduced to formally undo the propagation of the output field of the source to the target. The Schrödinger equation satisfied by the time-retarded density operator is

$$\dot{\chi}_{\rm ret} = \frac{1}{i\hbar}[H_{\rm ret}, \chi_{\rm ret}], \tag{19.34}$$

where $H_{\rm ret}$ differs from H through the changed interaction term

$$(H_{SR}^{\rm target})_{\rm ret} = \hbar\sqrt{2\kappa_b}\big[b\hat{\mathcal{E}}^\dagger(d_-, d/c) + {\rm H.c.}\big], \tag{19.35}$$

with

$$\hat{\mathcal{E}}(d_-, d/c) \equiv U_{\rm source}^\dagger(d/c)\hat{\mathcal{E}}(d)U_{\rm source}(d/c). \tag{19.36}$$

The field $\hat{\mathcal{E}}(d_-, d/c)$ is the formal solution, at retardation time d/c, to the Heisenberg equation of motion for the reservoir field operator at the location of the target $z = d$. It is important to note that the time evolution is generated by Hamiltonian $H - H_{SR}^{\rm target} = H_{\rm source} + H_R + H_{SR}^{\rm source}$; hence we write $\hat{\mathcal{E}}(d_-, d/c)$, rather than $\hat{\mathcal{E}}(d, d/c)$, and have (Eqs. 19.28–19.30)

$$\hat{\mathcal{E}}(d_-, d/c) = \hat{\mathcal{E}}_f(d, d/c) + \hat{\mathcal{E}}_{\rm source}(d, d/c), \tag{19.37}$$

where there is no self-field term $\hat{\mathcal{E}}_{\rm target}(d, d/c)$, contrary to what might be expected by looking at (19.28).

We may now use the fact that the reservoir field propagates freely from the source output ($z = 0_+$) to the target input ($z = d_-$) to write

$$\begin{aligned}\hat{\mathcal{E}}(d_-, d/c) = \mathcal{E}(0_+) &= \hat{\mathcal{E}}_f(0_+) - ie^{i\phi_R^a}\sqrt{2\kappa_a}a \\ &= e^{i\phi_R^a}\big[\hat{\mathcal{E}}(0) - i\tfrac{1}{2}\sqrt{2\kappa_a}a\big],\end{aligned} \tag{19.38}$$

where the first and second lines follow, respectively, from the first and second lines of (19.30a). Now, after substituting (19.38) into (19.35), the Schrödinger equation for the time-retarded density operator is to be solved using Hamiltonian

$$H_{\rm ret} = H_S^c + H_R + H_{SR}^c, \tag{19.39}$$

where

$$H_S^c = H_{\rm source} + H_{\rm target} + i\hbar\sqrt{\kappa_a\kappa_b}(e^{-i\phi_R^a}ba^\dagger - {\rm H.c.}), \tag{19.40}$$

and

$$H_{SR}^c = \hbar\big[(\sqrt{2\kappa_a}a + e^{-i\phi_R^a}\sqrt{2\kappa_b}b)\hat{\mathcal{E}}^\dagger(0) + {\rm H.c.}\big]. \tag{19.41}$$

Thus, by introducing the retardation, the system–reservoir interaction is recast in a form that allows the master equation to be derived from the standard prescription of Sects. 1.3 and 1.4; both subsystems now couple to the same reservoir operator—the field at the source $\hat{\mathcal{E}}(0)$—in (19.41). The only novelty is that the coupling is collective—i.e., made through the sum of fields $\sqrt{2\kappa_a}a + e^{-i\phi_R^a}\sqrt{2\kappa_b}b$. Introducing the reduced density operator

$$\rho_{\mathrm{ret}}(t) \equiv \mathrm{tr}[\chi_{\mathrm{ret}}(t)], \tag{19.42}$$

and following the steps in Sects. 1.3 and 1.4 (with $\bar{n} = 0$), we arrive at the *master equation for source subsystem, mode a, cascaded with target subsystem, mode b, with retardation of the source*:

$$\dot{\rho}_{\mathrm{ret}} = \frac{1}{i\hbar}[H_S^c, \rho_{\mathrm{ret}}] + (J_c\rho_{\mathrm{ret}}J_c^\dagger - \tfrac{1}{2}J_c^\dagger J_c\rho_{\mathrm{ret}} - \tfrac{1}{2}\rho_{\mathrm{ret}}J_c^\dagger J_c), \tag{19.43}$$

with

$$J_c \equiv \sqrt{2\kappa_a}a + e^{-i\phi_R^a}\sqrt{2\kappa_b}b. \tag{19.44}$$

Note 19.1. The density operator $\rho_{\mathrm{ret}}(t)$ describes the target subsystem at time t and the source subsystem at time $t - d/c$. Generally, the retardation d/c is negligible on the timescale for significant change of the system state—apart, of course, for changes due to the free evolution, which occur on the scale of the optical period. An alternate derivation of the master equation makes this observation at the outset and neglects all but the essential effects of retardation. In place of the transformation (19.32), which leads to (19.38), we simply write

$$\hat{\mathcal{E}}(d,0) = e^{i[\phi_R^a + (\omega_a/c)d]}\big[\hat{\mathcal{E}}(0) - i\tfrac{1}{2}\sqrt{2\kappa_a}a\big], \tag{19.45}$$

where only the phase change under propagation (for a quasi-monochromatic field) is taken into account when relating the Schrödinger operators $\hat{\mathcal{E}}(d,0)$ and $\hat{\mathcal{E}}(0,0) \equiv \hat{\mathcal{E}}(0)$. The different approach amounts to the replacement $e^{i\phi_R^a} \to e^{i[\phi_R^a + (\omega_a/c)d]}$, with the density operator now describing the source and target states at the same time. Thus, in the alternate approach we arrive at the *master equation for source subsystem, mode a, cascaded with target subsystem, mode b, without retardation of the source*:

$$\dot{\rho} = \frac{1}{i\hbar}[H_S^{c\prime}, \rho] + (J_c'\rho J_c'^\dagger - \tfrac{1}{2}J_c'^\dagger J_c'\rho - \tfrac{1}{2}\rho J_c'^\dagger J_c'), \tag{19.46}$$

where

$$H_S^{c\prime} = H_{\mathrm{source}} + H_{\mathrm{target}} + i\hbar\sqrt{\kappa_a\kappa_b}\big(e^{-i[\phi_R^a + (\omega_a/c)d]}ba^\dagger - \mathrm{H.c.}\big), \tag{19.47}$$

and

$$J_c' \equiv \sqrt{2\kappa_a}a + e^{-i[\phi_R^a + (\omega_a/c)d]}\sqrt{2\kappa_b}b. \tag{19.48}$$

19.2.4 Photoelectron Counting Unraveling

Master equation (19.43) is in standard Lindblad form. Its distinction as the master equation for a cascaded system is contained in the specific form of the Hamiltonian (19.40) and jump operator (19.44). Given the explicitly displayed Lindblad form, it is straightforward to write down any of the standard quantum trajectory unravelings. For example, performing a Dyson expansion, as in Sect. 18.2.1, but here in the Schrödinger rather than the interaction picture, gives

$$
\begin{aligned}
&\rho_{\mathrm{ret}}(t') \\
&= e^{[(\mathcal{L}-\mathcal{S})+\mathcal{S}]t}\rho_{\mathrm{ret}}(-\tau) \\
&= \sum_{n=0}^{\infty}\int_{-\tau}^{t'} dt'_n \cdots \int_{-\tau}^{t'_3} dt'_2 \int_{-\tau}^{t'_2} dt'_1 \\
&\quad e^{(\mathcal{L}-\mathcal{S})(t'-t'_n)}\mathcal{S}e^{(\mathcal{L}-\mathcal{S})(t'_n-t'_{n-1})}\cdots\mathcal{S}e^{(\mathcal{L}-\mathcal{S})(t'_2-t'_1)}\mathcal{S}e^{(\mathcal{L}-\mathcal{S})(t'_1+\tau)}\rho_{\mathrm{ret}}(-\tau),
\end{aligned}
\tag{19.49}
$$

with source superoperator

$$
\mathcal{S} \equiv J_c \cdot J_c^\dagger, \tag{19.50}
$$

jump operator defined by (19.44), and between-jump propagator

$$
\mathcal{L}-\mathcal{S} \equiv \frac{1}{i\hbar}(H_B^c \cdot - \cdot H_B^{c\dagger}) = \frac{1}{i\hbar}[H_S^c, \cdot] - [\tfrac{1}{2}J_c^\dagger J_c, \cdot]_+, \tag{19.51}
$$

where the non-Hermitian Hamiltonian is

$$
\begin{aligned}
H_B^c &= H_S^c - i\hbar\tfrac{1}{2}J_c^\dagger J_c \\
&= H_{\mathrm{source}} + H_{\mathrm{target}} + i\hbar\sqrt{\kappa_a\kappa_b}(e^{-i\phi_R^a}ba^\dagger - e^{i\phi_R^a}b^\dagger a) \\
&\quad - i\hbar\tfrac{1}{2}\big(\sqrt{2\kappa_a}a^\dagger + e^{i\phi_R^a}\sqrt{2\kappa_b}b^\dagger\big)\big(\sqrt{2\kappa_a}a + e^{-i\phi_R^a}\sqrt{2\kappa_b}b\big) \\
&= H_{\mathrm{source}} + H_{\mathrm{target}} - i\hbar\kappa_a a^\dagger a - i\hbar\kappa_b b^\dagger b - 2i\hbar\sqrt{\kappa_a\kappa_b}e^{i\phi_R^a}b^\dagger a.
\end{aligned}
\tag{19.52}
$$

Following the argument in Sect. 18.2.1, we may interpret the expansion in terms of an unraveling of the density operator for photoelectron counting records. In this instance the detector is located at some $z = z_d > d$, where it intercepts the output from the target subsystem (Fig. 19.3). Retardation from source to detector, $\tau = (z_d - d)/c$, has been included in (19.49); thus, in the adopted notation, $0 < t_1 < t_2 \ldots < t_n$ denotes the sequence of photoelectron count times at the detector, while $-\tau < t'_1 < t'_2 \ldots < t'_n$, $t'_k = t_k - \tau$ denotes the corresponding times of photon emission at the source.

Note 19.2. The photoelectron counting unraveling of master equation (19.46) follows in the same way. The jump operator is given by (19.48) and the non-Hermitian Hamiltonian is

$$H_B^{c\prime} = H_{\rm source} + H_{\rm target} - i\hbar\kappa_a a^\dagger a - i\hbar\kappa_b b^\dagger b - 2i\hbar\sqrt{\kappa_a\kappa_b}e^{i[\phi_R^a+(\omega_a/c)d]}b^\dagger a. \tag{19.53}$$

The one difference from (19.52) is the additional phase factor $e^{i(\omega_a/c)d}$, which compensates for the omitted retardation of the source subsystem.

Note 19.3. The Dyson expansion (19.49) provides the quickest and most transparent way to unravel the density operator. More fundamentally, though, the unraveling is again based upon the probability densities for photoelectron counting records (Sect. 18.1.1). A development along these lines shows how jump operators (19.44) and (19.48) arise from the Heisenberg operator for the detected field. The field operator appearing in expansion (18.20) of the record probability density is $\hat{\mathcal{E}}(z_d,\tau) = \hat{\mathcal{E}}(d,0)$. Using (19.28)–(19.30), it is expressed as the sum of a free field and two source fields:

$$\hat{\mathcal{E}}(d,0) = \hat{\mathcal{E}}_f(d,0) - ie^{i(\phi_R^a+\phi_R^b)}\big[\sqrt{2\kappa_a}a(-d/c) + e^{-i\phi_R^a}\sqrt{2\kappa_b}b(0)\big]. \tag{19.54}$$

Retardation of the source subsystem (Eq. 19.32) then takes us to the operator

$$\begin{aligned} &U_{\rm source}(d/c)\hat{\mathcal{E}}(d,0)U_{\rm source}^\dagger(d/c) \\ &= \hat{\mathcal{E}}_f(d,d/c) - ie^{i(\phi_R^a+\phi_R^b)}\big[\sqrt{2\kappa_a}a(0) + e^{-i\phi_R^a}\sqrt{2\kappa_b}b(0)\big]. \end{aligned} \tag{19.55}$$

Alternatively, with the retardation omitted (Note 19.1), we simply make the substitution $a(-d/c) = e^{i(\omega_a/c)d}a(0)$, to arrive at

$$\hat{\mathcal{E}}(d,0) = \hat{\mathcal{E}}_f(d,0) - ie^{i[\phi_R^a+\phi_R^b+(\omega_a/c)d]}\big[\sqrt{2\kappa_a}a(0) + e^{-i[\phi_R^a+(\omega_a/c)d]}b(0)\big]. \tag{19.56}$$

The free-field term may be dropped for the reasons given below (18.22), and the unimportant overall phase factors $-ie^{i(\phi_R^a+\phi_R^b)}$ and $-ie^{i[\phi_R^a+\phi_R^b(\omega_a/c)d]}$ may be dropped too. Hence jump operators (19.44) and (19.48), respectively, are obtained from (19.55) and (19.56).

19.2.5 Coherent Driving Fields

By considering a coherent input to the source subsystem we readily verify that the introduced formalism maps inputs to outputs with the correct amplitude and phase shift. The example also illustrates how the time retardation works. For simplicity, let us take the source and target cavities to be resonant, with $\omega_a = \omega_b = \omega_0$. To provide the coherent input, one mode of the reservoir is in the coherent state $|\alpha_{\rm in}\rangle$ and on-resonance with the cavities at frequency ω_0; thus, the free field has nonzero expectation value

$$\langle\hat{\mathcal{E}}_f(z,t)\rangle = \sqrt{\frac{c}{L'}}\alpha_{\rm in}e^{-i[\omega_0(t-z/c)-\phi(z)]}. \tag{19.57}$$

A straightforward analysis of the steady-state boundary value problem shows that the input field picks up the phase factor $-e^{i\phi_R^a}$ on reflection from the first cavity—for $0<z\le d$—and the additional factor $-e^{i\phi_R^b}$ on reflection from the second—for $z>d$. Thus, we expect to recover the expectation value for the total field

$$\langle\hat{\mathcal{E}}(z,t)\rangle = \sqrt{\frac{c}{L'}}\alpha_{\rm in}e^{-i\omega_0(t-z/c)}\begin{cases} e^{i(\phi_R^a+\phi_R^b)} & ct>z>0\\ -e^{i\phi_R^a} & z=0\\ 1 & z<0. \end{cases} \tag{19.58}$$

To verify this, we first calculate $\langle a(t')\rangle$ and $\langle b(t')\rangle$ from the quantum trajectory unraveling of the cascaded systems master equation, and then show that the result follows from expansion (19.28)–(19.30) of the reservoir field.

Through their coupling to the reservoir both subsystems are driven by the classical field (19.57). The non-Hermitian Hamiltonian (Eq. 19.52) is then given by

$$\begin{aligned} H_B^c &= \hbar(\omega_0-i\kappa_a)a^\dagger a + \hbar(\omega_0-i\kappa_b)b^\dagger b - i\hbar 2\sqrt{\kappa_a\kappa_b}e^{i\phi_R^a}b^\dagger a\\ &\quad+\hbar\sqrt{\frac{c}{L'}}\sqrt{2\kappa_a}(\alpha_{\rm in}e^{-i\omega_0 t'}a^\dagger + \text{H.c.})\\ &\quad+\hbar\sqrt{\frac{c}{L'}}\sqrt{2\kappa_b}(\alpha_{\rm in}e^{-i[\omega_0(t-d/c)-\phi_R^a]}b^\dagger + \text{H.c.}), \end{aligned} \tag{19.59}$$

where the second and third lines are the driving field interactions for source and target, respectively, taken from (19.24a) and (19.24b), with reservoir field expectations $\langle\hat{\mathcal{E}}_f(0,t')\rangle$ and $\langle\hat{\mathcal{E}}_f(d,t)\rangle$ (Eq. 19.57) in place of the operators $\hat{\mathcal{E}}(0)$ and $\hat{\mathcal{E}}(d)$. Note that in the case of the source subsystem, the time dependence of the interaction is expressed through the retarded time $t'=t-d/c$. This is required because the trajectory equations evolve the time-retarded state of the source (Eqs. 19.32 and 19.33); of course, this brings the driving of source and target subsystems into phase. If we now transform to the interaction picture, with

$$|\tilde{\psi}_{\rm REC}^{\rm ret}(t)\rangle = e^{i\omega_0 a^\dagger a t'}e^{i\omega_0 b^\dagger b(t-d/c)}|\psi_{\rm REC}^{\rm ret}(t)\rangle, \tag{19.60}$$

the non-Hermitian Hamiltonian is given by

$$\begin{aligned} \tilde{H}_B^c &= -i\hbar\kappa_a a^\dagger a - i\hbar\kappa_b b^\dagger b - i\hbar 2\sqrt{\kappa_a\kappa_b}e^{i\phi_R^a}b^\dagger a\\ &\quad+\hbar\sqrt{\frac{c}{L'}}\sqrt{2\kappa_a}(\alpha_{\rm in}a^\dagger + \text{H.c.}) + \hbar\sqrt{\frac{c}{L'}}\sqrt{2\kappa_b}(\alpha_{\rm in}e^{i\phi_R^a}b^\dagger + \text{H.c.}). \end{aligned} \tag{19.61}$$

A special and convenient feature of coherent driving is that the conditional state evolves out of the vacuum as a coherent state product, with $|\tilde{\psi}^{\rm ret}_{\rm REC}(t)\rangle = |\tilde{\alpha}(t)\rangle_a|\tilde{\beta}(t)\rangle_b$. The form is preserved by the quantum jumps as coherent states are eigenstates of the jump operator $J_c = \sqrt{2\kappa_a}a + e^{-i\phi_R^a}\sqrt{2\kappa_b}b$, and also by the nonunitary Schrödinger equation between jumps, which causes only the normalization of the state to change (Exercise 19.2). It follows that the steady state is the coherent state product that satisfies

$$\tilde{H}^c_B|\tilde{\alpha}\rangle_a|\tilde{\beta}\rangle_b = \lambda|\tilde{\alpha}\rangle_a|\tilde{\beta}\rangle_b, \tag{19.62}$$

with λ a complex constant. Thus, substituting for $\tilde{H}^c_B$ from (19.61), we may set the coefficients of $a^\dagger$ and $b^\dagger$ to zero to obtain a pair of coupled equations for the complex amplitudes $\tilde{\alpha}$ and $\tilde{\beta}$. The equations are

$$-\kappa_a\tilde{\alpha} - i\sqrt{\frac{c}{L'}}\sqrt{2\kappa_a}\alpha_{\rm in} = 0, \tag{19.63a}$$

$$-\kappa_b\tilde{\beta} - 2\sqrt{\kappa_a\kappa_b}e^{i\phi_R^a}\tilde{\alpha} - i\sqrt{\frac{c}{L'}}\sqrt{2\kappa_b}\alpha_{\rm in}e^{i\phi_R^a} = 0, \tag{19.63b}$$

with solutions

$$\tilde{\alpha} = -i\sqrt{\frac{c}{L'}}\sqrt{2/\kappa_a}\alpha_{\rm in}, \tag{19.64a}$$

$$\tilde{\beta} = i\sqrt{\frac{c}{L'}}\sqrt{2/\kappa_b}\alpha_{\rm in}e^{i\phi_R^a}. \tag{19.64b}$$

Inverting the transformation (19.60), the intracavity field expectations are then given by

$$\langle a(t')\rangle = -i\sqrt{\frac{c}{L'}}\sqrt{2/\kappa_a}\alpha_{\rm in}e^{-i\omega_0 t'}, \tag{19.65a}$$

$$\langle b(t)\rangle = i\sqrt{\frac{c}{L'}}\sqrt{2/\kappa_b}\alpha_{\rm in}e^{i\phi_R^a}e^{-i\omega_0(t-d/c)}, \tag{19.65b}$$

from which, using (19.28)–(19.30), we recover the anticipated result for the expectation of the output field (19.58).

Exercise 19.2. For coherent driving the complete time-dependent solution to the quantum trajectory equations can be found. Consider the more general form of coherent driving with the constant amplitude $\alpha_{\rm in}$ of (19.57) replaced by $\alpha_{\rm in}(t-z/c)$. Show that for an initial product of coherent states, the solution to the nounitary Schrödinger equation between quantum jumps is given by

$$|\tilde{\tilde{\psi}}^{\rm ret}_{\rm REC}(t)\rangle = A(t)|\tilde{\alpha}(t)\rangle_a|\tilde{\beta}(t)\rangle_b, \tag{19.66}$$

where $|\tilde{\alpha}(t)\rangle_a$ and $|\tilde{\beta}(t)\rangle_b$ are coherent states whose amplitudes satisfy

$$\dot{\tilde{\alpha}} = -\kappa_a\tilde{\alpha} - i\sqrt{\frac{c}{L'}}\sqrt{2\kappa_a}\alpha_{\text{in}}(t-d/c), \tag{19.67a}$$

$$\dot{\tilde{\beta}} = -\kappa_b\tilde{\beta} - 2\sqrt{\kappa_a\kappa_b}e^{i\phi_R^a}\tilde{\alpha} - i\sqrt{\frac{c}{L'}}\sqrt{2\kappa_b}\alpha_{\text{in}}(t-d/c)e^{i\phi_R^a}, \tag{19.67b}$$

and

$$A(t)/A(t_{\text{last}})$$
$$= \exp\left\{-\int_{t_{\text{last}}}^{t} dt'\left[\kappa_a|\tilde{\alpha}(t')|^2 + \kappa_b|\tilde{\beta}(t')|^2 + 2\sqrt{\kappa_a\kappa_b}\text{Re}\left[e^{-i\phi_R^a}\tilde{\beta}(t')\tilde{\alpha}^*(t')\right]\right]\right\}, \tag{19.68}$$

where t_{last} is the time of the last quantum jump.

Exercise 19.3. Verify that the total field expectation (19.58) can also be recovered from the quantum trajectory unraveling of the master equation without retardation (Notes 19.1 and 19.2).

To conclude, a few comments on quantum jumps for coherent driving are in order. For the situation of Exercise 19.2, where the system state is a product of coherent states, quantum jumps at times t_i, $i = 1, 2, \ldots$, change the state norm, multiplying it by the factor $\vartheta(t_i) = \sqrt{2\kappa_a}\tilde{\alpha}(t_i) + e^{-i\phi_R^a}\sqrt{2\kappa_b}\tilde{\beta}(t_i)$, the eigenvalue of J_c; otherwise, they play no role in the evolution. The quantum jumps play a substantial role, on the other hand, when the state is a *superposition* of coherent state products or if something other than empty cavities is involved. The former possibility, in particular, should be noted, as it is related to what we learned in Sect. 18.3.2 about how a quantum trajectory evolution can resolve an initial superposition of states, bringing about a dynamical localization, or wavefunction collapse, onto one or other component of the superposition. The localization mechanism arises out of an interplay between the norm-changing factor (19.68), which operates between the quantum jumps, and the norm change by $\vartheta(t_i)$ at the time of a quantum jump.

Note first that the integrand in (19.68) is simply one half the square modulus of the eigenvalue $\vartheta(t')$. Consider now a superposition of two coherent state products with corresponding eigenvalues $\vartheta_1(t)$ and $\vartheta_2(t)$. The evolution of the unnormalized state is linear. Therefore in-between the quantum jumps, the conditional state remains as a superposition of coherent state products, but with state amplitudes $A_1(t)$ and $A_2(t)$ that are decaying at different rates; thus, one component of the superposition—that with the *smaller* $|\vartheta_j(t')|$—grows at the expense of the other. Contrast the effect of a quantum jump: the state amplitudes are multiplied, respectively, by $\vartheta_1(t_i)$ and $\vartheta_2(t_i)$, so the component of the superposition with *larger* $|\vartheta_j(t_i)|$ grows relative to the other. In this way a competition between the components of the superposition is set up where the state amplitudes increase and decrease relative to one another in response to fluctuations in the number of occurring quantum jumps. Ultimately, a deficit (excess) of quantum jumps leads to the complete dominance of the component with smaller (larger) $|\vartheta_j(t)|$, and dynamical localization occurs. Recognizing the Poisson fluctuation in the number of quantum jumps, localization to a different component of the superposition is expected in each stochastic realization.

It should be noted, finally, that jump operator (19.44) assumes that the free-field expectation has been subtracted from the output before it reaches the detector, in the manner of the subtraction of the coherent input in Fig. 18.1. The more natural unraveling would use a jump operator derived from the full output field, i.e., the operator

$$J_c = \sqrt{\frac{c}{L'}}\alpha_{\rm in} - i\sqrt{2\kappa_a}a - ie^{-i\phi_R^a}\sqrt{2\kappa_b}b. \tag{19.69}$$

Then, to replace (19.61), the non-Hermitian Hamiltonian is given by

$$\begin{aligned}\tilde{H}_B^c = &-i\hbar\frac{1}{2}\frac{c}{L'}|\alpha_{\rm in}|^2 - i\hbar\kappa_a a^\dagger a - i\hbar\kappa_b b^\dagger b\\ &+\hbar\sqrt{2\kappa_a}a^\dagger\sqrt{\frac{c}{L'}}\alpha_{\rm in} + \hbar\sqrt{2\kappa_b}e^{i\phi_R^a}b^\dagger\left(\sqrt{\frac{c}{L'}}\alpha_{\rm in} - i\sqrt{2\kappa_a}a\right).\end{aligned} \tag{19.70}$$

A derivation of (19.69) and (19.70) can be given by modeling the input field as the coherent output of a third cavity and applying the same cascaded system ideas to the three-cavity cascade (Fig. 19.4). In this case, the Hamiltonian $H_{\rm system}$ is that of an empty cavity driven by a classical current, with driving strength and cavity decay rate set to produce an output field matching (19.57). The approach has been used by Carmichael [19.10] and Nha and Carmichael [19.11] to model a two-state atom driven by a coherent field. The physical origin of the unraveling is clear from the terms in the second line of (19.70). Both show the non-Hermitian (unidirectional) coupling of a cascaded system. The first describes the coupling of the output from the source into target cavity A, while the second describes the coupling of the output of cavity A into target cavity B. Of course, many examples of coherent driving involve only one target system, in which case the second term is absent.

Exercise 19.4. Consider cascaded cavities with vacuum input ($\alpha_{\rm in} = 0$) prepared in a superposition of coherent state products. Write a computer program to simulate quantum trajectories with non-Hermitian Hamiltonian (19.59) and jump operator (19.44). Study the process of dynamical localization as a function of the complex amplitudes of the initial superposition state components, first for $\kappa_b = 0$ (single cavity decay) and then for $\kappa_b = \kappa_a$ (cascaded cavity decay):

1. Does localization always occur? If not, determine the conditions under which it does occur.

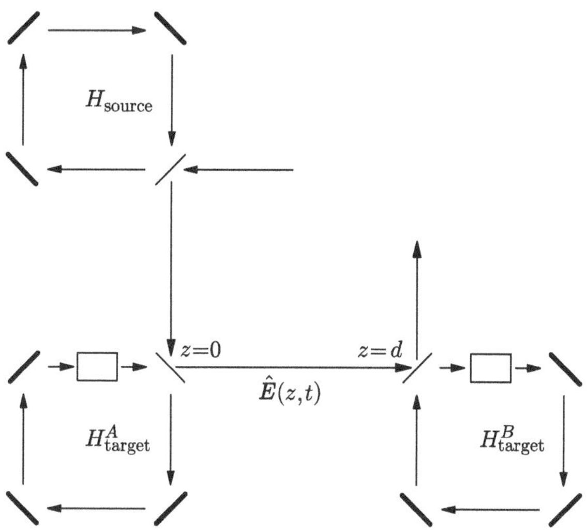

Fig. 19.4. Three-cavity cascaded system used to model the coherent driving of the two-cavity system of Fig. 19.3. The source outputs a coherent field that drives target cavities A and B, designated, respectively, as source and target in Fig. 19.3

2. Verify that the frequency of localization to different components of the initial superposition agrees with the usual quantum mechanical rule.

Exercise 19.5. By working backwards from the quantum trajectory formulation, show that (19.69) and (19.70) unravel the same master equation as (19.44) and (19.61).

19.2.6 Symmetric Irreversible Coupling

A natural extension of the cascaded system considered in Fig. 19.3 is the system with symmetric irreversible coupling depicted in Fig. 19.5. To obtain a master equation and its quantum trajectory unraveling in this case it is necessary to neglect retardation (Note 19.1). Then, as a generalization of (19.46)–(1948), adding cascaded system interactions in the two directions gives the modified system Hamiltonian

$$H_S^c = H_A + H_B + i\hbar\sqrt{\kappa_a\kappa_b}\big(e^{-i[\phi_R^a+(\omega_a/c)d]}ba^\dagger + e^{-i[\phi_R^b-(\omega_b/c)d]}ab^\dagger - \text{H.c.}\big), \tag{19.71}$$

and master equation

$$\begin{aligned}\dot\rho = \frac{1}{i\hbar}[H_S^c,\rho] &+ (J_\rightarrow\rho J_\rightarrow^\dagger - \tfrac{1}{2}J_\rightarrow^\dagger J_\rightarrow\rho - \tfrac{1}{2}\rho J_\rightarrow^\dagger J_\rightarrow)\\ &+ (J_\leftarrow\rho J_\leftarrow^\dagger - \tfrac{1}{2}J_\leftarrow^\dagger J_\leftarrow\rho - \tfrac{1}{2}\rho J_\leftarrow^\dagger J_\leftarrow),\end{aligned} \tag{19.72}$$

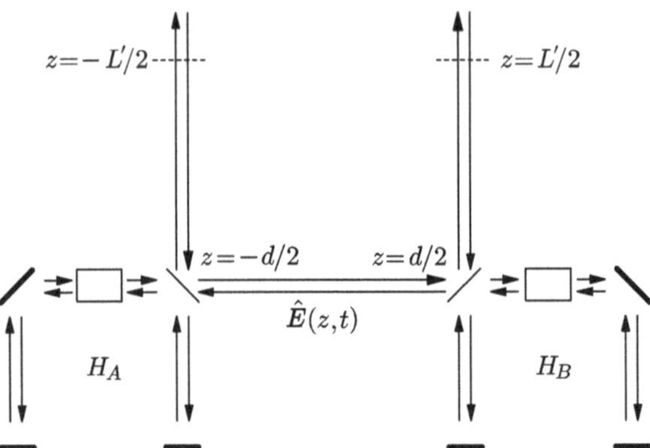

Fig. 19.5. Two-cavity system with symmetric irreversible coupling. The reservoir field $\hat{E}(z,t) = \hat{E}_{\rightarrow}(z,t) + \hat{E}_{\leftarrow}(z,t)$ is the sum of components traveling to the right and to the left

where now there are two jump operators, which correspond to a record of photoelectron counts collected now by two detectors, one intercepting the right- and the other the left-going output field:

$$J_{\rightarrow} = e^{i\frac{1}{2}(\omega_a/c)d}\sqrt{2\kappa_a}a + e^{-i\phi_R^a}e^{-i\frac{1}{2}(\omega_a/c)d}\sqrt{2\kappa_b}b, \tag{19.73a}$$

$$J_{\leftarrow} = e^{-i\frac{1}{2}(\omega_b/c)d}\sqrt{2\kappa_b}b + e^{-i\phi_R^b}e^{i\frac{1}{2}(\omega_b/c)d}\sqrt{2\kappa_a}a. \tag{19.73b}$$

The corresponding quantum trajectory unraveling has non-Hermitian Hamiltonian

$$\begin{aligned} H_B^c = {} & H_A + H_B - i\hbar 2\kappa_a a^\dagger a - i\hbar 2\kappa_b b^\dagger b \\ & - 2i\hbar\sqrt{\kappa_a\kappa_b}\big(e^{i[\phi_R^a + (\omega_a/c)d]}b^\dagger a + e^{i[\phi_R^b - (\omega_b/c)d]}a^\dagger b\big), \end{aligned} \tag{19.74}$$

the generalization of (19.53).

Time reversal for a near-perfectly reflecting mirror requires $\phi_R = \pi$; thus, let us set $\phi_R^a = \phi_R^b = \pi$. It is then apparent that the case of exact resonance, with $\omega_a = \omega_b = \omega_0$, is special, since the subsystem interactions cancel in Hamiltonian (19.71) and non-Hermitian Hamiltonian (19.74) may be written as

$$H_B^c = H_A + H_B - i\hbar\tfrac{1}{2}J_{\rightarrow}^\dagger J_{\rightarrow} - i\hbar\tfrac{1}{2}J_{\leftarrow}^\dagger J_{\leftarrow}, \tag{19.75}$$

with

$$J_{\rightarrow} = e^{i\frac{1}{2}(\omega_0/c)d}\sqrt{2\kappa_a}a - e^{-i\frac{1}{2}(\omega_0/c)d}\sqrt{2\kappa_b}b, \tag{19.76a}$$

$$J_{\leftarrow} = e^{-i\frac{1}{2}(\omega_0/c)d}\sqrt{2\kappa_b}b - e^{i\frac{1}{2}(\omega_0/c)d}\sqrt{2\kappa_a}a. \tag{19.76b}$$

Of course the two jump operators are effectively the same, since their difference in sign has no physical significance.

When written in this form, our model provides an elementary example of the sort of two-way scattering that underlies collective radiative phenomena such as superradiance and superfluorescence. A quantum trajectory unraveling of the superradiance master equation has been given by Carmichael and Kim [19.12]. It has non-Hermitian Hamiltonian

$$H_B = \sum_{j=1}^{N} \tfrac{1}{2}\hbar\omega_0\sigma_{jz} + \sum_{k\neq j=1}^{N} \hbar\Delta_{kj}\sigma_{k+}\sigma_{j-} - i\hbar\frac{1}{2}\int_{\Omega} J^{\dagger}(\theta,\phi)J(\theta,\phi), \quad (19.77)$$

where ω_0 is the common resonance frequency of the N atoms, and $\hbar\Delta_{kj}$ is the dipole-dipole interaction energy between atom j and atom k, $j \neq k = 1,\ldots,N$. This Hamiltonian has the form of (19.75), except there are now not two, but infinitely many jump operators,

$$J(\theta,\phi) = \sqrt{\gamma D(\theta,\phi)d\Omega}\sum_{j=1}^{N} e^{-i(\omega_0/c)\hat{\boldsymbol{R}}(\theta,\phi)\cdot\boldsymbol{r}_j}\sigma_{j-}, \quad (19.78)$$

associated with an infinity of point-like detectors—the general detector is located in the far field in direction (θ,ϕ) and subtends a solid angle $d\Omega$ at the source. The phase factors $e^{-i(\omega_0/c)\hat{\boldsymbol{R}}(\theta,\phi)\cdot\boldsymbol{r}_j}$, $j = 1,2,\ldots,N$, account for the different path lengths from the N scattering centers to the detector located in direction (θ,ϕ); $\boldsymbol{r}_j$ is the position of atom j and $\hat{\boldsymbol{R}}(\theta,\phi)$ is a unit vector in the direction (θ,ϕ). The factor before the sum in (19.78) determines the overall emission rate; γ is the Einstein A coefficient for the atomic transition and

$$D(\theta,\phi) \equiv (3/8\pi)\{1 - [\hat{\boldsymbol{d}}\cdot\hat{\boldsymbol{R}}(\theta,\phi)]^2\} \quad (19.79)$$

is the dipole radiation pattern, where $\hat{d}$ is a unit vector in the direction of the atomic dipole moment.

19.3 Optical Spectra

Quantum trajectory simulations provide us with a way to calculate things. Most directly they yield the density operator as the ensemble average (18.46); thus, an operator average $\langle\hat{O}(t)\rangle$ is computed as an ensemble average over the conditional operator expectation $\langle\hat{O}(t)\rangle_{\mathrm{REC}}$. The quantum trajectory method is more than a set of tricks for computing averages, though. It provides a connection to measurements—a connection to the averaged quantities that are constructed experimentally from data sets. In Fig. 18.5, for example, the squeezing spectra plotted were obtained as Fourier transforms of autocorrelation functions computed from simulated photocurrents.

In this section we set the squeezing example aside (see Sect. 9.3.3 for the connection between squeezing spectra and measurements) and consider measurements of the optical spectrum. We analyze two quantum trajectory unravelings, one corresponding to a measurement of the spectrum by a scanning interferometer and the other to a heterodyne measurement. Our aim is to see how a computation of the spectrum from simulated data sets is related to a standard calculation from the master equation and quantum regression formula.

19.3.1 Optical Spectrum Using a Scanning Interferometer

For the sake of definiteness, let us return to the example of the degenerate parametric oscillator treated in Sect. 18.1. The optical spectrum of the subharmonic mode is to be measured by coupling a small fraction of the output light into a scanning interferometer. The parametric oscillator (source) and interferometer (target) comprise the cascaded system depicted in Fig. 19.6, and we analyze its quantum trajectory unraveling with photoelectron counting records. The nonHermitian Hamiltonian in the interaction picture is given by

$$\tilde{H}_B^c = \tilde{H}_B + \hbar(\Delta\omega - i\Gamma)c^\dagger c - i\hbar\sqrt{\epsilon\Gamma}e^{i\theta}c^\dagger J_\rightarrow, \tag{19.80}$$

where $\tilde{H}_B$ is the Hamiltonian (18.32), ϵ is the fraction of the γ_{a2} output coupled into the interferometer, c and $c^\dagger$ are annihilation and creation operators for the interferometer mode, $\Delta\omega = \omega - \omega_C$ is its detuning from the parametric oscillator, and Γ is the interferometer half-width. The one-way coupling term is taken from (19.52), with $\sqrt{2\kappa_a}a \to \sqrt{\epsilon\gamma_{a2}}a = \sqrt{\epsilon}J_\rightarrow$, $\sqrt{2\kappa_b}b^\dagger \to \sqrt{\Gamma}c^\dagger$, and $e^{i\phi_R^a} \to e^{i\theta}$, where $\theta = \phi_R + \Phi(\ell + d) + \phi$ follows from the expression for the subharmonic mode output field given in Sect. 9.2.5 (Eqs. 9.115a and 9.117a).

We may assume, without loss of generality, that only a very small fraction of the available photon flux enters the interferometer—i.e., $\epsilon \ll 1$. Two convenient simplifications follow from this: (i) the interferometer outputs may be omitted from the making of records, and (ii) only zero- and one-photon states of the interferometer mode need be considered; thus, we expand the conditional state of the cascaded system as

$$|\bar{\tilde{\psi}}_{\mathrm{REC}}(\omega,t)\rangle = |\bar{\tilde{\psi}}^{(0)}_{\mathrm{REC}}(t)\rangle|0\rangle_c + e^{i\theta}|\bar{\tilde{\psi}}^{(1)}_{\mathrm{REC}}(\omega,t)\rangle|1\rangle_c, \tag{19.81}$$

where the record REC is made according to the photoelectric detection scheme of Fig. 18.1. For a particular record, up to time t, the conditional intracavity photon number expectation of the interferometer is

$$N_t(\omega|\mathrm{REC}) = \frac{\langle\bar{\tilde{\psi}}^{(1)}_{\mathrm{REC}}(\omega,t)|\bar{\tilde{\psi}}^{(1)}_{\mathrm{REC}}(\omega,t)\rangle}{\langle\bar{\tilde{\psi}}_{\mathrm{REC}}(\omega,t)|\bar{\tilde{\psi}}_{\mathrm{REC}}(\omega,t)\rangle}, \tag{19.82}$$

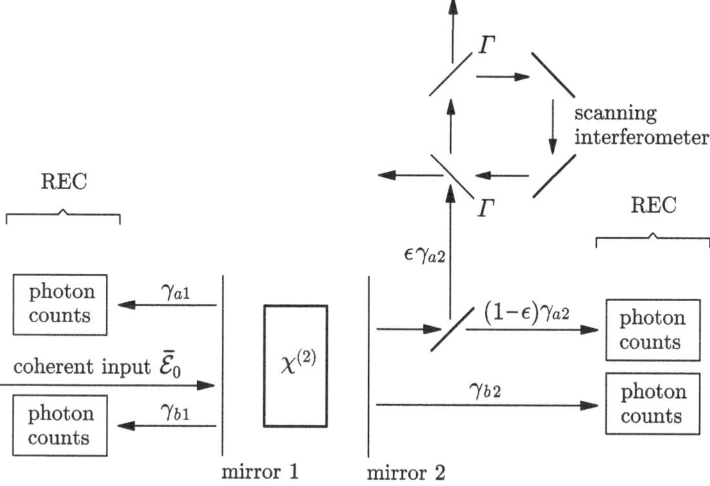

Fig. 19.6. Scattering scenario for the degenerate parametric oscillator with photoelectron counting records (Fig. 18.1) cascaded with a scanning interferometer to measure the optical spectrum of the subharmonic mode. The interferometer outputs may be omitted from the making of records when $\epsilon \ll 1$

with record probability $P(\text{REC}) = \langle \tilde{\bar{\psi}}_{\text{REC}}(\omega,t)|\tilde{\bar{\psi}}_{\text{REC}}(\omega,t)\rangle$ (Eq. 18.39). Then the average photon number is given by

$$
\begin{aligned}
N_t(\omega) &= \sum_{\text{REC}} N_t(\omega|\text{REC})\langle \tilde{\bar{\psi}}_{\text{REC}}(\omega,t)|\tilde{\bar{\psi}}_{\text{REC}}(\omega,t)\rangle \\
&= \sum_{\text{REC}} \langle \tilde{\bar{\psi}}^{(1)}_{\text{REC}}(\omega,t)|\tilde{\bar{\psi}}^{(1)}_{\text{REC}}(\omega,t)\rangle .
\end{aligned}
\tag{19.83}
$$

This expression defines the spectrum. Note that it is generally a time-dependent result. Our task is to solve for $|\tilde{\bar{\psi}}^{(1)}_{\text{REC}}(\omega,t)\rangle$ and perform the sum over records.

We begin by noting that the assumption $\epsilon \ll 1$ allows the zero-photon component of the conditional state to follow the trajectory evolution of the degenerate parametric oscillator alone—the presence of the cascaded interferometer may be ignored. Using (18.37) and (18.40), and omitting the retardation τ_R, we have

$$
|\tilde{\bar{\psi}}^{(0)}_{\text{REC}}(t)\rangle = \tilde{K}^{(0)}_{\text{REC}}(t)|\psi(0)\rangle, \tag{19.84}
$$

with

$$
\tilde{K}^{(0)}_{\text{REC}}(t) \equiv \tilde{B}(t-t_n)J_n\tilde{B}(t_n-t_{n-1})\cdots J_2\tilde{B}(t_2-t_1)J_1\tilde{B}(t_1), \tag{19.85}
$$

where

$$
B(t) \equiv \exp[-(i/\hbar)\tilde{H}_B t], \tag{19.86}
$$

and $J_k \in \{J_{\leftarrow}, J_{\rightarrow}, J_{p\leftarrow}, J_{p\rightarrow}\}$. The presence of the interferometer affects the evolution of the one-photon component of the conditional state, though. From non-Hermitian Hamiltonian (19.80), the nonunitary Schrödinger evolution between jumps is given by

$$\frac{d|\tilde{\bar{\psi}}^{(1)}_{\mathrm{REC}}(\omega)\rangle}{dt} = \left[\frac{1}{i\hbar}\tilde{H}_B - (\Gamma + i\Delta\omega)\right]|\tilde{\bar{\psi}}^{(1)}_{\mathrm{REC}}(\omega)\rangle - \sqrt{\epsilon\Gamma}J_{\rightarrow}|\tilde{\bar{\psi}}^{(0)}_{\mathrm{REC}}\rangle, \quad (19.87)$$

whose formal solution for the interval $t_k < t < t_{k+1}$ is

$$\begin{aligned} |\tilde{\bar{\psi}}^{(1)}_{\mathrm{REC}}(\omega,t)\rangle &= e^{-(\Gamma+i\Delta\omega)(t-t_k)}\tilde{B}(t-t_k)J_k|\tilde{\bar{\psi}}^{(1)}_{\mathrm{REC}}(\omega,t_k^-)\rangle \\ &\quad - \sqrt{\epsilon\Gamma}\int_{t_k}^{t} dt' e^{-(\Gamma+i\Delta\omega)(t-t')}\tilde{B}(t-t')J_{\rightarrow}|\tilde{\bar{\psi}}^{(0)}_{\mathrm{REC}}(t')\rangle, \end{aligned} \quad (19.88)$$

where $|\tilde{\bar{\psi}}^{(1)}_{\mathrm{REC}}(\omega,t_k^-)\rangle$ denotes the one-photon component of the conditional state immediately prior to the k-th quantum jump. Substituting $|\tilde{\bar{\psi}}^{(0)}_{\mathrm{REC}}(t)\rangle$ from (19.84) and (19.85), and defining

$$F_{t_a}^{t_b}(\omega,t) \equiv -\sqrt{\epsilon\Gamma}\int_{t_a}^{t_b} dt' e^{-(\Gamma+i\Delta\omega)(t-t')}\tilde{B}(t_b-t')J_{\rightarrow}\tilde{B}(t'-t_a), \quad (19.89)$$

this solution may be written as

$$\begin{aligned} |\tilde{\bar{\psi}}^{(1)}_{\mathrm{REC}}(\omega,t)\rangle &= e^{-(\Gamma+i\Delta\omega)(t-t_k)}\tilde{B}(t-t_k)J_k|\tilde{\bar{\psi}}^{(1)}_{\mathrm{REC}}(\omega,t_k^-)\rangle \\ &\quad + F_{t_k}^{t}(\omega,t)J_k\tilde{B}(t_k-t_{k-1})\cdots J_2\tilde{B}(t_2-t_1)J_1\tilde{B}(t_1)|\psi(0)\rangle. \end{aligned} \quad (19.90)$$

The result holds between quantum jumps number k and number $(k+1)$. It is at this stage only an implicit solution, due to the appearance of $|\tilde{\bar{\psi}}^{(1)}_{\mathrm{REC}}(\omega,t_k^-)\rangle$ on the right-hand side.

We now turn to the task of finding an explicit solution for times $t > t_n$. This is carried out by iterating backwards through the sequence of quantum jumps. Beginning from (19.90), with $k = n$, we write

$$\begin{aligned} |\tilde{\bar{\psi}}^{(1)}_{\mathrm{REC}}(\omega,t)\rangle &= e^{-(\Gamma+i\Delta\omega)(t-t_n)}\tilde{B}(t-t_n)J_n|\tilde{\bar{\psi}}^{(1)}_{\mathrm{REC}}(\omega,t_n^-)\rangle \\ &\quad + F_{t_n}^{t}(\omega,t)J_n\tilde{B}(t_n-t_{n-1})\cdots J_2\tilde{B}(t_2-t_1)J_1\tilde{B}(t_1)|\psi(0)\rangle. \end{aligned} \quad (19.91)$$

The expression requires us to substitute for the state $|\tilde{\bar{\psi}}^{(1)}_{\mathrm{REC}}(\omega,t_n^-)\rangle$ on the right-hand side. We use (19.90) once again, this time with $k = n-1$ and

$t = t_n^-$, i.e.,

$$
\begin{aligned}
&|\tilde{\bar{\psi}}^{(1)}_{\mathrm{REC}}(\omega, t_n^-)\rangle \\
&\quad = e^{-(\Gamma+i\Delta\omega)(t_n-t_{n-1})}\tilde{B}(t_n - t_{n-1})J_{n-1}|\tilde{\bar{\psi}}^{(1)}_{\mathrm{REC}}(\omega, t_{n-1}^-)\rangle \\
&\qquad + F^{t_n}_{t_{n-1}}(\omega, t_n)J_{n-1}\tilde{B}(t_{n-1} - t_{n-2})\cdots J_2\tilde{B}(t_2 - t_1)J_1\tilde{B}(t_1)|\psi(0)\rangle;
\end{aligned}
\tag{19.92}
$$

thus, with

$$
e^{-(\Gamma+i\Delta\omega)(t-t_n)}F^{t_n}_{t_{n-1}}(\omega, t_n) = F^{t_n}_{t_{n-1}}(\omega, t), \tag{19.93}
$$

the substitution brings us to

$$
\begin{aligned}
&|\tilde{\bar{\psi}}^{(1)}_{\mathrm{REC}}(\omega, t)\rangle \\
&\quad = e^{-(\Gamma+i\Delta\omega)(t-t_n)}\tilde{B}(t - t_n)J_n\tilde{B}(t_n - t_{n-1})J_{n-1}|\tilde{\bar{\psi}}^{(1)}_{\mathrm{REC}}(\omega, t_{n-1}^-)\rangle \\
&\qquad + \tilde{B}(t - t_n)J_nF^{t_n}_{t_{n-1}}(\omega, t)J_{n-1}\tilde{B}(t_{n-1} - t_{n-2})\cdots \\
&\qquad \cdots J_2\tilde{B}(t_2 - t_1)J_1\tilde{B}(t_1)|\psi(0)\rangle \\
&\qquad + F^{t}_{t_n}(\omega, t)J_n\tilde{B}(t_n - t_{n-1})\cdots J_2\tilde{B}(t_2 - t_1)J_1\tilde{B}(t_1)|\psi(0)\rangle.
\end{aligned}
\tag{19.94}
$$

The required explicit solution is reached by continuing to iterate in this way. It may be written as

$$
|\tilde{\bar{\psi}}^{(1)}_{\mathrm{REC}}(\omega, t)\rangle = \tilde{K}^{(1)}_{\mathrm{REC}}(\omega, t)|\psi(0)\rangle, \tag{19.95}
$$

where the propagator $K^{(1)}_{\mathrm{REC}}(\omega, t)$ is expanded in the series

$$
\begin{aligned}
&\tilde{K}^{(1)}_{\mathrm{REC}}(\omega, t) \\
&\quad = \sum_{k=1}^{n+1} \tilde{B}(t - t_n)J_n\tilde{B}(t_n - t_{n-1})\cdots J_kF^{t_k}_{t_{k-1}}(\omega, t)\cdots J_2\tilde{B}(t_2 - t_1)J_1\tilde{B}(t_1).
\end{aligned}
\tag{19.96}
$$

Note the simple relationship between $\tilde{K}^{(1)}_{\mathrm{REC}}(\omega, t)$ and $\tilde{K}^{(0)}_{\mathrm{REC}}(t)$: the general term in the sum defining the former is obtained from the latter by replacing $\tilde{B}(t_k - t_{k-1})$ by $F^{t_k}_{t_{k-1}}(\omega, t)$, while in the first and last terms of the sum the replacements are $\tilde{B}(t_1) \to F^{t_1}_0(\omega, t)$ and $\tilde{B}(t - t_n) \to F^{t}_{t_n}(\omega, t)$.

The first part of our task is now complete. It remains only to carry out the sum over records. Using (19.95), the expression for the average photon number (Eq. 19.83) reads

$$
N_t(\omega) = \sum_{\mathrm{REC}} \mathrm{tr}\big[\tilde{K}^{(1)}_{\mathrm{REC}}(\omega, t)|\psi(0)\rangle\langle\psi(0)|\tilde{K}^{(1)\dagger}_{\mathrm{REC}}(\omega, t)\big]. \tag{19.97}
$$

On introducing the explicit form of $\tilde{K}^{(1)}_{\mathrm{REC}}(\omega,t)$, it becomes convenient to work with superoperators rather than the operators; we write

$$\begin{aligned}N_t(\omega) = &\sum_{\mathrm{REC}}\sum_{k'=1}^{n+1}\sum_{k''=1}^{k'-1}\epsilon\Gamma\int_{t_{k'-1}}^{t_{k'}}dt'\int_{t_{k''-1}}^{t_{k''}}dt''e^{-\Gamma(2t-t'-t'')}e^{i\Delta\omega(t'-t'')}\\ &\times\mathrm{tr}\Big[e^{(\tilde{\mathcal{L}}-\mathcal{S})(t-t_n)}\mathcal{S}_n e^{(\tilde{\mathcal{L}}-\mathcal{S})(t_n-t_{n-1})}\dots\\ &\dots\mathcal{S}_{k'}e^{(\tilde{\mathcal{L}}-\mathcal{S})(t_{k'}-t')}\mathcal{J}_{\rightarrow}e^{(\tilde{\mathcal{L}}-\mathcal{S})(t'-t_{k'-1})}\dots\\ &\dots\mathcal{S}_{k''}e^{(\tilde{\mathcal{L}}-\mathcal{S})(t_{k''}-t'')}\mathcal{J}_{\rightarrow}^{+}e^{(\tilde{\mathcal{L}}-\mathcal{S})(t''-t_{k''-1})}\dots\\ &\dots\mathcal{S}_2 e^{(\tilde{\mathcal{L}}-\mathcal{S})(t_2-t_1)}\mathcal{S}_1 e^{(\tilde{\mathcal{L}}-\mathcal{S})t_1}\big(|\psi(0)\rangle\langle\psi(0)|\big)\Big]+\begin{pmatrix}\text{terms with}\\ k''\geq k'\end{pmatrix},\end{aligned}\tag{19.98}$$

with $\mathcal{S}_k\in\{\mathcal{S}_{\leftarrow},\mathcal{S}_{\rightarrow},\mathcal{S}_{p\leftarrow},\mathcal{S}_{p\rightarrow}\}$, where the source superoperators are defined in (18.23a) and (18.23b), and we have introduced

$$\mathcal{J}_{\rightarrow}\equiv J_{\rightarrow}\cdot 1,\qquad \mathcal{J}_{\rightarrow}^{+}\equiv 1\cdot J_{\rightarrow}^{\dagger}.\tag{19.99}$$

We proceed to carry out the summations. Our strategy is to compose the sums over k' and k'' as a double integral over time between 0 and t. We write the spectrum as a sum of two parts,

$$N_t(\omega)=N_t^{(+)}(\omega)+N_t^{(-)}(\omega),\tag{19.100}$$

where $N_t^{(+)}(\omega)$ is the expression displayed explicitly in (19.98), including all terms with $k''<k'$, and $N_t^{(-)}(\omega)$ is the similar sum over terms with $k''\geq k'$. Composing the sums as integrals, we have

$$\begin{aligned}&N_t^{(+)}(\omega)\\ &=\sum_{\mathrm{REC}}\epsilon\Gamma\int_0^t dt'\int_0^{t'}dt''e^{-\Gamma(2t-t'-t'')}e^{i\Delta\omega(t'-t'')}\mathrm{tr}\Big\{I_{t'}(\mathcal{J}_{\rightarrow})I_{t''}(\mathcal{J}_{\rightarrow}^{+})\\ &\quad\times\Big[e^{(\tilde{\mathcal{L}}-\mathcal{S})(t-t_n)}\mathcal{S}_n e^{(\tilde{\mathcal{L}}-\mathcal{S})(t_n-t_{n-1})}\dots\mathcal{S}_2 e^{(\tilde{\mathcal{L}}-\mathcal{S})(t_2-t_1)}\mathcal{S}_1 e^{(\tilde{\mathcal{L}}-\mathcal{S})t_1}\rho(0)\Big]\Big\},\end{aligned}\tag{19.101a}$$

with $\rho(0)=|\psi(0)\rangle\langle\psi(0)|$, and

$$\begin{aligned}&N_t^{(-)}(\omega)\\ &=\sum_{\mathrm{REC}}\epsilon\Gamma\int_0^t dt'\int_{t'}^{t}dt''e^{-\Gamma(2t-t'-t'')}e^{i\Delta\omega(t'-t'')}\mathrm{tr}\Big\{I_{t'}(\mathcal{J}_{\rightarrow})I_{t''}(\mathcal{J}_{\rightarrow}^{+})\\ &\quad\times\Big[e^{(\tilde{\mathcal{L}}-\mathcal{S})(t-t_n)}\mathcal{S}_n e^{(\tilde{\mathcal{L}}-\mathcal{S})(t_n-t_{n-1})}\dots\mathcal{S}_2 e^{(\tilde{\mathcal{L}}-\mathcal{S})(t_2-t_1)}\mathcal{S}_1 e^{(\tilde{\mathcal{L}}-\mathcal{S})t_1}\rho(0)\Big]\Big\},\end{aligned}\tag{19.101b}$$

where $I_{t'}(\mathcal{J}_{\rightarrow})$ denotes an operator that inserts the superoperator $\mathcal{J}_{\rightarrow}$ at location t' in the string of propagators to its right, and $I_{t''}(\mathcal{J}_{\rightarrow}^{+})$ makes the similar insertion of $\mathcal{J}_{\rightarrow}^{+}$ at t''. Written in this form, we see that the sum over records simply reconstructs the propagator $e^{\tilde{\mathcal{L}}t}$, reversing the expansion in (18.41). From (19.101a), we may write

$$\begin{aligned}
&N_t^{(+)}(\omega)\\
&= \epsilon\Gamma\int_0^t dt'\int_0^{t'} dt'' e^{-\Gamma(2t-t'-t'')}e^{i\Delta\omega(t'-t'')}\mathrm{tr}\left[I_{t'}(\mathcal{J}_{\rightarrow})I_{t''}(\mathcal{J}_{\rightarrow}^{+})e^{\tilde{\mathcal{L}}t}\rho(0)\right]\\
&= \epsilon\Gamma\int_0^t dt'\int_0^{t'} dt'' e^{-\Gamma(2t-t'-t'')}e^{i\Delta\omega(t'-t'')}\mathrm{tr}\left[e^{\tilde{\mathcal{L}}(t-t')}\mathcal{J}_{\rightarrow}e^{\tilde{\mathcal{L}}(t'-t'')}\mathcal{J}_{\rightarrow}^{+}e^{\tilde{\mathcal{L}}t''}\rho(0)\right]\\
&= \epsilon\Gamma\int_0^t dt'\int_0^{t'} dt'' e^{-\Gamma(2t-t'-t'')}e^{i\Delta\omega(t'-t'')}\mathrm{tr}\left\{e^{\tilde{\mathcal{L}}(t-t')}J_{\rightarrow}e^{\tilde{\mathcal{L}}(t'-t'')}\left[\tilde{\rho}(t'')J_{\rightarrow}^{\dagger}\right]\right\},
\end{aligned} \tag{19.102}$$

from which, noting that $e^{\tilde{\mathcal{L}}(t-t')}$ preserves the trace, we arrive at

$$N_t^{(+)}(\omega) = \epsilon\Gamma\int_0^t dt'\int_0^{t'} dt'' e^{-\Gamma(2t-t'-t'')}e^{i\Delta\omega(t'-t'')}\mathrm{tr}\left\{J_{\rightarrow}e^{\tilde{\mathcal{L}}(t'-t'')}\left[\tilde{\rho}(t'')J_{\rightarrow}^{\dagger}\right]\right\}. \tag{19.103a}$$

In a similar fashion, from (19.101b), we obtain

$$N_t^{(-)}(\omega) = \epsilon\Gamma\int_0^t dt'\int_{t'}^{t} dt'' e^{-\Gamma(2t-t'-t'')}e^{i\Delta\omega(t'-t'')}\mathrm{tr}\left\{J_{\rightarrow}^{\dagger}e^{\tilde{\mathcal{L}}(t''-t')}\left[J_{\rightarrow}\tilde{\rho}(t')\right]\right\}. \tag{19.103b}$$

Finally, adding the two pieces together, *the optical spectrum measured with a scanning interferometer of half-width Γ in time interval* $[0,t)$ is

$$N_t(\omega) = \epsilon\gamma_{a2}\Gamma\int_0^t dt'\int_0^t dt'' e^{-\Gamma(2t-t'-t'')}e^{i\Delta\omega(t'-t'')}\langle\tilde{a}^{\dagger}(t'')\tilde{a}(t')\rangle. \tag{19.104}$$

Exercise 19.6. It is usual to define the spectrum by the asymptotic limit $t\rightarrow\infty$, assuming a stationary correlation function $\langle\tilde{a}^{\dagger}(t'')\tilde{a}(t')\rangle$. Carry out the limit and show that (19.104) reduces to the form familiar from the Wiener–Kinchin theorem [19.13, 19.14]:

$$N_{\infty}(\omega) = \frac{\epsilon\gamma_{a2}}{2}\int_{-\infty}^{\infty} d\tau e^{-\Gamma|\tau|}e^{i\Delta\omega\tau}\langle\tilde{a}^{\dagger}(0)\tilde{a}(\tau)\rangle_{\mathrm{ss}}, \tag{19.105}$$

where $\langle\tilde{a}^{\dagger}(0)\tilde{a}(\tau)\rangle_{\mathrm{ss}} \equiv \lim_{t\rightarrow\infty}\langle\tilde{a}^{\dagger}(t)\tilde{a}(t+\tau)\rangle$.

Note 19.4. The normalization of spectrum (19.105) follows from its definition as the mean photon number inside the interferometer. As a check on this result, we might consider a monochromatic incident field with photon flux $\epsilon\gamma_{a2}\langle\tilde{a}^\dagger\tilde{a}\rangle_{\rm ss}$. In this case, on resonance ($\omega = \omega_C$), the transmitted photon flux should equal the incident photon flux. This requirement—$\Gamma N_\infty(\omega_C) = \epsilon\gamma_{a2}\langle\tilde{a}^\dagger\tilde{a}\rangle_{\rm ss}$—agrees with (19.105). An alternative normalization sets the integral over the spectrum to unity (Eq. 10.65) or to the incident photon flux (Eqs. 15.107 and 16.192). In the former case, for example, spectrum (19.105) is replaced by

$$T(\omega) = \left(\langle\tilde{a}^\dagger\tilde{a}\rangle_{\rm ss}\right)^{-1}\frac{1}{2\pi}\int_{-\infty}^{\infty} d\tau e^{-\Gamma|\tau|}e^{i\Delta\omega\tau}\langle\tilde{a}^\dagger(0)\tilde{a}(\tau)\rangle_{\rm ss}. \qquad (19.106)$$

19.3.2 Spontaneous Emission from a Driven Excited-State Doublet

The preceeding analysis does not merely provide a path to standard results for the optical spectrum. As an alternative to working from (19.104)–(19.106), one may simulate quantum trajectories for the cascaded system—source plus interferometer—to obtain the spectrum directly; Tian and Carmichael [19.15], for example, have computed optical spectra for a cavity QED system in this way. The method is time-consuming, however, since the simulations must be carried out for many settings of the interferometer frequency. Generally, it is faster to compute the source-field correlation function, $\langle\tilde{a}^\dagger(0)\tilde{a}(\tau)\rangle_{\rm ss}$, and obtain the spectrum by Fourier transform. Dalibard and coworkers [19.16] provide an algorithm to obtain the correlation function from quantum trajectories; alternatively, a method based on heterodyne detection trajectories can be used (Sect. 19.3.3).

Some situations, however, are especially well-suited to the scanning interferometer approach, particularly when it is possible to carry an analytical calculation through. The spontaneous emission spectrum for one atom in a cavity worked out in Sect. 13.3.1 provides a case in point. Though the derivation of this spectrum appears to follow the standard route—starting from expressions like (19.104), which require a correlation function to be calculated first (Eqs. 13.160a and 13.160b)—the method of solution takes it beyond this. Specifically, we are now in a position to interpret the decomposition of the density operator defined by (13.145)–(13.147). Although introduced merely as a solution technique, we can see now that it amounts to a quantum trajectory unraveling of the density operator for the simplest situation (spontaneous emission) where one and only one quantum jump occurs. The unraveling facilitates an analysis in terms of pure rather than mixed states; thus, to obtain the correlation function, and hence from it the spectrum, we need deal with nothing more complicated than a pair of coupled equations of motion for Schrödinger amplitudes (Eqs. 13.152a and 13.152b).

In this section we consider a similar example, working directly with the quantum trajectory equations for a source plus interferometer to derive a spon-

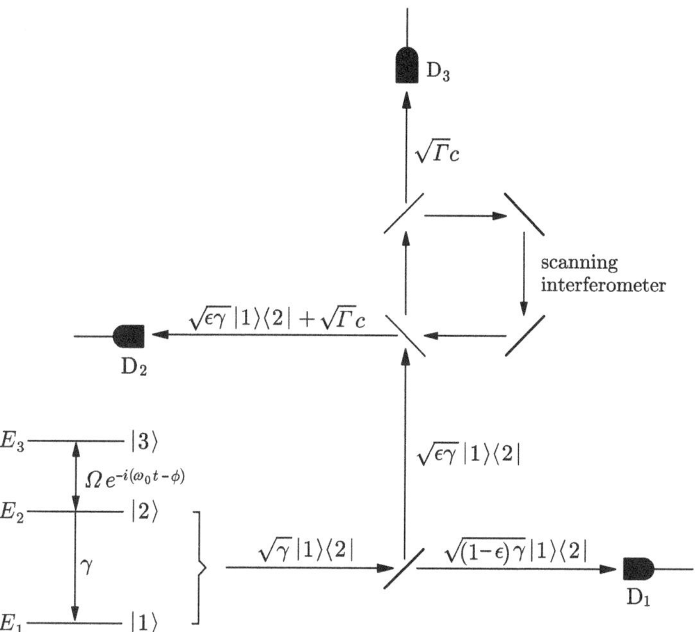

Fig. 19.7. Scattering scenario for measuring the spontaneous emission spectrum on the $|2\rangle \to |1\rangle$ transition of the three-level atom shown at lower left. The atom is driven on the $|2\rangle \to |3\rangle$ transition by a coherent field of Rabi frequency Ω. Output channels are labeled by their source fields, which are written in units of the square root of photon flux

taneous emission spectrum. The particular example is taken from the literature on quantum interference effects in spontaneous emission [19.17]. A three-level atom is prepared in its intermediate state $|2\rangle$ (Fig. 19.7) and interacts with a coherent external field, frequency ω_0, driving the $|2\rangle \to |3\rangle$ transition. It simultaneously undergoes spontaneous emission to arrive eventually in state $|1\rangle$. The spectrum of spontaneous emission is to be measured using a scanning interferometer. The scattering scenario is illustrated in Fig. 19.7. There are three distinct records, each corresponding to the recording of a single photoelectric count. Ultimately, a single count is recorded in one of the three detectors—D_1, D_2, or D_3. The spectrum is given by the probability, as a function of the frequency setting ω of the interferometer, that detector D_3 records the count.

The analysis begins from the non-Hermitian Hamiltonian for the cascaded system

$$\begin{aligned} H_B^c = {} & E_1|1\rangle\langle 1| + \Big(E_2 - i\hbar\frac{\gamma}{2}\Big)|2\rangle\langle 2| + E_3|3\rangle\langle 3| + \hbar(\omega - i\Gamma)c^\dagger c \\ & + \hbar\Omega\Big[|3\rangle\langle 2|e^{-i(\omega_0 t-\phi)} + |2\rangle\langle 3|e^{i(\omega_0 t-\phi)}\Big] - i\hbar\sqrt{\epsilon\gamma\Gamma}c^\dagger|1\rangle\langle 2|, \end{aligned} \qquad (19.107)$$

where Ω, ω_0, and ϕ are the Rabi frequency, the frequency, and the phase of the field driving the atom, respectively; E_1, E_2, and E_3 are the energies of the atomic states (Fig. 19.7); γ is the spontaneous emission rate for the $|2\rangle \to |1\rangle$ transition; and Γ is the interferometer half-width. Since only zero- and one-photon states are involved, unlike in Sect. 19.3.1, the coupling fraction ϵ into the interferometer need not be assumed to be very small. It is convenient to introduce detunings $\delta\omega$ and $\Delta\omega$, such that

$$E_3 - E_2 = \hbar(\omega_0 + \delta\omega), \qquad E_2 - E_1 = \hbar(\omega - \Delta\omega). \tag{19.108}$$

Then, working in the interaction picture, with

$$|\tilde{\psi}_{\rm REC}(t)\rangle \equiv e^{i\{[E_1|1\rangle\langle 1| + E_2|2\rangle\langle 2| + (E_3 - \hbar\delta\omega)|3\rangle\langle 3|]/\hbar\}t} e^{i[(\omega - \Delta\omega)c^\dagger c]t} |\psi_{\rm REC}(t)\rangle, \tag{19.109}$$

and absorbing the phase factor $-ie^{i\phi}$ into the definition of basis state $|3\rangle$, the non-Hermitian Hamiltonian is rewritten as

$$\begin{aligned}\tilde{H}_B = &-i\hbar\frac{\gamma}{2}|2\rangle\langle 2| + \hbar\delta\omega|3\rangle\langle 3| + \hbar(\Delta\omega - i\Gamma)c^\dagger c \\ &+ i\hbar\Omega\big(|3\rangle\langle 2| - |2\rangle\langle 3|\big) - i\hbar\sqrt{\epsilon\gamma\Gamma}c^\dagger|1\rangle\langle 2|.\end{aligned} \tag{19.110}$$

Our task is to solve for the probability, $P_3(\omega)$, that detector D_3 records a photoelectric count at some time in the interval $[0, \infty)$. With the conditional state expanded as in (19.81) (with $\theta = 0$), the probability is given by

$$P_3(\omega) = \Gamma \int_0^\infty dt \langle \bar{\tilde{\psi}}^{(1)}_{\{o\emptyset_t\}}(\omega) | \bar{\tilde{\psi}}^{(1)}_{\{o\emptyset_t\}}(\omega)\rangle, \tag{19.111}$$

where the state label indicates the no-count record up to the time t when detector D_3 fires. This expression is the equivalent of (19.83) multiplied by Γdt and summed over possible firing times t. No sum over records is required, since only one record—the no-count record—can preceed the firing of D_3. With the conditional state expansion

$$|\bar{\tilde{\psi}}^{(0)}_{\rm REC}(t)\rangle = \bar{\tilde{c}}_2(t)|2\rangle + \bar{\tilde{c}}_3(t)|3\rangle, \tag{19.112a}$$

$$|\bar{\tilde{\psi}}^{(1)}_{\rm REC}(\omega, t)\rangle = \bar{\tilde{c}}_1(\omega, t)|1\rangle, \tag{19.112b}$$

the required probability is

$$P_3(\omega) = \Gamma \int_0^\infty dt |\bar{\tilde{c}}_1(\omega, t)|^2. \tag{19.113}$$

We must determine the amplitude $\bar{\tilde{c}}_1(\omega, t)$, for which we turn to the nonunitary Schrödinger equation between quantum jumps.

From Hamiltonian (19.110), the Schrödinger amplitudes of the upper two atomic states obey the coupled equations of motion

$$\frac{d\tilde{\tilde{c}}_2}{dt} = -\frac{\gamma}{2}\tilde{\tilde{c}}_2 - \Omega\tilde{\tilde{c}}_3, \tag{19.114a}$$

$$\frac{d\tilde{\tilde{c}}_3}{dt} = -i\delta\omega\tilde{\tilde{c}}_3 + \Omega\tilde{\tilde{c}}_2. \tag{19.114b}$$

Amplitude $\tilde{\tilde{c}}_2$ then provides the source term in the Schrödinger equation of motion for $\tilde{\tilde{c}}_1(\omega)$:

$$\frac{d\tilde{\tilde{c}}_1(\omega)}{dt} = -(\Gamma + i\Delta\omega)\tilde{\tilde{c}}_1(\omega) - \sqrt{\epsilon\gamma\Gamma}\tilde{\tilde{c}}_2, \tag{19.114c}$$

with formal solution

$$\tilde{\tilde{c}}_1(\omega, t) = -\sqrt{\epsilon\gamma\Gamma}\int_0^t dt' e^{-(\Gamma+i\Delta\omega)(t-t')}\tilde{\tilde{c}}_2(t'). \tag{19.115}$$

Now, if the resolution of the interferometer is to be any good, Γ is required to be very small compared to Ω and γ; thus, we may replace the upper integration limit of (19.115) by infinity and set $e^{\Gamma t'}\tilde{\tilde{c}}_2(t')$ to $\tilde{\tilde{c}}_2(t')$. With this simplification, combining (19.113) and (19.115), the spectrum is

$$\begin{aligned} P_3(\omega) &= \frac{\epsilon\gamma\Gamma}{2}\left|\int_0^\infty dt' e^{i\Delta\omega}\tilde{\tilde{c}}_2(t')\right|^2 \\ &= \frac{\epsilon\gamma\Gamma}{2}\left|\tilde{\tilde{C}}_2(-i\Delta\omega)\right|^2, \end{aligned} \tag{19.116}$$

where $\tilde{\tilde{C}}_2(s)$ is the Laplace transform of $\tilde{\tilde{c}}_2(t)$.

Considering now the case where the atom is prepared in state $|2\rangle$, from the Laplace transforms of (19.114a) and (19.114b), we have

$$s\tilde{\tilde{C}}_2(s) - 1 = -\frac{\gamma}{2}\tilde{\tilde{C}}_2(s) - \Omega\tilde{\tilde{C}}_3(s), \tag{19.117a}$$

$$s\tilde{\tilde{C}}_3(s) = -i\delta\omega\tilde{\tilde{C}}_3(s) + \Omega\tilde{\tilde{C}}_2(s), \tag{19.117b}$$

with solution

$$\tilde{\tilde{C}}_2(s) = \frac{s + i\delta\omega}{(s+\gamma/2)(s+i\delta\omega) + \Omega^2}. \tag{19.118}$$

Thus, from (19.116) and (19.118), the *spectrum of spontaneous emission from a coherently driven atomic excited-state doublet and measured by an interferometer of negligible half-width* Γ is given by the probability of detection at detector D_3 (Fig. 19.7)

$$P_3(\omega) = \frac{\epsilon\Gamma(\gamma/2)(\Delta\omega - \delta\omega)^2}{[(\gamma/2)(\Delta\omega - \delta\omega)]^2 + [\Omega^2 - \Delta\omega(\Delta\omega - \delta\omega)]^2}. \tag{19.119}$$

Note how this result vanishes when $\Delta\omega = \delta\omega$. This is the so-called quantum interference, or interference of probability amplitudes for emission from the two atomic dressed states; see [19.17] for further discussion. It is readily shown that preparation of atomic state $|3\rangle$ rather than $|2\rangle$ results in the replacement of the interference minimum by an interference maximum.

Exercise 19.7. For an arbitrary interferometer half-width, the spectrum may be calculated as $P_3(\omega) = \Gamma\tilde{\tilde{\mathcal{C}}}_{11}(\omega, 0)$, where $\tilde{\tilde{\mathcal{C}}}_{11}(\omega, s)$ is the Laplace transform of $|\tilde{\bar{c}}_1(\omega, t)|^2$. Show that the *spectrum of spontaneous emission from a coherently driven atomic excited-state doublet and measured by an interferometer of half-width Γ* is

$$P_3(\omega) = \frac{\epsilon\Gamma(\Gamma + \gamma/2)(\Delta\omega - \delta\omega)^2 + \epsilon\Gamma^2[\Gamma(\Gamma + \gamma/2) + \Omega^2]}{[\Gamma\Delta\omega + (\Gamma + \gamma/2)(\Delta\omega - \delta\omega)]^2 + [\Gamma(\Gamma + \gamma/2) + \Omega^2 - \Delta\omega(\Delta\omega - \delta\omega)]^2}. \tag{19.120}$$

With Γ sufficiently small, the simpler expression (19.119) is recovered, while in the large Γ limit the Lorentzian response of the detector is obtained—i.e., $P_3(\omega) \to \epsilon\Gamma^2/(\Gamma^2 + \Delta\omega^2)$.

Note 19.5. We noted at the beginning of the section that this example shows similarities to spontaneous emission in single-atom cavity QED (Sect. 13.3.1). Indeed, if spontaneous decay from state $|3\rangle$ is added to the scheme of Fig. 19.7, the examples correspond to one another one-to-one; coupled equations (19.114a) and (19.114b) correspond to (13.152a) and (13.152b), and the spectrum given by (19.116)—i.e., for $\Gamma \ll \gamma, \Omega$—corresponds to that obtained from (13.169a). It follows that the earlier comment about a time-domain interpretation of the quantum interference (Note 13.14) also applies here.

19.3.3 Optical Spectrum Using Heterodyne Detection

The squeezing spectra displayed in the column to the right in Fig. 18.5 are Fourier transforms of the photocurrent autocorrelation functions shown in the column to the left. The latter were computed as time averages of the product $i(t)i(t+\tau)$, where $i(t)$ is a simulated homodyne current record. The optical spectrum can be computed in similar fashion, but from a simulated *heterodyne* current record. Our goal in this section is to prove this result—to show how the heterodyne current correlation function is related to the autocorrelation of the source field.

To this end, we return to the degenerate parametric oscillator example of Sect. 18.1 and the quantum trajectory unraveling of Sect. 18.2.3. From (18.91) and (18.92), the complex heterodyne current $\tilde{i}(t)$ satisfies the stochastic differential equation

$$d\tilde{i} = -\tau_d^{-1}(\tilde{i}dt - d\tilde{q}), \tag{19.121}$$

with incremental charge

$$d\tilde{q} = Ge|\mathcal{E}_{\rm lo}|\Big(\langle J_\to^\dagger\rangle_{\rm REC}dt + dZ\Big), \tag{19.122}$$

where dZ is a complex-valued Wiener increment and the conditional expectation $\langle J_\to^\dagger(t)\rangle_{\rm REC}$ is taken with respect to quantum state $|\tilde{\psi}_{\rm REC}(t)\rangle$, which satisfies the stochastic Schrödinger equation (18.90); jump operator $J_\to$ is defined in (18.35a). Solving (19.121) formally for $\tilde{i}(t)$ yields

$$\tilde{i}(t) = \tau_d^{-1}\int_0^t e^{-(t-t')/\tau_d}d\tilde{q}'. \tag{19.123}$$

Thus, the heterodyne current autocorrelation function is to be calculated as

$$\overline{\tilde{i}^*(t)\tilde{i}(t+\tau)} = \left(\frac{Ge|\mathcal{E}_{\rm lo}|}{\tau_d}\right)^2\int_0^t\int_0^{t+\tau} e^{-(2t+\tau-t'-t'')/\tau_d}\overline{(d\bar{q}')^*d\bar{q}''}, \tag{19.124}$$

with rescaled incremental charge

$$d\bar{q} \equiv (Ge|\mathcal{E}_{\rm lo}|)^{-1}d\tilde{q} = \langle J_\to^\dagger\rangle_{\rm REC}dt + dZ, \tag{19.125}$$

and from (19.122),

$$\begin{aligned}\overline{(d\bar{q}')^*d\bar{q}''} &= \overline{\langle J_\to^\dagger(t'')\rangle_{\rm REC}\langle J_\to(t')\rangle_{\rm REC}}dt'dt'' + \overline{\langle J_\to^\dagger(t'')\rangle_{\rm REC}(dZ')^*}dt''\\ &\quad + \overline{\langle J_\to(t')\rangle_{\rm REC}dZ''}dt' + \overline{dZ''(dZ')^*}.\end{aligned} \tag{19.126}$$

Our principal task is to evaluate the four stochastic averages on the right-hand side of (19.126).

Two of the terms are easily dealt with. The last term describes shot noise. It is simply the correlation function of the complex Wiener increment, for which we may write

$$\overline{dZ''(dZ')^*} = \delta(t'-t'')dt'dt''. \tag{19.127}$$

Substituting this result into the double integral of (19.124), we obtain the *shot noise autocorrelation function in heterodyne detection with detector half-*

width τ_d^{-1}:

$$\overline{\tilde{i}^*(t)\tilde{i}(t+\tau)}_{\text{shot}} = \left(\frac{Ge|\mathcal{E}_{\text{lo}}|}{\tau_d}\right)^2 \int_0^t dt' \int_0^{t+\tau} dt'' e^{-(2t+\tau-t'-t'')/\tau_d}\delta(t'-t'')$$
$$= \frac{(Ge)^2|\mathcal{E}_{\text{lo}}|^2}{2\tau_d} e^{-|\tau|/\tau_d}\left[1-e^{-2\min(t,t+\tau)/\tau_d}\right]. \tag{19.128}$$

For the second easily dealt-with term we observe that either the second or third term on the right-hand side of (19.126) vanishes. Note first that the Wiener increment is δ-correlated (Eq. 19.127) and through the stochastic Schrödinger equation, the conditional state—hence $\langle J_\rightarrow^\dagger(t)\rangle_{\text{REC}}$—depends on *past* Wiener increments only. It follows that

$$\overline{\langle J_\rightarrow^\dagger(t'')\rangle_{\text{REC}}(dZ')^*} = 0 \qquad t'' < t', \tag{19.129a}$$
$$\overline{\langle J_\rightarrow(t')\rangle_{\text{REC}}dZ''} = 0 \qquad t'' > t'. \tag{19.129b}$$

The two terms that remain require just a little extra thought.

To assist with their evaluation, let us divide the full heterodyne current record into two parts, with

$$\text{REC} = \text{REC1} \wedge \text{REC2}, \tag{19.130}$$

where REC1 covers the period of time from $t=0$ up to $\min(t',t'')$, while REC2 covers the period between $\min(t',t'')$ and $\max(t',t'')$. The record probability is

$$P(\text{REC2}\wedge\text{REC1}) = P(\text{REC2}|\text{REC1})P(\text{REC1}). \tag{19.131}$$

Consider then the first term on the right-hand side of (19.126) and the case where $t'' < t'$. The average over records is developed as a double sum over REC1 and REC2. We have

$$\overline{\langle J_\rightarrow^\dagger(t'')\rangle_{\text{REC}}\langle J_\rightarrow(t')\rangle_{\text{REC}}}$$
$$= \sum_{\text{REC1}} P(\text{REC1}) \sum_{\text{REC2}} P(\text{REC2}|\text{REC1})\text{tr}\left[J_\rightarrow|\tilde{\psi}_{\text{REC}}(t')\rangle\langle\tilde{\psi}_{\text{REC}}(t')|\langle J_\rightarrow^\dagger(t'')\rangle_{\text{REC1}}\right]$$
$$= \sum_{\text{REC1}} P(\text{REC1})\text{tr}\left\{J_\rightarrow e^{\tilde{\mathcal{L}}(t'-t'')}\left[|\tilde{\psi}_{\text{REC1}}(t'')\rangle\langle\tilde{\psi}_{\text{REC1}}(t'')|\langle J_\rightarrow^\dagger(t'')\rangle_{\text{REC1}}\right]\right\}, \tag{19.132}$$

where we have used the fundamental quantum trajectory expansion of the density operator (17.8) to write

$$\sum_{\text{REC2}} P(\text{REC2}|\text{REC1})|\tilde{\psi}_{\text{REC}}(t')\rangle\langle\tilde{\psi}_{\text{REC}}(t')| = e^{\tilde{\mathcal{L}}(t'-t'')}|\tilde{\psi}_{\text{REC1}}(t'')\rangle\langle\tilde{\psi}_{\text{REC1}}(t'')|. \tag{19.133}$$

Finally, the third term on the right-hand side of (19.126) is developed in a similar fashion:

$$
\begin{aligned}
&\overline{\langle J_{\rightarrow}(t')\rangle_{\mathrm{REC}} dZ''} \\
&= \sum_{\mathrm{REC1}} P(\mathrm{REC1}) \sum_{\mathrm{REC2}} P(\mathrm{REC2}|\mathrm{REC1}) \mathrm{tr}\Big[J_{\rightarrow}|\tilde{\psi}_{\mathrm{REC}}(t')\rangle\langle\tilde{\psi}_{\mathrm{REC}}(t')|\Big] dZ'' \\
&= \sum_{\mathrm{REC1}} P(\mathrm{REC1}) \mathrm{tr}\Big\{J_{\rightarrow} e^{\tilde{\mathcal{L}}(t'-t'')}\Big[|\tilde{\psi}_{\mathrm{REC1}}(t'')\rangle \overline{d\langle\tilde{\psi}_{\mathrm{REC1}}(t'')| dZ''}\Big]\Big\} \\
&= \sum_{\mathrm{REC1}} P(\mathrm{REC1}) \mathrm{tr}\Big\{J_{\rightarrow} e^{\tilde{\mathcal{L}}(t'-t'')}\Big[|\tilde{\psi}_{\mathrm{REC1}}(t'')\rangle\langle\tilde{\psi}_{\mathrm{REC1}}(t'')| \\
&\quad \times \Big(J_{\rightarrow}^{\dagger} - \langle J_{\rightarrow}^{\dagger}(t'')\rangle_{\mathrm{REC1}}\Big)\Big]\Big\} dt'',
\end{aligned}
\tag{19.134}
$$

where we note that dZ'' is uncorrelated with $|\tilde{\psi}_{\mathrm{REC1}}(t'')\rangle$, $d|\tilde{\psi}_{\mathrm{REC1}}(t'')\rangle$, and $\langle\tilde{\psi}_{\mathrm{REC1}} t'')|$, and we evaluate its correlation with $d\langle\tilde{\psi}_{\mathrm{REC1}}(t'')|$ from the stochastic Schrödinger equation using (18.90) and (18.91) [specifically using (18.98)].

We now collect our various results together. Substituting the four expressions (19.127), (19.129a), (19.133), and (19.134) on the right-hand side of (19.126), the autocorrelation of incremental charge, $t'' < t'$, is

$$
\begin{aligned}
&\overline{(d\bar{q}')^* d\bar{q}''}/dt'dt'' \\
&= \sum_{\mathrm{REC1}} P(\mathrm{REC1}) \mathrm{tr}\Big\{J_{\rightarrow} e^{\tilde{\mathcal{L}}(t'-t'')}\Big[|\tilde{\psi}_{\mathrm{REC1}}(t'')\rangle\langle\tilde{\psi}_{\mathrm{REC1}}(t'')| J_{\rightarrow}^{\dagger}\Big]\Big\} + \delta(t'-t'') \\
&= \mathrm{tr}\Big\{J_{\rightarrow} e^{\tilde{\mathcal{L}}(t'-t'')}\big[\tilde{\rho}(t'') J_{\rightarrow}^{\dagger}\big]\Big\} + \delta(t'-t''),
\end{aligned}
\tag{19.135a}
$$

where again the sum over records is taken using (17.8). A parallel calculation gives the autocorrelation for $t'' > t'$:

$$
\overline{(d\bar{q}')^* d\bar{q}''}/dt'dt'' = \mathrm{tr}\Big\{J_{\rightarrow}^{\dagger} e^{\tilde{\mathcal{L}}(t''-t')}\big[J_{\rightarrow}\tilde{\rho}(t'']\Big\} + \delta(t'-t''). \tag{19.135b}
$$

We are now in a position to prove the result asserted at the beginning of the section—that the heterodyne current correlation function (19.124) yields the source-field autocorrelation, $\langle\tilde{a}^{\dagger}(t)\tilde{a}(t+\tau)\rangle$, and hence, via a Fourier transform, the optical spectrum. First, we recognize the sequences of operators and superoperators on the right-hand sides of (19.135a) and (19.135b) as those appearing in quantum regression formulae (1.97) and (1.98); thus, using the definition (18.35a) of $J_{\rightarrow}$, the autocorrelation of incremental charge may be written as

$$
\overline{(d\bar{q}')^* d\bar{q}''}/dt'dt'' = \gamma_{a2}\langle\tilde{a}^{\dagger}(t'')\tilde{a}(t')\rangle + \delta(t'-t''). \tag{19.136}
$$

Combining this result with (19.124) and (19.128), we arrive at an expression for *the photocurrent autocorrelation function in heterodyne detection with detector half-width* τ_d^{-1}:

$$\overline{\tilde{i}^*(t)\tilde{i}(t+\tau)} - \overline{\tilde{i}^*(t)\tilde{i}(t+\tau)}_{\rm shot}$$
$$= \left(\frac{Ge|\mathcal{E}_{\rm lo}|}{\tau_d}\right)^2 \gamma_{a2} \int_0^t dt' \int_0^{t+\tau} dt'' e^{-(2t+\tau-t'-t'')/\tau_d} \langle \tilde{a}^\dagger(t'')\tilde{a}(t')\rangle. \tag{19.137}$$

Taking the limit of narrow detection bandwidth, $\tau_d^{-1} \rightarrow 0$, the right-hand side is proportional to the source-field autocorrelation. Its Fourier transform, assuming stationarity, yields the optical spectrum in agreement with (19.105).

References

Chapter 9

[9.1] R. J. Glauber: Phys. Rev. Lett. **10**, 84 (1963); Phys. Rev. **130**, 2529 (1963); **131**, 2766 (1963)

[9.2] P. L. Kelley and W. H. Kleiner: Phys. Rev. **136**, A316 (1964)

[9.3] L. Mandel: Proc. Phys. Soc. **72**, 1037 (1958); "Fluctuations of Light Beams", in *Progress in Optics*, Vol. II, ed. by E. Wolf (North Holland, Amsterdam, 1963) pp. 181–248

[9.4] L.-A. Wu, H. J. Kimble, J. L. Hall, and H. Wu: Phys. Rev. Lett. **57**, 2520 (1986)

[9.5] L.-A. Wu, M. Xiao, and H. J. Kimble: J. Opt. Soc. Am. B **4**, 1465 (1987)

[9.6] R. E. Slusher, L. W. Hollberg, B. Yurke, J. C. Mertz, and J. F. Valley: Phys. Rev. Lett. **55**, 2409 (1985)

[9.7] P. Grangier, R. E. Slusher, B. Yurke, and A. LaPorta: Phys. Rev. Lett. **59**, 2153 (1987)

[9.8] J. Mod. Opt. **34**, Special Issue on "Squeezed Light", June, 1987

[9.9] J. Opt. Soc. Am. B **4**, Feature Issue on "Squeezed States of the Electromagnetic Field", October 1987

[9.10] A. Yariv: *Introduction to Optical Electronics* (Holt, Rinehart and Wilson, New York, 1976) Chap. 8

[9.11] N. Bloembergen: *Nonlinear Optics*, Fourth Printing, 1982 (Benjamin, Reading, Massachusetts, 1985)

[9.12] E. C. G. Sudarshan: Phys. Rev. Lett. **10**, 277 (1963)

[9.13] J. R. Klauder and E. C. G. Sudarshan: *Fundamentals of Quantum Optics* (Benjamin, New York, 1968)

[9.14] B. R. Mollow and R. J. Glauber: Phys. Rev. **160**, 1076 (1967); **160**, 1097 (1967)

[9.15] L.-A. Wu and H. J. Kimble: J. Opt. Soc. Am. B **2**, 697 (1985)

[9.16] Reference [9.10] p. 205

[9.17] A. F. Kip: *Fundamentals of Electricity and Magnetism* (McGraw–Hill, New York, 1962) pp. 370–373

[9.18] C. Caves and B. L. Schumaker: Phys. Rev. A **31**, 3068 (1985); **31**, 3093 (1985)

[9.19] C. W. Gardiner and C. M. Savage: Optics Commun. **50**, 173 (1984)

[9.20] M. J. Collett and C. W. Gardiner: Phys. Rev. A **30**, 1386 (1984)
[9.21] H. J. Carmichael: J. Opt. Soc. Am. B **4**, 1588 (1987)
[9.22] H. P. Yuen and V. W. S. Chan: Opt. Lett. **8**, 177 (1983)
[9.23] M. J. Collett, D. F. Walls, and P. Zoller: Optics Commun. **52**, 145 (1984)
[9.24] M. J. Collett and D. F. Walls: Phys. Rev. A **32**, 2887 (1985)
[9.25] M. D. Reid and D. F. Walls: Phys. Rev. A **32**, 396 (1985); Phys. Rev. A **34**, 4929 (1986)
[9.26] H. P. Yuen and J. H. Shapiro: IEEE Trans. Inf. Theory **IT-26**, 78 (1980)
[9.27] B. Yurke: Phys. Rev. A **29**, 408 (1984)
[9.28] T. W. Marshall: Proc. R. Soc. Lond. A **276**, 475 (1963); Proc. Cambridge Philos. Soc. **61**, 537 (1965); Nuovo Cimento **38**, 216 (1965)
[9.29] T. H. Boyer: Phys. Rev. **182**, 1374 (1969); **186**, 1304 (1969); Ann. Phys. **56**, 474 (1970); Phys. Rev. D **11**, 809 (1975); **29**, 1096 (1984)
[9.30] E. Santos, Nuovo Cimento B **19**, 57 (1974); **22**, 201 (1974)
[9.31] L. de la Peña and A. M. Cello: J. Math. Phys. **18**, 1612 (1977); **20**, 469 (1979)
[9.32] J. L. Jiménez, L. de la Peña, and T. A. Brody: Am. J. Phys. **48**, 840 (1980)
[9.33] T. W. Marshall and E. Santos: Found. Phys. **18**, 185 (1988)
[9.34] A. Casado, T. W. Marshall, and E. Santos: J. Opt. Soc. Am. B **14**, 494 (1997); **15**, 1572 (1998)
[9.35] H. J. Carmichael and H. Nha: J. Opt. B **6**, S649 (2004)
[9.36] H. J. Carmichael and H. Nha: "Continuous Variable Teleportation within Stochastic Electrodynamics", in *Laser Spectroscopy*, eds. H.-A. Bachor and P. Hannaford (World Scientific, Singapore, 2004) pp. 324–333
[9.37] N. Bohr, H. A. Kramers, and J. C. Slater: Philos. Mag. **47**, 785 (1924); Z. Phys. **24**, 69 (1924)
[9.38] H. J. Carmichael: "Quantum Fluctuations of Light: A Modern Perspective on Wave/Particle Duality", in *Quantum Mechanics at the Crossroads*, eds. J. Evans and A. S. Thorndike (Springer, Berlin, 2006) pp. 183–212

Chapter 10

[10.1] C. W. Gardiner: *Handbook of Stochastic Methods for Physics, Chemistry and the Natural Sciences* (Springer, Berlin, 1983) pp. 299–302
[10.2] L. Mandel: Proc. Phys. Soc. **72**, 1037 (1958); "Fluctuations of Light Beams", in *Progress in Optics*, Vol. II, ed. by E. Wolf (North Holland, Amsterdam, 1963) pp. 181–248
[10.3] L.-A. Wu, H. J. Kimble, J. L. Hall, and H. Wu: Phys. Rev. Lett. **57**, 2520 (1986)
[10.4] G. Milburn and D. F. Walls: Optics Commun. **39**, 401 (1981)
[10.5] P. R. Rice and H. J. Carmichael: J. Opt. Soc. Am. B **5**, 1661 (1988)
[10.6] B. R. Mollow: Phys. Rev. **188**, 1969 (1969)

Chapter 11

[11.1] P. D. Drummond and C. W. Gardiner: J. Phys. A **13**, 2353 (1980)
[11.2] P. D. Drummond and S. J. Carter: J. Opt. Soc. Am. B **4**, 1565 (1987)
[11.3] S. J. Carter, P. D. Drummond, M. D. Reid, and R. M. Shelby: Phys. Rev. Lett. **58**, 1841 (1987)

[11.4] R. M. Shelby, P. D. Drummond, and S. J. Carter: Phys. Rev. A **42**, 2966 (1990)
[11.5] S. J. Carter and P. D. Drummond: Phys. Rev. Lett. **67**, 3757 (1991)
[11.6] P. D. Drummond, R. M. Shelby, S. R. Friberg, and Y. Yamamoto: Nature **365**, 307 (1993)
[11.7] M. J. Werner: Phys. Rev. A **54**, R2567 (1996); Phys. Rev. Lett. **81**, 4132 (1998); Phys. Rev. A **60**, R781 (1999)
[11.8] M. J. Werner and S. R. Friberg: Phys. Rev. Lett. **79**, 4143 (1997)
[11.9] P. D. Drummond and J. F. Corney: Phys. Rev. A **60**, R2661 (1999)
[11.10] J. F. Corney and P. D. Drummond: Phys. Rev. A **68**, 063822 (2003)
[11.11] J. F. Corney and P. D. Drummond: Phys. Rev. Lett. **93**, 260401 (2004)
[11.12] K. V. Kheruntsyan, M. K. Olsen, and P. D. Drummond: Phys. Rev. Lett. **95**, 150405 (2005)
[11.13] A. M. Smith and C. W. Gardiner: Phys. Rev. A **39**, 3511 (1989)
[11.14] P. D. Drummond, K. J. McNeil, and D. F. Walls: Optica Acta **28**, 211 (1980)
[11.15] D. F. Walls, G. J. Milburn, and H. J. Carmichael: Optica Acta **29**, 1179 (1982)
[11.16] B. Yurke: Phys. Rev. A **29**, 408 (1984); **32**, 300 (1985)
[11.17] C. W. Gardiner and C. M. Savage: Optics Commun. **50**, 173 (1984)
[11.18] M. J. Collett and C. W. Gardiner: Phys. Rev. A **30**, 1386 (1984); **31**, 3761 (1985)
[11.19] J.-M. Courty and S. Reynaud: Phys. Rev. A **46**, 2766 (1992)
[11.20] H. P. Yuen and P. Tombesi: Optics Commun. **59**, 155 (1986)

Chapter 12

[12.1] I. Prigogine, C. George, F. Henin, and L. Rosenfeld: Chem. Scripta **4**, 5 (1973)
[12.2] E. Helfand: Bell Systems Tech. J. **58**, 2289 (1979)
[12.3] H. S. Greenside and E. Helfand: Bell Systems Tech. J. **60**, 1927 (1981)
[12.4] W. Rümelin: SIAM J. Numer. Anal. **19**, 604 (1982)
[12.5] J. R. Klauder and W. P. Petersen: SIAM J. Numer. Anal. **22**, 1153 (1985)
[12.6] P. D. Drummond and I. K. Mortimer: J. Comput. Phys. **93**, 144 (1991)
[12.7] P. E. Kloeden and E. Platen: *Numerical Solution of Stochastic Differential Equations* (Springer, Berlin, 1992)
[12.8] M. J. Werner and P. D. Drummond: J. Comput. Phys. **132**, 312 (1997)
[12.9] C. J. Mertens and T. A. B. Kennedy: Phys. Rev. A **48**, 2374 (1993)
[12.10] L. I. Plimak and D. F. Walls: Phys. Rev. A **50**, 2627 (1994)
[12.11] O. Veits and M. Fleischhauer: Phys. Rev. A **55**, 3059 (1997)
[12.12] C. W. Gardiner: *Handbook of Stochastic Methods for Physics, Chemistry and the Natural Sciences* (Springer, Berlin, 1983) Chap. 9
[12.13] N. G. van Kampen: J. Stat. Phys. **17**, 71 (1977)
[12.14] M. Wolinsky and H. J. Carmichael: Optics Commun. **55**, 138 (1985)
[12.15] H. J. Carmichael, J. S. Satchell, and S. Sakar: Phys. Rev. A **34**, 3166 (1986)
[12.16] M. Dörfle and A. Schenzle: Z. Phys. B **65**, 113 (1986)
[12.17] A. M. Smith and C. W. Gardiner: Phys. Rev. A **39**, 3511 (1989)
[12.18] I. J. Craig and K. J. McNeil: Phys. Rev. A **39**, 6267 (1989)
[12.19] M. Wolinsky and H. J. Carmichael: Phys. Rev. Lett. **60**, 1836 (1988)

[12.20] Reference [12.12] pp. 146–148
[12.21] P. D. Drummond, K. J. McNeil, and D. F. Walls: Optica Acta **28**, 211 (1980)
[12.22] F. J. Dyson: Phys. Rev. **75**, 486 (1949)
[12.23] M. Abramowitz and I. A. Stegun: *Handbook of Mathematical Functions* (Dover, New York, 1965) p. 443
[12.24] R. Vyas and S. Singh: Opt. Lett. **14**, 1110 (1989); Phys. Rev. A **40**, 5147 (1989)
[12.25] R. Vyas and A. L. DeBrito: Phys. Rev. A **42**, 592 (1990)
[12.26] S. Schiller, G. Breitenbach, S. F. Pereira, T. Müller, and J. Mlynek: Phys. Rev. Lett. **77**, 2933 (1996)
[12.27] I. S. Gradshteyn and I. M. Ryzhik: *Tables of Integrals Series and Products* (Academic, New York, 1965) p. 481, §3.899 3
[12.28] P. Deuar and P. D. Drummond: Comput. Phys. Commun. **142**, 442 (2001); Phys. Rev. A **66**, 033812 (2002)
[12.29] P. D. Drummond and P. Deuar: J. Opt. B **5**, S281 (2003)
[12.30] P. D. Drummond, P. Deuar, and K. V. Kheruntsyan: Phys. Rev. Lett. **92**, 040405 (2004)
[12.31] G. J. Milburn: Phys. Rev. A **33**, 674 (1986)
[12.32] B. Yurke and D. Stoler: Phys. Rev. Lett. **57**, 13 (1986)

Chapter 13

[13.1] H. M. Gibbs: *Optical Bistability: Controlling Light with Light* (Academic, Orlando, 1985)
[13.2] L. M. Narducci and N. B. Abraham: *Laser Physics and Laser Instabilities* (World Scientific, Singapore, 1988)
[13.3] L. A. Lugiato: "Theory of Optical Bistability", in *Progress in Optics*, Vol. XXI, ed. by E. Wolf (North Holland, Amsterdam, 1984) pp. 69–216
[13.4] E. T. Jaynes and F. W. Cummings: Proc. IEEE **51**, 89 (1963)
[13.5] H. Paul: Ann. Phys. (N.Y.) **11**(7), 411 (1963)
[13.6] H. J. Kimble: "Structure and Dynamics in Cavity Quantum Electrodynamics", in *Cavity Quantum Electrodynamics*, ed. by P. R. Berman (Academic, Boston, 1994) pp. 203–266
[13.7] K. An, J. J. Childs, R. R. Dasari, and M. S. Feld: Phys. Rev. Lett. **73**, 3375 (1994)
[13.8] P. Münstermann, T. Fischer, P. W. H. Pinske, and G. Rempe: Optics Commun. **159**, 63 (1999)
[13.9] G. T. Foster, S. L. Mielke, and L. A. Orozco: Phys. Rev. A **61**, 053821 (2000)
[13.10] E. M. Purcell: Phys. Rev. **69**, 681 (1946)
[13.11] E. Goy, J. M. Raimond, M. Gross, and S. Haroche: Phys. Rev. Lett. **50**, 1903 (1983)
[13.12] D. J. Heinzen, J. J. Childs, J. E. Thomas, and M. S. Feld: Phys. Rev. Lett. **58**, 1320 (1987)
[13.13] D. J. Heinzen and M. S. Feld: Phys. Rev. Lett. **59**, 2623 (1987)
[13.14] P. W. Milonni and P. L. Knight: Optics Commun. **9**, 119 (1973)
[13.15] K. H. Drexhage: "Interactions of Light with Monomolecular Dye Layers" in *Progress in Optics*, Vol. XII, ed. by E. Wolf (North Holland, Amsterdam, 1974) pp. 165–232

[13.16] P. W. Milonni and J. H. Eberly: *Lasers* (Wiley, New York, 1988) pp. 484–490
[13.17] G. Rempe, R. J. Thompson, and H. J. Kimble: Opt. Lett. **17**, 363 (1992)
[13.18] C. J. Hood, M. S. Chapman, T. W. Lynn, and H. J. Kimble: Phys. Rev. Lett. **80**, 4157 (1998)
[13.19] M. Brune, F. Schmidt–Kaler, A. Maali, J. Dreyer, E. Hagley, J. M. Raimond, and S. Haroche: Phys. Rev. Lett. **76**, 1800 (1996)
[13.20] D. Kleppner: Phys. Rev. Lett. **47**, 233 (1981)
[13.21] G. Gabrielse and H. Dehmelt: Phys. Rev. Lett. **55**, 67 (1985)
[13.22] R. Hulet, E. S. Hilfer, and D. Kleppner: Phys. Rev. Lett. **55**, 2137 (1985)
[13.23] P. R. Rice and H. J. Carmichael: IEEE J. Quantum Electron. **QE 24**, 1351 (1988)
[13.24] H. J. Carmichael, R. J. Brecha, and P. R. Rice: Optics Commun. **82**, 73 (1991)
[13.25] R. J. Brecha, P. R. Rice, and M. Xiao: Phys. Rev. A **59**, 2392 (1999)
[13.26] G. Rempe, R. J. Thompson, R. J. Brecha, W. D. Lee, and H. J. Kimble: Phys. Rev. Lett. **67**, 1727 (1991)
[13.27] S. L. Mielke, G. T. Foster, and L. A. Orozco: Phys. Rev. Lett. **80**, 948 (1998)
[13.28] H. J. Carmichael, H. M. Castro–Beltran, G. T. Foster, and L. A. Orozco: Phys. Rev. Lett. **85**, 1855 (2000)
[13.29] G. T. Foster, L. A. Orozco, H. M. Castro–Beltran, and H. J. Carmichael: Phys. Rev. Lett. **85**, 3149 (2000)
[13.30] D. Meschede, H. Walther, and G. Müller: Phys. Rev. Lett. **54**, 551 (1985)
[13.31] G. Raithel, C. Wagner, H. Walther, L. Narducci, and M. O. Scully: "The Micromaser: A Proving Ground for Quantum Physics", in *Cavity Quantum Electrodynamics*, ed. by P. R. Berman (Academic, Boston, 1994) pp. 57–121
[13.32] O. Benson, G. Raithel, and H. Walther: Phys. Rev. Lett. **72**, 3506 (1994)
[13.33] F. De Martini and G. R. Jacobovitz: Phys. Rev. Lett. **60**, 1711 (1988)
[13.34] H. Yokoyama and S. D. Brorson: J. Appl. Phys. **66**, 4801 (1989)
[13.35] Y. Yamamoto and R. E. Slusher: Physics Today **46** (6), 66 (1993)
[13.36] P. R. Rice and H. J. Carmichael: Phys. Rev. A **50**, 4384 (1994)
[13.37] J. McKeever, A. Boca, A. D. Boozer, J. R. Buck, and H. J. Kimble: Nature **425**, 268 (2003)
[13.38] Yi Mu and C. M. Savage: Phys. Rev. A **46**, 5944 (1992)
[13.39] C. M. Savage and H. J. Carmichael: IEEE J. Quantum Electron. **24**, 1495 (1988)
[13.40] H. J. Carmichael, S. Singh, R. Vyas, and P. R. Rice: Phys. Rev. A **39**, 1200 (1989)
[13.41] S.-Y. Zhu, L. M. Narducci, and M. O. Scully: Phys. Rev. A **52**, 4791 (1995)
[13.42] J. J. Sanchez-Mondragon, N. B. Narozhny, and J. H. Eberly: Phys. Rev. Lett. **51**, 550 (1983)
[13.43] M. G. Raizen, R. J. Thompson, R. J. Brecha, H. J. Kimble, and H. J. Carmichael: Phys. Rev. Lett. **63**, 240 (1989)
[13.44] C. Weisbuch, M. Nishioka, A. Ishikawa, and Y. Arakawa: Phys. Rev. Lett. **69**, 3314 (1992)
[13.45] R. Houdré, C. Weisbuch, R. P. Stanley, U. Oesterle, P. Pellandini, and M. Ilegems: Phys. Rev. Lett. **73**, 2043 (1994)
[13.46] J. J. Hopfield: Phys. Rev. **112**, 1555 (1958)
[13.47] J. Gripp, S. L. Mielke, L. A. Orozco, and H. J. Carmichael: Phys. Rev. A **54**, R3746 (1996)
[13.48] L. Tian and H. J. Carmichael: Phys. Rev. A **46**, R6801 (1992)
[13.49] K. M. Birnbaum, A. Boca, R. Miller, A. D. Boozer, T. E. Northup, and H. J. Kimble: Nature **436**, 87 (2005)

Chapter 14

[14.1] H. M. Gibbs: *Optical Bistability: Controlling Light with Light* (Academic, Orlando, 1985)

[14.2] H. Seidel: "Bistable optical circuit using saturable absorber within a resonant cavity", U.S. Patent 3,610,731, filed May 19, 1969, granted October 5, 1971. The original proposal is summarized on pages 14 and 15 of [14.1]

[14.3] A. Szöke, V. Daneu, J. Goldhar, and N. A. Kurnit: Appl. Phys. Lett. **15**, 376 (1969)

[14.4] S. L. McCall, H. M. Gibbs, and T. N. C. Venkatessan: J. Opt. Soc. Am. **65**, 1184 (1975)

[14.5] H. M. Gibbs, S. L. McCall, and T. N. C. Venkatessan: Phys. Rev. Lett. **36**, 1135 (1976)

[14.6] L. A. Lugiato: "Theory of Optical Bistability", in *Progress in Optics*, Vol. XXI, ed. by E. Wolf (North Holland, Amsterdam, 1984) pp. 69–216

[14.7] I. S. Gradshteyn and I. M. Ryzhik: *Tables of Integrals Series and Products* (Academic, New York, 1965) p. 366, §3.613 1

[14.8] *Optical Instabilities*, Proceedings of the International Meeting on Instabilities and Dynamics of Lasers and Nonlinear Optical Systems, University of Rochester, June 18–21, 1985, ed. by R. W. Boyd, M. G. Raymer, and L. M. Narducci (Cambridge University Press, Cambridge, 1986)

[14.9] N. B. Abraham, P. Mandel, and L. M. Narducci: "Dynamical Instabilities and Pulsations in Lasers", in *Progress in Optics*, Vol. XXV, ed. by E. Wolf (North Holland, Amsterdam, 1988) pp. 1–190

[14.10] L. M. Narducci and N. B. Abraham: Laser Physics and Laser Instabilities (World Scientific, Singapore, 1988)

[14.11] K. Ikeda: Opt. Comm. **30**, 257 (1979)

[14.12] K. Ikeda, H. Daido, and O. Akimoto: Phys. Rev. Lett. **45**, 709 (1980)

[14.13] I. Stewart: *Does God Play Dice?: The Mathematics of Chaos* (Blackwell, Oxford, 1989)

[14.14] E. Ott: *Chaos in Dynamical Systems* (Cambridge University Press, Cambridge, 1993)

[14.15] G. L. Baker and J. P. Gollub: *Chaotic Dynamics: An Introduction* (Cambridge University Press, Cambridge, 1996)

[14.16] H. J. Carmichael: Phys. Rev. Lett. **52**, 1292 (1984)

[14.17] H. J. Carmichael: Optics Comm. **53**, 122 (1985)

[14.18] H. J. Carmichael: "Overview of the Theory of Optical Bistability", in [14.8]

[14.19] H. J. Carmichael and B. C. Sanders: Phys. Rev. A **60**, 2497 (1999)

[14.20] R. H. Dicke: Phys. Rev. **93**, 99 (1954)

[14.21] M. Gross and H. Haroche: Phys. Rep. **93**, 301 (1982)

[14.22] M. Gross, P. Goy, C. Fabre, S. Haroche, and J. M. Raimond: Phys. Rev. Lett. **43**, 343 (1979)

[14.23] J. M. Raimond, P. Goy, M. Gross, C. Fabre, and S. Haroche: Phys. Rev. Lett. **49**, 1924 (1982)

[14.24] Y. Kaluzny, P. Goy, M. Gross, J. M. Raimond, and S. Haroche: Phys. Rev. Lett. **51**, 1175 (1983)

[14.25] R. Bonifacio and L. A. Lugiato: Phys. Rev. Lett. **40**, 1023 (1978)

[14.26] L. M. Narducci, R. Gilmore, D. H. Feng, and G. S. Agarwal: Opt. Lett. **2**, 88 (1978)

[14.27] H. J. Carmichael, D. F. Walls, P. D. Drummond, and S. S. Hassan: Phys. Rev. A **27**, 3112 (1983)
[14.28] G. S. Agarwal: Phys. Rev. Lett. **53**, 1732 (1984)
[14.29] H. J. Carmichael: Phys. Rev. A **33**, 3262 (1986)
[14.30] C. J. Hood, M. S. Chapman, T. W. Lynn, and H. J. Kimble: Phys. Rev. Lett. **80**, 4157 (1998)
[14.31] F. Bernardot, P. Nussenzveig, M. Brune, J. M. Raimond, and S. Haroche: Europhys. Lett. **17**, 33 (1992)
[14.32] R. J. Thompson, G. Rempe, and H. J. Kimble: Phys. Rev. Lett. **68**, 1132 (1992)
[14.33] J. J. Childs, K. An, M. S. Otteson, R. R. Dasari, and M. S. Feld: Phys. Rev. Lett. **77**, 2901 (1996)
[14.34] R. J. Thompson, R. J. Brecha, H. J. Kimble, and H. J. Carmichael: Phys. Rev. Lett. **63**, 240 (1989)
[14.35] J. Gripp, S. L. Mielke, L. A. Orozco, and H. J. Carmichael: Phys. Rev. A **54**, R3746 (1996)
[14.36] Y. Zhu, D. J. Gauthier, S. E. Morin, Q. Wu, H. J. Carmichael, and T. W. Mossberg: Phys. Rev. Lett. **64**, 2499 (1990)
[14.37] C. Weisbuch, M. Nishioka, A. Ishikawa, and Y. Arakawa: Phys. Rev. Lett. **69**, 3314 (1992)
[14.38] R. Houdré, C. Weisbuch, R. P. Stanley, U. Oesterle, P. Pellandini, and M. Ilegems: Phys. Rev. Lett. **73**, 2043 (1994)

Chapter 15

[15.1] L. A. Lugiato: "Theory of Optical Bistability", in *Progress in Optics*, Vol. XXI, ed. by E. Wolf (North Holland, Amsterdam, 1984) pp. 69–216
[15.2] R. Gilmore: *Lie Groups, Lie Algebras, and Some of Their Applications* (Wiley, New York, 1974) pp. 149–181
[15.3] V. Benza and L. A. Lugiato: "Semiclassical and Quantum Statistical Dressed Mode Description of Optical Bistability", in *Optical Bistability*, ed. by C. M. Narducci, M. Ciftan, and H. R. Robl (Plenum, New York, 1981) pp. 9–29
[15.4] Reference [15.1] pp. 158–162
[15.5] P. D. Drummond: Phys. Rev. A **33**, 4462 (1986)
[15.6] P. Kinsler and P. D. Drummond: Phys. Rev. A **43**, 6194 (1991)
[15.7] H. J. Carmichael, J. S. Satchell, and S. Sarkar: Phys. Rev. A **34**, 3166 (1986)
[15.8] J. Gripp, S. L. Mielke, L. A. Orozco, and H. J. Carmichael: Phys. Rev. A **54**, R3746 (1996)
[15.9] R. Bonifacio, P. Schwendimann, and F. Haake: Phys. Rev. A **4**, 302 (1971)
[15.10] P. Schwendimann: Z. Phys. **265**, 267 (1973)
[15.11] J. P. Clemens and H. J. Carmichael: Phys. Rev. A **65**, 023815 (2002)
[15.12] R. Bonifacio and L. A. Lugiato: Phys. Rev. Lett. **40**, 1023 (1978)
[15.13] L. M. Narducci, R. Gilmore, D. H. Feng, and G. S. Agarwal: Opt. Lett. **2**, 88 (1978)
[15.14] H. J. Carmichael, D. F. Walls, P. D. Drummond, and S. S. Hassan: Phys. Rev. A **27**, 3112 (1983)
[15.15] M. G. Raizen, L. A. Orozco, M. Xiao, T. L. Boyd, and H. J. Kimble: Phys. Rev. Lett. **59**, 198 (1987)

[15.16] H. J. Carmichael, H. M. Castro–Beltran, G. T. Foster, and L. A. Orozco: Phys. Rev. Lett. **85**, 1855 (2000)
[15.17] G. T. Foster, L. A. Orozco, H. M. Castro–Beltran, and H. J. Carmichael: Phys. Rev. Lett. **85**, 3149 (2000)
[15.18] C. M. Savage and H. J. Carmichael: IEEE J. Quantum Electron. **24**, 1495 (1988)
[15.19] P. R. Rice and H. J. Carmichael: J. Opt. Soc. Am. B **5**, 1661 (1988)
[15.20] H. J. Carmichael, R. J. Brecha, M. G. Raizen, H. J. Kimble, and P. R. Rice: Phys. Rev. A **40**, 5516 (1989)
[15.21] B. R. Mollow: Phys. Rev. **188**, 1969 (1969)
[15.22] P. Kochan and H. J. Carmichael: Phys. Rev. A **50**, 1700 (1994) Appendix B
[15.23] H. J. Kimble, D. E. Grant, A. T. Rosenberger, and P. D. Drummond: "Optical Bistability with Two-Level Atoms", in *Laser Physics*, Lecture Notes in Physics, Vol. 182, ed. by J. D. Harvey and D. F. Walls (Springer, Berlin, 1983) pp. 14–40
[15.24] Y. Zhu, D. J. Gauthier, S. E. Morin, Q. Wu, H. J. Carmichael, and T. W. Mossberg: Phys. Rev. Lett. **64**, 2499 (1990)
[15.25] M. G. Raizen, R. J. Thompson, R. J. Brecha, H. J. Kimble, and H. J. Carmichael: Phys. Rev. Lett. **63**, 240 (1989)

Chapter 16

[16.1] H. J. Carmichael, R. J. Brecha, and P. R. Rice: Optics Commun. **82**, 73 (1991)
[16.2] G. Rempe, R. J. Thompson, R. J. Brecha, W. D. Lee, and H. J. Kimble: Phys. Rev. Lett. **67**, 1727 (1991)
[16.3] G. T. Foster, L. A. Orozco, H. M. Castro–Beltran, and H. J. Carmichael: Phys. Rev. Lett. **85**, 3149 (2000)
[16.4] P. R. Rice and H. J. Carmichael: IEEE J. Quantum Electron. **24**, 1351 (1988)
[16.5] S. L. Mielke, G. T. Foster, and L. A. Orozco: Phys. Rev. Lett. **80**, 3948 (1998)
[16.6] H. J. Carmichael, H. M. Castro–Beltran, G. T. Foster, and L. A. Orozco: Phys. Rev. Lett. **85**, 1855 (2000)
[16.7] C. W. Gardiner: *Handbook of Stochastic Methods for Physics, Chemistry and the Natural Sciences*, 2nd edition (Springer, Berlin, 1985) pp. 36–37
[16.8] R. Bonifacio and L. A. Lugiato: Optics Commun. **19**, 172 (1976); Phys. Rev. Lett. **40**, 1023 (1978); Phys. Rev. A **18**, 1129 (1978)
[16.9] H. J. Carmichael and D. F. Walls: J. Phys. B **10**, L685 (1977)
[16.10] G. S. Agarwal, L. M. Narducci, D. H. Feng, and R. Gilmore: "Dynamical Approach to Steady State and Fluctuations in Optically Bistable Systems" in *Coherence and Quantum Optics IV*, ed. by L. Mandel and E. Wolf (Plenum, New York, 1978) pp. 281-292
[16.11] L. A. Lugiato: Nuovo Cimento B **50**, 89 (1979)
[16.12] H. J. Carmichael: Z. Phys. B **42**, 183 (1981)
[16.13] L. A. Lugiato: Lett. Nuovo Comento **29**, 375 (1980)
[16.14] G. S. Agarwal: Phys. Rev. A **26**, 680 (1982)
[16.15] R. Louden: *The Quantum Theory of Light*, 3rd edition (Oxford University Press, Oxford, 2000) pp. 109–111

[16.16] P. R. Rice: "Quantum Theory of Atom–Field Interaction in a Resonant Optical Cavity", Ph.D. Thesis, University of Arkansas (1988)
[16.17] H. J. Carmichael and B. C. Sanders: Phys. Rev. A **60**, 2497 (1999)
[16.18] L. Horvath and H. J. Carmichael: "Atomic motion and density fluctuations in cavity QED with atomic beams", *Photonics, Design, and Packaging II*, Proceedings of SPIE, Vol 6038, ed. by D. Abbott, Y. S. Kishvar, H. H. Rubinsztein–Dunlop, and S. Fan, (SPIE, Bellingham, 2005) pp. 1–15
[16.19] J. McKeever, J. R. Buck, A. D. Boozer, A. Kuzmich, H.-C. Nägerl, D. M. Stamper-Kurn, and H. J. Kimble, Phys. Rev. Lett. **90**, 1336021 (2003)
[16.20] J. McKeever, A. Boca, A. D. Boozer, J. R. Buck, and H. J. Kimble: Nature **425**, 268 (2003)
[16.21] K. M. Birnbaum, A. Boca, R. Miller, A. D. Boozer, T. E. Northup, and H. J. Kimble: Nature **436**, 87 (2005)
[16.22] P. Alsing and H. J. Carmichael: Quantum Opt. **3**, 13 (1991)
[16.23] D. M. Meekhof, C. Monroe, B. E. King, W. M. Itano, and D. J. Wineland: Phys. Rev. Lett. **76**, 1796 (1996)
[16.24] M. Brune, F. Schmidt–Kaler, A. Maali, J. Dreyer, E. Hagley, J. M. Raimond, and S. Haroche: Phys. Rev. Lett. **76**, 1800 (1996)
[16.25] G. J. Milburn and P. Alsing: "Quantum Phase Transitions in a Linear Ion Trap", in *Directions in Quantum Optics*, Lecture Notes in Physics, Vol. 561, ed. by H. J. Carmichael, R. J. Glauber, and M. O. Scully (Springer, Berlin, 2001) pp. 303–312
[16.26] P. D. Drummond and H. J. Carmichael: Optics Commun. **27**, 160 (1978)
[16.27] P. D. Drummond: "Nonequilibrium Transitions in Quantum Optical Systems", Ph.D. Thesis, University of Waikato (1979); Phys. Rev. A **22**, 1179 (1980)
[16.28] H. J. Carmichael: J. Phys. B **13**, 3551 (1980)
[16.29] P. Alsing, D.-S. Guo, and H. J. Carmichael: Phys. Rev. A **45**, 5135 (1992)
[16.30] C. Cohen-Tannoudji and S. Reynaud: J. Phys. B **10**, 345 (1977)
[16.31] H. J. Carmichael, L. Tian, W. Ren, and P. Alsing: "Nonperturbative Atom–Photon Interactions in an Optical Cavity", in *Cavity Quantum Electrodynamics*, ed. by P. R. Berman (Academic, Boston, 1994) pp. 381–423
[16.32] Y. T. Chough and H. J. Carmichael: Phys. Rev. A **54**, 1709 (1996)
[16.33] S. Ya Kilin and T. B. Krinitskaya: Opt. i Spektrosk. **70**, 628 (1991); J. Opt. Soc. B **8**, 2289 (1991)
[16.34] H. J. Carmichael, P. Kochan, and L. Tian: "Coherent States and Open Quantum Systems: A Comment on the Stern–Gerlach Experiment and Schrödinger's Cat", in *Coherent States: Past, Present, and Future*, ed. by D. H. Feng, J. R. Klauder, and M. R. Strayer (World Scientific, Singapore, 1994) pp. 75–91
[16.35] P. Kochan, H. J. Carmichael, P. R. Morrow, and M. G. Raizen: Phys. Rev. Lett. **75**, 45 (1995)

Chapter 17

[17.1] P. D. Drummond and M. D. Reid: Phys. Rev. A **41**, 3930 (1990)
[17.2] P. Alsing and H. J. Carmichael: Quantum Opt. **3**, 13 (1991)
[17.3] J. Dalibard, Y. Castin, and K. Mølmer: Phys. Rev. Lett. **68**, 580 (1992)
[17.4] R. Dum, P. Zoller, and H. Ritsch: Phys. Rev. A **45**, 4879 (1992)

[17.5] R. Dum, A. S. Parkins, P. Zoller, and C. W. Gardiner: Phys. Rev. A **46**, 4382 (1992)

[17.6] E. B. Davies: *The Quantum Theory of Open Systems* (Academic, London, 1976)

[17.7] G. Lindblad: Commun. Math. Phys. **48**, 119 (1976)

[17.8] R. L. Hudson and K. R. Parthasarathy: Commun. Math. Phys. **93**, 301 (1984)

[17.9] K. R. Parthasarathy: *An Introduction to Quantum Stochastic Calculus* (Birkhäuser, Basel, 1992)

[17.10] C. W. Gardiner and M. J. Collet: Phys. Rev. A **31**, 3761 (1985)

[17.11] R. L. Hudson and K. R. Parthasarathy: Acta Appl. Math. **2**, 353 (1984)

[17.12] B. Kümmerer: J. Funct. Anal. **63**, 139 (1985)

[17.13] J. D. M. Maassen: "Quantum Markov Processes in Fock Space Described by Integral Kernels", in *Quantum Probability and Applications II*, Lecture Notes in Mathematics, Vol. 1136, eds. L. Accardi and W. V. Waldenfels (Springer, Berlin, 1985) pp. 361–374

[17.14] A. Barchielli: Quantum Opt. **2**, 423 (1990)

[17.15] A. Barchielli and V. P. Belavkin: J. Phys. A **24**, 1495 (1991)

[17.16] A. Barchielli and G. Lupieri: J. Math. Phys. **41**, 7181 (2000)

[17.17] L. Bouten, H. Maassen, and B. Kümmerer: Optics Spectr. **94**, 911 (2003)

[17.18] H. J. Carmichael, S. Singh, R. Vyas, and P. R. Rice: Phys. Rev. A **39**, 1200 (1989)

[17.19] H. J. Carmichael: *An Open Systems Approach to Quantum Optics*, Lecture Notes in Physics, New Series m: Monographs, Vol. m**18** (Springer, Berlin, 1993)

[17.20] K. Mølmer: Phys. Rev. A **55**, 3195 (1997)

[17.21] J. Javaneinen and S. M. Yoo: Phys. Rev. Lett. **76**, 161 (1996)

[17.22] J. I. Cirac, C. W. Gardiner, M. Naraschewski, and P. Zoller: Phys. Rev. A **54**, R3714 (1996)

[17.23] Y.-T. Chough: Phys. Rev. A **55**, 3143 (1997)

[17.24] H. J. Carmichael and K. Kim: Optics Commun. **179**, 417 (2000)

[17.25] J. Clemens, L. Horvath, B. C. Sanders, and H. J. Carmichael: Phys. Rev. A **68**, 023809 (2003)

[17.26] K. Mølmer, Y. Castin, and J. Dalibard: J. Opt. Soc. A. B **10**, 524 (1993)

[17.27] M. Planck: Verh. dt. Phys. Ges. **2**, 202 (1900); **2**, 237 (1900) [English translation in D. ter Haar: *The Old Quantum Theory* (Pergamon, Oxford, 1967) pp. 3–14]

[17.28] N. Bohr: Phil. Mag. **26**, 1 (1913)

[17.29] A. Einstein: Ann. Phys. **17**, 132 (1905) [English translation in D. ter Haar: *The Old Quantum Theory* (Pergamon, Oxford, 1967) pp. 91–107]

[17.30] A. Einstein: Verh. Dtsch. Phys. Ges. **18**, 318 (1916); Phys. Z. **18**, 121 (1917) [English translation in B. L. van der Waerden: *Sources of Quantum Mechanics* (North Holland, Amsterdam, 1967), Chap. 1]

[17.31] C. W. Gardiner: *Handbook of Stochastic Methods for Physics, Chemistry and the Natural Sciences* (Springer, Berlin, 1983) pp. 78–79

[17.32] N. Bohr, H. A. Kramers, and J. C. Slater: Philos. Mag. **47**, 785 (1924); Z. Phys. **24**, 69 (1924)

[17.33] H. J. Carmichael: Phys. Rev. A **56**, 5065 (1997)

[17.34] H. J. Carmichael: "Taming the Paradox of Lasing Without Inversion", in *Mysteries, Puzzles, and Paradoxes in Quantum Mechanics*, ed. R. Bonifacio (American Institute of Physics, New York, 1999) pp. 208–219
[17.35] N. C. Giri: *Introduction to Probability and Statistics* (Marcel Dekker, New York, 1993) p. 52
[17.36] L. Tian and H. J. Carmichael: Phys. Rev. A **46**, R6801 (1992)
[17.37] P. Kochan, H. J. Carmichael, P. R. Morrow, and M. G. Raizen: Phys. Rev. Lett. **75**, 45 (1995)
[17.38] M. D. Crisp and E. T. Jaynes: Phys. Rev. **179**, 1253 (1969)
[17.39] C. R. Stroud and E. T. Jaynes: Phys. Rev. A **1**, 106 (1970)
[17.40] D. Bouwmeester, R. J. C. Spreeuw, G. Nienhuis, and J. P. Woerdman: Phys. Rev. A **49**, 4170 (1994)
[17.41] E. Schrödinger: Naturwissenschaften **23**, 807, 823, 844 (1935) [English translation: Proc. Am. Phys. Soc. **124**, 323 (1980)]
[17.42] H. J. Carmichael: "Coherent States and Open Quantum Systems: A Comment on the Stern–Gerlach Experiment and Schrödinger's Cat", in *Coherent States: Past, Present, and Future*, eds. D. H. Feng, J. R. Klauder, and M. R. Strayer (World Scientific, Singapore, 1994) pp. 75–91
[17.43] H. J. Carmichael: "Stochastic Schrödinger Equations: What They Mean and What They Can Do", in *Coherence and Quantum Optics VII*, eds. J. H Eberly, L. Mandel, and E. Wolf (Plenum, New York, 1996) pp. 177–192
[17.44] P. Pearle: Phys. Rev. D **13**, 857 (1976)
[17.45] P. Pearle: Int. J. Theor. Phys. **18**, 489 (1979)
[17.46] N. Gisin: Phys. Rev. Lett. **52**, 1657 (1984)
[17.47] G.-C. Ghirardi, A. Rimini, and T. Weber: Phys. Rev. D **34**, 470 (1986)
[17.48] G.-C. Ghirardi, P. Pearle, and A. Rimini: Phys. Rev. A **42**, 78 (1990)
[17.49] N. Gisin and I. C. Percival: J. Phys. A **25**, 5677 (1992)

Chapter 18

[18.1] P. L. Kelley and W. H. Kleiner: Phys. Rev. **136**, A316 (1964)
[18.2] R. J. Glauber: Phys. Rev. Lett. **10**, 84 (1963); Phys. Rev. **130**, 2529 (1963); **131**, 2766 (1963)
[18.3] B. Saleh: *Photoelectron Statistics* (Springer, Berlin, 1978) Chap. 3
[18.4] B. R. Mollow: Phys. Rev. **168**, 1896 (1968)
[18.5] M. D. Srinivas and E. B. Davies: Optica Acta **28**, 981 (1981)
[18.6] M. D. Srinivas and E. B. Davies: Optica Acta **29**, 235 (1982)
[18.7] M. D. Srinivas: Pramana **47**, 1 (1996)
[18.8] L. Mandel: Optica Acta **28**, 1447 (1981)
[18.9] J. D. Cresser: "Ergodicity of Quantum Trajectory Detection Records", in *Directions in Quantum Optics*, Lecture Notes in Physics, Vol. **561**, eds. H. J. Carmichael, R. J. Glauber, and M. O. Scully (Springer, Berlin, 2001) pp. 358–369
[18.10] B. Kümmerer and H. Maassen: J. Phys. A **36**, 2155 (2003)
[18.11] H. P. Yuen and V. W. S. Chan: Opt. Lett. **8**, 177 (1983)
[18.12] N. Gisin and I. C. Percival: J. Phys. A **25**, 5677 (1992)

[18.13] N. Gisin and I. C. Percival: "Quantum State Diffusion: From Foundations to Applications", in *Experimental Metaphysics: Quantum Mechanical Studies for Abner Shimony*, eds. A. Shimony, R. S. Cohen, M. Horne, and J. Stachel (Kluwer, Dordrecht, 1977) pp. 73–90
[18.14] I. Percival: *Quantum State Diffusion* (Cambridge University Press, Cambridge, 1998)
[18.15] N. Bohr: Nature **121**, 580 (1928) [Reprinted as "The Quantum Postulate and the Recent Development of Atomic Theory", in *Quantum Theory and Measurement*, eds. J. A. Wheeler and W. H. Zurek (Princeton University Press, Princeton, 1983) pp. 87–126]
[18.16] H. J. Folse: *The Philosophy of Niels Bohr* (North Holland, Amsterdam, 1985)
[18.17] D. Murdoch: *Niels Bohr's Philosophy of Physics* (Cambridge University Press, Cambridge, 1987)
[18.18] H. J. Carmichael: "Coherent States and Open Quantum Systems: A Comment on the Stern–Gerlach Experiment and Schrödinger's Cat", in *Coherent States: Past, Present, and Future*, eds. D. H. Feng, J. R. Klauder, and M. R. Strayer (World Scientific, Singapore, 1994) pp. 75–91

Chapter 19

[19.1] G. Rempe, R. J. Thompson, R. J. Brecha, W. D. Lee, and H. J. Kimble: Phys. Rev. Lett. **67**, 1727 (1991)
[19.2] S. L. Mielke, G. T. Foster, and L. A. Orozco: Phys. Rev. Lett. **80**, 3948 (1998)
[19.3] G. T. Foster, S. L. Mielke, and L. A. Orozco: Phys. Rev. A **61**, 053821 (2000)
[19.4] H. J. Carmichael, H. M. Castro–Beltran, G. T. Foster, and L. A. Orozco: Phys. Rev. Lett. **85**, 1855 (2000)
[19.5] G. T. Foster, L. A. Orozco, L. A. Castro–Beltran, and H. J. Carmichael: Phys. Rev. Lett. **85**, 3149 (2000)
[19.6] G. T. Foster, W. P. Smith, J. E. Reiner, and L. A. Orozco: Phys. Rev. A **66**, 33807 (2002)
[19.7] H. J. Carmichael: "Continuous Variable Teleportation of Quantum Fields", in *Proceedings of the 1st Asia–Pacific Conference on Quantum Information Science*, ed. C. P. Soo (World Scientific, Singapore, 2004) pp. 11–26
[19.8] M. I. Kolobov and I. V. Sokolov, Opt. Spektrosk. **62**, 112 (1987)
[19.9] C. W. Gardiner: Phys. Rev. Lett. **70**, 2269 (1993)
[19.10] H. J. Carmichael: Phys. Rev. Lett. **70**, 2273 (1993)
[19.11] H. Nha and H. J. Carmichael: Phys. Rev. A **71**, 013805 (2005)
[19.12] H. Carmichael and K. Kim: Optics Commun. **179**, 417 (2000)
[19.13] N. Wiener: Acta Math. **55**, 117 (1930)
[19.14] A. Kinchin: Math. Annalen **109**, 604 (1934)
[19.15] L. Tian and H. J. Carmichael: Phys. Rev. A **46**, R6801 (1992)
[19.16] J. Dalibard, Y. Castin, and K. Mølmer: Phys. Rev. Lett. **68**, 580 (1992)
[19.17] S.-Y. Zhu, L. M. Narducci, and M. O. Scully: Phys. Rev. A **4791** (1995)

Index

Absorption
– cavity-enhanced 216
– coefficient 259
Adiabatic elimination
– "slow" and "fast" variables 88, 179
– for degenerate parametric oscillator
– – at threshold 89–90
– – of pump 135
– system-reservoir approach 141–144, 202–205
– through Heisenberg equation of motion 205–208
Amplification without inversion 425
Approximation
– Markov 144, 207, 211, 404, 418, 445, 486, 487
– parametric 8
– rotating-wave 234
– slowly-varying-amplitude 7, 256
– undepleted pump 5, 8
Autocorrelation function *see* Correlation function
Autocorrelation matrix 114, 116
– contracted 117

β factor *see* Laser, β factor
Baker–Hausdorff theorem 289
Bayes theorem 417
Bayesian inference 427, 449
Birth-death process 386
BKS proposal 425
Bloch equations *see* Optical Bloch equations
Bloch sphere 305, 370
Bloch vector 373
Bohr–Einstein quantum jump 410–412
– quantum trajectory generalization of 425
Born–Markov approximation *see* Approximation, Markov
Branching ratio 236, 238

Cavity *see* Optical cavity
Cavity Q *see* Optical cavity, Q
Cavity polaritons 242, 279
Cavity QED 133, 197
– antibunching of fluorescence (side-scattered light) 352–356
– breakdown of linear fluctuation theory 352
– coherent driving of atom and cavity mode compared 396
– coupled oscillator equations in 233, 281–282, 339, 345, 348, 362
– experiments 201, 209, 217, 222, 244, 279, 310, 316, 332, 344, 349, 369, 376
– forwards photon scattering 330–333, 345–349
– – including spatial effects 360–368
– laser 222
– perturbative and nonperturbative limits 197–198, 275
– scattering record for 407
– spectra of squeezing 357–360
– spectrum of the transmitted light 385–391
– strong coupling for many atoms 275

-- strong coupling for a single atom compared 356
- trapped ion analogy 376
Cavity-enhanced and -inhibited spontaneous emission *see* Spontaneous emission
Chaos 268, 298
Characteristic function 287
- analyticity of *see* Positive P representation, and analyticity of the characteristic function
- equation of motion for 62, 67, 103
- for a squeezed vacuum state 13
- for the electromagnetic field
-- in the positive P representation 98
Coherent state
- as a solution of quantum trajectory equations with coherent driving 494
- in a driven optical cavity 25
- role of in the modeling of projective measurements 476
- squeezed
-- defined 12
- superpositions 169, 176, 468, 471
-- resolved by quantum trajectory evolution 495
Collapse and revival *see* Quantum collapse and revival
Collapse of the wavefunction *see* Wavefunction collapse
Collective atomic operators *see* Operator, collective atomic
Collective atomic states 340
Commutation relation
- angular momentum 286
- boson 33
-- Fourier decomposition of for intracavity field 34
- for collective atomic operators 284, 286
- for Jaynes–Cummings ladder operators 392
- for quadrature phase amplitudes 9
- free field plus source field 53
Commutator *see* Commutation relation
Complementarity 467
Complex P representation *see* P representation, complex
Conditional expectation *see* Quantum trajectory theory, conditional expectation
Conditional homodyne detection *see* Homodyne detection, conditional
Conditional state *see* Quantum trajectory theory, conditional state
Contextuality *see* Quantum contextuality
Cooperativity parameter *see* Optical bistability, cooperativity parameter
Correlation function
- first-order 324, 388
- for quadrature phase amplitudes of intracavity field 37
- second-order 217, 331, 480
-- as conditional measurement 230
-- for cavity-enhanced resonance fluorescence 216, 220
-- for degenerate parametric oscillator 482–484
-- of forwards scattering for many atoms in a cavity 332, 352
-- of forwards scattering in cavity QED 347, 366
-- of the side-scattered light in cavity QED 355
-- quantum trajectory expansion of 481
-- related to waiting-time distribution 226
-- upper bound for 349
Correlations
- between free field and source field 30
- between free-field modes in thermal state 34
- like-atom and unlike-atom 317–320
-- difference between forwards and side scattering 353
Coupling constant
- dipole 198, 272
- for degenerate parametric amplification in a standing-wave cavity 7
- for degenerate parametric oscillator
-- and threshold photon number 66

-- classical model 15
-- quantum-mechanical model 21
- reservoir *see* Reservoir, coupling coefficient
Covariance matrix 114, 116, 123, 312
- for absorptive bistability 312–315
- for degenerate parametric oscillator
-- above threshold 84
-- at threshold 91–93
-- below threshold 75
Critical slowing down 146, 179

δ-function
- derivatives of 121
Davies process 402
Decay
- of unstable state 148
Degenerate parametric oscillator
- classical equations of motion 17, 73
-- derivation of 15–16
-- steady-state solutions 17–19
- counting distribution for subharmonic photons 175, 450–451, 468–470
- deterministic equations of motion in the positive P representation 124
-- solution to 152
- Fokker–Planck equation 74
-- above threshold 83
-- at threshold 89, 90, 180
-- below threshold 75
-- in the positive P representation 109, 115
-- with adiabatic elimination of the pump 135, 159
- optical spectrum below threshold 80
- phase-space equation of motion 73, 87
-- in the P representation 63
-- in the Q representation 68
-- in the Wigner representation 69
- photon number in the subharmonic mode 80, 122, 168
- pump parameter 17
- quadrature phase variances of intracavity field
-- at threshold 178
-- below threshold 36, 76
-- Fourier decomposition of 36–37
- quantum trajectory simulation of 468–473
-- with heterodyne-current records 463–464
-- with homodyne-current records 460
-- with photoelectron counting records 450–451
- second-order correlation function 482–484
- squeezing in 77–79, 84–86, 460
-- at threshold 180
- stochastic differential equations 113, 129, 135, 178
-- with adiabatic elimination of the pump 135
- stochastic Schrödinger equation
-- heterodyne-current records 462
-- homodyne-current records 459
- "turn on" of 148–149
- threshold behavior 19–20, 451
- threshold photon number 196
-- in pump mode 93
-- in subharmonic mode 93
-- of undepleted pump 66
-- outside the small-noise limit 146
Density operator (matrix)
- pure-state factorization of 218, 335
-- effect of nonradiative dephasing on 343
-- for many atoms 339–342, 358
-- for one atom 336–339, 354
-- limitations 357
-- physical basis for 343–344, 479–480
-- Schrödinger equation for 343, 361
- reduced 404, 408, 490
- time-retarded 489
- unraveling of *see* Quantum trajectory theory, unraveling of the density operator
Detection efficiency
- dependence of waiting-time distribution on 230
- in photoelectron counting 41
- in squeezing measurement 39, 44, 322

Diffusion
- non-positive-semidefinite 121–123, 285
-- and phase-space variables outside the Bloch sphere 305
-- consequences of 77, 96–98
-- for degenerate parametric oscillator 63–65, 69–71, 122
-- for optical bistability 303–306
- positive semidefinite 104, 189, 305
Diffusion matrix 107, 111, 113, 160, 293
- for optical bistability 304, 307
-- in the Q representation 297, 306
-- in the Glauber–Sudarshan P representation 295, 306
-- in the positive P representation 308
-- in the Wigner representation 297, 304
- in the positive P representation 108–109, 116, 136
- positive (semi)definite 64
Dipole radiation pattern 499
Dispersion
- related to vacuum Rabi doublet 279
Displacement
- of coherent state reservoir 141
Displacement operator *see* Operator, displacement
Distribution
- Gaussian
-- even-order moments of 122
Dressed Jaynes–Cummings eigenstates 383, 387
Dressed states 240, 245
- dressing of 245, 377
- self-consistent 374
Dyson expansion 171, 228
- of density operator 173, 421, 448, 491

Effective number of atoms, in cavity QED 273, 281, 362, 367
- and actual number of interacting atoms 275
Eigenvalue problem 270–271
Einstein A and B theory 410
Einstein A coefficient *see* Spontaneous emission, rate
Entangled state 426, 476
Entanglement
- between system and environment 405
- of source and target system states 407
Euler method *see* Stochastic differential equation, integration of
External field
- in quantum trajectory theory 406

Fabry–Perot cavity *see* Optical cavity, standing-wave
Field
- free and source 28, 201, 206, 487–488, 492
- nonclassical versus classical 3, 59, 96
-- and conditioning on photodetection 219
- TEM_{00} 247
First passage time 148
Fluctuations
- cavity-enhanced resonance fluorescence 213–216
- degenerate parametric oscillator 94
-- at threshold 91–93, 146–148, 178–181
-- in quadrature phase amplitudes above threshold 84
-- in quadrature phase amplitudes below threshold 75, 79
- linearized treatment of *see* System size expansion
- Poisson 457
- quantum
-- for small system size 164
-- size of 167
- regression of 336
- squeezed 10, 315
Fluorescence *see* Resonance fluorescence
Fokker–Planck equation
- consistency of in different phase-space representations 299
- diffusion matrix *see* Diffusion matrix

– drift matrix 116, 136
– drift vector 107, 111, 160, 292
– – for optical bistability 294
– first- and second-order moments, equations of motion for 293
– for cummulative charge in temporarily mode-matched heterodyne detection 476
– for degenerate parametric oscillator 74
– – above threshold 83
– – at threshold 89, 90
– – below threshold 75
– – in the positive P representation 109, 115
– – with adiabatic elimination of the pump 135, 159
– for optical bistability 303
– – in the Q representation 291
– – in the Glauber–Sudarshan P representation 289
– – in the positive P representation 308
– – in the Wigner representation 298
– generalized 105
– potential conditions 160
– solved by generalized function 121, 123

Gaussian distribution *see* Distribution, Gaussian
Gaussian moment theorem 351
Glauber–Sudarshan P representation *see* P representation
Green function 123

Hamiltonian
– anharmonic oscillator 190
– cascaded system 486
– – symmetric irreversible coupling 497
– coupled harmonic oscillator 246
– degenerate parametric amplifier
– – without pump depletion 8
– degenerate parametric oscillator 196
– Jaynes–Cummings 198
– – analogy with trapped ion and vibrational mode 376
– – dressed eigenstates of 376–383
– – driven 376, 392
– – eigenstates and eigenenergies of 245
– – related to $\sqrt{n}$ anharmonic oscillator 392–393
– many-atom 273, 284
– non-Hermitian 220, 233, 343, *see* Quantum trajectory theory, non-Hermitian Hamiltonian
– of the driven Jaynes–Cummings model
– – stationary states and quasienergies 382
– reservoir in cavity QED 199
– two-level atom interacting with intracavity field 256
Heisenberg equation of motion
– adiabatic elimination through 205–208
– for degenerate parametric amplifier 8
– for field in cavity QED 206
Heisenberg uncertainty relation
– and spectrum of squeezing 39, 162
– for quadrature phase amplitudes 9–10
Heterodyne current 462
– autocorrelation function from quantum trajectories 511–513
Heterodyne detection 460–466
– bandwidth 464
– measurement of the Q distribution 464, 473–476
– measurement of the optical spectrum 464, 510–514
– temporally mode-matched 471, 474
Homodyne current 459
Homodyne detection 40–44, 451–460
– balanced 453
– bandwidth 459
– conditional 219, 485
– – squeezing measurement with 322
– – violation of classical constraint 349
– measurement of the Wigner distribution 471
– temporally mode-matched 471
Homogeneous width 257

Ikeda instabilities 268
Indistinguishable particles
– contrasted with identical atoms 318
Input-output theory 118
Interference
– of probability amplitudes 242, 510
Inversion clamping (pinning) 225
Ito rule 458, 465

Jaynes–Cummings model *see* Hamiltonian, Jaynes–Cummings
Jump operator *see* Operator, jump
Jump process 227–228

Kerr effect 253

Laser
– β factor 222
– one-atom
– – photon antibunching in 230
– – pump parameter 224
– – waiting-time distribution of photon emissions 225–230
Lindblad form 404, 491
– and pure-state factorization in quantum trajectory theory 446
Linewidth narrowing
– squeezing-induced 328, 391
Local oscillator
– modeled as additional source in quantum trajectory theory 455
– shot noise 457
– temporally mode-matched 471

Markov approximation *see* Approximation, Markov
Master equation
– adiabatic elimination in 141–144, 202–205
– cascaded systems 490
– – with symmetric irreversible coupling 497
– dephasing term 273
– for cavity-enhanced resonance fluorescence 211
– for cavity-enhanced spontaneous emission 204
– for degenerate parametric oscillator 24, 437
– – with adiabatic elimination of the pump 136, 467
– for many-atom cavity QED 342
– – two-quanta truncation of 342
– for one-atom laser 222
– for optical bistability 286, 342
– for single-atom cavity QED 200, 210
– – as density matrix equation 336, 384
– – two-quanta truncation of 339
– for spontaneous emission 420
– for the driven damped harmonic oscillator 25
– – derivation of driving term 25–27
– for two-photon loss 137, 198
– in the Born approximation 143
– perturbation expansion for in powers of the driving field 336–342
– Scully–Lamb 410
– secular approximation 384
– source of mixed-state character in 343
Matrix
– diffusion *see* Diffusion matrix
– drift *see* Fokker–Planck equation, drift matrix
– Jacobian 113, 125, 308, 324
– – eigenvalues of in absorptive optical bistability 309–311
– orthogonal 107
– symmetric
– – decomposition of with quadratic form 107
Maxwell equation
– single-mode 268, 272
Maxwell equations
– coupled, for forwards- and backwards-wave amplitudes 256, 264–265
– for slowly-varying amplitudes 7
Maxwell–Bloch equations 271–273, 279, 281, 294, 302
– for “zero system size” 370, 373
– – stability of stationary states 374
– for standing-wave cavity 268
– – derivation of 263–268
– instabilities of 268
– linear stability analysis 269–271
Mean-field limit *see* Optical bistability, mean-field limit

Measurement
- null 417
-- equation of motion for conditional probabilities 428
-- in quantum trajectory theory 425
-- probability 227, 231, 427, 440, 459
- of Q distribution by heterodyne detection 464, 473–476
- of optical spectrum by heterodyne detection 464
- of Wigner distribution by homodyne detection 471
- projective 473–477
Micromaser 222
Mollow triplet *see* Spectrum, Mollow
Monte Carlo wavefunction simulation *see* Quantum Monte Carlo wavefunction method

Neoclassical radiation theory
- related to quantum trajectory theory 429
No-count probability *see* Measurement, null, probability
No-jump probability *see* Measurement, null, probability
Non-classicality
- of squeezed light 321
Nonclassical field *see* Field, nonclassical versus classical
Normal modes
- of linearized Maxwell–Bloch equations 270, 279
- related to vacuum Rabi doublet 279
Normal-mode resonances 242
Null measurement *see* Measurement, null

Operator
- cavity field 211
- collective atomic 286, 360
- displacement 12, 379
- free-field and source-field 404
- jump 228, *see* Quantum trajectory theory, jump operator
- ladder 283, 392
- nesting 138
- quadrature phase 380
-- defined 9
-- Fourier decomposition of 34–35
- rotation 378
- squeeze 12, 379
Operator average
- normal-ordered
-- for anharmonic oscillator 190
- two-time 114
Operator ordering
- and dropping of free fields 207
- and equations of motion for phase-space averages 293
- for two-time averages in the phase-space representations 81–82
- in the spectrum of squeezing 39, 53
Optical bistability 247
- absorptive 251
-- and cavity-enhanced emission 312
-- and spontaneous dressed-state polarization 397
-- eigenvalues of Jacobian matrix 309–311
-- like- and unlike-atom correlations in 318
-- lower and upper branch of 311
-- state equation 307
-- steady state unstable in regions of negative slope 311
-- steady states compared with the "zero system size" limit 372, 395
- cooperative branch 312, 317, 320
- cooperativity parameter 260, 274, 300, 317, 320, 362
-- threshold value of 262
- dispersive 253
- Fokker–Planck equation 303
-- in the Q representation 291
-- in the Glauber–Sudarshan P representation 289
-- in the positive P representation 308
-- in the Wigner representation 298
- independent-atom branch 312, 317, 320
- mean-field limit 254
- mean-field single-mode limit 263
- photon antibunching in 332
- single-atom 320
- squeezing in 315, 320–322

– state equation 250, 261–262, 279, 280, 307
– – negative slope regions unstable 271
– threshold condition 263
Optical Bloch equations 256
– for cavity-enhanced spontaneous emission 208
– for medium in standing-wave field 264–265
– for one-atom "laser" 223
– magnetic analogy 374
– of collective resonance fluorescence 375
Optical cavity
– bad-cavity limit 203, 237, 275, 312, 374
– – intensity fluctuations in 349
– – unlike-atom correlations in 319
– damping rate 15, 200
– driving of by a coherent field 25–27, 141
– free spectral range 264
– good-cavity limit 223, 312
– – vanishing of unlike-atom correlations in 319
– linewidth 200
– Q 239
– ring 251
– standing-wave 4, 198, 247
– – boundary conditions in 6, 249, 266
– – density of modes 27
– – equation of motion from round-trip field change 6, 15–16
– – mode volume 199
– – relationship between input and output 249
– – summing round trips 250
– TEM_{00} mode 198, 208, 272
– – traveling-wave 256
– ultra-high-finesse 209

P distribution 287
– as generalized function 122
P representation
– and non-classical fields 59, 96
– complex 106
– generalized 100
– positive *see* Positive P representation
– stochastic differential equations in 110–111
Period doubling 268
Perturbation expansion
– in number of photon emissions 480–484
Phase transition 298
Phase-space equation of motion *see* Quantum-classical correspondence, phase-space equation of motion
Phase-space interference 469
Phase-space methods
– limitations of for nonclassical fields 401
Photoelectric detection *see* Photoelectron counting
Photoelectron counting 403, 438–446
– assumption of freely propagating field 445
– criticism of Kelly–Kleiner by Srinivas and Davies 442–443
– detection efficiency 230
– distribution 41, 175
– – sub-Poissonian 230
– exclusive and nonexclusive probabilities 441–442
– normal- and time-ordering of operators in 441, 445
– – and dropping of free fields 444
– semiclassical simulation of 190
– theory of 41
– waiting-time distribution 225, 459
– – classical and nonclassical compared 229–230
– – for a Poisson process 230
– – for degenerate parametric oscillator, preceeded by even and odd photoelectron counts 469
– – for resonance fluorescence 434
– – quantum efficiency in 229
– waiting-time probability 440
Photon antibunching 212, 216, 220, 230, 245, 332, 348
– of fluorescence in single-atom cavity QED 352–356
Photon bunching 212, 221, 352, 484

Photon counting *see* Photoelectron counting
Photon flux
- coherent and incoherent 323
- conservation of in cavity-enhanced resonance fluorescence 213
- of one-atom and conventional laser compared 224–225
Photon pairs
- in parametric down conversion 177
Poisson distribution
- in photoelectron counting 41
Polariton *see* Cavity polaritons
Polariton frequencies 279
Polarization
- collective 281
- of two-level medium 256
Population grating 256
Positive P distribution
- conjugacy requirement 128
- construction of from Q distribution 100–101
- defined *see* Positive P representation, defined
- for a thermal state 102
- for degenerate parametric oscillator in steady state 160
-- above threshold 163
-- at threshold 162, 178, 179
-- below threshold 161
- nonuniqueness of 102, 190
Positive P representation 77
- and analyticity of the characteristic function 98, 104
- autocorrelation matrix in 116–117
- classical and non-classical phase space 152
-- divergent trajectories 152–156, 309
-- linearized evolution in 127
-- physical and nonphysical steady states 154
-- variables of 125
- classical and nonclassical phase space 124, 125, 188, 190
-- divergent trajectories 184–187
-- eigenvalues in 127
-- variables of 161
- defined 98–100, 106
- diffusion matrix in *see* Diffusion matrix
- Fokker–Planck equation in 109, 189
- normal-ordered averages evaluated in 99, 168
- physical and nonphysical steady states 309
- stochastic differential equations in 109
-- conjugacy requirement 128, 146
-- disagreement with master equation 182–189
-- for anharmonic oscillator 191
-- for degenerate parametric oscillator 113, 129, 135, 178
-- for two-photon damping 182
-- nonclassical character of noise 131–132
-- relationship to Glauber-Sudarshan P representation 111
-- "spikes" in simulation of 150, 156, 158, 164, 181, 309
-- trajectories of 129, 130, 145
- stochastic gauge 189
- two-time averages in 114
Probability
- conditional 228
-- expansion of unconditional in terms of 412, 416
- current 159
Probability amplitude
- conditional 424
Pump depletion 20, 137
- for degenerate parametric oscillator at threshold 146

Q distribution
- for a squeezed vacuum state 12
- measured by heterodyne detection 464, 473–476
Q representation
- normal-ordered, time-ordered averages in 82
Quadrature phase amplitudes
- defined 8
Quadrature phase operators *see* Operator, quadrature phase
Quantum collapse and revival 191
Quantum contextuality 467, 471

Quantum dynamical semigroup 402
- dilation of 403
Quantum jump *see* Bohr–Einstein quantum jump
Quantum Langevin equation 32, 402
Quantum Monte Carlo wavefunction method 402
Quantum regression formula 82, 217, 358, 405
- applied to phase-space and Maxwell–Bloch equations 325
- for joint photoelectron counts from quantum trajectory summation 481
Quantum state diffusion model 466
Quantum stochastic calculus 402
Quantum teleportation 485
Quantum trajectory theory 232, 402, 405
- applied to amplification without inversion 425
- applied to superradiance master equation 407
- conditional and joint probabilities compared 431
- conditional expectation 409
-- for dipole amplitude 425
-- of field amplitude in temporally mode-matched heterodyne detection 475
- conditional state 408, 414
-- in Dyson expansion of density operator 421, 448
-- in the presence of coherence 424
-- norm 459
-- normalized 448
-- unnormalized 447
- ergodicity of trajectories 451
- for cascaded systems 406
- idealized environment 467
- jump operator 423, 447
-- cascaded systems 491, 492, 496, 498
-- collective 426
-- for superradiance 499
-- in balanced homodyne detection 455
-- two-photon, for degenerate parametric oscillator with adiabatic elimination of the pump 468
-- without subtraction of the backscattered coherent driving 453
-- without subtraction of the coherent driving 496
- Monte Carlo simulation 431–433
-- based on waiting times 434
-- degenerate parametric oscillator 468–473
-- heterodyne-current records 463–464
-- homodyne-current records 459–460
- Monte-Carlo simulation
-- photoelectron counting records 448–451
- non-Hermitian Hamiltonian 422
-- cascaded systems 491, 493, 496, 498
-- for degenerate parametric oscillator 446, 453
-- for source cascaded with a scanning interferometer 500, 507
-- for superradiance 499
-- in balanced homodyne detection 455
-- unidirectional coupling in 453
- non-unitary Schrödinger equation 423, 446
-- with coherent driving, solved by coherent states 494
- optical spectrum in 499–514
-- construction from simulated data sets 464, 499, 510
- role of null measurement in 417, 425
- scattering record 406–409, 413
-- for degenerate parametric oscillator 438, 467
-- for Einstein stochastic process 411
-- for spontaneous emission 413, 416
-- no-count 482, 508
-- not generally Markov process 434
-- probability (density) for 408, 413, 419, 421, 423, 430–431, 438–447, 449, 458, 501
- second-order correlation function in 479–484
- source superoperator 444, 453, 455

– spectrum of squeezing in
– – construction of from simulated data sets 460, 510
– unraveling of the density operator
– – cascaded systems 491–492
– – defined 408
– – for initial pure state superposition 424
– – for spontaneous emission 414, 418
– – from Dyson expansion 421, 447–448
– – heterodyne-current records 460–466
– – homodyne-current records 451–460
– – nonuniqueness of 466–473
– – photoelectron counting records 447–451
– – sum over unnormalized states 431
– – without subtraction of the backscattered coherent driving 453
– – without subtraction of the coherent driving 496
Quantum-classical correspondence
– for two-time averages in normal order 324
– phase-space equation of motion
– – dephasing term in 288, 291
– – Fokker–Planck truncation of 289, 291–298
– – for degenerate parametric oscillator 63, 68, 71, 73–74, 87–88
– – for optical bistability 287, 290
– – generalized 105
– – third-order derivatives in 69
Quasienergies 377
– of the driven Jaynes–Cummings model 382
Quasimode 32, 202, 204, 206

R representation
– related to positive P representation 99
Rabi frequency 508
Radiation reaction 212
Random telegraph process 410
Rate equations 171, 226
– Einstein 410
– in the basis of dressed Jaynes–Cummings eigenstates 385, 388
Reduced state 217, 225
Regression formula *see* Quantum regression formula
Reservoir
– correlations for thermal state 34
– coupling coefficient
– – related to cavity damping rate 29, 486
– in coherent state 141
– mode density 33, 34
Resonance fluorescence 311, 403
– cavity-enhanced 210–216
– – and one-atom laser 223
– – forward- and backward-scattered fields 212
– – Rabi oscillation in 216
– – second-order correlation function 216, 220
– – spectrum of 328
– – weak-excitation limit 221
– collective 375
– for atoms in a cavity 352
– output field 286
Retardation
– effects for cascaded systems 490, 492, 493

Saturation intensity 257
Saturation photon number 196, 212, 275
– related to number of atoms 302
Schrödinger cat state 176, 430
Schrödinger equation 11, 220, 233, 488
– for time-retarded density operator 489
– nonlinear
– – for null-measurement conditioning 428
– periodic solution of 377, 383
– stochastic
– – heterodyne-current records 462
– – homodyne-current records 459
– – satisfied by the normalized conditional state 459, 465
– – temporally mode-matched heterodyne detection 474

Schwartz inequality
- for photon antibunching, generalized 348
Secular approximation *see* Master equation, secular approximation
Shot noise 43, 457, 460, 464
- and vacuum fluctuations 53
- autocorrelation function 512
Solid angle
- subtended by TEM_{00} mode 209
Spectral hole
- squeezing-induced 329
Spectrum
- Mollow 245, 352, 384
-- squeezing-induced linewidth narrowing in 330
- of photocurrent fluctuations in homodyne detection 44
- of transmitted light for many atoms in a cavity 322–330
- optical 79, 323, 388, 499–514
-- coherent and incoherent parts, defined 323
-- constructed from photocurrent autocorrelation 464
-- defined 501
-- effect of squeezing on 79–80
-- Lorentzian squared 80, 323, 328–330, 390
-- measured with a scanning interferometer 505
-- normalization of 506
- spontaneous emission 237
-- from driven excited-state doublet 506–510
- weak-probe transmission 277
Spectrum of squeezing 39
- in the classical limit 39
- at a single cavity ouput 45
- constructed from photocurrent autocorrelation 460
- for degenerate parametric oscillator 460
-- above threshold 85
-- at threshold 180
-- below threshold 49, 79
- for the forwards scattered light in cavity QED 360
- in the Wigner representation 56
- including free-field term 47
- related to optical spectrum 80
- role of detection and collection efficiencies 44
- source-field 44, 78, 85
- time ordering in 53
- with unit efficiency 47, 50
Spontaneous dressed-state polarization 374, 392, 395
- and absorptive optical bistability 397
- effect of spontaneous emission on 397–399
- Stern–Gerlach analogy 399
- threshold for 371
-- squeezing at 382
- two-atom generalization 400
Spontaneous emission
- cavity-enhanced 198, 202–210
-- and altered density of states 202
-- associated frequency shift 208
-- collective 276, 328
-- enhancement factor 204, 208–209, 221, 274, 328, 332
-- optical Bloch equations for 208
-- rate 208, 355
-- spectrum of 237–238
- cavity-inhibited 38, 210
- distinguishability of atoms through 318
- for collective (symmetrized) atomic states 340–342
- from driven excited-state doublet 506–510
-- related to nonperturbative cavity QED 510
- in cavity with spoiling of Q 239
- in nonperturbative cavity QED 232–239
-- and coupled oscillators 234, 237, 243
-- linewidth averaging 242
-- spectrum of 235–237
- master equation 420
- rate 200
-- subnatural 242
- related to weak-field absorption 259
Squeeze parameter
- defined 12

– for degenerate parametric oscillator 78
– for intracavity parametric amplification 17
– for single-pass parametric amplification 14
– in dressing of the Jaynes-Cummings eigenstates 382
Squeezed state *see* Vacuum (Coherent) state, squeezed
Stochastic differential equation
– for anharmonic oscillator 191
– for cummulative charge in temporarily mode-matched heterodyne detection 475
– for degenerate parametric oscillator 113, 129, 135, 178
– for filtered heterodyne current 462
– for filtered homodyne current 459
– for two-photon damping 182
– integration of
– – Euler method 145, 182
– – sampling error 145
– Ito in positive P representation 109
Stochastic electrodynamics 58
Stochastic Schrödinger equation *see* Schrödinger equation, stochastic
Superoperator 173, 203, 226, 232, 342
– associated 139
– commutator 139–140
– conjugate 139
– defined 138
– equation of motion for 140
– factorization in quantum trajectory theory 446–447
– roles in quantum trajectory theory 422
– source, in quantum trajectory theory 444, 453, 455, 491
Superradiance
– cavity-assisted
– – and decay rate in absorptive optical bistability 312
– quantum trajectory theory of 407, 499
Superradiance and superfluorescence 275, 499
Symmetrized atomic states 340, 360
– used for distinguishable atoms 340–342
Symmetry breaking 372, 395
System size expansion
– and Heisenberg uncertainty relation 162
– and many body quantum theory 148
– applied 71–74, 86–91, 299–306
– – to hierarchy of moment equations 119–120
– beyond lowest order 178–181
– limitations of 94, 133, 146–149, 273–275, 350–352, 395
– linearized treatment of fluctuations
– – divergence of fluctuations in 76, 86, 315
– – in operator form 118–120
– – recovered from exact results 160–163
– – using generalized functions 121–123
– scaling in
– – for degenerate parametric oscillator, at threshold 86–90
– – for degenerate parametric oscillator, away from threshold 72, 83
– – for optical bistability 299
System size parameter
– for degenerate parametric oscillator 72, 146, 164
– for optical bistability 300
– in good- and bad-cavity limits compared 320, 352
– limit of "zero system size" 370, 376
– related to coupling strength 133, 172, 196, 222

Teleportation *see* Quantum teleportation
Two-photon loss *see* Master equation, for two-photon loss

Uncertainty relation *see* Heisenberg uncertainty relation
Unraveling *see* Quantum trajectory theory, unraveling of the density operator

Vacuum fluctuations 237
– in spectrum of squeezing 39, 53–58
Vacuum noise *see* Vacuum fluctuations
Vacuum Rabi doublet 240, 281, 339
– many-atom 277, 310, 332
– – with squeezing-induced linewidth narrowing 330
– nonlinear extension of 279, 310
– of squared Lorentzians 323
– single-atom
– – with squeezing-induced linewidth narrowing 391
Vacuum Rabi oscillation 332, 339, 355
– frequency of 367
Vacuum Rabi resonances 242, 284
– and coupled oscillators 243
– squeezing at 321
– two-state behavior of 243–246
Vacuum state
– squeezed
– – defined 12
– – mean photon number in 12–13
– – Q and Wigner distributions for 12
Variance
– negative 122
Volterra-Lotka cycles 395
von Neumann measurement *see* Measurement, projective
Waiting-time distribution *see* Photoelectron counting, waiting-time distribution
Wavefunction collapse 466, 477
– mechanism for in quantum trajectory evolution 495
Wiener increment 145, 457, 462, 475, 511
– correlation function 511
Wiener process 183
Wiener–Khinchin theorem 505
Wigner distribution
– for a squeezed vacuum state 12
– for degenerate parametric oscillator 452
– – conditioned on even and odd photoelectron counts 469
– for spontaneous dressed-state polarization 396, 398
– from time-averaged density operator 451
– measured by homodyne detection 471
– stochastic 399
Wigner representation
– Fokker–Planck truncation of the phase-space equation of motion 289
– normal-ordered, time-ordered averages in 82
– visualization of vacuum fluctuations in 56–58, 166

Theoretical and Mathematical Physics

Statistical Methods in Quantum Optics 2
Non-Classical Fields
By H. J. Carmichael

An Introduction to Riemann Surfaces, Algebraic Curves and Moduli Spaces
2nd Edition
By M. Schlichenmaier

Concepts and Results in Chaotic Dynamics: A Short Course
By P. Collet and J.-P. Eckmann

The Theory of Quark and Gluon Interactions
4th Edition
By F. J. Ynduráin

From Microphysics to Macrophysics
Methods and Applications of Statistical Physics
Volume I
By R. Balian

From Microphysics to Macrophysics
Methods and Applications of Statistical Physics
Volume II
By R. Balian

Titles published before 2006 in *Texts and Monographs in Physics*

The Statistical Mechanics of Financial Markets
3rd Edition
By J. Voit

Magnetic Monopoles
By Y. Shnir

Coherent Dynamics of Complex Quantum Systems
By V. M. Akulin

Geometric Optics on Phase Space
By K. B. Wolf

General Relativity
By N. Straumann

Quantum Entropy and Its Use
By M. Ohya and D. Petz

Statistical Methods in Quantum Optics 1
By H. J. Carmichael

Operator Algebras and Quantum Statistical Mechanics 1
By O. Bratteli and D. W. Robinson

Operator Algebras and Quantum Statistical Mechanics 2
By O. Bratteli and D. W. Robinson

Aspects of Ergodic, Qualitative and Statistical Theory of Motion
By G. Gallavotti, F. Bonetto and G. Gentile

The Frenkel-Kontorova Model
Concepts, Methods, and Applications
By O. M. Braun and Y. S. Kivshar

The Atomic Nucleus as a Relativistic System
By L. N. Savushkin and H. Toki

The Geometric Phase in Quantum Systems
Foundations, Mathematical Concepts, and Applications in Molecular and Condensed Matter Physics
By A. Bohm, A. Mostafazadeh, H. Koizumi, Q. Niu and J. Zwanziger

Relativistic Quantum Mechanics
2nd Edition
By H. M. Pilkuhn

Physics of Neutrinos and Applications to Astrophysics
By M. Fukugita and T. Yanagida

High-Energy Particle Diffraction
By E. Barone and V. Predazzi

Foundations of Fluid Dynamics
By G. Gallavotti

Many-Body Problems and Quantum Field Theory An Introduction
2nd Edition
By Ph. A. Martin, F. Rothen, S. Goldfarb and S. Leach

Statistical Physics of Fluids
Basic Concepts and Applications
By V. I. Kalikmanov

Statistical Mechanics A Short Treatise
By G. Gallavotti

Quantum Non-linear Sigma Models
From Quantum Field Theory
to Supersymmetry, Conformal Field Theory,
Black Holes and Strings
By S. V. Ketov

Perturbative Quantum Electrodynamics and Axiomatic Field Theory
By O. Steinmann

The Nuclear Many-Body Problem
By P. Ring and P. Schuck

Magnetism and Superconductivity
By L.-P. Lévy

Information Theory and Quantum Physics
Physical Foundations for Understanding
the Conscious Process
By H. S. Green

Quantum Field Theory in Strongly Correlated Electronic Systems
By N. Nagaosa

Quantum Field Theory in Condensed Matter Physics
By N. Nagaosa

Conformal Invariance and Critical Phenomena
By M. Henkel

Statistical Mechanics of Lattice Systems
Volume 1: Closed-Form and Exact Solutions
2nd Edition
By D. A. Lavis and G. M. Bell

Statistical Mechanics of Lattice Systems
Volume 2: Exact, Series
and Renormalization Group Methods
By D. A. Lavis and G. M. Bell

Fields, Symmetries, and Quarks
2nd Edition
By U. Mosel

Renormalization An Introduction
By M. Salmhofer

Multi-Hamiltonian Theory of Dynamical Systems
By M. Błaszak

Quantum Groups and Their Representations
By A. Klimyk and K. Schmüdgen

Quantum The Quantum Theory of Particles, Fields, and Cosmology
By E. Elbaz

Effective Lagrangians for the Standard Model
By A. Dobado, A. Gómez-Nicola,
A. L. Maroto and J. R. Peláez

Scattering Theory of Classical and Quantum N-Particle Systems
By. J. Derezinski and C. Gérard

Quantum Relativity A Synthesis
of the Ideas of Einstein and Heisenberg
By D. R. Finkelstein

The Mechanics and Thermodynamics of Continuous Media
By M. Šilhavý

Local Quantum Physics Fields, Particles, Algebras
2nd Edition
By R. Haag

Relativistic Quantum Mechanics and Introduction to Field Theory
By F. J. Ynduráin

Supersymmetric Methods in Quantum and Statistical Physics
By G. Junker

Path Integral Approach to Quantum Physics An Introduction
2nd printing
By G. Roepstorff

Finite Quantum Electrodynamics
The Causal Approach
2nd edition
By G. Scharf

From Electrostatics to Optics
A Concise Electrodynamics Course
By G. Scharf

Geometry of the Standard Model of Elementary Particles
By A. Derdzinski

Quantum Mechanics II
By A. Galindo and P. Pascual

Generalized Coherent States and Their Applications
By A. Perelomov

The Elements of Mechanics
By G. Gallavotti

Essential Relativity Special, General, and Cosmological Revised
2nd edition
By W. Rindler

Zeitfracht Medien GmbH
Ferdinand-Jühlke-Straße 7
99095 Erfurt, Deutschland
produktsicherheit@kolibri360.de